Animal BEHAVIOR

Fifth Edition

Mechanisms · Ecology · Evolution

Lee C. Drickamer
Northern Arizona University

Stephen H. Vessey
Bowling Green State University

Elizabeth M. Jakob
University of Massachusetts

Boston   Burr Ridge, IL   Dubuque, IA   Madison, WI   New York   San Francisco   St. Louis
Bangkok   Bogotá   Caracas   Kuala Lumpur   Lisbon   London   Madrid   Mexico City
Milan   Montreal   New Delhi   Santiago   Seoul   Singapore   Sydney   Taipei   Toronto

McGraw-Hill Higher Education

A Division of The **McGraw-Hill** Companies

ANIMAL BEHAVIOR: MECHANISMS, ECOLOGY, EVOLUTION
FIFTH EDITION

Published by McGraw-Hill, a business unit of The McGraw-Hill Companies, Inc., 1221
Avenue of the Americas, New York, NY 10020. Copyright © 2002, 1996 by The McGraw-Hill
Companies, Inc. All rights reserved. No part of this publication may be reproduced or
distributed in any form or by any means, or stored in a database or retrieval system, without
the prior written consent of The McGraw-Hill Companies, Inc., including, but not limited to,
in any network or other electronic storage or transmission, or broadcast for distance learning.

Some ancillaries, including electronic and print components, may not be available to
customers outside the United States.

This book is printed on recycled, acid-free paper containing 10% postconsumer waste.

2 3 4 5 6 7 8 9 0 CCW/CCW 0 9 8 7 6 5 4 3 2

ISBN 0-07-012199-0

Executive editor: *Margaret J. Kemp*
Developmental editor: *Donna Nemmers*
Marketing manager: *Heather K. Wagner*
Project manager: *Joyce Watters*
Senior production supervisor: *Laura Fuller*
Design manager: *Stuart D. Paterson*
Cover designer: *Nathan Bahls*
Interior designer: *Jamie A. O'Neal*
Cover image: *Mark Moffet/Minden Pictures*
Senior photo research coordinator: *Carrie K. Burger*
Photo research: *Chris Hammond*
Executive producer: *Linda Meehan Avenarius*
Compositor: *Carlisle Communications, Ltd.*
Typeface: *10/12 Times Roman*
Printer: *Courier Westford*

The credits section for this book begins on page 403 and is considered an extension of the
copyright page.

Library of Congress Cataloging-in-Publication Data

Drickamer, Lee C.
    Animal behavior : mechanisms, ecology, evolution / Lee C. Drickamer, Stephen H.
Vessey, Elizabeth M. Jakob.—5th ed.
      p.   cm.
    Includes bibliographical references (p.) and index.
    ISBN 0–07–012199–0 (alk. paper)
    1. Animal behavior. I. Vessey, Stephen H. II. Jakob, Elizabeth M. III. Title.

QL751 .D73   2002
591.5—dc21   2001030656
                                            CIP

www.mhhe.com

For Judy Sellers, my very special partner, and for Donald A. Dewsbury, who has been a professional inspiration on many occasions.

LCD

For my students, who kept me on my toes and about whom I worried more than they will ever know.

SHV

For Adam Porter—thanks.

EMJ

BRIEF CONTENTS

Preface ix

CONTENTS

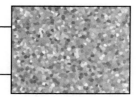

Preface ix

PART THREE
Mechanics of Behavior 78

PART FOUR
Finding Food and Shelter 218

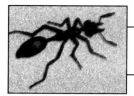

PREFACE

Animal behavior is a broad discipline with investigators and contributions from diverse perspectives, including anthropology, comparative psychology, ecology, ethology, physiology, and zoology. Our goal in this textbook is to use evolutionary principles as a unifying theme to provide students exposure to a number of approaches to the field of animal behavior. We also hope to demonstrate that the varied perspectives used to study behavior are complementary and often integrated; they are not mutually exclusive. The subtitle, "Mechanisms, Ecology, Evolution," reflects the broad themes that dominate the book.

We are very pleased to welcome Elizabeth Jakob as a third author for this fifth edition. She has brought a new perspective and a great deal of energy to our endeavors. Her influence on the book can be seen in a number of places, particularly with regard to current theory and research in behavioral ecology.

The text is designed for use by undergraduate students taking their first course in animal behavior, though over its 20+ years, it has also been widely used for upper level undergraduate and graduate level courses. Our approach, involving concepts, processes, and methods, makes the book readable and useful for all students, regardless of whether they are interested in the "gee-whiz" performances of animals so often observed on today's television nature programs, or in gaining a fuller understanding of the underlying mechanisms, ecology, and evolutionary biology behind what is observed. The 19 chapters cover all of the material encompassed by modern animal behavior. We have attempted to write each chapter so that it is self-contained. Thus, instructors can arrange the material to suit their personal approaches or the requirements of their particular courses. The length of the book and the degree of difficulty of the material are such that it should be completed in a semester or quarter with little difficulty.

We have assumed some knowledge of basic biology, but we provide some background material on evolutionary theory and related principles of genetics to ensure that all students have the same foundational knowledge with which to approach particular subareas of animal behavior. In each of the major parts of the book, and within each chapter, we have tended to follow a similar pedagogical pattern. We first define the concepts and processes that form the foundation for an area of investigation. Then, using appropriate research examples, we present the methods and techniques used to understand the problems that are explored in that subarea of behavior. By using this approach, we hope that students are introduced to a variety of viewpoints that have contributed to the richness and strongly integrative nature of the discipline.

The book is divided into five major parts. Our presentation begins with basic material on the types of questions posed and tested by animal behaviorists throughout the history of the discipline. Part One covers the background for the study of animal behavior, including history and approaches. Part Two deals with basic evolution, related principles of genetics, and behavior genetics, and the evolution of behavior. In Part Three, the mechanisms and processes that control behavior are presented and exemplified. In Part Four, the first of two parts on behavioral ecology, we deal with how an animal finds its way around, habitat selection, and aspects of feeding behavior. We conclude with Part Five, dealing with social and mating systems of animals and their evolution.

We have attempted to integrate proximate and ultimate perspectives throughout the book, though certain parts or chapters emphasize proximate questions and other parts emphasize ultimate questions. We begin with an evolutionary perspective, then examine events occurring inside the organism, followed by events involving ecological and social interactions. This provides a more complete, cumulative picture of the discipline of animal behavior than other possible plans of organization.

Changes for the Fifth Edition

Based on changes that have occurred in the past decade in animal behavior and in our thinking about the subject, we have made some organizational alterations, as well as many additions and deletions to the content of this new edition.

- The contents of Chapters 5 and 6 were reversed and reorganized. New material on behavioral genetics and the evolution of behavior has been added, including a discussion of phylogenetic analysis and the comparative method.

- We include many new neurobiological examples and a discussion of motor programs.

- We treat several topics in greater mathematical detail for enhanced clarity, while still keeping the text accessible.

- Material on animal cognition has been greatly expanded as part of the chapter on learning behavior.

- The glossary contains many new and revised definitions of all terms that appear in boldface in the text material.

- We have added more material on applied animal behavior in several chapters, reflecting the increased importance of this subfield of animal behavior.

- We have updated examples and references throughout the book and provided new illustrations and photos where appropriate. In doing so, we have attempted to retain the basic materials that provide an understanding of the history of the development of concepts, processes, and methods in animal behavior. In doing this, we avoid having the textbook become merely a catalogue of current research.

Acknowledgments

A special thanks is extended to Matthew M. Douglas, a science contributor to this edition, for both his words and helpful assistance.

Some reviewers beyond the "official" ones provided comments or assistance on various chapters. We thank Mitchell Baker, Russell Balda, Donald Dewsbury, Eben Goodall, Mike Henshaw, Chad Hoefler, Jeremy Houser, Christa Skow, Kristin Vessey, Elizabeth Wells, and especially Gordon Wyse and Adam Porter.

A hearty thanks to those individuals who have read all or parts of the manuscripts of the different editions and shared with us their helpful criticisms and suggestions. We especially acknowledge the reviewers of the fourth edition of this text:

Elizabeth Adkins-Regan
Cornell University

Christine R. B. Boake
University of Tennessee, Knoxville

Daniel A. Cristol
The College of William and Mary

Patricia DeCoursey
University of South Carolina

Perri Eason
University of Louisville

Jeremy J. Hatch
University of Massachusetts, Boston

Ann V. Hedrick
University of California

Norman A. Johnson
University of Massachusetts, Amherst

Christine R. Maher
University of Southern Maine

Catherine Marler
University of Wisconsin, Madison

Stephan J. Schoech
Indiana University

Peter Sherman
Transylvania University

Dan Weigmann
Bowling Green State University

Lee C. Drickamer
Flagstaff, Arizona

Stephen H. Vessey
Bowling Green, Ohio

Elizabeth M. Jakob
Amherst, Massachusetts

Animal BEHAVIOR

PART ONE

The Study of Animal Behavior

Part 1 of this book provides background information for examining the approaches and methods used to formulate and test questions in animal behavior. First, we examine the historical roots of the major approaches to the study of behavior (chapter 2). We then look at the problems with methods involved in designing experiments to test hypotheses about behavior (chapter 3). In the four major sections of the book that follow these introductory chapters, we will explore various internal and external factors that regulate and influence animal behavior.

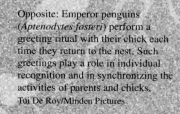

Opposite: Emperor penguins (*Aptenodytes fosteri*) perform a greeting ritual with their chick each time they return to the nest. Such greetings play a role in individual recognition and in synchronizing the activities of parents and chicks.
Tui De Roy/Minden Pictures

Introduction to Animal Behavior

WHY WE STUDY ANIMAL BEHAVIOR

All animals have a variety of complex relationships with members of their own species, with members of other species, and with the physical environment. The survival of a species depends on its individual members' ability to obtain food and shelter, to find mates and reproduce, and to protect themselves from parasites, predators, and the elements. In many ways, adaptive behavior ensures survival, and survival ensures evolutionary success. By studying animals, we learn much about the relationships between them and their environments, the physiological processes that determine their behavior, and some of the reasons for their abundance and distribution.

Behaviorists are interested in establishing general principles common to all animal behavior, including our own. There are researchers who study the behavior of animals in order to conserve and protect endangered species. Other behaviorists are interested solely in the actions of economically important predators, pests, and parasites. A few researchers study the behavior of domestic animals that provide for our well-being, while others study animals in captivity in order to preserve and exhibit them for the education of humanity. In short, the study of animal behavior is enormously important, both scientifically and economically.

We can illustrate the study of animal behavior by looking at the migratory and overwintering behavior of the Monarch butterfly (*Danaus plexippus* L.) in four general areas of inquiry: behavioral genetics and evolution; mechanisms of behavior; food and shelter; and social organization and mating systems. These four areas represent the four basic sections of animal behavior outlined in this book.

THE MONARCH BUTTERFLY: A CASE STUDY

The monarch butterfly is a large and beautiful orange butterfly with black wing veins and spots. This common species evolved from a large group of closely related danaid butterflies and has several subspecies throughout Mexico, Central America, and South America. All monarch butterflies have coevolved with a diverse group of plants called milkweeds, in the family Aescleapiadaceae. During World War II, the U.S. Army enlisted the help of civilians to make pillows for the troops by collecting the downlike projections that give the seeds their "parachutes." Many of these pillows contained seeds still attached to the down, and when the pillows were destroyed, some of the seeds germinated to produce milkweed plants in many parts of the world, including Hawaii, Australia, Europe, and Africa (Douglas, unpublished).

The monarch butterfly (a strong migratory species and an opportunistic stowaway on ocean-going ships) soon followed, and like its larval food plant, it too is now distributed widely over the earth. Interestingly, some of these new populations exhibit migratory behavior, whereas other populations do not. In North America (the United States and Mexico), there are at least two **disjunct** (separated) populations of monarch butterflies. One population on the West Coast breeds in the mountains of California, Oregon, and Washington, and in the fall migrates to form large communal roosts, containing thousands of butterflies, along the northern and central coastline of California. Here they overwinter in a state of near hibernation, their sexual organs held in reproductive dormancy or **diapause.** In the spring, they "turn on" reproductively, engage in stereotyp-

ical monarch mating behavior (figure 1.1), and disperse northeastward toward the mountains. The females lay their eggs on milkweeds then emerging from the warming soils, and the population expands, generation by generation.

A second population of monarchs inhabits the interior of North America between central Mexico and southern Canada, and between the Rocky Mountains and parts of the Atlantic Ocean. As summer progresses, this population of monarchs produces generations (broods) of butterflies as far north as central Canada. In the late summer and early fall, these butterflies migrate by the billions, in two or three massive but discontinuous waves (Douglas, unpublished), and ultimately form enormous overwintering roosts in the Transvolcanic Mountains of Mexico (figure 1.2) just northwest of Mexico City, at altitudes of nearly 3,000 meters (Brower 1985). There are at least seven major roosts, each only a few hectares in size, located in the same Oyamel fir stands from year-to-year. (Brower et al. 1977). Within these roosts hundreds of millions of butterflies spend the winter in a virtual refrigerator, instead of: relatively high in humidity and low in temperature, but never with the temperature extremes experienced in the higher latitudes of North America. Here's the mystery: The monarchs have never seen these roosts before and will never see them again, yet they find them unerringly, some traveling distances that exceed 5,000 kilometers.

The monarchs spend four to five months in these roosts, often completely covering the middle sections of the Oyamel fir trees. Rarely are butterflies found close to the ground, and rarely are they found near the tops of the trees (Calvert and Brower 1986). Here they lie dormant, butterfly hanging on

FIGURE 1.1   Mating monarchs.
Monarchs mate for extended periods, during which time the male transfers a large sperm- and nutrient-laden spermatophore to the female. If the mating pair is threatened, the male flies away, carrying the female passively behind him.
Courtesy of Karen Oberhauser

FIGURE 1.2   Roosting monarchs.
The transvolcanic range contains seven known permanent roosts at about 3,000 meters in altitude, just northwest of Mexico City.
Courtesy of Karen Oberhauser

FIGURE 1.3   Monarchs in mountain passes.
In late winter and early spring, the overwintering roosts break up, filling the skies with monarchs looking for nectar, water, and mates.
Courtesy of Tom Trow

butterfly, sometimes tens of thousands on a single branch. An individual butterfly weighs less than a paperclip, but enough weight is produced at times by the clustered butterflies to break massive limbs, sending them crashing to the earth below. If it is not too cold, the butterflies crawl back up into the roosts. (Calvert and Brower 1981). If it is too cold, they "shiver" (synchronously contract) the muscles that operate the wings to generate enough heat to move the legs (Douglas 1986).

In early March, the butterflies break their reproductive diapause. An incredible frenzy of mating butterflies then fills the mountain passes leading away from the volcanic peaks and flows down into the valleys below, where the milkweed is shooting up new leaves (figure 1.3). Rivers of slow-flying butterflies pour down the mountains to nectar at early spring flowers and to drink badly needed water before dispersing north and northeast to recolonize the central part of North America. We have a great deal of information about the "midwestern" population of monarchs, but many important and very basic questions remain about their migratory behavior.

A third, and much more poorly understood, population of monarchs inhabits the eastern seaboard of the United States (Walker 1991). A good percentage of these butterflies migrate from New England and the North Atlantic states into southern Georgia and Florida. However, unlike their western and midwestern cohorts, these butterflies remain in a state of potential reproduction and do not form roosts (Douglas, unpublished). Their offspring continue to breed throughout the year in these areas. In the spring, some of these butterflies disperse northward along the eastern seaboard to become the first monarchs to repopulate the North Atlantic and New England states. Much later in the spring, drifters from the central population of the Midwest may also make their way to these regions.

In all cases, females lay their eggs solely on milkweed plants (there are over 100 species of milkweed plants in North America). The next generation of monarchs hatches from their eggs, growing until the cuticle cannot accommodate further growth. They molt five times during their larval period

until they weigh thousands of times more than their original birth weight. At this point, the caterpillar is physiologically instructed to molt one last time into a chrysalis (pupa) from which the adult butterfly is assembled, literally from scratch, by liquefying and transforming the caterpillar parts into the adult parts of the butterfly. Unlike the overwintering generations, however, these generations of the summer are different from those that overwinter: They do not congregate, they do not store fat for migration, and they do not orient to the south and southwest. Furthermore, they are sexually active and mate soon after emerging from the pupae. In essence, two types of monarchs are produced during the year: the summer, nonmigratory form and the fall, migratory, overwintering form.

The three populations of monarchs are interfertile and produce viable offspring (Douglas, unpublished). Therefore, by the biological species definition, they must belong to a single species. However, each population has evolved different behavioral and physiological responses that adapt them to their respective environments. What questions might we ask ourselves concerning the different components involved in the study of monarch migratory behavior? Here are just a few possibilities, and a few potential answers—which in turn raise additional questions.

Behavioral Genetics and Evolution

1. Why do the three populations of monarchs in the continental United States respond differently to climatic changes in the fall? Why do the midwestern and western populations migrate, form colonies, and diapause, while the eastern population migrates but does not form colonies or enter reproductive diapause? Decreasing photoperiods and temperatures may somehow "turn on" a set of genes that induce migration in the midwestern and western populations. On the other hand, these same stimuli may "turn off" a set of genes that keep the butterflies in a reproductively active and nondiapausing state. Perhaps the eastern population lacks the ability to respond fully to environmental stimuli, or perhaps the genes regulating diapause are never completely turned off.

2. Why should the western population of monarchs migrate just to the coast of California while the midwestern population migrates all the way to Mexico? Why shouldn't the eastern population continue to Mexico by migrating across the Gulf of Mexico? We simply don't have enough information to understand the evolutionary genetics of this situation. All we know is what apparently has evolved, perhaps relatively recently. Migratory behavior is obviously adaptive to the survival of the adults, but wouldn't it have been easier to evolve a pupal diapause to survive winters in the temperate zone rather than migrate?

Mechanisms of Behavior

1. These different populations must share a common monarch ancestor, but why should the monarch's evolutionary "strategy" for surviving the winter months be so different from those of other butterfly species, which typically overwinter as larvae or pupae? The mechanism of the behavior inducing migratory behavior and diapause is not understood. We know that some butterflies hibernate and undergo diapause while others migrate, but the exact course of evolutionary events that led to

these different survival strategies is very difficult to decipher. How have neural transmitters, hormones, and physiological responses affected the evolution of this behavior?

2. How is it that monarch butterflies from Ontario can fly unerringly to the roosts of Mexico, a total distance of nearly 5,000 kilometers (assuming that butterflies rarely fly in a straight line)? The mechanism must involve following some cue or set of cues such as the arc of the sun across the sky (Taylor et al. 1997), the magnetic field of the earth (Etheredge et al. 1999; Taylor et al. 1998), or a combination of these and other cues. But what sort of physiological mechanism leads these butterflies to locations high in the mountains of Mexico—to occupy overwintering roosts that they have never seen before and will never see again? That is as difficult mechanistically as successfully sending a person (without a map) by foot from Chicago to Mexico City.

Food and Shelter

1. As with all animals, securing sufficient food and shelter is critical to the survival of migratory monarchs. Adults of the summer brood survive several weeks, but the migratory overwintering brood may survive nearly nine months. Why is there such a difference in lifespan between the two broods? In the summer, monarchs mate, use sugars as metabolic fuel, and senesce as do other butterflies. However, in the fall, monarchs enter reproductive diapause, store sugars as fat, and orient to the south and southwest. Is reproductive diapause somehow tied to the migratory response and lifespan? In years of drought, monarchs die by the tens of millions because of the lack of nectar-bearing plants. In the roosts, they must periodically descend from the mountains to the valleys below to imbibe at puddles and to take whatever nourishment they can. But by and large, their migratory and overwintering survival depends on their ability to store energy as larvae (from the milkweed host plant), and their ability to store enough energy as fat (derived from nectar resources) during the adult stage. Their survival is intimately tied to the successful location of food, both in the larval and adult stage.

2. Why should monarchs choose to shelter themselves in such specific locations within the transvolcanic mountains of Mexico year after year? They have never seen these roosts before, so why not spread out over the entire range—much of which superficially appears to be suitable for overwintering monarchs? Are there visual and olfactory cues from the Oyamel firs that direct the migrants to their roosting sites? Are these the only climate-protected spots in the transvolcanic mountains? There are no answers to these simple questions. Another question: Why do migrants roost in the middle of the trees and not the tops or bottoms? One possibility is that roosting in the middle of the trees protects them from the extremes of temperature, including radiational cooling and the cosmic cold of outer space possible at the tops of the trees (it is thermally much colder at the tops). The lack of butterflies at the bottoms might be explained by a number of factors, including increased predation by mice and several species of birds (which have evolved the ability to avoid the very toxic cardiac glycoside compounds incorporated in the cuticle and wings of the adult butterfly from the larval host plant). Whatever the answers, the roosting behavior is quite dramatic.

Social Organization and Mating Systems

1. How does monarch mating behavior during migration differ from that of the summer forms? Summer brood monarchs are reproductively active immediately after they emerge from the chrysalis. Monarch females may mate more than once. The males actively patrol for newly emerged females and literally tackle them in flight, bringing them to the ground. The males produce complex pheromones that settle the female and allow mating to take place. Over the course of nearly six hours, the male transfers a large spermatophore which contains both sperm and many nutrients for the female. If the female mates three times or more, a great deal of her available energy is given to her by the males' spermatophores. Compared to the human species, the monarch mating system is incredibly complex. Why should such a complex structural, physiological, and behavioral mating system have evolved when much more simplified mating systems suffice for most species?

2. Monarchs are social and form roosts only during the migratory season. Summer brood monarchs never migrate or form social roosts. Could it be just a simple genetic switch (cued perhaps by decreasing photoperiod or declining temperatures) that triggers aggregation, migration, and roosting behavior? If so, why is roosting behavior coupled with migration? Why not spend the winter alone, hiding from predators? Furthermore, what could be the selective advantage of grouping together in the roosts? Perhaps roosting originally offered a selective advantage because it promoted a greater probability of finding a suitable mate within the roost in the spring. Or perhaps the congregating behavior at the roosts (where butterflies hang on each other ten deep) prevents additional heat loss from the colony by increasing the size of the characteristic dimension of the colony and thereby reducing heat loss via convection and radiation. Could sociality be selected for and be adaptive for both reasons—or perhaps other reasons we cannot entertain at this point in our study of monarch behavior?

PROXIMATE VERSUS ULTIMATE CAUSATION

As you can see, even (apparently) simple behaviors such as monarch migration and overwintering are incredibly complex and difficult to study. Many animal behaviorists focus on "how" questions: those involving the ways in which behavior is directly produced and regulated. We call these structures and mechanisms **proximate factors.** For example, if we ask, How do male monarchs produce the "settling" pheromone, we could study hormonal and environmental cues that trigger the production and release of the pheromones. However, if we focus only on proximate factors, we may miss the important adaptive components of this behavior.

Both behavioral ecology and sociobiology are rooted in investigations using "why" rather than "how" questions. These are questions that investigate the **ultimate factors** that have influenced the evolution of behavioral patterns. For example, why should there be both migratory and nonmigratory forms of the monarch? Or, what are the evolutionary adaptive values of a female mating more than once? An evolutionary perspective generates important questions about the functions of behavioral patterns. However, a danger of focusing only on ultimate factors is that we may assume—without directly testing the assumptions—that certain mechanisms are adaptive. The study and testing of ultimate factors requires extra care and attention.

In the end, we must understand both proximate and ultimate factors whenever possible. Evolution by natural selection is the cornerstone of all subdisciplines of biology, including animal behavior. However, natural selection works at the level of the mechanisms that control and integrate behavior. To understand behavior more completely, therefore, we must have a solid grasp of both the internal mechanisms by which behavior is produced ("how" questions) and the external factors that influence whether those mechanisms will be passed on to future generations ("why" questions).

As you can readily see, a few observations of animal behavior can generate enough questions for a lifetime.

Throughout this book, we will explore case studies that examine both familiar and unfamiliar animal species and generate a range of questions, from the proximate to ultimate. In many cases, answers to initial questions have generated even more complex questions, and yet we will likely never know the entire story for even a single animal species. This is the mark of an exciting and dynamic discipline, and we hope that you are intrigued and challenged by the many questions that remain about animal behavior.

DISCUSSION QUESTIONS

1. Think through the seasonal changes experienced by the animals listed below. Then provide a series of questions about their behavior, questions similar to those presented in this chapter about monarch butterflies.

   a. tree squirrels

   b. turtles

   c. spiders

   d. red-winged blackbirds

2. Indicate which of your questions is concerned with proximate issues and which is concerned with ultimate issues.

3. Examine figure 1.1, which illustrates the copulatory behavior of monarch butterflies. From what you know at this point, which features of monarch mating behavior are likely to be similar to the mating behavior of birds? Which are likely to be different? How would these observations determine which questions you would first address if you were comparing the mating behavior of these species?

4. Based on your own daily experiences, perhaps with pets, from viewing nature programs on television, or your own natural history observations, make a list of 10 questions about the behavior of particular animals that have fascinated you. Using the categorization of four basic types of questions we have introduced in this chapter, first determine the categories that your questions best fit, and second, determine whether each question deals with proximate or ultimate causation.

SUGGESTED READINGS

Brower, L. P. 1985. New perspectives on the migration biology of the monarch butterfly, *Danaus plexippus* L. *Contrib. Mar. Sci Suppl.* 27:748–86.

Brower, L. P., W. H. Calvert, L. E. Hedrick, and J. Christian. 1977. Biological observations on an overwintering colony of monarch butterflies (*Danaus plexippus* Danaidae) in Mexico. *J. Lepid. Soc.* 31:232–42.

Calvert, W. H., and L. P. Brower. 1986. The location of the monarch butterfly (*Danaus plexippus* L.) overwintering sites in relation to topography and climate. *J. Lepid. Soc.* 40:164–87.

Douglas, J. Migrational orientation variation between different subpopulations of the monarch butterfly (*Danaus plexippus* L.). INTEL project report.

Douglas, M. M. 1986. *The Lives of Butterflies.* Ann Arbor: University of Michigan Press. 241 pages.

Etheredge, J., S. Perez, O. R. Taylor, and R. Jander. 1999. Monarch butterflies (*Danaus plexippus* L.) use a magnetic compass for navigation. *PNAS* (in press).

Taylor, O. R., S. Perez, and R. Jander. 1997. A sun compass in monarch butterflies. *Nature* 387:29.

Taylor, O. R., S. Perez, and R. Jander. 1998. Monarch butterflies (*Danaus plexippus*) are disoriented by a strong magnetic pulse. *Naturwissenschafen* 86 (3):140.

Tinbergen, N. 1963. On aims and methods of ethology. *Zeitschrift fur Tierpsychologie* 20:410–29.

Walker, T. J. 1991. Butterfly migration from and to peninsular Florida. *Ecol. Entomol.* 16:241–52.

History of the Study of Animal Behavior

H umans and their prehuman ancestors have left evidence—both deduced by us from archaeological explorations and drawn, sculpted, and written by them—of their interest in the natural world. We know that some of this interest originated in need. Animals were a primary source of food, clothing, and materials for tools and shelter; thus, knowledge concerning their behavior was necessary for successful hunting. During the course of history, interest in animal behavior has also stemmed from human curiosity about the natural world. In this chapter, we examine how and why people have studied animal behavior—from the early days of human evolution, through the emergence of animal behavior as a scientific discipline in the nineteenth century, to the experimental and theoretical approaches of the present.

INTEREST IN ANIMAL BEHAVIOR

Early Humans

For many thousands of years, humans and their ancestors were hunters and meat-eaters. The early hominids and the first *Homo erectus* practiced a crude variety of hunting techniques. Peking man, a form of *Homo erectus* that lived approximately 400,000 years ago, was an accomplished hunter, used fire, and made tools from animal bones.

L. S. B. Leakey (1903–1972), an anthropologist known best for his discoveries of early hominid remains in Tanzania, proposed and tested a hunting strategy that was based on knowledge of animal behavior—a strategy that early hunters may have used to capture rabbits or other small prey. Leakey suggested that, upon sighting the prey at about fifteen meters distance, the hunter should sprint directly toward the animal (a small animal often initially freezes in such a situation).

Within two or three meters of the prey, the hunter should turn sharply either left or right, because the typical escape behavior of the prey is to make a sudden dash in one direction or the other. If both prey and hunter go to the left, the hunter is upon the animal and can grab it bare-handed (as Leakey demonstrated), or he might use a club or stone to strike it. If the hunter guesses incorrectly, he should stop, turn, and wait for the animal to stop. The process is then repeated and perhaps results in a successful capture.

Early *Homo sapiens* must have been keen observers of animal habits and characteristics. They needed to be familiar with the behavior of animals, not only to know where and how to hunt their prey, but also to protect themselves from potential predators. Hunters of the Upper Paleolithic (35,000 to 10,000 years ago) probably used fire to drive animals over cliffs or into cul-de-sacs or bogs where they could be slaughtered with rocks or clubs (figure 2.1). A ravine with at least 100 mammoth carcasses has been located in Czechoslovakia, and the remains of thousands of horses that were stampeded over a cliff have been discovered in France.

Prehistoric cave paintings in France and Spain reveal other aspects of humankind's relationship to animals. These paintings realistically depict many types of game animals in ways that suggest close observation of the animals at various times in their life cycles. In addition, some of the drawings are symbolic representations of actual hunting scenes. However, while early people were aware of the animals in their environment, their knowledge of animal behavior was probably limited to mostly practical concerns.

Classical World

Interest in animal behavior in the classical world stemmed from curiosity about natural phenomena and a desire to record and categorize observations. For example, Aristotle (384–322 B.C.) wrote ten volumes on the natural history of animals, in which we note the first extensive use of the observational method. The following brief excerpts, translated from the original Greek, give us a flavor of what Aristotle's observations were like (the first two passages are true, the last is false) (Ley 1968, 36–37):

> They say that the cuckoos in Hellice, when they are going to lay eggs, do not make a nest, but lay them in the nests of doves or pigeons, and do not sit, nor hatch, nor bring up their young; but when the young bird is born and has grown big, it casts out of the nest those with whom it has so far lived.
>
> In Egypt they say there are some sandpipers that fly into the mouths of crocodiles and peck their teeth, picking out the small pieces of flesh that adhere to their teeth; the crocodiles like this and do them no harm.
>
> The goats in Cephallaria apparently do not drink like other quadrupeds; but every other day turn their faces to the sea, open their mouths and inhale the air.

The Roman naturalist Pliny (A.D. 23–79) made extensive observations of the natural world. A quote from his *Natural History* provides some insight into the anthropomorphism (ascribing human characteristics or attributes to nonhumans) that characterized Roman perceptions of animal behavior (Nordenskiöld 1928, 55):

> Amongst land animals, the elephant is the largest and the one whose intelligence comes nearest that of man, for he understands the language of his country, obeys commands, has a memory for training, takes delight in love and honour, and also possesses a rare thing even amongst men— honesty, self-control and a sense of justice; he also worships stars and venerates the sun and the moon.

We can see from these brief passages that early scholars were attempting to record what they observed in the world around them. Their perceptions of behavior were often colored by the lack of full knowledge about what was taking place, or by biases based on religion or philosophy. However, for many centuries these early observations served as the basis for human understanding of the natural world.

FOUNDATIONS OF ANIMAL BEHAVIOR

The rigorous scientific study of animal behavior did not begin until the latter part of the nineteenth century. We turn now to three major developments that contributed significantly to the study of behavior as it developed prior to 1900: (1) publication of the theory of evolution by natural selection, (2) development of a systematic comparative method, and (3) studies in genetics and inheritance.

FIGURE 2.1   Early Hunters.
Early humans practiced various hunting techniques that were based, in part, on their knowledge of the behavior of the prey animals. In some instances, they successfully drove individuals or groups of animals, such as the wooly mammoths shown here, into swamps or bogs, where the animals became trapped and could be killed.
© Granger Collection

Theory of Evolution by Natural Selection

For several centuries, European ships made voyages of exploration and discovery to all parts of the globe. Often scientists were officially attached to the voyages, as Charles Darwin (1809–1882) himself was. These scientists and other crew members made observations of exotic fauna and flora and brought live and preserved specimens to zoos and laboratories in Europe, where scholars could observe, record, and speculate about the anatomy, behavior, and interrelationships of these newly discovered species. The following passage from Darwin's account (figure 2.2) of the marine iguana of the Galápagos Islands illustrates the kind of observations he made on animals in their natural setting (Darwin 1845, 336):

> They inhabit burrows, which they sometimes make between fragments of lava, but more generally on level patches of the soft sandstone-like tuff. The holes do not appear to be very deep, and they enter the ground at a small angle; so that when walking over these lizard warrens, the soil is constantly giving way, much to the annoyance of the tired walker. This animal, when making its burrows, works alternately the opposite sides of its body. One front leg for a short time scratches up the soil, and throws it towards the hind foot, which is well placed so as to heave it beyond the mouth of the hole. That side of the body being tired, the other side takes up the task, and so on alternatively.

Like all major scientific paradigms, the theory of evolution drew upon contributions by and suggestions from the work of other scientists. In 1798, Thomas Malthus (1766–1834), in his *Essay on the Principle of Population,* hypothesized that humans have the reproductive potential to rapidly overpopulate the world and outstrip the available food supply. The inevitable result is disease, famine, and war. Malthus's theory was an important influence on Darwin's thinking about the competition for survival among members of a species. A contemporary and friend of Darwin's, geologist Sir Charles Lyell (1797–1875) was among those who made observations of rock strata and successions of fossils that gave evidence of a process of continuous change in living material through time, an idea that was at odds with the biblical suggestion of the simultaneous creation of all living things. This evidence of geological change led others to the idea that species themselves were not fixed entities. The artificial selective breeding of domesticated stocks by English farmers provided additional support for the thinking of both Darwin and A. R. Wallace (1823–1913).

Wallace's voyage to the Malay archipelago, Darwin's travels on the *Beagle* to South America and the South Pacific, and their other studies and the intellectual influences of the time, led each man independently to formulate the theory of evolution by natural selection. The original theory states that although each animal species has a high capacity for reproduction, the population size remains relatively constant over time. Thus, not all animals produce the maximum number of offspring. Heritable variation in traits exists within animals of one species. Because some traits are more advantageous than others, not all organisms produce an equal number of surviving offspring, and the operational process of natural selection occurs. Only those members of the species that are able to survive to produce more offspring contribute their characteristics to subsequent generations through their young.

Behavior, morphology, and physiology were all thought to be subject to the effects of natural selection. The following passage from *The Origin of Species* illustrates that Darwin clearly recognized the central role of animal behavior in determining the outcome of competition between animals (Darwin 1859, 94):

> Amongst birds, the contest is often of a more peaceful character. All those who have attended to the subject, believe that there is the severest rivalry between the males of many species to attract, by singing, the females. The rock-thrush of Guiana, Birds of Paradise, and some others, congregate; and successive males display with the most elaborate care, and show off in the best manner, their gorgeous plumage [figure 2.3]; they likewise perform strange antics before the females, which, standing as spectators at last choose the most attractive partner.

FIGURE 2.3   Male bower bird in display.
As in the passage quoted from Darwin, male birds of a variety of species display to attract females. The male bower bird builds a bower and adorns it with brightly colored objects.
© Patti Murray/Animals, Animals

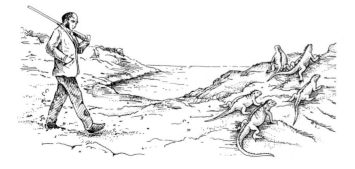

FIGURE 2.2   Charles Darwin investigating the unique marine iguanas of the Galápagos Islands.

Darwin concluded that species were not fixed entities. The theory of evolution by natural selection accounted for changes within a species through time and also for the gradual appearance of new species. Recent developments in other biological fields—genetics in particular—have modified the theory of evolution by natural selection proposed by Darwin and Wallace. Today, some evolutionary biologists believe that evidence from the fossil record and genetic mechanisms support the claim that rates of evolution vary through time (Stanley 1981). Change through evolution, in particular the appearance of new species, may occur more rapidly during some time periods than at other times. We will explore the theory of evolution and its consequences for animal behavior in chapters 4 and 5.

Comparative Method

George John Romanes (1848–1894) is generally credited with formalizing the use of the **comparative method** in studying animal behavior. For Romanes, the comparative method involved studying animals to gain insights into the behavior of humans. Romanes sought to support Darwin's theory with his proposal that mental processes evolve from lower to higher forms and that there is a continuity of mental processes from one species to another. He argued that although people could really know only their own thoughts, they could infer the mental processes of animals, including other humans, from knowledge of their own. For Romanes, the similarities between the behavior of humans and that of other animals implied similar mental states and reasoning processes in humans and in nonhuman species. He suggested that a sequence could be constructed for the evolution of various emotional states in animals. Worms, which exhibit only surprise and fear, were placed lowest on this scale; insects were said to be capable of various social feelings and curiosity; fish showed play, jealousy, and anger; reptiles displayed affection; birds exhibited pride and terror; and finally, various mammals were credited with hate, cruelty, and shame.

Romanes's theory relied largely on inferences rather than on recorded facts or direct observations of behavior; he made substantial use of anecdotes. A movement led by another Englishman, C. Lloyd Morgan (1852–1936), sought to counteract these faults by using the **observational method.** Morgan's basic tenet was that only data gathered by direct experiment and observation could be used to make generalizations and develop theories. Morgan is probably best known for his "law of parsimony," which is now axiomatic in animal behavior studies, "In no case may we interpret an action as the outcome of the exercise of a higher psychical faculty if it can be interpreted as the outcome of the exercise of one which stands lower in the psychological scale" (Morgan 1896, 53). This statement has been interpreted to mean that in the analysis of behavior, we must seek out the simplest explanations for observed facts. Where possible, we should reduce complex hypotheses to their simplest terms to facilitate the clearest understanding of the mechanisms that control behavior.

Theories of Genetics and Inheritance

The third development that greatly influenced research in animal behavior was the birth of the science of genetics and the development of modern theories of inheritance. In the 1860s, Gregor Mendel (1822–1884) reported his findings from breeding experiments using garden peas. These studies established key principles of the laws of inheritance of biological characteristics. Present-day behavioral biology is based on the combination of evolutionary theory, which explains how traits can change through time, and genetics, which explains how traits are passed from one generation to another.

We now know that, like morphological and physiological traits, an animal's behavior has a genetic component. Thus, behavior may change as a species evolves. This means that, as scientists, we can explore the genetic variation underlying various behavior patterns, just as others have investigated the effect of genetic inheritance on morphology and physiology. Behavior-genetic analysis had its beginnings in these early studies of inheritance and was then greatly expanded in the 1930s by the work of R. A. Fisher (1890–1962) and others. Behavior-genetic analysis (e.g., Boake, 1994) is a powerful tool used by many animal behaviorists; we will learn more about this in chapter 5.

EXPERIMENTAL APPROACHES

The ideas, methods, and theories established during the latter half of the nineteenth century form the foundation of today's experimental approaches to the study of animal behavior. (1) Comparative psychologists and physiologists have sought to determine the underlying causes of behavior—the control mechanisms. (2) Classical ethologists have been concerned primarily with the functional significance and evolution of behavior patterns but have also developed explanations for behavior mechanisms, including drives, innate releasing mechanisms, and similar concepts. (3) Behavioral ecologists and sociobiologists have explored the ways in which animals interact with their living and nonliving environments and have applied the principles of evolutionary biology to the study of social behavior and organization in animals. We should now briefly examine the historical development of each approach. From these varied approaches to the study of behavior has come the modern synthetic view of animals living and behaving in their natural environment. Though we examine these approaches here as separate entities, bear in mind that they did not develop entirely independently of one another, and that in recent decades, they have become melded into a single discipline. The modern approach to the study of animal behavior contains elements of all three approaches. As we can see by looking at the animal behavior courses offered at various colleges and universities and the titles of the textbooks used to teach such courses, those who work and teach in this area may call themselves ethologists, animal behaviorists, or comparative psychologists. However, they are all really pursuing similar goals using common theoretical frameworks, and practicing their craft using similar experimental techniques and methods.

Studies of Mechanisms

Comparative psychology is the study of different animals' behavior patterns in order to determine the general principles that explain their actions. Comparative psychology can best be understood by looking at the variety of approaches to behavior studies taken over the past century, which eventually led to comparative psychology's development. In today's world, comparative psychology has melded into the larger discipline that we call animal behavior or ethology.

Perceptual Psychology

Several distinct approaches to discovering the mechanisms underlying behavior emerged during the mid-nineteenth century. Researchers who were concerned with the mind/body dichotomy studied the relationships between physical and mental processes. Investigators were interested in separating the processes of sensation (body) and perception (mind). This usually involved the objective measurement of sensation (the reception of stimuli through the senses, such as sight and hearing) and the comparison of this direct measurement to objective interpretation (perception) of the sensations. Today's subdiscipline of psychophysics is an outgrowth of these early studies. These types of studies still impact our understanding of animal behavior in terms of what an animal makes of its world, both with regard to sensory systems and with respect to the animal's interpretations of its sensations.

Physiological Psychology

Modern physiological psychology developed from early attempts to relate behavior with the internal physiological properties and events of the organism. For example, Marie-Jean-Pierre Flourens (1794–1867) surgically removed portions of the brains of pigeons and recorded the resulting changes in the birds' behavior. Hermann von Helmholtz (1821–1894) studied the conduction speed of nerve impulses, and later, the physiology of vision. He ingeniously measured the speed of nerve conduction by experimenting on the frog motor neuron that triggers muscle contractions. First he stimulated the nerve at one point near the muscle, and then at a second point farther away from the muscle. The difference in amount of time elapsed between stimulus and muscular contraction in the two measurements is the conduction time for the distance between the two stimulus points. From this information, he calculated the speed of conduction.

Physiological psychology remains an important subdiscipline today, and work in this realm and in animal behavior are interconnected. Another classic study is Sperry et al. (1956), in which he surgically manipulated the position of the eyes in newts (*Notophtalmus viridescens*). Sperry removed the eyes and then replaced them so that they were upside down! Newts treated in this way behaved as if they saw the world upside down; they moved their eyes upward in response to the movement of an object downward in their visual field. This effect persisted even after several years. We learn from Sperry's work that in the visual system of the newt, the neurons in the optic nerve traveling from the retina to the brain are labeled for spatial orientation. Thus, even though the eye has been rotated, the message sent to the brain along the nerve remains the same as if the eye were in the correct, normal orientation. For his work on the nervous system, Sperry shared the Nobel Prize in 1981.

Functionalism

By the late 1800s and early 1900s, Europe was no longer the exclusive center of behavioral studies, and individuals were conducting research investigations in comparative psychology at a number of laboratories in the United States. Two major new theoretical and experimental points of view arose during this period: functionalism and behaviorism. The **functionalists,** among them, John Dewey (1859–1952), studied the functions of the mind and how the mind operates, in contrast to studying how the mind is structured. Functionalists attempted to answer three major questions: (1) How does mental activity occur? (2) What does mental activity accomplish? and (3) Why does mental activity take place? Functionalism employed objective observation rather than introspection as its primary method.

The functionalist approach was the introduction into psychology of **adaptive behavior,** a notion prevalent in biology that behavior functions in the animal's survival in its natural habitat. To these early psychologists, the concept of adaptive behavior implied that the response to a stimulus changes the sensory situation in such a way that the original conditions that produced the response are altered. For example, pain disappears when a sharp splinter is removed from the hand, and the original condition—the existence of a splinter—is also altered.

Behaviorism

John B. Watson (1878–1958) was the principal founder of a new approach to the study of behavior, **behaviorism.** The basic tenet of behaviorists is that animal behavior consists of an animal's responses, reactions, or adjustments to stimuli or complexes of stimuli. Thus, most activities of an organism are products of its past experiences. Behavior, rather than the mind, became the primary focus for study. To what degree can we predict and control behavior based on a knowledge of an animal's previous experiences? The methods utilized by Watson and his followers, for example, B. F. Skinner (1904–1990), were strictly objective. Reports of subjective feelings or emotions were, by definition, not acceptable as scientific data. This restriction forced the behaviorists to study human behavior in much the same way they studied the behavior of any other animal, without benefiting from their subjects' verbal judgments or reports of feelings and perceptions. (It is noteworthy that Skinner's earliest papers dealt with innate aspects of behavior; studying the history of our discipline provides many insights and surprises!)

FIGURE 2.4   Thorndike puzzle box.
A cat inside the cage can clearly see the reward, in this case a fish, placed outside. In order to obtain the reward, the cat must learn to manipulate a shuttle-lever system that raises the door of the cage.

Animal Psychology

Concurrent with the development of these viewpoints was the emphasis by Edward L. Thorndike (1874–1949) on the need for systematic, replicable experiments in comparative animal psychology. Thorndike used the puzzle box (figure 2.4) to perform a series of task-learning experiments, using cats as test subjects. A cat was placed in the box, which was fastened shut; by manipulating a shuttle-lever, the cat could open the door and obtain a reward placed outside the box. From these experiments, Thorndike concluded that much of animal learning takes place by trial and error and that rewards are a critical component of learning processes.

In 1950, Frank Beach stressed that the discipline of comparative psychology was devoting too much attention to the white rat as a test subject, while ignoring many other types of available vertebrate organisms. Others, notably Lockard (1971) and Hodos and Campbell (1969), called our attention to the lack of an evolutionary perspective in comparative psychology and to the incorrect use of the rat as a model for other organisms, especially humans. These critiques stimulated more truly comparative investigations, for example, the work of Dewsbury (1972, 1975) on reproductive behavior in rodents. More attention has also been given to the natural context and actual field investigation of animals (Lockard 1971; Barash 1973a,b, 1974a,b).

Animal psychology today is a diverse mixture of subdisciplines, both new and old. The study of comparative learning and learning theory is still quite important, as the works of Bitterman (1975), Seligman (1970), and Roitblat and Meyer (1995) exemplify. Ecological aspects of learning in a variety of animal species have been examined by investigators like Kamil and Sargent (1981), Davey (1989), Balda et al. (1998) and Dukas (1999). The development of behavior is also a subject of continued investigation by researchers like Oppenheim and colleagues (Oppenheim 1982; Oppenheim et al. 1992; Caldero et al. 1998), who examines aspects

of neural development; by Burghardt and colleagues, who are studying the development of feeding behavior in reptiles such as garter snakes (*Thamnophis sirtalis*) (Burghardt and Krause 1999); and by groups like the one developed at the Wisconsin Regional Primate Laboratory by Harlow and colleagues (Suomi and Harlow 1977) that explored primate behavior development. The study of the physiological processes underlying behavior has also diverged into several pathways: the relationship of hormones and behavior (Lehrman 1965; Crews 1980; Goy et al. 1988; Knapp et al. 1999; Strier et al. 1999); neural correlates of behavior (Hubel and Wiesel 1965; Brown et al. 1988; Glendinning et al. 1999) and brain chemistry; and psychopharmacology and behavior (Kelly et al. 1979; Ferris et al. 1999). The cross-fertilization between genetics and behavior also produced a new subdiscipline called behavior genetics, which is concerned with the hereditary bases of behavior and how the interactions of genetics and environment affect behavior (Hirsch 1967; Oliverio 1983; Miklosi et al. 1997; Kim and Ehrman 1998).

Studies of Function and Evolution

Ethology

The systematic study of the function and evolution of behavior, called **ethology,** is now a little over a century old. One of its most important principles is that behavioral traits, like anatomical and physiological traits, can be studied from the evolutionary viewpoint. For example, C. O. Whitman (1842–1910) made extensive observations of display patterns, which he termed instincts, in various species of pigeons. Whitman found that he could use displays (patterns of behavior exhibited by animals that function as communications signals) to classify animals according to similarities and differences in behavior. From its early beginnings, ethology developed into a separate science, with its own concepts and terminology, much as comparative psychology did. Today, as we noted previously, those working as ethologists are conducting the same sorts of studies as all others who study animal behavior.

The **ethogram,** an inventory of the behavior of a species, has been a starting point for many ethological studies. After making observations of an organism's behavior, ethologists then formulate specific questions about the adaptiveness and function of particular behavioral patterns. A student of Whitman's, Wallace Craig (1876–1954), defined two key categories of behavior patterns from his work with doves and pigeons. The first category includes the variable actions of an animal, such as its searching behavior to find food, a nest site, or a mate; these are called **appetitive behavior.** The second category includes stereotypical actions that are repeated without variation, such as the act of mating or the killing of prey; these are called **consummatory behavior.**

The ethological approach is used in another major area of inquiry: the determination of how key stimuli trigger specific behavior patterns. J. von Uexküll (1864–1944) demonstrated that animals perceive only limited portions of the

FIGURE 2.5   Courtship of male and female three-spined stickleback.
The enlarged belly of the female three-spined stickleback fish (top) is a sign stimulus for the male of the species (bottom) to court and to entice the female to enter the nest he has built.

total environment with their sense organs and central nervous systems. This sensory-perceptual world was termed the **Umwelt** by von Uexküll. Among the stimuli recorded by the sense organs, certain specific cues that ethologists call **sign stimuli** trigger particular stereotyped responses called **fixed action patterns (FAPs).** For example, the female three-spined stickleback fish's enlarged belly triggers courtship behavior in male sticklebacks (figure 2.5).

Credit for the synthesis of these early findings and for the further development of modern ethology belongs largely to two men, Konrad Lorenz (1903–1989) and Niko Tinbergen (1907–1988). Lorenz pioneered studies of genetically programmed behavior and investigated the importance of specific types of stimulation for young animals during critical periods of early development. Modern ethology's concern with four areas of inquiry—causation, development, evolution, and function of behavior—developed from a scheme proposed by Tinbergen (1963). (As psychologist Thomas McGill has noted, the first six letters of the alphabet can be used to remember these questions: Animal Behavior: Causation, Development, Evolution, Function.) Recognition for animal behavior as an independent discipline came in the fall of 1973, when the Nobel Prize for Physiology or Medicine was awarded to three ethologists: Konrad Lorenz, Niko Tinbergen, and Karl von Frisch (1886–1982). Von Frisch had conducted research on animal sensory processes and made important contributions to the study of bee behavior and communication.

Modern ethology is characterized by varied types of investigations ranging from more traditional observational studies in natural environments (Geist 1971; Joerman et al. 1988; Millesi et al. 1998) to experiments on the physiological bases of behavior (Bentley and Hoy 1974). The latter study is indistinguishable from those conducted by many physiological psychologists. Some ethologists work primarily with behavior genetics and the evolution of behavior

(Manning 1971; Gerhardt 1979; Ukegbu and Huntingford 1988) or explore the relationships between hormones and behavior (Hinde 1965; Truman, Fallon, and Wyatt 1976) or the nervous system and behavior (Nottebohm 1981; Rose et al. 1988; Oliveira and Almada 1998). Others work on research problems in the field or in a laboratory setting that resembles the natural habitat. By employing experimental manipulation to test specific hypotheses, Kummer (1971) investigated the effects that transplantation of individuals from troop to troop had on the social behavior of baboons, Wickler (1972) studied the significance of color patterns in fish, Gowaty and Wagner (1988) tested the aggressive behavior of eastern bluebirds, and Panhuis and Wilkinson (1999) investigated the effect of male eye span on contest outcome in stalk-eyed flies.

Since the mid-1950s, the distinctions between ethology and comparative psychology have been slowly disappearing. Several events have opened communication between scientists of the two approaches. These events include the biennial meetings of the International Ethological Conference, many cross-visitations between researchers in Europe and America, and the publication of a number of international animal behavior journals. A common approach, which started with the notion of species-typical behavior, has emerged from this exchange of information. **Species-typical behavior** involves actions and displays that are broadly characteristic of a species and that are performed in a similar manner by all its members. The autobiographical sketches of many leaders in animal behavior (Dewsbury 1985) include a variety of perspectives, and provide excellent insight into the way the various approaches have developed independently, and how they have recently coalesced into a unified and integrated approach to the study of behavior.

At least one major long-standing controversy in animal behavior has been largely resolved during recent years. Early ethologists believed that much of an animal's behavior was instinctive or preprogrammed and was not affected to any great extent by experience. Many psychologists claimed that learning and experience were the major determinants of behavior. Today, most animal behaviorists believe that neither of these viewpoints is entirely correct. Instead, as we shall see in chapter 10, the current focus is on the interaction of genotype, physiology, and experience as the determinants of behavior, and on how the relative contributions of genetic and environment effects differ among animal species.

Comparisons

Discerning whether a particular study has been conducted by an ethologist or a comparative psychologist may be difficult at first. If we understand the historical differences between these two approaches, we can better appreciate the synthetic approach that characterizes the behavior studies of the past several decades.

Ethology was developed, largely in Europe, by researchers trained in biology. Ethologists traditionally observed a wide variety of animals in nature and conducted experiments under conditions that mirrored the natural setting as closely as possible. They concentrated their efforts on exploring questions of ultimate causation—the "why" questions of the evolution and function of behavior.

Comparative psychology originated primarily in America. Until the past several decades, most psychologists generally worked under controlled laboratory conditions. Much of their research was carried out on small rodents, particularly the domesticated rat. Comparative psychologists placed primary emphasis on proximate issues—the "how" questions of the physiological and developmental mechanisms underlying observed behavior patterns. Dewsbury's history of comparative psychology (1984) provides many details regarding the development of concepts and theories in this field and many insights into the individuals responsible for the experimental and theoretical work. Dewsbury defines comparative psychology as the attempt to make comparisons across species in order to develop principles of generality regarding animal behavior. He examines the course of development that characterizes this field since 1900 and notes the many myths that have been associated with what scientists and nonscientists alike have come to believe a comparative psychologist is.

Animal behavior is now a unified discipline with a broad synthetic approach: much of the research conducted by animal or comparative psychologists today is indistinguishable from that of other animal behaviorists with different backgrounds. These research endeavors include explorations of the genetic aspects of food-searching behavior in blowflies (McGuire and Tully 1986), the effects of aversive conditioning on learning behavior of honeybees (Abramson 1986), the role of hormonal factors in infanticidal behavior in rats (R. E. Brown 1986), and the role of the brain in budgerigars' interpreting acoustic information from contact calls (S. D. Brown et al. 1988).

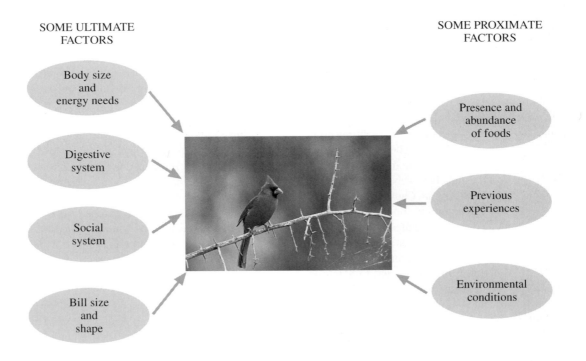

SOME ULTIMATE FACTORS

Body size and energy needs

Digestive system

Social system

Bill size and shape

SOME PROXIMATE FACTORS

Presence and abundance of foods

Previous experiences

Environmental conditions

FIGURE 2.6   **Factors affecting the feeding behavior of northern cardinals.**
Constraints that have arisen through evolution establish the limits on the dietary habits for the cardinal. Past experience and current environmental conditions influence the immediate choices made by the animal as it forages.
Source: Photo © Stephen J.Krasemann/Photo Researchers, Inc.

Behavioral Ecology and Sociobiology

In the past five decades, a third approach to the study of animal behavior has emerged. **Behavioral ecology** and **sociobiology,** with origins in zoology, examine the ways in which animals interact with their environments and the survival value of behavior (Morse 1980; Krebs and Davies 1993, 1997). "Environment" as used here includes animals of the same species (conspecifics), other animals within the same ecological community, plants, and inorganic physical features of the habitat.

Behavioral ecologists are concerned with both ultimate and proximate questions about behavior. Suppose we are interested in the feeding habits of the northern cardinal (*Cardinalis cardinalis*), living in a variety of places in the North American countryside (figure 2.6). Ultimate constraints affecting the cardinal would include its body size and related energy needs; the type of bill, which affects the foods it can consume; the digestive system, with regard to what foods the bird can process; and the social system of the species, which could influence the partitioning of available food resources. Proximate factors influencing feeding would include the presence and relative abundance of specific foods; past experiences of the bird in searching for and handling particular foods; and the season of the year, with particular regard to variations in energy needs due, for example, to reproduction or cold winter weather. Ultimate factors establish the limits, and proximate factors affect the behavior of an animal within those limits.

Behavioral ecologists, trained primarily in zoology, ecology, and related fields, are also greatly influenced by the methods of comparative animal psychology. Behavioral ecologists often begin a field investigation and define questions about, for example, population regulation or predator-prey relations. Does the predator maximize its energy intake by utilizing some form of optimal foraging strategy? Certain aspects of the overall investigation (what are the most important features of the prey for predator recognition and detection?) may require experiments more systematic than those that can be done in the field setting. Thus, as behavioral ecologists, we might bring specific, testable hypotheses into the laboratory or controlled outdoor setting where the experiments are conducted. Attempts can then be made to relate laboratory findings to what is known about the animal in its natural field setting.

For an example of an investigation using the behavioral ecology approach, consider the prey-catching behavior of the shore crab (*Carcinus maenas*). When these animals are given their choice of what sized mussel to consume (figure 2.7), they select the size that provides them with the highest rate of energy return (Elner and Hughes 1978). Notice that although the crabs do eat mussels in a variety of sizes, they may avoid the larger mussels because of the extra time and energy needed to crack open the shells. The wide size range that the crabs eat may represent a compromise: lots of time spent searching for just the right sized mussel would be inefficient. Thus, we see that both the time it takes to find food and the

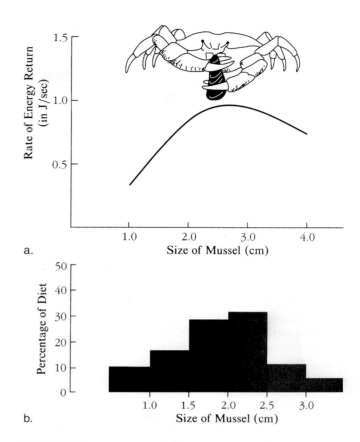

a.

b.

FIGURE 2.7   **Shore crabs select mussels for food.**
(a)Shore crabs select those sizes of mussels that provide the best rate of energy return more often than other sizes, though (b)they do eat mussels of a variety of sizes.

Source: Data from R.W. Elner and R.N. Hughes, "Energy Maximization in the Diet of the Shore Crab, *Carcinus maenas*," in *Journal of Animal Ecology,* 47:103–16, copyright 1978 by Blackwell Scientific Publications Ltd., Oxford.

ability of the predator to handle the prey can influence prey selection. We will explore these and other aspects of feeding in chapter 15. We'll look at similar factors that influence choices and selection of the most efficient pattern for animal habitat selection in chapter 14.

Among the investigations in behavior ecology, those of Emlen (1952a,b) on bird behavior and energy budgets, Davis (1951) on population biology of rats, and King (1955) on the relationship of prairie dog social behavior to habitat, were notable for the way they established topical areas for research work within the developing discipline. In more recent years, topics that have received particular attention by investigators include foraging strategy (Stephens and Krebs 1986; Vander Wall 1990; Bell 1991; Ydenberg and Hurd 1998), proximate mechanisms in behavioral ecology (Real 1994; Braude et al. 1999), predator-prey systems and predation risk (Haskins et al. 1997; Randall and Matocq 1997), the ecology of sex and strategies of reproduction (Askenmo 1984; Clutton-Brock 1988), and social systems in relation to ecology (Thornhill and Alcock 1983; Christenson 1984; Hill 1998). The last two topics are indicative of the joint nature of the approach involving behavioral ecology and sociobiology.

Sociobiology came of age in 1975 with the publication of E. O. Wilson's *Sociobiology: The New Synthesis.* Sociobiology applies the principles of evolutionary biology to the study of social behavior in animals. Sociobiology is a hybrid of behavioral biology (from the ethological perspective, with an emphasis on ultimate questions) and the study of social organization (with an ecological perspective) (Wittenberger 1981; Trivers 1985). As we shall see later, sociobiology relies heavily on the comparative method. Diverse groups of animals, living in a wide variety of habitats, are examined to find similarities and differences in their social systems. These examinations reveal if any general patterns explain the social behavior of a species. Thus, for example, we note that in some species of birds, young born in one year may not breed the second year, but help their parents rear the second year's brood. In other instances, adult birds that have lost their mate (or their clutch) may help close relatives rear young. These social systems that involve helping at the nest occur, for example, in Florida scrub jays (Woolfenden 1975), African white-fronted bee-eaters (Emlen 1984), acorn woodpeckers in the western United States (Koenig et al. 1984), and Seychelles warblers (Komdeur 1992; Komdeur et al. 1995). Investigations of these and similar social systems in birds reveal that nests with helpers are more successful—the number of young fledged is higher. Helping behavior would appear to be using energy to assist in rearing of offspring that are not the helper's own progeny. How can such behavior evolve? Sociobiologists are interested in exactly that question and also what the advantages are for the individual bird if it helps versus the advantages if it does not help.

The various theories and concepts that constitute sociobiology have their roots in many earlier works. Among the most significant are the writings of Williams (1966) on natural selection and the concept of adaptation, Trivers (1971, 1972) on the evolutionary aspects of altruism and parental behavior, and Hamilton (1964, 1971) on the genetic theory underlying the evolution of social behavior. Studies conducted under the general heading of sociobiology include, for example, those on altruism in ground squirrels (Sherman 1977), on strategies for reproduction in damselflies and other insects (Waage 1979; 1997), on parental investment in water bugs (Smith 1997), and on mate choice in American kestrels (Duncan and Bird 1989). In recent years, a major topic for investigators using the sociobiological approach has involved sexual selection and various factors influencing mate choice (Andersson 1994; Gowaty 1995; Eberhard 1996).

Since the mid-1970s, sociobiology has had a significant influence on research in animal behavior. Faced with the challenge of devising new research questions and new methods to test aspects of sociobiological theory, investigators have reexamined older data in light of new predictions. One area of prediction and hypothesis—and the source of considerable controversy—is the application of sociobiology to *Homo sapiens.* Some sociobiologists argue that the principles used to investigate the social behavior of animals can be applied to investigate the social behavior of humans. Other individuals argue that sociobiology is merely a form of biological determinism. A complete resolution of this controversy is probably impossible.

SUMMARY

Archeological evidence indicates that early humans had a practical knowledge of the behavior of animals, particularly of those animals that were potential food sources or predators. By Greek and Roman times, writers like Aristotle and Pliny recorded extensive observations about and deductions from natural phenomena.

Three developments of the last half of the nineteenth century contributed to the emergence of animal behavior as a scientific discipline. First, Darwin and Wallace, each working from his own data and from ideas of previous investigators, independently put forth the theory of evolution by natural selection. Second, Romanes pioneered the development of the comparative method and used it initially to study mental evolution. Third, with Mendel's work on inheritance and the rediscovery and development of his findings at the turn of the century, modern theories of genetics and evolution emerged.

Derived from these diverse beginnings, three major approaches characterize current studies of animal behavior. Investigations of the mechanisms controlling behavior have historically been conducted primarily by comparative animal psychologists and physiologists. Although much of the early psychological research relied heavily on introspection and inference, these methods were later replaced by systematic, objective observations and replicable experiments. Modern animal psychologists explore such areas as physiological control of behavior, sensation and perception, learning processes, and behavior genetics.

Ethology encompasses studies of the functional significance and evolution of behavior. Behavioral traits, like physical or physiological traits, are viewed in evolutionary terms and are thus subject to natural selection. Traditionally, ethologists have made many of their research observations in a natural setting. The research objectives and methods of modern ethologists range from observational studies and field experiments conducted to assess the function of behavior patterns, to investigations of the physiological bases of behavior.

Behavioral ecology and sociobiology generate studies that examine biological relationships between an organism and its environment and the evolutionary selection pressures that influence social systems. The questions are asked from an ecological and sometimes evolutionary viewpoint, and investigations conducted in both field and laboratory settings utilize systematic, controlled experimentation. Investigators using this approach are concerned with both proximate and ultimate factors influencing behavior.

DISCUSSION QUESTIONS

1. Select a bird, mammal, or insect that is easily visible during daylight hours and watch animals of that species for several hours. Record as many different behavior patterns as you can, their frequencies, and patterns of occurrence.

2. Consider yourself a prehistoric *Homo sapiens*. What characteristics and habits of the other animals in your environment would you want to know?

3. You have been asked to study the behavioral biology of the zinger (*Zingus zingu*), a snake that lives in the Arizona desert. Using each of the three major approaches to the study of animal behavior discussed in this chapter, formulate several questions you would want to answer about the zinger.

4. The animal's behavior, unlike the animal's skeleton, does not fossilize. What types of evidence would you look for in the fossil record to provide you with information concerning the behavior of animals?

SUGGESTED READINGS

Dewsbury, D. A. 1984. *Comparative Psychology in the Twentieth Century.* Stroudsburg, PA: Hutchinson Ross.

Dewsbury, D.A. 1985. *Leaders in the Study of Animal Behavior.* Lewisburg, PA: Bucknell University Press.

Dewsbury, D. A. 1989. A brief history of the study of animal behavior in North America. *Perspectives in Ethology* 8:85–122.

Morgan, C. L. 1894. *An Introduction to Comparative Psychology.* London: Scott.

Thorpe, W. H. 1979. *The Origins and Rise of Ethology.* London: Praeger Books.

3

Approaches and Methods

or any observation of animal behavior (for example, the annual reproductive cycle of monarch butter flies [*Danaus plexippus*] outlined in figure 1.2) we need to develop a framework in which to formulate testable hypotheses. What are the questions we want to ask about the lace bugs' mating system and egg-laying behavior? What about the internal events that take place in males and females during the changing seasons? Which questions are the most significant? In the first section of this chapter, we examine in detail how the various approaches to the study of behavior pose and answer experimental questions, and we consider the advantages and disadvantages of each approach.

Variation in how individual animals behave creates methods problems for the investigator. For example, some red-winged blackbirds perform territorial boundary displays repeatedly in the face of an intruder (figure 3.1), while others display much less frequently under the same circumstances. Nests constructed by redwings may vary in their height above ground and the types of plant material used in construction. How can we design experimental procedures to account for this variation? In the second part of the chapter, we examine some general principles for conducting behavioral research.

Finally, what are the problems and pitfalls of experimental research in animal behavior? An awareness of the limitations of the various methods and techniques will help us to evaluate the results of behavior investigations and to formulate better questions about behavioral patterns we observe. The principles for research and the problems and pitfalls of the experimental approach are presented in the final section of the chapter. Because these principles are part of a basic course in animal behavior, we believe that students of animal behavior must have some understanding of how to conduct research if they are to interpret properly what they are reading and if they are to participate in field and laboratory exercises.

FIGURE 3.1   Territorial display by male red-winged blackbird. Redwings display when faced with an intruder attempting to enter the territory. The wingspread display is often associated with the conkaree call.

Source: Courtesy Sarah Lenington.

APPROACHES

Ethology

We can define **ethology** as the biology of behavior: the exploration of functional and evolutionary questions, and the mechanisms underlying why an animal exhibits certain behavior patterns under certain circumstances. The first step in studying an animal is the compilation of an **ethogram,** an inventory of the behavior performed by animals of the species under investigation. Behavior can be divided either into broad descriptive categories (e.g., courtship, nesting, sleeping, and feeding), or into more restricted units (e.g., specific patterns shown during various phases of courtship [figure 3.2]). The ethogram serves as a basis for posing questions about the adaptive value, ecological importance, and regulation of the various behavior patterns.

An Ethological Study

By examining a study in detail, we can see how ethologists use observations from nature and field experiments to formulate and test questions. Niko Tinbergen studied the behavior of gulls for many years. Among the interesting problems

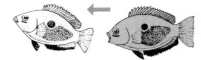

a. *Charging:* an accelerated swim of one fish toward another

b. *Tail Beating:* an emphatic beating of the tail toward another fish

c. *Quivering:* a rapid, lateral, shivering movement that starts at the head and dies out as it passes posteriorly through the body

d. *Nipping:* an O-shaped mouth action that cleans out the (presumptive) spawning site

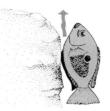

e. *Skimming:* the actual spawning movement whereby the fish places its ventral surface against the spawning site and meanders along it for a few seconds

FIGURE 3.2   Ethogram showing courtship pattern in orange chromide (*Etroplus maculatus*).

An ethogram can be compiled for all behaviors or for selected behaviors of a species.

he investigated was a curious phenomenon he first noted in his early observations of gulls (Tinbergen et al. 1962; Tinbergen 1963). After a young black-headed gull chick (*Larus ridibundus*) hatches and frees itself from the shell, the parents carefully pick up the remnant pieces of shell in their bills, fly off to some distant location, and drop them, a behavior also seen in many other bird species. Why do parent birds remove the eggshells? What selection pressures could have led to the evolution of such a behavior?

Tinbergen and his colleagues considered several explanations of the behavior. The sharp edges of the shells might injure the chicks, or the shells might clutter the nest and hamper the parents' attempts to brood and feed chicks. These explanations seemed unlikely, because the eggshells of black-headed gulls are thin and easily crushed. The investigators noticed that the outer surface of the eggshells was mottled—but the inside surface was white. Although the exterior camouflages the eggs during incubation, the bright white interior may attract potential predators, such as crows (*Corvus corax*) or herring gulls (*Larus argentatus*), which rely on visual cues for detecting prey. A more attractive hypothesis is that the parents remove the white eggshells to protect their black-headed young. The kittiwake (*Rissa tridactyla*), a related species, does not remove the shells after chicks hatch; the chicks are white and thus not camouflaged at all. However, kittiwakes nest on cliffs in regions where there are few predators.

Testable Hypotheses

A testable hypothesis arose from these observations: An eggshell left in a nest exposes the brood to a higher rate of predation. To test this hypothesis, Tinbergen and his co-workers performed field experiments in a large gullery in England. They scattered eggs of the black-headed gull over the dunes just outside the gullery. Some had the natural mottled coloration, and others were painted white. They watched the dune site from a blind and recorded the predation rates by crows and herring gulls for both egg types. The results, shown in table 3.1, clearly indicated that the white eggs were taken in greater numbers by both species of predators.

Tinbergen and his colleagues designed a second experiment as a further test of their hypothesis. They laid eggs out in the dune valley, and placed half-broken eggshells at varying distances from the whole eggs. The results revealed that

TABLE 3.1	**Predation of Black-Headed Gull Eggs**	
	Eggs Painted White	Eggs with Natural Mottled Coloration
Taken by crow	14	8
Taken by herring gull	19	1
Taken by others	10	4
Total taken	43	13
Total not taken	26	55

Source: Data from Tinbergen (1963).

TABLE 3.2	**Number of Eggs Taken and Distance Between Whole Gull Eggs and Broken Shells**		
	15 cm	100 cm	200 cm
Eggs taken	63	48	32
Eggs not taken	87	102	118

Source: Data from Tinbergen (1963).

the greater the distance between the whole egg and the half-broken eggshell, the lower the probability of detection of the whole egg (table 3.2). These and several additional experiments substantiated the idea that it is advantageous for adult gulls to remove the shells when the chicks hatch. Natural selection favors those gulls that remove the shells and thus eliminate a cue for predators to detect the nest.

Advantages and Disadvantages

Historically, ethologists proceeded from observations usually made in a natural setting to the formulation of hypothetical explanations of the functions and the evolutionary development of the behavior. The best explanations are those that lend themselves to experimental testing, such as the reason for eggshell removal by black-headed gulls. One major advantage of the ethological approach is that it is based on an evolutionary perspective: The functions and adaptive significance of behavior can be discerned best under field conditions with an understanding of how evolution affects morphology, physiology, and behavior. Another advantage of the ethological approach is that by working with the animals in a natural setting we are not restricting their behavior. Any answers that we obtain are not affected by caging or other artificial conditions that we might impose if we studied the animals in, for example, a laboratory setting.

There are two important disadvantages of the ethological approach: First, working in the natural setting, ethologists cannot control the environment and lack knowledge of the observed animals' past histories. Since factors like weather, habitat differences, or seasonal changes cannot be manipulated by the investigator, the functional significance of observed variations in behavior may be difficult to ascertain. (We should note, however, the investigator may turn certain of these situations to advantage by systematically gathering data under different weather or habitat conditions.)

Second, providing a hypothetical evolutionary explanation for a behavior is often easy, but devising and conducting an experiment that will thoroughly and convincingly test the hypothesis can be difficult. In particular, it is hard to distinguish current function from evolutionary origin. For example, some African wild dogs (*Lycaon pictus*) that prey on various ungulates (hoofed herd animals such as gazelles or zebras) exhibit cooperative hunting behavior. We can hypothesize that this behavior evolved as a means by which a group of dogs could succeed in isolating, running down, and killing a larger prey. A single dog could achieve this feat rarely, if at all. Given the right set of observational data, we might conclude that single dogs or pairs of dogs were less successful hunters than larger groups, but these data would tell us only about the current function of the behavior, and not necessarily why it initially evolved. Comparative studies based on phylogenetic information can sometimes address the question of the origin of a trait (see ch. 6).

Comparative Psychology

Comparative Analysis

Comparative psychology can be defined as the discipline devoted to comparative studies of behavior in animals. Early animal psychologists, working several decades before and after the beginning of the twentieth century, conducted their investigations on a wide spectrum of animals, ranging from invertebrates like protozoans and jellyfish to vertebrates like canids and primates. From about the 1930s until the early 1960s, comparative psychology was dominated primarily by theories of learning and secondarily by work on development. Since the 1960s there has been a resurgence of truly broad-based comparative work. Dewsbury (1984) characterizes the subject matter of current comparative psychology as involving the study of either behavior patterns that are closely tied to work on learning, motivation, or physiological psychology, or on species other than those commonly used for such studies (e.g., laboratory rats, pigeons), or both. This definition is more restrictive than many for the boundaries of comparative psychology. Thus, there have been and continue to be changes in what investigators call comparative psychology, and there is not likely to be firm agreement on the subject matter, even among those who work in this field. We should also add that, as detailed later in this chapter, the modern fields of ethology and comparative psychology are not far different from each other, certainly not as different as they once were, or as we once thought they were.

For comparative psychologists, the primary emphasis has been on "how" questions about the mechanisms that underlie observed behavior patterns. One way to begin research in comparative psychology is by identifying and characterizing the classes of behavior patterns in two or more species. A second way to begin is by selecting a species that is the most appropriate for investigation of a particular problem. However, too much emphasis on a single species limits the generality of the conclusions we may draw (Beach 1950; Hodos and Campbell 1969; Lockard 1971; Dewsbury 1984). The frequent use of the domestic rat has produced criticism of comparative psychology from both inside and outside the discipline.

A Psychological Study

By examining a specific research investigation, we can obtain a clearer view of the thinking and methods, and the advantages and disadvantages, of the comparative psychology approach. Begun and colleagues (1988) wished to test whether plains garter snakes (*Thamnophis radix*) could respond differentially to airborne odorants. In the first portion of the experiment, snakes were placed in a glass tank

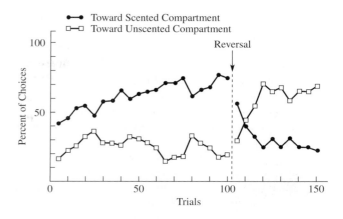

FIGURE 3.3   Choice behavior of garter snakes for airborne odor.
The location of the snake's head as being toward the scented or unscented side of the test apparatus is shown. For the first 100 trials the side with the lemon extract (scented) was correct, and for the remaining 50 trials the unscented side was considered correct.

with a false floor of wire mesh, several centimeters above a bedding of unscented pine chips. The tank was divided into three equal-sized compartments with removable partitions separating the sections. In the first part of the test, one end compartment had pine chips scented with lemon extract, while the middle compartment and remaining end compartment were left with unscented pine chips. The procedure involved placing a snake in the center compartment, waiting thirty seconds, and then removing the partitions. After two additional minutes elapsed, the position of the snake's head was recorded as being in one of the three sections of the tank.

Since the investigators wanted, initially, to determine whether a snake could detect the presence of the lemon extract, the section of pine chips scented with lemon was considered the correct side, and snakes with their heads in that section of the tank after two minutes were given a reward of a piece of earthworm. The results (figure 3.3, left side) revealed that over the course of 100 trials in the apparatus, the snakes shifted so that 75 percent of their responses were correct by the end of the tests.

To further test the learning ability of the snakes in this system, the researchers then reversed their definition of the correct side; now the snakes were rewarded for having their heads in the unscented section of the tank after a two-minute trial. The snakes clearly demonstrated what is termed reversal learning (figure 3.3, right side); after 50 additional trials, about 70 percent of the responses were now to the unscented (correct) side of the tank. The conclusion from this study is that snakes can discriminate airborne odorants. Snakes may use airborne cues under natural conditions for finding and trailing prey and for general orientation.

Advantages and Disadvantages

Many features of this study characterize investigations in comparative psychology and illustrate some of the advantages and disadvantages of this approach. First, the experience and environment of the subject animals can be controlled, and the investigator can draw exact or nearly exact conclusions about the effects of conditions on behavior. Second, one or two variables can be manipulated in a systematic and replicable experimental design to ascertain their effects without the potentially confounding influences of other uncontrolled variables. Finally, the investigation is concerned primarily with learning, in this instance, of a discrimination by the snakes for airborne odors.

Certain drawbacks are inherent in the comparative psychology approach. The animals, as those in the example, are often laboratory stock, housed in an environment that is simple compared with the natural habitat. How does domestication of laboratory animals affect behavior, and what are the effects of housing them in unnatural conditions? Comparisons of wild and laboratory rats suggest that some behavior patterns are affected but others are not (Boice 1972; Price 1972, 1984; Price and Huck 1976; Price and Belanger 1977). For example, wild rats are more active than their domestic counterparts, and laboratory strains exhibit different learning patterns than do those captured in the wild. Also, maternal behavior, including pup retrieval and nest construction, differs. Even the behavior of species is affected by rearing environment (Carducci & Jakob 2000). In general, it appears that during the process of domestication, selection may have produced a greater flexibility in behavior. Investigators must constantly be aware that laboratory confinement or housing in unnatural conditions may have changed their subject's behavior.

The comparative psychologists' approach may cause them to not view their experimental methods and results in a complete context involving ecology and evolution. We should be concerned with both physiological and evolutionary explanations for observed patterns of behavior in a species. For example, in most studies, knowing something about the natural environment of a species and its recent phylogenetic history is necessary. A species that has narrow diet preferences may not learn well when numerous types of foods are used as reinforcement. In the study just discussed, we might want to question the ecological relevance of lemon extract in the snake's natural world. As animal behaviorists, we must be constantly concerned about the *Umwelt*—the sensory-perceptual world of the species we are investigating.

Both the ethological and psychological perspectives have a great deal to contribute to the investigation and understanding of animal behavior. Because both "how" and "why" questions must be tested, an integrated approach that combines the methods of each discipline provides the best total analysis of behavior. During the last decade there has been a great deal of synergism between scientists approaching proximate (how) and ultimate (why) questions (Drickamer 1998; Dewsbury 1999). The integrated discipline that is emerging involves combining laboratory and field approaches to study simultaneously internal mechanisms of behavior, their ecological significance, and their evolutionary origin.

Behavioral Ecology and Sociobiology

In the past four decades, new frameworks and approaches for animal behavior study have emerged. Behavioral ecology and sociobiology can be characterized in similar ways; they both involve a great amount of field work with laboratory tests where appropriate; they both involve experimental manipulation under field conditions; and they are based on a reasoning process that includes both the ecological realm of the species under study, and concern with the way natural selection has shaped observed behavior. One may ask how behavioral ecology and sociobiology differ. Behavioral ecology deals with the habitat selection, feeding, and other aspects of a species' ecological niche, with particular reference to behavior. Sociobiology, considered by some to be a subdiscipline of behavioral ecology, is concerned primarily with the social system of a species and how and why their particular social organization has evolved. Both approaches make strong use of the comparative method, utilizing information on a variety of species to answer particular questions. An example of a recent study in behavioral ecology will provide some insights into these approaches.

A Behavioral Ecology–Sociobiology Study

How do behavioral ecologists and sociobiologists approach a research problem? Lindstrom (1998) tied together proximate and ultimate questions in a study of whether energy levels affect the mating performance of sand gobies (*Pomatoschistus minutus*). Sand gobies are small marine fish that exhibit male or paternal egg care. The study was conducted using artificial nest sites in a natural setting on the west coast of Sweden. The artificial nest sites consisted of flower pots cut in half. When male sand gobies adopted an artificial nest site, they were assigned either to a group that received additional food in the form of pieces of mussel (fed group) or they were in the unfed group, consuming only the natural foods in their environment. Each of the males in both groups were then followed through one nesting cycle. Daily, researchers observed the gobies by snorkeling and recording observations on waterproof paper. Each male's mating success was judged by the egg mass that had been laid by one or more females in his flowerpot artificial nest. When the nesting cycle was completed, Lindström and his colleages captured the fish and measured their total length and fresh weight. The fish were then killed and dried, and their dry weight and lipid content were measured.

There were no differences between males in the fed and unfed groups for total body length, nor was there any difference in the number of days that it took males from the two groups to establish nests at the artificial sites. Lindström found that males given supplemental feeding were able to spend more time (54% of the observations) at their nest sites compared to unfed males (38% of observations); the extra feeding meant that they did not have to spend as much time as the unfed males searching for food. This should provide males in the fed group with more time at the nest to mate with females and to tend the egg mass. This was indeed the case, as fed males mated sooner than unfed males, and there was a tendency for fed males to receive more eggs than unfed males. Body fat was greater in the fed group than in the unfed group. Thus, it appears that for sand gobies, energy availability does influence mating success.

Advantages and Disadvantages

One advantage of studies that use behavioral ecology approaches is that animals are generally studied in their natural habitat or in an environment designed to mimic certain natural features. A second advantage is that these studies generally provide specific data to test particular hypotheses by dissecting a larger problem into its component pieces for a better view of the behavior and its functional significance (as in the song repertoire study). The disadvantages of the approaches used by behavioral ecology and sociobiology are similar to those faced by ethologists: the lack of environmental control and the inability to regulate the animal's prior experiences. Sometimes it is also difficult to provide testable hypotheses for the functional and evolutionary aspects of behavior. One of the challenges of the past decade well met by behavioral ecology was to produce ways to ask and test such questions.

DESIGN FEATURES IN ANIMAL BEHAVIOR STUDIES

To conduct studies of animal behavior properly, it is necessary to have a basic understanding of the thought processes used to formulate hypotheses and design experiments to test these hypotheses. We now examine some principles and problems associated with conducting experiments in animal behavior. To provide a basis for examining these principles, we will consider the design of an investigation of the effect of testosterone on aggressive behavior in male Mongolian gerbils (*Meriones unguiculatus*) (figure 3.4). Our test animals will be normal intact male gerbils, castrated males, and castrated males given testosterone replacement treatments.

Definitions and Records

The first task is to agree on which of the many behavior patterns exhibited by gerbils we will consider to be "aggressive." We begin by observing several pairs of gerbils interacting. By watching and discussing the various behavior patterns exhibited, we may be able to decide which to record as aggression.

All observers must record similar behavior patterns in exactly the same way. (For an excellent discussion of the problems involved in describing behavior, read Drummond 1981.) To ensure that our observations and data collection remain consistent, we need a permanent record of the behavior. Each investigator can refer to the record when making observations, so that definitions of behavior will not change during the time period of the study, and so that other scientists can use the

FIGURE 3.4   Mongolian gerbil (*Meriones unguiculatus*). These animals, native to the Gobi Desert and nearby regions in Mongolia and China, have become widely used test subjects in laboratory investigations of behavior. Here two unfamiliar males meet, and one sniffs the other, which freezes.
Source: Lee C. Drickamer.

visual or written definitions to repeat the experiments or to make comparisons between gerbils and other rodents.

In addition to films, photographs, or drawings, we often provide written definitions of behavior patterns. Two books help us formulate written definitions for behavior (see McFarland 1981; Immelmann and Beer 1989). Both volumes contain definitions and descriptions for many behavior patterns and also various processes and concepts in animal behavior. Thus, for example, in the study of gerbils we will record three patterns as aggressive behavior:

1. *Attack:* One gerbil bites, kicks, claws, or pushes the other, possibly inflicting physical harm.
2. *Chase:* One gerbil vigorously pursues another in order to catch and attack it.
3. *Aggressive Groom:* One gerbil mouths the neck, flanks, or anogenital region of the other.

As rigorous as these definitions may be, we still need to check the agreement between observers about the behavior patterns. Therefore, it is necessary to conduct a series of **interobserver reliability** tests. Two observers simultaneously watch several pairs of male gerbils interacting, and each observer maintains a separate record of the frequencies of the three kinds of aggressive behavior. Then we compute the correlation between the two sets of data; a high positive relationship between the two data records indicates good agreement between what the two observers have recorded for the behavior patterns. Thus, during the course of data collection, several different individuals can observe and record, and we will not have to worry that each of them is recording something different for the same behavior pattern.

When the results of the experiment are reported, all aspects of the methods used in conducting the investigation must be specified. Thus, another scientist can independently **replicate** our experiment and verify, contradict, or modify our conclusions.

Design of the Experiment

Hypothesis Formulation

Once the agreement is reached on what to consider aggressive behavior, our next step is to design the experiment. First, we must formulate specific hypotheses (singular, hypothesis), or questions for testing. Two types of hypotheses are involved in behavioral research: experimental hypotheses and null (statistical) hypotheses.

Experimental hypotheses are ideas that are developed by combining the reported investigations of other scientists with our own ideas. In our present example we may want to know:

1. Is aggression in adult male gerbils dependent upon the hormone testosterone?
2. Do male gerbils begin to fight before or after puberty?

These hypotheses lead us to a series of predictions. For example, if testosterone is critical for adult male gerbil aggression, then eliminating testosterone by castration should lessen or eliminate aggression in castrated males. Further, based on hypothesis 2 we can predict that male gerbils will not fight before the age of puberty, when production of testosterone increases.

It is extremely important to understand that animal behavior research is probabilistic. If two animals engage in a contest for a particular feeding site on a particular day, animal *A* may be more successful than animal *B*. But, on another occasion *B* may be more successful than *A*. The reasons for this reversal are not readily apparent or easily explained. In another example, consider the problem of measuring how fast a springbok (*Antidorcas marsupialis*) can run a hundred meters when fleeing from a cheetah (*Acinonyx jubatus*). Some antelope will run faster than others. To generalize about the speed of all antelope of a particular species, we need to average the speed of many animals and determine how much variation in speed can be expected among members of the test group. Clearly, an exact determination of fighting ability or running speed cannot be made, so animal behaviorists must deal with probabilities and predictions about behavior. Although we will not deal directly with statistics in this text, we encourage a thorough knowledge concerning statistics and experimental design as necessary for research in animal behavior. For now, a few basic principles can provide us with a background for understanding and critiquing animal behavior research.

The **null hypothesis** proposes that the frequencies of animals' behavior patterns do not differ significantly when the animals are given different experimental treatments. The null hypothesis cannot be proved; it can only be accepted or rejected using a statistical test procedure. Our null hypothesis might be: There are no differences in the levels of aggression of normal intact male gerbils, castrated male gerbils, or castrated male gerbils given daily replacement injections of synthetic testosterone to replace the hormone normally secreted by the testes.

From the null hypothesis, we derive alternative outcomes for the experimental test and examine their interpretation. If indeed there were no significant statistical differences between all the treatment groups, our null hypothesis would be accepted. We would then return to experimental hypothesis 1 and its predictions. The major prediction would be incorrect: no basis exists from the test data to claim that testosterone plays a critical role in male gerbil aggression. However, if the intact males and those receiving daily injections of synthetic testosterone exhibited statistically significantly higher levels of aggression than castrated males, we could reject the null hypothesis. Returning once again to the experimental hypothesis, we would conclude that the test revealed a key role for testosterone affecting aggression in male gerbils. Regardless of whether we accept or reject the null hypothesis, there may be important biological conclusions to be drawn from our experimental finding. (For expansion of these ideas and a discussion of the distinctions between hypotheses and deductions, see Moore 1984.)

Designation of Variables

The next step in a research project is to establish the exact parameters to be investigated: the independent and dependent variables. **Independent variables** are the factors that the investigator has manipulated to define the treatment groups and conditions of the experiment. To test the first question regarding testosterone in male gerbils, at least three treatment groups, or independent variables, are necessary: (1) normal intact adult male gerbils; (2) castrated adult male gerbils; and (3) castrated adult male gerbils given daily injections of synthetic testosterone. The outline of the experiment might look like that shown in table 3.3. In addition to establishing the treatment groups, the investigator must systematically control many other aspects of the experiment, including the ages and previous experiences of the gerbils, the length of the observation period, and the size of the arena in which the males interact. The investigator should hold these factors constant for all experimental test groups, so that the *only* manipulated factor in the design is the presence or absence of the hormone testosterone. In the present study, we will pair each gerbil with another male from the same treatment group in an open circular arena one meter in diameter for a ten-minute observation period.

The measures of behavior patterns that are observed and recorded are the **dependent variables.** In our experiment, the frequencies of the three aggressive behavior patterns will be the dependent variables.

Control Groups

In addition to the design features discussed earlier, at least three other considerations arise in the course of setting up an experiment. Many experiments require control groups in order to provide a baseline for comparison with other experimental groups. A **control group** is an unmanipulated set of test subjects, used so the investigator can see what animals do *without* the experimental condition. In the gerbil experiment, the intact males can provide data on normal gerbil aggression. The frequency of aggressive behavior by the pairs of castrated males and by the pairs of castrated males

TABLE 3.3   **Outline of Experiment to Test Effects of Testosterone on Aggression in Adult Male Gerbils**
Groups 4 and 5 serve as controls for manipulations performed on gerbils in groups 2 and 3 to ascertain that the treatment, not the manipulation, affects aggressive behavior.

Treatment Groups	Behavior Patterns (Frequencies)		
	Attack	Chase	Aggressive Groom
1. Normal intact males			
2. Castrated males			
3. Castrated males given testosterone injections			
4. Sham-operated intact males			
5. Castrated males injected with oil			

given testosterone injections can be compared with that of the pairs of normal intact males of the control group.

In other instances, a control group may serve to demonstrate that some portion of the experimental treatment procedure has not produced an unwanted effect that confounds the results. The castration surgery performed on some test subjects involves anesthesia and an incision in the scrotum to remove the testes; the surgery itself may have altered aggressive tendencies. Castrated males' aggression levels may be different because of the decrease in blood testosterone levels, or because of the surgery. Thus, we must add a fourth treatment group, sham-operated males (table 3.3, group 4). All surgical processes except actual removal of the testes are carried out on these males. If the surgical processes have no effect on aggression, then we would expect sham-operated males to exhibit the same amounts of aggressive behavior as do normal, intact males. Unless we have a control group, we will not be able to distinguish whether any observed behavioral effects in the testosterone-replacement therapy group are the result of the hormone or of the injection process. Thus, we need to add a fifth treatment group of gerbils that are injected with peanut oil, which is the vehicle for the testosterone. For each manipulation in an experimental design, we need to determine what, if any, control groups are appropriate to the investigation.

Control groups are not necessary in all behavioral studies. For example, in research that simply describes the behavior patterns of various animal species, no control groups are needed. Nor would control groups be necessary in an assessment of a particular behavior pattern among groups of one species living in different habitats—for example, grooming behavior in groups of squirrel monkeys (*Saimiri* spp.). In this situation we would make comparisons between the monkeys in various habitats and draw conclusions about the relationship of grooming patterns to habitat features.

Independent Data Points

The need to gather numerical pieces of information that are independent of each other is probably the most difficult problem for behavioral biologists and is based on both biological and statistical arguments. Behavioral data independence in our gerbil experiment can be considered in several ways. For any pair of gerbils, the behavior of one male will definitely

influence the behavior of the other male. The principle of independent data points dictates that data from both males cannot be used as two distinct points in the same analysis. There are at least two solutions to this problem. We can combine the records of aggressive behavior for the two gerbils to produce one score for each pair. Or, data for "winners" and "losers" could be analyzed separately, but then we would need preestablished criteria for defining winners and losers.

In the gerbil investigation, we will use each animal only once, because the aggressive performance of the male the second time it is tested is partly a function of its experiences in the first interaction. After being severely attacked in the first encounter, some males become submissive during all subsequent interactions, and erroneously low estimates of aggression frequencies are a likely result. In contrast, other males become hyperaggressive after winning the first encounter; reusing these males would result in erroneously high estimates of aggression frequencies.

Another aspect of the problem of independent data points is related to animals living in groups. For example, in an investigation of adult female goose feeding rates, the total number of females and males living in the flock could influence the feeding rate of each individual female (figure 3.5). The independent units for data collection and analysis in this example are flocks; an average feeding rate for all females in a flock would provide data for making comparisons between flocks of different sizes or flocks resident in different geographical locations or habitats. Assuming that each of the females in a single flock represents a separate data point would be incorrect: the behavior of each is not necessarily independent. The improper assessment of what constitutes an independent data point or observation for analysis is perhaps the most frequent problem with studies of animal behavior. Repeated use of the same animals for multiple tests without proper statistical analyses or counting all members of an interacting group as independent of one another results in **pseudoreplication** and can lead to erroneous conclusions. If we wish to design and conduct an experiment properly, we must pay considerable attention to the difficult, but necessary, requirement of ensuring that the data points are independent.

The principle of independent data points is a derivative of one of the critical requirements of most statistical tests utilized by animal behaviorists—that is, each piece of information

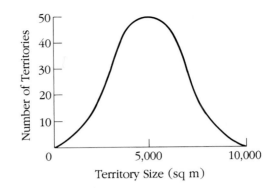

FIGURE 3.6   **Territory size in red-winged blackbirds.**
Many measurable behavioral traits show some form of continuous variation. Territory size in redwings provides such an example. This graph illustrates territory size for a large sample of over 500 birds.

FIGURE 3.5   **Canada geese feeding on ground beside pond.**
The behavior of each bird may be influenced by actions of the other birds in the flock. Information gathered on the behavior of an individual bird generally cannot be considered independent of similar data on another bird in the same flock.
Source: Courtesy Robert Gates.

used in the analysis must be independent of every other piece of information. Special statistical tests have been developed to handle data analysis when the same test procedures are repeated on the same animals over time. These tests do not require complete independence of data points for information gathered on the same animal.

Sample Size

A final factor in designing an experiment is the sample size—the number of animals that must be used in the investigation. Each experimental treatment group must contain enough animals to provide a complete and accurate assessment of the behavior. If the experiment is concerned with only qualitative descriptions of behavior, then the sample size should be large enough to permit an observer to record a thorough picture of the behavior, including estimates of average values for the dependent variables, and estimates of the amount of variation around the averages. In addition, most statistical tests require a certain minimum sample size for correct data analysis. Deciding the correct sample size to use for a particular experiment requires some practical experience with both the animal being studied and with the type of statistical test used in analyzing the results. Good animal behaviorists also take into consideration the fact that using too many animals may sometimes be wasting animal lives unnecessarily. Most animal behavior research organizations have ethical guidelines that dictate that minimal sample sizes are used when tests are harmful to the individual.

Variation and Variance

Animals exhibit variation in appearance or performance for most morphological, physiological, and behavioral traits. For example, we noted that red-winged blackbirds exhibit variation in display frequency and in nest-site selection.

Variation may be either discrete or continuous. **Discrete** variation involves those measures that can take on only certain values. For example, clutch size (number of eggs layed in a nest) in birds is a discrete measure; a species may lay 1, 2, 3, or up to 6 or more eggs per clutch, but they do not lay 1.5 or 3.6 eggs. Other measures are **continuous;** they can take on any value between some lower and upper limit. The nesting territories of red-winged blackbirds may vary from a few square meters to thousands of square meters (figure 3.6). Values assigned for different birds can be anywhere within this range and are limited only by how refined we wish to make our measurement.

To assess the amount of variation that occurs for a particular behavior or trait in a sample, we use two principal measures, the mean and the variance. We can calculate the mean and variance for any group of values for a trait regardless of whether the variable is discrete or continuous. The **mean,** $(\overline{x})$ or central tendency, is the arithmetic average. Thus, from a sample of four clutches of eggs from yellow-shafted flickers we might get values of 2, 4, 6, and 8—with a mean of 5 eggs per clutch. In a second sample of four clutches, the values of 4, 5, 5, and 6 eggs would also produce a mean of 5 eggs per clutch.

The **variance** is an estimate of the amount of deviation of the values in a sample around the mean of that sample; it is, by definition, the average squared deviation of the values from the mean of the sample. We calculate the variance S as,

$$S^2 = \frac{\Sigma(x - x_i)^2}{{}^nN - 1,}$$

where each x_i represents an individual value in the sample, x is the mean, and N is the total number of values in the sample. So, for our first sample of clutches the variance S is 6.7, whereas for the second sample, the variance S would be 0.7. The means for the two samples are identical, but the variances are quite different; there is much more variation around the mean in the first sample.

We will refer to the notions of mean and variance in both general and specific terms a number of times throughout the book. Two other measures of the variation around the mean, the standard deviation and the standard error of the mean, are also often encountered in the animal behavior literature. The **standard deviation** of a sample is simply the square root of the variance and is, therefore, a more useful number for describing variation around the mean. The **standard error of the mean** is an estimate of the standard deviation of a group of means from samples taken from a large population.

Capturing, Marking, and Tracking Animals and Animal Signs

Before discussing specific methods used for data collection in field settings, let us digress a moment to examine several related problems—issues that generally arise when we are conducting field studies, but that may also pertain to some laboratory work in animal behavior. These issues concern the methods of capturing animals, ways of marking animals for individual identification, procedures used for tracking animals, and signs left by animals that may assist us with our interpretation of their behavior. For additional techniques and details concerning the methods listed next, the student is urged to consult Lehner (1996) and Southwood (1978).

Capturing Animals

To study animals properly, it is often necessary to capture them, even if only briefly. We may need to mark the animals for identification. We may need to make measurements of size, sex, or age. Or, we may wish to bring them into the laboratory for certain portions of our work. There are a variety of methods that we can employ (Eltringham 1978); the exact capture method we use in a particular situation will depend upon the species and the nature of its habitat. The most common methods involve nets or traps. For many insect species we would use either sweep nets or dip nets. For birds we often use mist nets, nets of fine mesh set up using two poles, somewhat like a volleyball net. Mist nets are often set up to intersect the flight of the birds, for example, across a clearing in the woods or along the path of a stream. Larger nets can be used with small cannon to capture flocks of birds at a feeding station or mammals baited to come to a particular site. Live traps for vertebrates and invertebrates are available in an enormous variety of shapes and forms and can be used in the capture of almost all species of animals. Seine nets are often used to obtain samples of fish populations from a stream or lake. Pitfall traps, cans placed into the ground so their top rims are flush with the ground surface, are used for capturing a variety of invertebrates, reptiles, amphibians, and mammals. Insects can be removed from tree trunks via the use of brushes or small vacuums, and they can be removed from vegetation by shaking or beating the plant while holding a tray underneath. In aquatic habitats, electroshock can be used to stun large numbers of fish, which then float to the surface; they soon recover from their stun-

ning. Finally, for many mammals, particularly larger beasts such as bears or ungulates, we employ a CO_2 cartridge gun and a tranquilizer dart. Animals shot in this manner are anesthetized for varying periods, and they either wake up naturally, or they are given an antidote when the investigator has finished tagging or measuring them.

Animal Identification

In much of animal behavior research, we want to be able to identify individual animals (see Stonehouse 1978). Our research questions might include: Which animals are dominant? Which females are mating with which males? Are young animals cared for only by their mother, or do other members of a group share in this process? We also may often wish to avoid measuring the same animals twice by mistake. One way we can identify individual animals is by their natural marks (Pennycuik 1978). Examples of the use of this technique include coat color patterns in zebras (*Equus burchelli*) (Petersen 1972) and giraffes (*Giraffa camelopardalis*) (Foster 1966); differences in bill patterns in Bewick's swans (*Cygnus columbianus*) (Scott 1978); and variation in physical characteristics, natural mutilations, and scars in primates (Ingram 1978), lions (*Panthera leo*) (Pennycuik and Rudnai 1970), and bottlenose dolphins (Wursig and Wursig 1977).

A second technique for individual identification involves tagging animals; the nature and location of such tags will vary with the species (see Stonehouse 1978 for details). Among the various tagging techniques used are metal and plastic leg bands in birds (Spencer 1978; Patterson 1978); fin clips or punches and metal fingerling tags for fish (Laird 1978); toe clips, ear punches, dye marks, and tattoos for a variety of mammal species (Lane-Petter 1978); and small dots of paint or dye on invertebrates (Southwood 1994) (figure 3.7).

Another animal-marking technique involves radioisotopes, either in collars or bands, or in subcutaneous implants (Linn 1978). Because of potential problems with radioactivity and possible effects both on the animals bearing the radioisotope label and on the environment, the use of this technique has been curtailed considerably. More recently, Sheridan and

FIGURE 3.7 **Marking technique**
One way to mark animals such as the goldenrod leaf beetle (*Trihabda canadensis*) is to paint a number on the wings of each individual.
Photo courtesy of Patrice Morrow

Tamarin (1988) have developed the use of radionuclides for identification of individual meadow voles (*Microtus pennsylvanicus*) in order to study longevity and reproductive success; radionuclides are safer and not harmful to the animals carrying the label.

We can also examine the DNA of animals, using several techniques; the most widely used is DNA fingerprinting. DNA can be extracted from blood or tissue samples or, less invasively, from hair or feces.

Tracking Animals

Having marked animals for individual identification, we now might ask: How can the animal be followed over time to study its behavior? We can choose from among several methods. Probably the most popular of these techniques involves radiotelemetry. An animal is captured and fitted with a collar containing a radiotransmitter and battery or implanted with a similar device. Once released, the transmitter will send out signals that can be detected by a radio receiver and antenna. This technology has been adapted for use with animals ranging in size from small birds, lizards, and rodents, to whales and elephants. The size of the transmitter package and the range over which the receiver can pick up the signal vary with the size of the animal and the needs of the investigator. By following an animal for many days, it is possible to obtain an accurate picture of the area that it uses and the portions of the day when it is active. Radio signals can now be detected by satellites circling the earth; information from the satellite is then fed to a computer at a ground station and animal movements can be analyzed. This technique has been used effectively with caribou (*Rangifer tarandus*) in Alaska (Miller et al. 1975).

For some animals, particularly large mammals or birds in flight, we may use vehicles moving along the ground or airplanes from overhead to follow movement patterns. For many aquatic animals, including many vertebrates (especially fish and whales) and a variety of invertebrates, we can use boats and diving gear to observe both from the water's surface and from below. In all of these situations, photographs or videotapes may aid in gathering information and enhance the processes of identifying the individual and scoring the behavior patterns.

Another technique for following animals has been developed and is popular, especially for studying rodent movement. (Kaufman 1989). The procedure, sometimes called the "Shake-and-Bake" method, involves using a finely powdered fluorescent dye that comes in a variety of colors (e.g., magenta, lime green, blue, etc.). The animal to be tracked is captured and placed into a plastic bag with the powder. After gently shaking the bag for a few seconds, the pelage of the animal becomes covered with the dye. Once the animal is released, it will deposit a trail of dye for four to eight hours wherever it goes. On a night after the animal has been allowed to put down the dye trail, we go out with a special black light that causes the dye to fluoresce, and the trail of the animal can be followed quite easily. In this way, we obtain a record of the travels of the animal; the technique can be used with facility on both nocturnal and diurnal animals.

Animal Signs

Finally, are there any traces, signs, or constructions animals leave that can help us to interpret their behavior? Examples of animal signs that we might expect to find include tracks, fecal material, eggshells, and animal remains. Almost all animals leave tracks of some kind, except those in aquatic habitats. Impressions of feet can be found in mud, sand, or other types of surfaces. We may use tracks to determine the direction of movement of an individual or group, the composition and size of a group, and, depending upon whether there are age or sex differences in the nature of the tracks, something about the age and sex of the individuals. Examination of fecal material can tell us something of the diet of an animal. Eggshells may help us identify nest sites and indicate which species are nesting in a particular habitat. The remains of animals might hold clues to the cause of death or could even tell the tale of an act of predation. Other, possibly more active signs we might discover would be nests, egg cases, and spider webs. Nests of birds can provide considerable information about the site selection and construction of the nest, and when eggs or young are present, about the development of the nest.

Field Methods for Data Collection

The design features for studies of animal behavior apply equally in laboratory and field situations. One of the early pioneers in conducting animal behavior research, T.C. Schnierla (1950), used an experimental strategy that combined field investigation with appropriate laboratory tests of specific points. He was also a strong proponent of the comparative method and carried out behavioral observations under varied field conditions. Many of the pioneering ethologists, including both Tinbergen and Lorenz, conducted large portions of their research under field conditions.

In recent years, field studies of many animal species have focused considerable attention on research strategies and methods of collecting data under field conditions. Many of the techniques for behavioral observation that were first devised for use with primates have found broad application with many other animal species. We will mention only briefly topics like focal-animal sampling and instantaneous sampling; these and other related topics are discussed in detail by Hinde (1973), Altmann (1974), Dunbar (1976), and Lehner (1996).

Focal-animal sampling (Altmann 1974) involves recording all of the actions and interactions of one particular animal during a prescribed time period. Using this technique, an observer may watch a large number of animals, recording the behavior of each for a short period (e.g., five or ten minutes per animal), or the observer may record the behavior of fewer focal animals, each being watched over a longer time period (hours per animal). As an example of the use of this technique, consider the study conducted by

Bercovitch (1986) on dominance rank and mating activity in male olive baboons (*Papio anubis*). The question tested concerned the relationship between male dominance rank and access to sexually receptive females for mating. Intense observations were made on only one or two baboons each day. The females that were the subjects of the focal sampling were in behavioral estrus, exhibiting consort relationships with males. Using focal animal sampling, it was possible to record all of the male partners of the sexually receptive females. When the data were analyzed using information from only adult males that interacted with the receptive females, there was no clear relationship between dominance status and mating activity. However, if adolescent males were included in the analysis, then a relationship emerged, with dominant males engaging in more sexual activity with receptive females. One advantage of this observation technique is that intense samples can be gathered with a focus on particular animals.

A counterpoint to focal-animal sampling involves scan sampling; the technique is sometimes also called instantaneous sampling. An observer using this technique watches each animal for only a few seconds at periodic intervals and records the activity(ies) that the animal is performing only at the specific time marks indicated by the sampling scheme. The intervals between samples of the behavior of each individual can vary, but generally, they are a few minutes to a half hour. As an example of the use of this technique, consider a study of feeding behavior of two captive groups of lemurs (Ganzhorn 1986): the ring-tailed lemur (*Lemur catta*) and the ruffed lemur (*Varecia variegata*) housed in seminatural environments. Scan sampling was conducted on the groups of lemurs, using a thirty-minute interval between records of individual activity. The species and part of the plant being eaten were recorded when the lemur was eating—at the time the data were to be recorded. By recording the data in this fashion, the observer was able to make many records on all the lemurs in both groups during all seasons of the year. The analyses of these feeding data indicated that roughed lemurs spent less time eating during the day, had longer feeding bouts, and consumed more fruit than ringtails. One advantage of this technique is that it allows the observer to sample widely across all animals in the group and across a wide range of behavior patterns.

One-zero sampling involves recording the occurrence or nonoccurrence of each of a set of behavior patterns within a series of time periods (Renner and Rosenzweig 1986). This scheme is seen by some investigators as the best way to record a wide range of activities encompassing solitary actions, objected-directed behavior, and social interactions. This may also be a useful method for capturing the occurrence of behavior patterns that either occur with very low frequency or are of brief duration.

There are other sampling schemes, such as sequence sampling (Altmann 1974), that are beyond our present scope. Also not covered are other aspects of correct procedure design for collecting and analyzing data; these may be of interest to some students (see also Hinde 1973; Dunbar 1976).

We should note two final important points. First, despite the absence of rigid laboratory-type control, field studies should not be considered less accurate than laboratory investigations. Second, the same constraints and methods considerations apply whether we are working in the field or in the laboratory.

Observational Methods

In addition to the topics covered elsewhere in this chapter and at other appropriate points in the book, several selected issues should be discussed briefly in conjunction with the use of observational methods for studying animal behavior. Most of these issues have particular relevance to investigations conducted in the field, but many points also pertain to laboratory conditions. (For a thorough treatment of these and related topics, see Lehner 1996.)

Observing versus Watching

Most of us watch animals on a daily basis—pets, wildlife, or simply other humans. We may notice when something about the behavior of a dog is different from the usual or when a particular bird species flocks to the feeder in greater numbers. These visual and mental notes constitute "watching." Observing animal behavior involves systematic recording (writing, tape recording, or filming) of the activities of particular animals, usually with specific test questions governing the nature of the data recorded.

Making thorough, proper observations requires that the observer maintain an awareness of some critical considerations. Two of these—the problem of equal observability of all of the subjects under study and the establishment of interobserver reliability—are discussed elsewhere in this chapter. The latter is important both in terms of the agreement between the data recorded by two or more people on the same animals, and may act as a check against possible observer sex or age bias. One observer may be recording a disproportionate number of some actions for a particular age or sex.

Another key consideration for conducting observational research is scheduling the sessions to coincide with the time of day (and year) when the behavior we want to study is likely to occur with reasonable frequency, but not such that a distorted picture is obtained. For example, it may be erroneous to select only the peak periods for the occurrence of a particular class of behavior, such as aggression, as this would give a biased impression; but it would be equally unwise to study the same animals only during their resting period.

Proper observational research requires certain equipment. Generally, binoculars or spotting scopes are needed to identify animals and to see some aspects of their behavior accurately. For some species, it may be necessary to construct a blind or other unobtrusive vantage point in order to record the animals' activities without disrupting their normal patterns. We must always be aware that in both field and laboratory settings, the mere presence of an observer can influence the subjects' activities.

Finally, some form of reliable, accurate scheme of data collection must be devised. Construction of proper data sheets, ways of encoding behavior, and ways to keep track of the identities of subjects require practical experience. Lehner (1979) devotes a section of his book to this subject.

Identifying Subjects

At least two major problems should be mentioned in conjunction with marking animal subjects. First, some of the techniques just outlined may alter the behavior of the subject or may influence the appearance or actions of the subject, resulting in changes in the reactions of other animals to the subject (see Burley, Krantzberg, and Radman 1982). This could indirectly influence changes in the subject's behavior. Second, some investigators have adopted the practice by which animals are identified and recorded in the data by names such as "Swifty," "Old One," and the like. This practice should be avoided. When we name an animal based on its physical appearance or some behavioral trait, we attribute to it certain qualities that may result in subsequent observation bias. Even the practice of randomly assigning human names (such as "Ralph" and "Sue") to animals can result in the association of particular traits with names; this can also bias what we do and do not record.

Natural Variation versus Manipulation

We often begin a behavioral study by making general observations of the subjects. When we have decided on the hypotheses to be tested and the general approach, we arrive at several critical questions: Can the hypotheses be investigated without any manipulation—that is, can we rely on natural variation? Do we need to consider the possibilities for manipulating either the subject(s) or the environment? For some studies it will be necessary only to devise a usable scheme for recording the data. For instance, if we were interested in comparing the frequencies of a particular bird species' foraging behavior in deciduous versus conifer forests, no manipulations would be required, as the natural variation in the two types of habitats provides us with the basis for a testable hypothesis.

If, however, the hypothesis being tested involves the effects of some social variable—for example, flock composition on foraging behavior—it may be necessary to manipulate the number and sexes of birds in the foraging flocks by netting and removing some birds. Alternatively, we might be interested in whether provisioning the habitats with supplementary food would alter the foraging rates of the birds in either habitat. In this instance, we would be manipulating the environment.

Method Pitfalls

Several potential pitfalls may entrap the unwary investigator; we become aware of most of them through experience. Three pitfalls deserve special attention: (1) perceptual worlds, (2) correlation versus causation, and (3) differential observability of animals.

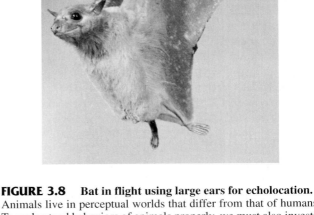

FIGURE 3.8   Bat in flight using large ears for echolocation. Animals live in perceptual worlds that differ from that of humans. To understand behaviors of animals properly, we must also investigate their varying perceptual worlds.
Source: ©Wildlife Conservation Society/Bronx Zoo.

Perceptual Worlds

One important task of animal behaviorists is to gain a thorough understanding of the perceptual world—the *Umwelt*, of each animal. Every animal lives in a different perceptual world. All too frequently, animal behaviorists assume that the animals they are studying live in the world of our human perceptions; this situation is only rarely true. For example, bats and many rodents emit sounds at very high frequencies, well beyond the range of human hearing (Griffin 1958; Noirot 1972). These high-frequency sounds are important, in bats for navigation and catching prey (figures 3.8 and 3.9), and in mice for communication between the young pups and the mother. Or, consider the visual world of insects—which is quite different from our own; bees and many other insects have eyes consisting of hundreds of individual cells, called omatidia, each of which sends signals to the insect's brain. Its visual world thus appears as a series of points, sort of like a photograph that has been taken with very large-grained film. To understand the behavior patterns of animals and to test various behavioral phenomena properly, we must know something about animals' sensory systems and how organisms interpret various stimuli in their environments. We will examine sensation, perception, and the *Umwelt* in more detail in chapter 7.

Correlation versus Causation

When explaining why a particular behavior occurs, we must be aware of a second problem area—confusing correlation with causation. For instance, if we are investigating songbird reproduction and find that the birds always mate at the same time of the year soon after the beginning of heavy spring

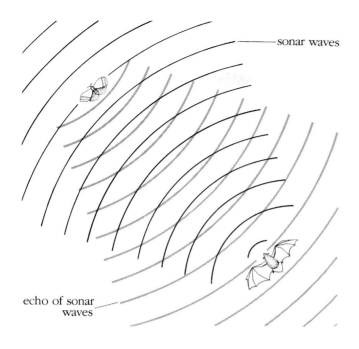

sonar waves

echo of sonar
waves

FIGURE 3.9   Bat's system of echolocation.
Bats navigate by echolocation whereby they emit high-energy
sounds that bounce off objects in the environment and return as
echoes to their sensitive hearing systems.

rains, we would probably find a good correlation between the
onset of rains and subsequent mating behavior. However, it
would be incorrect to infer that rainfall itself is a direct cause
of mating behavior. Increased rainfall may lead to changes in
both the quality and quantity of vegetation available to the
birds for food. Changes in diet may, in turn, produce hormonal changes, which may then lead to behavioral changes.
In addition, other factors may affect the mating behavior of
the birds, including external factors (such as number of hours
of daylight, availability of nesting materials, or social relations with other birds) and internal factors (such as biological clocks [see chapter 9] or previous mating experience).

Thus, explaining the timing of the onset of songbird
mating behavior involves considerably more than simple
correlation between environmental and behavioral changes.

A series of experimental tests might establish the links in the
chain between an external process, such as the increase in
rainfall, and the internal events that resulted in behavioral
changes. Correlation is an important tool for structuring our
thinking and for formulating hypotheses about the causes of
behavior, but we must not confuse correlation with causation. Any of several possible sequences can explain a particular behavior. Each step in any hypothetical sequence must
be analyzed experimentally to confirm the causal connections between events in the sequence.

Differential Observability of Animals

A third potential pitfall for investigators conducting studies of
animal behavior centers on how long and for what time of day
or season they observe different animals within the group. For
example, if investigators observe juveniles more frequently
than other group members, the conclusions of the investigation
could be significantly biased. A good illustration of this point is
the relationship between social-dominance status and mating
frequency of adult male rats at a garbage dump (table 3.4). To
test this relationship, we need the following information: (1)
the social rank of each adult male rat, (2) the observed frequency of mating by each male, and (3) the number and duration of times each male was observed.

We can determine social rank by watching the males at
the dump for several weeks. The top-ranking male, *B,* is the
rat that can defeat all six of the other adult male rats living at
the dump. The next highest male, *D,* can defeat the remaining five, and so forth down to the lowest-ranking rat, *M,* who
loses all his fights. We must also record how many times we
see each male copulating with an adult female (column 3).
On the basis of the frequency of copulation and rank of the
male rats, we may conclude that higher social status among
males is associated with a greater amount of mating activity.
However, this may not be true; what if the lower-ranking
male rats were not seen as often? The lower-ranking males
may be observed less frequently and recorded as mating less
frequently, but they could be engaging in frequent mating
behavior in areas not visible to the observer.

To complete the analysis, we need to know the number
of times and length of time each male rat is seen in a standard
time period—say, one month. If all the males are seen about

Male	Social Rank	Observed Matings per Month	Times Observed per Month (hr)	(Corrected) Number of Matings per Month per Times Observed (hr)
TABLE 3.4				**Social Dominance and Mating in Adult Male Rats at a Garbage Dump**
B	1	18	85	0.21
D	2	15	80	0.19
K	3	14	68	0.21
G	4	10	41	0.24
C	5	7	30	0.23
J	6	5	26	0.19
M	7	4	22	0.18

the same number of times and for the same total amount of time, then the correlation between higher social status and more mating activity is valid. If, however, as table 3.4 shows, males are seen more frequently or less frequently depending on social rank, then we must make a correction for observability. The data in column 5 of table 3.4 have been corrected for the **observability** of each male and are expressed as the number of matings per month per time observed. In this case we note that there are no differences in the mating rates of males of different ranks. While this finding does not mean that the original conclusion is invalid, the corrected results do indicate that further study will be necessary to determine what the lower-ranking, less-visible males are doing when the observer cannot see them.

APPLIED ANIMAL BEHAVIOR

One area of animal behavior research, applied animal behavior, had a relatively prominent place in the two decades after World War II. This is particularly evident if we examine the contents of *Animal Behaviour,* a primary journal in this field, published in Great Britain and now shared by the Animal Behavior Society in the Americas and the Association for the Study of Animal Behaviour in Great Britain and Europe. Many of the articles dealt with such domestic farm animals as sheep and poultry. During the past twenty years or so, this subfield of applied animal behavior has grown and expanded in scope. There are now several journals that provide coverage of applied animal behavior issues, most notably *Applied Animal Behaviour Science.*

What is applied animal behavior? The scope of applied animal behavior has grown in the recent past so that we now generally include investigations of behavior for domestic livestock, companion animals (pets), and the large variety of animals kept in zoological parks and aquaria, as well as studies of the care and well-being of animals housed in laboratory settings. Studies involving exotic animals in captivity have been quite important with regard to conservation efforts for a number of species, such as the California condor and the cheetah. There is a wide range of topics that are studied by those who conduct applied animal behavior research. These include the same types of investigations of behavior carried out by others doing animal behavior work (i.e., studies of physiological aspects of behavior, development, communication, aggression, etc.). In addition, there are some topics that are of more particular interest to applied animal behaviorists. Exploration of problems with companion animals has led to a relatively new profession involving pet psychology. Other individuals are interested in the effects of domestication on animals and the effects of captivity on their behavior. Medical science is heavily dependent upon animal models. In order to have healthy subjects for this sort of work, it is necessary that we study the effects of the conditions under which we house and breed these stocks of mice, rats, dogs, primates, and others.

The past decade has seen the emergence of a new subfield within applied animal behavior, the interface between conservation biology and behavior (Clemmons and Buchholz 1997). Many aspects of animal conservation require considerable knowledge concerning the behavior of the subject species. Traits such as foraging habits, social group composition, seasonal breeding patterns, territory size, and other behavioral phenomena are all important when investigators are attempting to understand why a species has diminished numbers living in the wild. These same traits can be critical to captive breeding programs wherein researchers are attempting to generate stocks of threatened or endangered species under captive conditions for reintroduction into natural habitats.

The methods and approaches used today in applied animal behavior are a mixture of those from the three approaches we outlined earlier in this chapter. Throughout the remainder of this textbook, we will endeavor to provide appropriate examples of behavioral phenomena that illustrate the increasing contributions of applied animal behavior to the larger discipline of which it is an active component.

USING ANIMALS IN RESEARCH

One issue that has become particularly important in the past decade for those who conduct research on animals involves the care and well-being of the subjects. This is particularly true for laboratory animals, though there are genuine concerns for the safety and health of animals in field pens and under natural, free-ranging conditions as well.

A major area of importance involving wild animals is the study of endangered or threatened species. The federal governments of the United States and in Canada, many of the individual states and provinces, and many other countries have passed laws governing the sale and importation of species that are endangered or threatened in the wild. Scientists who wish to conduct research on these animals must comply with the laws, and when they publish their findings, they generally are asked to certify that they have obeyed all applicable laws. Ironically, only intensive and extensive study of the behavioral and reproductive biology of these endangered species, whether they be mammals, birds, reptiles, or other taxa, can give us any hope of protecting the appropriate habitat, of providing proper breeding conditions, and of ensuring the continued reproduction of these important animals.

For those who work in the laboratory, at least two types of issues are important: matters concerning the actual care of the animals and matters concerning experimental research. Among the former are caging, lighting, food, water, and general health. These issues have been addressed by the "Guide for the Care and Use of Laboratory Animals" (1996) put forth by the National Institute of Health. Experimental research can range from simple observations of animals housed under specific conditions that, although not involving surgery, may cause the animal a certain amount of pain or distress, to manipulations that do involve surgery, such as removal of the testes, or brain operations. Guidelines put forward by groups like the Animal Behavior Society and the Association for the Study of Animal Behaviour (reprinted each year in the first issue of the journal), and others (American Society of Mammalogists 1987; American Ornithologists Union 1988), now serve to direct investigators in these animal welfare matters. Proper procedures must be followed; they often involve approval of the protocols by a committee of fellow scientists and a veterinarian. Investigators are urged to consider whether the procedures are necessary, what the significance is of the answers that will be obtained, and how the number of animals used can be minimized, consonant with the objectives of the experiment. It is the responsibility of everyone working with animals to be aware of these rules and guidelines, and to make every effort to comply on all issues. We think it is important to note that scientists have taken the lead on these matters, putting forth their own sets of guidelines, serving on committees that formulate national guidelines, and serving on thoughtful review boards for scientific proposals.

There are enormous benefits to be gained from research on animal behavior. A number of fields benefit from the investigations conducted on animals, including human medicine, conservation biology, wildlife biology, and anthropology, to name but a few. The benefits range from preservation of species to gaining a deeper understanding of the way in which all animals, including our own species, interact with their living and nonliving environment.

SUMMARY

An ethological study of animal behavior traditionally begins with an ethogram, an inventory of an animal's behavior, and seeks to answer "why" questions by determining the functions and evolution of behavior, as is exemplified by Tinbergen's studies of the significance of eggshell removal in gulls. The advantages of the ethological approach are that animals are studied in their natural habitat where functional relationships are more readily discerned and problems can be examined from an evolutionary perspective. However, the ethological approach sacrifices control over environmental parameters and over knowledge of an animal's history. In addition, devising a good test of the evolutionary significance of a behavior pattern is often difficult.

A study of animal behavior using the comparative psychology approach begins with the classification and comparison of the behavior of different species in order to discover relationships between them. This approach is exemplified by Begun et al.'s exploration of the discrimination of airborne odor cues in garter snakes. Experimenters have a high degree of control over both test subjects and environmental conditions; they can manipulate one or more variables while they hold all others constant. However, the breadth and strength of the conclusions of these studies are limited by the heavy reliance upon laboratory stocks for test subjects, by the influence of domestication on laboratory animals, and by an overemphasis on methods to the detriment of a more complete view of the context and functional importance of a behavior pattern.

Behavioral ecology and sociobiology attempt to dissect the ecological and social aspects of animal behavior from a functional and evolutionary perspective. These approaches have the advantage of studying the animal in a natural or natural-like habitat. Furthermore, when applied correctly, the resulting dissection of behavior patterns provides a bridge between the functional perspective and the more mechanistic perspective we have used to characterize comparative psychology. In addition to sacrificing some degree of control over the animals, studies using the frameworks of behavioral ecology or sociobiology sometimes suffer from oversimplification of past evolutionary processes that have resulted in the species' presently observed behavior patterns. There is a tendency to formulate "just-so" theories that often prove difficult to test experimentally.

Because animal behavior deals with probabilistic phenomena, correct experiment design is of great importance. The investigator must define the behavior patterns being investigated; must state the experimental hypotheses, or questions about observed biological phenomena, and the null hypotheses, or statements about expected differences between experimental treatment groups; and must designate the independent variables, or factors that are controlled and manipulated by the investigator, and the dependent variables, or behavior patterns that are measured as outcomes of the experiment.

In addition, the investigator must determine if control groups are necessary, either to provide a foundation for comparison or to demonstrate that the factors being manipulated, and not the manipulative processes themselves, are responsible for the observed results; must ensure the independence of data points; must use sufficient numbers (sample size) of animals in the experiment to provide proper analytical results; and must keep accurate and complete records of experimental procedures so that experiments can be replicated. Field investigations of animal behavior operate under the same design constraints and method requirements as do laboratory experiments.

Behavioral scientists must be aware of the need to understand the perceptual world of the animal being investigated and of the problem of differences in observability in animals of different ages, sex, or social status. Correlations between environmental factors and behavior should not be interpreted as causes of behavior.

Applied animal behavior involves the study of domestic livestock, companion animals, and animals that are housed in zoos and aquaria. The questions explored by individuals working in this emerging subfield of behavior include topics such as dealing with behavioral problems of pets, the effects of captivity on behavior, and the need for input concerning proper care and treatment of laboratory animals. The approaches and methods used by applied animal behaviorists are the same as those used by others who study animal behavior.

In recent decades, animal behaviorists, like all biological scientists, have become more cognizant of issues surrounding the health and welfare of the animals with which they work. Under both field and laboratory conditions, there are a variety of issues pertaining to the care and safety of research animals that are now governed by various guidelines and procedures. These procedures are designed to aid and facilitate research, not impede progress. There are enormous benefits to be gained from animal behavior research in human health and medicine, in conservation biology, and in our general understanding of biological phenomena.

DISCUSSION QUESTIONS

1. Animal behaviorists ask questions concerning causation, development, evolution, and function. Some species of gulls pick up shellfish, fly up a short distance in the air, and drop them on rocks to break them open. Describe the specific series of questions you might ask in each of these four areas about this behavior, and suggest some of the methods you would use to answer these questions. What specific hypotheses would you test? What experimental procedures would you use to test these hypotheses?

2. The effects of early feeding experiences on later food preferences were studied in guinea pigs. The animals were grouped five per cage, and each of the three groups was fed only one type of food—A, B, or C—for three weeks. Then the guinea pigs were tested individually to determine their food preferences in a "cafeteria" situation, with all three food types present. The results are shown in the table that follows. What can you say about the relative effects of early feeding experiences in guinea pigs? What principles of experimental design and methodology have been violated in conducting this experiment?

3. Suppose you were asked to conduct a study assessing the diet of a web-building spider living in a forested area. You have sufficient funds and spiders, and a natural setting in which to

Food Eaten in Cafeteria Test (g/24 hr)

		Type A	Type B	Type C
	Type A	9.6	2.8	5.7
Rearing	Type B	3.8	6.2	7.8
food	Type C	1.4	0.6	16.2

conduct both observation and manipulative experiments. What types of tests would you conduct to answer the question about diet? What experimental problems would you expect to encounter and how would you solve them?

4. To conduct sound experiments in animal behavior, an understanding of proper methods and experimental design is necessary. What other reasons can you cite for acquiring a firm knowledge of these principles for work in animal behavior?

5. Many research animals are housed in zoos throughout the world. What special concerns about caged animal behavior pertaining to health and welfare would you have if you were an administrator or veterinarian at a zoo?

SUGGESTED READINGS

Altmann, J. 1974. Observational study of behavior: Sampling methods. *Behaviour* 49:227–67.

Beach, F. A. 1960. Experimental investigations of species-specific behavior. *Amer. Psychol.* 15:1–18.

Dewsbury, D. A. 1973. Comparative psychologists and their quest for uniformity. *Ann. N.Y. Acad. Sci.* 223:147–67.

Drummond, H. 1981. The nature and description of behavior patterns. In *Perspectives in Ethology,* eds. P. P. Bateson and P. H. Klopfer. 4:1–34. New York: Plenum Press.

Hirsch, J., ed. 1987. *Journal of Comparative Psychology* 101 (3). This issue, marking the 100th anniversary of the journal, is dedicated to a review of comparative psychology, past, present, and future.

Immelmann, K., and C. Beer. 1989. *A Dictionary of Ethology.* Cambridge, MA: Harvard University Press.

Krebs, J. R., and N. B. Davies. 1987. *An Introduction to Behavioural Ecology,* 2d ed. Oxford, England: Blackwell Scientific Publications, Ltd.

Lehner, P. N. 1996. *Handbook of Ethological Methods.* New York: Cambridge University Press.

Martin, P., and P. Bateson. 1993. *Measuring Behaviour: An Introductory Guide.* New York: Cambridge University Press.

Tinbergen, N. 1963. On aims and methods of ethology. *Zeit. Tierpsychol.* 20:410–33.

Trivers, R. 1985. *Social Evolution.* Reading, MA: Benjamin Cummings.

PART TWO

Behavioral Genetics and Evolution

We began the book with a backward look at the discipline of animal behavior, and discussed many of the research approaches that behaviorists have developed over the decades. In part 2, we begin by reviewing the basic principles of genetics and evolutionary theory (chapter 4). Next, we'll take a close look at the study of behavioral genetics, with a review of both venerable and recent techniques (chapter 5). Finally, we'll look at the evidence for evolutionary change in behavior at both the microevolutionary and macroevolutionary scales (chapter 6).

A chimpanzee, which shares 98 percent of our DNA sequence.
© Macolm S. Kirk/Peter Arnold

4

CHAPTER

Genes and Evolution

arsh wrens (*Cistothorus palustris*) from the San Francisco area sing more than 2.5 times as many distinctly different songs as do marsh wrens from New York, and these songs are more complex in structure (figure 4.1). As an animal behaviorist, what sort of questions might you ask about this observation? As discussed in chapter 1, you may ask proximate questions, such as whether there is something different about the physiology or anatomy of the two populations that is likely to influence song development. In fact, western marsh wrens devote more nervous tissue in their brains to song than do eastern marsh wrens (Canady et al. 1984; see chapter 7 for further discussion of the nervous system and behavior).

You might also ask ultimate questions about the evolution of these differences. However, traits cannot evolve unless they are at least partially under genetic control. It is possible, for example, that the differences between these two populations can be entirely explained by differences in the environments of the birds. Perhaps, if you were to raise both wrens in the very same conditions, they might devote the same volume of their brain to singing, and sing the same number of song types. In fact, this is not the case: some differences are still there even when birds are raised in identical environments (Kroodsma and Canady 1985), indicating that they are at least partially genetically based. In more general terms, we say that the **phenotype,** or observable traits of an organism, may be influenced by both its **genotype,** or genetic makeup, and the environment. Much of behavioral genetics is concerned with exactly how, and to what extent, phenotypic differences across individuals are determined by genotypic differences.

In the first chapter of part 2, we begin with a review of fundamentals about genes and heredity. We cover some basic population genetics, with an emphasis on natural selection. Much of this material may be familiar to you from previous coursework, but it is a necessary base for understanding later

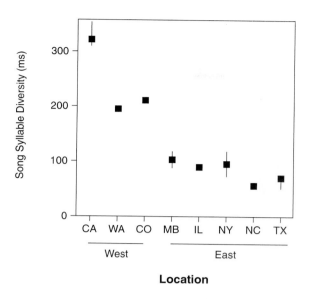

FIGURE 4.1   Diversity of song in western and eastern marsh wrens.
One measure of song is the syllable period, or the time between repetitions of a syllable. Song diversity was measured by taking the difference between the maximum and minimum syllable periods (in milliseconds) produced by a given bird. In western North America, in California (CA), Washington (WA), and Colorado (CO), songs had high syllable diversity. In more eastern populations, less-diverse songs were sung by birds from Manitoba (MB), Illinois (IL), New York (NY), North Carolina (NC), and Texas (TX). The squares indicate medians, and bars indicate range.
Source: Redrawn from Kroodsma 1989.

chapters. We end the chapter with a caveat about why we should be cautious about using evolutionary explanations uncritically.

BASIC PRINCIPLES OF GENETICS

Chromosomes, Genes, and Alleles

In 1859, when Darwin laid out the theory of evolution by natural selection in *The Origin of Species,* he knew that offspring tend to resemble their parents, but knew nothing about the mechanisms of inheritance. It was not until 1944 that DNA was discovered to be the molecule that stores genetic information, and not until 1953 that the helical structure of DNA was described. DNA is a molecule composed of a string of nucleotides (G, C, A, and T) that comprise the genetic alphabet. It is the sequence of these nucleotides that determines the function of any particular section of DNA. **Genes** are sections of a DNA molecule that have a particular function.

In eukaryotes, DNA molecules and associated proteins are organized into **chromosomes.** Each chromosome thus has many genes. The location of a gene on a chromosome is called its **locus** (plural loci) and locus is often used as a synonym of gene. An animal has two copies of each gene because chromosomes exist in homologous pairs. **Homologous** chromosomes have genes for the same traits in the same

order along their length. One chromosome of each pair is inherited from each parent (figure 4.2). An exception are the sex chromosomes: members of one sex (e.g., male mammals, female birds, female butterflies) usually have one pair of chromosomes that are not homologous, such as the X and Y chromosomes of humans (figure 4.2).

Genes can have different forms or **alleles.** The two copies of a gene received from the mother and father may be the same allele (so we say the organism is **homozygous** for that gene) or may be different alleles (so we say the organism is **heterozygous** for that gene). Different alleles of the same gene affect the same general trait, but different combinations of alleles at a locus may produce slightly different phenotypes. In Mendel's famous experiments with pea plants, for example, both alleles of a single gene affect the surface of peas. Plants that are homozygous for one allele have wrinkly peas, and those homozygous for another allele have smooth, round peas. In a behavioral example, researchers have identified a gene influences foraging behavior in *Drosophila melanogaster.* Larval flies with the *rover* allele move significantly more while eating than those homozygous for the *sitter* allele (Pereira and Sokolowski 1993). Alleles at the same locus may interact with one another. For example, one allele may be **dominant** or **recessive** to another allele for a particular trait. In a heterozygous individual with both types of alleles, only the dominant phenotype is expressed, or seen in the developed individual.

Genes can be divided into two categories. **Structural genes** are transcribed into messenger RNA molecules, which in turn are translated into proteins. Proteins have a myriad of effects on the phenotype. For example, pheromones are odor signals between individuals of a species. Female silkworm moths (*Bombyx mori*) release pheromone molecules that are detected by male moths because of proteins in the membranes of their antennae. Similarly, the proteins found in the membranes of nerve cells regulate how the cells respond to neurotransmitter substances (chapter 7), many of which are also proteins. In fact, all functions of nerve cells are regulated or strongly influenced by the structures of the numerous proteins that are found in their organelles and membranes. One particularly important category of proteins is enzymes. Enzymes play a critical role in the synthesis of biologically active molecules (e.g., the moth pheromone) and are necessary for the function of tissues and organs. The second category of genes are **regulatory genes.** These are transcribed into RNA molecules, which then function to control other genes by turning them on and off. They do so by binding to regulatory sites, DNA regions on the chromosomes near structural genes.

How does the genotype of an animal influence its behavior? In a broad sense, many regulatory and structural genes underlie behavioral traits. For example, behavior is intimately tied to morphology: a crab cannot crack open a mussel unless it has a large strong claw, and the development of the claw depends on a suite of both structural and regulatory genes. However, we are primarily interested in genes that are more directly linked to behavior. Even here the situation is complicated: genes do not directly affect behavior, but affect

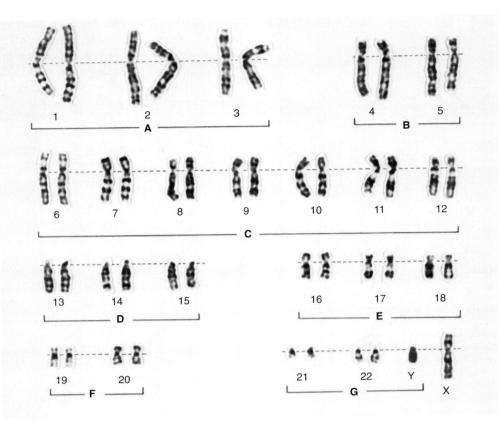

FIGURE 4.2   Human karyotype.
There are 23 pairs of homologous chromosomes in each body cell of the human ($2n = 46$).
Source: Courtesy J. S. Yoon.

the sensory and nervous system, which in turn causes behavior (chapter 7). Genes do not even provide a detailed blueprint for the nervous system. Thus, the exact relationship between a particular gene and a particular behavioral trait is often difficult to discern. We will look in detail at some of the best worked-out examples of the relationships between genes and behavior in chapter 5.

Mutation as a Source of Genetic Variation

As we will discuss in more detail later in this chapter, in order for natural selection to cause evolutionary change, there must be genetic differences among individuals. How does such genetic variation arise? The fundamental source of genetic variation is mutation, which results in a change in the structure of a DNA molecule. Changes in the DNA of gametes will be inherited by offspring. Mutations may be relatively minor changes in DNA structure, or larger changes. Recall that DNA is a long chain composed of sequences of four nucleotide molecules. Types of mutations include the addition or deletion of a single nucleotide or string of nucleotides, or the substitution of one nucleotide for another. Even small changes may result in a change in a gene's function (figure 4.3). Larger changes in genetic structure can also

occur, including chromosomal inversions (where two breaks occur in a chromosome, and the segment between them is rotated 180 degrees), and translocations (exchange of segments of two nonhomologous chromosomes).

Mutations can be beneficial, neutral, or deleterious, but most are deleterious. Mutations may even be lethal for the bearer. Mutation rates are also generally low, on the order of about one mutation per 100,000 copies of a gene (Strickberger 1996). In the face of odds such as this, how can beneficial mutations ever arise and spread in a population?

First, although the mutation rate is low, the sheer number of genes make it likely that at least some of them will be mutants. For example, human gametes carry an estimated 100,000 genes, so each new zygote is likely to have two new mutations (Strickberger 1996). Second, there is a reservoir of genetic variability in most populations. Neutral genes, and even genes that are mildly deleterious only in the homozygous form, can accumulate in a population over time. When the environment changes, these genes may be advantageous under the new conditions. Third, when a beneficial mutation does arise, and results in an offspring that is better adapted to the environment so that it survives and reproduces at a higher rate than other individuals, the frequency of the mutant allele will increase in the population. This is an example of evolution through natural selection, and will be discussed in

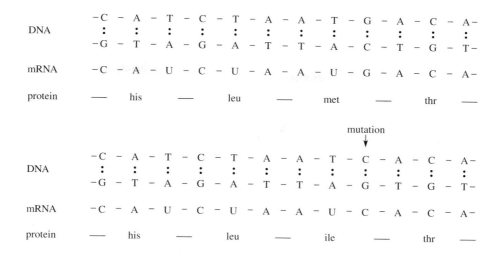

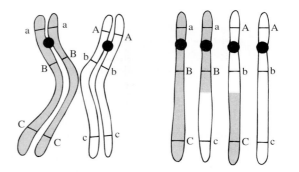

FIGURE 4.3   **Mutation in a structural gene.**
A very small portion of a DNA molecule is represented. The sequence of nucleotides in one strand (the bottom strand in both DNA molecules shown) codes for the formation of a specific sequence of nucleotides in a messenger RNA (mRNA) molecule. The sequence of nucleotides in the mRNA, taken three at a time, determines the amino acid structure of the protein. A substitution of one nucleotide (G) by another (C) is indicated by the arrow. This particular mutation causes the substitution of the amino acid isoleucine (ile) for methionine (met). The change in amino acid changes the structure and therefore the function of the protein that is synthesized.

greater depth in the next section. It is important to understand, however, that mutations occur at random: the chance that a specific mutation will occur is *not* determined by how useful that mutation might be (see Futuyma 1998 for further discussion).

Sexual Reproduction and Genetic Variation

Genetic variation in a population also arises during the process of sexual reproduction. The gametes (ovum and sperm) of each animal are termed haploid and contain *n* chromosomes, while the **somatic** (nongamete) cells are diploid because they contain 2*n* chromosomes, in homologous pairs. The value of *n*, the haploid number, differs for different species. In *Drosophila melanogaster, n* = 4. For humans, *n* = 23, so that most of your cells have 2 × 23 = 46 chromosomes (figure 4.2), while your gametes have 23 chromosomes, one from each homologous pair. At fertilization, the paternal chromosomes from a sperm cell join with the maternal chromosomes of an ovum to reestablish the diploid set of chromosomes that will be found in each of the new organism's somatic cells.

During the process of meiosis, the diploid gonadal cells that give rise to gametes divide so that only one chromosome copy from each homologous pair ends up in each ovum or sperm cell. Variability is introduced at each generation because the maternal and paternal homologues separate randomly and independently during the formation of gametes, producing new combinations of chromosomes that differ from those of either parent. Further variability is generated through the process of **recombination:** during

FIGURE 4.4   **Homologous chromosomes.**
Before homologous chromosomes (shown in gray and white) divide during meiosis, they are replicated. The replicates (chromatids) are held together by the centromere (black circle). When homologues pair up prior to division, they may "cross over" and exchange sections.

meiosis, homologous chromosomes come into contact with each other so that pieces of the two chromosome copies are likely to be exchanged. Recombination doesn't change which genes are on either homologue, but it may change the combinations of alleles at the different loci of the two "new" homologues (figure 4.4). Thus, sexual reproduction means that offspring are likely to have different sets of alleles than their parents, and therefore different phenotypes. In addition, interactions between alleles, whether of the same genes or between different genes, also can affect the phenotype (see chapter 5). This makes it even more likely that offspring will differ in appearance from their parents.

AN INTRODUCTION TO EVOLUTION

Darwin and the Theory of Natural Selection

The term **evolution** is often used in a general sense to connote change over time. However, we will use the term as it is defined by evolutionary biologists: evolution is a change in the frequencies of different alleles in a population or species over the course of generations. Evolutionary change comprises minor changes in a population (**microevolution**) as well as larger-scale changes that can eventually result in the formation of new species (**speciation**). There are four forces that cause change in allele frequencies over time: **mutation** (although it is so rare that its effects on gene frequencies are almost always extremely small), which we have already discussed; natural selection; gene flow; and genetic drift.

Charles Darwin developed the idea of evolution through the process of natural selection. The following account, summarized in figure 4.5, attempts to reconstruct the stages by which Darwin arrived at the theory of natural selection (adapted from Mayr 1977). Darwin made many of the following observations during the five years he spent as a naturalist aboard the *H.M.S. Beagle,* travelling throughout the world and especially along the coast of South America. Some of Darwin's "observations" also came from his reading of Thomas Malthus's 1798 essay on population growth as well as the writings of other individuals.

- *Observation 1: All species are capable of overproducing.* For example, one pair of houseflies, if unchecked, could produce more than six trillion progeny in just one year. Even elephants, which are not particularly fast breeders, could potentially produce nineteen million descendants in 750 years from just one pair.

- *Observation 2: Populations of species tend to remain stable over time.* We are neither overrun with houseflies nor elephants. In other words, the death rate tends to equal the birth rate, and most fluctuations are either temporary or repeated cyclically.

- *Observation 3: Resources are limited.* Populations do not go on increasing indefinitely because resources are limited, and eventually a growing population depletes some resource. We

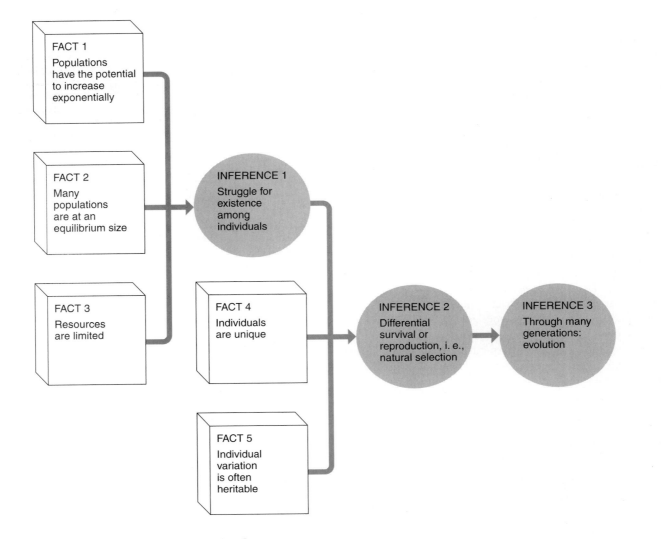

FIGURE 4.5   Components of natural selection theory.

usually think of food as most important, but other requirements, such as shelter or nesting space, may be limited for a population.

Inference 1, *following from Observations 1–3: There is a struggle for existence.* Exponential population growth, combined with a fixed supply of resources, results in competition among individuals. Although this idea was not new with Darwin, he emphasized that much of the competition is between members of the same species. Competition can be quite subtle, and does not necessarily entail combat or other obvious behaviors; in fact, the most aggressive animals may not produce the most offspring.

- *Observation 4: Individuals are unique.* From his studies of animal breeding, Darwin confirmed what others had observed: every animal in a group is different from every other, and the extent of variation in a population can be very large.

- *Observation 5: Individual variation is heritable.* Although the mechanisms of heredity were not understood in Darwin's time, it was well known that offspring inherited, to some degree, the traits of their parents. For example, animal breeders understood that they should carefully select which of their cattle to breed together so that the offspring would have the most desirable traits. In *The Origin of Species,* Darwin spends considerable time discussing the breeding of pigeons, because pigeon fanciers had accumulated a great deal of information about inherited differences among their birds.

Inference 2: *There is differential survival and reproduction, or natural selection.* The birth of more individuals than resources can support and the presence of individual variation set the stage for natural selection. Some individuals have traits that make them better competitors for food, mates, or other resources, and these individuals are more likely to survive and reproduce. Therefore, such individuals produce a disproportionate number of offspring that inherit those traits.

Inference 3: *Through many generations, evolution occurs.* Continued natural selection of individuals with heritable qualities over many generations leads to a change in the frequencies of those traits in the population or species.

In the 1930s and 1940s, Darwin's ideas were integrated with the new understanding of genetics in order to create the Modern Synthesis of evolutionary theory (e.g., Fisher 1930, Dobzhansky 1937, Simpson 1944). Since then, other fields, including the study of animal behavior, have been incorporated in the framework of evolutionary theory.

The term **adaptation** is important to understand in the context of evolutionary biology. Whereas physiologists often use the term to mean the process that an individual organism undergoes to adjust to environmental stress, evolutionary biologists use the term in two other ways. First, the verb "adapt" describes the process of genetic change that occurs during evolution by natural selection. Second, the noun "adaptation" describes a trait that has been shaped by natural selection (for example, see figure 4.6). Not all traits are adaptations, as we will see below.

Other Causes of Evolution

So far we have discussed two causes of evolutionary change in populations, mutation and natural selection. A third way that gene frequencies change is by **gene flow.** Gene flow occurs when animals move from one population to another. If the populations differ in their allele frequen-

FIGURE 4.6   Bill adaptations in six species of birds.
The bills of (a) the falcon, (b) the kingfisher, (c) the anhinga, (d) the pelican, (e) the chickadee, and (f) the flycatcher illustrate some of the variation in bill types adapted for each species' specialized techniques for obtaining and consuming food. Falcons use hooked bills to tear flesh; kingfishers use long slender bills to grab and hold fish; anhingas spear fish with their long slender bills; pelicans use their large bill and the pouch below it to dip into the water to net fish; chickadees have small short bills for gleaning insects from vegetation and for eating seeds; and flycatchers gather insects as they fly out from and back to their perches with their mouths wide open.

cies, those in the recipient population will change in the next generation if the migrants reproduce (figure 4.7). High levels of gene flow make populations very similar to one another, morphologically and behaviorally.

A fourth cause of evolutionary change is **genetic drift,** the variation in allele frequencies through random fluctuations. In a small population, chance fluctuations in breeding success (e.g., perhaps an animal that has a genotype with many beneficial alleles dies accidentally before reproduction) can have a large impact on the frequencies of alleles in the overall population. For example, imagine a population of ten mice, half of which carry an allele that gives them a black coat color. If, by chance, three animals that carry alleles for black coats should happen to die before they reproduce, the frequency of that allele would change dramatically in a single generation. In a larger population, say of 500 individuals, such random fluctuations in the reproduction of a few individuals would not have much of an impact on the frequencies of alleles. Genetic drift may also be important when a new population is established. If a population is founded by only a small number of individuals, only a fraction of the alleles present in the source population may be present in the new population (the **founder effect**). Genetic drift is of particular interest for conservation biologists, because small populations of endangered species are likely to be strongly

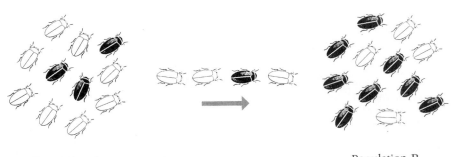

Population A   Population B

FIGURE 4.7   Evolution by gene flow.
Population *B*, with a low frequency of the white allele, gains members from population *A*,
which has a high frequency of the white allele. The frequencies of the white allele would be
expected to increase in future generations of population *B*.

a.   b.

**FIGURE 4.8   Cheetahs (A) have reduced genetic variation, which is correlated with deformed
sperm (B).**
© Johnny Johnson/Animals, Animals
Courtesy of Dr. JoGayle Howard/National Zoological Park

influenced by chance factors. The proportion of heterozygous individuals (the heterozygosity) in these populations tends to be reduced by genetic drift, as alleles may be rapidly lost (frequency = 0) or fixed (frequency = 1.0), so that all individuals become homozygous for the same allele. Cheetahs and Florida panthers both have extremely low heterozygosities (O'Brien 1994; figure 4.8). Genetic drift can eliminate even beneficial alleles. Reduced genetic variation is often correlated with a host of other problems, such as decreased sperm counts, abnormal sperm, and increased susceptibility to disease (O'Brien 1994).

It is important to understand the difference between the two processes of natural selection and genetic drift. With natural selection, individuals with certain heritable traits have more offspring than do those without such traits, and the difference in reproduction is a result of the traits they possess. With genetic drift, individuals possessing certain traits have more offspring than do those with other traits, but the difference in reproduction is due solely to luck and has nothing to do with the traits. In both cases, the genes of individuals that reproduce

more are represented at higher frequency in the next generation. However, a trait that increased in frequency in a population by chance (i.e., by genetic drift) is not called an adaptation. That term refers only to traits that evolve as a result of natural selection.

The Hardy-Weinberg Equilibrium

Using molecular techniques, it is possible to assay the genotypes of individuals in a population. These techniques are widely used to examine a range of questions, such as determining how closely related individuals are, and whether populations are genetically different from one another. For example, one might examine whether a population is undergoing evolutionary change through the use of the Hardy-Weinberg equilibrium. This allows us to determine what the expected genotype frequencies in a population would be if no evolutionary forces were acting. If we observe different frequencies in our study population, we can conclude that some evolutionary force is at work.

Suppose that there are just two alleles, A and a, for a given locus. Each gamete carries either A or a. We'll define p as the relative frequency (i.e., the proportion) of the A allele in a population, and q as the relative frequency of the a allele. These frequencies sum to 1:

$$p + q = 1. \qquad \text{(eq. 4.1)}$$

In other words, all a's and all A's together constitute 100 percent of the alleles at that locus.

Hardy and Weinberg showed how to use these allele frequencies to calculate genotype frequencies in the next generation. With these two alleles, there are three possible genotypes, AA, Aa, and aa. Since p is the frequency of the A allele in the parents, it is also the probability that a gamete produced by the parents will contain A. Likewise, q is the probability that a gamete will contain a. The probability of two independent events occurring together is found by multiplying their individual probabilities. If we assume that organisms mate randomly (that is, any organism is equally likely to mate with any of the others), we can calculate the probability of two A gametes coming together at mating, one from a sperm and one from an egg, as

$$p \times p = p^2 = \text{the frequency of genotype } AA \text{ in the offspring.}$$
$$\text{(eq. 4.2)}$$

Similarly,

$$q \times q = q^2 = \text{the frequency of genotype } aa \text{ in the offspring.}$$
$$\text{(eq. 4.3)}$$

To calculate the frequency of genotype Aa, we need to take into account that this genotype can be formed in two ways: the sperm might have A and the egg a, or vice versa. Thus, we need to count up both possible ways of forming Aa:

$$pq + pq = 2pq = \text{the frequency of } Aa \text{ in the offspring.} \qquad \text{(eq. 4.4)}$$

Since AA, Aa, and aa are the only possible genotypes, their frequencies must sum to 1:

$$p^2 + 2pq + q^2 = 1. \qquad \text{(eq. 4.5)}$$

This is the Hardy-Weinberg formula.

This is also called the Hardy-Weinberg equilibrium because, with no evolutionary forces acting, allele frequencies do not change from one generation to the next. To see this, let's calculate the frequency of A gametes produced by the three genotypes of the offspring generation. The AA genotypes will produce only A gametes. Because the frequency of AA genotypes in offspring is p^2, the frequency of A gametes produced by AA offspring is also p^2. The Aa genotypes will produce half A and half a gametes, so the frequency of A gametes that they produce is $2pq/2$, or pq. We can add together these numbers to discover the total frequency of A gametes that the offspring will produce: $p^2 + pq$. This factors to become $p(p+q)$, and because $p + q = 1$, we arrive at $p^2 + pq = p(p+q) = p$. Thus, the frequency of A gametes in the offspring generation will be p, the same as in the parents' generation. Similarly, the frequency of a gametes will again be q, as you might wish to confirm your-

self. This pattern holds for all subsequent generations: allele frequencies do not change from generation to generation under the conditions we have described. This is the Hardy-Weinberg equilibrium.

In a moment we'll do an example with data, but first let's more explicitly lay out the conditions. This equilibrium is relevant only for diploid organisms with sexual reproduction. The gene under consideration has two alleles only, and allele frequencies are identical in males and females. Mating is random. Population size is very large so that genetic drift has an imperceptibly small effect. Gene flow and mutation are so small as to be negligible. Finally, natural selection is not acting on the alleles under consideration. How can such a restrictive, unrealistic formula possibly be useful?

The Hardy-Weinberg principle is an example of a **null model**. Just as a null hypothesis is the hypothesis of no difference, this null model tells us what to expect when no evolutionary forces are at work. It is a very simple model, but it forms the basis for more realistic permutations.

The Hardy-Weinberg formula enables us to sample a population at one point in time and to determine whether it is at equilibrium for a set of alleles, without having to take repeated samples across several generations. Let us apply the formula to an example from chimpanzees (Morin et al. 1994). New techniques allow investigators to sample DNA noninvasively by using hair samples rather than blood or tissue. Chimps make fresh nests of leaves and branches every night, so the researchers watched known individuals build nests, and then returned in the morning to collect hair. Eight different loci were sampled. For this example, we will look at only one locus, *Rena4*. This locus has only two alleles, so calculations are easier than for other loci; researchers often favor loci with multiple alleles because they are more informative for many purposes. Of 32 individual chimpanzees sampled, carrying 64 alleles, 7 were homozygous for one allele (we'll call them AA) at *Rena4*, 11 were heterozygotes (we'll call them Aa), and 14 were homozygous for the other allele (we'll call them aa). Is this locus in Hardy-Weinberg equilibrium?

To find the expected frequencies in the next generation, we first calculate the frequency of the alleles A and a. Every AA individual has two copies of allele A, so we multiply the number of AA homozygotes by 2. Every Aa individual has one copy of allele A, so we add the number of Aa heterozygotes. Thus,

$$p = \frac{(\text{number of } Aa \text{ individuals}) + 2(\text{numbers of } AA \text{ individuals})}{\text{total number of alleles}}$$

$$p = \frac{11 + 2(7)}{64} = 0.39 \qquad \text{(eq. 4.6)}$$

Similarly, the frequency of allele a can be calculated as

$$q = \frac{(\text{number of } Aa \text{ individuals}) + 2(\text{numbers of } aa \text{ individuals})}{\text{total number of alleles}}$$

$$q = \frac{11 + 2(14)}{64} = 0.61 \qquad \text{(eq. 4.7)}$$

Now we can use those allele frequencies to calculate the proportions of the three genotypes that are expected if the population is in Hardy-Weinberg equilibrium, using formulas 4.2–4.4.

the frequency of genotype $AA = p^2 = (0.39)^2 = 0.15$

the frequency of genotype $Aa = 2pq = 2(0.39)(0.61) = 0.48$

the frequency of genotype $aa = q^2 = (0.61)^2 = 0.37$ (eq.4.8)

If we multiply each of these by 32, the number of individuals sampled, we can figure out how many chimpanzees of each genotype we would expect to find if the population was at equilibrium, and compare that to the numbers that were actually observed.

Genotype	Number Expected	Number Observed
AA	$0.15 \times 32 = 4.8$	7
Aa	$0.48 \times 32 = 15.4$	11
aa	$0.37 \times 32 = 11.8$	14

As you can see, it appears that fewer heterozygotes and more homozygotes were observed than we expected to find. However, a statistical test, such as a goodness-of-fit test, would tell us that these differences are not statistically significant. Thus, we can say that at the *Rena4* locus, this population of chimpanzees is not detectably out of Hardy-Weinberg equilibrium. However, other loci sampled by the researchers (with more alleles, and thus more difficult calculations) were not in Hardy-Weinberg equilibrium.

What does this tell us about chimpanzee biology? First, the fact that some loci are not in Hardy-Weinberg equilibrium tells us that some evolutionary force is acting on this population. One likely possibility is that dispersal may cause these differences. Males are philopatric (they do not disperse from the area in which they were born), whereas females disperse when they reach adolescence. Dispersal patterns such as this may cause deviations from Hardy-Weinberg equilibrium. Thus, when we detect deviations, we are often motivated to do further research. A second point from this study is that different loci may be affected differently by evolutionary forces. For example, some loci may be under stronger selection than others in the same genome, and thus might show different patterns.

Fitness and Adaptation

The basic tenet of Darwin's theory of natural selection is that some individuals possessing certain heritable traits are better able to survive, reproduce, and therefore pass on the genes for those traits than are individuals with different traits. The term selection is a metaphor that can be misleading, since it seems to imply that some force is making an active choice. For example, animal breeders perform artificial selection when they breed only animals with particular traits. However, in natural selection there is not a force that "selects" in that sense. Natural selection has no active goal, not even the survival of the species, as we shall see.

One of the key concepts of Darwin's theory was that of **fitness,** the number of offspring that an organism with a particular genetic and phenotypic makeup can be expected to produce. A particular fitness value is meaningless unless it is compared with the reproductive performance of other members of the population or species. Thus, one should think of relative fitness rather than fitness in an absolute sense (Mettler et al. 1988). Moreover, the adaptive value of certain genes or genotypes depends on existing environmental conditions. A genotype may have high fitness in one environment, but low fitness in another. Fitness is therefore not an unchanging characteristic of an organism, such as eye color, but is determined by both the organism's characteristics and the environment.

How should we measure fitness? We might be more interested in either short-term fitness (the potential of each genotype to be represented in the next generation) or long-term fitness (the potential of genotypes to be represented in gene pools in subsequent generations). There is no obvious point at which a researcher should assess the fitness of a particular genotype or phenotype. We generally test our hypotheses about the fitnesses of different genotypes and phenotypes by measuring the **reproductive success** of the organisms in question. Reproductive success is a measure of an organism's production of offspring. It may be measured in several ways, including the number of offspring born, the number that survive to weaning, or the number that survive to mating.

Fitness and reproductive success are similar, but distinct, ideas. For example, consider the popularized definition of natural selection as "survival of the fittest" (coined by one of Darwin's contemporaries, Herbert Spencer). If the fittest are simply those organisms that have survived and reproduced, this statement becomes a tautology: it is an empty statement because it is always true. Rather, the fittest are those that are expected to produce the most offspring, whereas reproductive success is a measure of the number of offspring that are actually produced. Fitness is a property of traits or genotypes, while reproductive success is a property of individuals.

The relative fitnesses of organisms are a function of a tremendously wide range of adaptations. In the rest of this book, we will discuss adaptations for survival, such as finding food and avoiding predation. We will also discuss adaptations related to social interactions, mating, and rearing offspring.

Units of Selection

What does selection act upon—single genes, chromosomes, individuals, social groups, or whole species? Natural selection "sees" the phenotype of the individual organism: that is, it is the organism as a whole that survives and reproduces, not its individual genes. Because of this, many people reason that it is most useful to consider the individual as the unit of selection. The fitness of an individual genotype depends upon the action and interaction of all its genes. For example, some individuals possessing a particular allele, may have well-adapted phenotypes (because of the other genes that they possess), while others possessing *a* may be less fit. The same holds true for individuals possessing allele *A*. The reason is that the fitness of the individuals is usually not determined by just the alleles of a single locus, and selection on a particular locus happens against the background of all the other genes. If, on average, the fitness of individuals possessing the *a* allele is lower than that of individuals possessing the *A* allele, then the frequency of the latter will increase in the population or species.

Another view is that the gene is best considered as the unit of selection. The DNA of diploid individuals, as we discussed earlier in this chapter, is reshuffled during the formation of gametes, so the genetic makeup of individuals changes from generation to generation. Dawkins (1976) has suggested that therefore it is more useful to think of the individual as a unique and temporary vehicle for pieces of DNA, which replicate themselves and are potentially immortal.

Groups may also be considered to be selective units. The concept of group selection has evolved since it was first introduced by Wynne-Edwards (1962). He proposed it to explain apparent cases of altruistic behavior—behavior that is performed, at some cost, to benefit another individual. Wynne-Edwards argued that groups with a high proportion of altruists would be more fit than groups without many altruists. Even though there is selection against altruistic individuals within a group, the group as a whole survives better than other groups because it would not suffer major crashes in size due to intense competition and overuse of resources. This is often popularized as a "good of the species" argument.

There are several serious criticisms of this formulation of group selection. First is the question of whether the extinction rate of altruistic individuals within populations can be compensated for by the disproportionate success of populations containing altruists. It has been argued that the success of altruists in pioneer populations would be unlikely to outweigh their rate of extinction in existing populations (Williams 1966). In addition, populations with a disproportionate number of altruistic individuals may be invaded by mutations back to the selfish condition, or by selfish animals moving in from other populations. Maynard Smith (1976) has argued that in order for the selfish mutation not to spread, the population in which it arose must go extinct at a rate so high that it is not likely to occur in nature. Thus, Wynne-Edward's ideas, and especially "good of the species" explanations, have been generally rejected.

In recent years, the phrase "group selection" has come to mean something different than in Wynne-Edwards original formulation. It now refers to differential survival of groups. For example, Wade (1976) randomly assigned flour beetles to different groups, and then measured the total number of offspring for each group. Groups varied tremendously in their reproductive output because of complex interaction of a number of traits, such as sensitivity toward crowding and cannibalism (Wade 1979, McCauley and Wade 1980). Some groups may thus be more likely to survive and pass on their genes than are other groups. In this formulation of group selection, individual and group selection may act in the same direction or in opposition to one another. Biologists disagree about the extent of the importance of group selection in the natural world, and this subject has been given recent attention (e.g., Sober and Wilson 1998).

Finally, another area in which selection is operating on a different level than the individual is kin selection. W. D. Hamilton (1964) presented his model of kin selection to account for the evolution of altruistic behavior. Hamilton argued that most altruistic acts are performed by individuals that are related to the recipients. Smaller units within populations are usually composed of related individuals who share a certain amount of their genetic material through common descent. Since related individuals may possess identical alleles at many loci, we can think of kin as genetic extensions of individuals. Hamilton's model introduced the concept of **inclusive fitness,** the sum of direct and indirect fitness. **Direct fitness** is based on the probability of reproductive success, measured by the number of one's own offspring, while **indirect fitness** is based on the additional reproductive success due to one's actions of nondescendant relatives. The implications of kin selection for altruism and social behavior will be addressed in detail in chapter 19. Group selection and kin selection can be expressed in identical mathematical terms, and some biologists consider kin selection to be a subset of group selection (e.g., Wade 1985, Sober and Wilson 1998, Frank 1998).

CAUTIONS ABOUT ADAPTIVE EXPLANATIONS

Darwin's theory of evolution by natural selection revolutionized the way people thought about organisms. It is the single best unifying explanation for the ultimate causes of the traits that organisms possess. However, we must show caution when thinking about the evolution of behavioral traits. It is important not to automatically assume that the traits that we see have been optimized by natural selection.

First, not all traits are heritable. Natural selection can act only on traits that are at least partly determined by the genotype; many traits may be caused entirely by the action of the environment. We must be particularly careful about studying behavioral traits in animals that are capable of learning (see chapter 11 for details).

Second, although it is easy to be persuaded by the logic of natural selection, it is important to remember that other evolutionary forces, as described above in the chimpanzee example, may be at work. In particular, it is often difficult to assess how genetic drift has influenced the evolution of traits. Simplistic thinking about the power of natural selection is apt to lead to unrealistic assumptions about the functions of the traits that organisms possess.

Third, selection pressures change over time. The environment may change, and some previously beneficial traits may become deleterious, or vice versa. An allele may be favored at one developmental stage, in one sex, or in a particular season, then may be selected against in another. Different genotypes can be favored in different habitats within a population. We must remember that our studies of animal behavior present only a snapshot of a species over evolutionary time, and that it is quite difficult to understand the past events that have influenced what we now see. We should guard against overextending hypothetical explanations of the behavior's adaptive significance and the natural selection pressures that led to each aspect of a complex behavioral sequence.

Fourth, because traits are not always independent of each other, not every trait can be optimized by natural selection. You might reason that if evolution by natural selection is prevalent, then over time, under a constant selection pressure, every organism would be homozygous for the most favorable alleles. Often, however, the situation is more complex than that. For example, some traits exhibit heterozygous advantage: heterozygotes can have greater fitness (at least under some circumstances) than either homozygote. In that case, natural selection would not act to favor one allele over the other. **Epistasis,** where genes at different loci interact to produce the phenotype, is common, so selection may act differently on genes in different genetic backgrounds. Many genes also exhibit **pleiotropy,** meaning that they affect more than one trait—one allele may produce a positive effect on one trait, but a negative effect on another. Genes can also be physically linked together on the same chromosome. If they are close together, they tend to be inherited together, and thus it is difficult for natural selection to optimize each trait independently. This is especially likely to be important over short time periods before crossing over has had a chance to break apart linkages.

Fifth, we must bear in mind that there is no predetermined goal in the evolution of any species. The only thing that is important is how much relevant genetic variability there is and which traits are best suited for reproduction in a particular environment. In addition, we should not assume that a species can evolve any particular trait that would suit it. It is quite possible that a species does not have the relevant genes in its gene pool to evolve certain traits that appear obviously adaptive to us. Mutations do not arise in order to fulfill a need. The evolutionary history of a species determines, in part, what is available for natural selection to work on in the future. We will discuss this concept in more detail in chapter 6.

Bearing in mind these precautions, it has proved very useful to use adaptation and natural selection as a framework in which to study animal behavior. The adaptationist approach is misused when, without supporting evidence traits are automatically considered to be the result of natural selection. However, generating testable hypotheses about the adaptive value of traits is a very powerful approach to the study of biology and has guided us to many important discoveries (Mayr 1983).

SUMMARY

The total set of genes that an individual possesses is its genotype. The genotype and the environment determine the phenotype, or the observable traits of an organism. Genes are located on homologous pairs of chromosomes, and may have different forms, or alleles. Genes can have either structural or regulatory functions.

Genetic variability in populations comes from several sources. When gametes (egg and sperm) are formed, homologous chromosomes exchange parts in a process called crossing over. After crossing over, during gamete formation the homologues separate independently. Hence, sexual reproduction shuffles the genetic makeup of each offspring relative to its parents. The ultimate source of new variability, however, is mutation, which is a heritable change in the genetic material.

Evolution is a change in the frequency of alleles in a population over generations. Darwin's theory of evolution by natural selection proposed that some organisms had heritable traits that made them better suited for survival and reproduction than organisms with different traits. Therefore, more offspring are born with those traits, and the traits increase in frequency in the population. Such traits are what give an animal its fitness, or probability of surviving and reproducing. Traits that have been shaped by the process of natural selection are called adaptations.

Evolution may result not only from natural selection, but also from genetic drift, the random change in frequencies of alleles, particularly in small populations. Mutations and gene flow also contribute to evolutionary change. If evolutionary forces are at work in a population, the genotype frequencies will usually not be in Hardy-Weinberg equilibrium.

Selection can work at the level of genes, individuals, kin groups, or unrelated groups. We should be careful about uncritically invoking adaptive explanations for the behavioral patterns that we see. Traits may be environmentally induced or result from other evolutionary forces besides natural selection. Selection pressures change over time. Traits are not always independent of one another. Finally, natural selection can act only on the genetic variation available. In spite of this, natural selection has provided a very useful framework for the study of behavior.

DISCUSSION QUESTIONS

1. Define evolution and describe how it can occur. Use the terms natural selection, genetic drift, fitness, and mutation in your answer.

2. Two common misunderstandings about evolution are that it acts for the good of the species, and that it is goal directed. Why are these faulty ideas?

3. A population is observed to have two alleles, *A* and *a,* for a particular locus. The three possible genotypes are found in the proportions *AA* = 37%, *Aa* = 22%, *aa* = 41%. Is the population in Hardy-Weinberg equilibrium?

4. There has been controversy among scientists about the level upon which selection acts. For example, some scientists feel it is more valuable to think about selection at the level of the gene, others emphasize the individual, and still others believe that this is a noncontroversy and both views can be accommodated. What, to your mind, are the advantages and disadvantages of these three perspectives?

SUGGESTED READINGS

Dawkins, R. 1976. *The Selfish Gene.* New York: Oxford University Press.

Futuyma, D. 1998. *Evolutionary Biology.* Sunderland, MA: Sinauer Associates.

Gould, S. J. and R. C. Lewontin. 1979. The spandrels of San Marco and the Panglossian paradigm: a critique of the adaptationist programme. Proceedings of the Royal Society of London, B. 205:581-98.

Hartl, D. L. and A. G. Clark. 1997. *Principles of Population Genetics.* 3rd edition. Sunderland, MA: Sinauer Associates.

Mayr, E. 1983. How to carry out the adaptationist program? *American Naturalist* 121:423-434.

Ridley, M. 1996. *Evolution.* Cambridge, MA: Blackwell Science.

Williams, G. C. 1966. *Adaptation and Natural Selection.* Princeton: Princeton University Press.

5
CHAPTER

Behavioral Genetics

 hroughout this book, we discuss the idea that many patterns of animal behavior are adaptations that have resulted from the process of natural selection. In order for natural selection to operate on behavior, behavioral patterns must be at least partly genetically based. This chapter begins with a discussion of the sorts of questions that behavioral geneticists ask. We then discuss cases in which single genes or small groups of genes influence behavior. Next, we will discuss cases in which many genes influence behavior. Throughout the chapter, we will focus on introducing techniques, both established and new, necessary for behavioral genetic research.

QUESTIONS IN BEHAVIORAL GENETICS

The study of behavioral genetics includes both proximate and ultimate questions. First, proximate questions concern how genes can determine behaviors. Genes, after all, are simply strings of nucleotides, and it is a long path between that and, for example, a behavioral pattern of complex predatory sequences. Some of the influences on the pathway connecting genes and behavior are sketched out in figure 5.1. This is a complex but exciting area of research that draws from many fields, including molecular biology, developmental biology, and neurobiology. This field is in a rapid phase of growth, and we will touch only the surface here.

The second set of questions are ultimate in nature. In order for behavior to evolve, it must be at least partly under genetic control, and behavioral geneticists often study the extent to which this is true. This is sometimes phrased as nature versus nurture, a phrase that originated with Shakespeare's *The Tempest:* a trait that is influenced by the genes is determined by "nature," while a trait under environmental control is determined by "nurture." However, it is naïve to

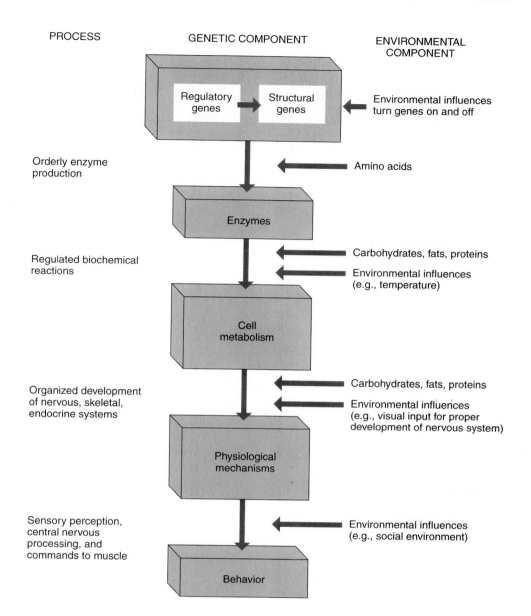

PROCESS   GENETIC COMPONENT   ENVIRONMENTAL
                                                COMPONENT

Regulatory genes → Structural genes ← Environmental influences turn genes on and off

Orderly enzyme production   ← Amino acids

Enzymes

Regulated biochemical reactions   ← Carbohydrates, fats, proteins
                                         ← Environmental influences (e.g., temperature)

Cell metabolism

Organized development of nervous, skeletal, endocrine systems   ← Carbohydrates, fats, proteins
                                         ← Environmental influences (e.g., visual input for proper development of nervous system)

Physiological mechanisms

Sensory perception, central nervous processing, and commands to muscle   ← Environmental influences (e.g., social environment)

Behavior

FIGURE 5.1   Model illustrating relationship between genes and environment in the control of behavior. There are many opportunities for environmental influences.

think that there is a black and white dichotomy. Nearly all traits are influenced by both genes and the environment, and one goal of behavioral genetics is to parse out their relative effects. Questions such as these are necessary for understanding the evolution of traits, but in order to answer them we do not need to know the details of exactly how every gene determines behavior. This area of behavioral genetics is also in a period of explosive growth.

STUDYING SINGLE GENE EFFECTS

We will begin by discussing cases in which we know that a single gene, or small group of genes, is known to affect a particular behavior. Although these cases are relatively rare,

they provide valuable model systems for studying the mechanisms by which information in the genes is translated into behavioral patterns.

Mendelian Crosses

Classical genetic crosses can sometimes be used to investigate simple patterns of inheritance for behavioral traits, much as Mendel did with his peas. A famous example is hygienic behavior in honeybees (*Apis mellifera*). Honeybees are afflicted with a bacterium, the American foulbrood (*Bacillus larvae*). Foulbrood can infect bee larvae and destroy the entire colony. Most strains of honeybees are resistant to the bacterium because the worker bees open diseased cells and remove the diseased larvae. These are called hygienic bees.

However, some strains are unhygienic, and allow diseased larvae to remain in their cells. When a true-breeding, homozygous hygienic strain was crossed with a true-breeding unhygienic strain, none of the hybrid offspring removed infected larvae (Rothenbuhler 1964a): the unhygienic trait is dominant. When heterozygotes are crossed with the two homozygote strains, four types of offspring were generated: the workers would either (1) uncap cells with dead larvae but not remove them; (2) remove dead larvae from cells that were uncapped for them but not uncap the cells; (3) not uncap cells or remove larvae; or (4) uncap cells and remove larvae. These results indicate that a minimum of two genes are responsible for hygienic behavior—one that controls the predisposition to uncap diseased cells and one that controls the predisposition to remove larvae. These behaviors are not wholly unaffected by the environment, however, as even unhygienic bees sometimes uncap and remove dead larvae, and hygienic bees sometimes fail to do so (Rothenbuhler 1964b). Be aware that Rothenbuhler's results do not mean that only two genes are responsible for generating these complex behaviors. It is likely that several genes turn suites of other genes on and off.

Mutations and Knockout Genes

Mutants are very useful in the study of the link between genes and behavior. Researchers can compare the behavior of animals with known mutations to those without in order to assess the function of the target gene. One research strategy is to cause mutations with one of several mutagenizing techniques, such as exposure to chemicals or x ray, and then screen the offspring of those animals for changes in behavior. Then, through experimental breeding regimes and analysis of offspring phenotypes, one can determine the number of loci that have mutated. However, this technique usually results in mutations that make animals sick and dysfunctional (Hall 1994). In addition, the mutations are random in nature because no specific region of the DNA is targeted, so expensive and time-consuming screening of many candidate mutants is necessary.

A new and potentially powerful technique involves the introduction of a genetically engineered mutant or **knockout gene** into an embryo. A knockout gene is one that has been targeted for disruption so that it no longer functions normally. In this case, the mutation affects a specific gene of interest, rather than randomly affecting the genome as has been the case with older mutagenizing techniques. The embryos are grown and animals are screened to find those that have the inactivated knockout gene in their DNA. Those individuals are bred to create a line that is homozygous for the inactivated gene. This allows investigators to study the animal without that particular gene being expressed. However, one of the potential problems with this approach is that by knocking out the function of a gene at the earliest stages of development, the role that gene plays in development is also removed. This role may involve the interaction between the knockout gene and other genes, so interpretation of results from animals with knockout genes must be made cau-

tiously. Often, researchers try to test the same knockout gene with a variety of genetic backgrounds in order to be more sure of its effect.

These techniques have yielded a number of examples of mutations at single loci that result in a major dysfunction of some aspect of behavior. For example, mutant strains of paramecium vary in their movement patterns (Kung et al. 1975); these have the evocative names of *sluggish, spinner,* and *paranoiac* (which moves backwards). Mutations that have been identified in *Drosophila* (reviewed in Lindsley and Zimm 1992, Hall 1994) include *spinster* (unreceptive females); *dunce, rutabaga,* and *amnesiac* (poor performance on learning and memory tasks); and *fruitless* (males court but never attempt to copulate). Simply identifying an association between a particular mutant and a phenotype is only the beginning, however. Researchers have been working to identify exactly which systems are disrupted by a given mutant. This is made difficult by the pleiotropic effects of many of the genes. For example, the mutant locus carried by *dunce* flies affects both development and the adult central nervous system (Dubnau and Tully 1998). Some of the effects of this mutation on memory have been traced to its disruption of the cyclic AMP pathway, which translates extracellular signals into intracellular responses (summarized in Belvin and Yin 1997).

Single gene effects have been found for some surprisingly complex behaviors, including aggression, sexual behavior, and anxiety and fear (reviewed in Nelson and Young 1998). For example, the *fosB* mutant in mice affects nurturing behavior. Mutant mice mothers do not retrieve their offspring or keep them warm (figure 5.2). Nurturing behavior in normal mice results both from changes in the hormone titers of the mothers at birth, as well as from responses that are induced over a period of several days by exposure to the pups. The hormonal profiles of *fosB* mutant mothers are normal; the effect of the gene appears to be on the induced response (Brown et al. 1996).

Mosaics

Some investigators have used mosaics, organisms whose tissues are of two or more genetically different kinds, to test aspects of the gene-to-behavior sequence. Mosaics permit us to observe both anatomical and behavioral anomalies combined in the same animal. The classic studies of Hotta and Benzer (1972, 1973) took advantage of an odd characteristic of a strain of *Drosophila* in order to develop this technique. In this strain, there is an unstable ring-shaped X chromosome. Female flies have the genotype XX. During development, the unstable X chromosome becomes lost in some cells, leading to mutant X cells that are male. Other cells retain both copies of the X, and are normal and female. The researchers produced mosaics that had some normal and some mutant tissues, and some normal and some mutant behaviors. Hotta and Benzer then determined what tissue parts must be mutant for the abnormal behaviors to be expressed. For example, one behavioral mutant, when placed in a tube with a light source above it, moves toward the light

FIGURE 5.2   A mutant allele in mice affects nurturing behavior.
fosB mutants do not retrieve their pups or keep them warm.
From Brown J., Ye H., Bronson R., Dikkes P., and Greenberg M. "A Defect in Nurturing in Mice Lacking the Immediate Early Gene *fosB*." *Cell* 86: 297–309. 1996. Reprinted by permission of Elsevier Science. Photo Courtesy of Dr. Jennifer R. Brown.

but in a spiral. This is a complex behavior, involving receptor cells, integration in the central nervous system, and motor signals. With the mosaic technique, Hotta and Benzer localized the cause of the mutant behavior in the eye itself: mutant eyes do not respond as well to light. Because flies move to equalize the input of light into both eyes, the mutant eye generates a spiral walking pattern (reviewed in Benzer 1973). More recently, mosaics have been used in the study of the courtship song of *Drosophila.* Male flies produce song when they are attempting to entice females to mate. Mosaic flies show that the focus of the song rhythm generator is apparently in the thoracic ganglia of the central nervous system, because a mutation in that area destroys rhythmicity (Konopka et al. 1996).

STUDYING MULTIGENIC EFFECTS: SIMPLER METHODS

Most behavioral traits are caused by multiple genes rather than single genes. Here, it is more difficult to examine the mechanisms by which gene action occurs. A great deal of research has focused on developing techniques that allow us to determine the extent to which a trait is influenced by genetics, the environment, or both. In this section, we will examine several methods that allow us to do this in a straightforward way. These techniques are conceptually simple to understand, although in practice they may be quite challenging to carry out.

Cross-Fostering Experiments

We can transfer neonatal animals from the parent female to another female of the same species or strain, or to a female of a different species or strain (if she will successfully rear the young). We can then compare genetically similar animals with different rearing environments and assess the relative importance of the effects of genotype versus the effects of maternal-care environment on certain behavioral traits. This technique of cross-fostering helps us differentiate species-specific behavior from environmentally influenced behavior. If genetically similar animals reared under different conditions exhibit similar behavior, then we can conclude that genetic control of that behavior is fairly rigid. However, if the behavior differs significantly, we can conclude that it is strongly influenced by environment. This technique has been extensively used in bird studies, as many species will accept foreign eggs as their own. For example, great egret (*Casmerodius albus*) chicks regularly attack and kill younger nestmates (siblicide). Great blue heron (*Ardea herodias*) chicks, in contrast, rarely do. Mock (1984) tested the hypothesis that the difference in behavior has to do with prey size: egret parents provide small fish that are easy for aggressive chicks to monopolize, whereas herons feed their young larger prey. Cross-fostered heron chicks raised on small prey by egret parents became siblicidal, showing that environmental conditions can induce siblicide in this species. In contrast, cross-fostered egret chicks were still very aggressive.

Twin and Adoption Studies

The study of the behavioral genetics of humans is obviously constrained, as many of the experimental techniques open to researchers on other animals are not available. Two tactics have been commonly used to separate environmental and genetic effects on human behavior: twin studies and adoption studies. In twin studies, researchers take advantage of the existence of two types of twins. Identical, or monozygotic, twins develop from the splitting of a single fertilized egg, and have identical genotypes. Fraternal, or dizygotic, twins develop from two separate fertilized eggs and share half their genes. We can compare the resemblance between identical twins to that between fraternal twins. If genetics is important in determining a particular trait, we expect the resemblance to be stronger between identical twins. In adoption studies, adopted children are compared to their adopted parents, with whom they share an environment, and birth parents, with whom they share genes.

These studies provide evidence for genetic effects on a number of behavioral traits. For example, both verbal and spatial performance appear to be under partial genetic control. Both are more strongly correlated between identical twins than between fraternal twins (Plomin et al. 1997). Adopted children do not significantly resemble their adoptive parents in these traits, but do resemble their birth parents (reviewed in Plomin and Craig 1997).

Any human studies must be treated with caution because it is extremely difficult to control environmental factors as well as one might like. For example, identical twins are often treated more similarly than fraternal twins, and therefore would experience more similar environments (Ehrman and Parsons 1981). Adoption agencies often try to place children in homes that are similar to those they came from, again making it difficult for researchers to distinguish between genetic and environmental effects. There is likely to be a strong interaction between genotype and the environment as well: genetically different traits in children elicit different responses from family and peers, thus creating differences in the children's environments (Plomin 1994). Interactions such as these make it very difficult to tease apart the effect of genes on traits. Additional problems arise from the fact that twins share the same environment in the womb: it is possible that we might ascribe similarities between twins to genetics, when it is in fact due to shared intrauterine environment (Devlin et al. 1997). It is likely, however, that there are significant genetic influences on at least some aspects of human behavior. This does not mean environmental influences can be ignored; quite to the contrary. Environmental influences are extremely important in determining human behavior, and interact with the genotype to have broad effects. Therefore, educational and other social opportunities play a critical role in human behavioral development despite a genetic influence on many psychological traits (Bouchard 1994).

Inbred Lines

One way to study the effects of environmental parameters on behavior is to hold the genetic component constant by using homogeneous strains of animals. One method of achieving genetic homozygosity (a population of animals with the same homozygous genotype) is by inbreeding, or allowing only brother-sister matings for many generations. In mice, after about thirty generations of inbreeding, virtually 100 percent of the allele pairs are homozygous. During the inbreeding process, many recessive alleles that are lethal or otherwise detrimental to successful reproduction may also become homozygous. Thus, to obtain a viable inbred strain

of mice, we must begin with numerous brother-sister pairs because many lines will die out before a high degree of homozygosity is achieved.

After a homozygous strain is created, we have an interesting genetic tool. Because we know that all the animals are genetically identical, we know that any phenotypic differences among them are a result of environmental differences. Conversely, we can expose two different inbred strains to the same environment, and know that any differences we see are due to genetic differences rather than environmental effects.

A classic experiment on maze learning in rats (Cooper and Zubek 1958) illustrates how both the environment and genetic makeup can affect behavior in an inbred strain. Two strains of rats were established by selective breeding (this technique will be covered in more detail in the following text). Animals in the "maze-bright" strain were adept at learning mazes and making relatively few errors, while "maze-dull" animals did not learn as fast. Young animals from each strain were raised under identical conditions until weaning, at which time some animals from each strain were reared in an enriched environment and some in a restricted environment. The enriched environment was brightly colored and contained many toys that the rats could manipulate. The restricted environment was grey and had no toys. The enriched animals from the maze-dull strain learned better than the restricted maze-dull animals (figure 5.3), demonstrating that the environment can have a strong influence in a genetically homogeneous strain. However, it was also clear from the results that the genetic differences between the two strains also affected their maze-learning abilities.

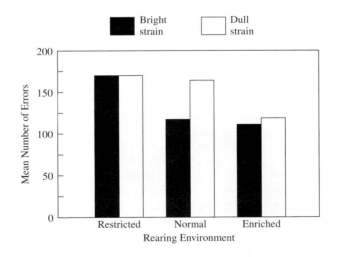

FIGURE 5.3   The effects of genes and environment on maze learning in rats.
The "maze-bright" strain of rats made fewer errors than the "maze-dull" strain in a normal environment. An enriched environment reduced the difference between the two strains significantly. In addition, a restricted rearing environment caused animals in the two strains to perform almost identically.

Source: Data from R.M. Cooper and J.P. Zubek, "Effects of Enriched and Restricted Early Environments on the Learning Ability of Bright and Dull Rats," *Canadian Journal of Psychology,* 12:159–64.

Other experiments have shown differences in the responses of different strains to more ecologically relevant problems. For example, paradise fish are small insectivores that live in shallow marshes and rice fields in east Asia. Several other fish species prey upon paradise fish, so being able to recognize predators and respond appropriately is important. When paradise fish see fish of other species, especially fish with large eyespots (markings that appear similar to eyes), they investigate them, and if the fish are dangerous predators, the paradise fish flee (Gerlai 1993, Csányi 1986). Two inbred strains of paradise fish, raised under identical conditions, were presented with model predators. One strain was more likely to flee or back away and less likely to approach the predator than was the other strain (Miklósi et al. 1997). The difference is attributable to genetic differences between the strains.

Conclusions from studies using inbred strains are limited to the particular strain and to the specific variables measured in an investigation. Numerous additional factors, such as differences in methods or procedures between laboratories, and the age, sex, and previous experience of test subjects, may affect the behavior and thus the interpretation of such studies. Because inbred strains are homozygous, we cannot use this technique to tease apart the effects of interactions among loci or between alleles at the same loci. The approaches described in the next section, however, offer this possibility.

STUDYING MULTIGENIC EFFECTS: QUANTITATIVE GENETICS

Quantitative Traits and the Sources of Behavioral Variation

Many of the behavioral traits that we see are not all or nothing, but rather a matter of degree: for example, the sprint speed of scorpions varies across individuals and encompasses a wide range of speeds. Variation without natural discontinuities is called continuous variation. Most traits like this depend on a large number of genes, each with a small effect, and are said to be under **polygenic** control (though not all traits under polygenic control show continuous variation, and not all traits that show continuous variation are under polygenic control). In order to understand the evolution of continuously varying traits, we can use the tools of **quantitative genetics.**

First, we need to understand the concept of variance. Variance is a statistical measure of variability. It takes into account how different each individual measurement is from the population average. Imagine for a moment a frequency distribution of the heights of a population of 500 different people. Curves such as this are often normal (bell-shaped) curves, with a lot of people of medium height, and fewer and fewer people out toward the very short and very tall ends. You can imagine that bell-shaped curves might take several shapes (figure 5.4). If most of the people are similar in height, then each person is close to the average and the variance is small (figure 5.4a). In contrast, if there is a broad range of heights, then many people

will be different from average and the variance is larger (figure 5.4b). We can therefore interpret the variance as describing the spread of points around the mean. Populations with a low variance have little spread around the mean, and those with a high variance have a lot of spread.

We can describe variance with a symbol, V. Traits that we measure for each individual are phenotypic traits, and we can describe a population's variance in phenotypic traits as V_P. In quantitative genetics, we are often interested in the source of this behavioral variation: does it result from genetic effects or environmental effects? We can phrase this as an equation:

$$V_P = V_G + V_E \qquad \text{(eq. 5.1)}$$

The total phenotypic variance is the sum of the genetic variance (V_G) and environmental variance (V_E) (this and the following discussion follows from Falconer and MacKay 1996). In other words, some of the phenotypic variation we see in a population results from differences in the genes, and some results from differences in the environment. Brain development, for example, is determined partly by genes and partly by environmental conditions during development, such as nutrition and the amount of stimulation in the environment. Each individual may differ from the population mean partly because of the genes they carry, and partly because of the environment they experience. This is the same principle that we have been discussing throughout this chapter; it is simply a formal statement in quantitative genetics terminology.

Sometimes there is an interaction between the genotype and the environment, which is also a part of the total phenotypic V. For example, a genotype that increases territorial aggression may be expressed only in habitats with few predators, whereas another genotype might do the reverse. In this case, we can add another component, V_{GxE}, or gene by environment interaction.

$$V_P = V_G + V_E + V_{GxE} \qquad \text{(eq. 5.2)}$$

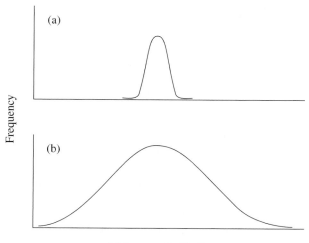

FIGURE 5.4   Variance in quantitative traits.
Different frequency distributions illustrating populations with (a) low variance or (b) high variance in a particular phenotypic trait.

Heritability

In order to understand whether a trait can respond to selection, we are interested in calculating the extent to which variance in the phenotypic trait is due to resemblances among relatives, i.e. the proportion of phenotypic variance due to genetic variance. This is called heritability. There are two ways to calculate heritability. The first way is to calculate it as V_G/V_P (the genetic variance divided by the phenotypic variance). This is called **heritability in the broad sense,** and is a fairly rough measure of the resemblance between relatives. It is possible to measure heritability more accurately, but we need to return to our equation again in order to understand the reasoning.

We can break down V_G into smaller parts. Alleles within a genome may interact with each other. We have already discussed dominance interactions between alleles at the same locus. For example, in the hygienic bees discussed previously, the unhygienic alleles were dominant to the hygienic alleles, so the heterozygous bees were unhygienic. We can call the portion of phenotypic variation that is due to dominance interactions V_D. Alleles at different loci can also interact, as was discussed in chapter 4. This interaction is called epistasis, or gene interaction. We can call the portion of phenotypic variation that is due to interactive effects V_I. The point to note about dominance interactions and epistatic interactions is that they are not transmitted intact from parent to offspring. For example, you have inherited only a haploid set of genes from each of your parents, so the dominance interactions at each of your genetic loci are not inherited from either parent. Similarly, you do not inherit the same sets of alleles across loci that either of your parents possess. Thus, in order to measure heritability, or the proportion of phenotypic variance due to the resemblance between relatives, as accurately as possible, we need to remove these two effects. The portion of phenotypic variation that is left over is called V_A, or additive genetic variance. This is what we are most interested in measuring. We can thus break down V_G into its component parts:

$$V_G = V_A + V_D + V_I \qquad \text{(eq. 5.3)}$$

Narrow sense heritability, which measures the relationship between relatives as accurately as possible, is written as V_A/V_P. For arcane historical reasons, this is also written as h^2. Narrow-sense heritability, like broad-sense heritability, ranges from 0 to 1: when there is no resemblance between relatives, $h^2 = 0$, and when the resemblance is complete, $h^2 = 1$.

It is important to remember that heritability describes populations, not individuals. A trait may be inherited, in that it is passed on in the genes from parent to offspring, yet if there is no genetic variation in a population for that trait ($V_G = 0$), its heritability will be zero. The lower the heritability, the less able the population will be to respond to selection. In the following text, we will discuss an example of calculations of heritability, as well as caveats about its use.

QUANTITATIVE GENETICS TECHNIQUES

Artificial Selection

Domesticated pets, livestock, and many ornamental and agricultural plants are the products of selective breeding. In behavioral genetics research, artificial selection can tell us about the genetic basis for a trait, including its heritability. Artificial selection requires that some degree of genetic variability exists for the trait in question. Commonly, animals at the distributional extremes of a large sample population are then bred through several generations under the same environmental conditions.

In figure 5.5*a*, we see the frequency distribution of a particular continuous trait in a particular population. The

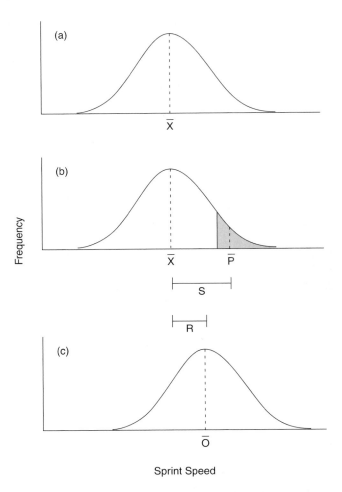

FIGURE 5.5   An illustration of a selection experiment.
(a) This is a frequency distribution illustrating how sprint speed varies in a population. Some quantitative measure of sprint speed is chosen, such as the number of centimeters run per second. The mean of the population is $\bar{X}$. (b) To create a fast line, only the fastest are chosen for mating. The mean of these parents is $\bar{P}$. The difference between the parental mean and the original population mean is S, a measure of the strength of selection. (c) The offspring generation is illustrated here. There is still variation in this group, but there is a new mean, $\bar{O}$. The difference between the original population mean and the new mean is the response to selection, R.
Source: Hugh Dingle.

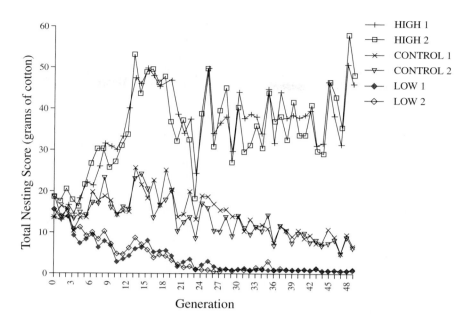

FIGURE 5.6   **The results of selecting for many generations on nesting behavior in mice.**

Shown are the results of an artificial selection experiment on the trait of the amount of cotton that mice use in constructing their nests. Two high, two low, and two control (randomly mated) lines were established.

Source: Lynch, C. B. 1994. Evolutionary inferences from genetic analyses of cold adaptation in laboratory and wild populations of the house mouse. In C. R. B Boake, ed. *Quantitative Genetic Studies of Behavioral Evolution.* Chicago: University of Chicago Press.

trait could be a morphological trait, such as the wing length of milkweed bugs, or a behavioral trait, such as sprint speed. In this population, the distribution follows a normal curve: for example, a few individuals are very fast or very slow, but most are somewhere in the middle. This is a typical distribution for quantitative traits. The mean of the original population is $\overline{X}$. We will now select for mating only the animals at one end of the distribution. Shown in the stippled area in figure 5.5*b*. These are the parents of the next generation, and their mean value for the trait under study is $\overline{P}$. The strength of selection is the difference between $\overline{X}$ and $\overline{P}$: if our selective regime is very stringent, and only the fastest individuals get to mate, S will be large, whereas if selection is weak, S will be small.

In figure 5.5*c*, we see the frequency distribution of the offspring generation, with a mean of $\overline{O}$. In this illustration, the offspring generation is faster, on average, than the original population. The difference between the offspring mean and that of the original population is R. We can then use R and S to calculate **realized heritability,** which is an estimate of narrow-sense heritability: $h^2 = R/S$. This makes intuitive sense: if the response to selection is very strong compared to the selection pressure, R/S will be close to one. If the response to selection is very weak, R/S will be close to zero.

Unfortunately for ease of research, the response to selection is often quite variable from one generation to the next. Therefore, usually large sample sizes and multiple generations of selection are needed to accurately estimate heritability. Researchers often establish selected lines in both extremes of

FIGURE 5.7   **Mice from lines selected for different levels of nest building.**
Source: Lee Drickamer.

the trait, as well as a control line with randomly mated individuals, and often high- and low-selected lines respond differently. An example of a long-term selection experiment is that on nest building in mice, illustrated in figure 5.6. Differences between the behavior of the selected lines are striking (figure 5.7). We must also be cautious when we interpret the results of selection experiments, as the targets of selection are not always obvious. For example, researchers have selected for changes in phototaxis, on response to light, in *Drosphila melanogaster.* A suite of genes, affecting vision, motor control, activity, etc. is involved in producing this phenotype. An analysis of individuals from the selected lines shows that artificial selection has produced eyes that are visually impaired (see Roff 1994 for details). Similarly, natural selection in the wild may act on any number of genes that influence a phenotype.

Parent-Offspring Regression

Another way to calculate heritability involves using the resemblance between relatives (Falconer and MacKay 1996; Hartl 1988), commonly parents and their offspring. In this method, the parental generation of animals is reared under standard conditions. They are randomly assigned mates; no selection or inbreeding takes place. The offspring are then reared under the same standard conditions. The trait of interest is measured in the parents and the offspring, at the same point in each animal's development. If parents have more than one offspring, either each offspring or a sample of several offspring per family is measured and all the offspring measurements in a family are averaged together. The offspring measurements are then plotted against the mean of the parental measurements, called the midparent. A statistical analysis technique called **regression** is used to fit a line to these points. The slope of the line equals the heritability of the trait under these conditions. If offspring are very similar to their parents, the slope of the line will be near 1 (figure 5.8a). If there is very little resemblance between offspring and parents, the slope of the line is near 0 (figure 5.8b). In some cases a trait is shown by only one sex, and the procedure is slightly modified. For example, elaborate tail feathers in many birds are produced only by males, so sons' tail measurements would be regressed against those of their fathers. In this case, the slope of the regression line equals half the heritability.

Here is an example. In field crickets (*Gryllus integer*), males attract females by calling with a pleasant trill. However, not all males trill the same way: trill bouts (periods of uninterrupted calling) can be of long, medium, or short length. These differences are obvious to human ears, but more importantly they are also obvious to female crickets. When presented with a choice between a speaker playing a long-trill call versus a speaker playing a short-trill call, 10 of 11 females walked toward the long-trill call (Hedrick 1986).

Is the trait of trill length heritable? The evidence suggested that the trait had a genetic basis, as crickets were not simply changing their trill bouts in response to temperature or some other environmental variable. In the laboratory, under constant conditions, individual crickets were consistent in their trills, even when they were tested three weeks later. Therefore, Hedrick (1988) estimated the heritability of trill length with a parent-offspring (father-son) regression. This is a simpler procedure to carry out than a selection experiment, as it is necessary to rear only one generation. Hedrick found evidence that call length was heritable: the estimate of h^2 was 0.75. This means that about 75 percent of the phenotypic variation in cricket trills in her study population was due to additive genetic variance. There was a large amount of variation around this estimate (the standard error was 0.25), as is often the case with estimates of heritability. However, heritability was significantly greater than zero, so we can conclude that this trait is at least partially under genetic control.

These findings present a puzzle. Here is a male trait that is highly preferred by females, and that appears to be highly heritable. However, substantial variation in this trait still exists in the population. Why is that? You might expect that, over generations, all the genes for the short trills would have disappeared, as those males would be passed over by the females. A number of explanations for the maintenance of genetic variation are possible. For example, short-trilling males might intercept females on their way to a long-trilling male. In addition, female crickets may not always choose the male that they find most attractive. Crickets are at risk of predation by birds and mammals, and they are especially at risk in the open. When given a choice between a preferred, long-trill male at the end of an exposed path, or a less preferable

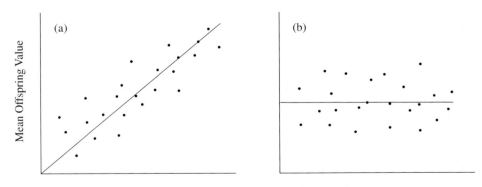

FIGURE 5.8   **Parent-offspring regression and the calculation of heritability.**
Offspring means are plotted against the mean of the two parents to generate these graphs. (a) In this example, there is a strong resemblance between parents and offspring, and the slope of the line (and therefore the heritability estimate) is near 1. (b) Here, there is little resemblance between parent and offspring, and the slope of the line is near 0. Other types of relatives may be used in calculations like these to estimate heritability.

short-trill male at the end of a covered path, females choose safety and the short-trill male (Hedrick and Dill 1993, Hedrick 1994).

Caveats about Heritability Estimates

Heritability estimates vary in different environments. This makes sense: the denominator of the formula for narrow-sense heritability is V_P, the total phenotypic variance, which includes genetic and environmental effects. It is very important to recognize that heritability estimates are valid only for the population in which they are measured, and only in the environment in which they are measured. Heritability estimates measured under uniform, controlled laboratory conditions will be higher than in more variable environments found in nature. It is not valid to think of a qualitative trait as being genetically determined to an exact extent, and environmentally determined to an exact extent, as if it were a fixed feature (Futuyma 1998). This restricts the uses to which we can confidently put numerical estimates of heritability. However, it does not undermine the larger point that heritability studies demonstrate: both genes and environment play fundamentally important roles in the expression of behavioral traits.

Quantitative Trait Locus Analysis

Behavioral traits are generally governed by more than one gene, which makes behavioral genetic analyses quite difficult. In particular, it is difficult to determine where the particular genes responsible for a behavior are located on the chromosome(s). Quantitative trait loci (QTL) are the set of loci that govern a trait (e.g., aggressiveness) that is not completely determined by any one gene acting alone (Takahashi et al. 1994). A QTL analysis involves crossbreeding animals that show different levels of a behavior (e.g., aggressive versus nonaggressive mice). Through a series of genetic crosses and analysis for specific DNA markers possessed by the original strains (figure 5.9), behavioral geneticists can determine which parts of different chromosomes have genes for a particular polygenic trait (Barinaga 1994). The actual process of locating such loci is quite complicated, but the technique of QTL analysis holds great promise for the identification of specific loci that play important roles in behavioral traits. For example, researchers have found two genomic regions in honeybees that affect foraging behavior (Hunt et al. 1995, Page et al. 1995). These regions affect both the amount of pollen that honeybees store and whether foragers will collect pollen or nectar.

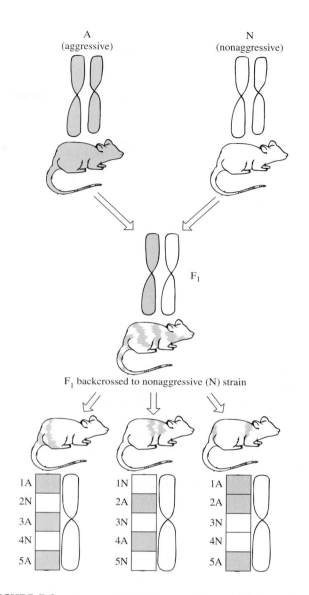

FIGURE 5.9   An example of the procedure used for quantitative trait locus (QTL) analysis.

Two inbred strains (e.g., aggressive and nonaggressive) are crossbred, which gives them one homologue from each pair of homologous chromosomes. The progeny from the crossbreeding (F_1) are then mated to individuals from one of the original strains (in this case, strain N). These backcrossed progeny have one homologue that is a combination of parts of the chromosomes from each strain (formed when the F_1 mice formed gametes) and one homologue that is from one of the inbred strains. The backcrossed mice are ranked for aggressiveness. The recombinant chromosomes of the these mice are then analyzed for genetic markers that are unique for each of the inbred strains. For each marker, the mice are sorted into those that have the A-type DNA and those that have the N-type DNA. If the A-type animals possessing a particular marker are more aggressive than the N-type, that particular section of the chromosome represents a QTL that may contain a gene(s) contributing to aggressive behavior.

Source: M. Barinaga, *Science* 264:1691 (17 June 1994).

SUMMARY

There are two categories of questions in behavioral genetics: first, proximate questions concern the role and relative importance of inheritance in development and regulation of behavior, and second, ultimate questions concern genetics and behavioral evolution.

Most proximate questions about the relationship of genes to behavior are studied by use of traits that are affected by a single gene or a small group of genes. Sometimes classical genetic crosses can be used to identify the pattern of inheritance for behavioral traits, such as hygienic behavior in bees. More frequently, mutations that are naturally occurring, caused by mutagenic agents, or genetically engineered are used as research tools. By comparing animals with normal genes to those with mutant genes, researchers have been able to trace the function of particular genes with increasingly greater precision. Another technique to localize the effects of mutants on the phenotype is the use of genetic mosaics.

In order for behaviors to evolve, they must have at least some genetic basis. Most behavioral traits are caused by multiple genes rather than single genes, and methods have been developed to study the degree to which particular traits are under genetic influence. For example, in cross-fostering experiments, newborn animals are trans-ferred from their parent to another member of the same species or strain. If genetically similar animals raised in different environments are similar, genetic effects are important. Similarly, in humans, studies of twins and adopted children enable us to separate, to some extent, the effect of genes and the environment. Inbred lines, where all animals have the same genotype, are also a useful tool.

The last major topic of this chapter is quantitative genetic techniques. Here we make use of the equation $V_P = V_G + V_E + V_{GxE}$, which says that the phenotypic variance in a population is due to variance due to genetics, to the environment, and to the interaction between genes and environment. We can use this equation to estimate broad-sense heritability, or the proportion of the total variance in the phenotype that is a result of genetic variance. We can also calculate a more meaningful measure, narrow-sense heritability, that measures the proportion of total phenotypic variance due to additive genetic variance. Narrow-sense heritability can be measured with selection experiments or with measures of the resemblances among relatives, such as parent-offspring regression. Heritability estimates must be used with caution, as they are good only for the populations and the environments in which they are measured.

DISCUSSION QUESTIONS

1. Riechert and Hedrick (1990, *Anim. Behav.* 40:679–687) stud-ied two populations of web-building spiders. One population is in a habitat where predation is rare, whereas spiders in the other population are commonly eaten by birds. Those in the risky habitat are very cautious and remain in the safety of their retreats for a long time, whereas those in the safer habitat read-ily move back out onto their webs after a disturbance. Design a series of experiments to determine whether there is a genetic basis for this behavior in this species. Find the article in the library. Is your design similar to that chosen by the authors of the study?

2. In selection experiments that are carried on for generations, it is quite common to see the response to selection decrease from one generation to the next, until finally the population stops responding to selection at all. Why might this be so?

3. Two populations of fish of the same species inhabit neighbor-ing lakes that have identical environmental conditions, includ-ing fauna and flora. The population in one lake is genetically homozygous; the population in the second lake is heterozy-gous. Chemical pollution from agricultural practices in nearby fields enters both lakes and results in rapidly shifting, unstable conditions, producing changes in the fauna and flora. What can you predict about the ability of the two populations of fish to respond to these changes? Explain your prediction(s).

4. Among the techniques for studying the genetics of animal behavior that we discussed in this chapter were use of inbred strains, artificial selection, and cross-fostering. What are the advantages and disadvantages of each of these approaches? For each technique, give a brief example situation in which you would use it to test a question about behavioral genetics.

5. The study of behavioral genetics in humans is controversial. Studies of the heritability of intelligence have been especially so. What are some of the reasons why we should be careful about conclusions that we draw from human studies? In your opinion, does the knowledge we gain justify investing time and money into this work?

SUGGESTED READINGS

Articles on genes and behavior by Hall; Crabbe et al.; Takahashi et al.; and Plomin et al. In *Science,* Vol. 264, 17 June 1994, pp. 1702–39.

Boake, C., ed. 1994. *Quantitative Genetic Studies of Behavioral Evolution.* Chicago: University of Chicago Press.

Futuyma, D. 1998. *Evolutionary Biology.* Sunderland, MA: Sin-auer Associates.

Plomin, R., J. D. DeFries, G. E. McClearn, and M. Rutter. 1997. *Behavioral Genetics,* 3d edition. New York: W. H. Freeman and Company.

Weiner, J. 1999. *Time, Love, Memory: A Great Biologist and His Search for the Origins of Behavior.* New York: Knopf.

The Evolution of Behavior

As we have seen, much of behavior has a genetic basis, even though the specific genes involved and their exact contribution to a behavior pattern are rarely understood. As discussed in chapter 4, traits with a genetic basis are capable of evolving over time. In this chapter, we will discuss the evidence for behavioral evolution. We include both **microevolution,** or genetic change within populations or species, as well as **macroevolution,** or the evolutionary patterns of behavior recognizable above the species level.

A common misconception about evolution, particularly among nonbiologists, is that it is progressive: that it proceeds in a predetermined, linear direction toward an optimum phenotype. Even the language that is sometimes used to describe evolution suggests this idea (for example, when traits or species are referred to as "lower" and "higher," or "primitive" and "advanced"). However, evolution is not necessarily directional or progressive. For example, environments may vary over time, and a once-optimal phenotype may no longer be favored. In some cases, ancient lineages, such as horseshoe crabs, have remained essentially unchanged over vast spans of time, presumably because they are in relatively constant environments where they are successful. Some have even become less complex: internal parasites that have evolved from free-living forms have adapted to a new environment where elaborate sensory structures and complex nervous systems are unnecessary. Organisms may also be unable to reach an optimum phenotype: as we saw in chapter 5, genetic variation is necessary for evolution by natural selection to occur, and variation is not always present. It may also not be possible to optimize every part of an animal's phenotype, as an adaptation in one trait may mean a trade-off or compromise in some other area. Nonetheless, many of the examples in this chapter are of cases where we clearly see adaptation in response to a particular known selective force, as these are the most compelling evidence of natural selection.

MICROEVOLUTIONARY CHANGES IN BEHAVIOR

Microevolutionary changes occur within species. We discuss evidence for microevolutionary changes in behavior from two sources: (1) domestication and (2) observations of natural selection in the field.

Domestication and Behavioral Change

Humans have had long associations with numerous species of animals. Humans control all aspects of the lives of many domesticated species, including housing, feeding, socialization, and breeding. Because we prefer animals with particular traits (such as hogs with more meat, or horses that obey their riders), in many species we have selectively bred animals with particular traits. This process of artificial selection has led to some remarkable changes in the behavior and morphology of domestic animals. These changes provided a big piece of the evolutionary puzzle for Charles Darwin: he understood that natural selection could produce similar changes in animal populations in the wild. He spent many years breeding pigeons, included a lengthy discussion of domestication in *The Origin of Species*, and wrote several volumes on domestication.

Dog behavior provides a striking example of evolution under domestication. Consider the morphological differences between, say, pugs and Great Danes (figure 6.1), and breed-specific behavioral characteristics, such as sheepherding dogs that can maneuver their charges through complicated paths, and hunting dogs that will carry prey back to their masters instead of eating it themselves. As discussed in chapter 5, effective artificial selection requires both the presence of genetic diversity as well as selective breeding. Let's examine these two factors for the case of artificial selection in dogs.

It has long been suspected that dogs arose from wolves, which were the major carnivore in the northern hemisphere (Scott and Fuller 1965, Morey 1994), but the fossil record is inadequate to give much detail about when and how this happened. Molecular techniques have been used to reconstruct the history of the group through genetic analysis. By comparing the strings of nucleotide sequences across a variety of groups, we can deduce the branching order in which mutations arose, and determine which groups are likely to be closely related. These studies suggest that either wolves were domesticated in several places and at different times, or that dogs mixed with wolves repeatedly after their first domestication (Vilà et al. 1999). This accounts for the great genetic diversity in the domestic dog, which provided the raw material for the evolution of different dog breeds.

When did selective breeding lead to the diversity of dogs we see today? Bones of wolves have been found in association with hominids from up to 400,000 years ago (Clutton-Brock 1995). However, it probably wasn't until 10,000 to 15,000 years ago, when hominids shifted from nomadic hunter-gatherer societies to more sedentary agricultural soci-

FIGURE 6.1
At first glance, these two dogs may appear to be an adult and juvenile of the same breed, but they are actually adults of two different breeds, pug and Great Dane. This illustrates the evolution of the remarkable diversity in domestic dogs in a short span of time.
© Jeanne White/Photo Researchers.

eties, that dogs began to diverge from wild wolves (Vilà et al. 1997). Even after their divergence from wolves, dogs did not immediately diversify. For several thousand years, dogs were likely to interbreed with each other: they probably accompanied travelling humans as companions or for trade, and thus gene flow among dogs was very high and differentiation among different groups was low. It wasn't until recently that modern breeding practices led to the breeds we see today (Vilà et al. 1997). Evidence of this evolutionary history is reflected in the high diversity of mitochondrial DNA within dog breeds, which would have been lost if dog breeds had been under selection for a long time.

Scott and Fuller (1965) concluded that much of the difference in behavior we see between dogs and wolves, and among dog breeds, has been the result of selection on agonistic (aggressive) and investigatory behavior. For example, the bulldog breed, in the interest of an English sport, was selected for its tendency to attack the nose of a bull and hang on, in contrast with the more typical slashing attack from the rear used by wolves. Terrier breeds have been selected for their tendency to attack prey relentlessly, regardless of any injury suffered; the usual wolf pattern is to snap and then withdraw to avoid injury. In other breeds, selection has been to reduce agonistic behaviors; scent hounds and bird dogs are examples of peaceable animals that can be kept in groups in a kennel.

LEARNING THE SECRECTS OF INTERSPECIFIC COMMUNICATION

Up on a stage in front of a packed auditorium, Patricia McConnell conducts a simple demonstration. Before her is a wriggling young dog who periodically jumps up to try to lick her hands. First Patricia shows the audience an ineffective way to teach a dog not to jump: she reaches toward the jumping dog with her hand to push it down. It's natural for primates to push, and it's a signal we readily understand. However, to a dog, outstretched front legs are an invitation to play. That's the message this dog is getting, and it wags its tail even harder and jumps again.

Next Patricia demonstrates how to make this cross-species communication more effective. Dogs that want some personal space use a shoulder slam or body block to get it. This time when the dog jumps, Patricia swings her hips into it, keeping her hands at her side. The dog quickly gets the message, and soon sits back on its haunches when Patricia simply leans toward it. The dog rapidly learned what is wanted because the signal was naturally understandable.

Patricia McConnell, Ph.D., is a Certified Applied Animal Behaviorist and an Adjunct Assistant Professor at the University of Wisconsin, Madison. She runs a highly successful dog training business and hosts a nationally syndicated radio show on applied animal behavior. Her success comes from this key insight: it is vital to take into account the natural behavior of an animal in order to communicate with it effectively.

Auditory communication has been one of her main interests. For an undergraduate honors project, she analyzed the voice and whistle commands handlers give to sheepherding dogs to tell them to walk toward the sheep, circle, lie down, etc. Handlers had told her that it's not important which whistle is used for which behavior, as long as the sounds are easy to distinguish. However, when Patricia sorted the sonographs (pictures of the whistles) into piles in categories based on what the command meant, her eyes showed her patterns that her ears had not caught: certain types of commands shared the same structure. As she was thinking about her data while riding horseback, she had another epiphany when she said "Whoooaa" to the skitterish horse: long, continuous notes are used to soothe many animals, and short, repeated notes (like clicking to a horse) are used to make them speed up. She remembers the joy of the discovery: "The clouds parted and the angels sang!"

In her dissertation, Patricia looked across cultures. She recorded more than 105 animal handlers that spoke 19 different languages in communicating with rodeo and draft horses, obedience and sled dogs, camels, yaks, guard geese, and cats, among others. Across these languages she found the same patterns: short, repeated notes meant speed up ("Kittykittykitty!"); slow, unmodulated notes were used to soothe; and single, modulated notes ("Whoa!") were used to stop animals that were already active. Could it be that mammals are predisposed to respond in particular ways to these sounds? Patricia tested this idea by training naive Border collie and beagle puppies to either four short tones with a rising frequency that meant "come," and one long tone with a descending frequency that meant "stay," or vice versa. As predicted, a command with four short tones was more effective than the long signal at eliciting approach and increasing motor activity.

Many people assume that they will automatically know the best way to communicate with their dog, but interspecific communication is more difficult than one might imagine. As Patricia says, "Imagine doing an ethogram on humans as an alien species, and figuring out what something as simple as a smile means: it could be joy, nervousness, tension, or something else." A dog faces a similar problem as it tries to understand its trainer.

The business Patricia started with $100 when she was fresh out of graduate school has grown into a success. She says it is more intellectually and emotionally challenging than she would ever have dreamed to apply her animal behavior degree in this way, and is as proud of her abilities as an animal trainer as she is of her Ph.D. Days when she can solve a behavioral problem and prevent a dog from being put down are satisfying indeed.

Wolves and dogs search for prey rather than lying in wait and therefore have a well-developed investigatory repertoire. Some dogs, such as scent hounds, are bred for the ability to follow a trail. Others, such as bird dogs, use their visual and olfactory senses much more equally: after visually searching the ground, they locate prey by scent only when they are a few paces away, and then they freeze. Dogs must be trained for all of these tasks, but in each case, artificial selection has produced a phenotype that is predisposed to specific behavioral traits (Scott and Fuller 1965).

An experimental approach to the study of the effects of domestication has been ongoing for over 40 years in Russia (Trut 1999). Dmitry Belyaev believed that the patterns of morphological change that occur in domesticated animals are the result of selection directly on behavior rather than on morphological traits. He and his collaborators tested this idea by using artificial selection on a species that had never been domesticated, the silver fox (*Vulpes vulpes*). In their experimental design, the only trait that was used in choosing which animals would reproduce was their friendliness toward humans in a standardized test. The results are fascinating. As expected, foxes became more and more friendly over generations of selection, so that today, after 30 to 35 generations of selection, 70 to 80 percent of the foxes whimper to attract attention and are eager to make contact with humans. Morphological changes are also apparent in the selected line, many of which are reminiscent of domestic dogs: the foxes are more likely to have a white pattern on their forehead, brown mottled fur, floppy ears, shortened legs and tails, shorter and wider snouts, and a tail curled up in a circle (figure 6.2). In wild foxes, many of these traits are characteristic of very young animals but disappear during ontogeny. It is likely that selection for friendliness has acted on the timing of development. These results suggest that Belyaev was right: selection on behavior alone can lead to striking morphological differentiation.

Undesirable behavioral traits may also appear as a side-effect of selection for other traits. Hatchery-reared trout, for example, are much more aggressive than wild fish, and may devote so much energy to unnecessary aggression that their survival in the wild is decreased (Deverill et al. 1999). These examples illustrate that morphological and behavioral traits can be intertwined in both artificial and natural selection.

FIGURE 6.2
This fox is from a lineage that has been selected for friendliness. Note the curled tail, a morphological trait that increased dramatically in the domesticated line compared to nondomesticated foxes (Trut 1999).

From Lyudmila N. Trut. "Early Canid Domestication: The Farm-Fox Experiment." *American Scientist* 87: March-April 1999. Reprinted by permission of *American Scientist*, magazine of Sigma Xi, The Scientific Research Society. Photo courtesy of Dr. Lyudmila N. Trut.

Natural Selection in the Field

Observing Change Over Time

It is extremely difficult to observe evolution in action in the field. The best chance of seeing natural selection at work occurs when there is a strong selection pressure acting on a population with a short enough generation time so that humans can track changes from one generation to the next. The most famous case is the work of Kettlewell (1965) on **industrial melanism,** or changes in color in the peppered moth, *Biston betularia.* This moth is active at night and rests on bark during the day. Moths that do not match the background upon which they rest are more likely to be picked off by birds. Selection pressures on the coloration of this moth changed with the advent of the industrial revolution, when pollution killed the light-colored lichen that covered the bark of many trees. Thus, initially moths were subject to selection for light color, but as the lichen disappeared darker moths were more likely to survive and reproduce, and the frequency of dark moths in the population increased. As pollution control laws went into effect, the frequency of the dark morph declined again as the air became cleaner (Cook et al. 1986, Cook et al. 1999).

Grant and colleagues have been studying the action of natural selection on finches in the Galápagos Islands (Grant 1986). On one of the islands, Daphne Major, regular rainfall throughout the early 1970s produced abundant seeds, and seed-eating finches thrived. In 1977, a drought occurred, and the ground finches (*Geospiza fortis*) failed to breed, declining in numbers by 85 percent. The corresponding decline in seed supply was nonrandom, with small seeds becoming scarce and large, hard seeds remaining relatively common. The result was strong selection for birds with large beaks that could handle the seeds, and within a short period of time, the average beak size increased dramatically (Boag and Grant 1981). When an El Niño year brought spectacular rains in

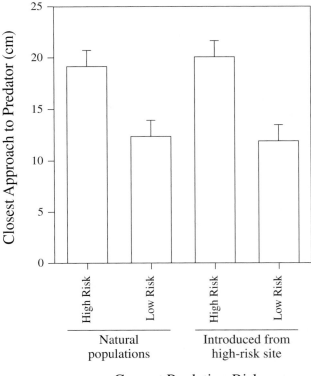

Current Predation Risk

FIGURE 6.3
Guppies were taken from a site with many predators and introduced into the Turare River, which has both high-risk and low-risk areas. After 100 generations, the introduced guppies behaved in an appropriate way: guppies in high-risk areas were cautious about approaching predators, and those in low-risk areas were not. They were similar to guppies in natural high- and low-risk populations.

Modified from A. Magurran. 1999. The causes and consequences of geographic variation in behavior. In S. Foster and J. Endler (ed.), *Geographic Variation in Behavior.* Oxford U. Press.

1983, selection pressures again shifted to favor small-billed birds that ate small seeds (Gibbs and Grant 1987).

An example of behavioral change in a short time period is described by Magurran (1999). In 1957, a researcher collected 200 guppies from a site on a river in Trinidad that also had predatory pike cichlids, and introduced them to a site that had neither guppies nor predators. Thirty-four years and 100 guppy generations later, fish were collected from this population and their offspring were reared in the laboratory. When tested as adults, these guppies showed reduced schooling behavior (schooling reduces predation on individuals), and were more inclined to inspect predators (a dangerous behavior which nonetheless provides useful information about the predator) than were guppies from sites where predators occurred (figure 6.3). This suggests that the descendents of the transplanted guppies had evolved to be less cautious.

Inferring Evolution from Geographic Patterns of Behavior

Often it is not possible to conduct long-term field studies that would enable us to measure evolution as it happens. Another useful approach is to compare populations of the same species that are found under different environmental conditions. In this way, we can identify possible ecological causes of adaptation (Foster and Endler 1999).

Agelenopsis aperta is a spider that lives in the southwestern United States. This species builds a web shaped like a flat sheet with an attached funnel, where it sits while waiting for prey (Fig 6.4). Susan Riechert (1999) has studied populations of this spider in two different habitats. The desert grassland is a harsh environment, very dry and hot with few insect prey. The riparian (streamside) woodland habitat is more tolerable: spiders live alongside a permanent spring-fed stream, with shelter from the sun and plenty of insects. However, birds also do well in this habitat, and the risk of predation by birds is very high.

These spider populations exhibit behaviors that appear to be adaptations to these environments. The grassland spiders attack a greater proportion of prey that hit their webs, fight more vigorously over web sites, and have larger territories than do the riparian spiders. Riparian spiders in an isolated Texas population show stronger antipredator behavior: they are more likely to retreat into a funnel when startled, and are slower to return to foraging after they have retreated.

Interestingly, in a riparian population of spiders in Arizona, not all behaviors of the spiders appear to be adaptive. For example, they escalate to high levels of fighting more often than would be optimal, and attack even prey that is not very profitable (thus unnecessarily exposing themselves to predation risk). A series of experiments established that gene flow from surrounding populations is probably the evolutionary force that keeps this population from being completely adapted to its environment. In chapter 4, we discussed how gene flow serves to homogenize populations, and how it can act in opposition to natural selection. *Agelenopsis* spiders move from populations in one habitat type to those in another, and thus prevent behavior from evolving to become perfectly adapted to the environment. If spiders are prevented from entering a population for just a single generation, a measurable shift in the behavior toward the optimum occurs (figure 6.4). This work provides a very clear example of the different evolutionary forces that interact on behavioral traits.

MACROEVOLUTIONARY CHANGES IN BEHAVIOR

Macroevolutionary change occurs above the level of the species. We discuss evidence for the macroevolution of behavior from the following sources: (1) fossil evidence; (2) phylogenetic reconstruction and the comparative method; (3) detailed studies of adaptive radiation; and (4) the role of behavior in speciation.

Behavior and the Fossil Record

The fossil record provides a great deal of data about the evolutionary history of organisms. Of course, behavior itself does not leave fossil remains. However, we can infer much about behavior from bones, teeth, horns, and tracks. Some species construct structures, or artifacts, that can be studied in fossil form (Hansell 1984).

One of the most fascinating fossil finds is that of dinosaur nests. Birds are descendents of dinosaurs, and researchers have wondered whether advanced parental care originated in the birds or in dinosaurs. Fossils show that some dinosaurs incubated their eggs much as modern-day birds do. Figure 6.5 shows a fossil *Oviraptor* in association with its eggs. The eggs are arranged in a circle in two layers, with the broad end toward the middle, suggesting that the parent manipulated the eggs into this configuration, much as is done by living birds (Norell et al. 1995).

Morphological structures can be good sources of information about behavior. For example, the horse lineage provides a particularly clear fossil record. Earlier horses were browsers, eating leaves and succulent plants. However, when the climate of North America became drier during the Miocene, grasslands became more widespread. Under this new selective regime, fast running and the ability to chew tough grasses became favored, and this is reflected in the foot and tooth structures (Stanley 1993).

Evidence of predation can be gained from drill holes in the fossilized shells of bivalve molluscs from the late Triassic (about 200 million years ago). These holes were made by carnivorous snails that penetrated the shells of their prey. Oddly, the habit seems to have disappeared, only to reappear some 120 million years later (Fursich and Jablonski 1984). Thus, what we might consider a real breakthrough in predatory technique did not persist; the presumed selective advantage of such a behavior pattern may not be a good predictor of its long-term survival. In further studies of snail predation on molluscan prey during the last 100 million years, Kitchell (1986) concluded that prey selection was nonrandom and governed by foraging rules similar to those used by present-day organisms.

Evidence about the behavior of extinct mammalian carnivores was discovered by Hunt et al. (1983), who excavated ancient burrow systems containing the skeletons of bear dogs from the early Miocene, 20 million years ago. These animals were about the size of a wolf or hyena and apparently used the dens much as do modern carnivores.

We can also glean information about mating systems by studying fossils. In mammals, the amount of sexual dimorphism in body weight and length is a good predictor of the degree of polygyny, the number of mates a male has (Alexander et al. 1979). Across a wide variety of modern species, the larger or heavier the male is relative to the female, the greater the number of females the male monopolizes. The degree of sexual dimorphism can sometimes be estimated from fossilized remains (Martin et al. 1994), which can give us an idea of the degree of polygyny in extinct species.

Studies of Adaptive Radiation

Adaptive radiation is the rapid evolution of new lineages. Isolated areas such as oceanic islands are particularly good places to look for examples of adaptive radiation. Because of their small size and large distance from a colonizing source, such places are likely to be invaded by few species (see

Arid Habitat

Extreme temperatures
Low prey levels
Low predation risk

Spiders are:
    Aggressive
    Not cautious

Well adapted to environment

Direction of
gene flow

Riparian Habitat

Moderate temperatures
Plenty of prey
High risk of predation

TX riparian: No gene flow

Spiders are:
    Not aggressive
    Cautious

Well adapted to environment

AZ riparian: Incoming gene flow

Spiders are:
    More aggressive and less
    cautious than predicted

Not completely adapted to environment

Experiment to Test the Effects of Gene Flow

Experimental Treatment

**Offspring of Spiders
Exposed to Treatment**

AZ riparian

Gene flow cut off
Predators excluded → Broad range of
phenotypes, both
aggressive and
cautious.

Shows importance of
predation as a selective force

Gene flow cut off
Exposed to predators → Not aggressive
Cautious

These are now better adapted
to their environment, and are
now similar to TX population

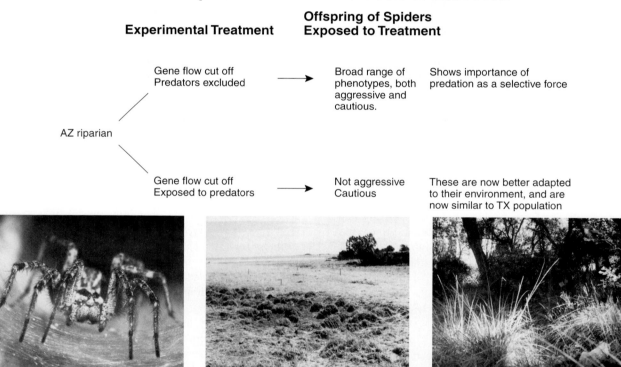

a.   b.   c.

FIGURE 6.4

Agelenopsis aperta spiders (a) are found in both arid habitats, (b) where conditions are poor but predators are few, and riparian habitats, (c) where conditions are good but there are many predators. Spiders from arid habitats are aggressive and will fight for the few good web sites, and are not cautious (they are quick to return to foraging, and will come out on their web even for low-quality prey). Spiders from Texas (TX) riparian habitats are not aggressive and are cautious, as predicted. However, spiders from Arizona (AZ) riparian habitats appear to not be well adapted, because of spiders immigrating from arid habitats. This hypothesis was tested by cutting off gene flow and manipulating exposure to predators.

6.4(a–c): From Riechert, S. E. 1998. Using behavioral ecotypes to study evolutionary processes. In S. Foster & J. Endler eds. *Geographic Variation in Behavior: Perspectives on Evolutionary Mechanisms.* Oxford University Press. Photo courtesy Dr. S E. Riechert.

FIGURE 6.5   **A fossil *Oviraptor* and its eggs.**
The arrangement of the eggs suggests that the dinosaur showed parental care.

Neg./Transparency no. 5789(3) Courtesy the Library, American Museum of Natural History.

chapter 14). In the relative absence of interspecific competition, new forms evolve rapidly. The best-studied case of such adaptive radiation is Darwin's finches, which we have already mentioned. Darwin was the first to study these birds on his 1835 voyage on *H.M.S. Beagle*. Biologists believe that a single species of finch (family Fringilidae) colonized the Galápagos and radiated into fourteen species in four genera (figure 6.6). Although adaptive radiation is evident in such traits as beak morphology and plumage, feeding and breeding behaviors also provide evidence of a common ancestry and adaptive radiation. The ancestral form was probably a heavy-beaked ground finch that fed on seeds, from which evolved: modern ground finches; cactus-feeding finches with long, decurved bills; insectivorous forms, including the famous woodpecker finch, which uses a tool in the form of a stick in its beak to probe underneath bark for insects; and slender-beaked warblerlike finches that feed on small insects.

Phylogeny and the Comparative Approach

Ethologists have long used the technique of comparing different species in order to understand the evolution of a trait,

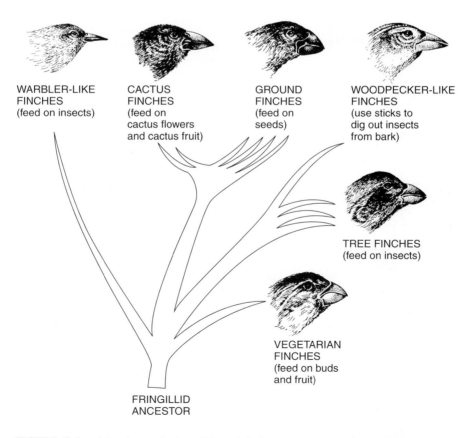

FIGURE 6.6   **Adaptive radiation of Darwin's finches on the Galápagos Islands.**
Small, isolated islands with low colonization rates serve as useful sites to study evolution and adaptation. It is likely that a single species of finch gave rise to all these types.

Source: Data from A. Sato, C. O'hUigin, F. Figueroa, P.R. Grant, B.R. Grant, H. Tichy, and J. Klein. 1999. Phylogeny of Darwin's finches as revealed by mtDNA sequences. Proceedings of the National Academy of Sciences USA

although the details of the approach have changed in recent years. First we will begin with a classic example of a comparative study, then we will discuss modern-day advances.

A Classic Comparative Study: Balloon Flies

According to Kessel (1955), in 1875, Baron Osten-Sacken was visiting the Swiss Alps, and he noticed groups of bright silvery flashes in the shadows of the fir forest. Believing that they were silvery insects, he captured some in a net but found instead that they were dull-colored flies in the family Empididae, or empidids. However, along with the flies, he had netted packets of filmy material (figure 6.7). At first, the function of these balloons was unclear: various people hypothesized that they acted as surfboards or perhaps as warning signals to predators. Instead, it appears that males use the balloons to attract mates. The balloons apparently are not valuable to the females in any way: they contain no nutrients. How could something like this have evolved?

One can gain insight into the possible evolutionary routes that a behavior has taken by examining other species that have similar behaviors. For example, Kessel (1955) described possible evolutionary stages in the origination of this behavior.

- *Stage 1:* In the majority of empidid species, both sexes capture insects independently, and no presentation of prey is associated with mating. These flies sometimes prey on conspecifics; when the male attempts to copulate, he may be eaten by the female.

- *Stage 2:* The male captures a prey item and presents it to the female as a nuptial gift. He copulates with her as she eats the prey instead of him. Several empidid species, as well as many other insect species, use nuptial gifts in mating. In an unrelated insect, the scorpionfly *Hylobittacus apicalis* (order Mecoptera), a male with a prey item advertises to females by means of a pheromone, a chemical signal. But, sometimes a male assumes the behavioral posture of a female; he approaches a male that has a prey item, lets the male try to copulate, then steals the prey item and either eats it or uses it to attract a female (Thornhill 1979).

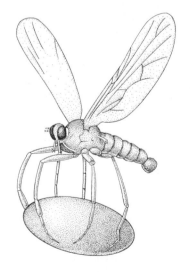

FIGURE 6.7   Male balloon fly (*Hilara sartor*) carrying balloon. Males with balloons fly in a swarm from which a female selects a mate. That male gives the balloon to the female prior to copulation.

- *Stage 3:* Rather than taking the prey and searching for a female, some male empidids join other males, each with a prey item, in an aerial dance. The prey, Kessel suggests, is now a stimulus for mating rather than a distraction to avoid mate cannibalism. The female enters the swarm and selects one of the males. Several empidid species represent this stage.

- *Stage 4:* In many species of the empidid genus *Hilara,* the male wraps the prey loosely with some silken threads, an action that seems to quiet the prey.

- *Stage 5:* In several species of the genus *Empis,* from the western United States, the male applies elaborate silken wrappings to the prey, which then resembles a balloon. When male and female meet in midair, the male transfers the balloon to the female and climbs on her back. The pair alights on a plant, and the female rolls the balloon about, probes it, and eventually consumes the prey item while the male copulates with her. The balloon itself stimulates mating.

- *Stage 6:* The male catches a small prey and may consume its fluid so it is no longer edible; he then constructs a complex balloon. The female accepts the balloon and plays with it during copulation but gets no meal from it. *Empis aerobatica* may represent this stage.

- *Stage 7:* The prey item is very small, of no food value to either sex, and pieces of it are plastered at the front end of the balloon. The balloon is now the sole stimulus for copulation. *Empimorpha geneatis* illustrates this stage.

- *Stage 8:* The *Hilara sartor* male gives the female a balloon that has no prey at all. Kessel suggested that this behavior was the final stage in the series.

Arranging existing species with different behaviors along a continuum (a transformation series) is a very simple way of doing a comparative study. It offers some insight as to how a complex trait may have evolved in stages. However, there are problems with this method. For example, we do not know whether the balloon species went through every stage during the course of its evolution. In addition, we may not have the order of the stages right. As discussed at the start of this chapter, traits do not always evolve from simple to complex. For example, maybe the balloons evolved first to make the male appear larger and more attractive, and later evolved into nutritional gifts.

Ideally, we would also like more information about the current functions of the traits in all of the species, as this can give us insight into the selection pressures that may have produced each behavior, as well as the selection pressures that currently act upon it. For example, for stage 2 species, Kessel suggested that the nuptial feeding of the female initially functioned to reduce the chances of the male's being consumed by the female during mating. Thornhill (1976) suggested an alternative function based on parental investment. Male insects may invest in their offspring by providing nutrition to their mates in the form of glandular secretion, prey captured by the male, regurgitative food offering (figure 6.8), or even the male himself. Females may select mates on the basis of the quality of food the males offer. In another example, recent work has shown that a balloon-bearing species, *Empis snoddyi* (stage 8), faces conflicting selection pressure. The males that are most likely to be successful at attracting mates are large males with intermediate-size balloons. Intermediate-size balloons represent a trade-off between long-range attrac-

FIGURE 6.8 Regurgitative food offering during copulation. Food offerings during copulation occur in many kinds of animals, such as in the stilt-legged fly (family Micropezidae) shown here. This behavior may be a form of parental investment by the male. By giving the female a protein-rich meal, the male may increase the survival of his offspring.
Source: © Edward S. Ross.

tion of females with larger balloons, and improved flying efficiency with smaller balloons (Sadowski et al. 1999).

The Use of Phylogeny in Studying Behavioral Evolution

Clearer answers about the evolution of traits can be gathered with the use of a **phylogeny,** the history of descent of a group of taxa. A phylogenetic approach is helpful with several of the problems that confront students of behavioral evolution. First, it can help us ascertain the order in which traits evolved. Second, it can help us to distinguish **homology** (similarities due to common ancestry) from **analogy** (similarities due to convergent evolution).

How do we reconstruct a phylogeny? It has long been recognized that closely related species are more likely to share traits than are distantly related species. Darwin introduced the metaphor of a tree to describe the relationships among taxa. The trunk, or the earliest ancestor, grows into separate limbs, limbs give rise to branches, which in turn give rise to twigs. Some branches have many twigs, or descendents, while others have few. Some branches fall off, representing extinction. The tips of the twigs represent species that are alive today. Twigs on the same branch should share characteristics because of common descent. Today, many scientists prefer the term bush instead of tree, as the many branches of a bush provide a better metaphor for evolutionary patterns than the single trunk of a tree; however, both terms are commonly used. The problem that faces modern-day systematists is to reconstruct the shape of the tree or bush from present-day species and fossils.

The first step in phylogenetic reconstruction is to identify particular **characters** of interest. A character is an inherited trait. The characters of all the specimens of interest are examined and their **character states** are described. For example, in spiders, the plane in which the jaws are moved is a character. There are two possible character states: jaws can either move from side to side or front to back. Morphological traits were historically the most often used characters, but in recent decades, new techniques have led to increased use of genetic characters. For example, a precise spot on a gene is a character, and the particular nucleotide in that spot is the character state. The assumption is that character states of more closely related species are likely to be similar, because they have been inherited from a more recent common ancestor.

After characters are scored, we can reconstruct a phylogeny. One well-known method is **cladistics,** which was developed in the 1950s by Willi Hennig. This method relies on identifying **monophyletic** groups, or groups of taxa that share a more recent ancestor with each other than they do other groups. **Polyphyletic** groups, in contrast, are taxa that arose from several different ancestors.

Using Behavior to Reconstruct Phylogeny

Beginning with Darwin and continuing with the early ethologists, students of behavior have recognized that phylogeny and behavior are intertwined. Scientists have used this association in several ways (Brooks and McLennan 1991).

Scientists have been interested in knowing whether behavioral patterns can be used like morphological or genetic characters in constructing a phylogeny. It is not obvious whether behavioral traits are useful for this purpose: as we have already seen, behavior is often under a great deal of environmental influence, and might be too variable to shed light on phylogenetic relationships. The ethologists were of the opinion that behavior could be useful. One of the cornerstones of their research, beginning in the early 1900s, was the idea that behavioral patterns could be treated like anatomical features. They "dissected" behavioral patterns to learn their "anatomy." They also demonstrated homologies in behavior, which are patterns shared by species through descent from a common ancestor. For instance, Van Tets (1965) studied birds of the order Pelecaniformes, scoring species for the presence or absence of behavioral traits, such as head wagging, and comparing the pattern with the phylogeny as inferred from morphology (figure 6.9). He concluded that behavioral data could be used to provide accurate assessments of phylogeny. Recently, his data were further analyzed by Kennedy et al. (1996), who constructed trees from the behavioral data and compared them to trees based on morphology and genetics, and found that all produced similar results. Similarly, Paterson et al. (1995) used 72 behavioral and life history characters to construct phylogenetic trees for 18 species of albatrosses, petrels, and penguins, and found that these trees were similar to those based on molecular data.

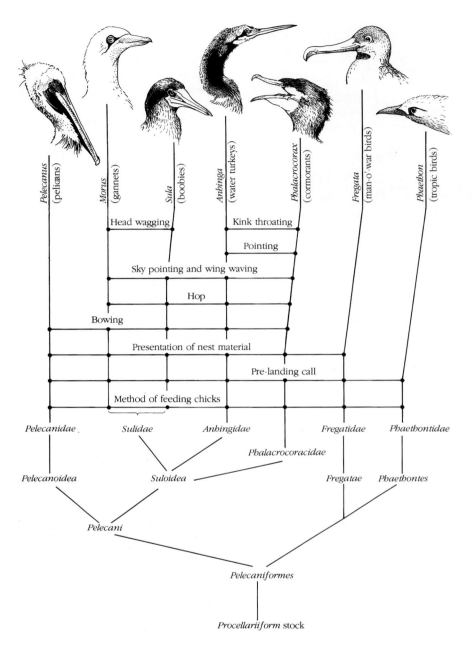

FIGURE 6.9   Behavioral traits and phylogeny in the order Pelecaniformes.
The top half of this figure shows the distribution of behaviors among genera. Dots indicate
that the behavior is present in that genus. In the bottom half of the figure is a phylogenetic
tree based on morphological traits. Note that genera that are closely related based on morphology also share the most behaviors, whereas those distantly related share the fewest.
Source: G. F. Van Tets, *"A Comparative Study of Some Social Communication Patterns in the* Pelecaniformes," in
Ornithological Monographs 2:1–88, copyright 1965 American Ornithologists' Union, Washington, D.C.

Using Phylogenies to Test Evolutionary
Hypotheses about Behavior

The examples in the previous section demonstrate that, at
least in some cases, behavior can provide data that are useful
in constructing phylogenies. Phylogenetic trees can also be
used to test hypotheses about the evolution of behavioral
traits. With this approach, a phylogenetic tree is constructed,
generally using morphological and/or genetic traits. Then,

behavioral characters are mapped onto the tree. (It is important not to construct the tree using the behavioral characters
under test, because that would be circular reasoning.) By
examining the tree, it is often possible to determine the direction of evolution of a behavioral trait. It is also possible to
identify selection pressures that are associated with the evolution of particular traits. In order to do this accurately, we
need to know how many times particular traits have evolved.
This is illustrated in figure 6.10.

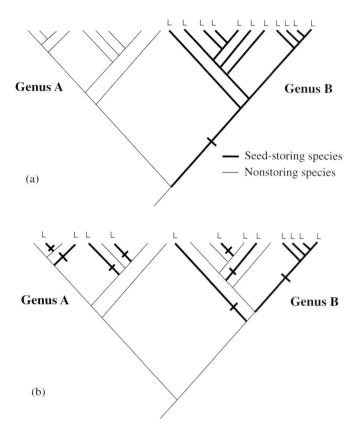

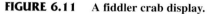

Seed-storing species
Nonstoring species

(a)

(b)

FIGURE 6.10

Imagine that you are interested in the relationship between food storing and spatial learning: You hypothesize that animals that depend on finding seeds that they store will perform better in laboratory tests of spatial learning. If you examine 10 seed-storing species and find that they all are good learners, and 10 nonstorers and find out they are poor learners, you might conclude that your hypothesis is correct. However, perhaps these are not independent samples: If all the seed-storers are closely related, and all the nonstorers are closely related, there is likely to have been only one evolutionary event (a), and your sample size is one, not 20 (L indicates species able to learn; crossbars indicate evolutionary events). A different phylogeny would generate another answer: In (b), there are eight evolutionary events, and that would be your sample size. (Modified from Rosenheim 1993.)

An example of the use of phylogeny in understanding the evolution of a trait comes from tiger moths. Some species use high-frequency sound in their mating displays. High-frequency sound is also used as a defensive display against bats and other predators with high-frequency hearing: these moths are distasteful to predators, and the sound deters them. A phylogenetic analysis shows that the use of sound in defense against predators came first, and sound subsequently took on a courtship function in some species (Weller et al. 1999).

Fiddler crabs provide an example where our initial perception of the direction of evolution was apparently incorrect. Fiddler crabs signal to one another in visual displays, often waving their claws and dancing in elaborate patterns (figure 6.11). Species vary in the complexity of their dis-

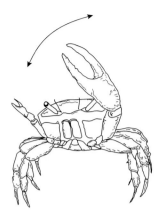

FIGURE 6.11   A fiddler crab display.

plays, and it was hypothesized that those with simpler displays, from the Indo-west Pacific, were ancestral to American species with more complex displays. However, species with the hypothesized "derived traits" of display behaviors appear to be phylogenetically ancestral. Instead of a simple evolutionary pattern from simple to complex, it is likely that behavioral complexity has arisen multiple times in the fiddler crabs (Sturmbauer et al. 1996).

Phylogenies have been used to examine the evolution of structures that animals make. Many species construct objects for prey capture, shelter, or mate attraction—for example, spider webs, caddisfly cases and nets, beehives, and bird nests. It is relatively easy to examine these semipermanent structures and to compare closely related species (Hansell 1984). Winkler and Sheldon (1993) constructed a phylogeny with DNA data for 17 swallow species, and then placed nest characters on it. The primitive nesting mode appears to be burrowing. Construction of mud nests originated once, and mud-nesting species have primarily diversified in Africa, where the climate is dry and mud is a feasible construction material. Mud nests seem to have increased in complexity from simple cups to fully enclosed nests.

Unfortunately, it can sometimes be very difficult to construct a reliable phylogeny for a particular group, and the exact structure of the phylogeny sometimes depends on the methods used to construct it (Losos 1999). However, the methods for constructing phylogenies from DNA data are improving rapidly. When robust phylogenies become available for diverse animal groups, we can address many additional interesting questions about behavioral evolution.

Behavior and Speciation

When an ancestral species gives rise to new species (**speciation**), the process usually starts with the interruption of gene flow by some sort of physical barrier. This is called **allopatric speciation** (Mayr 1963). If the populations diverge from one another while they are separated, they may not be able to interbreed again if they are reunited: they have become separate species.

Barriers to interbreeding between groups of organisms are called **isolating mechanisms,** and can be divided into two types. **Prezygotic isolating mechanisms** are those that occur before the formation of zygotes. **Postzygotic isolating mechanisms** occur after fertilization, and are generally physiological in nature (the zygote dies, or the hybrid has reduced viability or fertility).

For our purposes, prezygotic isolating mechanisms are more interesting, because many of these are behavioral. For example, members of different species may be active at different times of the day or the year, or may use different parts of the habitat, and would thus be unlikely to come into contact with one another.

In another type of prezygotic isolating mechanism, animals may contact each other but not mate: either one partner fails to court, or the other partner is not receptive. Which cues do animals use in species recognition? In many cases, morphological traits convey information. This is demonstrated by a recent study of cichlid fishes in Lake Victoria (Seehausen et al. 1997). Different species live sympatrically in the lake, but are sexually isolated by mate choice. Mate choice is determined on the basis of coloration. In recent years, human activity has led to eutrophication of the lake, and it is now much more turbid. The fish cannot see colors as well, and now mate choice is relaxed: fish are more likely to mate with members of other species.

Behavioral cues are also very important in courtship and species recognition. For example, there are several species of green lacewings (insects with long, delicate wings; figure 6.12) that look exactly alike to our eyes. The males and females sing to each other in a duet composed of low-frequency sound. Females respond most strongly to the songs of their own species (Wells and Henry 1992). Other examples of courtship behaviors are described in chapter 12.

We often see that closely related species that are **sympatric** (in the same location) are more likely to have divergent courtship signals than those that occur in separate locations. Littlejohn (1965) recorded the songs of two species of Australian tree frogs, *Hyla ewingi* and *Hyla verreauxi.* In zones of sympatry, the songs are quite distinct; in zones of allopatry, the songs are more similar (figure 6.13). In playback experiments, Littlejohn and Loftus-Hills (1968) gave gravid (egg-bearing) females a choice of two songs. They chose their own species' song when the choice was between the two species in sympatry, but showed no preference between songs of allopatric species. This pattern has been seen in numerous other animal taxa (reviewed in Wells and Henry 1998), including moths, crickets, birds, and fishes.

Reproductive isolating mechanisms do not necessarily evolve specifically for the purpose of keeping groups of animals apart, but rather are likely to occur as incidental by-products of change between groups (Paterson 1985, Templeton 1989). For example, if two groups become geographically separated, they may diverge in their courtship behaviors because of different selection pressures in their two environments. Then, if the groups come into contact

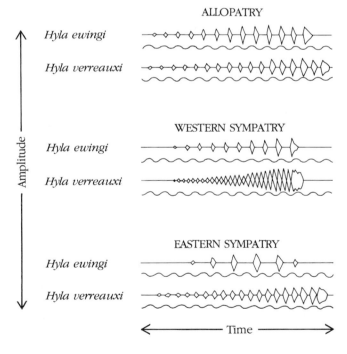

FIGURE 6.13   Oscillograms of songs of two Australian frog species.
The amplitude of each note of the song is shown on the vertical axis, and time is shown on the horizontal axis. Under each song recording is a 50-cycles-per-second reference line. The songs of the two species are similar when comparing individuals caught in areas where the two species do not overlap (allopatry). Where the two species do overlap (sympatry), song differences are noticeable.

FIGURE 6.12   A green lacewing.
Many species are morphologically indistinguishable, but differ in their courtship songs.
© Stephen Dalton/Photo Researchers.

again, they will fail to recognize each other as potential mates. However, in some cases, prezygotic reproductive isolating mechanisms may be under direct selection, in a phenomenon called **reinforcement of prezygotic isolation.** This idea has fluctuated in popularity since Dobzhansky elaborated upon it in the 1940s, and has gained favor in recent years with new theoretical and empirical work (Noor 1999). In reinforcement, hybrids may not be as fit as either parental species, so natural selection strengthens sexual isolation: animals that do not mate with the "wrong" species will have higher fitness, so behaviors that keep animals from interbreeding will increase in the population. Reinforcement may be one cause of the pattern that we saw previously, where sympatric species show greater discrimination in mate choice than allopatric species, but there are other possible causes (Noor 1999).

SUMMARY

Many sources of information about behavioral evolution are available to us. To understand microevolutionary changes that occur within species, we can examine domesticated animals. These provide a good demonstration of the power of selection to act on behavior: many morphological and behavioral traits have changed over a relatively short time via the process of artificial selection. For example, the vast differences among breeds of domestic dogs have come about very quickly.

Natural selection can occasionally be observed in action in a field population; the most well-known example is that of Darwin's finches, where weather-related selection pressures vary from year to year and result in measurable changes in beak morphology and diet. Given the difficulty of observing selection as it occurs, many researchers study geographic variation across populations of a single species. Here, the response of recently diverged groups to different selection pressures can be studied, and sometimes, such as in the case of the funnel-web spider *Agelenopsis aperta,* the opposing effects of gene flow and natural selection can be teased apart.

Macroevolutionary changes occur above the level of the species. Fossil evidence provides some clues about the behavior of extinct animals, such as the nesting habits of dinosaurs and the evolution of predation in snails. However, fossils that give us information about behavior are rare, so we must often rely on other methods for understanding the history of a group. Behavioral traits can be helpful in constructing a phylogeny, or the history of descent of a group of taxa. In addition, we can plot behavioral traits on phylogenies that have been constructed with morphological and DNA characters. This can tell us the evolutionary history of particular behavioral patterns.

DISCUSSION QUESTIONS

1. Compared to dogs, cats have changed relatively little over the course of domestication. What are three possible explanations of why that might be?

2. Imagine that you see two species that have similar behavioral patterns, but you are not sure whether these behaviors are analogous or homologous. How would you distinguish between these alternatives?

3. Tree frogs have both a mating call and an aggressive call. Which call do you believe would be more likely to differ between two closely related species that are sympatric? Why?

4. Here is a phylogenetic tree of sparrows based on genetic data (based on Irwin 1988). Indicated on the tree are the number of songs birds have in their repertoire. What does this tell us about repertoire size and the direction of evolution? Does this surprise you? Why or why not?

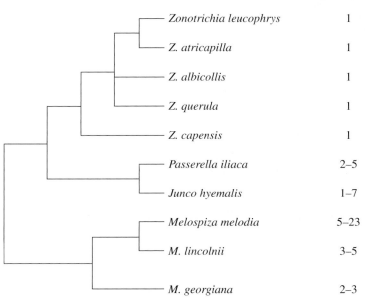

Repertoire size

Zonotrichia leucophrys	1
Z. atricapilla	1
Z. albicollis	1
Z. querula	1
Z. capensis	1
Passerella iliaca	2–5
Junco hyemalis	1–7
Melospiza melodia	5–23
M. lincolnii	3–5
M. georgiana	2–3

Based on figure 1 and table 1 of Irwin, R. E. 1988. The evolutionary importance of behavioural development: The ontogeny and phylogeny of bird song. *Anim. Behav.* 36: 814–24.

SUGGESTED READINGS

Brooks, D. R., and D. A. McLennan. 1991. *Phylogeny, Ecology, and Behavior: A Research Program in Comparative Biology.* Chicago: University of Chicago Press.

Foster, S. A., and J. Endler, eds. 1999. *Geographic Variation in Behavior: Perspectives on Evolutionary Mechanisms.* New York: Oxford University Press.

Futuyma, D. J. 1998. *Evolutionary Biology.* 3d edition. Sunderland, MA: Sinauer Associates.

Martins, E. P., ed. 1999. *Phylogenies and the Comparative Method in Animal Behavior.* Oxford: Oxford University Press.

Serpell, J. 1995. *The Domestic Dog: Its Evolution, Behaviour, and Interactions with People.* Cambridge: Cambridge University Press.

Weiner, J. 1994. *The Beak of the Finch.* New York: Alfred A. Knopf.

PART THREE

Mechanisms of Behavior

In the last section, we discussed the genetic bases of behavior. Now we look at the "black box" between the genes and the phenotype that is ultimately produced: what are the mechanistic underpinnings of behavior? In chapter 7, we discuss how stimuli are perceived by the sensory system, and how the nervous system coordinates behavior. In chapter 8, the focus is on the role of hormones in organizing and releasing behavior. Next we examine the control of behavioral cycles by biological clocks (chapter 9). The development of behavior throughout the lives of animals is discussed in chapter 10. Finally, we address how behavior is modified through experience during the course of an animal's life (chapter 11). By the end of this section, we hope that you will have a good understanding of proximate causes of animal behavior.

Opposite: ©PhotoDisc

7

CHAPTER

The Nervous System and Behavior

E very living organism is continually bombarded by environmental stimuli. Animals have sensory receptors to receive information, some type of nervous system to sort out and interpret the stimuli, and effector (e.g., muscle and endocrine) systems to produce appropriate behavioral responses. In this chapter, our first topic is the basic unit of the nervous system, the neuron, and how it transmits action potentials. Following that, we consider different types of nervous systems and their evolution. We next examine sensory systems, perception, and the filtering mechanisms that allow animals to sort out irrelevant stimuli and respond to only important stimuli. We'll examine the nervous control of movement. We'll look at research techniques used to study the nervous system. Finally, we consider some in-depth examples of the relationships between the nervous system and behavior.

THE NERVE CELL

The nervous system of an organism is composed of specialized cells called nerve cells, or **neurons** (figure 7.1). A typical neuron consists of the nerve cell body, which contains the nucleus; a long **axon** with fingerlike **synaptic processes** at the end; and **dendrites** emanating from the cell body. Information moves through the dendrite toward the cell body, and away from the cell body through the axon. Nerve cells vary tremen-

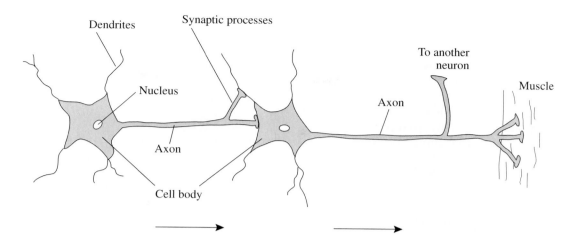

FIGURE 7.1   A generalized pathway in the nervous system.
Neurons are the structural building blocks of nervous systems. The arrows indicate the direction in which information travels.

dously in form, including differences in the position of the cell body, the length of the axon, and the number and kinds of dendrites. Several types of **glial cells** insulate and provide metabolic support for the nerve cells. The nervous system can be divided into the **central nervous system (CNS),** consisting of the brain and nerve cord, and the **peripheral nervous system,** consisting of all the nerves that connect to the CNS.

There are three keys to understanding the transmission of information within the nervous system. First, a neuron at rest is negatively charged compared to its surroundings. Second, stimulating a neuron causes a wave of change in its charge, called an **action potential,** that is conducted along its length. Finally, neurons communicate with one another at specialized junctions called **synapses.** We'll look at each of these in turn, through the example of a male moth that is searching for a mate.

The Neuron at Rest

Many animals are faced with the problem of locating mates that are a long distance away, and many species rely on chemical cues called pheromones in mate finding. Female silk moths release a pheromone called bombykol from a special gland in their abdomen. Male moths are extremely sensitive to this pheromone, and have feathery antennae equipped with sensory cells specialized for detecting it.

Imagine a male silkworm moth (figure 7.2) at rest on a leaf without any bombykol molecules near him. If we examined the neurons associated with the sensory cells on his antennae, we would find that there is a different concentration of ions (atoms and molecules with an electrical charge) inside and outside of the cell. The membrane of a neuron is semipermeable: some ions or molecules can pass easily in and out, others can never pass, and still others pass only when special channels open. A neuron at rest contains relatively low concentrations of sodium ions (Na^+) and relatively high concentrations of potassium ions (K^+) compared

FIGURE 7.2   Male silkworm moth.
The male moth, here the giant silkworm moth, *Hyalophora euryalis*, has specialized receptors on its antennae for detecting chemical attractants, or pheromones.
Source: © Edward S. Ross.

to its surroundings. Specialized active transport (energy requiring) proteins imbedded in the membranes of neurons constantly pump Na^+ out of each neuron and K^+ into each one. Some of the Na^+ and K^+ "leaks" back along its concentration gradient into and out of the cell, respectively, but at different rates. In addition, some large negatively charged proteins are trapped within the cytoplasm of the neuron. The end result of the unequal distribution of ions is that the inside of a neuron at rest (i.e., one that is not being stimulated), such as the bombykol receptor cell of our male moth, is negatively charged (about–70 millivolts) relative to the outside. This is called the **resting potential** of the neuron, and the neuron is said to be **polarized.** The resting potential sets the stage for the movement of ions across the membrane of the moth's neuron when it is stimulated.

Action Potentials

Now imagine that air currents containing bombykol molecules emitted from a nearby female waft past the male moth. When the molecules make contact with the receptors on the sensory cells of his antennae, the neuron becomes slightly less polarized (**depolarization**). If the membrane potential reaches a certain threshold polarity, **voltage-gated sodium channels** in the membrane open. This allows Na^+ to rush in through the open channels across the membrane, down the concentration gradient. The Na^+ that enters will cause that localized region inside of the neuron to become positively charged with respect to the outside (figure 7.3). At that point, **voltage-gated potassium channels** open and allow K^+ to rush out of the cytoplasm until the original resting potential is established. In fact, in many neurons, more K^+ than is necessary to reestablish the resting potential leaves the cell so that very briefly in that location the inside becomes **hyperpolarized;** that is, the inside of the cell is even more negatively charged with respect to the outside than during the resting potential. After a brief period of hyperpolarization, the resting potential is reestablished. The Na^+/K^+-pump constantly works to maintain the unequal ion concentrations.

The action potential is transmitted down the axon. When the Na^+ ions initially rush into the neuron through the voltage-gated channels, some ions diffuse laterally within the cytoplasm and depolarize the neighboring regions of the axon. This causes the adjacent sodium voltage-gated channels to reach the threshold polarity and open. Therefore, the action potential is a self-propagating wave of localized depolarization that sweeps along the membrane. An action potential can spread in either direction from the point of origin, but they normally travel from the cell body to the terminals of the axon. This is because the point of origin is typically near the origin of the axon at the cell body. Once a region of a membrane has experienced an action potential, it goes through a brief refractory period (when the membrane is hyperpolarized) and the voltage-gated sodium channels cannot open. This keeps action potentials from propagating themselves repeatedly back and forth along a neuron.

Synapses

Now the signal must be transmitted from the one neuron to the next. This occurs in a region called a synapse. There are several types of synapses. **Chemical synapses** use a chemical intermediary called a **neurotransmitter** to communicate between neurons. Here we will describe an excitatory chemical synapse, where the presynaptic neuron (the neuron before the synapse) causes the postsynaptic neuron (the neuron after the synapse) to be more likely to fire. When the action potential reaches the end of the axon in the presynaptic neuron, it causes vesicles filled with neurotransmitter molecules to dump the neurotransmitter into the synaptic space between the two neurons (figure 7.4). The neurotransmitter molecules diffuse across the synaptic space and attach to specific receptor proteins on the postsynaptic neuron. Channels in the membrane of this neuron open, and positively charged molecules flow in. These positive ions spread out from the point of origin near the synapse. If enough positive ions flow into this area so that the depolarization exceeds the threshold for an action potential, a self-propagating action potential is initiated and it sweeps down the postsynaptic axon. Not all chemical synapses cause the postsynaptic neuron to fire; some cause it to be less likely to fire.

In this fashion, the male moth's receptor cells send action potentials to the brain in response to stimulation by bombykol. The brain integrates these action potentials with signals from other neurons to generate a behavioral response. At the end of any particular nerve "path," the axon of the last neuron synapses with an **effector,** which is a muscle cell, gland, or organ that gives the final response to the stimulus. In a nerve-muscle junction (figure 7.5), the transmitter sub-

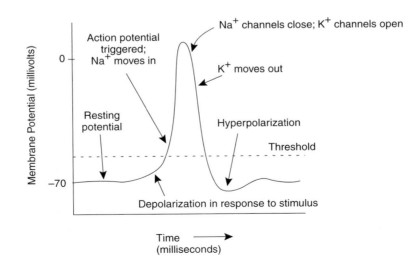

FIGURE 7.3   **The change in membrane potential over the course of an action potential.**

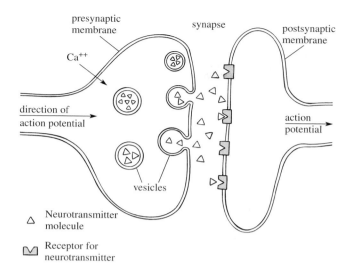

△ Neurotransmitter
    molecule

ᙯ Receptor for
    neurotransmitter

FIGURE 7.4   An excitatory chemical synapse between two neurons.

When an action potential arrives at the end of a synaptic process of an axon, the permeability of the presynaptic membrane to calcium ions increases, and the ions enter the cytoplasm. The increase in Ca^{++} causes membrane-bound vesicles filled with neurotransmitter (e.g., acetylcholine) to move to the end of the process, fuse with the membrane, and dump the neurotransmitter molecules into the synaptic space between the two neurons. The neurotransmitter molecules attach to specific receptors on the postsynaptic membrane and open ion channels. If sufficient numbers of positive ions (e.g., Na^+) cross the postsynaptic membrane to depolarize it, a self-propagating action potential may result. An enzyme (e.g., acetylcholinesterase) present in the synaptic space close to the postsynaptic membrane breaks down the neurotransmitter so that its effects are relatively short-lived. Many insecticides work by inhibiting the enzymatic activity of acetylcholinesterase so that acetylcholine builds up in the synapse and the insects lose control of their muscle contractions.

stances from the synaptic processes at the end of the axon cause an action potential to sweep across the muscle membrane, resulting in a contraction, and usually movement by the animal. In this case, the moth's response is to begin flying upwind, where he will be likely to locate and mate with the female that is emitting the bombykol.

A variety of neurotransmitters have been identified (see Delcomyn 1998 for more detail). Some are amino acids (the building blocks of proteins), such as glutamate and aspartate. Others are amines, which lack the acid group that amino acids carry. These include histamine, octopamine, dopamine, serotonin (which mediates alertness in vertebrates), and norepinephrine and epinephrine (which are important in the "fight or flight" behavior exhibited by animals in danger). Other neurotransmitters are peptides (short chains of amino acids), that work in conjunction with other transmitters. Still others are acetylcholine, an amine, found at the neuromuscular junction in vertebrate muscles as well as in the central nervous system of insects.

At an **electrical synapse,** neurons are even closer together, and the currents generated by an action potential are strong enough to depolarize an adjacent neuron above threshold. Electrical synapses are exceptionally fast, and do not require a chemical intermediary. These are often found between neurons that trigger an escape reaction such as when an earthworm snaps back into its burrow when touched. Chemical synapses, on the other hand, are easier to modulate. By changing the nature of the receptor protein with which the neurotransmitter binds, a neuron can vary the way it responds to a particular input. Such changes are thought to underlie learning and memory.

Some synapses are excitatory, as in the bombykol example, and make the postsynaptic neuron more likely to fire (undergo an action potential). Other synapses are inhibitory, and make the postsynaptic neuron less likely to fire by causing it to become hyperpolarized (more negative than

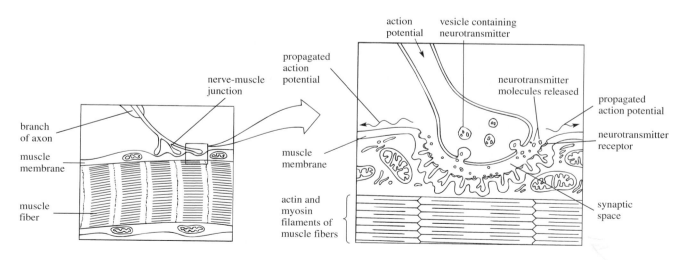

FIGURE 7.5   A nerve-muscle junction.

A nerve-muscle junction is similar to a synapse between two neurons. When neurotransmitter molecules attach to receptors on the muscle membrane, they cause an action potential to sweep down that membrane. The action potential causes the actin and myosin filaments that form the muscle fibers to interact so as to cause a contraction.

normal). A single neuron may receive many inputs from a number of different neurons. Often a signal from only one synapse is not enough to make a neuron fire. However, signals can be summed from many synapses, or even from a single synapse firing multiple times. In addition, synapses can have immediate, rapid effects on the postsynaptic neuron, or they may alter the neuron's response to further input. The latter, called **neuromodulation,** occurs over a longer time. The integration of all these signals is modified by the electrical properties of the neuron, which can also change over time. All of this adds up to a complex but very flexible system that forms the basis for animal behavior.

PHYLOGENETIC VARIATION ACROSS NERVOUS SYSTEMS

In this section, we'll examine broad patterns in nervous system structure and function across different taxa, and discuss some of the evolutionary implications of these patterns. Our generalizations are, of necessity, rather broad.

Simple Systems

Unicellular organisms do not possess a formal structure that could be called a nervous system, but they do exhibit irritability, the wavelike passage of a response to a stimulus from one point of a cell to another. In multicellular animals, increased complexity and longer distances between parts of the organism necessitate a more efficient form of communication. Over evolutionary time, specialized tissues and structures developed for receiving and transmitting information within the organism. These changes in the nervous system were critical for the evolution of complex, flexible, and integrated behavior.

The **nerve net** represents one of the earliest specialized neural organizations that conducts messages between cells. The nerve net appears to be a random arrangement of nerve fibers. It is characteristic of members of the phylum Cnidaria, including sea anemones, hydra, and jellyfish. Synapses occur at junctions between the neurons, but unlike synapses in the nervous systems of most other organisms, those in cnidarian nerve nets are nonpolarized and conduct impulses in all directions. An impulse that begins at one point thus spreads out and is passed on to most of the other neurons in the net. However, in spite of this seemingly disorganized nervous system, animals with nerve nets can perform simple behaviors. For example, anemones can capture food with their tentacles and transfer it to their mouths, and (slowly) contract and stretch their bodies to climb upon rocks and shells. Jellyfish swim by rhythmically contracting their bell. However, there is no centralized processing of information.

Some animals with other forms of nervous systems retain some form of nerve net—for example, peristaltic contraction in the gut wall of vertebrates is controlled by a nerve net. Starfish (phylum Echinodermata, class Asteroidea) have small pincers on their backs that help keep small organisms from settling there, and the actions of these pincers are controlled by local nerve nets.

Radially Symmetrical Nervous Systems

Two trends marked the evolution of slightly more complex nervous systems. First, different types of neurons became specialized to serve different functions, and second, arrangements of neurons became more ordered, forming nerve tracts and integrative centers. One interesting arrangement is that of the echinoderms (phylum Echinodermata), the starfish, brittle stars, and sea urchins; these are not closely related to other major groups. Adults have a radially symmetrical nervous system with a circular nerve ring in the central portion of the body, and nerve tracts (collections of axons) extending into each arm (figure 7.6). Within each arm is a network of sensory and motor neurons. The radially symmetrical nervous system permits more flexible and more complex behavior than does the nerve net, but control is still decentralized.

In starfish, coordinated movements of the five arms are partly controlled by impulses that travel through **sensory neurons** and **interneurons,** through the central ring, to **motor neurons.** Alternation of excitation and inhibition of muscle tissue by motor neurons produces extension and retraction of the arms and results in locomotion. Starfish are capable of righting themselves if placed upside down and of wrapping themselves around a bivalve mollusc (such as a clam) to open the shell and reach the meat inside; the interneurons and central ring mediate these coordinated movements.

Animals with radially symmetrical nervous systems also evolved specialized types of sensory receptors, including systems for receiving contact, taste, and general chemical signals. The combination of a refined system for sensing the environment and coordinated responses means that starfish and related organisms possess a greater degree of flexibility and diversity of behavioral responses than do animals with nerve nets, although there are still limitations.

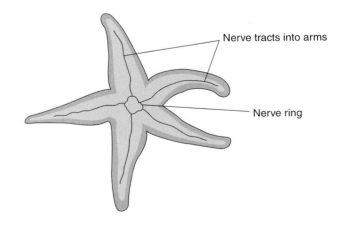

FIGURE 7.6   **The nervous system of an echinoderm.**

Bilaterally Symmetrical Nervous Systems

Most animals that exhibit radial symmetry are sessile or drifting, but bilaterally symmetrical animals, with bilaterally symmetrical nervous systems, generally have directional movement. These include several phyla of worms (including annelids and flatworms), molluscs, arthropods (including insects and crustaceans); and chordates. Associated with the evolution of a bilateral body plan was **cephalization,** or the evolution of a distinct head region where the sense organs are concentrated. The head is generally at the anterior end of the animal so it is first to detect changes in the environment as the animal moves along. In conjunction with cephalization, there has been a refinement and diversification of sensory organs; we will see these in more detail later in this chapter.

The ventral surface of bilaterally symmetrical animals generally became specialized for movement. Motor control also became more centralized, with coordination of different areas of the body. For example, flatworms (phylum Platyhelminthes) have two anterior **ganglia,** or clusters of neurons and synapses. Earthworms (phylum Annelida) have a single ventral nerve cord that runs the length of the body (figure 7.7). Both sensory and motor neurons innervate each segment, and the ventral nerve cord coordinates the movements of different segments. This enables earthworms to be very effective burrowers. (To gain an appreciation of the difficulty of the burrowing lifestyle, imagine how difficult it would be to burrow into a soft beach with your hands tied behind your back and your ankles tied together.) In phyla in which articulated exo- or endoskeletons evolved, we also see very precise motor control. For example, arthropods (phylum Arthropoda) have a jointed exoskeleton capable of controlled movement, and generally have larger ganglia and better sense organs than the wormlike phyla.

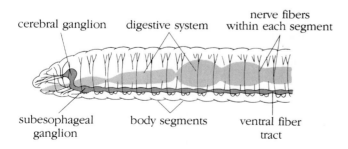

cerebral ganglion   digestive system   nerve fibers within each segment

subesophageal ganglion   body segments   ventral fiber tract

FIGURE 7.7   Earthworm nervous system.
Earthworms are bilaterally symmetrical organisms, with bodies divided into many segments. The earthworm nervous system has several important features: the concentration of neurons in ganglia toward the anterior or head region, the ventral nerve tract, and sensory and motor neurons that connect to the ventral nerve cord within each body segment allowing for coordinated movement.

The Vertebrate Nervous System

Vertebrate nervous systems are especially complex. The evolutionary changes in vertebrate nervous systems were characterized by further concentration and enlargement of the central nervous system, including further cephalization, and development of many interconnections between nerve cells. Many behavioral phenomena characteristic of vertebrates are intimately tied to these changes in the nervous system. First, the complexity of behavior patterns and the general flexibility of behavioral responses of vertebrates exceed that of most other animal groups. Second, because of changes in the structure of the nervous system and in the morphology of neurons themselves, vertebrates' responses are generally faster than those of invertebrates, though some invertebrate responses occur as fast as any vertebrate response (e.g., escape responses of cockroaches). A third characteristic of vertebrates is, in general, a greater capacity for information storage. This feature of the vertebrate nervous system has enormous ramifications for behavior: the retention of past experiences may influence future behavior. Although many invertebrates are capable of learning, it is most developed in vertebrates.

Throughout vertebrate evolution, the spinal cord has changed little in structure, but the brain has changed dramatically, increasing in both size and complexity. In fish, for example, the ratio between the weight of the brain and the spinal cord is close to 1:1, but in humans it is 55:1 (Hickman et al. 1998). Similarly, other areas of the brain vary greatly in size and complexity across the vertebrate classes (figure 7.8). Functions of the major areas of the brain are shown in table 7.1. The hindbrain includes the medulla, which controls breathing and heart rate. This is an evolutionarily ancient part of the central nervous system. Another part of the hindbrain is the **cerebellum,** which controls posture and balance. It is small in amphibians and reptiles but highly developed in birds and mammals, as might be expected from the lifestyle of these groups. The cerebellum also is involved in learning motor tasks. The forebrain includes the **cerebrum,** which is subdivided into the paleocortex and neocortex. The paleocortex is the main site of the **limbic system,** which is associated with behavior related to drives such as sex and feeding, as well as emotion. The neocortex (or cerebral cortex) is especially large in mammals. The cerebrum interprets sensory input and organizes motor output.

SENSORY RECEPTORS

Now we will examine separate processes of the nervous system in more detail, beginning with sensation. The term **sensation** refers to the process of transducing stimuli (e.g., sound, light, heat, mechanical forces, or molecules) into action potentials. The stimuli may be internal or external to the animal. Sensory receptors within muscles and organs provide information on muscle tension, body position, and

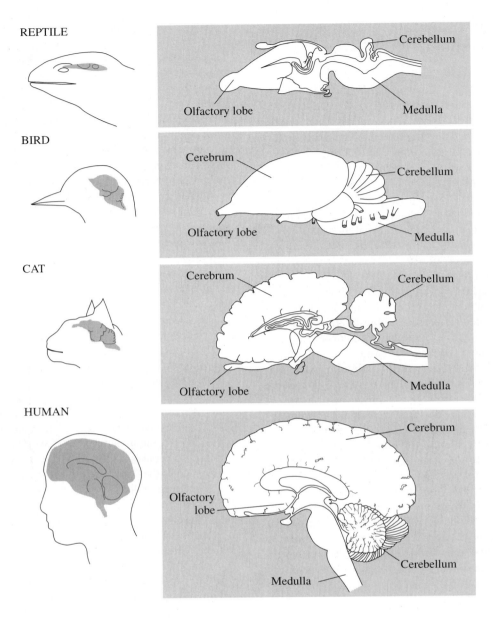

REPTILE

BIRD

CAT

HUMAN

FIGURE 7.8   Vertebrate nervous systems.
Nervous systems of vertebrates are characterized by centralization, cephalization, and the development of many interconnections within the nervous system. Differences among vertebrate species in the size of the brain and cranial capacity are evident here in the drawings of reptile, bird, cat, and human nervous systems. Note the variation across taxa in the size of the olfactory lobe (used in the sense of smell) and the size of the cerebellum (important in balance).

internal conditions, whereas other sense organs provide information about the environment.

Sensory processing begins in the specialized membranes of receptor cells. In most cases, the specific stimulus that the cell is sensitive to causes a conformational change (change in shape) in a specific receptor protein within the membrane. This conformational change usually results in a change in the permeability of the membrane to an ion(s) and, therefore, a change in polarity of the receptor cell. If the change is great enough, it generates an action potential and excites or inhibits neighboring neurons.

Sensory receptors are classified according to the type of stimulus to which each responds, and we will discuss, in turn, chemoreceptors, mechanoreceptors, electroreceptors, thermoreceptors, and photoreceptors. As we will see, animals vary tremendously in the type, number, arrangement, and sensitivity of their sense organs. The information available in the environment is perceived in dramatically different ways by different animals. This idea is encapsulated in the ethologists' word Umwelt, or the sensory view of the world. It is vitally important that we keep this in mind as we design experiments in animal behavior.

TABLE 7.1	The Major Divisions and Functions of Parts of the Vertebrate Brain	
Divisions	**Main Component in Adults**	**Primary Function**
Forebrain		
Telencephalon	Cerebrum	Motor coordination of voluntary muscle movements
		Sensory perception and integration
Diencephalon	Thalamus	Integrates sensory information
	Hypothalamus	Appetitive drives and homeostasis
Midbrain		
Mesencephalon	Optic lobes (tectum)	Integrates visual information with other sensory inputs
		Relays auditory information
Hindbrain		
Metencephalon	Cerebellum	Equilibrium, muscle, posture
	Pons	Links cerebellum with other brain centers
Myelencephalon	Medulla	Regulates heart rate and respiration

Modified from Hickman et al. 1998.

Chemoreceptors

The receptors that are designed to sense specific molecules are generally categorized as either **gustatory** (taste) or **olfactory** (smell) receptors. Gustatory receptors allow animals to identify appropriate and inappropriate foods. In insects, gustatory receptors are generally housed in hairlike structures, such as the chemosensory hairs on the feet of blowflies, whereas in vertebrates gustatory receptors are generally in the mouth and tongue. In gustation, different chemical substances are transduced via a variety of mechanisms. For example, sour or salty substances directly influence ion channels in the membrane of the receptor cell, whereas the effect of sweet substances is usually mediated by a pathway with a second messenger, a cascade of chemical reactions that results in the opening of an ion channel.

Olfaction often plays a role in animal communication, both within and between species. Olfaction is often used in long-distance mate finding, as we saw in the moth example. Male moths may have as many as 75,000 receptor cells per antennae, and are able to detect the pheromone even when the concentration is as low as several molecules per million (Schneider 1974). Males of other moth species and most other animals are apparently not affected by the chemical attractant and do not respond to its presence in the air. These species-specific pheromones can be used as bait to capture pest insects (e.g., Japanese beetle traps, figure 7.9). Animals also frequently use olfaction to determine the reproductive status of conspecifics. For example, some ungulates exhibit a behavior called flehmen, where they lift their upper lip and draw air into a special vomeronasal organ in the nasal region. The vomeronasal organ is highly sensitive to chemicals in the urine of conspecifics. Olfaction is also used to distinguish kin from nonkin. Golden hamsters (*Mesocricetus auratus*) apparently learn their own odor, and match it with the odor of strangers they meet to judge relatedness (Mateo and Johnston 2000); this has been dubbed the "armpit effect" (Dawkins

FIGURE 7.9   A Japanese beetle trap that uses pheromones as bait.
Beetles collide with the flat fins on the top of the trap and then drop into the capture vial below. These traps are very effective at attracting beetles; the disassembled trap is shown spilling beetles onto the ground. In fact, the pheromone is so effective that traps may attract more beetles into an area than they kill, so it is much better to place traps away from valuable plants.
© Scott Camazine/Photo Researchers.

1982). In addition, many predator and prey species are able to recognize each other by odor cues. For example, foraging storm petrels (seabirds) are attracted from long distances to dimethyl sulphide, a chemical released by zooplankton (Nevitt et al. 1995).

In olfaction, the sensory receptor is protected by a layer of either fluid (insects) or mucus (vertebrates), and odor-causing chemicals must first be dissolved in this fluid. A diversity of biochemical pathways are involved in transduction; in both invertebrates and vertebrates, most involve a second messenger system. Decoding the signals from olfaction is a complex process: animals may have many different receptor proteins with a range of response properties, and the pattern of response across many neurons can provide the animal with information about the chemicals that it detects (reviewed in Delcomyn 1998).

Mechanoreceptors

Mechanoreceptors are sensitive to touch, pressure, stretching, sound, and gravity, and monitor both internal and external stimuli. Mechanoreceptors vary greatly in their complexity. For example, many touch and pain receptors are simple undifferentiated nerve endings in the connective tissue of vertebrate skin. Other touch receptors are more complex, such as Pacinian corpuscles that detect vibration. Invertebrates also have mechanoreceptors; leeches, for instance, have touch and pain receptors in their skin.

Other mechanoreceptors include the more specialized **sensilla,** hairlike extensions on the exoskeleton of arthropods that contain a nerve cell, sensitive to pressure or wind. At the end of the chapter, we will discuss an example of escape responses in insects in more detail. Even more specialized are **hair cells** of vertebrates (figure 7.10). Hair cells are found in the inner ear organs that regulate equilibrium,

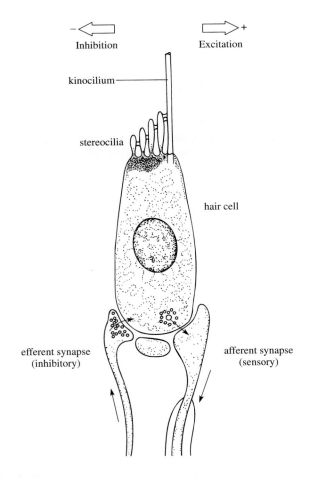

FIGURE 7.10   Hair cell of a vertebrate.
If the stereocilia of a hair cell are displaced toward the kinocilium, the hair cell is depolarized, and causes an increase in action potential frequency down the afferent nerve to the central nervous system. If they are displaced away from the kinocilium, the cell is hyperpolarized, and action potential frequency decreases. In addition, signals from efferent neurons can hyperpolarize the cytoplasm of the hair cell (and many other sensory cells). This reduces the frequency of action potentials, and therefore the sensitivity to the stimulus.

and in the lateral line systems of fishes and amphibians that provide information about their own movements and other animals in the vicinity. In addition, hair cells are important in vertebrate hearing.

The mammalian auditory system is particularly complex (figure 7.11). Sound energy enters the auditory canal, where it causes the eardrum (tympanic membrane) to vibrate. The eardrum vibrates three small bones in the middle ear, which in turn vibrate the membrane of the oval window of the **cochlea.** When the oval membrane vibrates, the waves are transmitted to the fluid in the cochlea, and depending on the frequency of the waves, specific parts of the **basilar membrane,** that bears the hair cells, vibrate. The hair cells in that region of the membrane are stimulated when the membrane vibrates. The hair cells have specialized membranes at their tips called **stereocilia.** When the stereocilia are bent, ion channels in the hair cell either open or close, depending on the direction of bending, and the polarization of the hair cell changes accordingly. Hair cells are nonspiking, so instead of an action potential they generate a graded signal that correlates with the intensity of the stimulation. Thus, a signal is sent from that part of the cochlea to the brain that indicates a particular intensity of sound.

In addition to mechanoreceptors that respond to external stimuli, there are several types of internal receptors that monitor stimuli within the body. For example, **stretch receptors** respond to the lengthening of muscles and connective tissue. Signals from these receptors protect animals from stretching tissues to the point of damaging them. In addition, various **proprioreceptors,** such as those found in the inner ear, are also found in muscles and tendons and relay information about the position and motion of the body.

Electroreceptors

Some fishes have hair cells on the surface of their bodies that have lost their cilia and are modified for sensing electric currents. Some species (for example, certain sharks) are able to detect the weak electric currents produced by the musculature of prey fishes.

Other species can generate a stronger electric field around their bodies using modified nerve or muscle tissue. Weakly electric fish generate electrical signals that are used in both electrolocation and communication. For example, the electric fish *Gymnarchus niloticus* inhabits muddy rivers in parts of tropical Africa, where visual navigation is difficult. Special cells near the rear of the fish emit a continuous stream of weak impulses, and porelike structures on the head contain sensory receptors that are stimulated by extremely small changes in the electrical field around the fish (Machin and Lissman 1960; Hopkins 1974). The sensory receptors detect moving or stationary objects in the environment as distortions in this electrical field. The electric fish can thus navigate in its murky environment with little difficulty.

A comparative study (see chapter 6 for details about this approach) shows that the ancestral form of the discharge of

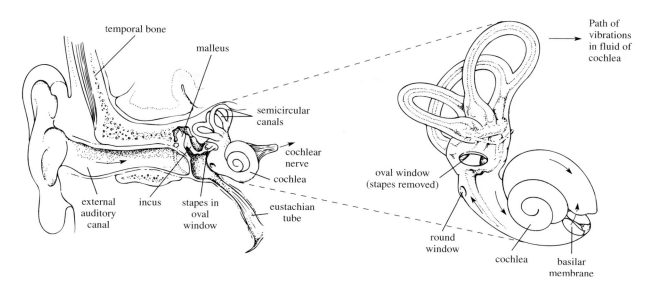

FIGURE 7.11   The inner ear of a human.
Auditory signals are conveyed by three small bones (malleus, incus, and stapes) from the eardrum to the oval window at the base of the cochlea. The vibrations in the cochlear fluid move through one side of the coiled cochlea and then back out along the other side of the cochlea where their energy is dissipated at the round window. Depending on their frequency, the vibrations cause specific sections of the basilar membrane to vibrate, thus stimulating hair cells that send signals to the CNS via the cochlear nerve.

the electric organ in a particular group of electric fish was an intermittent, simple pulse. Several modern-day groups have evolved a biphasic (two part), pulsed waveform. Which selection pressures might have led to this change? Stoddard (1999) trained a predator, an electric eel, to approach any playback of an electric field. It was half as likely to approach a normal biphasic waveform as one in which the second phase was experimentally deleted, suggesting that biphasic signals may have evolved to reduce predation risk. Another twist to this story is that in many species, there are sex differences in the second part of the signal: males extend the duration of the second phase, especially during the peak hours of reproduction (Franchina and Stoddard 1998). Thus, this signal may have initially evolved for crypsis, and then was further modified by sexual selection (see more on sexual selection in chapter 17).

Thermoreceptors

Some nerve-cell endings are specialized for signaling temperature either on the skin or internally. Many animals have specialized thermoreceptors that fire action potentials when warmed ("warm" receptors) and some that fire when cooled ("cold" receptors). For example, a rattlesnake has infrared receptors in specialized pits on its face. These receptors are sensitive to remarkably small changes in the surrounding temperature and allow the snake to detect prey. Because of the orientation of the facial pits, the snake also is able to detect the direction of the prey. In addition, internal thermoreceptors allow animals to maintain the proper body temperature, metabolically and/or behaviorally.

Photoreceptors

Photoreceptors use excitable **photopigments** that are associated with the receptor membrane. The energy of the light causes a structural change in the pigment molecules. This change initiates a cascade of biochemical reactions that results in a neural signal. This basic process is similar across animals, but there is a tremendous amount of variation in how the photoreceptors are arranged, their sensitivity, and how the information they receive is processed.

The simplest photoreceptors have no lenses to focus images, but simply detect the presence or absence of light. For example, the familiar flatworm, the planaria *Dugesia,* has large eyespots that do not form images but allow their owner to avoid bright light.

Many insects and crustaceans have a more elegant type of eye, or **compound eyes,** composed of numerous **ommatidia** that each have a lens and a retina (figure 7.12). Each ommatidium is oriented in a slightly different direction. Animals with compound eyes are very good at picking up motion, as a moving object crosses the visual fields of many ommatidia and stimulates many sets of retinal cells. (In fact, many arthropods are rather poor at seeing stationary objects. Flying insects often zip back and forth in front of an object that they are investigating—by moving their bodies relative to the object they can produce the same effect as if the object were moving.) Insects vary tremendously in the number of ommatidia they have. For example, dragonflies (order Odonata) depend heavily on vision for feeding, territory defense, and mating. They have up to around 9,000 ommatidia, arranged so as to get a wide view around them. However, they cannot see equally well in all directions and are

COMPOUND EYE

DETAIL OF
OMMATIDIUM

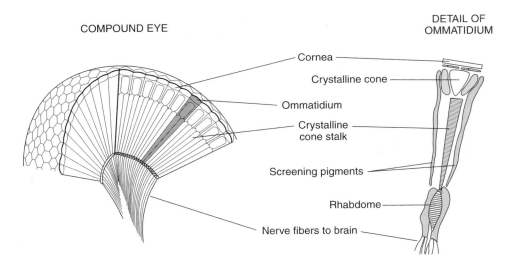

FIGURE 7.12   **Compound eye and ommatidium of an insect.**
The compound eye of many insects is comprised of many individual photoreceptive cells called ommatidia. Each ommatidium "sees" one point of the visual field, and together these points form an image.
The corneal and the crystalline cone focus light on the rhabdome, which contains visual pigments. Light-
sensitive screening pigments disperse under bright light so that all the light that reaches the rhabdome
enters through only one ommatidium. In dim light, the screening pigments contract so light from many
ommatidia strike each rhabdome. This makes the eye more sensitive, but the image quality worsens.

more likely to detect territorial intruders coming from the
front rather than other angles (Switzer and Eason 2000).
Their visual system thus has implications for their behavior,
such as where they spend time on their territories. Other
insect species have far fewer ommatidia, and this is correlated with a lesser dependence on vision.

Not all arthropods have compound eyes. Spiders (class
Arachnida) have simple eyes, with only one visual unit (in
contrast to compound eyes), but in abundance; most species
have four pairs. Many spider species have very poor vision,
and rely mostly on vibration (mechanoreception) for information about their environment. In contrast, jumping spiders
(family Salticidae) have one pair of very good eyes in the
front of their cephalothorax. These form images and, in some
species, are reported to have the acuity of primate eyes.
Other pairs of eyes detect motion but do not form images,
and we see this in the spider's behavior: when something
moves behind a jumping spider, it quickly turns its body to
orient its image-forming eyes toward the stimulus.

Members of the phylum Mollusca are a wonderful
example of adaptive radiation. They vary tremendously in
their lifestyles and sensory systems, ranging from burrowing
clams with very little in the way of sense organs, to
cephalopods (squid and octopuses) that have remarkably
good vision, with eyes that are functionally similar to human
eyes. Cephalopods have complex visual communication.
They have chromotaphores, or cells in the skin that contain
pigment molecules. These are under nervous control, so
color change is very rapid. Different color patterns are used
in aggression, courtship, and other behaviors. An individual
can simultaneously give one signal to, say, a rival on one
side, and a different signal to a receptive female on the other.

Vertebrates have photoreceptors called **rods** and **cones**
(figure 7.13). Rods are more sensitive to light, and are useful
in night vision. Cones are less sensitive, but are tuned to particular wavelengths of light, and are responsible for color
vision. The photopigments are in the outer segment of the rod
and cone cells. Under dark conditions, sodium channels in
the rod or cone are open, and sodium flows into the cell while
potassium flows out. When a photon of light hits the photopigment, it initiates a cascade of chemical events that result
in the closing of some sodium channels. Sodium can no
longer enter the cell at the same rate as before, but potassium
still leaves the cell. This leads to a hyperpolarization of the
cell. This is a graded response: the brighter the light, the
more hyperpolarized the cell becomes. The rods and cones
send synaptic outputs to other rods and cones and to horizontal and bipolar cells. Horizontal cells connect rods and
cones in different regions of the retina. Bipolar cells send
information to amacrine cells. All these cells produce graded
responses with the exception of amacrine cells, which generate action potentials that are sent to ganglion cells, which
in turn communicate with the brain. This complex network
of cells does more than passively transmit signals from the
retina. The information that the brain receives is shaped by
the specialized ways in which these cells respond to sensory
input, as we will see in an example later in the chapter.

In vertebrates, we again see patterns in the placement
and sensitivity of photoreceptors that relate to the animal's
ecology. The eyes of nocturnal animals have rods rather than
cones for better night vision. Prey species (such as rabbits)
generally have their eyes on the sides of their heads, which
enables them to spot potential danger from a number of
directions. Eyes arranged side by side, such as primate and

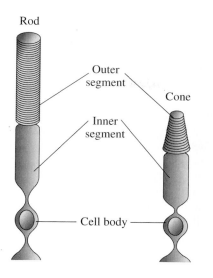

Rod

Outer
segment

Cone

Inner
segment

Cell body

FIGURE 7.13   Rod and cone cells from a vertebrate eye.

cat eyes, allow for better depth perception because both eyes see the same object but from slightly different angles.

It is extremely important when designing experiments to keep in mind that what an animal can see is not necessarily what we see—in fact, it's likely to be very different. This is true of any sensory system, but we are particularly likely to fall into error with the visual system because we are such visual animals ourselves. For example, many animals can see ultraviolet light, although we cannot. When it was first discovered that bees see in the UV range, excited researchers immediately gathered up many seemingly plain white flowers and put them under UV light. Lo and behold, many flowers had petals that were patterned in the UV; "plain white flowers" often appear, to a bee, more like a landing strip with patterns pointing to the nectar and pollen. Birds can also see in the UV, so it is important to keep this in mind when designing experiments; for example, painting feathers to investigate the effect of feather color on mate choice may also change UV-reflecting characteristics.

Processing and Filtering Sensory Information

An action potential is an all-or-none phenomenon: a nerve cell either fires or it does not (Hodgkin 1971). The propagation process appears to be nearly identical for all neurons. If nerve impulses are all similar, how do neurons transmit any information about the nature of a stimulus or its strength? Information about the nature of a stimulus lies in the wiring of the nervous system: different types of environmental information are picked up by different sensory receptors and carried along separate neural paths to particular areas within the central nervous system for decoding and interpretation. The separate but integrated systems for different types of information (for example, information from the eyes or the ears) prevent confusion about whether the stimulus was orig-

inally photic (relating to light) or auditory (relating to sound). The strength of stimulation may be communicated in at least two ways: (1) the frequency of spikes may vary with signal intensity, or (2) several neurons carrying the same type of information may fire simultaneously. Thus, intensity information may be encoded by an individual neuron's firing frequency and by the number of neurons firing.

Once information has been coded into action potentials, it may be conveyed throughout the animal's nervous system. Perception is the analysis and interpretation of sensory information. This decoding process is the function of the ganglia, bundles of neurons found in many invertebrates, or in the central nervous system in vertebrates. In the simplest nervous systems, little or no decoding takes place; a stimulus-response system is "wired" into the basic plan of the animal. The way an animal perceives a particular stimulus is a function of the type of sensory information it receives, the structure of its nervous system, and the past experiences permanently encoded in its nervous system.

How can we tell what an animal perceives? We can use both physiological and behavioral measurements. For example, if we place electrodes in nerves that connect a sensory end organ to decoding centers, we can record whether action potentials are initiated when we provide stimulation. We can study the natural behavior of animals that is elicited by particular stimuli (for example, we can test whether a male moth can detect a particular concentration of pheromone by watching whether it begins to fly upwind). Sometimes animals can be taught to perform a particular behavior when they detect a stimulus. For example, Goldsmith et al. (1981) trained hummingbirds (*Archilochus alexandri*) to go to feeders illuminated with particular wavelengths of light. By presenting pairs of feeders with different wavelengths and monitoring the choices of the birds, the researchers were able to determine the birds' sensitivity to small differences in light.

A major problem faced by the sensory system of animals is not just picking up information, but editing the available data. Every animal is continuously bombarded by an enormous number of diverse stimuli. Receiving, processing, and responding to all of these stimuli would be impossible, and natural selection has acted on sensory systems so that they respond best to the most pertinent stimuli. Ethologists developed the concept of **sensory filters,** one peripheral and the other central. Peripheral filters function at the level of the sensory receptors. As we have seen, each species receives information from a limited number of sensory modalities, and within each modality, species differ strikingly in their range of sensitivity. These limitations of the peripheral sensory system act as an initial screen of the environmental stimuli available to an animal.

Central-filtering processes within the nervous system sort out incoming information, select relevant or important stimuli for further action, and eventually produce a particular response. Central filters are components of what are called **innate releasing mechanisms,** or IRMs, by ethologists. Evidence for the existence of these processes comes from several sources. When we present an animal with a variety of stimuli,

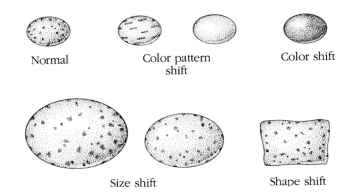

Normal   Color pattern   Color shift
                shift

Size shift   Shape shift

FIGURE 7.14   Eggs used to test retrieval behavior in herring gulls.
The eggs shown here, which vary in size, shape, color, or color pattern were used by Tinbergen to test the stimulus qualities that are most important in eliciting egg retrieval behavior in herring gulls. Tinbergen found that size and shape were more critical than other factors.

all of which are detected by its sense organs, it responds selectively: its behavior is based on only one or a few of the stimuli presented. For example, when an egg rolls out of its nest, a brooding herring gull (*Larus argentatus*) retrieves it with its bill, pulling it back into the nest. Tinbergen (1953) tested herring gulls to determine which stimulus qualities of the egg— its size, shape, color, or pattern (figure 7.14)—might trigger the retrieving response. Tinbergen measured the bird's preferences by placing pairs of model eggs on the edge of the nest and recording which egg the bird chose to retrieve first. Then, using eggs that varied in size, shape, color, and color pattern, he found that for retrieval, the most critical sensory cues were egg size and shape. The gull's visual system is quite capable of receiving incoming stimuli from the eggs and transducing these stimuli into messages sent to the central nervous system, but the gull's behavioral response was directed only at specific stimuli. Thus, Tinbergen concluded that some type of central-filtering process must be in operation during the decoding of the visual input.

Our everyday experience demonstrates how the central-filtering process operates on the auditory mode. In a crowded college dormitory, a stereo is playing loudly and many conversations are taking place. Two students are deep in a discussion about campus politics. Although they are receiving a large quantity of auditory input, they can hear one another and can concentrate on the topic at hand, because central-filtering processes act on the incoming nerve impulses from the ear and sort out only the relevant stimuli for further processing.

THE CONTROL OF MOVEMENT

Perceiving stimuli is only half the job of a behaving animal: now it must respond appropriately. How does the nervous system control movement? Motor neurons usually innervate a large number of muscle fibers, called a motor unit. In

insects, a motor unit may be an entire flight muscle, but in vertebrates most muscles are composed of more than one motor unit. The motor neuron synapses with the muscle at the neuromuscular junction. When stimulated, muscles contract; they do not actively lengthen, but are rather pulled out again by an antagonistic muscle. For a more detailed description of muscle function, see Robinson (1998).

Simple, stereotyped responses to particular stimuli are called **reflexes.** For example, the sea slug *Aplysia* withdraws its gill in response to touch. A reflex is controlled by, at a minimum, a sensory neuron that conducts a signal away from a sense organ, a motor neuron that controls the muscle, and an interneuron that links the two. Many reflex arcs are more complicated, however, involving many more neurons. For example, in some cases motor neurons inhibit the firing of other motor neurons that would pull the intended limb in the wrong direction.

More interesting than simple reflexes are the neurons controlling repetitive behavior. Walking, scratching, chewing, etc., are all repetitive movements. A puzzle for early neuroethologists was whether these movements required feedback from the environment, or whether they were self-sustaining. Interestingly, it is now clear that many repetitive movements are controlled by **central pattern generators.** Sensory input is not necessary to keep the repetitive movements going, although it is often capable of modifying the basic pattern. Central pattern generators have been found in both vertebrates and invertebrates. For example, they are involved in singing in crickets and grasshoppers, swimming in leeches, and walking in cats, rats, and birds (Pearson 1993). We'll look at a more detailed example later in the chapter.

Motor systems are hierarchically organized in both vertebrates and invertebrates: often, pattern generators are controlled by higher-level command neurons (reviewed in Pearson 1993). In some cases, a single neuron commands an entire motor program, such as in the tail-flip escape of a crayfish. More commonly, a population of neurons commands a behavior. In animals with a highly developed brain, the brain acts at the executive level and has organizational control over behavior.

METHODS OF INVESTIGATING THE NERVOUS SYSTEM

A wide range of techniques are available to explore the neural mechanisms that control behavior. Here, we briefly review a selection of commonly used techniques. In the following section, we will examine case studies that employ them.

Recording Neural Activity, Neural Stimulation, and Self-Stimulation

One direct way we can discover relationships between neural mechanisms and behavior is by implanting electrodes in an organism's nervous system and recording the neural

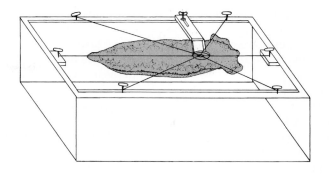

Tritonia suspended in apparatus for electrode stimulation

(a)

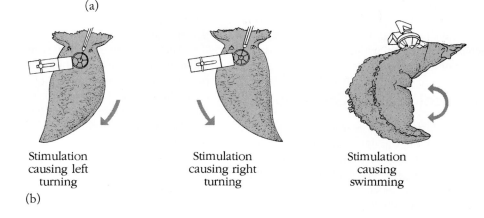

Stimulation
causing left
turning

Stimulation
causing right
turning

Stimulation
causing
swimming

(b)

FIGURE 7.15   **Testing *Tritonia* through stimulation.**
Use of stimulation through electrodes to evoke distinct behavior patterns is one method for mapping which nerves and decoding centers affect particular behavior patterns. An apparatus (a) used to suspend the marine nudibranch *Tritonia* in seawater permits placement of electrodes into specific ganglia or neurons. Turning and swimming movements (b) can be produced when specific neurons are stimulated.

impulses (action potentials) directly. By recording impulses and simultaneously observing behavior, we can draw correlations between patterns of neural discharges and patterns of behavior, and map the neural pathways in the brain. The general pattern of activity of the outer region of the brain can also be measured noninvasively with electroencephalograms, or EEGs, which are disk-shaped electrodes placed on the scalp.

Another method we can use to investigate neural control of behavior is neural stimulation—that is, we can place electrodes directly into specific areas of the peripheral nervous system or the brain and pass an electrical current through the electrodes. If the stimulation evokes a recognizable sequence of behavior that is normally part of that species' repertoire, then the electrode is probably in or near a nerve pathway or a brain area that controls or generates the observed behavior (figure 7.15).

A variation of the electrode stimulation technique permits the subject animal to provide self-stimulation by pressing a lever or some other device. When the subject presses the lever frequently, we infer that the stimulation is rewarding or pleasurable.

Transection, Lesions, and Transplantations

These are "cut and paste" techniques—normal neuronal pathways are either disrupted or removed, or new nervous tissues are added. In **transection,** particular nerves are cut and the resulting loss of function is observed. In **lesions,** electrical currents, chemicals, or microknives are used to destroy specific brain areas. It is also now possible to make reversible lesions, which allows elegant controls as well as reducing ethical concerns about animal use. In **transplantation,** neural tissue is transferred from one organism to another, or occasionally from one location to another within the same organism.

Functional Neuroanatomy and Imaging

Functional neuroanatomy is the study of the size, structure, and arrangement of cells within the nervous system, including the brain. Teasing apart where a particular neuron goes is not always clear or easy, and often staining is useful. For example, a region of the nervous system might be injected with a chemical such as horseradish peroxidase, an enzyme that is transported throughout the nerve. The tissue is then fixed and the area treated with chemicals in order to make the staining visible. The tissue is then thinly sliced with a microtome and examined under a microscope.

Noninvasive imaging techniques are now also available, which are especially useful for human studies or in studies when researchers do not wish to kill the test animal. The structure of the brain can be examined with magnetic resonance imaging (MRI), which is based on the absorption and emission of energy in the radio frequency range of the electromagnetic spectrum. Hydrogen atoms, found in water molecules in tissue, produce a small magnetic field that is detectable by MRI imaging. CT scans, or computerized tomography, use x-rays projected through the body section by section, which can be later assembled to construct a three-dimensional picture of the body.

Psychopharmacology and Cannulation

The identification of various brain neurotransmitters (e.g., acetylcholine, serotonin) and the development of synthetic drug compounds chemically related to these neurotransmitters have permitted investigators to explore relationships between the nervous system and behavior. This area of brain research, called **psychopharmacology,** incorporates chemistry, physiology, and psychology.

Hollow electrodes or fine tubes (cannulas) or needles can be inserted into specific areas of the brain in order to introduce substances, such as androgens. They can also be used to collect secreted transmitters.

Metabolic Activity of Neurons

One area of research in neurobiology involves the use of autoradiography to measure the rate of metabolic activity of localized regions of the nervous system, particularly the central nervous system (Sokoloff et al. 1972; Gochee, Rasband, and Sokoloff 1980). The technique is based on the fact that neurons metabolize glucose for energy, as do all cells (Gallistel et al. 1982). Deoxyglucose (2-DG) labeled with ^{14}C can be injected into the test animal. This compound is taken up by cells differentially, depending upon their rate of activity and energy utilization. Once inside the cell, the 2-DG is converted to 2-deoxyglucose-6-phosphate (2-DG6P). The 2-DG6P is not metabolized further and is trapped within the cell. The test animal is then killed, its brain tissue sectioned, and the sections placed directly on x-ray film (autoradiography). When the film is later developed, the remaining images show varying intensities of gray that depend upon rate of uptake and utilization of 2-DG. Neuronal activity can thus be measured using the labeled compound and autoradiography (Gallistel et al. 1982).

A less invasive procedure uses positron emission tomography (PET). The subject is given water or oxygen that has positron-emitting, radioactive elements, and then is scanned with a special apparatus. Blood flow and neuro-transmitter metabolism can then be measured and displayed on a TV screen. A variation on MRI imaging, called functional MRI or fMRI, makes use of the proportion of nonoxygenated to oxygenated blood in order to identify which areas of the brain are undergoing metabolic activity.

Potential Problems

All the techniques that we have mentioned here have limitations and assumptions. For example, it might be difficult to be certain that the nerve targeted for lesion is really destroyed, and that no unintended damage occured to other areas. Thus, researchers often strive to use a combination of techniques that have nonoverlapping assumptions in order to get an unambiguous result.

Another problem is that of interpretation: care must be taken not to confuse correlation with causality. That is, just because the disruption of part of a nervous system leads to the disruption of a behavior, it is not necessarily the case that one has identified with certainty the part of the nervous system that is solely responsible for a behavior. Often, a series of experiments using a range of methods are needed to determine whether a particular structure is needed for a behavior to be produced (the behavior cannot be produced without it) as well as whether it is sufficient (other structures are not also necessary in order to produce the behavior).

EXAMPLES OF NEUROBIOLOGY AND BEHAVIOR

The Mechanics of Feeding

All animals need to feed. This seemingly simple behavior generates many questions. What prompts an animal to begin feeding? What signals them to stop eating? This was studied in the blowfly (*Phormia regina*) in a series of classic experiments by Vincent Dethier and his associates (Dethier and Bodenstein 1958; Dethier 1962; Dethier 1976; Dethier and Gelperin 1967).

When a blowfly is hungry, and lands on a potential food source, chemoreceptors on its legs are stimulated. It extends its proboscis, thereby stimulating its oral taste receptors, and it begins to feed. Dethier and his colleagues tested a series of hypotheses about the initiation of feeding. First, they thought that blood sugar levels might drive hunger and satiation. They injected hungry flies with sugar solutions, but the flies were not satiated. They then hypothesized that hunger and satiety might be directly controlled by the fullness of the hindgut, the major portion of the fly's alimentary canal. They tested this idea by delicately giving the flies "enemas," thus filling the hindgut. Again, the flies remained hungry.

Dethier and his colleagues then considered another part of the digestive system—the foregut, where the initial step in digestion takes place. By using very refined, miniaturized surgical techniques (involving a paraffin block to hold the fly, forceps, and a razor blade) and a dissecting microscope, they were able to transect a small nerve that connects the foregut and the brain. They hoped to find out whether this nerve carries the message of foregut fullness, which in turn stops feeding behavior. When hungry blowflies with this nerve transection were given an opportunity to feed, they too began to eat—but they did not stop eating. From these data and related information, Dethier and his colleagues concluded that as long as there is food in the blowfly's foregut, stretch receptors located there send messages to the brain that inhibit further eating behavior. Severing the nerve makes the fly permanently hungry. Additional research revealed that stretch receptors located in the body wall also send messages to the brain that act to inhibit further feeding when the fly is satiated.

Repetitive Movement

Consider animal motion. Much movement—walking, flying—is essentially repetition of the same patterns of muscle contraction. Earlier in this chapter, we discussed briefly the role of central pattern generators in generating movement. The most famous example of a central pattern generator is Donald Wilson's (1960) work on locust flight. The four wings of the locust beat in a rhythmic pattern. In order to test whether or not this results from of a series of reflexes triggered by environmental stimuli, Wilson removed sense organs and, in some experiments, even the wings themselves. He found that the motor pattern of flight was still apparent (this is called "fictive flight"), even when the only parts remaining in the preparation were the head and ventral side of the thorax and the thoracic nerves. The flight pattern does not require sensory input to be maintained. However, the locust does adjust its flight in response to sensory stimulation. For example, hairs on the locust's head are sensitive to wind, and enable the locust to compensate if it begins to roll in flight. This is often the case with central pattern generators: sensory input often modifies basic motor patterns in both vertebrates and invertebrates (Pearson 1993).

Escaping Danger

When an animal is suddenly confronted with danger, its survival depends on fast and appropriate action, such as darting away, pulling back into a shell or a hole, or flying evasively. We expect that natural selection has operated very strongly on escape responses, because an animal that fails to evade predators will not have the chance to pass on its genes. In fact, many escape responses have evolved to be very fast and are often under the control of a very straightforward neural arrangement. This makes them effective for the animals, as well as easy for us to study.

Many insects have cerci (singular: cercus) projecting posteriorly off their abdomen (figure 7.16). These are mechanoreceptors, covered with fine hairs that are highly sensitive to the movement of air currents. When cockroaches sense the approach of a crushing foot or a predatory toad, they scurry quickly away. This behavior has been studied with a number of different techniques. To verify that the cerci

FIGURE 7.16
Cerci (indicated by arrow) are sensitive to air currents and are crucial to the speedy escape behaviors of many insects, such as this cricket.
© Grant Heilman Photography.

were the wind detectors, researchers covered them in vaseline; roaches with covered cerci were always eaten by toads, whereas control roaches escaped half the time (reviewed in Levi and Camhi 1996). A series of neurological experiments determined that the escape response is primarily managed by a system of giant interneurons, first described by Roeder (1967). The diameter of an axon is correlated with the speed that an impulse travels, and these giant interneurons quickly transmit a message of approaching danger to the leg muscles in the thorax. Roaches do more than simply run, however. High-speed video shows that roaches first turn away from the stimulus; this seems to get them out of the "line of fire." A network of ganglia in the thorax take input from the cerci about the direction of the stimulus and change the position of the legs in order to orient the roach properly. Interestingly, there is a second, albeit slower, system for mediating escape behavior in the roach. If the giant interneurons are transected, the roach can still escape (Stierle et al. 1994).

Some predators can eradicate or manipulate the escape response to their own advantage. For example, some wasps capture roaches to serve as food for their offspring. Often, the roach will not be immediately killed but only paralyzed. Some wasp venoms interfere at the neuromuscular junction, either by blocking neurotransmitter release or by blocking the receptors. Other wasp venoms raises the threshold for running so that the roach is less likely to escape (Fouad et al. 1996; Libersat et al. 1999).

Another well-studied escape response is that of moths (family Noctuidae) that evade approaching bats. Bats detect prey with the use of ultrasonic radar in a process called echolocation. They emit a pulse of sound, then listen for its echo as it bounces off of nearby objects. (This sound is not audible to humans, but it is possible to buy a "bat detector" that clicks when it detects a bat's echolocation signal.) Noctuid moths have special hearing receptors that consist of tympanic membranes located on each side of the body in the thoracic region (Roeder and Treat 1961; Roeder 1970). These ears can detect ultrasound.

Two neurons, A_1 and A_2, connect each "ear" to the central nervous system. Roeder placed small electrodes into each of these neurons and recorded the nerve impulses produced as a result of bat vocalizations and artificially created sounds. By varying the frequency and intensity of these sounds and recording the nerve impulses, he discovered several things. First, when sound frequency is varied, there is no change in pattern from either neuron: the moth cannot detect differences in frequency. When the bat is far away and cannot get an echo from the moth yet, the sounds the moth hears are soft. At this point, the A_1 neuron responds but the A_2 does not. When the A_1 neuron is stimulated, the moth flies away from the sound of the bat by comparing input from the A_1 neurons in the left and right ears. When the neurons are responding equally, the moth is flying directly away.

When the sounds are loud (i.e., the bat is closing in), both A_1 and A_2 neurons respond, and stimulation of the A_2 neuron causes desynchronization of the moth's wingbeats so that the moth tumbles through the air in a random path, making it much more difficult for the bat to capture it. Thus, different patterns of firing in the two neurons encode the stimuli for sounds that are passed on to the central nervous system.

The moth faces additional problems when trying to determine whether a bat is approaching. There is extra "noise," or irrelevant stimuli, in the system. First, the wingbeats of some species of noctuid moths produce ultrasound, and this causes the A_1 neurons to fire (Waters and Jones 1994). In addition, the A_1 neurons spontaneously fire even when no ultrasound is present (Waters 1996). Both of these phenomena may reduce the ability of the moth to discriminate bat signals from background noise. If a moth responds to every action potential of its A_1 neurons as a bat, it will spend much of its time escaping from nonexistent predators, but it will pay with its life for not detecting a real bat. One way to get around this problem is to be more responsive to repeating signals, such as the feeding buzzes produced by many bats.

Still other moths (such as *Cycnia tenera,* the dogbane tiger moth) apparently jam the echolocation signals of bats (Fullard 1994). As a bat approaches, just before it reaches its target, *C. tenera* emits high-frequency clicks that cause the bat to break off pursuit. These clicks seem to provide "phantom echoes" to the bat and confuse its radar system.

A lesson about the necessity of paying attention to an animal's Umwelt in designing research projects comes from bat research. Researchers often mark bats with plastic and metal rings on the forearm to provide them with individual identification codes. Rings, however, clink together, and this sound is detectable by the A_1 neurons of moths (*Noctua comes*), which then initiate escape behavior (Norman et al. 1999). Thus, marking bats with two rings on the same forearm might reduce their foraging success.

Detecting Prey

Now let's examine feeding from a predator's point of view. How do animals distinguish prey from other objects? Clearly, selection will favor a predator that does not attack a dangerous animal or a conspecific by mistake. A famous set of experiments have been carried out by Ewert on toads in the genus *Bufo* (1980, 1984, 1985). Toads are voracious predators, and eat a variety of insects and worms. Ewert showed toads cardboard models, and watched their reactions to them. Some models elicit prey capture behavior. The toads orient to it (and will continue to orient to it as it moves around) and then approach it. Ewert demonstrated that small square stimuli elicited only a small response, but long wormlike rectangles, moving past the toad in the direction of the long axis, produced the greatest number of turning and prey capture movements. Vertically oriented rectangles (the "antiworm") elicited no response. However, toads responded to an antiworm that moved vertically (figure 7.17).

How does the toad's nervous system make these distinctions? In vertebrates, neurons in the retina pick up visual stimuli, and pass it on to bipolar cells, which in turn connect to ganglion cells. Ganglion cells connect to the optic tectum, a region of the midbrain, and to the thalamus in the forebrain. At each level, the information from the external world is edited and shaped. Each neuron has a **receptive field,** an area of the sensory surface to which it responds (figure 7.18). For example, each rod and cone cell responds to stimulation of a particular part of the retina. Similarly, each bipolar cell responds to a particular set of rods and/or cones. Some bipolar cells become hyperpolarized when the center of their receptive field is stimulated (off-center bipolar cells), whereas others become depolarized (on-center bipolar cells). At the next level, ganglion cells show similar patterns. Further processing occurs in the optic tectum and the thalamus. For example, in the toad, recordings of T5(2) neurons in the optic tectum show that they exhibit selective firing activity when wormlike stimuli are presented. There is a good correlation between the pattern of neuronal firing in this group of cells and the behavioral response profile when the length, velocity, and other attributes of the stimulus presented to the toad are varied. This is called a **feature detector:** neurons respond only to stimuli with particular characteristics. Other areas of the brain, besides the T5(2) neurons, are also involved in identifying prey (Ewert et al. 1999).

FIGURE 7.17   **Prey capture behavior in toads.**
Toads respond strongly to wormlike shapes that move in the direction of their long axis (a), but not to "antiworms" that move in the direction of their short axis (b).
© Leonard Lee Rue III/Animals, Animals.

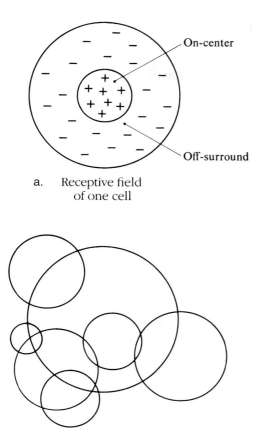

a.   Receptive field
of one cell

b.   Series of receptive fields

FIGURE 7.18   Receptive fields of mud puppies.
(a) When light is shown on the retina, the activity of ganglion cells is excited if the light strikes on-center and inhibited if it shines on the off-surround neurons. (b) Mapping a series of receptive fields of different retinal cells results in an overlapping pattern, with fields of the same or different sizes, depending on the organism.

Feature detectors have also been found in other animals. For instance, dragonflies in the genera *Aeshna* and *Anax* rely on excellent vision and fast flight to capture airborne prey. A group of eight feature detectors in the dragonfly are tuned exclusively to moving, contrasting objects. Some of these neurons respond only to targets of particular sizes, whereas other neurons help to steer the dragonfly during prey tracking (Frye and Olberg 1995; figure 7.19).

Not every neuron is a feature detector, however. Schiller (1996) emphasizes that most species of vertebrates have a combination of feature detectors and neurons that act as general-purpose analyzers. He likens a feature detector to having a specific tool for a particular job: it is very effective, but there might be a limit on how many tools you can carry. A general-purpose analyzer is like a multipurpose tool that will do a passable job but require less dedicated space in the brain compared to many specific tools. Neurons become increasingly multifunctional in more central areas of the nervous system. The cortex is the site of many multifunctional neurons. Nonetheless, the existence of feature detectors demonstrates how the mechanics of perceptual systems fundamentally determine how animals perceive the world.

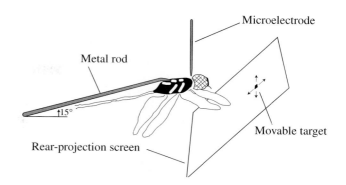

FIGURE 7.19   A method for testing the feature detector in dragonflies.
A dragonfly is suspended from a metal rod and shown images projected onto a translucent screen, as a microelectrode records neural activity.

Communication

Communicating animals face two problems: how to produce signals, and how to detect signals produced by conspecifics. The communication of crickets is our first example. Males of most species of crickets (*Teleogryllus* spp., *Gryllus* spp.) emit species-specific sounds via **stridulation,** a process that involves rubbing together a portion of the cuticle of one wing (the scraper) against a series of ridges on the other wing (Bentley and Hoy 1974). The songs serve to attract females, which orient to and move toward the calling males (**phonotaxis**).

From early work, we know that the central nervous system of some adult crickets (e.g., *Teleogryllus commodus*) produces stereotyped motor output patterns that develop during the nine to eleven molts after hatching. The neurons involved in sound production are present at hatching, but they undergo changes in size, and new synaptic connections are made during the molts that lead to the adult stage. By examining specific nymphal crickets' neurons known to be involved in production of calls in adults during various instars (stages between two molts), Bentley and Hoy (1970) demonstrated that the neural processes underlying the production of sound develop progressively with each molt. Only at the last instar is the neural machinery complete for calling.

The neuronal events involved in sound production have been investigated using a series of recording and stimulating electrodes placed at designated locations in the cricket nervous system (figure 7.20). A nerve impulse begins when the command interneurons connecting the head region and thoracic ganglion are stimulated (Bentley and Hoy 1974). This leads to a muscle contraction, followed by changes in wing position that result in stridulation, producing the sounds we hear from crickets.

The motor program that produces the stridulation is an example of a central pattern generator. Hedwig (2000) used intercellular recordings to study the effects of other neurons on this pattern generator. An interneuron in the brain acts as a command neuron: when it fires at a rate of 60 to 80 action potentials per second, the cricket stridulates. The more it fires, the faster the rate of stridulation.

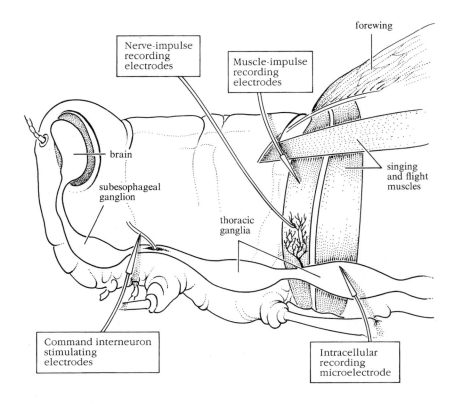

FIGURE 7.20   **Production of cricket song.**
A series of microelectrodes can be used to study the neural bases for production of cricket song. The electrodes can be used either to track the impulses moving through the system or to stimulate particular structures in the system. The song of the cricket is elicited when the command interneurons, located in the group of neurons between the head region and the thoracic ganglion, stimulate thoracic interneurons and eventually the motor neurons leading to the muscles involved in song production. The vertical muscle fibers are involved in wing closure for sound production, and the horizontal muscle fibers open the wings during sound production.

Hedwig also studied the effects of a wind stimulus on calling behavior. You may have noticed that, as you walk through a field of calling crickets, they go silent upon your approach. In the laboratory, a puff of air on the cerci caused the chirping to stop, but intracellular recordings indicate the command neuron keeps firing at the same rate. Therefore, the firing of the command neuron is necessary for stridulation to occur, but not sufficient: other neural pathways can override it.

Considerable work has also been done on the neurophysiology of sound reception in female crickets (see Huber 1978, 1983a, 1983b). The ears of crickets are located on the tibiae (on the forelegs) near the animal's body. Intracellular electrodes in a group of about sixty receptor cells at each ear revealed that some neurons are tuned specifically to the frequency of conspecific males' courtship songs. Other neurons appear to be tuned to frequencies of various intraspecific calls and to sounds from potential predators. Female crickets follow the rule "Turn towards the side with the ear that is most strongly stimulated" (reviewed in Horseman and Huber 1994). The neurons originating at the auditory receptors project to specific locations in the prothoracic ganglion of the cricket.

Many animals, including humans, partition continuously varying stimuli into categories. This is especially well known from human speech: we don't always detect every possible gradation between two stimuli, but are likely to divide the continuum into discrete categories. Crickets (*Teleogryllus oceanicus*) also categorize sound (Wyttenbach et al. 1996). Crickets call at 4 to 5 kilohertz, whereas bats call at 25 to 80 kilohertz when they are echolocating. A cricket must identify a sound and either fly toward a conspecific, or fly away from a potential predator. When crickets are presented with the range of frequencies, they apparently categorize all sounds of less than 16 kHz as attractive, and all sounds of more than 16 kHz as repulsive.

Now consider the calling behavior of anurans, or frogs and toads. From early spring until well into the summer months, male frogs and toads call (figure 7.21) in the evening hours to attract females. Females follow the sound to the males. The pair form a bond, termed amplexus, in which the male grasps the female. As the female lays a cluster of eggs, the male fertilizes them.

In many locations, there are a variety of frogs and toads of different species and genera calling on the same evening. Female frogs are capable of making species-specific discrim-

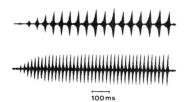

FIGURE 7.21   Representative frog calls.
Oscillograph of representative mating calls of *Hyla versciolor* (upper tracing) and *Hyla chrysoscelis* (lower tracing). The small bar represents 100 milliseconds. Notice that the temporal form along the axis as the song proceeds, and the emitted energy of the call (deviations from the central line) differ dramatically for these two species of tree frogs.

Source: H.C. Gerhardt, "Sound Pattern Recognition in Some North American Treefrogs," *American Zoologist* 22:581–595, copyright 1982 by the American Society of Zoologists.

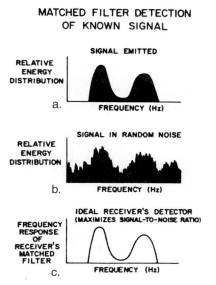

FIGURE 7.22   Frog matched filter system.
Depiction of how a matched filter may work. The emitted signal has a particular spectral pattern, in this instance bimodal (a). However, in nature, the call would normally be embedded in noise (b). If the receiver has a matched filter system, then we would expect the frequency response in the sensory input system of that receiver (c) to be very similar to the pattern emitted by the sender.

inations and can selectively use phonotaxis to find mates of their own kind from the cacophony of noise (e.g., Gerhardt 1974 a, b). This is evolutionary advantageous: males and females of the same species will find each other and mate; the calls serve as a premating isolation mechanism. Without such discrimination, much energy would probably be expended on costly reproduction attempts that do not lead to successful egg fertilization. Furthermore, in some anuran species (e.g., *Hyla* spp., *Rana catesbeiana*), it may be possible for females to make distinctions among conspecific males (Howard 1978; Gerhardt 1982). That is, female anurans may exhibit some degree of mate choice based upon the calls' features. This could lead to differential reproductive success for some females, and, for those species where this type of discrimination occurs, could be a selective force in the evolution of calling behavior and mate selection processes.

How then do females discriminate among calls? The anuran's peripheral auditory system contains two organs: an amphibian papilla and a basilar papilla. The amphibian papilla is more sensitive to low to midrange sound frequencies, and the basilar papilla is more sensitive to higher frequencies (Feng, Narins, and Capranica 1975; Capranica and Moffat 1977). Different species are sensitive to different frequencies of sound. There is some evidence for a matched filter system (figure 7.22): the auditory system acts as a peripheral filter when it is tuned to be especially sensitive to calls of its own species (e.g., Capranica and Moffat 1983). However, the match is often quite crude, and there is evidence of mistuning in some species (reviewed in Schwartz and Gerhardt 1998).

More processing occurs in the brainstem and the auditory centers of the thalamus. For example, in the leopard frog (*Rana pipiens*), neurons in the posterior thalamic nucleus are highly selective, and respond only to the simultaneous presentation of both low and high frequency tones. These correspond to two frequency peaks in the advertisement call of this species (reviewed in Hall 1994). Central processing also is responsible for distinguishing among temporal patterns (such as repetition rate and duration) of songs. For example, phasic units of neurons another are those that respond only

when a stimulus first begins, and tonic units are those that respond throughout the duration of a stimulus. This combination of types of neurons allows the animals to encode information about the timing of stimuli (Hall 1994).

Birds produce and interpret sounds differently from insects and amphibians. (Detailed descriptions are in Bradbury and Vehrencamp (1998) and Hauser (1996)). Birds produce song during exhalation, and they require muscular action to completely exhale: this gives them great muscular control over their vocalizations. The junction of the two bronchi from the lungs is modified into a specialized **syrinx.** Muscles surrounding the syrinx control the tone and timing of the song.

Nottebohm and his colleagues have worked on aspects of birdsong, especially in canaries (*Serinus canarius*), for nearly 30 years. They were first to discover that the avian brain has discrete nuclei responsible for the production and processing of song (e.g., Nottebohm 1975, 1980, 1981; Nottebohm and Nottebohm 1976; DeVoogd and Nottebohm 1981; Paton and Nottebohm 1984; Nottebohm, Nottebohm, and Crane 1986). The neural pathway for song production originates in the HVC, or higher vocal center. There is a pathway from the HVC to the RA (robustus archistriatum). The RA sends projections to the DM (dorsomedial nucleus) and to the hypoglossal nerve. DM also projects to the hypoglassal nerve, which in turn innervates the syringeal muscles. Other centers control breathing and the timing of signals. Lesioning any connection distorts or eliminates song output. The HVC is also involved in a second loop that deals with song perception and development.

Bird Song in the Lab and Field

Charles Darwin himself routinely fields questions about evolutionary biology in Steve Nowicki's introductory biology class at Duke. A Phil Donahue clone, giant frogs, mad scientists set on pithing those frogs, and the orca whale from *Free Willy* have also made appearances. It's street theater with a point: explaining biological principles in terms of everyday experiences. Even his lectures in front of professional societies sometimes incorporate plastic light sabers. Clearly, this is a person who enjoys his work.

Steve also has a lively approach to research, which leads to him being a "moving target," he says, for people trying to pin down what he does. He emphasizes an integrative approach, moving from the field to the lab and back again, as one question leads to another.

A series of experiments was launched by Steve's observations of winter flocks of chickadees. The familiar *chick-a-dee-dee-dee* call of these birds may sound very simple, but its acoustic structure varies across flocks. In fact, birds can tell which flock another bird belongs to by its call, as Steve showed with playback experiments in the field. But how do members of a flock all come to sound alike? Flocks disband every spring, and membership changes through the years. Steve addressed this question by capturing birds from different flocks and housing them together in an aviary. Over the course of three weeks, their calls converged to a common theme.

This song convergence was interesting because it demonstrated a level of ongoing vocal plasticity in adult birds that was not previously suspected from learning studies that had focused on sensitive periods in the acquisition of song. In addition, the way that the birds converged (that is, the particular acoustic features that showed convergence) presented a new puzzle that could not be accounted for by established theory on how birds produce sound. This led Steve and his collaborators to investigate the proximate factors in song production by transecting the two motor nerves leading to either side of the syrinx, the unique sound-generating organ in birds. The two sides of the syrinx function somewhat independently, but it is their interaction that produces the characteristics of one particular note in the chickadee call. Neither side acting alone can generate the call, nor can the call be explained by simply adding together their output. Thus, by varying the relative tuning of the two sides of the syrinx, chickadees have control over part of their calls, and this control enables flockmates to converge rapidly on a common call. Another experimental technique was to record birds singing in air in which nitrogen is replaced by helium, a less dense gas. This demonstrated that the vocal tract of birds acts as an acoustic filter, and is more analogous to the control of human vocalization than had been previously thought. Later, Steve and his colleagues returned to studying the whole bird, and found that different parts of the call are subject to different amounts of learning.

Steve favors this multilevel approach in other studies as well. For example, current projects include testing the "nutritional stress hypothesis" developed by Steve and his colleagues Susan Peters and Jeffrey Podos: the idea is that growing birds face nutritional conditions that influence their brain development and song learning, and that females can use aspects of song as an indicator of male quality. Stay tuned to see the results.

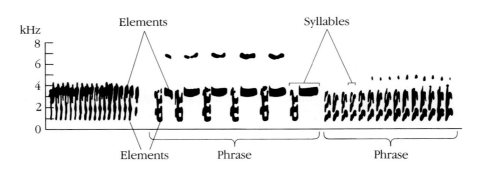
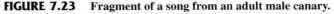

FIGURE 7.23   **Fragment of a song from an adult male canary.**
Canary song includes 20–30 syllable types, each repeated several times forming a phrase. This figure shows three phrases. Each syllable in the first and second phrase is composed of two elements, whereas syllables in the third phrase consist of a single element. The vertical axis represents sound frequency (in kHz), and the horizontal axis is in seconds.

Male canaries sing (figure 7.23); females generally do not sing. When a young canary matures at one year of age, it learns a song repertoire. Each succeeding year, the bird learns a new repertoire. Most songbirds show increased testosterone around the time when they are singing. When birds are experimentally treated with testosterone, song nuclei show increased dendritic growth. In males, the HVC nucleus is 99 percent larger in the spring (when they sing) than in the winter (when they do not sing), and the RA nucleus increases by 76 percent. Female birds treated with testosterone show anatomical changes similar to those observed in males, and they start to sing. Experiments with adult canaries using radioactively labeled thymidine (a marker for DNA synthesis) indicate that new neurons are being formed in one of several brain nuclei involved in song control. Electrodes detected impulses resulting from auditory stimuli and confirmed that these new neurons are parts of functional circuits.

Thus, neuroanatomical changes occur on a seasonal basis that corresponds to the annual behavioral cycle of the birds. As a new song repertoire is learned each spring, testosterone increases and new connections involving newly generated

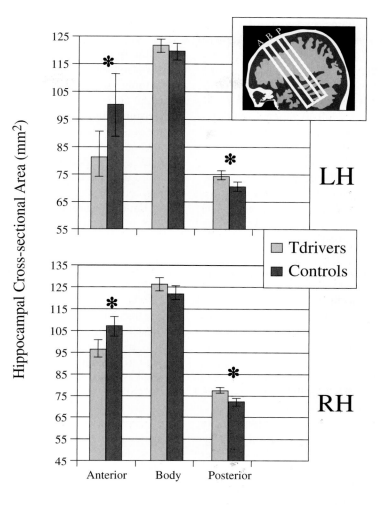

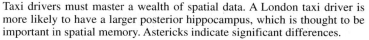

FIGURE 7.24   Hippocampal differences in London taxi drivers relative to controls.
Taxi drivers must master a wealth of spatial data. A London taxi driver is more likely to have a larger posterior hippocampus, which is thought to be important in spatial memory. Astericks indicate significant differences.

neurons are formed in brain areas associated with song control. These new neurons appear in response to a protein called brain-derived neurotrophic factor (BDNF): BDNF receptors are found in both sexes, but BDNF increases in response to testosterone, so it is normally found in males. When BDNF is infused into the HVC of adult females, the number of new neurons triples (Rasika et al. 1999). At the end of the breeding season, there is a reduction in the size of the nuclei responsible for controlling the canary's song (Nottebohm, Nottebohm, and Crane 1986), and this corresponds to the birds' forgetting the songs of that year. Neuronal genesis in adults, once thought unlikely, has now been found in a variety of taxa (Tramontin and Brenowitz 2000).

Learning and Memory

The neurobiology of learning and memory is a hot topic, and we have space to discuss only a few examples. First, we'll discuss spatial memory. Many animals benefit from learning

spatial information, such as the location of food, home, and areas of safety. The neurobiology of spatial learning has been studied in both vertebrates and invertebrates. Fascinating work has come from the study of food-storing birds, including chickadees, titmice, magpies, nutcrackers, jays, crows, and nuthatches (reviewed in Clayton and Lee 1998). These birds hide hundreds to thousands of seed caches, and then find them again by memory, often long after they have hidden them. For example, Clark's nutcrackers (*Nucifraga columbiana*) can remember the location of cache sites for over 285 days (Balda and Kamil 1992). A particular section of the brain called HF, which includes both the **hippocampus** and the parahippocampus, is involved in spatial memory. If the HF region is lesioned, birds continue to store food but cannot remember its location (e.g., black-capped chickadees; Sherry and Vaccarino 1989). Across species, HF size is generally positively correlated with the reliance on stored food.

The hippocampus is important in spatial learning in other animals as well, including rodents (O'Keefe and Nadel

1978) and homing pigeons. Homing pigeons use a navigational map and a compass sense to find their way back to their loft from a release site. Young homing pigeons given a hippocampal lesion before they have had a chance to learn a map are never able to learn one. Adult, experienced birds transferred to a new location where they have to learn a new map are also impaired. However, adult, experienced birds on familiar territory can continue to navigate even after hippocampal lesions, suggesting that the hippocampus is involved in map learning but not in its operation.

In mammals, mapping seems to occur via "place cells" in the hippocampus that tend to fire when their owner is in a specific site in a space that it has had a chance to explore. The mapping is so precise that researchers can record the firing patterns of a rat that is moving around in a familiar space and make a good guess about its physical location during the recording. Place cells also appear to be sensitive to information about the animal's movement, above and beyond just sensory inputs about its location (Sharp et al. 1995).

The hippocampus is also important in the spatial learning of humans. London taxi drivers were used as examples of humans with very good spatial skills. These drivers undergo rigorous training to learn to navigate among places in the city. The training is known as "The Knowledge," and takes, on average, about two years to master. MRI imaging showed that the posterior hippocampal region of taxi drivers, the site of spatial memory in mammals, was significantly larger than a control group (figure 7.24). Perhaps it is possible that humans with naturally good spatial skills (and larger posterior hippocampi) become taxi drivers. However, the amount of time spent as a taxi driver also corresponded with posterior hippocampal volume (Maguire et al. 1997).

In insects, the site of spatial learning appears to be the mushroom bodies, a structure in the central nervous system. Worker honey bees (*Apis mellifera*) have to find their way repeatedly between flowers and the hive. A bee learns the location of the hive during an orientation flight (described in Capaldi et al. 1999): as she leaves the hive, she turns and looks back at the entrance, and hovers back and forth while turning in ever-growing arcs. She spirals out of sight, then returns to the hive without nectar or pollen. Only older workers forage outside the hive. Associated with this developmental change is an expansion of the mushroom bodies, as determined by sectioning the brain and measuring the volume of different regions. Researchers addressed the question of whether the size of the mushroom body increased as a result of the experience of flying around the hive by gluing plastic disks to the backs of young bees. These "big-back" bees could not get out of the hive, but they had the same amount of mushroom body growth as did control bees. This demonstrates that neither flight outside the hive nor foraging experience was essential for the growth of the mushroom bodies (reviewed in Capaldi et al. 1999). This contrasts to the avian hippocampus, where changes are more likely to occur in response to experience.

The molecular details of memory formation are beginning to be understood. Memory is thought to be produced when two connected neurons are simultaneously activated in such a way as to strengthen the synapse. This strengthening known as long-term potentiation, requires the activation of NMDA receptors on the cell membranes of the postsynaptic neuron. NMDA receptors have been called "coincidence detectors" because they help the brain to associate two events. Scientists have investigated this molecular machinery by producing a smarter mouse, dubbed "Doogie mouse" after the television series about a boy genius (Tsien 2000). These mice have been genetically engineered to make more than the usual amount of a key subunit of the NMDA receptor. Doogie mice remember an object they have previously investigated, remember a mild unpleasant shock, and learn a spatial task (finding a platform submerged in murky water) better than normal mice. Mice with the gene for part of the NMDA receptor knocked out (made inactive) were worse at memory tasks. Of course, other molecules are also sure to be involved in learning and memory.

SUMMARY

Animals face the challenge of transducing energy from different environmental stimuli into neural impulses, interpreting them, and acting upon them. Nerve cells are specialized for carrying signals from one area of the body to another via action potentials, or waves of changes in electrical charge. Neural functioning is based on the nerve membrane's differential permeability to different ions, especially sodium and potassium ions. Signals move from one neuron to the next in specialized regions called synapses.

Taxa vary greatly in the details of their nervous systems. For example, radially symmetrical animals have decentralized nervous systems that suit their ecological lifestyle. Bilaterally symmetrical animals generally move unidirectionally through their environment. They tend to have increased numbers and types of sensory receptors, the formation of specific neural pathways through aggregations of individual neurons into fiber tracts, and cephalization, or the concentration and expansion of neural tissue near the organism's anterior end.

The sensory and perceptual worlds of animals differ from our own in three general ways. Some organisms (e.g., electric fish sensing electrical currents) have completely different modes of sensation. Other organisms (e.g., silkworm moths sensing pheromones) may share the same mode as humans, but the range over which they can transduce stimuli from the environment may differ. Finally, some organisms (e.g., herring gulls pulling eggs back into their nests) may transmit the same sensory information to the central nervous system but may differ in their interpretation of this information.

A variety of techniques are available for studying the nervous system, including recording neural activity, disrupting nervous tissue via transections or lesions and studying its effect on behavior, and transplanting neural tissue. Researchers study the anatomy of the nervous system through a variety of techniques from staining to imaging of the function of intact animals. Chemicals can be introduced to the nervous system as well as sampled via tiny cannulas.

We explore a number of case studies of the nervous system and behavior. Classic work on feeding in blowflies shows us how hunger and satiety work in a fairly simple system. Repetitive movement, characteristic of much animal behavior, is under the control of central pattern generators. It has been particularly well-studied in locust flight. The speedy response of roaches and moths to an approaching predator is an example of a finely honed system shaped by natural selection. The neurobiology of communication of crickets, frogs and toads, and birds has been examined from both the sender and the receiver's points of view. Finally, researchers are beginning to understand the processes underlying learning and memory, especially spatial memory.

DISCUSSION QUESTIONS

1. What is meant by the phrase "an action potential is all-or-none"? Given this fact, how is the nervous system capable of encoding information about the intensity of a stimulus?

2. Imagine that you are initiating a research project on two species of little-known insects. One is attracted to a particular species of plant, whereas the other is not. You are interested in the proximate reasons underlying this difference. Design a series of experiments. Remember that the sensory and perceptual systems of various organisms may differ in at least three ways: (a) organisms may have different sensory equipment and may be capable of receiving different types of stimuli; (b) organisms may have the same sensory equipment but different thresholds, so that they receive a particular stimulus over different ranges; or (c) organisms may interpret the same stimulus in different ways. Attempt to tease these apart in your design.

3. A number of changes—including changes in numbers and types of sensory receptors, the aggregation of neurons into fiber tracts, and cephalization—characterize the evolution of nervous systems. However, even animals with simple nervous systems may be well adapted to their environment. What advantages and disadvantages do the following animals have: (a) unicellular organisms, (b) worms, (c) insects, (d) amphibians, (e) mammals?

4. Complementary techniques are often used to address the same neurobiological question. Why is that? Describe two or more techniques you might use to address the question of whether a particular brain region is responsible for producing attack behavior in fish. Defend your choices.

SUGGESTED READINGS

Balda, R., I. Pepperberg, and A. Kamil, eds. 1998. *Animal Cognition in Nature.* New York: Academic Press. (Some chapters deal with neurobiological questions.)

Delcomyn, F. 1996. *Foundations of Neurobiology.* New York: W. H. Freeman and Company.

Fenton, M. Brock, and J. Fullard. 1993. Moth hearing and the feeding strategies of bats. In *Exploring Animal Behavior: Readings from* American Scientist. P. Sherman and J. Alcock, eds. Sunderland: Sinauer Associates. Pp. 87–96.

May, M. 1993. Aerial defense tactics of flying insects. In *Exploring Animal Behavior: Readings from* American Scientist. P. Sherman and J. Alcock, eds. Sunderland: Sinauer Associates. Pp. 74–86.

Shettleworth, S. 1998. *Cognition, Evolution and Behavior.* New York: Oxford University Press.

Simmons, P., and D. Young. 1999. *Nerve Cells and Animal Behaviour,* 2d edition. Cambridge: Cambridge University Press.

Tsien, J. 2000. Building a brainier mouse. *Scientific American* 282:62–68.

Hormones and Behavior and Immunology and Behavior

 Hormones are chemical substances produced either by specialized ductless glands in various parts of the body, by specialized individual cells, or by neurons called neurosecretory cells within the nervous system. In the last case, they are called **neurosecretions.** Both are carried by the circulatory system. Both are messengers to various target organs, and they influence such processes as growth, metabolism, water balance, and reproduction. In this chapter we consider how the endocrine system acts as a behavior-regulating mechanism through the production of hormones in both invertebrates and vertebrates, and how behavior can influence the endocrine system.

Let's begin by considering the types of questions we will examine in this chapter. A male rat placed with a sexually mature female rat will mount the female within a matter of seconds and begin a sequence of behavior that ends with copulation (figure 8.1). A castrated male rat takes much longer to initiate mounting (figure 8.2). Which hormonal mechanisms underlie this behavior? The testes are a source of androgens, hormones that affect reproductive behavior. Injections of synthetic androgens can replace the hormone normally produced by the testes and restore mount latencies to near normal levels in castrated male rats.

Many insects, like grasshoppers (*Melanoplus* spp.), hatch from eggs and reach adult form after going through a series of nymphal stages (figure 8.3). Between the stages, the insect molts, shedding its exoskeleton and undergoing other external and internal changes. What internal processes regulate this sequence? What behavioral changes are associated with these morphological shifts? If, when the insect is in an early nymphal stage, we remove the prothoracic gland, which secretes **ecdysone** (the hormone that controls molting), the normal developmental sequence stops. If, however, a prothoracic gland from another grasshopper is transplanted, or if ecdysone is injected, the sequence of nymphal stages resumes.

FIGURE 8.1   **Male rat mounting female rat.**
The male may perform only a mount or a mount with intromission. The sexual performance and timing of sexual behavior patterns in male rats are influenced by the presence of testicular hormones.
Source: Courtesy Ronald J. Barfield.

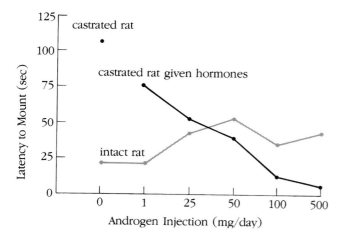

FIGURE 8.2   **Sexual behavior in castrated rats.**
The average latency to the first mount varies in intact rats, castrated rats, and rats given daily injections of doses of synthetic androgen.
Source: F. A. Beach and A. M. Holz-Tucker, "Effects of Different Concentrations of Androgen upon Sexual Behavior in Castrated Rats," *Journal of Comparative Physiology and Psychology,* 42:433–453, copyright 1949 by the American Psychological Association.

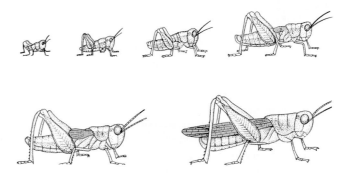

FIGURE 8.3   **Grasshopper development.**
Hoppers, young grasshoppers that hatch from eggs, proceed through a series of five to seven molts, changing in body size, and in internal and external characteristics with each molt, until they attain the adult stage in several months. Molting and other changes in physiology and behavior are controlled largely by the neuroendocrine secretions produced by the brain and various specialized glands within the insect.

The hormonal changes are also correlated with certain behavior transitions relating to diet, activity rhythms, and choice of habitat, for example. Both the dispersal of grasshoppers—individually or in large numbers—and the mating behavior of grasshoppers are partially under hormonal control.

We take a general look at the endocrine systems of invertebrates and vertebrates in the first part of this chapter. The major portion of the chapter explores the ways that hormones influence behavior; by examining various studies we gain an appreciation of the effects of hormones on behavior and look closely at the techniques that are used to explore these relationships. Our primary focus is on the mechanisms of the hormonal system, but we will also note some functional aspects of the hormone-behavior relationship. Later, in chapter 18, we will discuss more thoroughly the ecological and evolutionary significance of endocrine systems.

In the past decade there has been an increased emphasis on the relationship between the immune system and animal behavior. Aspects of the immune system can influence traits such as mate choice and sexual selection. All evidence indicates that this subfield of animal behavior will continue to expand. It is likely that in a future edition of a textbook like this there will be a separate chapter on the immune system and behavior. For now, we present some basic information on the immune system at the conclusion of this chapter on hormones and behavior. We will also place new materials relating the immune system to behavior throughout other pertinent chapters in the book.

GENERAL FEATURES OF ENDOCRINE SYSTEMS

A general feature of endocrine systems is that they either involve neurons directly, as do the neurosecretory cells, or they are closely tied to the nervous system by nerves that connect them with the brain and with one another. The glands are also interconnected by the circulatory system. This dual relationship of the neuroendocrine glands and the nervous system is an important feature of hormonal systems in both vertebrates and invertebrates. Thus, the mechanisms that exert controlling influences on behavior have apparently evolved in close harmony.

A neurosecretion or a hormone does not necessarily exert a direct influence on behavior. Instead, the neurosecretion or hormone may have another endocrine gland as its target. These secretions are termed **trophic neurosecretions** or **trophic hormones** and they occur in vertebrates as well as invertebrates.

Invertebrate Endocrine Systems

Different types of invertebrates—for example, echinoderms (e.g., starfish), segmented worms (e.g., leeches), molluscs (e.g., clams), and crustaceans (e.g., lobsters)—possess different endocrine systems. Perhaps the role of hormones in affecting behavior is most clearly understood for insects. Let's examine more closely one such system, that of the grasshopper. Neurosecretions or hormones that affect behavior are released from five major locations: (1) The **neurosecretory cells** in the brain (figure 8.4); (2) the paired **corpora cardiaca** just behind the brain, near the aorta; (3) the paired **corpora allata** alongside the esophagus; (4) the elongated **prothoracic gland** at the rear of the head, near the neck membrane; and (5) the male and female **gonads** in posterior body regions (not shown). Most insect species have similar structures, with some variations and exceptions. These five locations do not act independently: neurosecretions from the brain of many insects exert a trophic effect on the prothoracic glands, which in turn produce and secrete the hormone ecdysone that controls molting. Ecdysone, in turn, controls molting from one developmental stage to the next (see figure 8.3). Other trophic hormones influence dispersal and mating behavior.

The specificity of action of all hormones is dependent upon the specificity of receptor sites at the target tissues. This is true for both peptide and steroid hormones, whether the targets are other endocrine glands or nonendocrine tissues. At any particular time, the bloodstream may be circulating many "messages" in a variety of hormones, but a hormonal effect on physiology and behavior occurs only when the hormone makes contact with the appropriate receptor sites. Thus, the testes secrete androgens into the bloodstream, and the hormones are carried throughout the body. The androgens affect seminal vesicle growth, changes in secondary sex characteristics, and behavior through target tissues in the brain. In each instance there are appropriate receptors for the androgens located on or inside cells of the target tissues.

Vertebrate Endocrine Systems

What are the characteristics of vertebrate endocrine systems, and how do they differ from those of invertebrates? During the course of evolution, many changes have taken place in animal endocrine systems. In particular, the endocrine system of vertebrates has evolved into two major components. (1) The **hypothalamo-pituitary system,** located on the ventral side of the brain, has close connections to several central nervous system structures (figure 8.5). It is composed of the closely connected pituitary gland and the hypothalamus; together they form an important bridge between the nervous and the endocrine system. This link between the two systems is the key to integration of the two control systems. Two connections exist: the posterior pituitary is mainly composed of neurons that originate in the hypothalamus; the anterior pituitary is linked to the hypothalamus by the hypothalamic-pituitary portal system. The pituitary produces trophic hormones, which affect other endocrine glands, and direct-acting hormones. (2) A series of endocrine glands, including the **thyroid, pineal gland, adrenals, pancreas,** and **gonads,** are located throughout the body. The trophic hormones released by the various parts of the pituitary are peptides (proteins). The peptides' basic structure consists of amino acid chains. The hormones from the adrenal glands, testes, ovaries, and placenta include steroid hormones, organic compounds with a common carbon atom ring-structure and varying additional side chains.

Endocrine Gland Secretions

What are the major endocrine glands of vertebrates, and what behavioral processes do they influence? Several pituitary gland secretions exert controlling effects on behavior and related physiological mechanisms in vertebrates (table 8.1). Oxytocin and vasopressin are produced by hypothalamic neurons and are stored by nerve terminals in the posterior pituitary (pars nervosa); they are then released as neurosecretions into the bloodstream. Oxytocin acts to stimulate uterine contractions that may facilitate sperm movements in the female genital tract after copulation and help expel the fetus during parturition. Oxytocin is also intimately tied to nursing behavior: it stimulates the ejection of milk from the mammary glands. Vasopressin influences the physiology of the kidneys; it alters urine

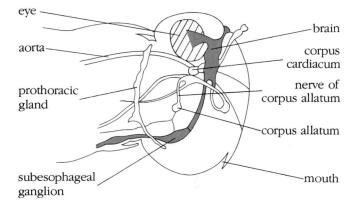

FIGURE 8.4 **Brain neurosecretory cells and glands of the head region in the grasshopper.**

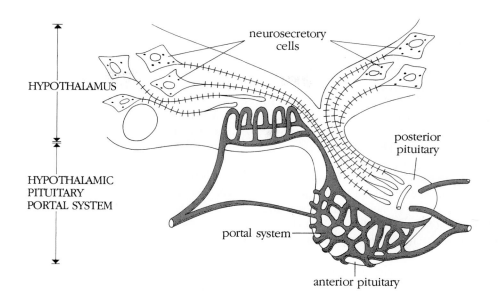

FIGURE 8.5   Hypothalamus and pituitary of a generalized vertebrate neuroendocrine system.

The products of various portions of the pituitary gland regulate other endocrine glands and secrete hormones that directly influence physiology and behavior. Posterior pituitary cells are extensions of nerve cells in the hypothalamus; these cells store and release neurosecretions. Other neurosecretions released in the hypothalamus are carried by blood vessels to the anterior pituitary, where they directly influence hormone production and release.

TABLE 8.1   Sources of hormones that affect behavior, hormones produced, and regulatory and physiological effects

Source	Hormone	Regulatory and Physiological Effects
pineal gland	melatonin	annual reproductive cycle
posterior pituitary gland	oxytocin, vasopressin	milk ejection; parturition; water balance
anterior pituitary gland	luteinizing hormone (LH)	corpora lutea formation; progesterone secretion; Leydig cells stimulation (androgen secretion)
	follicle-stimulating hormone (FSH)	follicle development (♀); ovulation (with LH and estrogen); spermatogenesis
	prolactin	milk secretion; parental behavior in birds
	adrenocorticotrophic hormone (ACTH)	regulates adrenal glands, which release corticosteroids
intermediate pituitary gland0	melanophore-stimulating hormone (MSH)	color change
adrenal cortex	steroids	water balance; metabolism; electrolyte balance; blood sugar level; stress reaction
adrenal medulla	adrenaline (epinephrine), noradrenaline (norepinephrine)	blood sugar level; stress reaction
testis	androgens	testis development; spermatogenesis; secondary sex characteristics; sexual activity
ovary and placenta	estrogens	uterine growth; mammary gland development; sexual activity
	progestogens	gestation

concentration and thus helps regulate water balance. For example, excretion of highly concentrated urine and retention of body water are physiological adaptations of many desert mammals, such as camels, kangaroo rats, and gerbils.

The intermediate pituitary (pars intermedia) secretes a hormone important in communication. Melanophore-stimulating hormone (MSH) affects the concentration or dispersion of pigment granules in chromatophores, or color cells, found in many vertebrates, particularly fish, reptiles, and amphibians. In the absence of MSH, pigment granules remain clumped; MSH stimulation leads to dispersion of the granules and a color change. For example, adult male three-spined stickleback fish (*Gasterosteus aculeatus*) normally have pale-colored sides. But when two males engage in a territorial boundary display, release of MSH triggers dispersion of pigment granules, and the fish's sides take on a bright blue color. Color changes in vertebrates may serve as communication signals, or they may produce coloration patterns that render an animal inconspicuous against a particular background (camouflage).

The anterior pituitary (pars distalis) secretes four hormones that indirectly affect behavior; three of these are trophic hormones that exert their effects on other endocrine glands. In females, follicle-stimulating hormone (FSH) and luteinizing hormone (LH) affect the cycle of egg maturation in the ovaries, sexual receptivity, conception, and pregnancy. In males, FSH and LH control sperm production and secretion of male hormones, or androgens. A third trophic secretion, adrenocorticotrophic hormone (ACTH), affects the production and secretion of adrenal cortex steroid hormones. Prolactin, a pituitary hormone present in birds, mammals, and fish, is important for parental behavior. It influences milk production in mammals, crop milk accumulation in pigeons and doves, and possibly parental behavior in fish. Prolactin is also found in amphibians; in certain species, it stimulates migration to water to breed.

The pineal gland within the brain secretes several hormones that are indoleamines, proteins, and polypeptides. In the study of behavior, the most important of these is probably melatonin. Melatonin functions in the modulation of reproduction and annual breeding activity patterns in mammals (Reiter 1980), among other things. You may be familiar with it, as it is sometimes recommended to combat jet lag and other problems in (daily) biological rhythms. We will postpone further discussion of the role of the pineal gland and melatonin until chapter 9, where we discuss biological rhythms.

Both sexes produce androgens, estrogens, and progestins but in different amounts. The testes produce androgens, primarily testosterone, which directly acts on tissues to produce male sexual behavior. Testosterone is also converted to estradiol, which also affects sexual motivation in several species. The ovaries produce estrogen and progesterones, which influence mating behavior in most mammals. Estrogen primes progesterone receptors. In addition, the ovaries release small amounts of androgens; these may be important for sexual receptivity in humans and other primates. During pregnancy, progestogens secreted by the placenta play a key role in maintaining gestation. Adrenal hormones are involved in maintaining the body's water balance, metabolism, and electrolyte balance. Adrenaline and noradrenaline, from the adrenal medulla, play important roles in emergency stress reactions.

Endocrine Gland Interactions

Feedback Loops   How do endocrine glands interact with each other and with target tissues? Many endocrine gland secretions are characterized by feedback loops; an example is diagrammed in figure 8.6. Pituitary secretion of FSH and LH is controlled by a releasing hormone, Gonadotrophin (GnRH), that travels via the hypothalamic-pituitary portal system from the hypothalamus. The trophic hormones FSH and LH are then carried in the blood to the testes, where they stimulate the gonadal processes of spermatogenesis in the seminiferous tubules and the production and release of testosterone from the interstitial cells. Testosterone, in turn, is carried in the bloodstream to other locations, including the sex accessory glands (e.g., the seminal vesicles, where semen production is stimu-

lated) and the hypothalamus. Specialized hypothalamic sensory cells are part of the body's homeostatic machinery and continuously monitor blood levels of various chemicals, including testosterone and its metabolites. Thus, when we castrate an animal, the amount of testosterone present decreases, but at the same time, releasing factors are secreted into the hypothalamic-pituitary portal system, travel to the anterior pituitary, and cause an increase in output of FSH and LH. To what should we then attribute any observed changes in behavior—to the lower levels of testosterone or to the higher levels of FSH? This last point illustrates one of the difficulties encountered in the study of behavioral endocrinology behavior.

The levels of circulating testosterone influence the secretion of hypothalamic-releasing factors to the pituitary in a negative feedback relationship. As the testosterone level in the blood increases, the secretion of hypothalamic-releasing factors decreases. Conversely, as the testosterone level in the blood decreases, more releasing factors are secreted into the hypothalamic-pituitary portal system, travel to the anterior pituitary, and cause an increase in output of FSH and LH.

There are also feedback loops involving FSH, LH, and the female gonadal hormones, estrogen and progesterone, as we can see by tracing the sequence of events in the estrous cycle of a female mammal. At the start of the cycle, the hypothalamus stimulates release of FSH and LH from the pituitary; FSH predominates at this time. FSH and LH stimulate the growth of follicles (or of one follicle in some species) within the ovary and the production of estrogen from the follicles. When the estrogen level in the blood reaches a peak, indicating the follicles are mature, the estrogen has a negative feedback effect on the pituitary, and it decreases the release of FSH. However, estrogen has a positive feedback effect on LH; the pituitary releases more LH when the estrogen level rises, and LH becomes the predominant pituitary hormone. In those mammals that ovulate spontaneously, the follicles rupture and the ova are released at about the time of the transition to LH dominance. The ruptured follicles secrete estrogen and progesterone under the continuing influence of LH and prolactin. However, rising blood levels of progesterone exert a negative feedback influence on the pituitary that cause it to diminish its release of LH. In mammals where ovulation is induced, the stimulation received during copulation triggers this sequence of hormonal events and results in the release of ova.

Estrogen and progesterone are also responsible for changes in behavior. Species vary in their reproductive physiology and endocrinology, but for mammals in general, the estrous phase of the cycle, when the females are receptive to males for courtship and mating, is influenced by the two ovarian steroid hormones. The complex physiological and behavioral changes are the product of several endocrine pathways that involve positive and negative feedback systems.

Similar feedback loops exist for ACTH and adrenal steroid hormones. The feedback principle is important for a complete understanding of hormone interactions and their effects on behavior. Feedback loops can also be influenced by factors in the environment (e.g., daylength) to alter or set hormone levels.

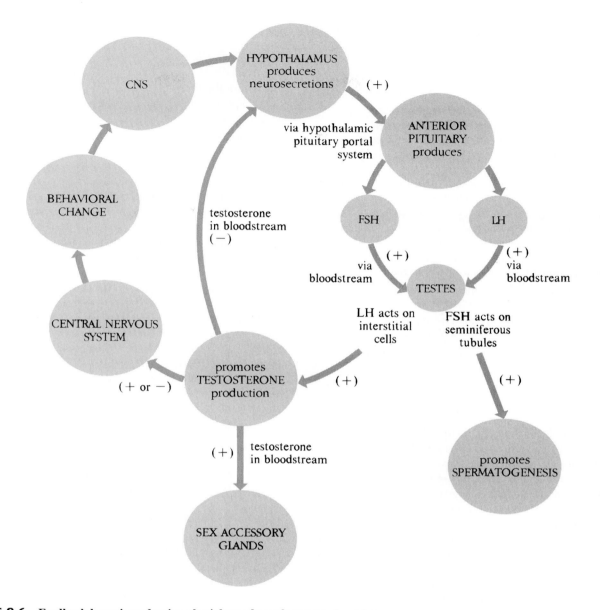

FIGURE 8.6 **Feedback loops in endocrine physiology of a male mammal or bird.**
Feedback loops are a key feature of hormone-behavior relationships. Some of the pathways involve direct effects (e.g., FSH and LH act to stimulate the testes). Other pathways involve negative feedback effects. (e.g., testosterone in the bloodstream passing through the hypothalamus is monitored by specific receptors. High levels of testosterone may result in reduced output of the releasing factors, which affect the anterior pituitary.) (+) beside an arrow indicates a stimulatory effect and (−) indicates an inhibitory effect.

Synergism and Antagonism   Two key endocrine interactions involve synergism and antagonism. Female estrogens and progesterones are often in circulation and simultaneously act together to affect sexual behavior in vertebrates. These effects are termed **synergism.** For example, as mentioned previously, estrogen must be present to prime females to respond to progesterone. Sexual receptivity in female rats depends on the synergistic action of the two hormones (Beach 1976). In contrast, some hormones, when they occur in circulation together, have **antagonistic** actions; the effects of the two hormones appear to be in opposition to one another. Male pigeons initially act aggressively toward a female that is a potential mate, yet eventually their aggression is inhibited as a result of the antagonis-

tic interactions of testosterone and progestogens.

EXPERIMENTAL METHODS

What types of techniques have investigators developed to explore the relationships between hormones and behavior? In the sections that follow, we will explore each technique in more detail as we examine the various ways that hormones can affect behavior. These techniques include the following:

1. Extirpation, or removal, of a particular endocrine gland to assess the absence of a specific hormone on behavior

2. Hormone replacement therapy, in which the investigator injects a specific hormone into an animal, or transplants a gland from another animal to replace one previously removed by surgery

3. Augmenting the animal's own hormone supply with exogenous hormone

4. Antagonist or competitive chemical provision, which involves giving the animal a chemical that antagonizes a particular hormone or a chemical that competes with the hormone and blocks its receptor sites, interfering with the effectiveness of the hormone

5. Blood transfusions to transfer the "hormonal state" of one animal to another in order to observe possible behavioral effects

6. Bioassays to assess circulating hormone levels indirectly by measuring a secondary characteristic that is dependent on a particular hormone such as skin gland size

7. Radioimmunoassay to measure directly circulating levels of a hormone from a blood, saliva, or fecal sample

8. Autoradiography and cytoimmunochemistry to localize the sites where hormone uptake occurs

9. Immunoneutralization by adminstering antibodies against a hormone to effectively remove it from circulation

Several techniques that offer exciting new possibilities for exploring the relationships between hormones and behavior have been developed in the past two decades. Knockout mice (see chapter 6) provide an opportunity to explore more directly the connections between genetics and hormones influencing behavior (Nelson 1997). When a particular gene sequence has been identified, it is disabled by substituting a mutated sequence of DNA for that gene. This enables scientists to study mice that are missing a gene, for example, that codes for production of a particular hormone or for cell receptors that receive particular hormonal messages (Lydon et al. 1995). Another relatively new procedure involves implanting hormones directly into particular areas of the brain. Estrogen implants in the medial pre-optic area of the brains of male rats induces them to become more maternal (Rosenblatt and Ceus 1998). Last, new techniques involving immunocytochemistry have been developed. In these procedures an antibody to a particular hormone is developed. It is then linked with a marker such as a fluorescent dye. When the antibody is injected into a test subject, it binds to the hormone and provides us with a picture of where that hormone is located in the body. This technique has been used, for example, to locate the estrogen receptors in specific areas of the brains of quail (Balthazart et al. 1990).

ORGANIZATIONAL EFFECTS

We can divide hormonal influences on behavior roughly into two categories: organizational and activational effects. The organizational effects of hormones are manifested during an organism's development. **Organizational effects** involve hormonal influences during critical periods in behavioral development that produce relatively permanent changes in the organism's nervous system and other tissues.

For example, sex differentiation and patterns of growth for body tissues are partly under hormonal control. In activational effects, hormones act as triggering influences on the expression and performance of behavior patterns. Direct **activational effects** are occurring when hormone secretion or inhibition of secretion leads to a relatively rapid response. Indirect activational effects require more complex stimulation and hormone secretion sequences. Let us first consider some examples of organizational effects on behavior and related processes, and then some examples of activational effects.

In birds and mammals, sex is determined by the sex chromosomes. However, the sex chromosomes do not directly give rise to phenotype differences between males and females: these come about because of sex hormones secreted by the gonads.

Sexual and Aggressive Behavior

Studies of quail, zebra finches, guinea pigs, mice, rhesus monkeys, musk shrews, and other animals, including several reptiles, provide clear evidence that certain hormones exert critical effects on sex differentiation during early development (Toran-Allerand 1978; Adkins-Regan 1987; Crews 1998; Freeman et al. 1998). Investigations of both mammals and birds have demonstrated the effects of gonadal hormones on organizational processes as they affect later adult sexual and aggressive behavior (Young et al. 1964; Davidson 1966a; Luttge and Whalen 1970; Baum et al. 1990; see also review by Feder 1981; Adkins-Regan and Ascenzi 1987; Schumacher et al. 1989). Let's consider some specific examples of the ways hormones exert organizational effects on behavior and related processes.

If we castrate a male rat within the first four or five days of his birth, he will not show normal sexual behavior as an adult. If we give a neonatally castrated male rat estrogen and progesterone as an adult, he will exhibit female sexual responses, such as the lordosis posture that a receptive female adopts to permit mounting and intromission by a male (see figure 8.1): he shifts her tail to one side, raises his hindquarters, and lowers his abdomen. If we give a male rat that was castrated as an adult estrogen and progesterone, he will not show female sexual behavior.

In an experiment, neonatal male rat pups were injected with estrogen. Histological examination after sacrifice as adults revealed some degeneration of the seminiferous tubules where sperm are produced. Although these males showed mounting behavior when near a receptive female, the behavior was irregular: the mounts were often incorrectly oriented, and no ejaculation occurred.

Female rats treated within the first four or five days after birth with either artificial androgen (TP) or artificial estrogen (EB) did not show normal estrous cyclicity as adults and did not respond with lordosis when injected, as adults, with estrogen and progesterone. When these neonatally injected females were given injections of TP as adults, they exhibited malelike sexual behavior. In fact, the neonatal injection of androgen alone will produce malelike sexual behavior in adults of some inbred strains of mice (Manning and McGill 1974). Females given estrogen injections neonatally sometimes had irregular estrous cycles; females injected with TP neonatally exhibited permanent vaginal cornification. (Vaginal cornification, the sloughing off of the epithelial lining, is produced by the action of estrogens on the lining of the vaginal tract and is a sign of estrus.) Masculinized females also exhibit higher propensities and shorter latencies to fight, compared to those of adult males (Edwards 1968; see review by vom Saal et al. 1992).

Similar studies of the organizational effects of hormones on female behavior have been conducted on guinea pigs (Phoenix et al. 1959) and rhesus monkeys (Goy 1970; Goy, Bercovitch, and McBrair 1988). Female progeny of females treated with androgens during pregnancy have external genitalia that are masculinized (smaller vaginal opening and hypertrophied clitoris), and they exhibit malelike sexual behavior.

These data demonstrate that there is a critical period during which hormone injections must occur for certain behavior patterns to be affected. In rats and mice, this critical period occurs during the initial four or five days after birth; in guinea pigs and rhesus monkeys, the critical time period occurs prenatally. Neonatal injections do not have the same effects on guinea pigs and rhesus monkeys that they do on rats and mice. Prenatally, rhesus monkeys and guinea pigs attain the stage of development that rats and mice attain postnatally. The effects of the injected hormones occur at about the same relative stage of development in these different species of animals. There are also critical periods for the organizational effects of hormones on the physiology and behavior of quail (Schumacher, Hendrick, and Balthazart 1989) and zebra finches (Adkins-Regan and Ascenzi 1987). For quail, the critical time period for organizational effects of estradiol occurs during the nestling phase, in the first two weeks after hatching (figure 8.7).

One interesting aspect of the organizational effects has emerged in recent years: not all aspects of physiology and behavior are influenced in the same manner. Thus, for example, different groups of quail were treated with three different doses of estradiol (0, 5, or 25 μg) on either the 9th or 14th day of embryonic development (Schumacher, Hendrick, and Balthazart 1989). The birds were then castrated at day 4 after hatching to avoid the confounding effects of post-hatching hormone release, and then were given exogenous testosterone as adults. Male birds treated at day 9 during embryonic development, but not those treated on day 14, exhibited demasculinized sexual behavior and increased cloacal gland growth. However, males treated at either day 9 or day 14 exhibited diminished crowing activity and lowered levels of plasma LH levels. Thus, the process of demasculinization may not be restricted entirely to a specific critical period; different behavior patterns may be affected at different times during development.

Similarly in rhesus monkeys, the timing of androgen treatment of females determines the extent of masculiniza-

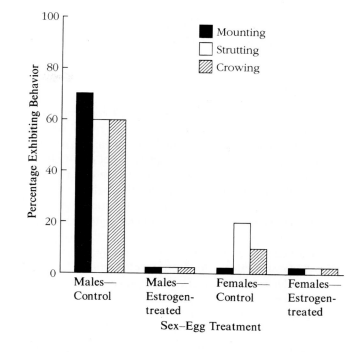

FIGURE 8.7   Organizational effects of estrogen treatment on male and female quail.
Estradiol benzoate was injected into the eggs on the tenth day of incubation. As adults, the birds were held in short day-length conditions to produce regressed gonads. They were then injected with testosterone and tested for three behavior patterns when placed with normal female partners.

tion. Early, but not late, androgen treatment results in masculinization of the female genitalia (Goy, Bercovitch, and McBrair 1988). In a series of behavioral test measures that included mounting, initiation of play, rough play, and grooming, androgen treatment at neither time resulted in masculinization of all four types of behavior. Early androgen treatment produced females that exhibited more mounting behavior (with mothers and peers) and less grooming, whereas females treated with androgen later in gestation exhibited more rough play and more grooming. Thus, it appears that in those animals where the phenomenon has been tested, organizational effects may occur over longer periods than were previously thought; the timing of the effects and the sensitivity of various systems to the organizing hormones vary.

Research on reptiles from three different orders indicates that the organization of behavior via hormonal effects occurs in alligators (*Alligator mississippiensis*), turtles (*Trionyx spiniferus*), and geckos (*Eublepharis macularius*) (see Adkins-Regan 1981). Eggs were obtained from each species and were injected with estradiol at an appropriate age before hatching. The artificial estrogen caused female development in all of the treated animals. Thus, estradiol apparently feminizes the gonads of a variety of reptiles.

Additional Examples of Organizational Effects

In addition to organizational effects on male and female sexual behavior and aggression, we know there are organizational effects on other behavior patterns. Meany and colleagues (1982) investigated the effects of glucocorticoids on play-fighting behavior of Norway rat pups (*Rattus norvegicus*). Glucocorticoids administered prenatally to male rat pups suppressed the normally high levels of play-fighting found in males. Similar treatments given to female rat pups did not influence the level of play-fighting behavior. Thus, there appears to be a sex-dependent, organizational effect of glucocorticoids on the development of play-fighting in rats.

Organizational effects of hormones on sexual behavior are suggested by some data for humans (Money and Ehrhardt 1972). Two such effects involve masculinization of genetically female fetuses. Mothers given progestogens to help maintain pregnancy often gave birth to females whose appearance was masculinized. (This treatment has since been discontinued for pregnant women.) Among nonhuman mammals, some newborn females have a hereditary disease that results in high levels of adrenal gland androgen output—this disease in a pregnant woman could masculinize a female fetus. Genetic males may have target tissues that cannot respond to the androgens secreted from their own testes, and this results in a more feminine appearance. It is important to note that in humans, social experiences often predominate in shaping behavior; individuals usually respond to the sex of rearing regardless of the genetic sex.

In animals that bear multiple young at the same time, fetuses may be influenced by the hormones of their siblings. In cattle, a female with a male twin is called a freemartin and is unable to produce her own offspring in adulthood. In the early part of this century, scientists recognized that hormone action was probably involved; presumably at some time during early embryonic life, a circulatory link connected the blood flow of the two fetuses of opposite sexes. This process could result in the transfer of some androgens to the heifer, resulting in a sterile individual (Cole 1916; Lillie 1916). We now know that this hypothesis is substantially correct.

The intrauterine position of a female fetus in rats and mice can influence her genital morphology and sexual behavior (Clemens 1974; vom Saal and Bronson 1978; vom Saal 1989; vom Saal et al. 1992). Rats and mice release multiple eggs at each cycle and bear litters of sizes ranging from one to twenty pups. Both species have bicornate uteri; fetuses are arranged sequentially in each arm of the uterus (figure 8.8). Thus, fetuses of one sex can be positioned between two other fetuses of the same sex, between two of the opposite sex, or between two fetuses of opposite sexes. In rats and mice, there is a brief period late in gestation when testicular androgens are produced and released. Clemens (1974) proposed that females in utero could be masculinized by exposure to testosterone from neighboring littermates. It turns out that anogenital distance, a measurement of the distance between the anal opening and the genital opening, is a reliable measure of androgen exposure.

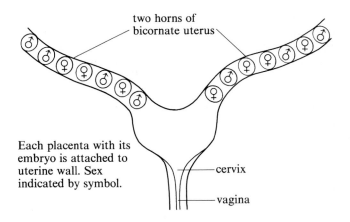

FIGURE 8.8   Rat uterus.
Both rats and mice have bicornate uteri. The often numerous fetuses of a pregnancy are arranged sequentially in each of the two horns of the uterus as shown here for the rat. Pups of one sex may be positioned between two pups of their own sex, between two of the opposite sex, or between fetuses of opposite sexes. The location of a female next to a male fetus results in masculinization of the genetic female.

For rats and mice, anogenital distances for genetic females positioned between males in utero are larger than for females positioned between two females in utero. In addition, certain behavioral and physiological traits are affected. In rats, females that have been masculinized due to in utero position exhibit more mounting behavior (Clemens, Gladue, and Coniglio 1978), while in mice the estrous cycles are longer, levels of aggressive behavior are increased, senescence occurs at earlier ages, daily activity levels are lower, and the tendency to exhibit malelike sexual behavior is increased in females between two males compared with females between two females or between a male and a female (Gandelman, vom Saal, and Reinisch 1977; vom Saal and Bronson 1980a,b; vom Saal 1989). Interestingly, male fetuses that are positioned between two female fetuses have higher concentrations of estradiol and show organizational effects as well; such males have altered genital and brain morphology, higher levels of activity, less sexual behavior, a tendency to be infanticidal toward pups, and less responsiveness to doses of testosterone used to induce intermale aggression than do males positioned between two other males in utero (vom Saal 1989).

Some species have alternative male phenotypes. For example, in tree lizards (*Urosaurus ornatus*) males of different phenotypes engage in different types of mating tactics (Hews and Moore 1995; Moore et al. 1998). Some males are territorial, whereas others sneak into territories to obtain matings with females. Depending upon the frequencies of these various mating types, it may be advantageous to switch to a different phenotype. One process by which these phenotype arise may involve both organizational influences of steroid hormones in the traditional pattern of organizational influence followed by activational effects later in life. The other process, termed the relative plasticity hypothesis, entails hormonal mechanisms that can switch phenotypes on and off during adult life.

From these studies we can reach some tentative conclusions about the effects of gonadal hormones on the organization of behavior in animals. Gonadal hormones (either estrogens or androgens) secreted at the appropriate critical time during early development affect the undifferentiated brain by their production of a malelike adult organism, or one exhibiting malelike behavior, regardless of the actual genetic sex. Early gonadal hormone secretion apparently sensitizes the brain to hormones that circulate in the blood later. In the absence of gonadal hormone secretion, a typical cyclical female pattern develops (see reviews by Arnold and Gorski 1984; vom Saal et al. 1992). Again, the absence of hormones establishes certain patterns of sensitivity in critical areas of the brain, such as the hypothalamus, so that later hormone circulation produces particular physiological and behavioral responses. The general sensitivity of particular areas of the brain, and these areas' sensitivity to different chemicals' temporal patterns or blood levels of different chemicals are affected (Dörner and Kawakami 1978).

Thyroid and Adrenal Glands

Are there other endocrine glands whose products can have organizational effects on behavior? Hormones from the thyroid and adrenal glands have been implicated in organizational processes. Thyroidectomized rats exhibit characteristics of cretinism, slower growth, delayed sexual maturation, and retarded development of the nervous system. Their actions are slower, and learning is slower and occurs with great difficulty.

Young rats that are given a few minutes of handling each day during early infancy exhibit less extreme responses to stressful situations as adults (Levine 1967, 1968). When adult rats that were handled as pups are placed in a stressful situation, they secrete less adrenal steroid hormones and show less fear response to presentations of novel stimuli. When they were pups, they secreted more adrenal steroid hormones in response to shock or ACTH injections. Early in development, these higher levels of early adrenal steroid hormones may have affected brain mechanisms related to stress responses and thereby established permanent changes that continued into adulthood. These findings exemplify another principle regarding organizational effects on behavior: Some of the effects organize behavior in a manner that does not require hormones for activation in adulthood, whereas other effects do require hormones later in life for activation of the behavior.

Invertebrates

All of the preceding material has dealt with vertebrates. Are there organizational effects of hormones in invertebrates? There are, in fact, several processes in invertebrates that involve organizational effects of hormones on behavior and related physiological processes (Truman and Dominick 1983; De Loof and Huybrechts 1998), though additional studies in this area are needed. The processes involved in insect metamorphosis from pupa to adult, and in the molting that takes place periodically as the insect forms a new exoskeleton and sheds its old one, involve organizational effects of specific hormones. The last stage of metamorphosis is called eclosion, and shedding of the old skin is called ecdysis. There are apparently some close parallels involved in the hormonal control of these processes. For the molting sequence, there are three behavior patterns associated with ecdysis: the finding of a suitable perch, the stereotyped movements involved in removing the old exoskeleton, and the expansion of the new cuticle to fit the body of the animal. Many of the steps in this sequence depend upon the level of eclosion hormone present in the insect. Changes in hormone titers and the timing of these changes are activational hormone effects that influence the behavior sequence. However, it also appears that for some insects (e.g., the tobacco hornworm [*Manduca sexta*]), certain hormone changes in the sequence prime or organize the system for later effects. Thus, the decline of eclosion hormone level at one stage early in the sequence is necessary to trigger certain preparatory behavior, (e.g., perch finding), and also to make the system responsive to increases in the eclosion hormone titer at a later stage, inducing the next step in the sequence. If an exogenous hormone that mimics the actions of eclosion hormone is given to the hornworms at the time that they would be experiencing a decline in the natural levels of the hormone, then the proper preparatory stages are delayed. This is a dose-dependent effect, with larger doses of exogenous hormone leading to longer delays before the next step in the sequence occurs.

Two invertebrate hormones, MH (molting hormone, also called ecdysone, or ecdysteroid) and JH (juvenile hormone), interact in the control of growth and metamorphosis in insects. When levels of JH in the blood are high, and MH is also present, the insect continues to grow and differentiate, but will not molt to the adult stage. However, if MH acts alone, molting is induced and the insect will metamorphose and differentiate into the adult form. There is additional evidence from studies of some insect species that the differentiation and maturation of sexual organs depend partly on gonadal hormones (Gilbert 1974; de Wilde 1975; Fraenkel 1975).

Recent studies have shown that there are also organizational effects in insects with respect to sexual differentiation (De Loof and Huybrechts 1998). Early work by Naisse (1966a,b,c) revealed that in the firefly (*Lampyris noctiluca*) there are differences between males and females that begin during the fourth larval instar. These differences occur in males and involve the presence of a hormone that eventually leads to differentiation of normal male sex organs. The absence of this hormone results in the development of a female. More recently an androgenic hormone when isolated from a crustacean amphipod (*Orchestia gammarella*) appears to play a role in sexual differentiation. Finally, investigators have now isolated a series of compounds from invertebrates called ecdysteroids; questions are now being tested about the equivalency of these compounds with estrogens and androgens in vertebrates (De Loof et al. 1981; Girardie 1995; Lageaux 1981). Progress in this area when working with invertebrates is slower than comparable work on vertebrates because many of the experiments take longer periods of time and glands are quite small, meaning that many animals (up to 20,000) are needed to successfully extract, isolate, and characterize the hormones. With modern molecular techniques, we can expect additional progress with understanding effects of organizational hormones and related processes in invertebrates.

ACTIVATIONAL EFFECTS

Aggression and Sexual Behavior Patterns

When male ring doves (*Streptopelia risoria*) are castrated, they exhibit decreased levels of aggression, courtship, and copulation behavior. When castrated birds are treated with implants of crystalline testosterone propionate in specific sites in the hypothalamus, the normal levels of these behavior patterns are restored (Hutchinson 1978; Barfield 1971). Such experiments clearly demonstrate the activational effects of testosterone on sexual and aggressive behavior and suggest that specific brain sites may be stimulated by testosterone—sites that influence sexual behavior. Seasonally higher levels of testosterone are recorded for male snow buntings (*Plectophenax nivalis*) in conjunction with increased territorial behavior (Romero et al. 1998). Similarly, territorial behavior and testosterone levels are higher on a seasonal basis in male northern fence lizards (*Sceloporus undulatus hyacinthinus*) (Klukowski and Nelson 1998). In groups of spotted hyenas (*Crocuta crocuta*), males that are natal to the group have lower testosterone levels, are less aggressive, and engage in far fewer sexual encounters than immigrant males that have joined the group (Holekamp and Smale 1998).

Results from other researchers indicate that the presence or absence of testosterone influences aggressive behavior in birds and mammals, such as ring doves, roosters, mice, rats, and domestic cats (Bennett 1940; Tollman and King 1956; Guhl 1961; Barfield et al. 1972; Leshner 1975). Intact male birds and mammals show more aggression, have shorter latencies to initiate fighting behavior, and fight more frequently than do castrated subjects of the same species. Castration is used with some domestic livestock, such as swine and cattle, to alter behavior patterns, to reduce aggression, and thus to make the animals easier to handle and less likely to injure one another.

Presence or absence of testosterone influences sexual behavior in a variety of animals other than birds and mammals, including South African clawed frogs (*Xenopus laevis*) (Kelley and Pfaff 1976) and the Carolina anole (*Anolis carolinensis*) (Crews 1974). Harvey and Propper (1997) showed that the presence of testosterone is necessary for both amplexus, a key feature of sexual behavior in frogs in which the male clasps the female, and the development of darker coloration in the thumb pads of the males, a secondary sexual characteristic. Testosterone influences the electric organ discharge of various species of electric fish (Dunlap et al. 1998; Dunlap and Zakon 1998).

An interesting series of studies has been done on the hormonal control of newt reproduction. Newts are amphibians that live on land most of the year but migrate to water for breeding. Prolactin induces migration, and also influences the morphological expansion of the tail fin in males, which they use to fan a stream of water towards a prospective mate (Chadwick 1941). Toyoda et al. (1992), using intraperitoneal injections of hormones, found that prolactin resulted in increased performance of the tail-fanning behavior. Moreover, injections of both prolactin and growth hormone produced even greater amounts of the behavior pattern. This demonstrates both the hormonal control of a sexual behavior pattern and also the synergism that occurs when two or more hormones act in concert to produce an additive or multiplicative effect on behavior.

Both male and female Syrian golden hamsters (*Mesocricetus auratus*) have paired sebaceous flank glands that they use to mark objects in their environment by depositing sebum from the gland (figure 8.9). The flank glands are androgen dependent; measurement of their size and pigmentation can be combined in an index that is a bioassay measurement of relative levels of circulating androgens (Vandenbergh 1971, 1973; Drickamer and Vandenbergh 1973; Drickamer, Vandenbergh, and Colby 1973). When groups of four intact male hamsters that had been isolated since weaning and that were equal in body weight were allowed free social interaction in a large pen, there emerged a significant positive correlation (r = .77) between the outcomes of encounters and the index of gland size and pigmen-

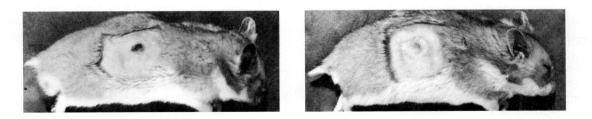

FIGURE 8.9   Hamster flank glands.
Hamsters have paired sebaceous flank glands that are used to mark objects in the environment with sebum. The size and pigmentation of the glands are androgen dependent. A normal, intact male (left), and (right) a castrated male.
Source: Courtesy John G. Vandenbergh.

tation. The key feature of this experiment was that the investigators measured the glands before they placed the hamsters together and could predict the outcomes of social encounters on the basis of the bioassay (gland size) of relative levels of circulating androgens. In a related experiment, the investigators gave injections of testosterone propionate to castrated male hamsters in four groups. Each hamster in a group was given a different dose, and when the four were allowed to interact, the investigators could again predict the outcomes of social encounters (r = .81) on the basis of dose level and corresponding gland index. Interestingly, investigators could also predict the outcomes of social interactions among female hamsters on the basis of the measurement and pigmentation of their flank glands (Drickamer and Vandenbergh 1973).

Whether we are studying the relationships between hormones and aggression in vertebrates and whether we conduct our investigations in laboratory or field, there are some constraints. This is particularly true when we remove the animals from their natural context to study hormones and aggression (Monaghan and Glickman 1992). Aggression in nature may be about food, mates, living space, etc. Without a clear knowledge of what conditions are important for aggression in the particular species under study, we risk misinterpretation of what we observe. In fact, it is possible that animals may not exhibit aggression under some conditions where we have manipulated their environment, whereas those same conditions could normally result in aggressive interactions. Thus, if we supply extra food, we may reduce aggression, or we may crowd animals too much by confining even a single test pair of animals in an area that is too small, and aggression would be reduced or increased accordingly. We may also attempt to record relationships between hormones and aggression at an inappropriate time of the year. Many male vertebrates exhibit heightened levels of aggression (and correlated with this, higher levels of testosterone) when they breed, due to competition either for space or mates. Testing such animals at the nonbreeding period of the year could have strong consequences for our results and their interpretation.

Hormones and Cognition

In addition to its organizational and activational effects on reproductive behavior, estrogen has been implicated in brain functions related to cognitive processes (Williams 1998). Studies on rats have revealed that depriving the brain of estrogens results in a reversible learning deficit (Singh et al. 1994). Fugger and colleagues (Fugger et al. 1998) tested the effects of the absence of the gene for an estrogen receptor in mice on their ability to learn a water maze task. Mice are placed in a tank with water and a platform to which they can escape. Female mice with this knockout gene treated with estrogen did not learn the maze task, whereas male mice and controls of both sexes did. These findings indicate a role for estrogen in spatial learning tasks. Learning and memory are modulated, in part, by actions of estrogens. Similar findings in humans suggest that estrogen replacement therapy can protect postmenopausal women from several aspects of functional decline in cognitive processes (Kimura 1995).

Secondary Sex Characteristics

In addition to their direct activational effects on behavior, hormones can affect secondary sexual characteristics. The characteristic cock's comb of the male chicken decreases greatly in castrated roosters. Male cats, which spray urine probably as a marking behavior, often cease to spray after their testes are removed. In these two examples, a change in a secondary sex characteristic (cock's comb) and in a sex-related behavior (urine spraying) both affect the process of communication.

Sexual Attraction

In many insect species, females release pheromones that act as sex attractants for males. A **pheromone,** as discussed in chapter 7, is a chemical produced by an animal and released in the external environment where it may affect conspecifics' behavior or physiology. If a female cockroach's corpora allata are surgically removed after she has molted to the adult state, she does not produce pheromones when she becomes reproductively active (Barth 1965). Females treated in this way are incapable of attracting males and do not mate. However, if a corpus allatum from another adult female is transplanted into the test subject, she is again capable of pheromone production and release. Apparently a hormone that is produced by the corpus allatum acts on peripheral glandular tissues to stimulate production of the sex-attractant pheromone. Interestingly, if a female is gonadectomized as an adult, the processes of pheromone release, mate attraction, and mating are not affected.

Why has this system of mate attraction evolved in some moths, cockroaches, and other insects? There are several possible answers to this question. First, emission of a species-specific sex-attractant chemical may be a species-isolating mechanism. If several closely related species are reproducing at the same time, it would be advantageous—to avoid gamete wastage and to allow insects to locate potential mates of their own species. Thus, when each species has a different sex-attractant pheromone, individuals can easily locate and mate with partners of their own species. Second, some pheromones may serve to synchronize the reproductive activities of the two sexes of a given species so that males and females are ready to mate at the same time. Third, these pheromones may set the stage for hormonal and behavioral changes accompanying the nest-building and parenting behavior. Finally, of course, the chemical attractant ensures that a male will find a female and mate with her.

Eclosion

The process whereby the adult form of an insect emerges from the pupa after metamorphosis is called **eclosion** and is another activational effect controlled largely by hormones.

Many moth species eclose at a species-specific time of day. The eclosion hormone, produced by neurosecretory cells in the brain, plays a critical role in this process (Truman and Riddiford 1970; Truman 1971). If the eclosion hormone is injected into pupae that are near the end of metamorphosis, eclosion behavior, such as abdomen movements and wing spreading after emergence, can be activated at any time of day. Moths that have had their brains removed usually emerge successfully; therefore the presence of the eclosion hormone is not an absolute requirement for eclosion to take place. The process, however, is not as coordinated in brainless subjects, and some activities (e.g., wing spreading) are usually absent. Thus, although the hormone may not be necessary for eclosion, it does appear to be necessary for proper coordination of the sequence.

Life Stages

Adult male desert locusts (*Schistocera gregoria*) fail to exhibit sexual behavior when the corpora allata have been removed. When corpora allata from other adult males are transplanted into allatectomized males, sexual behavior is restored (Lohrer 1961; Pener 1965). However, similar investigations have revealed that the corpora allata are not needed for sexual behavior in certain grasshoppers (Barth 1968).

Several locust species exhibit both solitary and gregarious phases; young hoppers (see figure 8.3) reared in isolation exhibit moderate activity levels and do not engage in sustained flights; hoppers reared under crowded conditions do engage in long flights (Johnson 1969). Some evidence supports the conclusion that this difference has a hormonal basis. First, solitary hoppers have larger prothoracic glands than do gregarious forms (Carlisle and Ellis 1959). Second, adult locusts that develop from solitary hoppers retain the prothoracic glands, but the glands are absent in adults that develop from gregarious hoppers. Third, if homogenates of prothoracic glands are introduced into gregarious hoppers, general activity is reduced (Carlisle and Ellis 1959), and if prothoracic glands from solitary adults are transplanted into gregarious adults, sustained flight activity is diminished (Michel 1972).

Molting

Studies of hormones and neurosecretions in invertebrates other than insects have revealed common patterns of effects. We know that crustaceans, molluscs, and some worms, have one or more endocrine glands, the secretions of which affect sexual differentiation, gamete maturation, and reproduction stimulation (Wells and Wells 1959; Charniaux-Cotton and Kleinholz 1964; Golding 1972). In some crustaceans, several behavior-related phenomena are under partial endocrine control. Many crustaceans (as well as other arthropods) **molt,** shed their exoskeletons, periodically as they grow. Removal of both eyestalks in these animals shortens the interval between molts. If the crustaceans are given extracts of a particular neurosecretory gland, molting is prevented. Clearly the gland produces a molt-inhibiting factor. It is also possible that some cells in the eyestalks are involved in the production and secretion of a substance that shortens the interval between molts (Kleinholz 1970).

Color Change

As we noted earlier, a melanophore-stimulating hormone (MSH) affects color changes in fish. We can also see MSH action in short-tailed weasels (*Mustela erminea*), which undergo seasonal changes in **pelage,** or coat color, during spring and fall molts (Rust 1965; Rust and Meyer 1969). In the spring, MSH secretions increase, and new brown hairs replace the white winter coat. During the fall months, MSH release is inhibited by the action of melatonin, another hormone that is secreted by the pineal gland. The developing hairs at this time of year are not pigmented, and the weasel's coat returns to its white winter color. These seasonal shifts in coat color may be both behaviorally and functionally significant because the result matches the general background color of the weasel's environment. Camouflage coloration serves a possible dual function because it enables the weasel to hunt inconspicuously and to be "hidden" from potential predators.

MSH has also been implicated in the color changes that occur in fish, amphibians, and reptiles. For example, the color of two fish competing at a territory boundary may change or deepen during the course of the interaction. One interesting feature of the way MSH affects color changes in fish, amphibians, and reptiles is the rapidity of the physiological changes—some take only seconds.

BEHAVIOR-ENDOCRINE EFFECTS

In addition to the ways in which endocrines influence behavior, we should also consider the manner in which behavior and the outcomes of behavioral interactions can influence hormone levels. Two rather different examples, one involving domestic sheep and the other humans, serve to illustrate this pathway.

Previous sexual experience, presence of particular mating cues, and season of the year were tested with regard to their influence on endocrine responses in domestic male sheep (rams) (Borg et al. 1992). Testosterone and several other hormones were measured via radioimmunoassay on rams that were (a) sexually experienced or inexperienced, (b) provided complete access to estrous ewes or given only restricted access permitting courtship but not mating, and (c) tested during summer and fall seasons. All experienced rams achieved ejaculation when permitted to do so, but only 33 percent (summer) and 67 percent (fall) of the previously sexually inexperienced rams achieved ejaculation. Levels of testosterone were greater in the experienced rams than in the inexperienced rams. Testosterone levels were elevated more during restricted access to ewes, com-

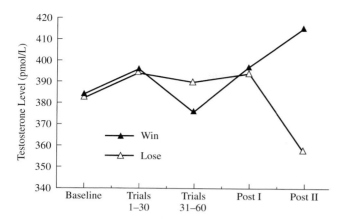

FIGURE 8.10   Human male testosterone levels.
Corrected testosterone levels for human males as winners or losers measured by radioimmunoassay based on saliva for a noncompetitive situation. Measurements were taken at five separate times for both winners and losers.

pared to the levels when males were given complete access to the ewes. Thus, behavior and related events can have consequences for levels of circulating hormones; these hormonal events may, in turn, affect subsequent behavior.

That particular situations can affect hormone levels for male humans was recently tested by McCaul, Gladue, and Joppa (1992). Other investigators have previously reported relationships between testosterone levels and aggression (Christiansen and Knussman 1987) and sexual behavior (Knussman, Christiansen, and Couwenbergs 1986). Conversely, aggressive behavior in a competitive situation can influence testosterone levels (Elias 1981) as can sexual behavior (Fox et al. 1972). What might the consequences for testosterone levels be in noncompetitive situations in which males can win or lose? Male college students were used to test this question. Saliva was used to assay testosterone levels (Gladue, Boechler, and McCaul 1989) before and after the subjects participated in a random coin-flipping experiment in which they could win or lose $5. Testosterone levels for winners were significantly higher than for losers (figure 8.10). The winners also reported positive mood changes relative to the losers. Again, behavioral and environmental events apparently can influence endocrine levels, possibly mediated, in this instance by a change in mood.

ENDOCRINE-ENVIRONMENT-BEHAVIOR INTERACTIONS

Some activational effects of hormones involve complex interactions between behavior, hormones, and environmental stimuli. We shall now discuss in some detail four examples that illustrate these interrelationships: reproductive sequence in ring doves, parturition and maternal behavior in rats, reproduction in lizards, and reproduction in house sparrows.

Reproductive Sequence in Ring Doves

Figure 8.11 illustrates the reproductive sequence in ring doves. A male begins courtship display shortly after being placed with a female. The failure of castrated males to court females indicates the importance of a continuous supply of androgens for the initiation of the cycle. Male courtship stimulates pituitary release of FSH in the female dove; FSH in turn stimulates follicle development in the ovaries. The follicles secrete estrogen, which affects uterine growth and development. Within a day or two, the birds begin nest construction; during this phase of the cycle, they copulate and continually add to their nest. The presence of a nest stimulates the production and secretion of progesterone in females. Progesterone in both sexes promotes incubation behavior after eggs are laid. Egg laying is activated partially by secretion of LH by the female's pituitary.

Incubation, maintained by progesterone secretion, lasts 14 days; the male and female take turns on the nest. Under the influence of the presence of eggs in the nest and as a result of stimulation from incubation behavior, both the male's and the female's pituitary glands secrete prolactin. Prolactin acts to inhibit FSH and LH secretion, and all sex behavior ceases. Prolactin also stimulates crop development and the production of crop milk (a nutrient-rich fluid secreted in the gullets of males and females) and may also help to maintain incubation behavior. When young squabs hatch after 2 weeks, the parents immediately feed them with crop milk. During the next 10 to 12 days, both parents continue to feed the young with crop milk; however, feeding behavior wanes toward the end of this period, in part because the secretion of prolactin decreases. As prolactin decreases, the pituitary secretes FSH and LH, the same pair of doves resumes courtship, and the sequence begins again.

At each stage of this sequence, the internal state of each bird interacts with external variables to produce the observed behavior patterns. The variables consist of (1) the hormonal state of both the male and female dove, including the feedback loops; (2) the behavior of each member of the pair that stimulates changes in the hormonal levels and behavior of its mate; and (3) environmental cues, such as nests and eggs, that influence hormonal and behavioral changes in both (figure 8.12).

Daniel Lehrman and his colleagues conducted many ingenious experiments to clarify the interactions and changes that together comprise the reproductive sequence in ring doves. A description of a few of their experiments gives us a glimpse of the logic and experimental method they employed (see Lehrman 1955, 1958a,b, 1959, 1961, 1964, 1965; Lehrman, Brody, and Wortis 1961; Erickson and Lehrman 1964; Cheng 1979).

To determine whether the presence of a mate or of nesting material affects incubation behavior of female ring doves, they used three experimental groups: (1) control females housed alone, (2) females housed with a male only, and (3) females housed with a male and nesting material. They assessed the results of these pairings in terms of the percentages of test females in each group that exhibited incubation

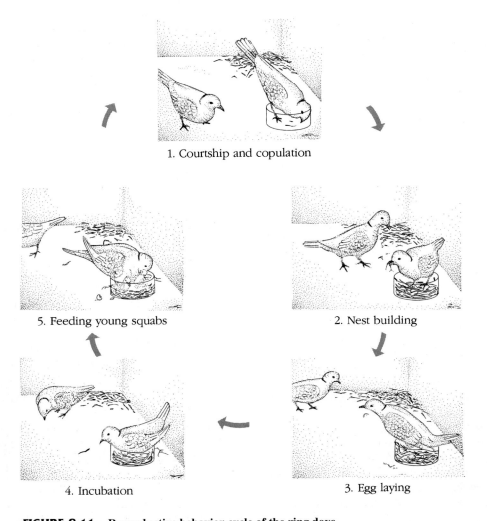

1. Courtship and copulation

5. Feeding young squabs

2. Nest building

4. Incubation

3. Egg laying

FIGURE 8.11   Reproductive behavior cycle of the ring dove.
This cycle provides an example of indirect environmental determinants of behavior. The sequence involves (1) courtship and copulation, (2) nest building, (3) egg laying, (4) incubation, and (5) feeding crop milk to the young squabs after they hatch. The cycle then repeats.

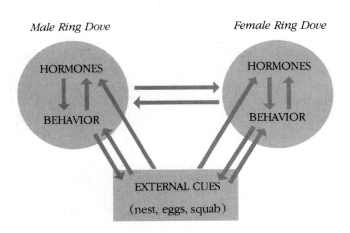

FIGURE 8.12   Interrelationship between hormones and behavior in ring doves.
Hormones and behavior patterns of each individual, interactions between individuals, and external cues affect the synchrony of reproductive behavior. Bidirectional arrows indicate feedback relationships, and unidirectional arrows indicate direct effects.

behavior when presented with a nest containing eggs (figure 8.13). Control females never incubated eggs. By days 6, 7, and 8 after pairing, increasing percentages of females in the second group (those housed with a male only) incubated eggs; by day 8, 100 percent of females caged with a male and nesting material incubated test eggs. Thus, we can conclude that the presence of the male and the nesting material are necessary for complete incubation behavior in the female ring dove. Similar research with the male ring dove shows that the presence of both the female and the nesting material are necessary for males to show complete incubation behavior.

A combination of several techniques has been used by Silver and her colleagues to explore the stimuli that affect males and possible corresponding hormonal changes during the early portion of the ring dove reproductive cycle (Silver 1978; O'Connell et al. 1981a,b). Male doves were presented with varying stimuli, their behavior was observed, and radioimmunoassay measurements were made of plasma hormone levels. Males presented with intact courting females had higher levels of testosterone than males paired with

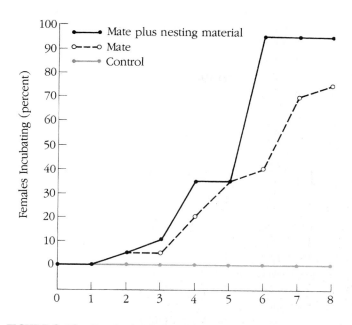

FIGURE 8.13   Incubation behavior in ring doves.
Development of incubation behavior in female ring doves is affected by association with a mate or with a mate plus nesting material. Each point on this graph is derived from tests of twenty different birds.

FIGURE 8.14   Ring doves courting.
Female ring doves separated from their male partners by a glass partition can still induce both behavioral and hormonal changes in the males. Males in this situation begin to court the female and exhibit increased levels of plasma testosterone.
Source: Rae Silver.

ovariectomized females. Males exposed to females caged behind glass partitions had testosterone levels comparable to those for males given free access to females (figure 8.14). Thus, it appears that the female's gonadal condition can influence the male's response. When surgically deafened males are exposed to intact females, their testosterone levels were lower than those of normal males given the same exposure. Some social and auditory stimuli appear to be important early in the reproductive sequence and result in high male testosterone levels and stronger behavioral responses to females.

By about the eighth day after pairing, when incubation begins, the male's testosterone levels have declined to the precourtship baseline level. Courting males were exposed to either an incubating female or another courting female. Those in the first group exhibited reduced testosterone levels within a few days, whereas testosterone levels remained high in the second group. Also, the effects of nests and nest material on testosterone levels were tested. Males whose nests were destroyed each day, or who were given no nest material at all, had higher testosterone levels than did males who were given nest material and left undisturbed.

Both external stimuli and female dove behavior appear to influence male behavior and male hormone levels at several early stages in the sequence. In the male's behavior, transitions appear to be strongly mediated by context stimuli; whereas for females, transitions appear to be influenced more by hormonal changes, as we noted previously.

In another study, male ring doves were allowed to court and mate with females and to participate in nest construction. After the eggs had been laid and incubation had begun, each male was placed behind a partition so that he could see his mate sitting on the eggs, but he could not incubate the eggs.

These males, stimulated only by the visual cues from an incubating mate, underwent normal crop development, and when permitted, they fed the young squabs normally after hatching. Males that were not permitted to watch their mates incubate eggs failed to develop crops or to feed the young. Although participation in nest building is necessary for crop development, direct involvement in incubation is not. Cheng (1992) provided another twist concerning behavioral effects on endocrine levels. For many years, the interpretation of the behavioral sequence included male ring doves performing a bow and cooing to the female at an early stage as part of the courtship. These actions stimulated endocrine processes, leading to nest building by the pair and changes in the female, leading to ovulation. Cheng has now determined, through additional experiments that involved devocalizing female ring doves, that they would not undergo follicular development unless they were exposed to their own coos during the courtship. Thus, it now appears that both the male courtship and the female hearing her own coos are necessary for the endocrine stimulation resulting in ovulation.

Why has such a finely tuned system of complex interactions between behavior, hormones, and external cues evolved in ring doves? The almost lockstep system that characterizes the reproductive cycle of ring doves—where each stage in the sequence is dependent on certain interactions and cues from preceding stages—ensures that as the cycle proceeds, both members of the pair are in synchrony and both are in the proper "frame of mind" to perform the required behavior as needed. Since the system involves the participation of two birds and necessitates that their activities be coordinated throughout, the reproductive success of the pair is guaranteed only if both perform certain actions in synchrony. For example, females that laid eggs before a nest was completed would contribute little to future generations. If the male developed a crop or began to produce crop milk just as the female laid the eggs, long before any squabs had hatched, and ceased to produce crop milk during the two weeks they are normally fed by both parents, only the female would be capable of supplying the squabs with nourishment. Her food supply might be insufficient, and both of their reproductive energies would have been wasted because some or all of the squabs would die before fledging.

Parturition and Maternal Behavior in Rats

In some ways, maternal behavior in the laboratory rat is similar to the cyclical reproductive behavior of ring doves: The internal state of the female rat is partly a response to external stimulation of the presence of a nest and pups. Three major events of the maternal cycle will illustrate the interaction between internal state and external stimulation: (1) parturition and the events surrounding the birth of a new litter, (2) nest building, which occurs both before and after parturition; and (3) the period of lactation (Lehrman 1961; Zarrow 1961; Rosenblatt and Lehrman 1963; Richards 1967; Lott and Rosenblatt 1969; Rosenblatt 1970; Lubin et al. 1972; Rosenblatt, Siegel, and Mayer 1979).

First, let's examine events at the time of parturition. Progesterone, sometimes called the pregnancy hormone, helps maintain proper internal conditions prior to parturition. Internal changes in hormone secretions and shifting external cues occur in the days just before birth and at the time of birth. The previously low level of estrogen begins to increase, and progesterone level may decrease; the overall effect is a shift in the estrogen/progesterone ratio. These hormonal changes help trigger parturition, and after birth, to trigger the retrieving and nursing behavior that characterizes a mother's treatment of her newborn pups. Estrogen acts synergistically with oxytocin released by the pituitary to promote the secretion of milk from the mammary glands. Prolactin from the pituitary acts to promote milk production (Masson 1948; Lott 1962; Grota and Eik-Nes 1967).

To demonstrate that hormonal changes occurring around the time of birth are at least partially responsible for producing changes in maternal behavior, Terkel and Rosenblatt (1968, 1972) used a system of chronically implanted heart catheters in rats. They mounted catheters on the rats' necks and connected them to a swivel pump that permitted the shunting (transfusion) of blood from one rat to another. When blood was shunted from newly parturient females to virgin females provided with young pups, the virgin females exhibited maternal behavior, such as retrieving pups and crouching over the pups to nurse, after 14 to 15 hours of exposure to the pups.

Some of the events that occur during parturition do not have to take place for the female to exhibit maternal behavior, for example, the birth process itself and the consumption of birth fluids or the placenta. If a female is prevented from experiencing the events associated with normal birth through the removal of the fetuses by Cesarean section and then is shortly thereafter presented with newborn pups, she still exhibits characteristic maternal behavior (Moltz, Robbins, and Parks 1966).

What about the hormonal relationships during the period of nest building? Observations of preparturient female rats tell us that nest-building behavior begins to increase in intensity 4 to 6 days before parturition. The construction of the first nest, called a prepartum nest, is apparently a behavior that is partially under hormonal regulation, although investigators have obtained conflicting results. Several experiments have demonstrated that the change in the estrogen/progesterone ratio is an important nest-building trigger. When pups are present, a female builds a litter nest with higher sides (figure 8.15); the trigger for this behavior seems to be the pups' presence.

Finally, consider the events that occur during lactation. Maternal behavior during lactation and up to the time of weaning (3 to 4 weeks of age) depends on close behavioral synchrony between mother and pups. Most of the interactions between them operate through nonhormonal channels. For example, the retrieving responses and the nursing crouch have been induced in virgin females and males. The test subjects exhibited maternal behavior when they were presented with stimulus pups each day for up to several weeks; they did not exhibit maternal behavior at the first presentation of pups. In addition, a female that has been presented with a replacement litter at any time up to the midpoint of her lactation period will often continue to lactate (Nicoll and Meites 1959). As the development of a normal litter proceeds, pup behavior interferes with maternal behavior—that is, the pups make it difficult for the female to perform maternal behavior. Maternal rats retrieve pups frequently up to the second week of lactation, and then this behavior declines. At first, pups are nursed only in the nest and with the mother's assistance; but later, nursing can be initiated by pups and can take place away from the nest, wherever pups and mother encounter each other. After 3 to 4 weeks of nursing, FSH increases again in the female, and the estrous cycle resumes (Bruce 1961; Rosenblatt 1967; Terkel and Rosenblatt 1971).

FIGURE 8.15   Female rat with pups in litter nest.
Prior to parturition, the female rat constructs a prepartum nest, a flat mat of material. After giving birth, she builds a better nest with higher sides, partially due to stimulation provided by the presence of her pups.
Source: Courtesy Ronald J. Barfield.

Reproduction in Lizards

Using primarily the green anole lizard (*Anolis carolinensis*), Crews (1975, 1977, 1979, 1980, 1983; Crews and Greenberg 1981) and his colleagues conducted field and laboratory investigations of the behavioral endocrinology of reproduction and the significance of various endocrine-behavior events in environmental adaptation. To understand these hormone-behavior interactions better, we first need to make note of the annual sequence of events in these lizards' life cycle. Louisiana populations of the green anole, which investigators have studied most extensively, exhibit a four-part annual cycle:

1. From late September to late January, the lizards are quiescent and live under tree bark and rocks.

2. In February, the males emerge from dormancy and establish breeding territories.

3. In March, the females become active; mating begins in late April, and by May, they are laying one egg every 10 to 14 days, a pattern that they continue for several months.

4. In August, both sexes enter a refractory period for about one month; during this time, the same environmental and social cues that in the spring bring about gonadal recrudescence are no longer effective.

Combined laboratory and field techniques have led to significant conclusions about reproduction in the green anole lizard. Let's examine a sequence of observations and see what each can tell us about hormones and behavior. The seasonal rise in temperature in the spring (Licht 1973) and the male's courtship behavior (Crews, Rosenblatt, and Lehrman 1974) affect ovarian development and egg laying in the female. The extension of the male's dewlap during displays is the key to selection of males as mates by the females and to the promotion of ovarian activity (figure 8.16). Females are receptive to advances by males only when conception is most likely to occur. This receptivity is partially regulated by the estrogen secreted by the developing follicle. Mating activity inhibits further receptivity by the female—the inhibition begins within 24 hours after mating and lasts for the duration of that cycle.

Several points regarding the adaptive significance of this reproductive pattern emerge from the data. The mating inhibition in females is adaptive, because copulation in this species of lizard is prolonged and usually takes place in the open—for example, on tree limbs—where the mating pair is vulnerable to predation. Field data support the contention that more lizards are captured when they are mating than when they are engaged in other activities. The restriction of receptivity and mating to the time when fertilization is most likely to occur is thus evolutionarily adaptive (Valenstein and Crews 1977). Cessation of reproduction during the refractory period that precedes the winter dormancy period is adaptive in at least two ways. First, it ensures that young do not hatch at a time of diminished food resources and poorer environmental conditions. Second, the female is able to build up fat reserves for the winter rather than devote energy toward reproduction.

Sexual behavior in male and female lizards depends on gonadal hormones, a pattern that we have noted in many other vertebrates. Investigators found that when a female lizard was given an injection of progesterone and 24 hours later was given an injection of estrogen, a synergistic effect—induction of female receptivity—resulted. We noted a similar synergistic effect on receptivity in rats with these same two hormones (see Beach 1976, for review).

Data obtained by autoradiography on the uptake of sex steroids by specific areas of the brain augment these parallel findings in lizards and rats. Experimenters inject the animal with radioactively labeled estrogen; several hours later they sacrifice the animal; section the brain with a freezing microtome; and place the sections on special glass plates that have been treated with a photographic emulsion. They leave the plates in the dark for periods ranging from days to months depending upon the exact procedure used; and then they develop the plates to find the specific sites where uptake occurred. The results from rats and lizards are strikingly similar. Concentrations occur in the septum and preoptic regions of both species (Morrell, Kelley, and Pfaff 1975; McEwen et al. 1979). These and related findings of comparative investigations point toward an exciting avenue for future research on common patterns of endocrinological events and interactions between hormones and behavior in different classes of vertebrates.

Further investigations of the reciprocal relationships between males and females in lizards have been conducted with the whiptail lizard (*Cnemidophorus inornatus*). Like the anoles, these lizards exhibit highly seasonal patterns of reproduction, and the gonads regress at other times of the year (Lindzey and Crews 1988). The presence of a male whiptail facilitates ovarian recrudescence in females. Conversely, the presence of a reproductively active female will stimulate testicular development in males. This is one of the few species

FIGURE 8.16   Dewlap display of male *Anolis* LIZARD.
The male *Anolis* lizard extends the dewlap toward the female as part of his courtship display. The dewlap display patterns are important for species-specific mate selection. This behavior pattern also appears to be critical for stimulating ovarian activity in the female.
Source: Courtesy David Crews.

in which either sex can directly affect the coordination of the reproductive condition of the opposite sex. Two additional findings augment our understanding of the degree of reciprocation that occurs in this species. Female whiptail lizards can stimulate some, but not all, of the male courtship behavior sequence in castrated lizards; exogenous androgens are necessary for reappearance of the entire courtship sequence. Female whiptail lizards are responsive to male courtship only at a specific time in the process of egg formation; they will actively reject males that approach too early or after ovulation has occurred. The high degree of coordination is extremely important for successful reproduction in a seasonally reproducing species, but with few exceptions, such as these studies on whiptail lizards, the complete nature of the male-female interaction has not been demonstrated.

Reproduction in House Sparrows

For our final example, consider another way in which hormones and behavior interact during reproduction in the house sparrow (*Passer domesticus*). In a pair of studies, Hegner and Wingfield (1987a,b) demonstrated the interactions of hormones and environmental clues that affect reproduction in free-living house sparrows. The first study involved manipulation of male sparrows' testosterone levels to determine testosterone's effect on the amount of parental investment exhibited by the sparrows or on their breeding success (number of nestlings fledged) (Hegner and Wingfield 1987a). The investigators hypothesized that testosterone levels are somewhat elevated early in the breeding season as males compete for establishment and maintenance of territories. However, for males to exhibit normal parental behavior, including assistance with nestling feeding, the levels of testosterone must drop as the reproductive sequence proceeds. Males with high levels of testosterone when they would normally be collecting food for the nestlings might spend too much time defending the territory, and thus would not provide sufficient food for the nestlings. As the nestling period ends, testosterone levels again rise as they prepare for another reproductive sequence.

To test the foregoing hypothesis, Hegner and Wingfield captured breeding male house sparrows in the spring and implanted a small capsule beneath the skin. The capsule contained one of three substances: (a) testosterone, (b) flutamide, an androgen antagonist that inhibits testosterone uptake and binding at receptor sites, or (c) nothing, as a control procedure. They then monitored the behavior of the birds during reproduction. Blood samples were taken from the birds periodically to assess levels of circulating hormones. The changes in circulating levels of testosterone in the birds are shown in figure 8.17. Overall, there was a significant decrease in reproductive success in the testosterone-treated birds (mean of 2.6 young fledged per nest) compared to the birds that were treated with flutamide (3.8 young fledged) or those receiving an empty capsule (4.2 young fledged). One major contributing factor to this difference was that males treated with testosterone made only about half as many feed-

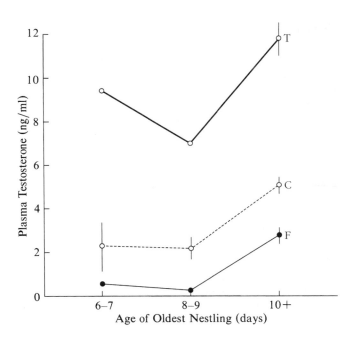

FIGURE 8.17   Hormone levels in sparrows.
The diagram depicts the relationship between the age of the oldest nestling and circulating testosterone level in male house sparrows under three different treatment conditions. The top line represents males given implants of testosterone, the middle line consists of data from males given an empty capsule implant as a control, and the bottom line represents males given flutamide, a substance that blocks the actions of testosterone. The vertical bars indicate $\mp$ 1 standard error of the mean.

ing visits to the nest during the nestling phase as birds given flutamide or an empty capsule, a significant difference. Indeed, males given testosterone were actively defending their nests two to six times more than the other birds, which detracted from the time they could have been providing food.

In a companion investigation, these same investigators (Hegner and Wingfield 1987b) studied the relationships between the number of nestlings present in the nest and parental investment, reproductive success, and endocrinology. They manipulated the number of nestlings present in house sparrow nests by (a) adding two nestlings to some nests, (b) removing two nestlings from some nests, or (c) leaving nests alone so that they contained the normal numbers of nestlings, ranging from three to six birds. They then recorded the number of feeding visits by the parents, the number of young fledged, and by obtaining blood samples late in the nestling phase of the cycle, the levels of androgens, corticosterone, and luteinizing hormone. House sparrows were able to rear the larger clutches successfully; those clutches where nestlings were added resulted in a greater number of fledged young than the other treatments, though the average weight per fledgling was less in the larger clutches. Also, parent birds spent much more time feeding larger clutches than smaller clutches. Those birds that reared larger clutches had longer intervals until the next brood, and the next brood was reduced in size. These latter findings may be an outcome of the extra energy cost of rearing the larger brood, particularly for the female, and the extra time needed to rear a larger brood of nestlings to the stage where they become independent. Male parents with larger broods had higher levels of dihydrotestosterone circulating in their blood, but all other hormone measurements were similar across the three treatments. The results indicate that there is not any stress for the parents related to rearing the larger broods, though there are consequences for the reproductive output of the next brood. When the number of nestlings is greater, parent birds of this species apparently invest more in the current brood and less in the subsequent brood.

While hormone levels play a significant role in affecting reproductive success within a brood as demonstrated by the first study, it is not clear that hormone levels have any important influence on reproductive success between broods as shown by the second study. However, additional hormone measurements taken from blood samples obtained at various times during both the first and second clutches in the second experiment would be needed to further clarify this conclusion.

HORMONE-BRAIN RELATIONSHIPS

It is important to understand that the nervous system and the endocrine system are feedback systems, that both systems are key parts of the body's mechanisms for interfacing with the environment, and that both systems are critical for adaptation. In general, the nervous system provides faster and more specific responses to external and internal events; the endocrine system provides slower responses and more general effects. The nervous system regulates the fine-tuning responses to specific external conditions or stimuli. The endocrine system regulates more generalized responses to external and internal conditions.

In the previous chapter and in this one, we have mentioned the ongoing research concerning the mechanisms by which the endocrine and nervous systems interact with one another. A few additional comments will provide an appropriate finale to our examination of these behavior control systems.

We now know there are regions of the brain with cells that possess specific receptor sites for many of the hormones that influence behavior. The experimental procedures involve a combination of autoradiography (to pinpoint the receptor locations precisely), and electrodes (to measure electrophysiological activity from the same receptor sites) (Morrell and Pfaff 1981; Pfaff 1981). Thus, for example, when radioactively labeled sex steroids are given to rats (Pfaff and Keiner 1973), frogs (Kelley, Morrell, and Pfaff 1975), or chaffinches (Zigmond, Nottebohm, and Pfaff 1973), the label becomes concentrated in similar brain areas in all three animals. The preoptic area and portions of the limbic system and hypothalamus are among the sites that commonly have the highest amounts of the labeled steroid.

When electrode recordings are made from cells in these same preoptic and limbic locations, some important relationships between hormones, behavior, and neural activity emerge. Firing rates for neurons in the highly labeled portions of the limbic system and hypothalamus vary during the course of the rat estrous cycle; generally, they are highest shortly before ovulation (Terasawa and Sawyer 1969; Kawakami, Terasawa, and Ibuki 1970). When ovariectomized female rats are given injections of estrogen, firing rates in these same sites are elevated (Cross and Dyer 1972).

As we have noted earlier, progesterone facilitates sexual behavior in female rats. Pleim and Barfield (1988) demonstrated that intracranial administration of progesterone, but not estradiol, facilitates sexual receptivity. Small cannulas were placed into either the midbrain or ventromedial hypothalamus of adult female rats. Progesterone, estradiol, cholesterol, or nothing was given to the female via the cannula; the latter two treatments served as control procedures. When females were placed with sexually active adult male rats at 1 hour and at 4 hours after treatment, only females that had been treated with progesterone in the ventromedial hypothalamus, and that were tested 4 hours after hormone administration, exhibited sexual receptivity. Studies such as this further our understanding of the location of key hormonal receptor areas that trigger behavioral output.

Several behavior patterns typical of male house mice are elicited by appropriate intracranial implants of testosterone (Nyby, Matochik, and Barfield 1992). The behavior patterns involved were the 70 kHz ultrasonic vocalizations given by males during mating when presented with estrous females or their urine, urinary marking in response to stimulus males or females, mounting of estrous females, and male-male aggression. Test subjects were given testosterone in the following brain regions; septum, medial preoptic area, anterior hypothalamus, or ventromedial hypothalamus. Control mice were given subcutaneous silastic capsules of testosterone or empty capsules. For the controls, the testosterone-implanted mice performed all of the behavior patterns at rates typical of normal responsive males. The control mice with empty capsules did not respond. Males with implants in the median preoptic area exhibited increased levels of ultrasonic vocalizations; implants at the other brain locations resulted in few of these vocalizations. Males with implants in the median preoptic area and the two regions of the hypothalamus urine marked more than the males given empty capsules, but not as often as males with the testosterone in the silastic capsule implants. The only males that mounted were those given testosterone in capsule implants and those given testosterone in the median preoptic area. Aggression occurred infrequently in all male mice given brain implants of testosterone. Together, these findings demonstrate the functional connections between specified brain regions, which likely have receptors for the testosterone, and the processes of triggering particular brain regions.

In the lobster, various hormones and neurosecretions act upon the nervous system to influence aggressive behavior (Kravitz 1988). In particular, various neurotransmitter substances (e.g., serotonin, octopamine) are found throughout the nervous system. These amines apparently regulate the

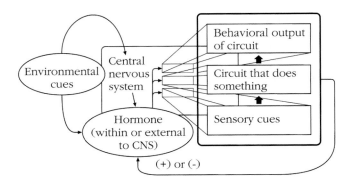

FIGURE 8.18   Hormone sensitization.
Schematic diagram of the manner in which hormonal sensitization of various tissues can occur. The responses of certain neurons can be primed by doses of hormones or neurosecretions that precede the input from particular environmental stimuli.

Reprinted with permission from Edward A. Kravitz, "Hormonal Control of Behavior: Amines and the Biasing of Behavioral Output in Lobsters" in *Science*, 241:1775–1781. Copyright 1988 American Association for the Advancement of Science.

sensitivity of the nervous system and effector organs (e.g., muscles) to environmental stimuli (figure 8.18). In lobsters, injections of serotonin resulted in stereotypical postures seen in dominant animals, and injections of octopamine produced postures characteristic of subordinate animals. Further investigations led to the conclusion that a key effect of the aminergic neurotransmitter substance is the priming of receptors in exoskeletal muscles to respond in particular ways when appropriate stimuli occurred (e.g., another lobster). Serotonin directed the muscle program toward flexion of the postural muscles involved in more threatening, dominant postures. Octopamine produced the opposite effect, inhibiting flexion and producing postures characteristic of a subordinate animal. Through this regulatory process, the input to the nervous system, the processing of incoming information, and the behavioral output are biased toward certain stereotyped responses.

Thus, there appear to be some clear connections between hormonal and neural activity. We are now nearing the time when we will understand mechanisms that relate hormones, nervous system, and behavior at the level of the individual cells and the biochemistry within and between those cells (Komisaruk 1978).

HORMONES AND ECOLOGY

As our knowledge about mechanisms involving hormones has increased (particularly, our concern for the sorts of behavior-hormone-environment interactions sequences just discussed), so too has our understanding of behavioral ecology for some of these animals. There are now some systems (e.g., the foregoing analyses of hormones and behavior for lizards and sparrows), where the linkage between internal hormonal processes and knowledge concerning ecological and evolutionary biology underlying particular behavioral

phenomena has provided a strong sense of the likely causal links between environmental and physiological events. Let us consider another example of this linking process involving hormone levels in red-winged blackbirds (*Agelaius phoeniceus*). This work has been carried out primarily by a group that includes an endocrinologist (John Wingfield), a behavioral ecologist (Gordon Orians), and their students (see Wingfield et al. 1987; Beletsky and Orians 1987, 1989, 1991; Dufty 1989; Beletsky, Orians, and Wingfield 1989, 1990a, 1990b, 1992).

Male red-winged blackbirds establish seasonal breeding territories in suitable habitat. The aforementioned investigators explored a variety of relationships in these male redwings, involving the habitat, the mating system (polygyny), reproductive success, and possible seasonal and annual variations in hormone levels for testosterone and corticosterone. The methods used for these studies include behavioral observation, recording hatching and fledging success, and, of particular importance to our interest here, taking blood samples from the male birds on which radioimmunoassay techniques were used to assess testosterone and corticosterone levels. Also noteworthy in this instance is that one of the studies (Beletsky, Orians, and Wingfield 1992) involved measuring hormone levels at the same point in the breeding season for four consecutive years.

The findings can be summarized in the following statements. Testosterone levels peak in most male redwings during the 2-week period at the height of the nesting season, when territory defense is frequent and males are likely engaged in mate guarding. Testosterone levels measured during that limited period were higher for males holding territories in high-density areas when compared to males in low-density areas. Males holding territories had higher levels of testosterone than floater males without territories. There was a significant positive correlation between testosterone level and the number of females nesting in a male's territory (harem size). There was a trend for males with high levels of testosterone during this breeding period to fledge more offspring. There were no significant relationships between testosterone and male age or breeding experience. There were significant differences within the same males measured over several years. There were no differences across the 4 years with regard to mean testosterone level in the males (table 8.2). For corticosterone, there were no relationships between levels of this hormone and age or breeding experience or with reproductive success, and no year-to-year variation in mean levels was observed. However, there was a high correlation within individual males from year-to-year with regard to corticosterone level.

What do these findings tell us? One key conclusion is that the data support the hypothesis that males facing more frequent challenges, such as those living in high-density areas, will have, on average, more elevated levels of testosterone. Therefore, increased levels of testosterone may relate to heightened aggressiveness, which, in turn, could result in higher numbers of mates and fledglings for particular males.

TABLE 8.2   **Mean testosterone and corticosterone levels for male red-winged blackbirds measured at the onset of the breeding season in each of 4 years**

Year	Testosterone ($\bar{X} \mp$ SD, ng/ml plasma)	n	Corticosterone ($\bar{X} \langle$ SD ng/ml plasma)
1987	4.21 ∓ 2.42	18	7.66 ∓ 4.49
1988	4.29 ∓ 4.30	37	7.24 ∓ 5.98
1989	4.38 ∓ 3.63	30	10.41 ∓ 6.10
1990	2.89 ∓ 1.44	28	8.03 ∓ 6.88

Source: Beletsky, Orians and Wingfield, 1992, "Year-to-year Patterns of Circulating Levels of Testosterone . . ." in *Hormones and Behavior*, vol. 26, p. 425.

For corticosterone, the conclusion seems to be that it may be regulated endogenously and is thus independently controlled relative to testosterone, or it may be that there is an approximately equal amount of stress on all males, regardless of breeding density, age, etc. Additional work on the costs and benefits of higher levels of testosterone are certainly needed, but this type of study is a solid beginning to relating proximate mechanisms and ultimate consequences.

IMMUNE SYSTEM AND BEHAVIOR

Immune System

In vertebrates several body defense mechanisms protect the organism from pathogens, such as bacteria, parasites, or viruses. The skin and lining of the digestive tract act as barriers, providing a first line of defense. Other defense mechanisms are mediated by various aspects of the immune system. The immune system consists of glands, the thymus for example, bone marrow, and a series of lymph nodes. One type of cell produced in the immune system, macrophages, are attracted by chemicals released by foreign cells or damaged tissue. **Macrophages** function to immobilize and destroy foreign materials.

A second major group of cells characteristic of the immune system is the **lymphocytes,** or white blood cells. There are a variety of lymphocytes, some of which perform specialized functions in relation to recognition and disabling of pathogens. Foreign materials, called antigens, may stimulate certain lymphocytes to produce antibodies which attach to the surface of the antigen to immobilize it. Then other lymphocytes, including macrophages, destroy the antigen. The **immunocompetence** of an organism refers to its capacity to recognize and destroy pathogens of various types. Invertebrates also have defenses against pathogens that include an immune system, but these processes are not as well understood in invertebrates and little work has been done to relate their immune systems to behavior.

Implications for Behavior

The health of an organism, and thus, for example, its suitability as a possible mate, may be related to its immunocompetence. Natural selection may favor those organisms that are better able to withstand the challenges of a variety of pathogens, both in terms of their own survival and with respect to their reproductive fitness. Zuk (1994) has summarized the relationships between immunology and the evolution of behavior. Some specific behaviors such as mate choice may be strongly influenced by the immune system. The major histocompatibility complex (MHC) is a segment of DNA (genes) found in all vertebrates. The MHC codes for several aspects of the reactions to pathogens described in the previous section. House mice (*Mus musculus*) demonstrate mate preferences for members of the opposite sex with an MHC genotype that is different from their own (Yamazaki et al. 1976; Potts et al. 1991a,b). The rationale is that mice selecting a mate of different MHC type would produce progeny with more heterozygosity with respect to the ability of their immune system to fend off pathogens (Brown 1997).

Hamilton and Zuk (1982) first proposed parasites might play a role in sexual selection. The traits that are used by members of the opposite sex to select mates, such as body ornaments, might reflect the condition of the animal with respect to its load of parasites. Presumably a partner with fewer parasites would be a better mate. Zuk (1994) has further proposed that the condition of the animal with respect to parasites could be directly reflective of the organism's immune system. In addition, there are sex differences in disease susceptibility and immunosuppressive effects of testosterone affecting the secondary sex characteristics of males that could be reflected in the behavior of animals (Zuk 1990; Braude et al. 1999). The interrelationship of the immune system with hormones has been nicely demonstrated in voles (*Microtus* spp.) (Klein et al. 1997). The results indicated both species and sex differences in the responses of the immune system to different social housing conditions and with correlated differences in sex steroids. The study of the relationships between the immune system and behavior is an area ripe for additional investigation and new results should appear rapidly in the coming years. We have cited additional examples of the relationship between the immune system and behavior in later chapters in this book.

SUMMARY

Hormones are chemical substances produced by specialized duct-less glands or by neurosecretory cells. They influence the processes of growth, metabolism, water balance, and reproduction. The endocrine system has evolved in concert with the nervous system; the two systems are closely interrelated in their regulation of behavior. Most endocrine structures in invertebrates, such as the insects discussed in this chapter, involve neurons directly, or they are closely tied to the nervous system. Trophic hormones in both invertebrates and vertebrates affect the production and secretion of other hormones. Feedback loops, particularly in vertebrates, regulate the release of hormones and neurosecretions and aid in the process of monitoring hormonal levels. Synergism and antagonism between hormones are important determinants of the effects of those hormones on behavior.

Three major regions of the vertebrate pituitary gland produce hormones that affect behavior. The posterior pituitary secrets oxytocin, which affects uterine muscles and milk ejection, and vasopressin, which plays a key role in regulating water balance. The central portion of the pituitary secretes MSH, a hormone that affects pigments, and thus skin and hair color. The anterior pituitary secretes three trophic hormones—FSH and LH, which affect the gonads and reproduction, and ACTH, which influences production and release of hormones from the adrenal cortex. The anterior pituitary also produces prolactin, a hormone that affects maternal behavior, milk production, and related processes in birds and mammals. The gonads and adrenals are endocrine glands secreting hormones that affect behavior.

Some hormones exert organizational effects because these influences are manifested during development. In particular, sexual differentiation is an organizational effect in some mammals, birds, and reptiles because it is under some degree of hormonal regulation. Other hormones exert activational, or triggering, effects on behav-ior. Examples of activational effects include coat color changes in the weasel, the release of pheromones and the process of eclosion in insects, molting in crustaceans, sexual and courtship behavior in rats and ring doves. Some hormones can exert both activational and organizational effects.

There are complex environment-hormone-behavior interactions, as exemplified here by the reproductive sequences in ring doves, maternal behavior in rats, seasonal reproductive patterns of lizards, and reproduction in house sparrows. The hormonal effects within individual animals, the effects of behavior on hormones, and the interactions between the hormonal/physiological state of the animal and features of its environment combine to produce observed behavior patterns. The hormonal state of the animal can affect nestling production in house sparrows, and the number of nestlings can influence levels of circulating hormones.

In the past decade, we have witnessed an emerging synthesis involving proximate hormonal mechanisms integrated with ecological and evolutionary aspects of behavior for several animal systems. Among the best delineated systems to date are those for green anole lizards, house sparrows, and red-winged blackbirds.

The research techniques investigators use to explore hormone-behavior relationships include the removal of endocrine glands, hormone replacement therapy, blood transfusions, administration of hormone substitutes or use of chemicals that inhibit particular hormone effects, autoradiography, and assays for hormone levels. Using several of these techniques, investigators have explored the mechanisms by which the endocrine and nervous systems are interrelated in regulating behavior. There is an increasing interest in the relationship between the immune system and behavior. An understanding of some basic principles of immunology and knowledge of the manner in which the immune system can influence behavior provide the basis for examples presented in later chapters.

DISCUSSION QUESTIONS

1. Both male and female golden hamsters possess sebaceous flank glands that produce sebum, a substance that is used to mark objects in their environment. Suggest three or four experimental questions (hypotheses) you would want to ask in studying the hormonal regulation and control of this scent-marking behavior. For each question, identify the treatment groups you would use for testing your hypotheses and justify the need for each.

2. The following summary table is from a review of early organizational effects of hormones on behavior by Adkins-Regan (1981). A number of reptiles have been tested using various techniques to manipulate the early hormone environment. What general conclusions can you make about organizational effects on sex structures in these reptiles? How do these results compare with those discussed in this chapter on mammals?

3. The reproductive sequences of ring doves and lizards and the maternal behavior of rats illustrate feedback loops involved in internal hormonal conditions, behavior, and environmental cues. Can you describe or explain other behavior sequences that have similar complex feedback interactions? How would you use available techniques to investigate the various steps in the hormones-environment-behavior sequence?

4. The testes of 25 males of a bird species were weighed in each of four seasons. To organize the data, the birds were divided into groups of five on the basis of their social dominance interactions; the five top-ranking birds were placed in category I,

the next five in category II, and so forth. The following table shows average testes weights in milligrams for the birds in each category and in each of the four seasons. What general conclusions can you draw from these data? What conditions of interpretation would you place on these conclusions?

Average weight of bird testes (milligrams)

| | Male dominance category | | | | |
	I	II	III	IV	V
Spring	7.6	7.0	6.8	6.4	5.8
Summer	8.0	7.7	7.2	6.9	6.2
Fall	4.1	3.9	3.8	3.9	3.8
Winter	2.6	2.5	2.6	2.4	2.5

5. Hormonal mediation of environmental stimuli and the role of the endocrine system in regulating behavior in response to external stimuli may affect reproduction and the reproductive success of organisms. Using the work of Wingfield and his colleagues described in this chapter as a model system, can you devise an experimental scheme to explore the relationship between hormone levels and reproductive success in (a) red-winged blackbirds, (b) garter snakes, and (c) woodchucks?

Summary of experiments on sex differentiation in Reptilia

Species	Age at Treatment	Treatment	Gonads	Gonaducts	Other Sex Structures
				Effect on[a]	
Several	Embryos	Gonadectomy		M F	
Several	Embryos	Estrogens	m f		
Lacerta vivipara	Embryos	Gonadectomy		M F	F M
	Embryos	Parabiosis			M F
	Embryos	Androgens			f m
	Embryos	Estrogens	m f	M F	M F
Anguis fragilis	Embryos	Testosterone		F M	
Anolis carolinensis	Juveniles	Estrogen			M F
Sceloporus sp.	Juveniles	Estrogen			M F
Alligator mississippiensis	Juveniles	Testosterone		m f	f m
Crocodylus niloticus	Juveniles	Testosterone	f m	m f	
Thamnophis sirtalis	Embryos	Testosterone		m f	f m
	Juveniles	Estradiol	m f	m f	
Chrysemys marginata	Embryos	Testosterone	f m		
Emys orbicularis	Embryos	Estradiol or Testosterone	m f		
Emys leprosa	Recently hatched	Androgens		m f	f m
	Recently hatched	Estrogen	m f	m f	
Testudo graeca	Embryos	Estradiol	m f	m f	

From E. Adkins-Regan, "Early Organizational Effects of Hormones," in *Neuroendocrinology of Reproduction*, ed. by N. Adler, p. 188. Copyright © 1981 Plenum Press, New York. Reprinted by permission.

[a]FM indicates extensive or complete masculinization of the character; M→F indicates extensive or complete feminization. f→m indicates slight or partial masculinization; m→f indicates slight or partial feminization.

SUGGESTED READINGS

Adkins-Regan, E. 1981. Early organizational effects of hormones: An evolutionary perspective. In *Neuroendocrinology of Reproduction,* ed. N. Adler. New York: Plenum.

Beach, F. A. 1965. *Sex and Behavior.* New York: John Wiley and Sons.

Becker, J. B., S. M. Breedlove, and D. Crews, eds. 1993. *Behavioral Endocrinology.* Cambridge, MA: MIT Press. *BioScience.* 1983. Vol. 33 (October Issue). This issue of the journal *BioScience* contains summary articles on hormones and behavior in invertebrates and vertebrates.

Brown, R. E. 1994. *An Introduction to Neuroendocrinology.* Cambridge, England: Cambridge Univ. Press.

Colborn, T., and C. Clement, eds. 1992. *Chemically-Induced Alterations in Sexual and Functional Development: The Wildlife/Human Connection.* Princeton, NJ: Princeton Scientific Publications.

Crews, D. 1980. Interrelationships among ecological, behavioral, and neuroendocrinological processes in the reproductive cycle of *Anolis carolinensis* and other reptiles. *Adv. Stud. Behav.* 11:1–74.

Hutchinson, J. B. 1976. Hypothalamic mechanisms of sexual behavior, with special reference to birds. *Adv. Stud. Behav.* 6:159–200.

Nelson, R. J. 1995. *An Introduction to Behavioral Endocrinology.* Sunderland, MA: Sinauer Associates.

Truman, J. W., and L. M. Riddiford. 1974. Hormonal mechanisms underlying insect behavior. *Adv. Insct. Physiol.* 10:297–352.

9

CHAPTER

Biological Rhythms

 Many birds produce their melodious songs in the early morning hours just before and after sunrise. Moths and butterflies generally emerge from their cocoons at dawn when the still, somewhat moist air provides the best conditions for the slow drying of their unfolding wings. Many animal behavior patterns occur daily. Many songbirds of the north temperate deciduous forest fly south during autumn, overwinter in tropical or subtropical zones, and begin their return flight north as spring approaches. Ground squirrels emerge from their winter dens in spring, remain active for 4 to 6 months, and return to hibernation in autumn. During their months above ground, they mate, produce offspring, rear the young, and prepare for another winter. Certain animal activities occur on an annual cycle. We can observe other animal behavior patterns that occur with frequencies that approximate cyclic features in the environment, for example, tides and lunar phases. **Biological rhythms** occur when animal activities and behavior patterns can be directly related to distinct environmental features that occur with regular frequencies. Biological rhythms are the external manifestations of biological clocks and are regulated by them. **Biological clocks** are internal timing mechanisms that involve both self-sustaining physiological pacemakers and environmental cyclic synchronizers (*Zeitgebers*).

In this chapter we first examine the characteristics of biological rhythms. Next, we look at pacemakers and their physiology: the mechanistic, or proximate, questions. We then ask questions about the functional significance of biological rhythms within ecological contexts: the ultimate questions. Natural selection should favor individuals whose peak activity periods of feeding, for example, coincide with peak activity periods of their prey. Finally, we cover the significance of biological clocks with particular emphasis on hibernation and migration.

BIOLOGICAL RHYTHMS

Let us first consider the terminology and concepts for describing and characterizing biological rhythms. Each biological rhythm is composed of repeating units called **cycles.** The length of time required to complete an entire cycle is the rhythm's **period;** 24 hours is the period in the example shown in figure 9.1. The magnitude of the change in activity rate during a cycle, the difference between peaks and troughs, is the **amplitude.** Any specified recognizable part of a cycle is called a **phase;** the designated portion of the cycle shown in figure 9.1 would be an active phase.

Biological rhythms may be characterized by several properties. First, whereas temperature changes alter the rate of most chemical reactions and cellular processes, biological rhythms are **temperature-compensated.** Generally, the rate of a chemical reaction doubles for each 10°C increase in temperature. However, biological rhythms are relatively insensitive to change in temperature (Sweeney and Hastings 1960; Pittendrigh 1993). This fact is significant, for if biological rhythms were sped up or slowed down by ambient temperature changes, they would not help the organism keep accurate time.

Second, biological clocks are generally unaffected by metabolic poisons or inhibitors that block biochemical pathways within cells. We might expect that application of a metabolic poison such as sodium cyanide would alter the period of a biological rhythm, yet this does not happen.

A third property is that periods of biological rhythms occur with approximately the same frequency as one or more environmental features. The Latin word *circa* (about or around) usually is part of the name of most biological rhythms.

Fourth, biological rhythms are self-sustaining, maintaining approximately their normal cyclicity even in the absence of environmental cues.

Fifth, and last, most biological rhythms can be entrained by environmental cues. The self-sustaining, internal pacemaker mechanism(s) may be set and adjusted according to input from the external environment.

Types of Rhythms

Epicycles or Ultradian Rhythms

Different organisms exhibit a variety of biological activities with varying frequencies and periods (table 9.1). Some of these cycles are of short duration and are termed **epicycles** or **ultradian rhythms.** Lugworms (*Arenicola marina*) that live in burrows on sand flats in the intertidal zones feed every 6 to 8 minutes. Some small mammals like meadow voles (*Microtus pennsylvanicus*) that are active primarily during daylight hours show short cycles of activity bursts followed by periods of rest that vary from 12 minutes to 2 hours (Madison 1985). Ultradian rhythms have been identified in *Euglena gracilis,* a unicellular aquatic organism; individuals held in constant darkness develop identifiable rhythms of variable lengths, for example, 8 minutes to 1.5 hours (Balzer and Hardeland 1992). The brooding cycle of some birds, such as junglefowl (*Gallus gallus*), which consists of a behavior sequence involving brooding, preening, feeding, exploring, dustbathing, and brooding again, involves epicycles of varying duration (Hogan et al. 1998). The occurrence and importance of ultradian rhythms are still being explored; we will not treat them further here, but additional information is likely to emerge on this subject in the next decade.

Tidal Rhythms

A primary environmental feature of seacoasts is the ebb and flow of the tides. Tidal rhythms affect activity periods in many organisms that inhabit the zone. Conditions in the intertidal zone vary tremendously over the tidal cycle—salinity, temperature and wave action all change over wide ranges. Tides, the result of unequal gravitational forces of the sun and the moon, exhibit about a 12.4-hour period from one low-tide phase to the next (though there are some variations in time period from one location to another). Many species of small crabs (e.g., *Uca minax, Uca crenulata*) inhabit these seashore regions; they time their activity cycles so that they feed on the sandy shore when the tides are out, but return to burrows when the tide flow returns (figure 9.2). In this way they avoid being stranded above the tide line where they might desiccate or be swept out to sea by the inrushing tide (Barnwell 1966; Palmer 1990).

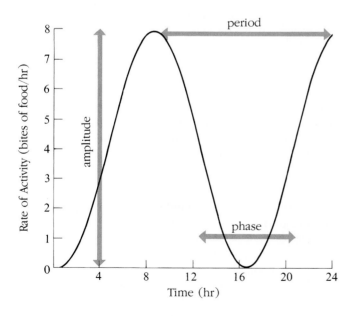

FIGURE 9.1   Biological rhythm.
The characteristics of a biological rhythm include period, phase, and amplitude. Phase refers here to the quiet period in terms of activity. Each of these characteristics can vary over time for the same animal, between animals of the same species, or between animals of different species.

TABLE 9.1   **Summary of biological rhythms**

Type of Cycle	Organism	Behavior
Ultradian (variable)	lugworm	feeding (every 6–8 min.)
	meadow vole	feeding/resting (every 15–120 min. during daylight)
	euglena	motility
Tidal (12.4 hours)	oyster	opening of shell valves
	fiddler crab	locomotion/feeding
Lunar (28 days)	midge (marine insect)	mating/egg laying
	grunion (marine fish)	egg laying
Circadian (24 hours)	deermouse	drinking/general activity
	fruit fly	emergence of adults from pupa
Circannual (12 months)	woodchuck	hibernation
	chickadee	reproduction
	robin	migration/reproduction
Intermittent	desert insect	reproduction (triggered by rain)
(variable—several days to several years)	lion	feeding (triggered by hunger)
	shiner (river fish)	reproduction (triggered by flooding)

FIGURE 9.2   Tidal rhythms affect fiddler crabs.
Fiddler crabs inhabit shorelines where they forage along the beaches. They must do their foraging between the tides and return to their burrows with each new incoming tide. Without a biological clock to warn them, they would either be swept out to sea by the water, or be left high on the beach away from their burrows, where they would desiccate in the sun.
Source: © Robert and Linda Mitchell.

Lunar Rhythms

Based on the 29.4-day cycle of the moon, lunar rhythms are clearly related to the tidal rhythm. Some marine insects, like the midge (*Clunio marinus*), coordinate eclosion, mating, and egg-laying activities with the lunar cycle. They lay their eggs at very low tide, thereby ensuring that the larvae will hatch in the proper marine environment (Neumann 1966). Grunion (*Leuresthes* spp.), a marine fish, spawns during the spring tides on the sandy beaches of California and uses the moon as a timing cue for these activities (Walker 1949).

Circadian Rhythms

Many organisms exhibit biological rhythms of about 24 hours' duration that are governed by self-sustaining internal pacemakers. These we call **circadian rhythms.** In their daily cycle, some animals exhibit peak activity during the daylight hours **(diurnal);** some are active primarily at night **(nocturnal);** and still others exhibit peak activity around dusk and/or dawn **(crepuscular).** Activity periods may shift seasonally. Many bird species that are year-round residents of northern temperate zones are primarily crepuscular through late spring and summer but shift to a more diurnal pattern during the cold winter months; thus they avoid the very cold temperatures of early winter mornings. Circadian activity periods may also show age-dependent shifts. Young woodchucks (*Marmota monax*) restrict most of their activity to early evening hours, whereas adult woodchucks are more diurnal in their pattern. Similarly, young dragonflies fly during the hours just after dawn, but older adults fly mostly in the middle of the day (Corbet 1960).

Circannual Rhythms

As the term implies, **circannual rhythms** are behavioral and physiological patterns that are governed by self-sustaining internal pacemakers and that occur within a period of about one year (Gwinner 1986b). Some mammals enter **hibernation,** a condition of deep sleep and reduced metabolic activity, during the winter months. By doing so they avoid the harsh conditions of winter. Many bird species escape the rigors of winter in northern and temperate climates by migrating to southern latitudes. The annual life cycle of many insects that live where there are seasonal climatic shifts incorporate a **diapause phase,** a period of dormancy, during the more rigorous portions of the climatic cycle. For example, silkworm moths (*Bombyx* spp.) and mosquitos (*Aedes* spp.) lay eggs that are dormant during the winter (Kogure

1933; Vinogradova 1965); nymphs of the dragonfly (*Tetragneura cynosura*) overwinter as larval forms and complete development the following spring (Lutz and Jenner 1964); both the parasitic wasp (*Nasonia vitripennis*) and its host, the flesh fly (*Sacrophaga argyrostoma*) enter diapause as larvae (Saunders 1978); still other insects go through diapause in the pupal stage or as adults.

Endogenous Pacemaker

One of the critical characteristics of biological rhythms is the existence of an internal self-sustaining pacemaker. What is the evidence for the existence of **endogenous** clock mechanisms? As we examine the following sequence of information in support of the existence of internal chronometers, bear in mind that what we are measuring are generally the overt, observable manifestations of the clock—the periodic changes in physiology and behavior—not the mechanism itself.

Free-Running Rhythms

First, let us consider what happens when we place an animal in constant environmental conditions; for example, in constant darkness. When this is done, many organisms exhibit activity rhythms, called **free-running rhythms,** with a period different from that of any known cyclic environmental variable. This provides indirect evidence for an endogenous pacemaker. The pattern of activity of a flying squirrel (*Glaucomys volans*) housed in constant darkness for several weeks is shown in figure 9.3 (DeCoursey 1960, 1961). Since the animal's activity has a clearly rhythmic pattern in the absence of any obvious cyclic cue, evidence is in favor of an endogenously based system of timekeeping.

For most species studied, the free-running periods follow what has become known as **Aschoff's Rule** (Aschoff 1960, 1979). When animals are kept in constant darkness, their activity rhythm continues with a period of nearly 24 hours, but it drifts slightly, becoming somewhat shorter (as in the flying squirrel) or somewhat longer each day. Aschoff's Rule states that the direction and rate of this drift away from the 24-hour period are a function of light intensity and of whether the animal is normally diurnal or nocturnal. For nocturnal animals like the flying squirrel, housing under constant dark conditions results in a free-running rhythm period shorter than 24 hours; the activity begins slightly earlier each day. Conversely, for a diurnal animal housed in the dark, the free-running period is slightly longer each day, and the activity begins slightly later each day. There are some exceptions to this rule; but in general, the data show that a variety of animal species conform to the pattern.

Isolation

What about "learning" or other similar influences as a mechanism for biological rhythms? Birds and some reptiles that hatch from eggs can be kept under constant conditions in an incubator from just after they are laid until after they hatch. If

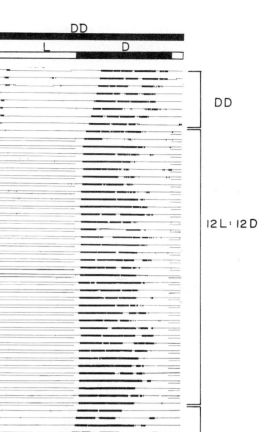

FIGURE 9.3   Free-running periods and entrainment.
When a flying squirrel is placed in constant darkness, it adopts a free-running rhythm with a period different from any known cyclic environmental variable. When the squirrel later is placed in an environment with a varying cue, for example, a light-dark cycle, the squirrel's activity onset becomes locked onto the lights-off signal.
Source: Courtesy of Patricia DeCoursey.

newly hatched organisms exhibit circadian rhythms, a major component of biological rhythms would appear to be inherited and endogenous. Hoffman (1959) maintained lizard eggs under one of three conditions: (1) 18-hour days consisting of 9 light hours and 9 dark hours; (2) 24-hour days, with 12 light hours and 12 dark hours; and (3) 36-hour days with 18 light hours and 18 dark hours. Lizards from all three groups, hatched and maintained under constant conditions, exhibited free-running activity periods of 23.4 to 23.9 hours. Therefore, we can conclude that a component of the biological clock mechanism in these lizards is inherited, is unaffected by various rearing regimes, and is thus endogenous.

Genetics

Another piece of evidence for an endogenous pacemaker comes from reports of mutations in the gene(s) that regulate the basic program of biological clocks in various invertebrates, for example, *Drosophila* spp. (Konopka 1979). A mutant golden hamster (*Mesocricetus auratus*) with a free-running period of 22.0 hours, considerably shorter than the species' norm of 24.1 hours, has been found (Ralph and Menaker 1988). Further work revealed that the mutant gene in the hamster was at a single locus. In hamsters with the heterozygous condition, such as the animal in which the phenomenon was first discovered, the dominant mutant gene produced the 22.0 hours free-running period. In hamsters with the homozygous condition, the free-running period decreased to about 20 hours. Animals with mutant alleles exhibited abnormal patterns of entrainment to a 24-hour period, or would not entrain. Mutations such as these provide direct evidence for the inherited endogenous machinery for circadian rhythms and for the basic underlying program of activity.

In recent years, genetics has become a powerful tool for examining various aspects of biological clocks and, in particular, their control mechanisms (Dunlap 1993). The foregoing example involving the hamster strain with a shortened circadian period is a good example of **mutational analysis.** The use of similar techniques, involving **genetic screening,** has resulted in isolation of strains of *Drosophila* spp. with characteristics that aid in understanding biological clocks. In some mutants, the clock itself has been altered, and in others, the clock remains intact, but the period has been altered. In other strains, mutations can result in uncoupling of the clock, resulting in organisms with arrhythmic patterns (Dunlap 1990; Hall and Kyriacou 1990a). This latter process is akin to distinguishing between the clock itself and the hands of the clock (Feldman and Hoyle 1976; Dunlap 1993).

Translocation

Additional support for the hypothesis of endogenous control of biological rhythms comes from **translocation** experiments. Honeybees (*Apis* spp.) visit particular feeding sites at the same time each day. This behavior is functionally significant because many flowers that provide bees with food are open only at a specific time of day. Renner (1959, 1960), working in a specially designed room with constant conditions, trained bees to leave the hive to forage at a specific time each day. He trained the bees in Paris and then flew them to New York during the night, where he placed them in a similar enclosed room with constant conditions. On their first day in New York, the bees began to forage at the same time as would have been expected had they remained in Paris—twenty-four hours after their last foraging. In related studies, bees that inhabited outdoor hives were translocated from Long Island to California. Although the bees initially foraged at the new site at the same time as they had foraged at the original site, they gradually adjusted to local time by foraging later and later each day (figure 9.4).

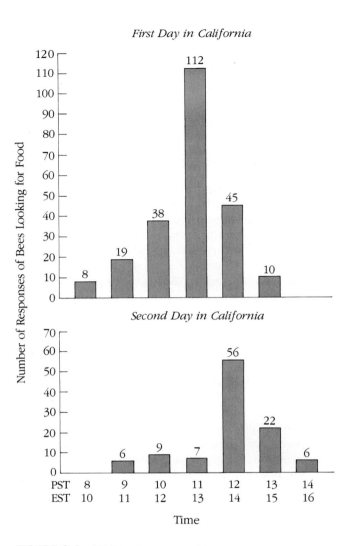

FIGURE 9.4   Visiting frequency of bees.
Bees were trained to forage at a particular time period (1:00 P.M. to 2:30 P.M. EST) in New York and were then flown to California at night. During the first two days at the new site, the bees gradually shifted the time of their feeding. These data support the notion of an endogenous biological rhythm that shifts in response to local conditions.

Source: Data from M. Renner, "The Contribution of the Honeybees to the Study of Time-Sense and Astronomical Orientation," *Cold Spring Harbor Symposia on Quantitative Biology* 25:361–367, 1960.

A key aspect of current biological clock models illustrated by these experiments is that while the clocks are endogenous, external conditions can reset the clocks of many biological rhythms. Similar relocation studies of circadian skin color changes in fiddler crabs and shell-valve-opening time in oysters reveal that these animals also react initially as if they were still in the original location, but gradually their clocks become reset to local time. The jet lag experienced by humans making long-distance airline flights is another example of this phenomenon. Smith and colleagues (1997) have shown how this effect can provide an advantage in terms of sports teams. They determined that, for the Monday night games in the National Football League, there appears to be an advantage for teams from the West Coast. When games are in the West, the West Coast team is at home and, with a mid-evening starting time, the athletes may be within the daily time period for peak athletic performance, while teams from the East Coast playing in the West will have passed that peak time period by the time the games start. Conversely, teams from the West Coast playing in the East will still be competing at a time that is within the daily peak period of athletic performance so that they compete on a more even basis with their East Coast hosts, who are also still playing at a peak time when the game starts.

Variation in Period

A final source of evidence for the endogenous nature of biological rhythms comes from the variation that exists in the natural activity periods of most organisms. For example, if we measure the activity periods of two rattlesnakes (*Crotalus* spp.) under constant conditions, we find circadian rhythms of 23 hours, 10 minutes and 24 hours, 33 minutes. These rhythms do not match the period of any known environmental variable. The animals must have internal chronometers that, due to individual differences, lead to deviations from a 24-hour rhythm.

Zeitgebers

As we have seen, many organisms exhibit circadian and circannual periodicities that appear to be closely adjusted to patterns of daylength, tidal rhythms, or other environmental features. However, the endogenous rhythm generally does not exactly match the environmental cycle; under constant conditions, the endogenous cycle is somewhat longer or shorter. Endogenous rhythms must be sychronized with the external stimulus. This process is called **entrainment.** As we noted in the introduction, cues that provide information to animals about periodicity of environmental variables are called *Zeitgebers* ("time givers"). **Zeitgebers** are the entraining agents defined as those cyclic environmental cues that can entrain free-running endogenous pacemakers. We know that Zeitgebers can influence rhythms by affecting both the phase and the frequency.

Return for a moment to the example in figure 9.3. When the flying squirrel is provided with a daily cycle of light and dark, its activity rhythm generally conforms to the light cycle, and the onset of activity occurs at about the same time each evening (DeCoursey 1961). The squirrel is thus said to be entrained in the daily light-dark cycle. The light-dark cycle is the critical cue for entrainment in most endothermic vertebrates tested so far. Further, in most terrestrial organisms, the daily light-dark cycle is the Zeitgeber. It is noteworthy that the animal does not need to be exposed to the entire light-dark regime to have its clock set or adjusted. In fact, 15-minute pulses of light given to flying squirrels as they awaken from their sleep and enter the lighted portion of a 12-hour light/12-hour dark regime will induce a phase shift in their activity pattern, causing a nap followed by a delay in starting the activity phase (DeCoursey 1986). Similarly, light

pulses of 15 minutes duration are sufficient to reset the daily biological rhythms of the field mouse (*Mus boodgua*) (Sharma and Chandrashekaran 1997).

A question arises from the finding that photoperiod is a Zeitgeber: Why should photoperiod be the primary cue that sets and alters biological clocks? Because, for most animals, the cue with the greatest degree of predictability is photoperiod. This is true whether we are looking at circadian or circannual rhythms; day in and day out, year in and year out, the daily and seasonal patterns of changes in photoperiod will be extremely consistent. Thus, during the course of evolution, animals that developed biological timekeeping systems that used photoperiod to set and reset their clocks would be most likely to have "the correct time." As we shall see in the paragraphs that follow, other cues may interact with or modulate the major reliance on photoperiod as a Zeitgeber.

What other cues can serve as Zeitgebers? A major exception to the photoperiod rule is the group of organisms that exhibit tidal rhythms. The most predictable and most important cue in these organisms' environment is the ebb and flow of the tides. We would expect natural selection to favor development of timekeeping processes that are tuned-in to the tides, rather than to the photoperiod. For animals in the intertidal zone, the ebb and flow of the tides may be the prime Zeitgeber. Other cues, such as salinity and hydrostatic pressure, which also vary with circatidal rhythms, may influence these rhythms. For shore crabs (*Carcinus maenas*) tested in a system where artificial tidal flows, variations in salinity, and shifts in pressure could all be manipulated, Warman and Naylor (1995) demonstrated that the crabs would exhibit three peaks during each circatidal period. The three peaks occurred at the times expected for the peak values for each of the three tide-related variables tested. Further, for the intertidal crab (*Hemigraspus sanguineus*) there are interactions between tidal rhythms and photoperiod that regulate activitiy rhythms such as larval emergence (Saigusa and Kawagoye 1997).

Ectotherms such as lizards and insects that cannot fully control their own body temperatures may use temperature or light cues as Zeitgebers. When entrained, these ectotherms exhibit increased activity under warmer conditions and decreased activity during the cooler portions of the daily cycle (Refinetti and Susalka 1997). To measure this effect in tsetse flies (*Glossina* spp.), the number of animals that are caught during each hour of the day was sampled with nets (Crump and Brady 1979; Brady 1988). During cooler seasons, they were active only at midday when ambient temperatures reach 20°C; they tended to be active only when conditions were above about 10°C. During hotter seasons, they were active during the early morning and evening hours when the temperature drops to about 20°C, and they avoided the peak heat (40°C) during the midday hours.

Other environmental cues that may, in some organisms, entrain biological rhythms are summarized in Moore-Ede et al. (1982). Cycles of food availability, but not water availability, can entrain activity rhythms in several species of small mammals (Moore 1980; Richter 1922). Goldfish (*Carassius auratus*) were tested with varying light and feeding regimes

(Sanchez-Vazquez et al. 1997). When light and food cues were placed in opposition to one another, the investigators found that the fish shifted their daily rhythm according to their feeding schedule. Goldfish appear to have either two clock mechanisms that can be entrained by light and food, or a single clock that can be trained by both light and food. Food is also a circadian Zeitgeber for house sparrows (*Passer domesticus*) (Gwinner and Hau 1996).

For developing organisms, particularly mammals, cues from the internal maternal environment can influence the rhythms of the progeny when they are born. The activity rhythm of the mother is transferred to her progeny—her own rhythm is, in effect, a Zeitgeber for her developing young. This occurs, for example, in Siberian hamsters (*Phodopus sungorus*) (Duffield and Ebling 1998).

In humans, social cues may be involved in the entrainment of biological rhythms. Two groups, each with four volunteer subjects, were isolated in separate identical chambers (Vernikos-Danellis and Winget 1979). Each individual's activity rhythm was synchronous with those of the others in that room. However, the average free-running periods for the two rooms differed; 24.4 hours for one room and 24.1 hours for the other. To avoid any confounding effect from self-selected light-dark cycles, both groups were constantly illuminated. The possible role of social cues in synchronization of biological clocks within the rooms, but not between the rooms, was tested further by moving one subject from one group to the other. That subject soon went through a phase shift and became synchronized with the activity rhythm of the new group. The exact nature of the social cues involved in the entrainment of circadian rhythms in humans is not yet known.

Social cues may act in concert with other environmental cues to entrain biological rhythms. When individuals who have just flown by jet across six time zones are required to remain in their hotel rooms, they entrain more slowly to the new local time than do individuals who are permitted to leave their rooms and who thus obtain more social-environmental cues from their new surroundings. One method for overcoming jet lag is to spend extra time in the daylight (sunshine) in the new location (Daan and Lewy 1984).

The human circadian pacemaker can be reset using exposure to bright light (Czeisler et al. 1989). Subjects were assessed to determine the normal pattern of circadian rhythmicity. They were then subjected to 5-hour periods of bright light. The phase shift effects produced by the bright light were dependent upon when during the daily cycle the treatment was provided. Preliminary evidence from medical studies indicates that melatonin may help to alleviate jet lag and related problems (Arendt et al. 1986). In a test, eight subjects were given melatonin for 3 days before and 4 days after a San Francisco–London flight, and nine subjects were given a placebo before the flight. None of the subjects who had taken melatonin had jet lag effects, whereas six of the nine who had been given the placebo had appreciable jet lag effects.

Daylength and social cues also act together to entrain certain aspects of the annual rhythms in various animals, as, for example, in the house sparrow (*Passer domesticus*). In

this species, both hormone levels and behavior are influenced by a combination of photoperiod duration and conspecific competition (Hegner and Wingfield 1986).

If we bring an animal such as a skunk (*Mephitis mephitis*) into the laboratory and reverse the light and dark portions of its natural cycle with artificial lighting, the animal's activity phase will shift to the light portion of the cycle within a few days. The ability to manipulate the activity phase of animals' daily rhythms has been used by zoological parks to display some nocturnally active animals. By having special buildings that are kept darkened when the zoo is open to visitors and illuminated when the zoo is closed, the public is able to watch them be active, instead of watching them sleep. As investigators, we can manipulate the rhythms of nocturnal animals, and under darkened conditions, conduct behavioral observations during our own diurnal activity phase.

Models

Using information obtained on biological rhythms, we can depict what we know in a model (figure 9.5). The model system shown accounts for the two major elements important for biological timekeeping: an endogenous self-sustaining pacemaker and a system for entrainment to environmental Zeitgebers. In the example illustrated here, we note that some form of environmental input (e.g., light) is picked up from the environment by the appropriate receptor system and transmitted to the circadian oscillator (e.g., the suprachias-

Environmental input
(e.g., light)

ENTERTAINMENT
PATHWAY +
CLOCK

Receptor system
(e.g., photosensitive cells)

Nervous system
(e.g., brain or ganglion)

Circadian oscillator generates
endogenous self-sustained rhythm
(e.g., suprachiasmatic nucleus or SCN)

OUTPUT
PATHWAY

Message carried to other
neural structures via neurotransmitters
(e.g., serotonin)

Gene expression affected
in target cells

Output system via messenger
to effector tissue and organs
(e.g., feeding behavior)

Overt observed rhythm
(e.g., circadian pattern of activity)

FIGURE 9.5   Model of a mammalian pacemaker system portraying the various component parts and how they may function together, resulting in the overt behavioral and physiological rhythms that can be observed.

matic nucleus in the brain). From there a message, which may be neural, hormonal, or both, is sent to various tissues and organs and results in behavioral output (e.g., feeding), thus producing the rhythm we observe and record. The same model may be used with other types of rhythms involving entrainment to tidal, annual, or other cues. There may be a variety of different clock mechanisms in the same organism, each regulating a different function, and there may be links between these various oscillator systems.

Location and Physiology of the Pacemaker

Having described biological rhythms and having examined how they appear to be controlled, we might now wish to ask: Where is the pacemaker device responsible for doing the time-keeping? During the past several decades, great progress has been made on locating the pacemakers and studying their physiology and genetics. Much of this work has been summarized by Aschoff (1981), Brady (1988), DeCoursey (1983), Menaker and Takahashi (1995), and Wilsbacher and Takahashi (1998).

Cockroaches

Are clock mechanisms primarily neural or hormonal? The early work by Harker (1960, 1964) suggested that hormonal and neurosecretory products play key roles in the cockroach's biological clock mechanism. Harker made one cockroach (*Leuccophaea moderae*) arrhythmic by keeping it in continuous light. She then interconnected its blood system with that of a cockroach kept in a normal light-dark cycle; that animal showed regular circadian periodicity. The arrhythmic recipient cockroach adopted the regular circadian periodicity of the donor. However, other studies (Roberts 1966; Brady 1967a,b) have failed to replicate Harker's work and instead have shown that cockroaches kept in constant conditions can maintain regular circadian periodicities; the recipient cockroach in Harker's studies may simply have been reverting to the free-running rhythm.

Decapitation of a cockroach results in **arrhythmic activity;** the surgery apparently removes the source of the key oscillator, but not the ability to exhibit rhythmic locomotor patterns. Experimenters have used these headless organisms in implantation and transection experiments to test various organs and tissues as candidates for endocrine or neural clocks. Brains, corpora allata, and corpora cardiaca transplanted from hosts exhibiting rhythmic activity to headless cockroaches have all failed to produce regular rhythmic activity in the recipients. Brady (1969) found that the optic lobes play a key role in cockroach circadian rhythms and that periodicity is communicated to effector organs via the nervous system.

Further studies of the optic lobes of the cockroach lead us to the tentative conclusion that the critical area for the pacemaker lies primarily in cell bodies in the inner region of the optic lobes (Roberts 1974; Sokolove 1975). Another series of experiments (Page et al. 1977) revealed that separate pacemakers in each optic lobe interact with one another.

In addition, the presence of either compound eye alone can serve to entrain the pacemakers in both optic lobes.

Page (1982) extended our understanding of cockroach circadian pacemakers one step further. The optic lobes of a cockroach were surgically removed and replaced by the optic lobes of a second cockroach with a different circadian rhythm. After a 4- to 8- week lapse, the recipient cockroach again exhibited a circadian activity pattern — and the restored rhythm's free-running period matched that of the donor cockroach. These findings support the hypothesis that the optic lobes contain a circadian pacemaker, and that after a lapse of some weeks, neural connections between the optic lobes and the brain regenerate in the recipient cockroach — resulting in the reestablishment of a recognizable pattern of circadian activity.

Sea Hares

Experiments on the sea hare (*Aplysia*) provided further evidence of the relationships between neural structures, optic system, and biological rhythms (Jacklet 1969, 1973; Jacklet and Geronimo 1971). When sea hares' eyes are isolated and kept in total darkness in seawater, their optic nerve impulses exhibit a circadian rhythm (figure 9.6). Eyes taken from sea hares that have been under the 12-hour light/12-hour dark regimen before eye excision show a peak firing frequency at "dawn" on the day after removal. Eyes from sea hares kept in constant light before surgery exhibit regular optic nerve impulse rhythms. Thus, some type of oscillator, possibly related to neurosecretory processes, must be present in the eye. The phase of the eye rhythm is readjustable at each dawn, and the oscillator in the eye may control other sea hare rhythms. A clock mechanism that uses a pacemaker in the eye to adjust to day-to-day changes in daylength could be adaptive for an organism that lives in a variable environment. The cockroach appears to have two coupled pacemakers in each optic lobe; sea hares have pacemakers

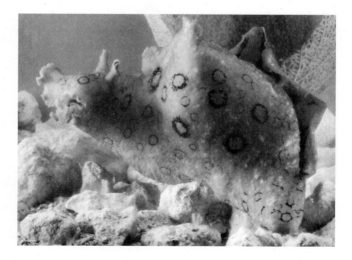

FIGURE 9.6   Sea hare.
The sea hare's eyes are embedded in the integument of its head. When we remove the eyes and place them in seawater, they continue to exhibit a circadian pattern of optic nerve impulses.
Source: © Runk/Schoenberger/Grant Heilman.

in their eyes as well (Hudson and Lickey 1977). A more recent study on crickets (*Gryllus bimaculatus*) demonstrated a similar coupling of two pacemakers located in the cricket's optic lobes; the coupling occurred via information transmitted to the contralateral optic lobe in the brain (Tomioka et al. 1994).

Further work on *Aplysia* has resulted in an increased understanding of its circadian oscillator's biochemical events (Eskin and Takahashi 1983). Using isolated eye preparations from sea hares, these investigators determined that the amount of cyclic adenosine 3′, 5′-monophosphate (cAMP) formed in homogenates of the eye was affected by the amount of (exogenous) serotonin, but not other neurotransmitters. The action of serotonin is mediated by the activity of adenylate cyclase, an enzyme critical for synthesis of cAMP. The investigators also demonstrated that forskolin, a specific activator of adenylate cyclase activity, produced both advance and delay phase shifts in the timing of the circadian rhythm of the isolated *Aplysia* eye. From these findings we can conclude that cAMP is important in mediating phase shifts in the pacemaker located in the sea hare eye.

Drosophila

Recent studies have utilized several species of the fruit fly, *Drosophila*, to explore the genetics of biological clocks.

Drosophila have a diurnal activity phase circadian rhythm that is entrained by photoperiod. Two types of behavioral rhythms have been studied most closely, locomotor activity and eclosion, the process when the adult fly emerges from the pupa. Much of this work is based on the discovery of a gene called period (*per*); mutations at this locus can lengthen, shorten, or possibly abolish circadian rhythms of locomotion or eclosion (Konopka and Benzer 1971; Hall and Kyriacou 1990b). The *per* locus is active in the photoreceptor cells of the compound eye, in glial (support) cells in the brain, and in two groups of brain neurons, one in the dorsal cortex and the other near the optic lobes (summarized by Helfrich-Förster 1995). Circadian cycling of mRNA from *per* genes occurs in these cells and there is a neuropeptide that is produced (coded for) by the *per* gene, which also exhibits a cyclical activity pattern (Helfrich-Förster 1995).

Recently, a second mutation has been discovered called timeless (*tim*); flies with this mutation do not exhibit circadian rhythms (Sehgal 1995). Glossup et al. (1999) recently demonstrated that for the circadian oscillator in *Drosophila* there are two interlocked negative feedback loops involving gene expression (figure 9.7). The two negative feedback loops involve gene pairs with activation or repression of a particular locus by the presence or absence of proteins that are produced from the other member of the gene pair. Further evi-

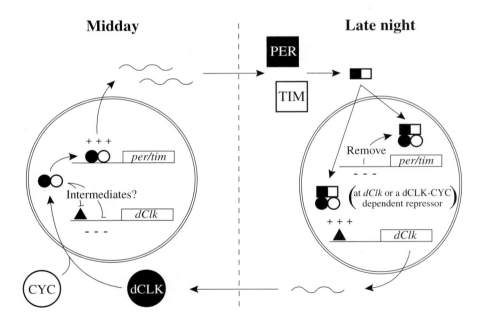

FIGURE 9.7   **Model for gene regulation within the *Drosophila* circadian oscillator.**
During the late evening (right side of diagram), PER-TIM dimers (closed and open squares, respectively) enter the nucleus and bind dCLK-CYC dimers (closed and open circles, respectively), thereby repressing *per-tim* activation. Concurrently, the binding of PER-TIM dimers to dCLK-CYC releases dCLK-CYC-dependent repression of *dCLK*, thus enabling the *dCLK* transcription via a separate activator or activator complex (triangle). By midday (left side of diagram), high levels of dCLK-CYC (in the absence of PER-TIM) serve to activate *per-tim* transcription and repress *dCLK* transcription (either directly or through intermediate factors). As the circadian cycle progresses, PER-TIM dimers accumulate and enter the nucleus during the late evening to start the next cycle. Dashes denote maximal repression; plus signs denote maximal activation; wavy lines denote mRNA. Dimers are combinations of two molecules.

Reprinted with permission from N.R.J. Glossop, L. C. Lyons, and P.E. Hardin, "Interlocked feedback loops within the Drosophila circadian oscillator," in *Science*, 286:766-768. Copyright 1999 American Association for the Advancement of Science.

dence suggests that a neuropeptide (PDF) acts as a transmitter to mediate the pathway between the specific neurons in the *Dropshila* brain and other brain regions that then control various behavioral manifestations of the rhythm (Renn et al. 1999). These studies indicate a key role for specific genes in establishing and regulating biological rhythms and they help us to further elucidate the mechanistic pathways for both the entrainment of rhythms and the effector pathways by which the clock mechanisms exert their effects on behavior.

Vertebrates

What about pacemaker mechanisms in vertebrates? Investigations of biological clock mechanisms in rats (*Rattus norvegicus*) led to the finding that a direct neural connection exists between the retina and the hypothalamus, a pathway that terminates in the suprachiasmatic nuclei (SCN), a region of the hypothalamus (Hendrickson et al. 1972; Moore and Lenn 1972). Lesions to cut connections to these nuclei resulted in rats that lost circadian rhythms of drinking behavior and wheel-running activity (Moore and Eichler 1972; Stephan and Zucker 1972).

Ralph et al. (1990) and Matsumoto et al. (1998) transplanted SCNs into host hamsters where their own SCN had previously been ablated. In about 80 percent of the cases, recipient hamsters exhibited activity patterns of a circadian nature; that is, their rhythmicity was restored. Moreover, the period of the resumed rhythm matched that of the donor hamsters. Animals receiving a SCN from a normal hamster exhibited a period of about 24 hours, whereas those given a transplanted SCN from a mutant hamster of the type we mentioned earlier in the chapter, with a period of about 22 hours, resumed their activity with the shortened period, like the mutants (Ralph et al. 1990). In a further study of this phenomenon, Liu et al. (1997) have demonstrated that individual neurons within the cells that comprise the clock have widely varying periods, but the observed rhythm is a mean of these individual rhythms of the neurons that make up the clock within the SCN.

A number of researchers have shown that SCNs are essential for circadian rhythms in mammals and that pacemakers located in these neural nuclei drive a variety of rhythms, including sleep-wake cycles, rhythms of various hormones, and feeding behavior (Rusak and Zucker 1979; Rosenwasser and Adler 1986; Meijer and Rietveld 1989). However, there is ample evidence that there must be other pacemakers as well. For example, bilateral SCN lesions in monkeys, resulting in arrhythmicity of the animal's activity-rest cycle, do not eliminate the circadian rhythm for body temperature (Fuller et al. 1981). Thus, there must be at least one additional pacemaker somewhere else in the body. Entraining rats to a particular feeding time followed by lesioning of the SCN does not alter the anticipation of feeding time in these animals (Stephan et al. 1979b; Clarke and Coleman 1986). This, too, suggests the existence of another clock outside of the SCN. There is some evidence that one location for such a secondary clock could be the ventromedial hypothalamus (VMH). Krieger (1980) demonstrated that circadian rhythms retained after destruction of the SCN are lost when the VMH is destroyed.

Investigations of rhythms in birds and mammals have implicated the pineal gland as a probable receptor of light stimuli that entrain and affect circadian patterns (Gaston 1971; Menaker and Zimmerman 1976). Existing evidence supports the hypothesis that both neural processes (Block and Page 1978) and hormone-neuroendocrine products (Zucker et al. 1976; Starkey et al. 1995) are involved in the mechanisms underlying biological rhythms in vertebrates.

The pineal gland in mammals lies within the brain, near the midline; but in some earlier terrestrial vertebrates, this structure was positioned on the top surface of the brain and served as a third or median eye. (We have all seen the mythical movie monsters with the third eye on the forehead!) In today's reptiles, birds, and amphibians, the pineal gland is located just under the skull — indeed, it is still sensitive to light in many of these organisms. We should not find it surprising that the pineal gland plays a key role in regulating certain rhythms based on photoperiod. The role of the SCN as a location for the primary circadian oscillator has been confirmed for reptiles through tests on the ruins lizard (*Podarcis sicula*); when the SCN was lesioned in these lizards they became arrhythmic (Minituni et al. 1995).

Melatonin, an indolamine (protein) is produced by the pineal gland (Reiter 1980). Daily subcutaneous injections of melatonin given to pinealectomized male hamsters induce regression of the gonads, mimicking the effect of shortened photoperiods in these animals (Reiter 1974a; Tamarkin et al. 1975). A series of investigations using hamsters, rats, mice, and other mammals have repeatedly demonstrated the antigonadotrophic effects of melatonin. In these mammals, the pineal normally receives photoperiod information via neural circuitry from the eyes. Reiter (1974b) proposed that one key function for the pineal may be the partial control of the annual rhythm of reproduction. Seasonal changes in photoperiod are translated into physiological effects by the pineal and its endocrine products. Using radioactively labeled melatonin and autoradiography, Reppert et al. (1988) determined that for humans, specific sites in the hypothalamus differentially bind melatonin. In particular, the site where the most binding of the labeled melatonin occurs is the suprachiasmatic nucleus. Further evidence for the role of melatonin and also its interrelationship with the SCN comes from work on Djungarian hamsters (*Phodopus sungorus*). When a strain of these animals that is not responsive to photoperiod was given daily injections of melatonin, it soon exhibited activity patterns and neuronal firing patterns that resembled those of normal photosensitive hamsters (Margraf and Lynch 1993).

As an example of the molecular approach to biological clock mechanisms, consider again the sea hare (*Aplysia;* figure 9.6). Serotonin, a neurotransmitter, appears to be involved in the pacemaker located in the eye. Studies on sea hares involve the use of light to shift the phases of activity. Immunocytochemical studies confirm the role of serotonin-mediated pathways involving the eye and the capacity of the eye to synthesize serotonin (Corrent and Eskin 1982). Serotonin, whose production would be increased by appropriate external stimulation, could thus be a potential messenger from the main oscillator to other locations in the animal's nervous system. In addition, cyclic-AMP (cAMP) acts as a second messenger within cells, receiving messages from the eye cells, and triggered by the serotonin (Eskin et al. 1982), results in potential alteration of gene expression within the target cells. Further, the use of pharmacological agents (e.g., forskolin or phosphodiesterase) results in elevation of the cAMP levels, mimicking the presence of serotonin. Thus, we are moving ever closer to having an understanding of the molecular mechanisms by which at least some biological clocks operate.

Several recent studies have extended our knowledge of the genetics and biochemistry of the vertebrate clock mechanism. Several genes, including CLOCK and PER1, that are involved in the biological clock mechanism have been found in mice. The Per1 gene exhibits a circadian periodicity of expression in terms of RNA and protein production in the SCN (Xiaowei et al. 1999). Takumi et al. (1998) have isolated another mouse gene, Per2, which exhibits a strong circadian period of expression in the mouse SCN. Interestingly, the amino acid sequence for the protein produced by Per2 is similar to the sequence of Per from *Drosophila* mentioned earlier.

Additional information regarding control mechanisms of circannual rhythms, primarily in vertebrates, has also accumulated. The data collected so far largely concern correlated responses that are probably several steps removed from the actual clock mechanism. In thirteen-lined ground squirrels (*Spermophilus tridecemlineatus*), selected mixtures of dialysates (materials that pass through the membrane in dialysis) and their residues obtained from the blood of other ground squirrels or from woodchucks (*Marmota monax*) accelerate or impede the induction of hibernation (Dawe and Spurrier 1972). The brown fat found in some animals, which was thought for many years to contain chemicals responsible for inducing hibernation (Johansson 1959), has been shown instead to contain a substance that produces arousal (Smith and Hock 1963). When turtles (*Testudo hermanni*) are housed outdoors, the levels of two compounds from their pineal gland—serotonin and melatonin—exhibit both circannual and circadian rhythms (Vivien-Roels et al. 1979). The turtles synthesize serotonin during the day and melatonin at night; this pattern of synthesis disappears entirely during hibernation. During the breeding season, concentrations of both chemicals and the amplitude of circadian fluctuations increase. Additional investigations are needed to determine the relationship between the concentrations of these chemicals and the observed circannual and circadian rhythms.

SIGNIFICANCE OF BIOLOGICAL TIMEKEEPING

Ecological Adaptations

Pittendrigh (1960) and Daan and Aschoff (1982) summarized the ecological and evolutionary significance of biological rhythms. Physical factors of the environment, such as light, temperature, and humidity, can be critical for some organisms: those whose integument loses water to the atmosphere, those who cannot fully regulate their own body temperature, and those who live in temperate, arctic, or desert environments where physical factors undergo dramatic seasonal or daily variation.

Many insect species live in environments where the photoperiod (daylength) differs from the periods of physical factors in the environment that may have either positive or deleterious effects, such as temperature, moisture, and chemical substances. Photoperiod exerts no direct beneficial or harmful influence but may serve as an adaptive timing cue, or a predictor of environmental conditions critical for survival (Beck 1968).

An example comes from woodlice, terrestrial isopod crustaceans that live in damp places. Field observations and laboratory experiments indicate that woodlice exhibit a circadian rhythm of activity that is regulated by photoperiod: they are active at night and quiescent during the day. At night, temperatures are lower and humidity is higher than during the day. Cloudsley-Thompson (1952, 1960) found that during the day, woodlice respond negatively to light and positively to moisture; thus, during the daytime, they tend to remain in dark, damp areas of the test chamber. At night woodlice are photonegative, but their positive reactions to humidity are not as pronounced. Under conditions of extremely low relative humidity, the woodlice become weakly photopositive.

Together, these behavioral reactions can be seen as adaptive traits related directly to water conservation and nocturnal habits. Since response to moisture is weaker at night, woodlice can move into and through dry places they would never go during the day. The woodlouse's stronger photonegative response at night ensures that when daylight comes, the insect will go into hiding. Finally, weakly photopositive responses when humidity is very low enable the woodlouse to move out of its resting place to seek a moister environment, even during daylight hours when prevailing conditions dictate such a move to ensure survival, for example, when remaining in a dry location could lead to death by desiccation.

These examples beg the question, however, of why animals have endogenous rhythms at all — why not just respond directly to external cues? The answer is that with an endogenous rhythm an animal is prepared for changes in its environment. An endogenous rhythm allows it to anticipate potentially dangerous conditions and prepare for them through physiological and behavioral change. Woodlice, for example, will not be "caught out" under hot sun if they anticipate the change.

Diurnality

For animals that have an impervious integument—birds, mammals, and organisms that live in environments where conditions remain relatively constant, like the ocean—a number of biotic factors, including reproduction, feeding habits, interspecific competition, and the possibility of predation, help to determine the timing of daily activities.

FIGURE 9.8 **Loris and hamadryas baboon.**
Today's primates have evolved from insectivore-like predecessors. The loris (left) is a descendant of one of the earliest forms to evolve; note the large eyes for night vision. The hamadryas baboon (right) has returned to a largely terrestrial existence, possibly aided by its large size and social defense mechanisms against predation.
Source: Photo by Ron Garrison, © Zoological Society of San Diego.

Whether the animal is nocturnal, diurnal, or crepuscular appears to depend on its responses to various biotic factors. It is sometimes difficult and complex for us to make determinations of the ecological significance of circadian rhythms in birds and mammals.

For example, let us consider the evolution of primates from primitive insectivores, an evolutionary sequence that is represented today by some species of primates in Africa and Asia (Simons 1972)—lorises, tarsiers, and some lemurs (figure 9.8). Primates usually exhibit a diurnal activity phase in their circadian rhythms; the shift from nocturnal to diurnal activity during primate evolution can be traced in part to the effects of certain biotic factors.

Rodents probably evolved concurrently with early primates from insectivore-like predecessors and rapidly became dominant in the terrestrial habitat. The resulting strong competition between rodents and early primates for food, nesting sites, and living space may have brought about two adaptive changes that characterized primate evolution: a shift by primates to arboreal habitats and a phase shift in the timing of their activity peak to the daylight hours. In addition, if the first change involved a move to the trees (the evidence supports this contention), then there may have been additional selection pressures for diurnal activity, because arboreal locomotion is safer in daylight.

An alternative explanation of the evolution of diurnality in primates is that the insectivore-like stem group, from which the primates evolved, was originally composed of small, diurnal, arboreal animals. Predation pressure might have led to natural selection for a nocturnal activity phase,

since adequate cover during the daylight hours exists only in the forest canopy. Later, as the true primates evolved, the food habits of some species shifted to a diet containing more fruits and leaves, and body size increased. With increased body size, the primates were better able to defend themselves or escape from predators, and some species adopted a terrestrial life-style. Having at least two equally plausible explanations (there may be others) for the diurnal activity of primates illustrates the problem of the use of post-hoc reasoning to account for the evolution of observed present-day behavior.

Several studies on insects provide some insight into the type of thinking and experimental testing that may be needed to demonstrate possible fitness consequences related to circadian rhythms. Daily task performance of honeybees (*Apis mellifera*) varies in an age-related manner (see figure 10.20). Activities within the hive, such as tending the brood and cell cleaning, are carried out during all periods of the daily cycle, but when the bees become foragers, they adopt a pattern with a diurnal activity phase (Moore et al. 1998). This is an adaptive shift in the daily activity rhythm that is related to the functional role(s) that the bees are engaged in at different times in their life. Insects that lay their eggs in the eggs or pupae of other insects are called insect parasitoids. Pompanon and colleagues (1995) demonstrated that adult parasitoids of the genus *Trichogramma* emerge primarily in the morning hours. This is adaptive in two ways. First, conditions are more humid in the early morning hours, meaning that there will be less stress in terms of drying out at this time of day as the adults emerge and their wings expand. Second, for this diurnal species, a morning emergence is better because it shortens the time between emergence and locating a host on which to lay eggs; emergence in the evening would necessitate waiting until the next morning to begin the search for a suitable host. Males emerge prior to females. The earlier emergence of males could optimize mating opportunities for males; males that emerge first could have more chance to mate and thus higher fitness. Further, the delay between emergence and first locomotion is shorter for males than for females. Male activity reaches its maximum level when females are emerging and becoming active. These strategies by males and females enhance the opportunities for females to be located by and mated with males. The highest level of male activity coincides with the emergence of the females, helping to insure that males find females for mating at the emergence site. Mated females are much more active than virgin females, while mated males do not alter their activity. Mated females are more active as they start to disperse to locate oviposition sites, while males remain active in an effort to seek additional females for mating.

Applied Aspects of Circadian Clocks

An example of the application of knowledge about circadian activity rhythms involves attempts to control rodent populations. Rodents can carry diseases, and they can be pests with respect to infestation of and consumption of grain stores; therefore, their control or eradication is important. Tkadlec and Gattermann (1993) reported that there was significant variation in terms of effectiveness when, during the daily cycle, two species of rodents, voles (*Microtus arvalis*) and golden hamsters (*Mesocricetus auratus*), were given different rodenticides. With the use of zinc phosphide, both species exhibited stronger susceptibility to the poison when it was provided during the light phase of the daily cycle. Interestingly, for a second poison, crimidine, hamsters were most susceptible at the beginning of the dark phase of the cycle, and there was no circadian effect for voles with regard to the poison's success. Variations in effectiveness for the rodenticides can be attributed to differences in their routes of physiological action (zinc phosphide has depressive effects on a number of body functions, whereas crimidine stimulates the central nervous system) and to the differences in activity phase. Voles are more likely to be active in the daylight hours, whereas hamsters are nocturnal. Thus, an understanding of circadian rhythms as well as physiological effects of specific poisons are potentially required by the rodent-control biologist.

Hibernation

Circannual rhythms in invertebrates and vertebrates have clear ecological significance. Some mammals, such as ground squirrels (*Spermophilus* spp.), chipmunks (*Striatus* spp.), and jumping mice (*Zapus hudsonicus* and *Napeozapus insignis*), exhibit annual patterns of hibernation (Kayser 1965; Pengelley and Asmundson 1974). From mid-fall until mid-spring, these mammals enter a physiological state in which their normal endothermic body temperature (ca. 37°C) falls to within a few degrees of the environmental temperature. Hibernators begin preparation for the winter season well in advance of the arrival of colder conditions. Some hibernating mammals, such as woodchucks, alter their diet by late summer and add a layer of body fat, which provides extra insulation and a food supply for the winter. Other animals, such as chipmunks and hamsters, which inhabit shallow burrow systems during the spring and summer, extend their underground system deeper or dig special burrows with a hibernation chamber. These winter sleeping sites are always located deep beneath the ground surface, usually below the frost line, so there is no danger of being frozen during the coldest periods of winter.

Some of these observations about hibernation must be examined in light of the finding that body temperatures in hibernating arctic ground squirrels (*Spermophilus parryii*, figure 9.9) are as low as −2.9°C (Barnes 1989). Other deep hibernators, such as the Richardson's ground squirrel (*Spermophilus richardsoni*), may also be able to survive these extreme climatic conditions (Wang 1978; Lyman 1982). How can the animal survive such conditions? Two ways that this might occur involve the ground squirrel's having suffi-

FIGURE 9.9   Arctic ground squirrel.
The arctic ground squirrel lives in a harsh environment where conditions are suitable for above-ground activity only 2–4 months out of each year. The squirrels spend the remainder of the year in hibernation with body temperatures as low as −2.9°C. These animals are able to tolerate temperatures below freezing because of supercooling of the blood and other body fluids without crystallization.
Source: © Leonard Lee Rue/Photo Researchers, Inc.

cient solute materials in its vascular system to depress the freezing point, or its having antifreeze molecules in its system. Studies indicate that neither of these occurs for arctic ground squirrels. In this instance, a phenomenon known as supercooling is involved; the temperature of the liquid (blood and other body fluids) can be cooled below the freezing point without crystallization. The ability of arctic ground squirrels to hibernate with core temperatures below 0°C may be related to the prolonged hibernation period (up to 8–10 months) and the extreme ambient conditions of their habitat in the far northern latitudes. The physiology of hibernation and some closely related behavioral phenomena have been reviewed by Lyman et al. (1982).

The most obvious adaptive value of a circannual rhythm of hibernation is that hibernating creatures avoid many problems of survival: they do not have to find and possibly compete for limited food supplies, and they do not have to maintain a warm and suitable nest site during adverse winter conditions. However, other mammals in the same regions do not hibernate but have adopted other strategies for survival. Some mice, for example, produce great numbers of progeny during the summer months; a portion of these will survive the rigors of winter to breed the next spring. Other mice live in communal nests during the winter and gain warmth from each other. Some mammals, like raccoons, do not actually hibernate, but during the coldest periods of winter, they may remain in their dens for several days in a state of quiescence or torpor.

A very important aspect of the circannual rhythms in mammals that do hibernate is that they are able to anticipate the coming of winter. During late summer and fall, their behavior changes; they shift their dietary habits and prepare hibernation burrow systems. If these animals waited until the food supply disappeared or the ground was frozen or covered with snow, it would be too late.

Hibernating mammals must mate soon after emerging from the winter den so that their progeny can develop before the onset of the next winter season only a few months away. How is this possible? When the animals emerge, both males and females are in nearly prime reproductive condition; some of their physiological systems have "awakened" before the end of hibernation. For example, thirteen-lined ground squirrels (*Spermophilus tridecemlineatus*) are in estrus within 1 to 2 days after emergence from the winter burrows (Landau and Holmes 1988), and they mate successfully the first week out of hibernation. The internal changes accompanying the observed behavioral phenomena are apparently triggered by an internal clock—an endogenous timekeeping mechanism that is set when the animal goes into hibernation in the fall.

Migration

The annual rhythm of migratory behavior exhibited by birds that breed in north temperate and polar regions, but winter in more southern latitudes, offers interesting parallels to hibernation. By flying south for the winter, these birds escape severe winter conditions and likely food shortages; the lack of sufficient physiological adaptation generally precludes their remaining in northern climates for the entire year. When they return to the north in the spring, these birds, like the hibernating mammals, must reproduce soon after their arrival at summer breeding locations to ensure that their progeny are fully developed by the time of their return flight south in the fall. As with hibernation, we might ask: How is this possible? Possession of a circannual clock mechanism permits these birds to anticipate the coming of fall in time to prepare physiologically for and to initiate migration. At the birds' wintering grounds to the south, where fewer dramatic seasonal changes in the environment occur, the same clock mechanism "tells" the birds when it is time to start their northbound flight in order to arrive back at the breeding area at the best time for reproducing. Bird species are entrained by light, and possibly in some cases by temperature Zeitgebers. Furthermore, as we shall see in chapter 13, biological clocks also play a role in the orientation and navigation of many birds.

Some marine organisms exhibit circannual rhythms. Longhurst (1976) studied vertical distribution and daily vertical migration in plankton and found that daily vertical movements are true circadian rhythms controlled by an internal clock mechanism, but that the nature and timing of the vertical migration vary with species, sex, age, and even location. Light is the principal, but not the only, cue that can affect these rhythms. Three ecological and evolutionary explanations of vertical migration have been proposed (McLaren 1963; Longhurst 1976). First, migrating upward at night (a common, but not universal, feature of the vertical movement of plankton) may be a means of avoiding predation. Marine fish that are potential predators are active in the upper, better-lighted zone of the ocean during the day and move to a lower zone at night. Thus, the plankton are in the upper zone at night when the fish that would most likely consume them are not. However, many of these plankton species are found at a depth during the day where enough light still penetrates to permit some predators to be active. Moreover, some of the plankton that rise to the upper level or to the surface at night are bioluminescent and thus announce their presence. Second, by migrating vertically between two levels of water, plankton may use the current that is generated as a means of dispersion. Third, the plankton may derive bioenergetic or nutritional benefits from these daily vertical movements.

Combined Clocks

The adaptive value of both circadian and circannual rhythms has been explored in several animals. Hydrophones were used to record the daily and seasonal cycles of sound production in weakfish (*Cynoscion regalis*) (Connaughton and Taylor 1995). Male weakfish make a drumming noise using certain sonic muscles that contract. Seasonal variation in drumming involved maximal levels from mid-May until August, when drumming ceased. On a daily basis, drumming occurred least frequently during the early and mid-morning hours, increasing to maximal levels in the early evening and remaining high until the following morning. Both the seasonal and daily rhythms of drumming behavior are adaptive in terms of their occurrence at times when mating is most likely. The drumming may serve to attract females, as part of the courtship, or as a means of stimulating reproductive readiness in the females.

In a similar manner, there are both annual and daily variations in the rhythms of singing and song-flight behavior in bluethroats (*Luscina svecia*) (Merilae and Sorjonen 1994). Singing and song-flight activity peaked for males soon after their arrival in the spring on the breeding area. The diurnal pattern of singing and song-flight activity were maximal in the early morning, 0300 to 0900 hours, with a secondary peak for some males at about 2200 hours. Once most males had secured a mate and eggs were layed, the singing and song-flight behavior ceased for those males. Only males that had not acquired mates or that had failed breeding attempts continued to sing. In addition, the speed with which males paired was highly correlated with song-flight activity. These data support the hypothesis that singing and song-flight activity in bluethroats are important for successful attraction of females.

SUMMARY

Animals exhibit patterns of activity and inactivity that recur at regular intervals. The timing of these behavior patterns is apparently controlled by biological pacemakers, regulated by circadian, lunar, tidal, or circannual rhythms. Biological rhythms involve two elements: an endogenous self-sustaining pacemaker and external cues that can entrain the pacemaker. Evidence for the existence of an internal chronometer comes from several sources. When an animal is placed in an environment with constant conditions, it exhibits an activity rhythm with a period different from that of any known environmental variable; these are called free-running rhythms. Animals hatched and reared in isolation and under constant conditions exhibit circadian (daily) free-running rhythms. Animals translocated from one time zone to another will, initially, exhibit an activity phase matching that of their original environment. After some period of time in the new location, they exhibit a phase shift and adjust the time of their activity to the new local time. Last, the activity rhythms for several animals of a species differ from one another; there are individual differences in the chronometers.

The activity rhythms of animals can be entrained to a variety of environmental cues called Zeitgebers. The most prominent and consistent environmental variable is the light-dark cycle. It is therefore not surprising that many daily and annual rhythms are entrained to the light-dark cycle in a wide variety of animals. Other rhythms may be entrained to tidal or lunar cues that are prominent in some environments. The availability of food, social cues, temperature, and maternal cues have also been demonstrated as Zeitgebers for rhythms in some organisms.

Research on cockroaches, sea hares, fruit flies, birds, and rodents provides evidence on the location of the pacemakers and on their molecular mechanisms. This research also helps to elaborate how the clocks function. Most biological clocks are relatively unaffected by either temperature changes or metabolic inhibitors.

Evidence to date implicates neural, neuroendocrine, and hormonal events in the pacemaker systems of different organisms. For both cockroaches and sea hares, the experimental results indicate that there are pacemakers in the optic lobes and eyes. Further data, from *Drosophila,* have revealed some of the genetic bases for biological clocks. In mammals, the evidence suggests the existence of at least two pacemaker systems, one in the suprachiasmatic nuclei (SCN) and possibly another in the ventromedial hypothalamus. The pineal gland, and in particular, the melatonin produced there, have been implicated in the regulation of annual cycles of reproduction in some vertebrates. Recent studies have elucidated some aspects of the genetics of vertebrate biological clocks and the associated cell molecular events that link the genes with protein products.

The ecological and evolutionary significance of biological timekeeping can be assessed best with respect to physical variables such as daylight, temperature, and moisture and to related biotic factors, like competition for limited resources, feeding habits, and predation. For example, the evolution of diurnality among primates may have resulted from their competition with nocturnal rodents. An alternative theory is that as primates evolved larger body sizes, they may have been better able to cope with potential predators, and thus could adopt a diurnal activity phase that enhanced feeding opportunities. Hibernation by some mammals and migration by some birds in temperate and polar climates are evolutionary adaptations to avoid the rigors of the winters in those regions.

There are also several important applied aspects resulting from an understanding of biological clocks. Examples include reversing the light cycles for nocturnal animals so they can be observed during the normal waking hours of humans, as, for instance, in many zoos; and the need to be cognizant of possible biological rhythm effects with regard to such practices as application of rodenticides.

DISCUSSION QUESTIONS

1. Inventive research techniques, such as translocation and molecular techniques to explore clock mechanisms, have been employed in the study of circadian biological rhythms. If you were about to embark on the study of the clock mechanisms of circadian rhythms and the factors that influence these rhythms, what new techniques might you employ?

2. Let us suppose that a new species of small mammal, the yellow-striped scamp, has just been discovered in the woodlands of central Ontario, Canada. Scamps are both arboreal and terrestrial; they are about 10 cm long, including the 4 cm tail; and they eat seeds and insects. What types of laboratory and field observations and experiments would you conduct to describe the biological rhythms of this species? In your answer, consider behavior patterns and underlying physiological mechanisms and the adaptive significance of the observed rhythms.

3. Certain behavior patterns (e.g., feeding) are regulated by circadian clocks, while other behavior patterns (e.g., hibernation) may be regulated by circannual clocks. What other behavior

patterns that may be regulated by circadian or circannual rhythms can you name? What is the probable adaptive significance of each of the behavior patterns you have named and how does the rhythmic pattern of the behavior relate to its functional significance?

4. As was noted in the chapter, birds that inhabit the north temperate zones migrate to warmer climates for the winter months and return the following spring to breed. In recent decades, some birds of certain species, for example, American robins, have started to remain throughout the year in the more northern zone where they spend the summer. How can this sudden change in behavior be explained? What are the consequences for the birds? Is it possible that some form of natural selection will change the overall annual behavior pattern of these birds?

5. The figure represents the singing behavior of a wood thrush over a 14-day period. For the first 7 days, the bird was housed in constant darkness. For the second 7 days, the bird was subjected to a daily light cycle with the lights on at 0600 hours and off at 1800 hours. What is the period of the activity rhythm in constant darkness and in the light-darkness cycle? What changes has the light-dark cycle brought about in the frequency and phase of the activity period?

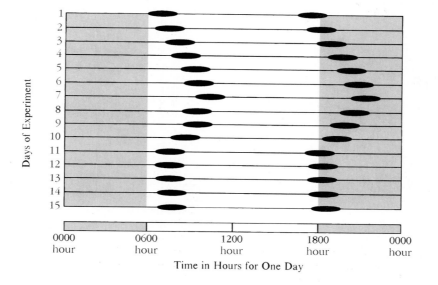

Actogram for singing of a thrush; note the animal has two active phases. Shaded area indicates darkness; black oval denotes singing.

SUGGESTED READINGS

Aschoff, J., ed. 1981. *Handbook of Behavioral Neurobiology.* Vol. 4, *Biological Rhythms.* New York: Plenum.

Brady, J. 1982. *Biological Timekeeping.* Cambridge England: Cambridge University Press.

DeCoursey, P.J. 1983. Biological timing. Biology of the Crustacea, Vol. 7, 107–62. New York: Academic Press.

Gwinner, E. 1986. *Circannual Rhythms.* Berlin: Springer-Verlag.

Hall, J. C., and C. P. Kyriacou. 1990. Genetics of biological rhythms in *Drosophila. Adv. Insect Physiol.* 22:221–98.

Pittendrigh, C. S. 1993. Temporal organization: reflections of a Darwinian clock-watcher. *Annual Review of Physiology* 55:17–54.

Development
of Behavior

hus far in this section of the book, we have been examining the mechanisms of behavior. We turn next to the process of behavioral development. We begin with an historical approach concerning an issue that has been at the heart of understanding behavior development for many decades and an area where some new critical thoughts have resulted in a refined understanding of what actually takes place. Following that, the major portion of the chapter is an examination of key factors and events that influence behavior development taken up in chronological sequence, beginning with the embryology of behavior, followed by postnatal events, juvenile events, play behavior, and comments on the continued development of behavior into adult life.

NATURE-NURTURE-NICHE

A key question that has been at the heart of our knowledge about behavior development for centuries concerns the relative importance of the genetics of the organism (nature) and environmental influences (nurture). Indeed, as West and King (1987) have chronicled, references to this issue appeared as early as the late sixteenth century. What then is involved in the "nature-nurture" controversy? What lines of reasoning have been applied to development by animal behaviorists to explain the appearance and performance of behavior patterns? What evidence exists to support these explanations? Animal behaviorists use several terms interchangeably when referring to events in behavior development. These include development of behavior, ontogeny of behavior, and, with regard to events before birth or hatching, embryology of behavior. We will use all of these during the course of the chapter.

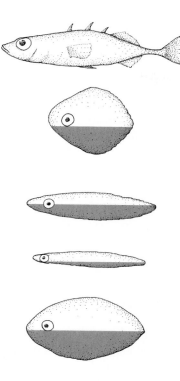

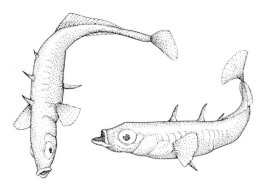

FIGURE 10.2 **Aggressive threat displays of male stickle-backs.**
These postures of the adult male three-spined stickleback fish are stereotypical, are performed in a similar manner by all males of this species, and are called fixed action patterns (FAPs).

FIGURE 10.1   **Fish models used to test aggression in male sticklebacks.**
The first model of fish resembles the normal three-spined stickleback except that it lacks the red belly characteristic of males of this species; the other four crude models all have red bellies. When these models are presented to a male, he attacks the last four models more than the first.

Genetic Influences

Proponents of the idea that behavior development is genetically programmed (innate) developed a special terminology for describing the manner in which behavior was thought to be controlled:

- **Sign stimulus:** an external signal that elicits specific responses from conspecifics.

- **Innate releasing mechanism (IRM):** a neural process, triggered by the sign stimulus, which preprograms an animal for receiving the sign stimulus and mediates a specific behavioral response.

- **Fixed action pattern (FAP):** an innate behavior pattern that is stereotyped, spontaneous, and independent of immediate control, genetically encoded, and independent of individual learning (Tinbergen 1951).

Ethologists suggest that we analyze behavior that is under genetic control as a sequence of events: sign stimulus, innate releasing mechanism (IRM), and then fixed action pattern (FAP). As an example of the use of this terminology applied to behavior, consider the adult male three-spined stickleback fish (*Gasterosteus aculeatus*). These fish establish territories and engage in aggressive displays and fights at territorial boundaries (Tinbergen 1948, 1951). One way to discover which characteristics (sign stimuli) elicit aggres-

sive responses is to present fish models (figure 10.1) to a male stickleback and see which one he attacks. The males display various standard, easily recognizable threat and aggressive attack postures (FAPs) in response to sign stimuli (figure 10.2). Crude models that have a red belly are attacked more often than are normal-looking stickleback models that lack the red underside. The aggressive posture exhibited by male sticklebacks appears rigidly stereotyped. This similarity of display patterns by all male sticklebacks may have some strong evolutionary advantages. If each male performed the threat behavior differently, the male being threatened might not be certain of the meaning of the posture. Confusion over an actual threat could lead to misinterpretation, and in turn, to an attack involving physical damage to one or both males.

The hypothesis of genetic control of behavior and the experimental methods used to test various aspects of this hypothesis were criticized on several grounds (Lehrman 1953, 1970; Moltz 1965). The term "innate" can have at least two different meanings to many animal behaviorists. First, it may refer to variations in a trait among individuals in a population. For example, human eye colors are blue, brown, green, and mixtures of these colors, with the color genetically based. In this instance, we might say that differences in a trait are inherited or innate but that external influences during ontogeny may still affect the development of that trait. Second, some ethologists have used the term innate to refer to the notion of fixed development of a specific behavior pattern—that is, the organism exhibits behavior that is preprogrammed in the genes, as is the case in the fixed action patterns of sticklebacks. Unfortunately, these different meanings create a great deal of confusion, and some animal behaviorists use the term indiscriminately. Many psychologists, zoologists, and ethologists prefer to use **innate** to refer only to genetically based differences between individuals or populations.

We know that genes carry information that codes for proteins and not directly for behavior patterns, morphological structures, or physiological processes. Biochemists and geneticists have discovered that the sequences of molecules in DNA, the chemical that carries genetic information, are codes for the production of specific protein molecules (see chapter 5) that are involved in the structures and processes within the cells and tissues of the body. To the extent that the genome provides the basic framework for ontogeny, genes have a critical role in determining the eventual behavior patterns exhibited by an organism.

If we define **experience** to include the effects of all interactions between an organism and its environment that influence behavior, then we must consider when a developing animal is capable of receiving internal or external stimuli that affect the organization of brain and the body structures involved in behavior. Lorenz (1937) argued that the neural structures, as encoded in the genes, appear first and control both the reception of external and internal stimulation, and the behavioral responses of an animal. In support of his ideas, reports on neurobiological development (see review by Jacobson 1978) have indicated that experience probably does not affect basic neural structure in some organisms. But the work of Oppenheim, Chu-Wang, and Maderut (1978), concerning the death of motor neurons in chick embryos, and the studies of Nottebohm and his colleagues reported in chapter 7 on canary song provided evidence that neural structures can be altered by ongoing experiences. Also, Moltz (1965) and Lehrman (1970) argued that the developmental processes involved in the establishment of the neural circuitry for receiving stimuli and producing responses are a function of both the genome and the experiences of an organism, beginning at conception, and this idea has stood the test of time.

Schleidt (1974), who carefully examined aspects of the stereotyping of FAPs, concluded that more quantitative studies of stereotypy were needed. He utilized data from the measurement of gobbling calls of male turkeys (*Meleagris gallapvao*) to illustrate the variability that exists in an FAP (figure 10.3). The fixity of an FAP is a relative judgment that depends on the population of animals being sampled, the method of assessing the variation, and the limits of our own perception.

In response to criticisms of the concept of FAPs, Barlow (1968) developed the concept of **modal action patterns (MAPs)**. Barlow defined the MAP as a spatiotemporal behavior pattern that is common to members of a species; different individuals tend to perform the pattern in a recognizably similar or modal fashion. This sounds very much like the species-typical behavior pattern concept proposed by Beach (1960). The MAP concept takes into account the possibility of variation around the modal pattern, the necessity of possessing some flexibility in behavior for individual adaptation, and the possibility that environmental input or the sign stimulus can result in variation in the action pattern.

We should also make note of the fact that in a general sense, different animals apparently possess varying degrees of genetic preprogramming of behavior. Invertebrates appear to have more of their behavior under a higher degree of genetic control: they possess more stereotyped responses and less flexibility. This is in contrast to vertebrates, where more flexibility is apparent in many behavior patterns. Even within the vertebrates, some investigators would argue that there are differences in the degree of genetic programming; generally, fish show more stereotyped patterns than amphibians; amphibians exhibit more stereotyped patterns than reptiles; etc. Thus, as we noted in chapter 7 on the nervous system, there are changes in the degree of behavioral programming and flexibility of the responses accompanying changes in the nervous system; these changes are reflected in the evolutionary sequences, from the invertebrates to the vertebrates.

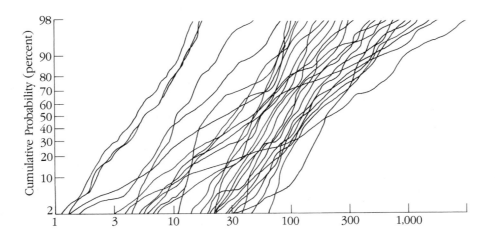

FIGURE 10.3   Inter-gobble intervals of turkeys.
The inter-gobble intervals (IGIs) for 35 samples of 100 gobbles each are plotted as the cumulative distribution in logarithmic probability coordinates. Data were obtained from 9 different male turkeys. Data such as these that show variation in a particular parameter of an FAP are useful in assessing the variation of such fixed action patterns.

Environmental Influences

Researchers have attempted to clarify the roles of environmental influences and experience on the development of behavior with several types of investigations. In the first half of this century, many researchers emphasized the role of experience and practice in behavioral development. Among these investigators, E. L. Thorndike, R. M. Yerkes, and J. B. Watson emphasized the study of animal intelligence and learning processes. Watson (1930) helped to develop the approach called **behaviorism.** This theoretical and experimental approach operates with the premise that much of behavior results from previous experience. Later, B. F. Skinner (1938, 1953) espoused and supported with experiments the idea that most animal actions can be analyzed functionally in terms of combinations of stimulus and response. According to Skinner, these S-R combinations are learned by the organism. Since learning, by its very definition, implies an important role for experience and practice (see chapter 11), it is likely that we have historically ascribed a heavy environmental emphasis to animal psychologists that investigate learning behavior, when many of them actually did not hold such an extreme position (see Dewsbury 1984). Watson, for example, was very interested in the development of instincts in young animals and the improvement of such behavior with practice. Yerkes actually was a proponent, for much of his career, of heredity as a controlling influence on behavior, as evidenced by his first book, *The Dancing Mouse* (1907). Thorndike also had strong hereditarian views and was, later in his career, active in the eugenics movement.

Many comparative psychologists have tested the effects of early experience on later behavior. The sensory, motor, and social aspects of experience have all been manipulated to ascertain the resultant effects on behavior. Some of these investigations involved enrichment, and others utilized deprivation, including isolation. The results often, but not always, demonstrated key roles for experience and environment in the development process. This is not surprising when we realize that these were exactly the aspects the investigators were manipulating. The very nature of what is being tested provides the impression that those conducting such studies are, in fact, heavily biased toward an environmental point of view. Many studies of experiential effects also involved genetics, in the form of testing different species or strains.

One criticism of studies of learning and experience as environmental influences on the development of behavior is that too often the work has been conducted only in laboratory settings, and using only laboratory rats as a test. Beach (1950) presented a numerical analysis of research published in the *Journal of Comparative and Physiological Psychology* during the period from 1930 to 1948. He found that between 60 and 70 percent of the studies had used laboratory rats as subjects, and less than 10 percent had used invertebrates or nonmammalian vertebrates as test subjects. Hodos and Campbell (1969) have also decried the lack of a truly comparative psychology.

Epigenesis and Ontogenetic Niche

In the past several decades, our understanding of what transpires during behavior development has been greatly enhanced by two related conceptual frameworks for this process. The first of these is epigenesis, and the second is ontogenetic niche.

Epigenesis is the integrated process of behavior development that involves both the genome and environmental influences. According to the epigenetic approach, expression of genetic material, which leads to the synthesis of tissues, organs, and thus to behavior patterns, is dependent on the environmental context. Therefore, in different environmental conditions, the same genes may be expressed differently. For example, individuals of a species of fruit fly (*Drosophila*) reared at different temperatures develop wings capable of normal flight, irregular weak flight, or no flight at all (Harnly 1941).

The general structure of the organism, which develops through the processes of maturation, is dictated in large measure by its genome, but the development of various structures and behavior is influenced by experience. Genes set the limits, and through interaction with the environment, the final product or phenotype is determined. What an animal inherits—that is, what is dictated by the genome—consists of a range of possible expressions of each measurable physical, physiological, and behavioral trait; genes set the limits on phenotypic expression of traits. For some morphological, physiological, or behavioral traits, the prescribed limits of expression may be quite flexible, as dictated by the genetic makeup; whereas for other traits, the limits of potential expression may be quite narrow. Also, as noted previously, within the animal kingdom, there are some distinct patterns of the degree of genetic control of behavior; invertebrates are less flexible than vertebrates.

The theory of epigenesis involves elements of both genetic and environmental viewpoints (see Miller 1988). Today, most scientists interested in the development of behavior are exploring questions about mechanisms that underlie particular behavior patterns. This often involves working closely with embryologists investigating gene expression and with physiologists and anatomists investigating relationships between structure and function in embryos and neonates. Others interested in the development of behavior are concerned with questions about life history strategies, the evolution of rates of development, and the relationships between patterns of development and the ecological habitats of various species.

As an augmentation of and improvement on the idea of epigenesis, West and King (1987) have proposed that we alter the nature-nurture dyad into a triadic expression—nature-nurture-niche. Much the same view, but using slightly different terminology, has also been put forth by Hall and Oppenheim (1987). **Ontogenetic niche** involves the multitude of ecological traits that are, in effect, passed on from generation to generation and that play integral roles, in concert with genetic inheritance, influencing behavior development. A major effect of this conceptualization is that it places environmental and genetic influences on equal footing (West and King 1987). Thus, nature and nurture should be viewed as acting together and not in opposition as has been the conceptual framework for many generations of scientists. Hall and Oppenheim (1987) state that this interactive process involves selection pressures operating on particular stages of development, resulting in particular behavioral capacities and their accompanying neurobiological substrates that we observe and record. A slightly different interpretation of this same process has also been posited. In this interpretation, ontogenetic refers to the fact that as an organism develops it encounters changing environmental situations. In this manner, changes in behavior and related morphology and physiology during the course of development would be viewed as integrated with the changes in the environmental surroundings in which the organism finds itself. This notion would apply to behavior development in the egg or in utero as well as after the organism has hatched or is born.

Many of the earlier constructs that were used to view the role of nature in the developmental process and, as a result, the experimental designs used to test such ideas, promulgated the idea that nature was a negative force, acting upon or constraining the developmental process. Environment has often been viewed as a harsh influence from which the animal should be "protected or buffered." West and King (1987) reformulate this idea to view nature as a habitat in the broad sense, in which behavioral ontogeny occurs. Thus, ontogeny is fully dependent upon and acts in concert with the developing animal's surroundings.

Recognizing that science is a cumulative process, we will, in the remainder of this chapter, build upon the traditional study of both behavior development and the notion of nature-nurture-niche. Much of this modified approach will involve attempting to interpret the earlier work in light of the new conceptual framework. Since much of the research that led to the new framework concerned early events in behavior development, particularly with regard to the nervous system, we will begin there.

EMBRYOLOGY OF BEHAVIOR

When does behavior development begin? Some integral processes important to behavior development take place before hatching or birth. Investigators have documented a variety of pre- and postnatal influences on sensory and motor development, and more general effects on subsequent behavior. Developmental biologists are accumulating considerable information about chemical and physical processes involved in the epigenesis of the tissues and organs of the body, including the nervous system. The integration of information from studies of embryology and of behavior development is leading to a new synthesis of what takes place during development and to new frontiers in research.

Nervous System Development

During development, continual changes take place in the nervous system, including cell proliferation, migration of neurons, and differentiation of cell types (see Jacobson 1978 for a review). We are concerned here specifically with how these embryological events are related to the ontogeny of discrete, observable behavioral events. According to Wolff (1981), the neural developmental process can be divided into two major categories of events: (1) the production and distribution of neurons, and (2) the establishment of appropriate synapses between neurons. In the critical developmental process, not only neuron proliferation is important, but also their selective death. Cell death is partially a consequence of another process characteristic of nervous system development: overproduction of neurons. This overproduction leads to competition for appropriate connections, and ultimately to either death or survival of neurons, depending on the interactions between the neurons and their targets.

Oppenheim and his colleagues explored the development of motor neurons that innervate the limbs of embryonic chicks (Oppenheim et al. 1978). When chick embryos are immobilized with neuromuscular blocking agents for varying periods between days 4.5 and 9 of incubation, the chicks have more motor neurons in the spinal cord motor columns than untreated chicks (Pittman and Oppenheim 1979). Apparently, the number of motor neurons undergoing cell death during this period is related to the embryo's muscular activity. Treatment started after day 12 of incubation did not influence the rate of cell death. Also, the immobilization treatment was stopped on about day 10, the cell death rate in treated chicks was merely delayed; they had the same total number of cells as control chicks by days 16 through 18. Apparently, some functional interactions that determine cell death or survival occur at the developing neuromuscular junctions.

In another study, the developmental effects of removing the limb bud on one side were compared with the same processes in the intact limb bud, which served as a control. The nerve cell death rate was greatly enhanced on the side where the limb bud was ablated. However, the characteristics of the developing motor neurons that grow outward from the central nervous system were no different from those of normal motor neurons; the neurons exhibited normal cell morphology, axon growth, and synaptic enzyme systems. Thus, it appears that the motor neurons of the chick spinal cord motor column have the capacity to initiate differentiation. Neurons deprived of their target, the limb bud, are no different from those not deprived. Further, the number of degenerating neurons can be increased by electrical stimulation (Oppenheim and Nunez 1982). Apparently, certain groups of nerve cells appear at stages of development, serve some function, and then disappear. Their appearance and disappearance seem to be regulated by various factors, including the expression of particular genes (turning on and off protein production) and the context of the developing embryo in which the neurons are located. West and King (1987) refer to this phenomenon as an ontogenetic adaptation. Perhaps a more familiar example of a trait that appears, serves an important function, and then disappears is the egg tooth. In many avian species, a tooth develops on the bill, is used by the chick to break out of the shell at hatching, and then is resorbed. Also, play behavior, which we will discuss later in the chapter, appears in greater intensity in young animals and may disappear in adults of many species (West and King 1987). As we shall see, play can serve a variety of functions.

Because of studies such as these of Oppenheim, West, and colleagues, there has been a change in philosophy regarding how we view the ontogeny of behavior. Formerly, investigators adopted the perspective that behavior seen in prenatal or prehatching organisms was an imperfect reflection of what was later to be seen in the adult (Hall and Oppenheim 1987). The investigations concerning cell death and related experimental studies have revealed that the connections between early developmental events and adult behavior are not nearly as linear as we once conceived. Further, where we formerly tended to operate with the notion that behavior began at birth or hatching, we now know, based particularly on the studies of the developing nervous system, that many events prior to hatching or birth have significant consequences for the ontogeny of behavior.

Sensory and Motor Stimulation

Can sensory and motor responses be evoked and measured prenatally? The sensory and neural systems of an embryo at various developmental stages can be tested to see if they are mature enough to respond to appropriate stimuli. In this way, we can ascertain whether particular stimuli are critical for the developmental processes occurring at that moment, as well as whether they have later consequences. It is this examination of the immediate import of the stimulation that is different from the earlier examination of only the potential future consequences of the input.

For example, we can measure an embryo's reflex responses to tactile stimulation. Hamburger (1963, 1973), and others studying various avian species, produced evidence that both general and specific embryonic movements can be induced by stimulating an embryo with a fine brush or blunt probe. The embryos of some bird species begin to respond to stimulation as early as the first week of development. Mammalian embryos, which are much more difficult to study because they are isolated in the uterus, respond with reflex-type movements to stimulation of different body zones. Touching the snout of a 16-day-old rat fetus produces head movements; touching the side of the face of an 8-week-old human fetus produces general movements of the body and limbs. For humans, Prechtl (1984) summarized much of what we know about prenatal behavior. Fetal movements are, as in other vertebrates, at least partly endogenous, being generated spontaneously. However, there is also ample evidence that sensory feedback plays key roles in the ongoing ontogenetic sequence. Of importance here is that we can relate some fetal movements to their immediate significance;

they need not be thought of only as precursors of later movements (practice). Rather, the frequent changes of position may be critical for avoiding adhesions and could aid in preventing stoppage of circulation in the fetal skin. Structure and function are certainly closely related as development progresses.

Some of the difficulties of studying mammalian embryos have been overcome by two techniques employed by Smotherman, Richards, and Robinson (1984) for studies of pregnant female rats. In one procedure, chemomyelotomy (Basmajian and Ranny 1961; Narayanan et al. 1971), permanent paralysis of the lower abdomen and legs, was induced using ethyl alcohol. The second procedure involved spinal transection in the region of the thoracic and lumbar junction. These surgical techniques were performed on pregnant females late in gestation with observations conducted on days 17–20. Additional surgery was then performed to expose the rat pups in utero (figure 10.4), and discrete behavior patterns were observed and classified into eight categories: head movement, opening and closing of the mouth, foreleg movement, hind leg movement, twitching of the trunk, body curling (figure 10.4), body stretching, and movements by the mother. There were some differences, particularly in frequency of occurrence, between fetuses in dams subjected to the two initial procedures (figure 10.5); activity was generally greater with the spinal transection technique. Fetuses also tended to be more active with additional time in the saline solution used to bathe them after they were exposed in the uterus. Also, the extra freedom of movement permitted by the removal from constraints of the uterus may have contributed to observed rates of behavior. These techniques offer promise for providing data on phenomena and events similar to those recorded and tested in chick embryos.

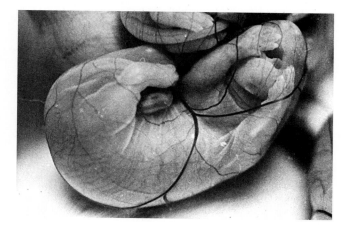

FIGURE 10.4   Foreleg and body movements shown in photograph of 20-day-old rat fetus.
Use of one of two procedures for permanently paralyzing the lower abdomen of pregnant female rats, followed by opening the uterus, permits a view of the behavior patterns exhibited by the developing fetus.
Source: William P. Smotherman.

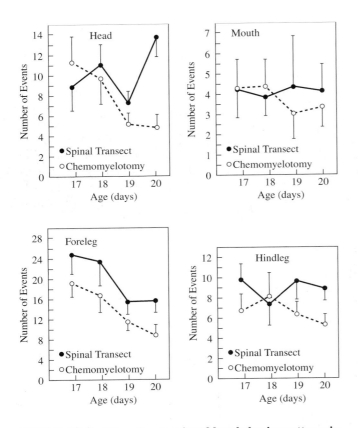

FIGURE 10.5   Mean frequencies of four behavior patterns in rat fetuses.

Frequencies per 5-minute observation period are shown for four behavior patterns observed in rat fetuses at 17–20 days of gestation. Note in particular where the two surgical techniques produce similar and different results.

From W.P. Smotherman, et al., "Techniques for Observing Fetal Behavior in utero ..." in *Developmental Psychobiology,* 17:661–674. Copyright © 1984 International Society for Developmental Psychology. Used by permission.

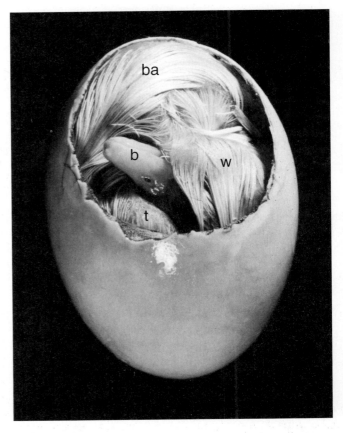

FIGURE 10.6   Bird embryo after removal of portions of eggshell.

By using a delicate technique, we can observe development in avian embryos without disturbing the normal ontogenetic processes. (ba = back, b = bill, w = wing, t = thigh)

Source: Courtesy R.W. Oppenheim.

In studies of the development of the response to maternal signals, Gottlieb (1968) showed that several days before hatching, Peking duck embryos (*Anas platyrhynchos*) respond selectively to sounds of their species' maternal call. Recordings of other species' maternal calls produce a quickening of the heart rate, which is a general activation response; but only the species-specific maternal call induces a heightened frequency of bill-clapping responses.

A slightly different, but related, set of experiments revealed that vocalizations made by prehatching birds may play an important role in synchronizing the hatching time of the birds in a clutch for some avian species (Vince 1964, 1966, 1969, 1973). Developmentally advanced bobwhite quail (*Colinus virginianus*) embryos accelerated the development of less-advanced embryos in a clutch. In particular, a "clicking" vocalization made by the embryos is important for hatching stimulation and synchronization in bobwhite quail. Vince has also provided evidence that in some instances, slower-developing embryos may retard those that are more advanced. Thus sensory-motor systems that operate prenatally in certain animal species may have important functional significance.

The motor development of prenatal organisms has been investigated descriptively and experimentally. In a study of duck and chick embryos, Oppenheim (1970, 1972) recorded general activity levels and body movement sequences from the embryo's first weeks of incubation to its hatching. His method (see Kuo 1967; Gottlieb 1968) was to remove a portion of the eggshell, exposing the developing embryo (figure 10.6), which remained viable and continued to develop normally. Figure 10.7 illustrates activity and inactivity of embryos at different stages of development. Oppenheim also analyzed the individual movements of the head and various parts of the body, and related the motor development sequences to pipping of the shell and hatching of the duckling or chick. His work demonstrated the functional relationships between embryonic motor movements and successful hatching.

Prenatal development of the motor system is not limited to general body, limb, or head movements. Duck and chick embryos are capable of vocalizing several days before hatching (Gottlieb and Vandenbergh 1968). Three types of calls—distress, contentment, and brooding—can be elicited by prodding the embryonic bird's oral region.

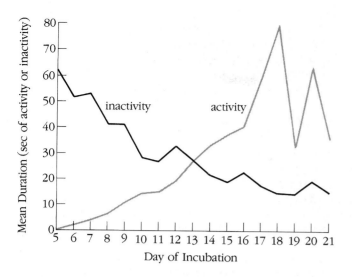

FIGURE 10.7   Measurement of activity in duck embryos. The mean duration in seconds of activity (gray line) and inactivity (black line) varies for each day of incubation in duck embryos.

All of these calls resemble vocalizations made by birds shortly after hatching that are used in communication with the mother.

Other investigators have reported generalized prenatal stimulation affecting later behavior. Subjecting pregnant female rats (*Rattus norvegicus*) to stress can affect the activity of the young at postnatal ages of 30 to 140 days (Thompson 1957; Ader and Conklin 1963). Offspring of females that were stressed during pregnancy were less active than those of unstressed mothers.

Studies of human mothers and neonates (Sontag and Wallace 1935; Ferreira 1965) have revealed differences between infants born to mothers who experienced stress during pregnancy and infants born to mothers who did not experience stress. During the weeks immediately after birth, neonates of stressed mothers were more active, cried more, and gained weight slower than did neonates of unstressed mothers.

Maternal Experiences Affect Offspring Behavior

Knowing that an organism's sensory and motor systems are functionally capable of some responses before birth or hatching, we might then want to know: What are the effects of maternal experience on later behavior of progeny? Denenberg and Whimbey (1963) conducted an experiment in which young female rat pups were handled by the experimenters during their first 20 days after birth. The pups were removed from the dam by hand, placed on clean wood shavings in a small can for 3 minutes and then returned to the mother by hand. At maturity, a group of these handled females, and another group of nonhandled females, were bred with colony males. The litters from these dams were all weaned at age 21 days and housed with like-sexed littermates until they were 50 days of age. Starting at that age, the young rats of both

sexes were tested for activity rates and defecation scores in a 45-in² open-field arena that consisted of a circular or square pen with lines marked off on the floor. Activity was measured by the number of lines crossed and/or the number of different squares entered by the test subject in a standard time period. Defecation scores were the number of fecal boluses dropped in the same time period. (Open-field tests are designed to measure emotionality; the more emotional animals cross fewer lines, enter fewer areas of the pen, and defecate more often. For a review of this method, see Archer 1973.)

The offspring of handled mothers were less active and defecated more than the offspring of nonhandled mothers. Thus, treatments administered to female rats before they were bred had a strong influence on their pups' behavior. These effects may be mediated both by the prenatal mother-fetus relationship and by the interaction between mother and pup after birth. There could also be some effects attributable to interactions between littermates.

In a related investigation, Denenberg and Rosenberg (1967) tested whether these effects could be extended a second generation to the grandpups of the original handled females. The same general procedures were employed as in the previous experiment, except that the grandpups were tested in the open field at 21 days of age. Female grandpups of handled females were significantly less active than female pups descended from nonhandled mothers, but male grandpups were only slightly affected. Thus, the effects of handling female rats in infancy may have effects on behavior two generations later. It should be noted that in this second experiment, the mother rats were also provided with a modified housing experience involving objects present on the sawdust-covered floor of the cage for the period from 21 to 50 days of age; this alternative caging may have enhanced the carryover effects to a second generation. The observed effects, particularly on second-generation female rats, could result from a variety of mechanisms, including physiological influences on the fetus, alterations of the milk provided during lactation, or extrachromosomal inheritance.

EARLY POSTNATAL EVENTS

What do we know about events that occur immediately after hatching or birth? The young of some species are born relatively helpless, often with little or no fur or feathers, and are generally incapable of locomotion or ingestion of solid foods; these we call **altricial young.** The young of other species are born in a more advanced stage, are capable of locomotion and other behavior patterns, and are often able to consume at least some solid foods; these we call **precocial young.** A variety of events at or near the time of birth have important consequences for behavioral processes immediately and later in life. Among these are social attachments and feeding behavior. We now examine each of these in more detail using examples of precocial and altricial young.

Imprinting

Imprinting occurs when an animal learns to make a particular response to only one type of animal or object. There are two kinds of imprinting: filial imprinting and sexual imprinting. The study of imprinting is an interesting story with recent research that extends our conceptualization of this process to include the notion of ontogenetic niche.

Filial Imprinting

Filial imprinting is the process by which animals develop a social attachment to a particular object. The phenomenon of imprinting has been observed for centuries. One of the earliest scientific descriptions of imprinting is that of Spalding (1873), and the first thorough investigations of the process were carried out by Lorenz (1935). Originally, this effect was thought to be both instantaneous and irreversible, but we now know that although the process occurs in many animals, it is not necessarily instantaneous, and it can be reversed or otherwise altered (Salzen and Sluckin 1959; Salzen and Meyer 1967; Hinde 1970).

The basic experimental method we use to study imprinting in precocial birds is to expose them to an imprinting stimulus soon after hatching, and then to test them later (usually several days later) using the same stimulus and a different stimulus. Measures of following and approaching behavior are used to assess the success of imprinting. However, to test the possibility that birds may have been predisposed to prefer the imprinting stimulus, a second test group is necessary: one of birds that are initially exposed to the second stimulus. If the majority of the responses are again directed at the initial stimulus, then imprinting is judged to have been successful, and both objects are usable imprinting stimuli. For imprinting stimuli, we can use live or stuffed birds of the same species, flashing lights, colored spheres or cubes, and stationary colored lights—many animate and inanimate objects work. The same objects are used to test the effectiveness of imprinting. A common apparatus used for imprinting and testing birds is a circular arena with a rotating boom from which the stimulus object can be suspended (figure 10.8).

For clarity, we will define the time the animal can develop an attachment as the **sensitive period**, and the part of the sensitive period that the attachment response performance and reinforcement are greatest, the **critical period**. In mallard ducks the sensitive period occurs from about 5 to 24 hours after hatching (Hess 1959); the critical period when imprinting is most successful encompasses the interval from 13 to 16 hours after hatching (figure 10.9).

One characteristic of imprinting, most often referred to as the **locomotion-fear dichotomy**, may be closely related to the timing of the critical period (figure 10.10). Immediately after hatching, a young duck or chick shows little fear

FIGURE 10.8 Circular arena apparatus used for filial imprinting and testing the following response of young precocial birds.
Note the stuffed bird on the boom, the young duckling, and the lines on the arena floor that can be used to measure the movement of the test bird as well as the proximity of the duckling to the moving object. Here a mallard duckling is following a mallard maternal call in preference to following a moving visual replica of a mallard hen.
Source: Courtesy Gilbert Gottlieb.

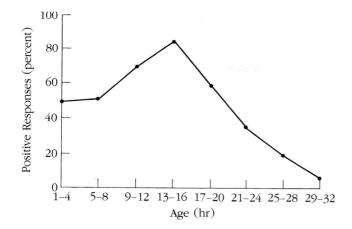

FIGURE 10.9   Period of sensitivity for imprinting in mallard ducklings.
At each 4-hour age interval the percentage of ducklings that show successful imprinting is plotted. The peak percentage occurs at 13–16 hours after hatching; we call this the critical period.
Source: From E. H. Hess, "Two Conditions Limiting Control Age for Imprinting," in *Journal of Comparative & Physiological Psychology,* 52:516, copyright 1959 by the American Psychological Association.

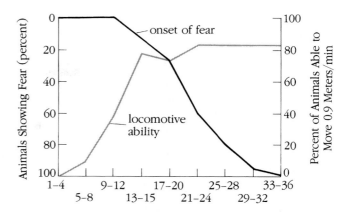

FIGURE 10.10   Locomotion-fear dichotomy.
As a young bird develops a fear of novel objects (left axis), it also improves in locomotive ability (right axis). In the middle region of the graph, where the fear response is still low but locomotive ability is relatively high, imprinting is most likely to occur and to have the highest degree of success.
Source: E. H. Hess, "Imprinting," *Science,* 130(1959): 133–41, copyright 1959 by the AAAS.

of strange objects, as measured by its avoidance and vocalizations, but it is also not yet very mobile. The bird gradually develops locomotive abilities, but at the same time, its fear of strange objects increases. Thus, there is a period when locomotion is high and fear response is low; and the interval between 13 and 16 hours after hatching is the time that imprinting in ducklings is maximally successful.

Another aspect of filial imprinting concerns the importance of different sensory modalities. In his original investigations of imprinting, Lorenz (1935) found that ducklings responded to both auditory and visual stimuli, but the strongest responses were to the mother's call. The work of Gottlieb (1971) and his colleagues with precocial birds has revealed that either visual or auditory stimuli used alone are effective in producing successful imprinting in some species. However, when both auditory and visual cues are provided, the degree of successful imprinting exceeds that achieved with either used alone. Stimuli of several sensory modes presented simultaneously may represent natural conditions more than do some single-mode experimental regimes.

Social context can also influence imprinting. Johnston and Gottlieb (1985) reported that for young Peking ducklings, the social experience of the chicks after imprinting, but prior to testing, can affect performance in the test situation. Birds that are isolated after imprinting exhibit a discrimination preference for a mallard over a pintail model. However, ducklings that are housed in groups between imprinting and testing are not successful with this discrimination.

In a related set of experiments involving responses to alarm calls rather than imprinting, Miller, Hicinbothom, and Blaich (1990), and Miller (1994) have explored further the importance of social context for responses of young birds. Mallard ducklings respond to a maternal alarm call by freezing. This response depends upon auditory experiences at about the time of hatching. We know this because ducklings that are devocalized prior to hatching and reared in isolation show diminished responsiveness to the alarm call. But, providing sound recordings of other chicks to devocalized ducklings reinstates the normal level of responsiveness (Miller and Blaich 1987). Miller, Hicinbothom, and Blaich (1990) extended this work by socially rearing devocalized ducklings and then testing them for alarm call responsivity. Interestingly, these socially reared ducklings exhibited normal response levels to maternal alarm calls. So, for some devocalized birds, playing recorded calls reinstates the response, whereas for others, social rearing suffices to produce the same outcome. The investigators conclude that there must be multiple pathways in behavioral development; different ontogenetic trajectories result in similar end points.

Miller (1994) has continued this line of investigation, adding an additional contextual dimension relating to the frequency of call notes/second. In earlier work, Miller (1980) demonstrated that a key characteristic of the freezing response given to the maternal alarm call in domestic mallards (Peking ducks) was the fact that calls pulsed at rates of 0.8 to 1.8 notes/second were most effective. In his recent study of this phenomenon (Miller 1994), ducklings reared individually or in social groups were exposed to alarm call rates of 0.2 notes/second and 2.6 notes/second, both outside the range previously determined to be most effective. The incidence of freezing when presented with calls of 0.2 notes/second was higher in socially reared than in individually reared ducklings (figure 10.11). In fact, the frequency of freezing in the socially reared ducklings was comparable to that recorded in wild mallards at nests under field conditions. There were no differences in freezing response frequency for ducklings of either rearing condition when presented with call rates of 2.6 notes/second. These findings help to further demonstrate the importance of contextual cues, here in the

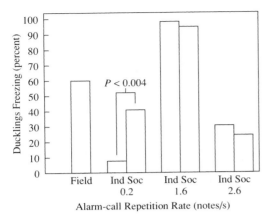

FIGURE 10.11   Responses of ducklings to maternal alarm calls.
The percentage of ducklings exhibiting the freezing response is shown for seven different treatment regimes. Three different call rates were used in conjunction with individual- or social-rearing conditions for the ducklings. The freezing response rate for ducklings in the wild is shown for comparison.

form of consideration of the natural setting and call rates, when exploring the ontogeny of behavior.

Research conducted in the last decade has begun to provide a picture of the neural structures and pathways involved in imprinting. An area of the forebrain called the intermediate hyperstriatum ventrale (IMHV) is involved in visual and auditory association processes. Studies involving lesions, recording from neurons in this brain area, and measurement of metabolic activity of IMHV cells all provide supporting evidence for involvement of this brain area in imprinting (Bredenkoetter and Braun 1996; Gruss and Braun 1996; McCabe and Nicol 1999; Tsukuda et al. 1999). The dorsal neostriatal complex (dNC) has also been implication in the association processes involved in imprinting (Metzgar et al. 1998; Bock and Braun 1999; Braun et al. 1999). Additional studies like these, involving neural structures, cellular processes, and related aspects of the DNA in the cells in these brain areas will eventually close the gap between the external, observed behavior we call imprinting and the underlying bases for such behavior patterns.

What about imprinting phenomena in mammals? In goats (*Hircuscapra*), the offspring are imprinted on the adult female parent, and parents learn the characteristics of their offspring. This imprinting is based primarily upon olfactory and possibly visual cues (Klopfer et al. 1964). The sensitive period for formation of the attachment between the doe and kid may last for several hours after the kid's birth, but the critical time period for formation of the bond is during the initial hour after parturition. During the first postpartum hour, a foster kid can be substituted for the neonate, and the doe accepts the alien kid as her own (Klopfer and Klopfer 1968). A similar imprinting phenomenon occurs in sheep (*Ovis aries*). This has been used by Price et al. (1984) and

Martin et al. (1987) to study cross-fostering in sheep providing useful data for sheep ranchers in many western states. In one experiment (Price et al. 1984), alien sheep were placed into stockinettes (a body sock made of synthetic material that fits over the head and limbs of the lamb) that had previously been on lambs belonging to a particular ewe. The alien lambs were then placed with the ewe; 84 percent of the alien lambs were adopted. Alien lambs placed with ewes while retaining their own stockinettes served as controls; all were rejected by the foster ewes. Thus, the odor cues that impregnated the stockinettes served as a critical cue for this fostering process. In a follow-up experiment, Martin et al. (1987) demonstrated that the stockinette technique could be used to transfer a second lamb to ewes that already had single lambs.

What is the functional significance of filial imprinting? The imprinting or following response is important in some species, particularly in ducks, which may build nests located at some distance from water, the eventual home of the birds. Within several days of hatching, the mother leads the ducklings to the water. Imprinting with visual and/or auditory species-specific cues may enable young birds to follow the mother successfully through dense vegetation or other obstacles between the nest and the water. Gottlieb (1963, 1971) has shown the importance of auditory communication in wood ducks (*Aix sponsa*). Shortly after hatching when the young are ready to leave the nest, which is in a tree hole several feet above the water or other suitable site, the mother descends to the water and calls to the ducklings. In response to the call, the young jump up to the nest opening and drop down into the water to follow the mother.

Miller and Gottlieb (1978) demonstrated that incubating female mallard ducks emit species-specific vocalizations. These vocalizations are similar to those of the mother when she calls her young from the nest. In this instance, the imprinting process actually begins before hatching.

Both visual and auditory imprinting may play an important role in predator avoidance. The ability to follow the mother's lead quickly to safety in the seconds between the sighting of a predator and its attack may mean the difference between life and death. Alternatively, young birds may respond to the maternal alarm call by freezing and remaining motionless until danger has passed (Miller 1982; Blaich and Miller 1986; Miller and Blaich 1987). In addition, the recognition of conspecifics that may result from imprinting may be significant in the young birds' socialization process, and in conspecific cooperation in a social organization. Associating with conspecifics may be important for surviving, for locating food and shelter, and for migrating.

Sexual Imprinting

A different type of imprinting, in which individuals learn selectively to direct their sexual behavior at some stimulus objects, but not at others, is termed **sexual imprinting.** Sexual imprinting may serve as a species-identifying and species-isolating mechanism. By appropriate exposure to an imprinting stimulus and later testing, the sexual preferences of birds have been shown to be imprinted to the stimulus to which they were exposed. Sexual imprinting generally involves longer periods of exposure to the stimulus than filial imprinting. Both precocial birds like turkeys (Schein and Hale 1959) and ducks (Schultz 1965), and altricial species like zebra finches (*Taeniopygia guttata*) (Immelmann 1965, 1972) and doves (*Columba* spp.) (Craig 1914), have exhibited mate preferences based on early rearing experiences. Most of these studies have involved cross-fostering young birds to another species or rearing them with models and postpubertal testing for mate selection preference.

It is interesting to note that male and female sexual imprinting differs for mallard ducks and zebra finches. Both species are sexually dimorphic: males have morphological (color, plumage) and behavioral (display) characteristics that trigger sexual responses; females lack these traits. When male mallards are cross-fostered with other ducks and zebra finches are cross-fostered with other finches, the resulting males court females of the foster species and not of their own; however, mate selection by female mallards and zebra finches is not affected by the rearing experience. In a review of this subject, Bateson (1978) presented a model for testing sexual imprinting that comprises four parameters: (1) the age of the animal, (2) the length of exposure to the foster species, (3) the actual imprinting stimulus value of the foster species, and (4) the exposure to other species before and after exposure to the foster species.

Most birds in nature have limited contact with members of other species during the first few days or even weeks after hatching. This ensures that young birds will imprint on members of their own species and that they will court only conspecifics when they engage in reproductive activities as adults. Thus, we can see that imprinting as a species-isolating mechanism helps guarantee that the investment of reproductive energy and gametes is not wasted on unsuccessful mating activities.

Cross-fostering has also been used to study the effects of early exposure to particular stimuli on rodents' sexual preferences (Lagerspetz and Heino 1970; Murphy 1980; Huck and Banks 1980; see also summary by D'Udine and Alleva 1983). These studies have involved a variety of rodents, including house mice (*Mus domesticus*), Norway rats (*Rattus norvegicus*), lemmings (*Lemmus trimucronatus* and *Dicrostonyx groelandicus*), and hamsters (*Mesocricetus auratus* and *Mesocricetus brandti*). In all of these studies, we find that beginning shortly after birth, a period of residence with a dam from another species resulted in the adult foster pup's reduced preference for conspecifics of the opposite sex and enhanced preference for the foster species. Clearly, some type of imprinting process or similar early experience phenomenon is occurring that strongly affects sexual preferences and potential reproduction. These early established preferences may serve as species-isolating mechanisms and/or may play critical roles in mate-seeking and mate-selection behavior.

Development of Feeding and Food Preferences

Effects of Lactating Mother's Diet and Conspecific Odors on Offspring Food Preferences

Is it possible that the food ingested by a lactating female influences the food preferences of her progeny? To test this question, Galef and Henderson (1972) experimented with 32 pups born to 4 female rats (*Rattus norvegicus*). After the birth of the pups, the investigators removed the lactating mothers from their home cages and placed them in separate compartments during three 1-hour feeding periods each day. They gave two of the females food prepared by Purina, and two food made by Turtox. When untested rats are given a choice test, they normally prefer the Turtox diet; the foods differ in taste, texture, and color. Galef and Henderson measured the food preferences of the pups for 7 days, beginning at 17 days of age. Figure 10.12 shows that the young rats reared by Purina-fed lactating mothers ate proportionately more Purina chow than Turtox; and pups reared by lactating mothers fed Turtox preferred Turtox chow.

A rat pup may acquire information about the food its mother is consuming in at least three ways. Young rats may consume feces dropped by the mother; they may ingest or smell particles of food adhering to the mother's fur or oral region; or the flavor or odor of the mother's food may be transmitted in her milk. Further experiments conducted by Galef and Henderson confirmed that information about diet

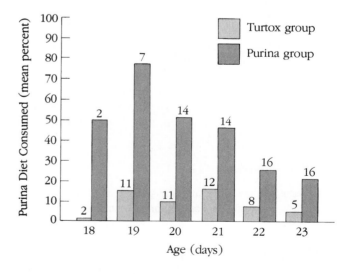

FIGURE 10.12   Feeding preferences in rat pups.
The mean amount of Purina chow eaten as a percentage of total food intake by pups reared by mothers fed either a Purina or a Turtox diet varied. The numbers above each bar indicate the sample size for that day.

Source: Data from B. G. Galef and P. W. Henderson, "Mother's Milk: A Determinant of the Feeding Preferences of Weaning Rat Pups," in *Journal of Comparative and Physiological Psychology*, 78:213–19, copyright 1972 by the American Psychological Association.

is transmitted through the mother's milk. The results of these experiments have some interesting implications for our understanding of the development of food habits in young rodents as they leave the natal homesite to fend for themselves. Prior to this work, it was assumed that young rats develop their own preferences of solid foods without any assistance from adult rats (Barnett 1956). However, the results outlined here and in other work by Galef and his colleagues confirm that young rodents may obtain some of this information from their mothers.

At the age of weaning, the feeding site selection of young rats is influenced by interactions with adults, and in particular, by olfactory cues associated with conspecifics (Galef and Clark 1971; Galef and Heiber 1976; Galef 1977, 1981). When presented with a choice, rat pups select a feeding site associated with either conspecifics or their excreta in preference to a clean site. If pups are reared without contact with conspecifics, they select a feeding site without regard to whether it is clean or has conspecific cues. When pups are reared away from conspecifics, five days of conspecific exposure before testing is sufficient for them to prefer feeding sites associated with conspecific stimuli.

Turtle Food Preferences

Can early experience with foods affect dietary habits in turtles? Burghardt (1967) investigated the feeding preferences of snapping turtles (*Chelydra serpentia*). He fed separate groups of newly hatched turtles either horsemeat or worms for one day. One week later, each turtle was given its choice between the two types of meat. The results revealed that (table 10.1) the turtles exhibited a clear preference for the food they had eaten immediately after hatching. We conclude from these results that the initial feeding experience may be a critical factor in later diet preferences in snapping turtles. Further testing would be needed to confirm and extend these conclusions to the retention into adulthood of the initial feeding experience effect, but some type of food imprinting may be occurring. The turtle's food preference may have become partially fixed, even though the turtle will still consume other foods.

Feeding Behavior of Gull Chicks

The feeding behavior of laughing gulls (*Larus atricilla*) has been studied in both field and laboratory settings (Tinbergen

FIGURE 10.13   Normal feeding behavior of laughing gull chick.
Chicks engage in two separate types of pecking. Here the chick aims an accurately coordinated peck at the beak of a parent, which prompts the parent to regurgitate food. In other instances the chick may peck at food (e.g., fish) the parent has regurgitated.

and Perdeck 1950; Hailman 1967, 1969). In the field, Tinbergen and Perdeck made observations and found that hungry chicks use their bills in a pecking and stroking pattern directed at the bill of a parent, which induces the adult bird to regurgitate food for the chick (figure 10.13).

Hailman examined how this food-begging behavior in gull chicks developed. His results indicate that the behavior resulted from an interaction of genes and environment. The genome of the young gull provides the information necessary for the correct maturation of the bird's sensory and motor systems. As it matures, the bird learns to peck at the parent's red bill, which contrasts with its black head. By using cardboard models of a gull's head to test pecking behavior, Hailman found that it took the chicks several days to attain 75 to 90 percent accuracy in hits (figure 10.14). Thus, we can conclude that full development of the begging behavior requires genetically programmed development of sensory-neural systems to receive and interpret the stimuli, motor capacities to respond to the stimuli, and experience to learn the pecking behavior.

TABLE 10.1	**Food preferences of snapping turtles**					
		First Meal		**Second Meal**		
Group	N	Food	Pieces Eaten	Food	Pieces Eaten	Number Choosing First-fed Food
1	12	Horsemeat	26	Worm meat	19	12
2	13	Worm meat	28	Horsemeat	34	8
Totals	25		54		53	20

Source: Data from Burghardt (1967).

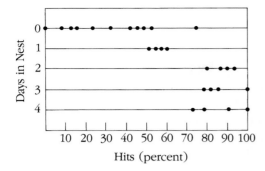

FIGURE 10.14   Accuracy of chick pecking.
Gull chicks' improving accuracy in pecking ability is illustrated here. By the time the chicks have been in the nest 2 to 4 days, they exhibit 75 to 90 percent accuracy in directing their pecks at the bill portion of a painted card representation of the parent's head.
Source: Data from J.P. Hailman, "How an Instinct is Learned," in *Scientific American,* 221(1969):100, copyright 1969 Scientific American, Inc.

JUVENILE EVENTS: BIRDS AND MAMMALS

The period that lasts roughly from fledging (birds) or weaning (mammals) until the animal's full independence on the way to maturity can loosely be termed the juvenile stage. Development is a continuous process that is usually marked by various identifiable events, such as birth, or fledging. For convenience, we as investigators often divide the continuum into stages or periods. Because animals of different species—and indeed, animals of the same species—often proceed through development at varying rates, the timing and length of these periods will vary. Several pivotal events usually occur during the juvenile period, as defined here, including natal site dispersal, puberty, appropriate communication signal learning, various experiences that influence later behavior patterns, and for birds, the first migration toward the equator. Play behavior, which for some species constitutes an important activity during the juvenile period, will be discussed in more detail later in the chapter.

Deprivation/Enrichment Experiments

Historically, one method for exploring the possible significance of a particular experience or stimulation during development has involved depriving the organism of one or more of these experiences or stimuli (**deprivation experiments**) or providing some form of enriched environment (**enrichment experiments**) and measuring the effects on subsequent behavior. Some test paradigms have involved treatments starting before or during the juvenile stage. The effects may be detectable immediately or later in life. Deprivation and enrichment may be social, sensory, motor, or some combination of these. With the new notion of an ontogenetic niche and the consideration of conditions like deprivation or enrichment as having immediate contextual consequences for the ontogeny of behavior as well as manifestations later in life, we can expect that while deprivation and enrichment experiments will continue to have importance for deciphering what happens

FIGURE 10.15   Young rhesus monkey with surrogate mother.
When given a choice, most infant rhesus monkeys preferred a cloth surrogate mother to a wire surrogate mother, even when the nursing bottles were attached to the wire mother.
Source: Courtesy Harlow, Primate Laboratory, University of Wisconsin.

during behavior development, there will be new interpretations and some new experimental techniques and designs. A more traditional approach is represented here by the studies of Harry Harlow and colleagues on maternal deprivation effects in primates. Following that, an examination of the ontogeny of bird song provides an opportunity to explore newer approaches to behavior development as they have grown out of older approaches. A final example, concerning puberty in house mice, illustrates the importance of contextual cues influencing reproductive physiology and related behavior.

Social Deprivation in Monkeys

From the 1950s onward, Harlow and his colleagues conducted a series of studies on the effects of **social deprivation** on the behavior of young, adolescent, and adult rhesus monkeys (*Macaca mulatta*) (Harlow and Zimmerman 1959; Mason 1960, 1961a, 1961b; Harlow 1962; Harlow and Harlow 1962a, 1962b, 1969; Arling and Harlow 1967; Mitchell 1970). Similar studies have also been conducted by Hinde and his associates in Great Britain (Hinde and Spencer-Booth 1967, 1970, 1971; Hinde and Davies 1972; Hinde and McGinnis 1977; Hinde et al. 1978). Throughout these investigations, young rhesus monkeys were reared under various treatment conditions for periods of time during the first two years of life. The conditions included the following:

1. Rearing in total isolation in chambers that remove the infant from all social contacts and most external stimulation

2. Isolate rearing, but with a cloth or wire surrogate mother (figure 10.15)

3. Peer-group rearing with other monkeys of similar age

4. Rearing with the mother only

5. Rearing with the mother, but with varying periods of separation at specified age intervals

6. Rearing in small social groups in the laboratory (often used as a control condition)

Both the age of the monkey at treatment initiation and the duration of treatment are critical parameters of social deprivation. This list of rearing treatments is not intended to be exhaustive; researchers have employed others as well as gradations or combinations of these.

As we might expect, the severity of the observed effects due to deprivation rearing varies with the degree of social isolation and treatment duration. The first five treatment conditions are listed in the order of approximate degree of deprivation. Behavioral deficits (e.g., failure to perform complete behavior patterns, performing abnormal behavior patterns, lack of responsiveness to conspecifics) are produced in varying intensities, depending upon the severity of the isolation. These effects include withdrawal, engagement in fewer social interactions, deficiency in understanding communication, and inability to perform normal sexual behavior. Among the better-known effects of social deprivation are rocking and swaying, self-clasping and other self-directed actions, huddling behavior (exhibited by monkeys reared in peer groups), and poor maternal behavior by surrogate-reared females, who are abusive toward their infants (figure 10.16); hence the name "motherless mother." Results of another type of effect investigated—the reestablishment of the bond between the mother and infant after deprivation or periods of separation—also depend upon the severity of the separation treatment but indicate there is a great deal of individual variation.

Studies performed by Harlow, Hinde, and their associates have taught us a great deal about the nature and development of bonds between mothers and their infants and between maternal members of conspecific groupings of primates. These studies show that several qualities of the rhesus mother, such as contact (even that provided by cloth surrogates), warmth, and some types of movement, are important for normal infant development and development of affection. Young rhesus monkeys spent more time clinging to cloth surrogate mothers than wire mothers, even when the source of milk was associated with the wire surrogates. The effects from experiments using brief separations can last up to 2 years. These investigations have provided information helpful for bettering primate rearing conditions in zoos and laboratories, and insight regarding the importance of parent-offspring contact during the human rearing process. However, today these studies would provoke ethical questions and would be less likely to be approved by animal care.

We should note two additional points in connection with these studies of social deprivation. First, some of the behavior that Harlow and others have recorded as "abnormal" occurs occasionally in rhesus monkeys reared under "normal" conditions (Erwin et al. 1973); self-directed aggression has been observed on a number of occasions in nonisolate-reared subjects. Another study reported that in free-ranging rhesus monkeys living on an island off the coast of Puerto Rico, only 50 percent of the young born to primiparous females (those giving birth for the first time) survive to the age of 12 months (Drickamer 1974b). Harlow and his coworkers have also noted that "motherless mothers" exhibit much better maternal behavior with their second infant. Part of the deficiency in maternal care recorded in "motherless mothers" may result from their ignorance of how to be good mothers, and not strictly to the emotional abnormalities associated with isolate rearing.

Second, and quite significant, Harlow and his associates (Harlow and Suomi 1971; Suomi 1973; Suomi et al. 1974) succeeded in socially rehabilitating isolate-reared monkeys. They accomplished this by exposing 6-month-old social isolates to 3-month-old normal monkeys, called "therapy monkeys," for 2 hours per day, 3 days per week, for 1 month. The effects produced by some types of early social isolation are not irreversible, as was once thought, though even rehabilitated monkeys continue to exhibit some behavior deficits.

Bird Song

In the past several decades, an increasing amount of animal behavior investigators' attention has been given to bird song pattern—its control, development, and evolutionary significance. We have already explored control of bird song in chapter 7, and some of its functional significance will be explored in chapter 12. There is a diversity of developmental strategies that led to the variety of songs and calls we are familiar with (Kroodsma 1978, 1981; Irwin 1988; Slater 1989; Marler 1990; Nowicki et al. 1998). Comprehensive analyses of many bird species show that there are at least two major strategies for song development: (1) imitation of the songs of others, particularly of adult conspecifics; and (2) invention or improvisation. Underlying both strategies are the questions: What kind

FIGURE 10.16   Socially deprived rhesus monkey mother.
Rearing young rhesus monkeys under varying conditions of social deprivation produces several types of deviant and abnormal behaviors. When mature, the socially deprived female may be a very poor mother, at least with her first infant.

Source: Courtesy Harlow, Primate Laboratory, University of Wisconsin.

of templates exist to provide some genetic basis for the song-learning process and to help shape or guide it? and Does a sensitive phase for development of the song repertoire exist? Let's consider several examples that should help explain song development.

The song learning of the marsh wren (*Cistothorus palustris*) has been studied extensively. In nature, male marsh wrens sing more than 100 types of songs; neighboring males generally sing identical song types; they often interact by countersinging with one another using the same song type. To test their song-learning development, males were reared in special housing conditions where the songs they heard could be completely controlled. Males were played specially prepared tutor tapes containing nine different songs each. Males were exposed to one tape for age 15 to 65 days, a second tape for age 65 to 115 days, and a third tape the following spring. By imitating them, the males learned the nine songs on the tape they were exposed to before 65 days of age, but they did not learn the songs on the subsequent two tapes. The males never sang any invented songs. Further investigation revealed a more refined estimate of the peak sensitive period, which falls between about age 35 and 55 days (Kroodsma 1978). In addition, males learned song types played for either 3 days or 9 days. Some males improvised some songs in these additional tests, presumably from elements of the songs on the tutor tapes.

The learning situation used for these marsh wrens involved only tutor tapes played over loudspeakers. In another test, young wrens were exposed to a number of song types on tutor tapes and then were given a period of social interaction with adult males with varied song repertoires. The period of social interaction occurred in the fall of the first year for some birds, and not until the following spring for others. Data on song repertoires for these birds indicated that song learning can occur early from the tapes, or at either of the periods of exposure to adult males. Hence, song learning is somewhat flexible, and social interaction may be an important feature of the process. These findings also make it clear that we should be careful when using only artificial stimuli like loudspeakers (see Kroodsma and Pickert 1984). Social interaction has also been shown to be critical for language acquisition in human children (Jerison 1973; Freedle and Lewis 1977).

A longitudinal study of song development has been done on male swamp sparrows (*Melospiza georgiana*) that were hatched in the wild and brought into the laboratory. Beginning at age 16 to 26 days, the birds were given song training with species-specific songs twice daily for 40 days (Marler and Peters 1977; Peters et al. 1980). During this initial phase of song learning, birds memorize the songs that they hear during a sensitive period. Weekly recordings of each bird's songs began shortly after training when they were just over 3 months old, and continued until they were older than 1 year. When these recordings were analyzed, a seven-stage sequence of song development was discernible (figure 10.17); taken together these stages form the motor phase of song development. The syllables used in training are shown

at the top of the figure. Young birds began by singing what Marler and Peters call subsong (stage VII) at an average age of 272 days. They progressed through subplastic song (stages V and VI) and began to sing the **plastic song** (stages II to IV) at an average age of 299 days.

Crystallized song (stage I) began at an average age of 334 days. During the course of this developmental sequence, the duration of the song decreased. Syllabic structures began to emerge during subplastic song. Syllable analysis revealed that about 30 percent were imitations of the training songs and 70 percent were inventions or improvisations. By the time the crystallized song emerged, the number of syllables sung was only 23 percent of the potential repertoire; both imitated and improvised syllables were included in the crystallized song of most birds (Marler and Peters 1981, 1982). During development, it was not unusual to record songs characteristic of more than one stage on the same day. However, once the crystallized song singing began, the birds rarely reverted to an earlier stage. Marler and Peters (1982) hypothesize that the pattern of song development noted for the swamp sparrow may be characteristic for many songbird species.

What internal events can be related to the song development process? The neural bases for song development have been explored in a series of studies using white-crowned sparrows (*Zonotrichia leucophrys*). Birds of this species learn their songs from hearing adult birds and have a sensitive period from about age 50 to 100 days. Birds that are isolated during this period develop abnormal song (Marler 1970). Further, white-crowned sparrows that live in different regions exhibit dialects in the basic song pattern (Marler and Tamura 1962). A series of neurons that are critical for song production have been located in a brain region called the pars caudale in the hyperstriatum ventrale (termed the HVc) (Margoliash 1983, 1986, 1987). These neurons respond selectively when the bird's own song, when songs from within the same dialect, or when sequences of synthetic song with specific properties are played to the bird. This responsiveness may be important during song development—modification of these neurons may occur during the sensitive period. Song recognition may be important for white-crowned sparrows' territory definition and mating partner recognition and could thus contribute to greater reproductive success.

The relationship of ontogeny and phylogeny to bird song has been examined by Irwin (1988). In sparrows, Irwin notes that the ontogeny of song parallels phylogeny: song development proceeds from general stages to more specialized stages. At the earliest ontogenetic stage, song is continuous, the repertoire size is undefined, syllable structure is minimal, and the number of syllable types per song is not defined. By the middle stages of ontogeny, the song is still continuous and the song repertoire remains undefined, but the syllable structure is now stereotyped and each song contains many syllable types. Still later in development, the song has become discrete and the repertoire contains many songs. In the adult stage, there are species variations. These varia-

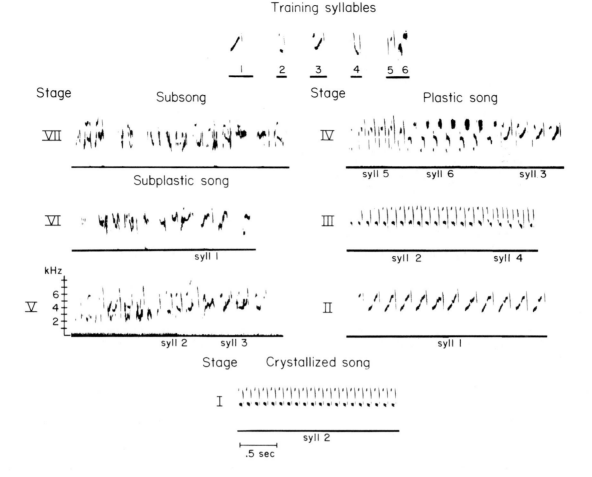

FIGURE 10.17   Song development in a swamp sparrow.
Samples of each of the seven stages of developing song in a male swamp sparrow. The stage is indicated at the left and major stages above. Imitations of the six training syllables portrayed at the top of the figure are identified by number. Training syllable 6 is a song sparrow syllable; the remainder are from swamp sparrows. Syllables were presented in one-part songs, except for syllables 5 and 6, which appeared together in the same "hybrid" song. In nature, swamp sparrows typically have a repertoire of three to four song types, and, although there is some sylla-ble-sharing in local populations, most individual repertoires are unique. The experimental male shown here had a final repertoire of two song types, shown in stages I and II. This crystallized song is virtually identical to the song of the local wild male that served as the original source of the training syllable.

tions differentiate this stage from the previous stage where patterns were generally common for all species in a group. For example, within the three species of sparrows studied, the song sparrow (*Melospiza melodia*), the swamp sparrow (*Melospiza georgiana*) and white-crowned sparrow (*Zonotrichia leucophrys*), all have discrete songs and stereo-typed syllable structure as adults. However, song and swamp sparrows have many songs in their repertoires; the white-crowned sparrow has only one. Song and white-crowned sparrows have many syllable types in each song, but the swamp sparrow has only one. Thus, general patterns in song development follow phylogenetic lines that lead to differen-tiation in adult song types. We should note, however, that Irwin's findings for sparrows are not necessarily applicable to all birds.

In a review of bird song development, Slater (1989) made two key points. The first is that bird song has different functions in different species. The second is that bird song probably has more than a single function in many species. Taken together, these two statements provide a backdrop for understanding what has been discovered in recent years about the manner in which birds learn their songs.

Much of the work on the ontogeny of bird song has been predicated on the idea that hearing other birds, often adult conspecifics, is a key feature of the process. Indeed, that is likely the case, but in many instances, we have ignored or forgotten the fact that there are many other elements in the overall context (niche) in which the bird is developing. It turns out that some of these other elements are also impor-tant for song development. Two examples should help illus-trate this point, one involving brown-headed cowbirds

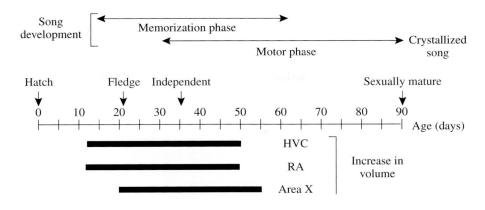

FIGURE 10.18   Bird song development.
Key life history events in the early post-hatching development of a zebra finch male are shown on the time line (age in days). The two phases of song development are depicted above the time line. Below the line, the black bars indicate the increases in volume for three areas (nuclei) of the zebra finch brain that occur during this time period. HVC = high vocal center; RA = *robustus archistriatalis;* Area X = nucleus key to song acquisition.

(*Molothrus ater*) and the other concerning a species we have discussed previously, the marsh wren.

The key finding of the work on cowbirds is that different song repertoires are developed by male cowbirds (females don't sing) housed in different social contexts (King and West 1983a, 1983b; West and King 1985). Cowbirds are brood parasites and must avoid imprinting on their hosts as potential mates. Young, acoustically naive male cowbirds were housed with either females of the same (*M. a. ater*) or different (*M. a. obscurus*) subspecies. Males reared under these two conditions developed songs that differed with respect to structural elements, and these differences were functionally significant. Since the females do not sing, the differences must be attributed to something other than vocal imitation, using the females as vocal tutors. Apparently, male cowbirds can modify their singing based on some type of social stimulation. A follow-up series of laboratory experiments and observations at a winter roost of cowbirds (King and West 1988) determined (a) the effects of differential social housing on song development can be traced to very early in song development and (b) male cowbirds living in a natural roost had ample opportunities to interact with females, and they shared the same structural categories of song during ontogeny as their laboratory kin. Song development in this, and possibly other, species is therefore a dynamic and interactive process that includes social as well as vocal components. Social factors have also been implicated in the song development of the white-crowned sparrow. As compared to recordings, live tutors extended the sensitive phase for song development (Baptista and Petrinovich 1986).

Further, an earlier study by Kroodsma and Pickert (1980) provides evidence for additional factors that form part of the context in which song development takes place. The investigators found that the duration of the sensitive period for song development in marsh wrens was influenced by at least two environmental factors. Both photoperiod

(amount of daylight/day) and the amount of adult song heard by the young birds in the form of tutor tapes played to them affected not only the sensitive period for song development during the hatching year, but also the ability of the birds to learn further songs the next spring and the dates at which the males developed their adult songs. Late-hatched young (e.g., from second broods) hear fewer songs from adults during that first summer and fall. These birds may also disperse farther to establish their first breeding territories the next year. It appears that the system has evolved to provide these late-hatching birds with some needed flexibility in terms of developing their song more during that first spring, when they have dispersed and they are exposed to a particular dialect of the species-typical song pattern. As a conclusion to this section, it is worth repeating a sentiment presented in a review of bird song development by Marler (1983). This is the notion that while a phenomenon like bird song has provided an excellent medium to study behavior development and elucidate many principles, these ideas are likely applicable in various forms to many other behavioral ontogeny phenomena.

Nowicki and colleagues (Nowicki et al. 1998) have recently provided a summary of what is known about song development and its neurological bases in zebra finches set against the background of their developmental life history (figure 10.18). The three brain regions noted in this figure are all important to the control of bird song. During the memorization phase all three areas increase in volume, indicating cell proliferation. Data from neuroanatomical studies indicate that during the period from about 20 to 40 days posthatch, new connections are also being formed between the HVC and the RA, as well as between other brain nuclei involved in bird song (Bottjer et al. 1985; Konishi and Akutagawa 1985; Nordeen and Nordeen 1988). Together, these findings provide an excellent picture of the close relationships between ongoing life history events, song development, and the underlying neural processes.

Puberty in Female House Mice

Puberty is a critical event in the lives of most organisms. For many, sexual maturation marks the onset of reproductive behavior and the production of progeny. Puberty is also important in the population biology of many species (see Drickamer 1986; Vandenbergh and Coppola 1986). House mice's (*Mus domesticus*) deme structure generally involves one to several adult males, three to seven females, and their young offspring. Some juvenile females may disperse from the natal site, whereas others may remain within the deme, but virtually all juvenile males disperse (Crowcroft and Rowe 1957; Bailey 1966; DeLong 1967; Crowcroft 1973). The timing of puberty in female mice can be affected by various social and environmental factors within this social context. The age of puberty is measured by the occurrence of first vaginal estrus; the heightened levels of estrogen associated with ovulation result in cornification of the cells that line the vagina. This phenomenon can be detected by microscopic examination of a vaginal smear.

The presence of a mature male mouse or daily exposure to urine from mature males accelerates the onset of puberty in young female mice (table 10.2), as does daily exposure to urine from lactating females and urine from individually caged females in estrus. Daily exposure to group-caged females or their urine, regardless of the age of the grouped females, results in delays in the onset of puberty. Female mice exposed daily to clean bedding or to urine from singly caged diestrous females reach puberty at ages that are intermediate between the acceleration and delay effects produced by exposure to urine from the other sources (table 10.2). These effects have been demonstrated in both laboratory and wild stocks of house mice (Vandenbergh 1967, 1969b; Vandenbergh et al. 1972; Drickamer 1974a, 1979, 1982a, 1983; Colby and Vandenbergh 1974; Drickamer and Murphy 1978; Drickamer and Hoover 1979); and several of the effects have been replicated in stocks of wild *Mus domesticus* maintained in the cloverleaf islands of superhighways (Massey and Vandenbergh 1980, 1981; Coppola and Vandenbergh 1987).

When mice are exposed simultaneously to urine from two or three sources, the outcome depends on which sources were used and the relative proportions of urine from each source (Drickamer 1988). If treatment involves equal proportions of urine from different sources and any exposure to urine from grouped females, puberty is delayed in young test females—regardless of what other urine sources are used. When the proportion of grouped female urine mixed with acceleratory chemosignal increases to a ratio of about 10:1 relative to urine that delays puberty, then the acceleration effect overrides the delay. If all donor sources involve urine that accelerate puberty, then the combination treatments result in earlier maturation, but not any earlier than using one of the sources alone (Drickamer 1982b). Diet and daylength have also been shown to affect the timing of puberty in female house mice (Vandenbergh et al. 1972; Drickamer 1975).

Hence, a variety of contextual cues from the developing mouse's environment influence its internal physiological events. These effects, in turn, have important consequences for the onset of reproductive behavior in the mouse. The chemosignal effects will also influence its **generation time,** the length of the interval between birth and onset of reproduction. Generation time is a key element in determining the rate of growth or decline in the numbers of young mice entering the population over a given length of time. Young female mice may be able to exhibit some degree of behavioral control over their exposure to the puberty-influencing urinary chemosignals (Drickamer 1992; Drickamer and Brown 1998). Females avoided exposure to male-soiled bedding or traps that had previously contained adult males, reducing their exposure to urinary cues that that would accelerate their puberty.

JUVENILE EVENTS: INSECTS AND FISH

Much of the research on the ontogeny of behavior in animals has concentrated on birds and mammals as exemplified by the material we have just covered. What about other types of animals? A number of excellent studies of behavior development have been conducted using insects, fish, and other animals. In light of what we have learned earlier in this chapter concerning the concept of an ontogenetic niche, it should be interesting to see how that concept might be applied to the examples that follow.

TABLE 10.2   Mean ages and ranges for sexual maturation of female house mice (*Mus domesticus*) given various treatments with urinary chemosignals.

Each young test female was treated starting at 21 days of age, and all test females were housed under the same conditions of photoperiod, food and water availability, and individual caging.

Treatment	Mean Age at Puberty (Days)	Age Range (Days)
Control females—exposed to water treatment daily	35	29–42
Females caged with an adult male	27	24–31
Females exposed daily to urine from adult males	31	27–36
Females exposed daily to urine from grouped females	41	36–45
Females exposed daily to urine from estrous females	30	26–36
Females exposed daily to urine from diestrous females	36	30–41
Females exposed daily to urine from lactating females	30	27–35
Females exposed daily to urine from pregnant females	31	26–36

Insects

Drosophila

Fruit flies (genus *Drosophila*) have been used in a wide variety of investigations of the development of behavior; studies utilizing field and laboratory conditions have been done both on larvae and on adults. The primary activity of insect larvae is feeding. As a larva moves across the food surface, it probes with its mouthparts and ingests food with each cycle of extension and retraction of its body. The larvae of *D. melanogaster* go through three developmental stages, or **instars,** and then pupate before emerging as adult flies. The rate of feeding reaches a peak early in the third instar and then declines during that instar (Burnet and Connolly 1974). Before pupation, the larva may migrate away from the food source to a pupation site, or it may remain at the food source during pupation (de Souza et al. 1970); the pattern varies for different species and sometimes even within a species. Some species may burrow into the soil to pupate, probably to decrease the risk of predation.

The adults of many *Drosophila* species, particularly females, have been studied more extensively than have larvae, because adults perform a considerably greater number of behavior patterns. Two major behavior patterns related to reproduction, sexual receptivity and oviposition, undergo developmental changes in females. Manning (1966, 1967) studied sexual receptivity in *D. melanogaster.* As the corpora allata, which release juvenile hormone (JH), and ovaries grow larger, the female becomes receptive. This generally occurs about 48 hours after the eclosion of the adult fly from the pupa. Immature females reject courting males that attempt copulation. One hypothesis that we can suggest relates the physiological and behavioral events: JH may affect the brain directly or indirectly, lowering the threshold for sexual receptivity (Schneiderman 1972). If actively functioning corpora allata are implanted in females at the pupal stage, the emergent flies exhibit earlier sexual receptivity and have larger ovaries than nonimplanted controls.

After mating, females become unreceptive again, and soon oviposition behavior increases. The turning off of receptivity appears to be due to at least three factors: the presence of sperm in the females' receptacles, the act of copulation itself, and a secretion from the paragonial gland of males (Manning 1967; Burnet et al. 1973). The increase in rates of oviposition is probably due to the increased size of the ovaries, which provides information to the nervous system via stretch receptors, and to the presence of male paragonial gland secretion (Merle 1969; Grossfield and Sakri 1972).

Ringo (1978) discusses the development and maturation processes in females and males of a group of *Drosophila* species inhabiting the Hawaiian Islands. Development takes a longer time in these species, continuing for days or weeks. For males of *D. grimshawi,* a series of eight behavior patterns were observed and recorded at five ages during a 1-month period following eclosion (figure 10.19). In general, the diversity of behavior observed increased with age, and the relative frequency of each behavior increased with age. Males of this species form leks, in which groups of males display communally and mate with females attracted to the lek. For days 15 and 22, there were pronounced increases in sexual and agonistic behaviors—courting, jousting, and abdomen dragging (see Spieth 1966, and Ringo 1976, for detailed descriptions of the behavior). The increases in behavior correspond to the time when the males are most likely to be competing for copulations.

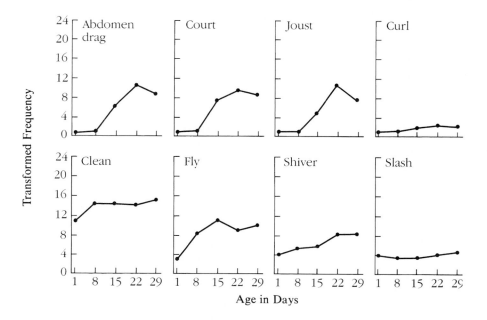

FIGURE 10.19 **Behavior of *Drosophila*.**
Frequencies of eight behaviors in *Drosophila grimshawi* males at ages of 1, 8, 15, 22, and 29 days, post-eclosion. Diversity and frequency of behavior increase with age.

Bees

Many bee species live in large colonies in which there are castes—physiologically, behaviorally, and often morphologically different forms occurring together (Wilson 1971; Michener 1974). For example, in the honeybee (*Apis mellifera*) and other related species of social bees, there are at least three castes: queens, drones, and workers. Queens are larger, they mate, lay eggs, eat proteinaceous food, and often do not forage or defend the colony. In contrast, workers generally are smaller, do not mate or lay eggs, and actively engage in foraging, defense of the colony, and nest building. The eggs of the queen(s) hatch into larvae, and after these pupate, they emerge as adults.

One key question concerning bee development is: How and when does the determination of caste occur? Several factors appear to be important in this process (Wille and Orozco 1970). Which factors are important varies between major groups of bees, but in general they include: size of the brood cell, amount of food mass provided to the developing larva, quality of the food supplied to the larva, and possibly some chemical cues transmitted with the food. For most species studied, these effects begin immediately upon hatching or very early in larval life (Ribbands 1953; Jung-Hoffman 1966; Weaver 1966; Darchen and Delage 1970). Conditions both within the hive (e.g., loss of a queen) and outside the hive (e.g., changing seasons) can influence the numbers of workers and queens produced.

Most worker bees of these social species undergo changes in behavior as they develop that correspond to changes in their functional roles within the colony or hive (Free 1965). The activity changes for a worker honeybee are shown in the histograms in figure 10.20. Resting and patrolling occurred throughout the 24 days surveyed. Many other activities occur in a sequential pattern, beginning with high levels of cell clearing in the early days after emergence. This is followed by activities related to the comb and tending the brood. For the last portion of their life span, the workers become foragers; the average worker bee living in a temperate zone climate survives to the age of 6 weeks. For many of the activities, there is some overlap: Bees shift among behavior patterns of all types on a given day, up to the start of foraging activity. In the last phase, they concentrate primarily on foraging for food for the hive.

Several physiological changes are correlated with the shifting patterns of functional roles that we have just noted. Young bees have enlarged hypopharyngeal glands. These produce a major component of the bee milk, part of the larvae's diet. Levels of secretion for invertase, an enzyme involved in the conversion of nectar to honey, are highest during the middle portion of the life span (Simpson et al. 1968). Wax glands are small in the youngest workers, but there is a gradual increase to maximum size by about days 16 to 18, and then the wax glands decline rapidly (Snodgrass 1956). These changes correspond with the higher levels of comb-related activities during this particular age range. Many other traits follow similar patterns of correlation

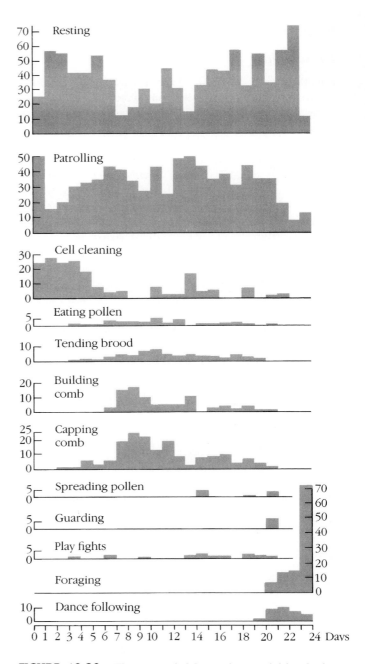

FIGURE 10.20   Time occupied by various activities during the first 24 days of life of a single marked worker honeybee living in a colony.
The columns of figures at the left and right represent percentages of its time during which the bee engaged in each activity, out of the 2–10 hours of observation daily.

between developmental events throughout the life of the bee and its changing functional activities in the hive.

Most social bees actively guard and defend their colonies. How do they recognize nestmates and discriminate them from other conspecifics? Work by Breed (1983) indicates that bees will use environmental odor sources, or if the environmental cues are controlled, they will use genetically based cues to discriminate nestmates from non-nestmates (see also review by

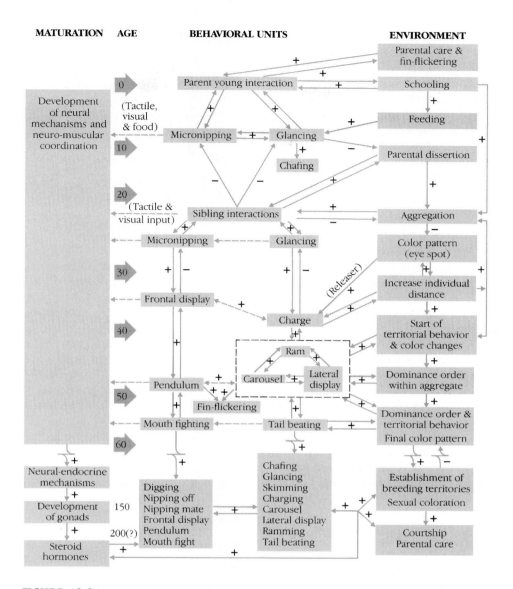

MATURATION   AGE   BEHAVIORAL UNITS   ENVIRONMENT

FIGURE 10.21   **A model representing the ontogeny of behavior in *Etroplus maculatus*, the orange chromide.**

The arrows marked "+" indicate facilitation; the solid arrows marked "−" indicate inhibition. The dotted arrows, for diagrammatic simplicity, represent environmental feedback to the organism. The environmental factors implied by the dotted arrows are those indicated feeding into the model from the right. The dotted arrows marked "+" are proposed routes of facilitation between behavioral units.

From R. L. Wyman and J. A. Ward, "The Development of Behavior in the Cichlid Fish *Etroplus maculatus* (Bloch)," *Zeitschrift für Tierpsychologie*, 33:461–91, 1973 Paul Parey Verlagsbuchhandlung, Berlin.

Hölldobler and Michener 1980). Investigators have shown that the cues necessary for making these discriminations are acquired prior to emergence as adults.

Fish

While considerably fewer studies of the ontogeny of fish behavior have been conducted than on some other vertebrate groups, recently there has been some progress (Noakes 1978; Huntingford 1986). Perhaps the most extensive investigations were those of Ward and his colleagues (Quartermus and

Ward 1969; Cole and Ward 1970; Wyman and Ward 1973). After observing the orange chromide (*Etroplus maculatus*), they proposed a theoretical model for the ontogeny of numerous behavior patterns (figure 10.21). All subsequent behavior patterns of this species are developed from two initial movements: glancing and micronipping. Data from a number of other cichlid species, and from some salmonid species, seem to fit this model, though extensive additional testing is needed.

Other research on salmonids (*Salmo* spp. and *Oncorhynchus* spp.) reveal some general patterns of behavior development in these genera, though there is considerable variation at each stage (Abu-Gideiri 1966; Jacobssen and Jarvik 1976; Dill 1977; see Huntingford 1986 for a summary of fish development). The eggs begin their development buried in the gravel of a stream several months after fertilization. The first signs of life are contractions of the heart muscle. Soon after, the body muscles begin contracting. Initially, these contractions do not occur in coordinated or regular patterns, but they develop into characteristic undulations that look like swimming movements. Shortly before hatching, the jaws and fins also show coordinated movements. Hatching results from swimming motions that are rigorous enough for larvae to break out of the egg. Soon after hatching, the yolk sacs are fully absorbed, and larvae rest in an upright position on the gravel bottom of the stream. At this early stage, fish are photonegative, and they tend to swim into the current. Making movements to flex the tail and push off the substrate, the fish succeed in leaving the gravel and entering the stream; at about this time they switch to being photopositive. Beginning before emergence from the egg and increasing in frequency after emergence, movements, such as darting toward and biting small objects in front of them begin to occur; the fish often direct some of these same biting actions toward conspecifics. Young fish of most salmonid species exhibit one other general trait during development: they form a school in the presence of a potential predator, and if attacked, they flee and then freeze, remaining motionless on the substrate, or occasionally near the surface of the stream. Salmonids in later stages of ontogeny show greater variability across the various species, though apparently some species temporarily maintain a form of territoriality before leaving the natal stream for larger bodies of fresh or salt water (Kennleyside and Yamamoto 1962).

PLAY BEHAVIOR

Many mammals, some birds, and possibly a few other animals exhibit play behavior during their course of development. What is play behavior and how does it function in the developmental process? **Play behavior** has been defined and characterized in many different ways. Fagen defines play as "an inexact term used to denote certain locomotor, manipulative and social behavior characteristic of young (and some adult) mammals and birds under certain conditions in certain environments" (Fagen 1981, p. 21). We can characterize play in various animal groups according to the actions of the animals and the contexts in which the play behavior is observed. Most investigators recognize at least three types of play that have some overlap. The first type is social play, exemplified by wrestling, chasing, and tumbling activities of the young of many species (Bekoff 1978) (figure 10.22). A second type provides exercise for developing muscles, locomotor patterns, and other movements (Byers 1998). The third type is often labeled diversive exploration or object play (Hall 1998). This type of play generally involves sensory inspection of an object followed by extensive repeated manipulation of the object and can be observed in both juveniles and adults of some species.

FIGURE 10.22   Canid play behavior.
Source: Dr. Mark Bekoff.

By reviewing the available literature, we find that one or more of these types of play behavior have been recorded in mammals ranging from rodents and bats, to bears, cats, elephants, marsupials, and whales. Play has also been recorded in a large number of avian species, including raptors, passerines, aquatic or oceanic birds, and parrots. To date, only anecdotal bits of evidence exist regarding play behavior in other vertebrate or invertebrate groups. Notable among these findings is the work of Burghardt (1982, 1998a,b), who has recorded playlike behavior in some reptiles and who has speculated on the importance of homeothermicity for the evolution of play behavior. We should gather additional observations before reaching any firm conclusions that limit play to birds and mammals.

Functions

Investigators have attributed a wide range of functions to play behavior (Loizos 1967; Bekoff 1974b; Bekoff and Byers 1981; Fagen 1981; and see Bekoff and Byers 1998). In general, we have tended to claim that the primary function of play behavior relates to practice for adult activities. Indeed, play behavior in many animals does involve performing actions that contain elements similar to those seen later in adult life. Perhaps aggressive behavior is the best example. As young cats or primates engage in various forms of social play, their mock attacks, chases, and mild, noninjurious bites are practice for "real life"—they may use these same patterns a few months or years later (Thompson 1998).

In the more recent interpretation of play, two other functions may be even more critical than the historical prescription that play was practice for adult life. First, play behavior may perform critical roles in growth and development in the process of maturation. As young foals or lambs cavort about, alone or in small groups, they use their muscles and develop coordinated movements. Additional studies will be necessary to obtain more exact measurements of the manner and degree of the influence of play actions on maturation of specific muscle groups, as well as the sensory and motor nervous systems. Some observational studies provide preliminary evidence in this regard. Investigators have reported on the energetics of

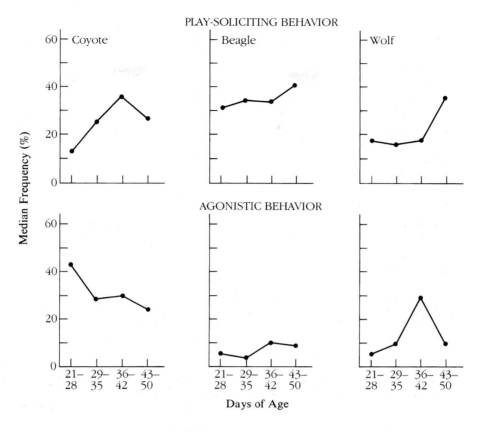

FIGURE 10.23   Play-soliciting and agonistic behavior in canids.
The median frequency of occurrence (percent) of action patterns observed during both play-soliciting and agonistic interactions in relation to the total number of actions performed during the stated time periods.

play behavior in the pronghorn antelope (*Antilocapra americana*) (Miller and Byers 1990; Bekoff and Byers 1992; Byers 1997). Play behavior, calculated in one way, accounts for 2 percent of the total time budget but actually occupies about 8 percent of the active time. In addition, for this species, the active nature of play means that well over 20 percent of the daily energy budget expended in excess of basal metabolic rate was devoted to play behavior. These data and similar findings for other animals, particularly mammals, underscore the likely importance of play behavior during development.

This leads to another function of play—gaining information about the environment. This would be particularly true of diversive or object play. By exploring and manipulating objects found in their environment, young animals accumulate information that may prove useful later in life.

Finally, a function of play can be to establish social relationships with peers and adults. Some of these may be pure affiliations, whereas others could be related to the establishment of dominant-subordinate relationships (Thompson 1998). Others have argued against this interpretation of play behavior (Pellis and Pellis 1998). The aggressive play observed in young juveniles of many primates and some canids gradually becomes more intense and adultlike. The patterns of dominance established in play encounters as juveniles can be retained into adult life. Let us consider several exam-

ples of the nature of play behavior and its functional significance during development, for better understanding.

Canids

One animal group in which play behavior has been studied extensively is the canids (Zimen 1972; Bekoff 1974a,b,c; Moehlman 1979). Bekoff (1974a) studied social play and play soliciting in coyotes (*Canis latrans*), wolves (*C. lupus*), and beagles (*C. familiaris* now interpreted by some mammalogists to be *C. lupus* as well). His observations indicated some species differences and some age-specific trends for several aspects of play behavior; for example, play soliciting and agonistic behavior (figure 10.23). Bekoff notes that the onset of play soliciting in the beagles was early and that the beagles exhibited very little agonistic behavior during play—no fighting occurred at all, only mild threats. Wolves showed moderate levels of play soliciting, which increased during the last age interval recorded in the sample. Like beagles, wolves displayed low levels of agonistic behavior that consisted primarily of threats. In contrast, coyotes exhibited high levels of agonistic behavior throughout the observation period and a correspondingly low rate of play-soliciting actions. Coyotes generally establish dominance relationships through fights at an early age (Fox and Clark 1971); this may account for the differences between these animals and the other two species. When all play behavior was summarized, the beagles were seven times more playful than the coyotes, and three times more playful than the wolves. It is interesting and worthwhile to speculate on the possible correlation between differences in observed play behavior and differences in social structures of the species tested. Wolves are generally social, group-living animals, though some individuals may lead a solitary existence (Mech 1970). Beagles are somewhat social animals, though we rarely observe domestic canids in seminatural or natural situations where their feral social structure can be fully recorded. Coyotes, on the other hand, are generally more solitary in their social organization, with each animal roaming a large home range.

In a review of play, Price (1984) suggested that neoteny explains why some domesticated breeds of animals are more playful than their wild ancestors were. Humans have selectively bred domestic animals so they would retain their juvenile characteristics in adulthood. Thus, the beagles in the foregoing example may be more playful than the wolves because of the effects of our selection.

Keas

Keas (*Nestor notabilis*) are large parrots that inhabit parts of New Zealand. A variety of play behavior has been observed and reported for this species (Derscheid 1947; Jackson 1963; Keller 1975, 1976). The most extensive observations were those of Keller on captive keas in zoos. The young keas perform a variety of acrobatic maneuvers that include somersaulting, sliding on snow-covered slopes, hanging by their bills from tree branches, and hanging by their feet, upside down. Social play in groups is also quite common and involves wrestling and activities that anthropomorphically resemble "hide-and-go-seek" and "king of the hill." In social situations, the solicitation to play may involve tactics like assuming a "defensive" posture—head lowered and one foot raised, or walking stiff-legged, as is commonly seen in some mammals and a few other birds. Last, keas engage in a great deal of object play. They manipulate objects of all sorts using feet and bills, they toss things into the air and fly at them, and they play with snow when it is present.

Human Children

Numerous studies have been conducted both by developmental psychologists and by ethologists on play behavior in young *Homo sapiens* (reviewed by Bruner et al. 1976; Fagen 1981). Hutt and Bhanvani (1972) conducted a follow-up study of earlier work by Hutt (1966, 1967a, b, 1970a, b) to predict differences in play behavior based on longitudinal sampling. A young child confronted with a new toy will first investigate and inspect it (specific exploration) and then play with the toy (diversive exploration). Three- to five-year-old nursery school children can be classified into three mutually exclusive categories: (1) nonexplorers who may visually inspect a new toy but do not handle or play with it; (2) explorers who thoroughly investigate a new toy but fail to do more than that with it; and (3) inventive explorers who investigate a new toy and then play with it in a variety of innovative ways.

Do these individual differences in exploratory behavior provide any insights into predicting traits of the children at a later age? To test this question Hutt and Bhanvani (1972) obtained data for about 50 children from the original sample of 100 used to generate the three categories above. The children were 7 to 10 years of age at the time of the second sampling procedures. Each child was given a series of tests to measure creativity and a personality questionnaire. Each child was rated by parents and teachers on behavior, adjustment, and development. The results provided support for several suggestive conclusions (which were really more like hypotheses for further testing):

1. Lack of exploration was related to later lack of curiosity and adventure in young boys and to difficulties in personality and social adjustment in young girls.

2. Children who had been more creative and imaginative in their early play behavior were more likely to be creative in the later ages; this was particularly true for boys.

Hutt and Bhanvani note that some of these observed effects may be attributable to early childhood differences between the sexes; boys are more exploratory in their play activity for a longer period of their life, and girls are socially and linguistically more advanced than boys at nursery school age. We might also note that longitudinal follow-up assessment of longer-term effects of early differences would be both appropriate and necessary to strengthen, extend, or refute these conclusions.

DEVELOPMENT INTO ADULT LIFE

The development of behavior, as we noted earlier, is a continuum. Many key events take place early in an organism's life. However, this does not mean that developmental processes cease when the organism reaches some particular chronological age. Rather, depending upon the species and other conditions, continuing developmental changes in some traits occur throughout most or all of the organism's life. Examination of developmental phenomena on into adult life represents an interesting and potentially productive extension of the idea of ontogenetic niche expressed by West and King (1987). We have already discussed in some detail the worker honeybee's chronological sequence of changes in physiology and behavior that continues throughout life. In the preceding section, we noted that both the young and the adults of many species engage in play behavior. Some plasticity for particular traits may be important to the ecology and ultimately to an organism's reproductive success, as the following example illustrates.

White-footed mice (*Peromyscus leucopus noveboracensis*) primarily occupy a woodland habitat, but they are ubiquitous in a wide range of habitats within their geographical range (midwestern and eastern United States). Although they are most abundant in a variety of types of forests, they also live in fields, in croplands, in marshes, and often in human dwellings. Prairie deermice (*P. maniculatus bairdi*) live in grasslands—usually open fields or croplands—and are almost never captured in wooded areas, brushland, marshes, human-made structures, and so on.

Adult mice of these two species were caught and brought into the laboratory, where they were presented with a cafeteria arrangement of seeds and grains from the various habitats where the mice had been caught (Drickamer 1970). Mice from both species ate large amounts of corn, but there were species differences in preferences to the other foods. The *P. leucopus* ate more elm and maple seeds than *P. maniculatus;* the latter species ate more bush clover seeds and wheat. Could there be a relationship between the mice's occupancy of particular habitats and their food preferences or feeding strategies?

In a further test, Drickamer (1972) explored the effects of dietary experience on subsequent food choices both for young mice (21 to 90 days of age) and for adult mice (over 90 days of age) of these two species. To circumvent the problems of inadequate nutrition that occurred when the mice were given only diets of seeds, laboratory mouse chow was used—but three different flavors were generated by placing an odor source (a drop of an essential oil such as anise or pine on a small piece of cotton) beneath the food in each dish. Mice were provided with a 2-week training experience either as young or as adults and were tested either immediately after the training or 1 month later. The mice were tested two

ways to determine whether the training experience influenced their choice of food-odor combination. The results of these tests are summarized in table 10.3. Preferences for particular food-odor combinations were influenced by prior training experience for young *P. maniculatus* and for both young and adult *P. leucopus*. Adult *P. maniculatus* were not affected by the training experience. Further tests assessed the patterns of visitation to various food sources by young and adult mice of both species. Adult *P. leucopus* changed feeding sites more frequently than young *P. leucopus,* and more frequently than either young or adult *P. maniculatus*. If the feeding site arrangements were shifted about, the young and adult *P. maniculatus* demonstrated a strong position preference, whereas *P. leucopus* of both ages either followed the foods to new locations or changed to another food.

How may these observed differences in feeding habits be related to the ecology of these species? Clearly, the adult *P. leucopus* are more flexible in shifting their feeding preferences and, apparently, in using a more diversified strategy, in comparison with the adult *P. maniculatus*. This suggests that *P. leucopus* may successfully occupy and utilize a wider variety of habitats because of more flexible feeding habits as adults. In contrast, *P. maniculatus* with age became more rigid in their food habits and feeding strategy. This correlates with their occupancy of a much more limited range of habitat types. The young of both species exhibit some flexibility in feeding, but by the time the animals are about 90 days of age, there has been a change for *P. maniculatus*. Thus, we can see that developmental processes can continue into adulthood and that the results of such processes may have important consequences for the life-history patterns of various species.

TABLE 10.3   **Effect of food-odor training on mice**

Results from tests on two species of *Peromyscus* to determine the effects of training with various food-odor combinations on the selection of diet from a series of food-odor combinations presented cafeteria style or on the chewing of balsa wood pegs to obtain a preferred food-odor combination. "+" indicates that the testing resulted in significant preferences for the food-odor combination with which the mice were trained. "−" indicates no significant effects of the training experience on food-odor combination preference.

	Appetitive	Consummatory
Peromyscus leucopus		
Young mice—tested immediately	+	+
—tested 1 month after training	+	−
Adult mice—tested immediately	+	+
—tested 1 month after training	+	−
Peromyscus maniculatus		
Young mice—tested immediately	+	+
—tested 1 month after training	+	+
Adult mice—tested immediately	−	−
—tested 1 month after training	−	−

SUMMARY

Historically, a paramount question with regard to behavior development has concerned the nature-nurture issue: Is behavior a function of genetic inheritance and/or the environment and experiences of the developing organism and to what degrees? Proponents of both viewpoints developed their own terminology (e.g., sign stimuli and fixed action patterns) to describe behavior and the developmental process. In more recent years, the idea of epigenesis, an interactive process involving both genetics and the environment has been prominent. Throughout much of this work, behavior development was viewed in terms of the nature of effects of early stimulation or experience on later behavior. Most recently, an exciting new idea has emerged, the ontogenetic niche. There are two key features of this new thinking. First, an organism inherits not only its genes from its ancestors, but also a host of ecological and social features. It is in this total context that behavior development occurs. Second, instead of viewing most of behavior development as a series of influences that are manifested later in life, we now view the process with respect to the immediate importance of ongoing events for the behavioral ontogeny of the organism as well.

Behavior development involves many events that begin long before hatching or birth—the embryology of behavior, which begins at conception. We now know that certain processes that occur during these early stages are critical for the period in which they occur, and, in fact, they may disappear as development proceeds. Examples include the considerable amount of cell death that occurs within the nervous system during ontogeny and the appearance of morphological structures such as the egg tooth in birds that disappears once it has served its particular function. A variety of forms of sensory and motor stimulation that occur early in development can have important consequences both in the immediate context, as with the development of motor reflexes, and as development proceeds, for instance, with regard to the synchronization of hatching in some birds. For some mammals, experiences of the mother (e.g., stress) that occur during pregnancy can affect behavior development.

A variety of events that occur just after birth or hatching have significant consequences for social behavior, in the immediate and later, more adult, contexts.

Sexual imprinting involves learning to direct sexual behavior at particular stimulus objects and may serve as a species-identifying mechanism, important for species isolation. Various environmental factors can influence the imprinting process and the related response to maternal alarm calls, particularly social factors, such as the presence of conspecifics. This illustrates the important principle that there may be multiple pathways to the same end point in the course of behavior development.

Though ontogeny of behavior is a continuous process, we divide it into segments for our convenience. Juvenile development commences shortly after birth or hatching and concludes at about the time of fledging (birds) or weaning (mammals). Studies involving enrichment and deprivation (e.g., the studies of monkey social

development) have provided insights regarding events during this period. The severity of the behavioral deficits can be related to the degree of deprivation.

The diet consumed by a female rat may affect the food preferences of her offspring. Also, young rats may be influenced in their selection of feeding sites during weaning by social cues from interactions with conspecifics. Young snapping turtles are strongly influenced in their selection of diet by their first feeding experiences after hatching. Young gulls learn to peck at the bills of their parents to obtain food through a combination of genetic predisposition and practice over time.

Birds learn to sing based on both a genetic template that varies with the species and learning from conspecifics, which may take the form of imitation or improvisation. Longitudinal studies of song development indicate that some birds progress through a series of stages in song acquisition: subsong, subplastic song, plastic song, and, finally, crystallized song. Various environmental factors can influence the song development process, including social factors, such as the presence of conspecifics, and photoperiod, in relation to the time during spring or summer when the birds hatch.

Puberty in house mice is influenced by a variety of social and environmental cues. Some social cues (e.g., urinary chemosignals) accelerate sexual development, whereas others retard the process. Both diet and daylength also influence puberty. The timing of puberty is important both for the onset of reproductive behavior and for the population biology of the mice.

Behavior development has been explored in a variety of insects and fish. For fruit flies there are clear relationships between internal physiological events and changes in sexual receptivity after eclosion, and later, the tendency to oviposit. After honeybees emerge from pupae, they progress through a series of functional roles in the colony—again, there are clear developmental correlations between the roles and physiological and morphological changes in the bees. One model for the development of behavior in fish demonstrates how the complex and varied behavior patterns of the young adult derive from a few simple patterns in the newly hatched animal.

Play behavior is an important component of the developmental sequence in many mammals and birds. Studies on canids, keas, and children illustrate the varied types of play: social, exercise/maturation, and exploration. The functions of play behavior include aiding the process of growth and development, learning about the environment, practicing for adult activities, and establishing social relationships.

Development does not end when the young animal becomes independent of its parent(s), but rather continues into adult life. Possessing some capacity to remain flexible in certain behavioral traits may have important consequences for the organism. For example, differences in habitat occupancy of two species of deermouse may be partly a function of the differential dietary flexibility of the adults of these two species.

DISCUSSION QUESTIONS

1. Careful definition of terms is important in the study of animal behavior. Write out your own definitions of the seven terms important in behavior development listed in the following text. Locate the definitions of these terms in this book and in other books on behavior. How do the definitions differ? (a) epigenesis, (b) ontogenetic niche, (c) play behavior, (d) critical period, (e) ontogeny, (f) filial imprinting, and (g) species-isolating mechanism.

2. Using the technique provided by Smotherman and his colleagues for viewing young rats in utero and the information you now have concerning the embryology of behavior and importance of ontogenetic niche in birds, what experimental questions can you devise on small mammals to test the ideas developed using birds?

3. Pratt and Sackett (1967) raised three groups of young rhesus monkeys with different degrees of contact with their peers. They allowed one group no contact; the second, only visual and auditory contact; and the third, full normal contact with peers. Next they allowed animals of the three groups to interact socially; then they gave each individual a three-choice preference test where the alternative choices were conspecifics—one from each of the three different treatment conditions. The data in the table represent the preferences of the test subjects. What conclusions can you draw from these data?

   The numbers in the table are the mean number of seconds spent by a test monkey with each of the three possible partners in a choice test.

4. Many animals we use for studies of behavior are from domesticated laboratory stocks or are derived from wild animals that have been kept for varying periods of time in the laboratory. What differences in developmental processes, if any, might you expect to find between these laboratory-reared animals and their counterparts in the wild?

5. Behavior development actually is a continuing process that occurs throughout the life of an organism. Birds of a number of species migrate to the equatorial regions for the winter months and return to the temperate zones to breed in spring and summer. What types of ongoing developmental processes may be taking place in these birds that survive and make this annual journey two, three, or more times? What types of experimental tests can you propose to explore these lifelong developmental processes?

6. Insect behavior development is an area that is likely to receive considerably more research attention in the next several decades than it has to date. Given what you know about physiological processes in insects from earlier chapters in this textbook, what types of experiments would you propose concerning behavior development in each of these three species?—(a) cockroaches, (b) blowflies, (c) grasshoppers?

7. Several of the studies cited in this chapter concerning the development of bird song and the responsiveness of young birds to maternal alarm calls shed new light on the importance of a variety of social and environmental contexts influencing ongoing behavior development. With this background information, what other contextual cues do you think might be important for song development in birds? What sorts of similar tests can you conceive that would demonstrate the importance of various contextual cues for the ontogeny of (a) odor cue detection in salamanders and (b) food preferences in ungulates such as deer or elephants?

Rearing Condition of Test Animal	Rearing Condition of Stimulus Animal		
	No Contact	Visual and Auditory Contact	Normal Contact
No contact	156	35	29
Visual and auditory contact	104	214	103
Normal contact	94	114	260

Source: Data from Pratt and Sackett (1967).

SUGGESTED READINGS

Bateson, P. P. G. 1978. How does behavior develop? In *Perspectives in Ethology,* ed. P. P. G. Bateson and P. H. Klopfer, 1998 Vol. 3, 55–66. New York: Plenum.

Bekoff, M., and J. A. Byers. 1998 *Animal Play.* New York: Cambridge University Press.

Fagen, R. 1981. *Animal Play Behavior.* New York: Oxford University Press.

Hall, W. G., and R. W. Oppenheim. 1987. Developmental psychobiology: Prenatal, perinatal and early postnatal aspects of behavioral development. *Ann. Rev. Psychol.* 38:91–128.

Immelmann, K., G. W. Barlow, L. Petrinovich, and M. Main. 1981. *Behavioral Development.* New York: Cambridge University Press.

Oppenheim, R. W. 1982. Preformation and epigenesis in the origins of the nervous system and behavior: Issues, concepts, and their history. In *Perspectives in Ethology,* ed. P. P. G. Bateson and P. H. Klopfer, Vol. 5, 1–99. New York: Plenum.

West, M. J., and A. P. King. 1987. Settling nature and nurture into an ontogenetic niche. *Develop. Psychobiol.* 20:549–62.

11
CHAPTER

Learning

 n this chapter we explore learning behavior in terms of its underlying processes, evolution, and functional significance. We begin by defining learning and examining what that definition means. We next explore the various forms of learning. This is followed by exploration of learning in invertebrates and vertebrates, then comparative learning. Our coverage of learning continues with an examination of the idea of variation in preparedness for learning, the constraints that must be remembered when we study learning, and several in-depth examples of the interaction of learning and ecology in animals' lives; the functional significance of learning. This leads to an examination of animal cognition, an exciting new subfield of animal behavior that has arisen in the past two decades. We conclude with brief examinations of memory and motivation.

Learning can be defined as a relatively permanent change in behavior or potential for behavior that results from experience. Several aspects of this definition need further consideration. The phrase "relatively permanent change" refers to some internal processes that change, in the form of memory, but that do not always guarantee recall and the ability to use the information. (We all forget things!) "Potential for behavior" refers to the fact that some things can be learned that are not used immediately or may never be used; that is, they are latent or unexpressed learning. Finally, we define "experience" in terms of all aspects of the environment, including genetic inheritance interacting with ontogenetic niche, that define and surround an organism beginning soon after its conception and lasting throughout its lifetime.

FORMS OF LEARNING

Our knowledge about learning in animals has accumulated from the work of psychologists and ethologists. A thorough historical review is beyond the scope of this text (see Thorpe

1956; Hilgard and Bower 1975; Kamil 1988; Rescorla 1988a,b). Two major types of association learning are recognized by animal behaviorists: classical (Pavlovian) conditioning and instrumental learning (operant conditioning). There are many similarities and some differences in the key characteristics of these two major types of learning. Though there are some who would disagree, we also recognize habituation and sensitization as distinct types of learning.

Habituation

We define **habituation** as the relatively persistent waning of a response that results from repeated presentations that are not followed by any form of reinforcement. **Reinforcement,** a term we will use often in this chapter, is defined as anything that alters the probability of behavior. Reinforcement can be positive (e.g., a food reward) or negative (e.g., the jerk on a choke chain used with a dog). There are other reasons why a response might wane besides habituation. **Fatigue** may cause a response to decline when it is repeated in rapid succession. **Sensory adaptation** generally occurs at the level of the peripheral sensory receptors and consists of a reduction in, or cessation of nerve impulses transmitted to the central nervous system. Both fatigue and sensory adaptation can be distinguished from habituation because they last a relatively short time, whereas habituation is a persistent central nervous system process involving changes in the brain or spinal cord.

Consider the following example of habituation. A group of students is listening to a lecture. If a classroom radiator clanks loudly, everyone will be momentarily startled. If after a short spell, the radiator again makes its offending noise, the degree of response will be considerably decreased. After several additional clanks and bangs, almost no one will be startled by each new episode of noise, and the class will go on as usual.

Habituation can be a functionally important aspect of an animal's behavior in its natural surroundings. Young ducklings scurry for cover when any shadow passes overhead, an adaptive response to avoid predators. Gradually the ducks learn, partly through habituation, which types of shadows signal potential danger and which are harmless. Ground squirrels, fiddler crabs, and marine worms all live in burrows. When danger threatens, these animals dash for their burrows for protection. They habituate to specific nonharmful stimuli in their environment but retain escape reactions to threatening or unusual stimuli. The benefits are clear: without habituation, animals would spend too much time responding to meaningless stimuli rather than foraging or engaging in some other useful behavior.

Classical Conditioning

In its original form, **classical conditioning** (or **Pavlovian conditioning**) involves a stimulus, the **unconditioned stimulus (US),** that elicits a specific response, the **unconditioned response (UCR).** At approximately the same time as the US, a second, neutral stimulus is presented that does not customarily elicit the UCR. When the neutral stimulus and US are

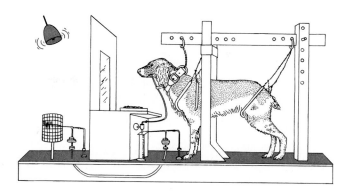

FIGURE 11.1   Conditioning test apparatus.
Ivan Pavlov discovered classical conditioning through his work on the salivary reflex in dogs using an apparatus similar to this. The dog in the restraining apparatus is ready to be tested using the classical conditioning paradigm.

paired for a number of trials, the response will eventually be elicited by the neutral stimulus. At this point, the neutral stimulus (in figure 11.1, the bell) is termed the **conditioned stimulus (CS)**, and the response is the **conditioned response (CR)**. The best known example of classical conditioning is that of Pavlov and the salivary responses of dogs (figure 11.1).

Our understanding of classical conditioning has been modified and modernized in the last 25 years (Bolles 1985; Rescorla 1988a,b). The general principle of the modern view is that conditioning involves the learning of relations (associations) among events. This broader perspective means that an animal actually has a very rich representation of its environment, as contrasted with the more restricted view of its world associated with a reflex tradition, as in the earlier formulation of classical conditioning. There are four major points that distinguish and characterize the modern view (Rescorla 1988a,b).

Contiguity refers to the association in time of events: How close in time must two events be in order for an association to be formed? In our current view of classical conditioning, the time interval between CS and US events can be more variable than originally thought. An examination of the variety of studies of conditioning represented in figure 11.2 reveals that across a broad range of species, using various forms of classical conditioning, there is, in all cases, a nonmonotonic relationship between the CS and US interval and the success of the conditioning. Animals that are poisoned by food often become ill some time after ingestion of the food, yet they learn, in many cases very rapidly, to avoid the food in the future. Thus, the general rule would be that the time interval between events where an association is learned will depend on the exact context and nature of the events. Evolution has resulted in a range of possible time intervals, flexible for whatever associations are adaptive for the organism.

Information refers to the fact that for successful conditioning to occur, there must be some intuitive informational relationship between the CS and US. One type of experiment to explore this phenomenon involves a failure or variable

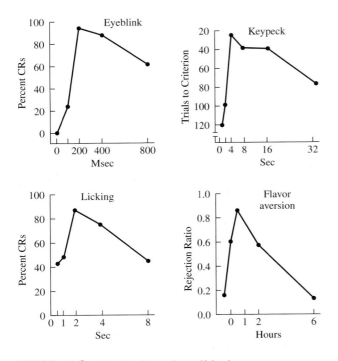

FIGURE 11.2   Contiguity and conditioning.
Data from four different experiments demonstrate the timing variation that can occur between CS and US with successful association: (a) eyeblink responses in rabbits (Smith et al. 1969), (b) the number of trials to an acquisition criterion during autoshaping in pigeons (Gibbon et al. 1977), (c) proportion of trials where a lick response occurred during an acquisition paradigm in rats (Boice and Denny 1965), and (d) proportion of rejection of a distasteful flavor during conditioned aversion training (Baker and Smith 1974).
Source: Modified from R. Rescorla, 1988a.

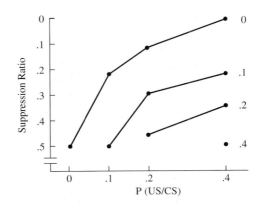

FIGURE 11.3   Fear conditioning in rats.
The probability of the US-CS connection is plotted against the effectiveness of conditioning, here represented in terms of suppression ratio in fear conditioning of rats with electric shock. Rats freeze in response to aversive stimuli and stop pressing a bar that gives them a food reward. The suppression ratio compares the number of presses during a conditioned period versus in a baseline period. A ratio of 0.5 indicates a lack of conditioning, and a ratio of 0 indicates excellent conditioning. The parameter at the right of each line graphed is the likelihood of a shock being administered in the absence of the CS.
Source: From Rescorla (1968).

success in establishing CS-US conditioning. Even if the CS and US often occur together, it will be difficult for an animal to learn to associate them if the US occurs frequently by itself (Rescorla 1968) (figure 11.3). Several theories have been proposed to account for the importance of information in CS-US conditioning (Rescorla and Wagner 1972; Mackintosh 1975), but detailed treatment is beyond the scope of our present examination of learning.

Inhibition, in the context of modern formulations of conditioning theory, refers to learning negative relations between the CS and US. This could be contrasted with learning positive CS-US connections, sometimes called excitatory conditioning. The latter has historically received the most attention.

Salience refers to the fact that animals can make some associations, dependent upon which stimuli are involved, more readily than others. An example from the work of Garcia and Koelling (1966) serves to illustrate this point. If we pair an auditory-visual stimulus (light and sound) with footshock, rats learn this association much more readily than if a distasteful flavor is paired with footshock. Conversely, the rat learns an association of the distasteful flavor and an ensuing gustatory illness rather than an association of the light and sound with the gustatory illness. Specific reinforcers are not equally effective with particular categories of stimuli. We can conclude that natural selection has resulted in a "match-

ing" of cues, such as gustatory and olfactory cues, with internal illness, providing the animal with the ability to make key associations pertinent to its survival and reproduction.

Sensitization

Sensitization refers to several related ideas in the literature on learning behavior. In one sense, the term refers to enhanced responsiveness to a repeated stimulus, particularly a noxious or intense stimulus (Groves and Rebec 1988). A loud noise may startle an animal and cause it to move away from the sound. Repeated occurrences of the loud noise could strengthen this response, leading the animal to eventually leave the area. In this sense, the term sensitization does not imply an association between the stimulus and a particular response.

In a second sense (Dewsbury 1978), sensitization refers to a strengthening of a response that was initially produced via a CS resulting from pairing with a US and UR. The enhancement of the response is appropriate to the CS and not the US. This would be the case, for example, in the Pavlovian salivary-conditioning procedure, if some response appropriate to the tone (CS) (e.g., the dog perking up its ears) was enhanced by the presentation.

In a third way, the term may be used with regard to a stimulus that primes the organism to pay particular attention to what follows (Immelmann and Beer 1989). As an example, consider that when a female mouse is presented with a live mouse pup before she is given a dead mouse pup, she will exhibit stronger parental behavior toward the dead pup than if she is presented with only the dead pup. In reality, these sensitization phenomena may be similar, and the underlying neural processes may also be similar.

FIGURE 11.4   Skinner box.
The interior of the box contains a lever, a light, a food bin, and a grid floor. Additional apparatus for automation and for monitoring the rat's behavior is housed behind the back panel and on the left side of the cage.
Source: Courtesy of the B. F. Skinner Foundation.

Operant Conditioning

Operant conditioning, or as it is sometimes called, **instrumental learning,** involves a wide variety of procedures. In each instance, the animal learns to associate its behavior with the consequences of that behavior—that is, the sequence of events is dependent upon the behavior of the animal. Usually some type of reinforcement is involved. The task performed by the animal may be relatively simple, as in a rat trained to press a lever in a Skinner box (figure 11.4). Pressing the bar is reinforced with food or water. Another example is when an animal has learned to run a maze to a goal box to receive the reinforcement. In nature, a weasel may learn to associate mouse odor with locating and catching a meal.

A large variety of different experimental paradigms has been used by investigators to study aspects of operant conditioning. Lever pressing by rats or pecking at colored discs by pigeons to produce reinforcement, often in the form of food, are among the most common. Two types of avoidance learning fall into the category of operant conditioning. To test **active avoidance learning,** we require an animal to move in order to avoid some noxious consequence of not moving. A shuttle box, used most often with rats, involves a cage with two compartments, each having a wire grid floor through which shock can be delivered. The rat must move between the two compartments when a signal (e.g., a light or buzzer) commences and the door between the two halves of the shuttle box is opened. The rat has, say, 15 seconds to make its move, or it receives a shock. This same procedure has been used successfully

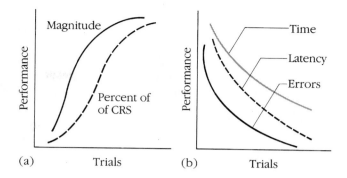

FIGURE 11.5   Characteristic forms of learning curves.
(a) The percentage of CRs and response magnitude increase with practice. (b) Latency measures and other time measures decrease, as do errors.

with goldfish in an aquarium with two compartments (Behrend and Bitterman 1963).

In **passive avoidance learning,** the animal does not make an overt response, but rather, an already existing response (e.g., a feeding behavior) is altered by providing a noxious or negative stimulus each time the response occurs. Brower and Brower (1962) demonstrated this phenomenon using toads (*Bufo terrestris*), which flick out their tongues to capture prey. In this instance, the prey were drone flies (*Eristalis vinetorum*) and meal worms (*Tenebrio*); drone flies are mimics of honeybees (*Apis mellifera*). After permitting a toad to consume several drone flies and meal worms, Brower and Brower presented toads with honeybees, and the toads were stung when they attempted to consume the bees. After just a few trials with bees, the toads continued to eat meal worms when they were presented, but now they avoided flicking their tongues out to capture drone flies if that prey was offered.

In addition to these examples of instrumental learning paradigms, there are many others that have been employed. These include escape learning and discrimination learning; the latter of these will be discussed later in the chapter when we cover vertebrate learning.

Classical and Operant Conditioning Compared

Many who study learning phenomena today postulate that these processes underlying classical and operant conditioning are similar, while others argue that they are different processes. Let's compare them.

Acquisition

The act of first developing a response is termed **acquisition.** Several different response measures, some common to both types of learning, are used. A graphic presentation of the response measures is a **learning curve** (figure 11.5). A sequence of trials given over time often serves as the scale for the horizontal axis. The vertical axis is usually some measure of performance or strength of CR (conditioned response). Among the common dependent measures used are the percentage of conditioned or correct responses, the magnitude of the response, the time required to complete a response, the error rate, or the latency (time interval) between some signal (e.g., the CS) and the response. For several of these (figure 11.5*a*), the curve increases over the trials, whereas for other measures, the curve decreases (figure 11.5*b*). We must emphasize that these are merely measures of performance, and do not tell us about the underlying mechanism(s) of learning.

Schedules of Reinforcement

In classical conditioning, the US provides the reinforcement for a behavior. In operant conditioning, the reinforcer is presented when the subject gives the appropriate response. The most basic schedule is continuous reinforcement, where reinforcement is given for each trial or response. A variety of other types of schedules provide partial or **intermittent reinforcement.** Some examples of each of these different reinforcement schedules from our own lives should help to understand these variations (with thanks to Michael Renner). Two types of schedules involve a predictable pattern. Fixed ratio schedules require the test subject to make a fixed number of responses to receive reinforcement. Examples include being paid by the job rather than an hourly wage, also known as piecework, and giving yourself a rest break after each time you have taken so many steps while climbing a hill. Fixed interval schedules provide reinforcement for the first response after the passage of a prescribed time interval following the last reinforced response. Examples include checking for the mail delivery (providing the mail delivery comes at the same time each day) or checking the oven at fixed periods to see if the cookies are done. Two types of schedules involve an unpredictable pattern. A variable ratio reinforcement program involves reinforcement occurring after a variable number of responses. Examples include playing slot machines, fly casting for fish, or being a door-to-door salesperson. Variable interval schedules involve reinforcement for the first response after a variable time interval. Examples of this schedule include studying for unannounced quizzes in a class and redialing a phone number when it is busy.

Extinction

Extinction refers to the decrease of response rate or magnitude with lack of reinforcement. If we eliminate the US for a period of time in classical conditioning or eliminate the reward (or punishment) in operant conditioning, the response is extinguished. The number of trials or responses that must occur without reinforcement before extinction occurs is a function of several variables. In general, the longer the conditioning procedure has been (the more well learned the response), the harder it is to extinguish that response. Interestingly, responses that have received partial reinforcement during acquisition are the most difficult to extinguish. Animals trained on partial reinforcement schedules "resist" extinction for several reasons: they learned to persist in responding when faced with some nonreinforced responses during training, and there is less difference between partial reinforcement and no reinforcement than between continuous reinforcement and none. Human gambling is an excellent example of a behavior affected by partial reinforcement schedules that is resistant to extinction.

Spontaneous Recovery

If a conditioned response has been extinguished and is then followed by a rest interval of several minutes or up to a day or more, depending upon the species and experimental conditions, the animal may exhibit **spontaneous recovery** upon reintroduction to the test situation. For example, when a rat, trained to press a bar for food reinforcement, is no longer given the reward, the response will be extinguished. After a rest interval of one hour, the rat will exhibit spontaneous recovery, pressing the bar many times without receiving a reward, until the response is again extinguished.

Generalization

We find that if an animal has been conditioned to respond to a certain stimulus, the response will usually also occur to stimuli similar to that used in the original acquisition trials; this is called **generalization.** In classical conditioning, we can vary the pitch of the bell tone; dogs will respond by salivation to tones similar to that of the original bell, but not to tones with quite different pitches. For operant conditioning in humans, consider the learned association between certain types of music accompanying various actions and scenes during a radio or television program. We learn to associate certain cadences and musical motifs with particular effects, such as danger, sadness, or mystery.

Differences

Several key differences exist between classical and operant conditioning. First, the basic paradigms used to demonstrate the two types of learning differ. In classical conditioning, the situation is thought by many to involve a stimulus-stimulus pairing; the CS and US are paired. In contrast, operant conditioning involves pairing the stimulus and response. A second difference has already been noted; in classical conditioning, the subject does not control the sequence of events, whereas in operant conditioning, the sequence is contingent upon the responses of the test animal. Another way we can state this is that in classical conditioning, the responses are elicited, but in operant conditioning, the responses are emitted by the subject. A third difference involves the **shaping** process characteristic of operant conditioning. In this process, the experimenter reinforces successive approximations of the desired response. For example, as a rat is trained to press a lever for food reinforcement, the experimenter may at first control the process by successively rewarding the rat for being in close proximity to the lever, for exploring the lever, and then for touching the lever, until the rat begins to press the bar to obtain the food.

Other Aspects of Learning

In addition to habituation, sensitization, and the two major types of association conditioning, several other similar processes should be noted and briefly explained. These additional types of learning phenomena are generally considered to be categories of operant conditioning (instrumental learning).

Latent Learning

Associations made with neither immediate reinforcement or reward, nor with particular behavior evident at the time of learning have sometimes been labeled **latent learning.** An example comes from the predatory digger wasps (*Philanthus triangulum*), which inhabit burrows in sandy soil. The wasps must leave their burrows in order to capture prey some distance away: how do they relocate their burrows? Tinbergen provided landmarks around the wasp's nest holes; upon emergence, the wasp reconnoitered the area and then flew off. If Tinbergen removed or rearranged some landmarks, the returning wasps became disoriented. A series of studies demonstrate that the entire configuration of landmarks is used by the returning wasp as a guide to the location of its burrow (Tinbergen and Kruyt 1938; Tinbergen 1958).

Observational Learning

The tendency to perform an appropriate action or response as the result of having observed another animal's performance in the same situation may involve either classical conditioning or operant conditioning and has been called **observational learning.** For example, Klopfer (1957) demonstrated that ducks (*Anas platyrhynchos*) can learn a discrimination task by observation. He conditioned a group of ducks to feed from one of two dishes by placing the incorrect dish on a wired shock grid. During the conditioning, he restrained observer ducks nearby, where they could see the subject ducks learning the discrimination but could not participate. He then released the observer ducks and placed them in the test situation alone with both food dishes present. The observer ducks avoided the incorrect dish with its wired shock grid.

Imitation occurs when an animal immediately copies the actions of another while they are both in each other's immediate presence (Thorpe 1963). The act performed is one that would not normally be expected in the species' behavioral repertoire. Consider, for example, the manner in which a group of Japanese macaques (*Macaca fuscata*) of Koshima Island learned some of their food habits. Imo, an inventive young female monkey in the group, introduced two new techniques that appeared to be copied by other group members through imitation. One new technique was taking sweet potatoes, which the monkeys were fed, to the water to wash them before eating (figure 11.6). Washing removed the gritty sand adhering to the skins of the potatoes. Also, the salt water may have added some flavor. Imo also took handfuls of wheat, another food given to the provisioned monkeys, to the

FIGURE 11.6   Japanese macaques washing sweet potatoes. A young female macaque initiated the practice of washing potatoes in water before consuming them—a process that was learned by other macaques through imitation.
Source: © M. Kawai/Kyoto University Primate Research Institute.

water and allowed a little at a time to fall into the water; the sand mixed with the wheat grains sank, and the wheat floated on the surface of the water. Imo then picked up and ate the individual wheat grains (Itani 1958; Kawai 1965).

Reexamination of the findings concerning imitation and the passage of a tradition such as the potato washing suggests that this may not be imitation (Galef 1976, 1990, 1996; Heyes 1996). First, the time scale for imitation of the potato-washing behavior is relatively slow. After Imo learned the trait, one other juvenile began washing potatoes the next month. Two other members of the troop, another juvenile and Imo's mother, started washing potatoes some 4 months after Imo. Within the next year, another monkey acquired the trait; two more began potato washing 2 years after Imo; and, finally, two more started 3 years after Imo initiated the behavior. If this were truly imitation, we might have expected that more macaques would have started sooner. Tied to this is a second issue. Normally, when learning is enhanced by imitation, we might expect that the process would spread in a cumulative or even almost exponential manner. If we plot the rate of increase of this potato-washing pattern in Imo's group, the curve is almost flat (Galef 1990). Thus, on these bases alone, we would question whether this is actually imitation. There are other reasons to wonder whether imitation occurred in this instance, having to do with patterns of human feeding of potatoes to specific monkeys (Green 1975) and with the fact that this trait, and the others, such as wheat placer-mining, all involve foods that are not naturally occurring in the diet of these monkeys.

Thus, we can learn that varied interpretations of such phenomena are possible. To confirm that such occurrences do involve imitation or the learning of a tradition would require various forms of experimentation. These experiments would likely include laboratory tests to determine what types of social interactions could facilitate imitation learning. A more parsimonious explanation might involve a young monkey being near its mother as she fed on the potatoes used as provisioning food. Picking up scraps of potatoes that fell into the water could account for the acquisition of potato washing by Imo. The others may also have acquired the trait in this manner. Many similar phenomena that suggest socially facilitated learning require careful scrutiny and are a challenge with respect to dissecting two central questions (Galef 1976): What is being transmitted? What mechanisms are involved in the transmission process?

Imprinting

We have already discussed imprinting as a process of attachment formation and following behavior and in terms of sexual imprinting, in chapter 10. Imprinting is sometimes classified as a separate, special type of learning.

Learning Sets

One general phenomenon important for an understanding of learning was first elucidated by Harlow and his associates (Harlow 1949, 1951; see also Warren and Barron 1956; and Shell and Riopelle 1957). A **learning set** is defined as the acquisition of a learning strategy by the animal or, in other words, interproblem learning. We find that, given a series of problems, an animal will transfer some of what it has learned about solving the first problem to the solution of subsequent problems in a series. The learning curves of monkeys given blocks of test problems (figure 11.7) demonstrate this effect. As the series proceeds, the animal scores an increasingly high percentage of correct responses by the second or third trial on each new problem. In effect, the formation of learning sets occurs as the animal "learns how to learn." Learning sets may also

restrict the behavior of an animal. Negative consequences may result if an animal has formed a learning set regarding a particular problem; it may not be able to shift strategies as readily as an animal without the learning set. The speed at which some animals learn the interproblem strategy varies (Warren 1965), with the most rapid learning restricted to primates (figure 11.8).

INVERTEBRATE LEARNING

We cannot retrace the steps in the evolution of learning, but a brief survey of the learning capacities and capabilities of organisms from various animal phyla should provide some insight into the possible generality of any learning principles and a sampling of the differences in learning capabilities. We do this in two sections: first, we explore learning processes in invertebrates, and then, we turn to vertebrates. Before proceeding, we must note four cautions. First, there has been a disproportionate examination of learning across the various phyla. More studies have been conducted on learning processes in vertebrates than in invertebrates; and within the vertebrates, much of the attention has been focused on mammals. Second, there are valid disagreements among scientists regarding the proper criteria for demonstrating various types of learning. Third, exploring the physiological and behavioral aspects of learning in many animals requires more refined techniques and objective, bias-free methods. Fourth, we measure performance on a particular task, and cannot directly measure a genetically determined ability. Many factors (notably, test design) can influence performance.

Protozoa and Cnidaria

Protozoans exhibit habituation responses, but there is no fully accepted evidence for any type of associative learning (Corning and von Burg 1973). For example, Patterson (1973) demonstrated that for *Vorticella convallaria* the contraction

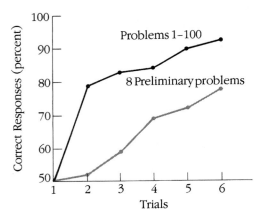

FIGURE 11.7 Discrimination learning curves.
Each dot on the graph represents the average percentage of correct responses on a particular trial for the 8 problems or 100 problems given to a rhesus monkey. Note the improvement that occurs with successive trials and particularly the large improvement by the second trial in the 100-problem set compared to the small improvement in the set of 8 preliminary problems.

Source: Data from H. Harlow, "The Formation of Learning Sets," in *Psychological Review*, 56:51-56, copyright 1949 by the American Psychological Association.

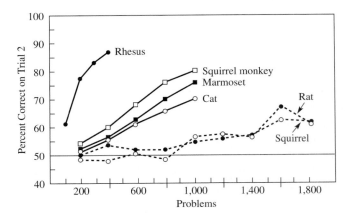

FIGURE 11.8 Learning set formation in mammals.
The rate at which learning set formation occurs is compared for six species of mammals, including three species of primates. Rapid learning occurs most readily in rhesus macaques in this sample, and all of the primate species exhibit learning set formation more readily than the other species shown.

response exhibited to both mechanical and electrical stimuli habituates with repeated stimulus presentations. Gelber (1965) claimed that *Paramecium aurelia* learned to associate presentations of a fine wire in their environment with food (bacteria). After presenting the wire coated with bacteria on a number of trials, the wire alone seemed to elicit an approach response by more paramecia than presentation of the wire alone to untrained animals or to populations trained with an uncoated wire. An alternative interpretation of these findings was provided by Jensen (1965). He tested the notion that the unicellular paramecia were merely responding to a food-rich zone in their environment, created by the presence of the bacteria-coated wire. The debate over these alternative interpretations has not been resolved.

For cnidarians, there is also evidence for habituation, but little evidence for associative learning. In the early years of this century, investigators generally assumed that cnidarians exhibited only certain involuntary, stereotyped reflexes. We now know that there are endogenous neural rhythms in many cnidarians and that, rather than being totally "passive," many cnidarians are capable of spontaneous, active responses in the course of interacting with their environment (Rushforth 1973). One possible demonstration of a form of association learning (Ross 1965) is based on the fact that sea anemones (genus *Stomphia*) respond to chemostimulation from certain starfish (e.g., *Dermasterias imbicata*) by stretching their bodies, detaching from the substrate, and "swimming" away (figure 11.9). The animal soon lands and eventually reattaches to the substrate. Chemostimulation is paired with gentle pressure applied near the base of the anemone. With repeated trials, the application of the pressure stimulus alone leads to some reduction in the "swimming" response.

Platyhelminthes and Annelida

Among the flatworms, considerable research has been conducted with planaria, but little is known about the other groups, some of which (e.g., tapeworms, flukes) are internal parasites. Conclusive evidence exists for habituation and several types of association learning in planarians, which have a relatively simple bilateral nervous system. Thompson and McConnell (1955) first reported classical conditioning in these organisms, but it was not until the report by Block and McConnell (1967) that satisfactory control procedures were employed. Among the issues involved was whether there was classical conditioning or whether the observed effects were due instead to either sensitization or pseudo-conditioning (Dyal and Corning 1973; Corning and Kelly 1973). Sensitization, as used here, is an increase in the strength of a response originally evoked by a CS as the result of pairing with an US and a response. **Pseudoconditioning** is the strengthening of a response to a previously neutral stimulus by repeatedly eliciting the response with another stimulus without pairing the presentation of the two stimuli. In general, researchers today agree that planaria are capable of exhibiting some form of conditioned learning.

Two of the most striking conclusions that arose from these experiments on learning in planaria were: (1) Learning, or some component thereof, was passed on to the two regenerated planarians that resulted from severing a pretrained worm in half. The two regenerated planaria learned a T-maze choice task faster than naive worms or naive regenerates (McConnell et al. 1959). (2) Studies in which naive worms were permitted to cannibalize worms that had previously been trained on a learning task exhibited much faster learning of that task than planaria that consumed naive conspecifics (Corning and Kelly 1973). One potentially serious problem with some of the studies of learning in planarians is that they leave a slime trail as they move along. It is thus possible that using the same apparatus for repeated trials, without cleaning, confounds the results.

Annelids have a segmented nervous system with ganglia in each body segment and some concentration and coalescing of ganglia in the anterior body segments. Among annelids, habituation has been demonstrated for earthworms (*Lumbricus terrestris*) using the backward movements that occur in response to puffs of air (Ratner and Gilpin 1974). Test paradigms with worms use either aversive conditioning or some

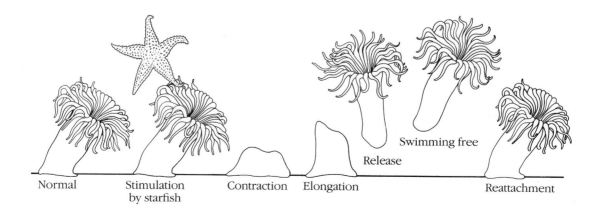

FIGURE 11.9   Interaction between starfish and sea anemone.
When a starfish (*Dermasterias*) makes contact with the sea anemone (*Stomphia cocinea*), the anemone will release from its attachment, "swim" free for a time, and eventually reattach itself to the substrate.

form of punishment training such as mild shock. Classical conditioning has been shown by pairing a light stimulus and mild vibrations (Ratner and Miller 1959). Operant conditioning has been demonstrated for annelids in a variety of experiments using a two-choice T-maze apparatus and mild shock for incorrect choices. As with the flatworms, the problem of mucous trails containing chemical cues after repeated trials is a potential problem for investigations using annelids.

Mollusca

Learning capabilities of two major groups of molluscs have been investigated: gastropods (snails, slugs) and cephalopods (squids, octopodes). One great advantage scientists have in working with these organisms is that the nervous systems of many species are readily accessible and contain large neurons. Thus, it is possible for us to explore some of the neurophysiological processes accompanying learning.

Habituation has been studied extensively in snails of the genus *Lymnaea*. Both visual and mechanical stimuli result in a withdrawal movement: the snail's shell is drawn downward and forward (Cook 1971). The response habituates with repeated stimulation. Habituation of the gill withdrawal reflex in *Aplysia* occurs with repeated stimulation from a jet of water (Pinsker et al. 1970). Kandel and coinvestigators (see Hawkins et al. 1983) have demonstrated differential facilitation of excitatory postsynaptic potentials in the neuronal circuit for this withdrawal reflex. Activity at the

synapses between the sensory and motor neurons involved in this reflex arc is facilitated more by tail shock (US) if the shock is preceded by spike activity in the sensory neuron than when the spike activity and shock occur in an unpaired pattern, or with the shock treatment alone.

Association learning has been reported in various species of both gastropod and cephalopod molluscs. Much of what we know about this group comes from studies conducted with octopodes (reviewed by Sanders 1973). Recall the material in chapter 3 regarding tactile discrimination in the octopus; members of this group can learn to discriminate cylinders based on the amount of grooved surface, but they fail to learn to separate cylinders based on the pattern of grooved surface when the amount of grooved area on different cylinders is roughly the same. Through convergent evolution, the peripheral aspects of the octopus' visual system are quite similar to the visual system found in vertebrates. However, examination of the results of visual discrimination tests (figure 11.10) reveals that the interpretations of the stimuli must be quite different than for most vertebrates; octopuses fail to make successful choices when the paired items are mirror images or when there are strong similarities between the arrangements of horizontal and vertical lines of the two stimuli. Last, via conditioned learning, octopuses can be trained to discriminate between various food items (Boycott 1965). If the octopus is presented with a fish and a geometric figure such as a disk and, upon approaching the fish, the octopus is given a mild shock, the animal learns to avoid

Problem	Discriminanda	Horizontal extent	Vertical extent	N	Percent correct responses
1				6	81*
2				8	71*
3				8	65.5*
4				6	50
5				6	59*
6				7	56*

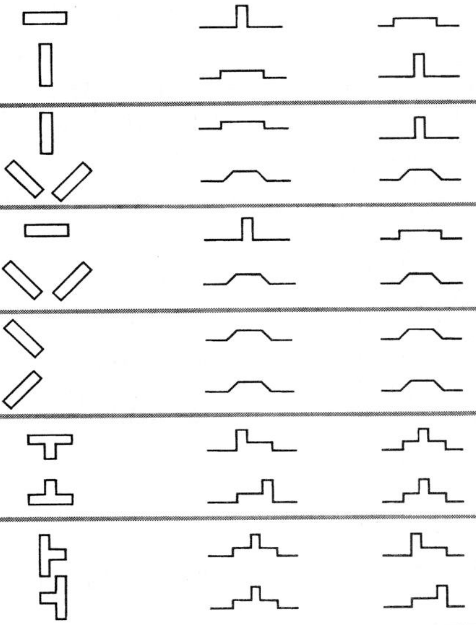

*Better than chance level of performance $P < 0.05$.

FIGURE 11.10 **Octopus performance over 60 trials on the discrimination of pairs of shapes differing in orientation only.**
When an octopus is given discrimination trials with pairs of outline shapes, it can successfully discriminate those that differ in orientation, but not those with the same or nearly the same horizontal and vertical extents.

the fish. When the disk and fish are presented together, the octopus will retreat. If the fish alone is presented, the octopus will attack. Such techniques can be used to train octopuses to approach one type of food (e.g., crabs), while avoiding fish, or vice versa. Octopuses are also trainable in spatial tasks (Boal et al. 2000).

Alkon (1980, 1983) demonstrated conditioning in the marine snail (*Hermissenda crassicornis*). These snails are normally positively phototactic during the daylight portion of their daily cycle. Snails were trained in glass tubes filled with seawater and mounted radially on a turntable. Prior to training, the snails were measured for their velocity of movement toward the lighted center portion of the turntable. Then the turntable was rotated, producing centrifugal force for the snails and possibly simulating the water turbulence encountered near the ocean's surface under stormy weather conditions. Thus, the light source and rotation were paired. After training, the velocities were again recorded; trained snails moved toward the center of the turntable with about one-third of the pretraining velocity. Care was taken to perform a number of control procedures to ensure that the observed effect was in fact due to some form of association learning by the snails. These control procedures included exposure to light alone, rotation alone, alternating light and rotation, and presenting light and rotation at random. By mapping the neural pathways for reception of information from the light source and from the rotational forces, Alkon (1983) produced a wiring diagram for the pertinent portions of the nervous system of *Hermissenda*. Further, he showed that one of the consequences of training is that particular receptors and cells in the nervous system become either more excited or inhibited, depending upon the conditioning.

Arthropoda

The arthropods are a large and diverse group comprising 80 percent of all animal species. Only the vertebrates have been studied more with respect to learning behavior. The arthropods have evolved a wide variety of types of sensory receptors and have a nervous system characterized by an aggregation of ganglia, termed a "brain," located in the anterior segment of the body, with a pair of ventral nerve cords passing to the thorax and abdomen. Learning has been studied more in three groups of arthropods than the others: the subphylum Chelicerata (e.g., *Limulus,* scorpions, spiders) and the classes Crustacea and Insecta. For each of these groups, habituation, classical conditioning, and operant conditioning have been demonstrated in a variety of species. Consider, for example, bees can learn to discriminate colors by associating different hues with sugar solutions or with nonsugared water (von Frisch 1967a,b; Wells 1973). Spiders that have lost a leg to a scorpion attack are more likely to avoid the scent of scorpions than those that have not encountered one (Punzo 1997). Various species of crabs and crayfish have been successfully trained in T-mazes (Gilhousen 1927; Datta et al. 1960). Krasne (1973) reported that there may be some transfer of learning from a simple maze to more complex mazes for crabs.

Probably the most thoroughly studied example of insect learning is the honeybee. Honeybees are easy to keep in large numbers, and can be easily trained to visit drops of sucrose set out in a dish. By manipulating cues such as color and odor that are associated with the dish, as well as reinforcement schedules, the same sort of learning processes can be tested in bees that have been tested in vertebrates.

For example, we can examine the effects of different reinforcement schedules on learning rates. Skinner (1938) found that if rats are trained with a partial reinforcement schedule, the task that had been learned (e.g., bar pressing) was more difficult to extinguish. Sigurson (1981a,b) and Bitterman (1988) report a similar effect in honeybees.

Bees were trained to enter an enclosed chamber where there was a Plexiglas tube surrounded by a Plexiglas disk that could be illuminated. Sucrose solution could be pumped into the tube in tiny amounts. The bees' approaches to the tube and dips into it were recorded electronically. The illumination of the Plexiglas disk and tub served as the conditioned stimulus.

Bees were given a period of training. Each trial commenced with the bee in darkness for a variable time period, after which the disk was illuminated and the bee could respond by dipping into the tube. One group of bees was reinforced on each of 15 training trials. A second group was always reinforced on the first trial, but then reinforced on only half of the subsequent trials (randomly determined). This was followed, for both groups, with 15 extinction trials during which there was no reinforcement. The results (figure 11.11) show that, during pretraining, both groups exhibited rapid acquisition of the task. Then, the bees with consistent reinforcement responded at higher rates than those receiving partial reinforcement. Finally, bees that had received partial reinforcement were more resistant to extinction. These results are reminiscent of those well known in vertebrates.

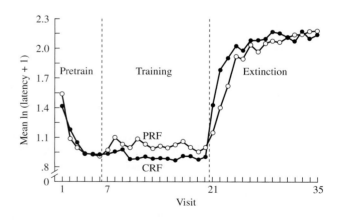

FIGURE 11.11   Learning in honeybees.
Demonstration of learning and extinction in honeybees. The x-axis represents trials or visits by the bee to the response tube (see text). The y-axis is the mean natural logarithm of the latency to respond plus 1 second. During the pretraining phase, all responses were reinforced. During training, the CRF group was reinforced on every trial (consistent reinforcement treatment), and the PRF group was given reinforcement on only half of the trials (partial reinforcement treatment). During extinction, none of the trials involved reinforcement.

Other sorts of tasks have also been presented to honeybees. For example, researchers have examined how bees respond to compound cues, such as color (e.g., orange, yellow) presented together with odor (e.g., jasmine, violet); whether bees are better able to learn some cues rather than others, and how bees learn landmarks that tell them where they are. There has been controversy over whether the results from bees differ from vertebrates, and, if one admits to differences, whether they are really fundamental. Some argue that learning processes in bees appear to be quite similar to what is known in vertebrates (e.g., Bitterman 1988). Others argue that learning in general, and honeybee learning in particular, is specialized and inherently biased, probably in an adaptive way. For example, although bees are able to learn about color, odor, and shape, the information from each of these modalities is not equally valued (odor has precedence over color, and color over shape). The differences in viewpoint might be at least partially ascribed to differences in methods; these issues are discussed in Gould (1993).

VERTEBRATE LEARNING

We know considerably more about the complex processes that underlie learning phenomena for the phylum Vertebrata (reviewed by Masterton et al. 1976) than for most invertebrates; this is particularly true for the mammals. Throughout this chapter, many of the examples we have used have come from the literature concerning learning in vertebrates. Our treatment here will be divided into two parts: an examination of work on two particular species of mammals and a synopsis of current thinking on vertebrate learning and intelligence. Rats and rhesus monkeys have been used in a large number of studies of vertebrate learning. We will consider an example for each species to provide some flavor of the research and how it is conducted.

Rats and Monkeys

We've focused on examples where animals successfully learn the task they are given, but clearly all animals are limited in what, and how much, they can learn. Rats can be trained in an operant conditioning apparatus to press a lever to receive food. They can also be trained in an apparatus with two levers (designated L for left and R for right) to press the levers alternately, LR or RL, for a food reward. Can rats be conditioned to press the levers in a LLRR or RRLL sequence, called double alternation? Investigators tested this question by conditioning rats first on the single lever task, either R or L, followed by the single alternation task, RL or LR (Travis-Niedeffer, Niedeffer, and Davis 1982). They then rewarded only double-alternation performances. The rats learned the double alternation task at a level exceeding chance expectations, and their performance improved over days. It was necessary, however, in some instances, to permit the rats to give extra responses to the first lever before pressing the second lever. Thus, LLLRR was rewarded the same

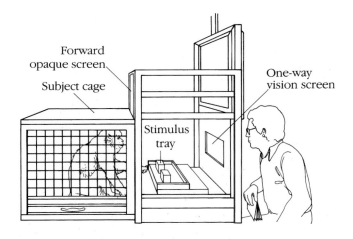

FIGURE 11.12   Wisconsin General Test Apparatus.
An early version of the Wisconsin General Test Apparatus (WGTA) used to test aspects of learning in primates and, with some modifications of the apparatus, cats and raccoons.

as LLRR. When the investigators attempted to condition the rats to a sequence of LLRRLLRR or RRLLRRLL (double alternation with a fixed ratio schedule of two repetitions), no rat could successfully perform at better than a chance level.

One particular apparatus, the **Wisconsin General Test Apparatus (WGTA)** (figure 11.12), has been very useful in studying learning behavior in rhesus macaques (*Macaca mulatta*) (see Meyer et al. 1965 for the history of this and related techniques). One type of learning problem studied with this apparatus is the **delayed response problem** (Fletcher 1965). The basic procedure starts with the test tray out of reach but in full view of the test animal. Food is placed in one food well, and then two identical objects are placed over the wells. After a prescribed delay, the test tray is moved closer to the monkey, which then responds by lifting one object or the other. A trial ends when the monkey picks up one of the objects, uncovering either the correct food well, containing a reward, or the empty well. A number of variables can be investigated with this procedure, including length of the delay phase, the nature and size of the reward, the nature and the similarity or dissimilarity of the objects used to cover the food wells, and whether the animal is permitted to watch the test tray during the delay phase or, instead, has an opaque screen lowered in front of the tray (Meyer and Harlow 1952). Other animals, such as cats and raccoons, have been tested with slight modifications of the apparatus.

These tests show that rhesus monkeys are capable of learning basic discriminations with delays of up to 30 seconds or more; more food reward leads to better performance; and the presence of the opaque screen during the delay phase increases error rates by up to 50 percent. One interesting and striking finding in these studies is that the behavior and performances of individual monkeys differ markedly. In general, monkeys that exhibit hyperactivity in the test situation and those that are more easily distracted during the delay phase exhibit lower levels of performance. Clearly, in studies of learning behavior, we must consider the significance of individual differences in performance, regardless of the species being tested or the task being performed (Warren 1973).

Vertebrate Intelligence

To begin this subsection, we need a definition of intelligence. However, this is a difficult task. Some have used the term animal intelligence as essentially synonymous with animal learning. This definition is probably too narrow. **Intelligence** may best be defined, for our purposes, as a collection of capacities (Immelmann and Beer 1989), including imagination, problem-solving ability, memory, and the ability to utilize information gained from past experiences, perceptiveness, and behavioral flexibility. Kamil (1988) defines animal intelligence as encompassing those processes by which animals obtain information about their environment, retain it, and use that information to make decisions during their behavioral activities. Given this range of capacities, it is unlikely or impossible that any single test could be designed to measure intelligence. It is for precisely this reason that much of the work that has been done on comparative intelligence can be criticized. Usually, only one, or at most two, of the aforementioned sorts of capabilities are tested and possibly compared across species. After extensive examination of studies of comparative intelligence, Macphail (1982, 1985, 1987) hypothesized that there was no real basis to assume that there are quantitative or qualitative differences in the mechanisms of intelligence among nonhuman vertebrates. Macphail puts forward two other alternative explanations for the fact that we, thus far, have been unable to demonstrate any clear differences in intelligence among nonhuman vertebrates but discards both in favor of the null hypothesis stated above.

The first rejected hypothesis is that data gathered to date have been misinterpreted. Macphail notes that some would argue that contextual variables, such as the value of particular food rewards, could account for any species differences with regard to intelligence. He rejects this argument based, in part, on the fact that for any particular learning phenomenon, original claims of the absence of particular forms of learning (e.g., reversal learning) have generally been countered once sufficient careful systematic testing has been conducted. The second hypothesis is that the questions that have been posed have been inappropriate. In particular, experimenters have not taken into account the ecological niches of the organisms under study, and this factor might account for any failure to find species differences in intelligence. Macphail also rejects this notion by arguing that, while some would claim the failure is due to the fact that animals have species- or niche-specific and problem-specific methods for problems, this line of reasoning does not account for (a) the fact that learning psychologists have not been able, with their artificial tasks, to demonstrate any species-specific differences (although, as we shall see, others argue that such differences exist) and (b) that arguments concerning biological constraints have been well accommodated by general learning theory (see Domjan 1983).

Mackintosh, Wilson, and Boakes (1985) suggest that a better way to proceed, in order to explore possible species differences in intelligence among vertebrates, is to compare fewer species, but to use a broader range of learning paradigms on which to base comparisons. Their general position, somewhat different than that of Macphail, is that additional types of testing and a different perspective are needed if we are to truly ascertain whether there are interspecific differences in intelligence among vertebrates. They used two types of learning paradigms to provide comparisons; reversal learning, where an animal uses the outcome of one trial to predict the outcome of the next trial; and transfer of matching and oddity discrimination, where matching involves choosing the stimulus that is correct in relation to a reference or comparison cue, and oddity involves selecting the stimulus that does not match the reference cue. While both rats and goldfish (*Carassius auratus*) are capable of reversal learning, rats show more rapid improvement, which indicates that there must be some differences in the mechanisms of intelligence between these two vertebrates (Mackintosh and Holgate 1969; Mackintosh and Cauty 1971). Both pigeons (*Columba livia*) and European jays (*Garrulus glandarius*) learn matching and oddity problems with little difficulty (Zentall and Hogan 1974; Wilson et al. 1985). Yet, when the problem involves transfer of matching or oddity to a new set of stimuli, the jay is much superior to the pigeon (Wilson et al. 1985). Again, a plausible explanation for the observed difference is that there is at least a quantitative difference and possibly a qualitative difference in the mechanisms of learning and intelligence for these species.

One aspect of vertebrate intelligence that may be critical is exploratory behavior (Renner 1990). **Exploratory behavior** (sometimes termed curiosity) can be defined as a spontaneous search for and active investigation of objects, situations, or other organisms in the absence of any homeostatic need. In this sense, exploration serves the function of information gathering for the animal; if that information is later used, then exploration can be seen to be closely related to latent learning. Given our definition of intelligence earlier in this section, exploratory behavior could be a large component of several of the capacities listed with that definition. Indeed, the functional significance of exploratory behavior has been demonstrated under laboratory conditions for both white-footed mice (*Peromyscus leucopus*) and meadow voles (*Microtus pennsylvanicus*). In both tests (Metzgar 1967; Ambrose 1972), rodents with prior exploratory experience in a habitat were better able to avoid predation than conspecifics placed in the same artificial habitat with no prior exploratory experience in those surroundings.

Many, though not all, tests of exploratory behavior use an apparatus called the open-field test, consisting of an open area, usually marked off in sections, used to assess the animal's behavior in terms of locomotion. Often the responses of test animals in an open field involve other actions (e.g., escape behavior) (Welker 1957; Suarez and Gallup 1981). As Renner (1990) notes, many sophisticated devices are now available that make measurement of activity more precise and less prone to potential observer problems, but even these merely extend the measurement of the locomotion (spatial) component of exploratory behavior.

If we are to fully investigate the exploratory behavior of animals of various species, then additional, more inventive sorts of studies are needed, particularly those that involve examination of other components of our definition of this phenomenon. Some earlier work, notably studies by Harlow (1950; Harlow et al. 1950) with providing rhesus monkeys mechanical puzzles, such as latches or hinges, and by Butler (1953, 1954), wherein monkeys could gain visual access (for exploration) to a conspecific or an electric train, provide an insight into the type of thinking that will be necessary to gain a fuller understanding of exploratory behavior. More recently, Hopf, Herzog, and Ploog (1985), studying squirrel monkeys (*Saimiri sciureus*), examined developmental changes with regard to exploration of a social surrogate object. In virtually all of these studies, the measures recorded involve rates of interaction or contact with the object. Other investigators, looking more at the nature of investigatory behavior in terms of how objects are manipulated, have shown that movement in space and object exploration in rats develop according to different calendars (Renner and Pierre 1991). Avenues for future work on exploratory behavior should include more thorough descriptions of what exactly it is that animals are doing when we say that they are exploring. An expanded operational definition of exploratory behavior will necessitate coming up with methods to make objective measures of what we are observing. The end result, however, will be a better understanding of the nature of the stimuli that are important to the animal that is engaging in such behavior rather than just a picture of the rate of exploration.

COMPARATIVE LEARNING

Over the past several decades, animal behaviorists have attempted to compare learning abilities of various phyla. Most notable of these are the reports by Bitterman (1960, 1965a, 1975) and Dewsbury (1978). We will examine first the work of Bitterman and then that of Dewsbury.

Bitterman's approach was to look at certain organisms (e.g., fish, rats, turtles, pigeons, and monkeys) and their learning capabilities and to classify organisms according to whether their learning abilities are like those of his test species. Basically, Bitterman believes different processes occur in different species; different learning phenomena may result from the same process in different organisms; and, conversely, what appear to be similar learning phenomena in different organisms may result from basically different underlying mechanisms. He argues strongly against the principle of **equation**—the attempt to equate situations and procedures for learning in different species. He favors instead the concept of control by **systematic variation**—systematically varying each potential variable that may adversely affect a species' performance. Bitterman's work is significant because it cautions animal behaviorists who would

plunge into comparative learning studies without a close look at the capacities of the various animals under study and the variety of methods used to test learning phenomena. However, control by systematic variation can be, in practice, very difficult to carry out. Kamil (1988, 1998) suggests that a more tractable approach is to test each species with a battery of tests of the same cognitive ability.

Dewsbury (1978) classified attempts to investigate comparative learning into two categories: quantitative comparisons and qualitative comparisons. Dewsbury first examines the cross-species comparisons for acquisition of learning (see also Brookshire 1970); there appear to be no general patterns across the various phyla for rates of acquisition or for avoidance learning. The development of learning sets for particular conditioned learning responses in different species has been widely used to compare learning in vertebrates. Again, the data do not support any clear relationship between phylogeny and the learning set phenomenon. Dewsbury summarizes the problems of making such quantitative comparisons under several headings, including (1) individual differences, which make it difficult to utilize mean performance levels as species-typical (see Rumbaugh 1968); (2) motivational differences, for example, in animals that are food-deprived and then given learning tasks; and (3) species differences that reflect variations in biological constraints on learning (see subsection later in this chapter). There is little evidence to suggest the possibility of constructing any sort of "scale of intelligence" for vertebrate organisms, let alone comparisons that attempt to include the invertebrate phyla.

PREPAREDNESS AND CONSTRAINTS ON LEARNING

Over time we have discovered that two general types of constraints are operative in studies of learning in general and of comparative learning in particular. These are the biological constraints brought to the learning situation by the animal, termed "preparedness" by Seligman (1970), and the methods constraints imposed by the investigator and test situation (see Hinde and Stevenson-Hinde 1973).

Biological Constraints—Preparedness

Seligman (1970) introduced the notion of **preparedness**—the genetically based predisposition to learn. According to this notion, an animal faced with a particular learning task, whether in nature or in the laboratory, may be prepared, unprepared, or contraprepared to perform that task. The constraints and limitations imposed by inheritance affect an animal's relative preparedness to learn (Garcia et al. 1972; Hinde and Stevenson-Hinde 1973; Warren 1973). In the evolutionary process, each species has been provided with tools and capacities that limit its range of potential responses to specified inputs.

For example, rats (both wild and laboratory strains) are prepared to learn to associate the taste of certain foods with subsequent illness, even when the onset of illness occurs up to several hours after they ingest the food (Rozin 1968; Garcia et al. 1976). This lack of a close timing relationship between the stimulus and response is an interesting and important contradiction of the previously established view that there had to be continuity between the presentation of a stimulus and the positive or negative reinforcement. Rats learn to avoid foods that make them ill after one or, at most, a few exposures; the evolutionary advantage of rapidly learning to avoid ingesting potentially harmful or poisonous foods should be obvious. Humans also make rapid associations between illness and a food they have ingested. Even though people have actually contracted an illness not connected with a food (e.g., the flu), they may associate something they ate shortly before becoming sick with the illness and may maintain the negative association for some time.

In other instances, it appears that even though an animal is capable of learning a particular task, we find its system is generally unprepared—that is, the animal requires training to complete the learning task. Rats do not naturally press levers for food, but they can be trained (conditioned) to do so. Chimpanzees normally communicate by vocalizations and gestures (van Lawick-Goodall 1968). They can, however, be trained to use some American Sign Language of the Deaf (Gardner and Gardner 1969; Fouts 1973). Chimps learning sign language require a great deal of time and training to become successful at even rudimentary communication with this system (figure 11.13).

A third possibility is that an animal may be contraprepared to perform a particular task—that is, even with many attempts or a great deal of training, the animal appears to be incapable of performing the task. In an investigation of avoidance or defensive reactions, an animal's responses must be selected from species-specific patterns (Bolles 1970). Thus, pressing a lever to avoid an electric shock is an avoidance task for which the rat is contraprepared, whereas it is prepared to avoid the shock by running away, a more natural tendency. For any particular learning task, the capacities of a specific animal species are somewhere along the continuum formed by preparedness, unpreparedness, and contrapreparedness. The predispositions that an animal brings with it to a test situation or to an experimental situation under natural conditions must be accounted for when we attempt to explore, interpret, and understand its learning behavior (Timberlake 1990; Timberlake and Lucas 1989).

Methods Constraints

As we have seen in earlier chapters, restrictions or constraints operate in many laboratory and field situations in which learning studies are undertaken (Shettleworth 1972); in fact, the apparatus and the situation being used to test the animal may alter the results. Stimulus cues must be related to the animal's sensory capacities and perceptual world, or Umwelt. For instance, when studying an animal's defense reactions through avoidance learning, some attempt must be made to relate the particular avoidance response measure to

FIGURE 11.13   Chimpanzee signing.
Using sign language is not part of the normal behavioral repertoire of chimpanzees. They can, however, be trained to learn a vocabulary of more than 100 words in sign language. This chimp has learned a variety of signs.
Source: © H. Terrace/Anthro-Photo.

the species-specific defense reaction (Bolles 1970). A decrease in an animal's observed performance may result not from its inability to display a certain action, but rather from its being presented with an inappropriate stimulus—one the animal cannot interpret.

We must be certain that the response tasks an animal is asked to perform are part of its potential repertoire. In addition, our experimental design must take into account the possibility of sex differences in learning capacities and the probability that animals of different ages may learn at different rates, may possess different amounts or types of prior experience, and may have different sensory/perceptual capabilities.

If we state that a form of learning behavior is present in most animal types, it is tempting to assume **commonality;** that is, we infer a phyletic, or evolutionary, relationship from the common exhibition of a type of learning. As noted earlier in this chapter, assumptions about the commonality of evolutionary relationships can lead to potential problems and limitations (Hodos and Campbell 1969). Too often we merge data on animals of one phylogenetic lineage (a group of related species) with data gathered on animals of a different lineage. To obtain information about the evolutionary history of learning, we have to follow the same procedures we use in studying the evolutionary history of any behavior, as described in chapter 6.

LEARNING AS ADAPTIVE BEHAVIOR

Up to this point in the chapter, we have explored various concepts and procedures involved in the study of learning, and we have been on a brief phylogenetic tour exploring learning capacities in various animals. What about the ways in which learning serves as adaptive behavior for animals in everyday life? Some instances of such phenomena have already been presented, but we need now to examine specific examples of ways in which learning plays a role in feeding, predator-prey relations, and reproduction and parenting. Each of these topics is covered in greater detail in the chapters that follow. The adaptive value of learning can be evaluated in both field and laboratory settings; the use of both locations for the study of learning may prove to be the most viable approach (Shettleworth 1984; Miller 1985; Dukas 1998).

Feeding

In the western United States, you might see corvids (members of the crow family) searching for seeds. At times, they consume the seeds immediately, but at other times, they place the seeds in caches, usually holes or crevices in trees or other vegetation (Balda 1980, 1987; Tomback 1980; Vander Wall and Balda 1981; Kamil et al. 1986; Balda and Kamil 1998). The three species that are most conspicuous in their seed-storage behavior differ in their general habitat ecology and also the degree to which they rely on stored seeds. Clark's nutcrackers (*Nucifraga columbiana*) (figure 11.14) live at higher elevations than their cousins and have fewer alternatives to seeds for food, particularly in the winter months when they may consume nearly 100 percent stored pine seeds; each of these birds may cache as many as 33,000 seeds each year. Pinyon jays (*Gymnorhinus cyanocephalus*) live at intermediate elevations, have some alternative foods available, and consume 70–90 percent pine seeds in winter months; they cache some 20,000 seeds each year. Scrub jays (*Aphelocoma coerulescens*) inhabit much lower elevations, do not face severe problems with food location in the winter, and each bird stores only about 6,000 seeds annually. These birds must somehow be able to remember their cache locations. We will explore several facets of the spatial memory of these corvids in the next section on animal cognition.

What about spatial memory and foraging behavior in other species? Many nectar-feeding birds, such as hummingbirds, sunbirds, and Hawaiian birds such as the amakihi (*Loxops virens*), use some form of spatial map to guide their feeding patterns (Gill and Wolf 1975; Kamil 1978; Gass and Montgomerie 1981). Nectar is a renewable resource, but one that, once depleted, requires some time for a new supply to be generated. Studies of the foraging patterns in the aforementioned species reveal that they generally follow a forag-

FIGURE 11.14   Clark's nutcrackers.
Learning and spatial memory play critical roles in the feeding habits of Clark's nutcrackers. These birds cache pine seeds during periods of abundance and then return to relocate and consume them during periods (e.g., winter) when food is hard to find. The caching behavior has been studied under laboratory conditions, as shown here, in a chamber constructed so that birds can cache seeds in holes in the floor.
Source: Courtesy Dr. Russel P. Balda.

ing pattern, as they visit clusters of flowers, that follows the renewal time for the various species they are utilizing. That is, their behavior is efficient in that they avoid returning to flowers they have visited previously until about the time that the nectar supply should be renewed. To accomplish this, the birds must have some type of spatial memory or mental map of their feeding territory and the ability to store information about where on that map they have been in some sort of temporal sequence. A similar phenomenon has been reported for Clark's nutcrackers, with respect to visiting clusters of seed caches; they do not revisit those that they have depleted, conserving their energy resources (Kamil et al. 1993).

Spatial memory has also been tested extensively in rats. The radial-arm maze (figure 11.15) has been used to examine whether rats forage in an optimal way (see chapter 15 for a detailed discussion of foraging strategy and optimal foraging). In an eight-arm radial maze with food reinforcement at the end of each arm, we can test whether a rat makes return visits to any arm before it has successfully been to all of the arms (Olton and Samuelson 1976; Olton et al. 1977; Olton 1979). If, as is necessary, we control for the possible use of odor trails, the conclusion is that rats can accomplish their high level of performance on this radial-arm-maze task by remembering where they have and have not been. In fact, they can remember for several hours. Roberts (1981) demonstrated that when rats were allowed to retrieve food from four arms of a maze and were then placed on a second eight-arm

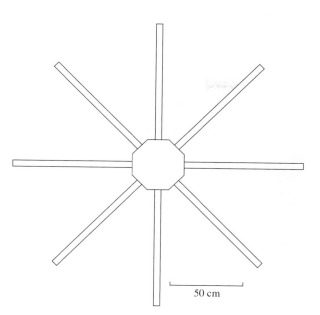

FIGURE 11.15 Radial-arm maze.
Diagram of Olton's elevated radial-arm maze with a start chamber in the center and eight arms, each of which would have some form of reward (e.g., food) at the end.

maze and allowed to collect from four arms before being returned to the original maze, they made excellent scores retrieving the food from the remaining four arms there, performing with the same low-error rate as rats that had no intervening experience. This can be extended to two intervening mazes without a measurable performance decrement, but when three mazes are used between the two trials on the original maze, errors increase significantly.

The radial-arm maze can be used to study several key properties of learning that relate well to studies of foraging behavior (Olton 1985). These include (1) primacy, which is having a more accurate memory for items that occur at the beginning of the list (e.g., places visited early in a foraging sequence); (2) recency, which is having a better memory of those items near the end of the list (e.g., places visited closest to the present); (3) proactive interference, which is the detrimental effect of some previously learned information affecting the ability to learn current information (e.g., the use of multiple radial-arm mazes with the rats); (4) retroactive interference, which is the detrimental effect of information being learned currently on information learned in the past (e.g., forgetting places you have been as new ones are visited); (5) decay, which refers to the progressive deterioration that occurs for memory as time passes from the point at which it was learned (e.g., imposing some delay interval between a learning experience and the test of that learning). Separating decay from types of interference is likely impossible. Consolidation is measured by the fact that memory is less subject to disruption as time passes from the point of learning.

Another series of learning studies has focused on aspects of foraging behavior, using rats in a laboratory setting that simulated the costs and benefits associated with food finding (Collier and Rovee-Collier 1982; Johnson and Collier 1987, 1989). Foraging costs in this test situation are simulated by bar-pressing requirements when a rat is placed into a series of test chambers that are much like Skinner boxes (see figure 11.4). Individual rats are given choices between test chambers where the size of the food reward and the cost, in terms of bar presses, can be varied. Interestingly, rats in this situation tend to take in a relatively constant amount of food each day, regardless of the variations in the test scheme. Rats did tend to take more meals in the more profitable patches, whether because of larger food pellets for the same effort or because of a lower effort required to obtain the food. These types of studies and those with the radial-arm maze represent novel approaches to questions concerning food sampling behavior and foraging patterns in animals, combining psychology, ecology, and animal behavior.

Social learning may also play a key role in the dietary selection and preferences of animals like rats. Learning about foods starts with young rat pups acquiring information about diet as we discovered in chapter 10 (Galef and Clark 1971). Additional studies by Galef and his colleagues have revealed that rats can learn to prefer or avoid foods based on watching a demonstrator rat (Galef et al. 1985), can transfer information about a distant food source (Galef et al. 1984), and will follow another rat to a food source more readily if the followers know the food source is palatable rather than toxic (Galef et al. 1987). In a most interesting development, Galef and collaborators (Galef et al. 1988; Galef 1996) determined that a chemical in the breath of rats, carbon disulfide (CS_2), acts as an enhancer during the process of learning about unfamiliar foods. For rats living in the wild, all of these processes could be critical for the development of food habits, acquiring new food habits, and avoiding potentially poisonous foods. Since rats are generally opportunistic feeders and have a relatively catholic diet, we can be fairly certain that social mediation of preference and avoidance is a significant feature in their overall dietary patterns.

Predator-Prey Relations

Both predators and prey may exhibit patterns of behavior that illustrate the importance of learning. In nature, animals probably learn a great deal about their surroundings during the course of their daily activities. Some of this information does not appear to have immediate functional value but may be important for survival later. Consider an example we mentioned earlier involving an owl as the predator and mice as the prey. Metzgar (1967) demonstrated how this process might work for the white-footed deermouse, *Peromyseus leucopus*. Mice in one test group were each given time to explore and live in a room containing features such as logs and trees. Mice in a second test group were held in laboratory cages without the experience in the test room. Over time, mice from each group were placed in the room with an owl present as a predator. Only two of twenty mice that had prior experience in the room were caught by the owl, whereas eleven of twenty mice that had no prior exposure to

FIGURE 11.16   Learning to be predators.
Young African hunting dogs begin obtaining knowledge about prey species while they are still living at the den. Later, they accompany groups of adults on hunting trips, and eventually they learn to participate in running down and capturing various prey such as the wildebeest shown here.
Source: © Harvey Barad/Photo Researchers, Inc.

the artificial habitat were captured. Spending time in the artificial habitat provided the mice in the first group with more knowledge of the habitat, enabling them to avoid becoming prey for the owl.

Wild hunting dogs (*Lycaon pictus*) in Africa live in social units involving several adults of each sex, juveniles, and pups (van Lawick-Goodall 1970). The dogs of a pack hunt cooperatively (figure 11.16), enabling them to run down and capture larger prey than any single individual could take on its own. Hunting in this manner also enables the pack to provide food for their young and other dogs that may remain at the den area during the reproductive season. As young hunting dogs develop, they can be observed joining first in the process of tearing apart and consuming prey animals that have been captured by other dogs from the pack and brought back to the den site. As the dogs grow a bit older and stronger, they venture out with the adult dogs and start to participate in the hunting process itself. Initially, this participation by younger dogs involves moving out with the adult dogs from the den area to locate possible prey. Later, the juveniles start running with the adults for portions of the chase, learning techniques such as selection of the prey, cutting a potential prey animal off from the herd, and running down the quarry by having dogs approach it from several directions. Eventually, the younger dogs become full participating members of the hunting process, and they learn to take portions of the kill back to the den area or to regurgitate food consumed at the site of a kill for animals that were not on the hunt. Similar learning sequences may occur in other related species that hunt cooperatively (e.g., wolves [*Canis lupus*] and spotted hyenas [*Crocuta crocuta*]) and in some cats (e.g., lions [*Panthera leo*]).

Reproduction and Parenting

A key process pertaining to reproduction for many animals involves the ability to locate conspecifics of the opposite sex. One way this can happen has already been explored in some detail in chapter 10, the phenomenon of sexual imprinting, a special form of learning. The cross-fostering technique has been used to test whether young rodents of various species learn cues that are important for species recognition at early ages. In many instances, when young rodents are transferred at or near the time of birth to a foster species and are later tested for species preference, they exhibit an enhanced selection of the foster species and a reduced selection of their natural species (reviewed by D'Udine and Alleva 1983). Among the species pairs that have been tested in this manner are southern grasshopper mice (*Onychomys torridus*) and white-footed mice (*Peromyscus leucopus*) (McCarty and Southwick 1977); montane voles (*Microtus montanus*) and gray-tailed voles (*Microtus canicaudus*) (McDonald and Forslund 1978); house mice (*Mus musculus*) and Norway rats (*Rattus norvegicus*) (Lagerspetz and Heino 1970); and pygmy mice (*Baiomys taylori*) and house mice (Quadagno and Banks 1970). Results from these studies are mixed; in some instances, the fostered animals prefer their own species when tested as adults, and in others, they prefer the foster species. In general, there are, in all instances, some effects of the fostering, but the degree to which preferences are affected varies.

Evidence from both primates and birds suggests that being a good parent and increased success in rearing offspring to independence can depend, in part, upon learning experiences. For some birds, for example, white-fronted bee-eaters in Africa (*Merops bulockoides*) (Emlen 1984; Emlen and Wrege 1988), and Florida scrub jays (*Aphelocoma coerulescens*) (Woolfenden and Fitzpatrick 1984), birds other than the biological parents may participate in the processes associated with rearing of a clutch. The helping behavior may include assisting with territory maintenance, guarding the nest, incubation, brooding, removal of fecal sacs, and feeding the young. Providing assistance means that the birds delay their own reproduction by one to several years. This topic is discussed in more detail in chapter 19, where we treat the evolution of social behavior.

The point here is that through the various learning experiences obtained by providing help at the nest, many birds gain experience that contributes to their own abilities to be good parents and to a higher rate of reproductive success when they become parents. The process of providing help may have evolved under a variety of selection pressures, but one outcome is that young birds have the opportunity to practice some aspects of parenting behavior before attempting to nest on their own.

In a similar way, a number of primate species exhibit behavior patterns in which individuals (usually females, but sometimes males) other than the biological mother provide temporary care for a developing individual (Quiatt 1979; McKenna 1981). This phenomenon has been termed "aunting" behavior or allomothering. Through handling and carrying an infant, other monkeys, often juvenile females or females nearing the age of first reproduction, will acquire some of the skills necessary for being a good parent. There is also evidence that, even with practice acquired through aunting behavior, for at least some primate species where data are available, infant mortality is high for first-or even second-time births. For example, in rhesus macaques, first-born infants have only a 74 percent chance of being alive through the first month and 55 percent chance of remaining alive to 6 months of age (Drickamer 1974b). These percentages increase only slightly for second infants. However, by the time the females are delivering and caring for offspring for the third time, the survival rate to 6 months of age has risen to 78 percent and by the fourth year to 91 percent. Thus, learning how to be a good, successful parent has a relatively high cost even in a social primate.

The foregoing are but a small selection of the many examples where learning plays a critical role in the survival and reproduction of various animal species. In chapter 14 on habitat selection, we will see in more detail how learning influences the habitat preferences of deermice (*Peromyscus maniculatus bairdi;* Wecker 1963) and chipping sparrows (*Spizella passerina;* Klopfer 1963). As part of their mode of orientation during migration (chapter 13), many birds use star patterns in the night sky. Studies using birds reared under star patterns that have been altered (e.g., in a planetarium) will orient differently from conspecifics reared under the natural night sky (Emlen 1970). Finally, many of the communication postures and signals to be discussed in chapter 12 are learned during the course of development.

ANIMAL COGNITION

The preceding section on learning as adaptive behavior leads us to the question of animal cognition. Cognition is presently used in two somewhat different ways by animal behaviorists. In one sense, **cognition** has been defined as a general term for mental function, including perception, thinking, and memory (Immelmann and Beer 1989) or as the study of the minds of organisms (Roitblat 1987). These notions encompass a more traditional definition and one that developed primarily from a psychological perspective; this has been called the anthropocentric approach to animal cognition (Kamil 1998). In recent years, cognitive ethology has grown tremendously as a subdiscipline within animal behavior (Balda et al. 1998; Dukas 1998; Shettleworth 1998). Investigators using this approach have adopted a perspective that the mental experiences and consciousness that characterize nonhuman animals can be explained by examining the impact of evolution on cognitive processes as well as the ontogeny and underlying neural bases for these processes; this has been termed the ecological approach (Kamil 1998). Examining a series of studies involving animal cognition will provide a better understanding of this "hot" topic.

Pigeons

An early example of this approach involves exploring the ability of pigeons to categorize or classify objects, based on the objects themselves or photographs (slides) of the objects (Herrnstein et al. 1976; Cerella 1979; Herrnstein and de Villiers 1980). The work begins with the premise that many of the objects used in discrimination learning are not natural to the normal world of the wild pigeon. At issue is whether an organism, like the pigeon, can classify objects that may be from its natural environment, a classification process like humans do when we place objects into open-ended categories at various levels, such as people, leaves, etc., or, on a more refined scale, into subcategories, such as gender of the person, type of tree from which the leaf came, etc.

The investigators devised a modified pigeon chamber in which pigeons were trained to peck a key when shown slides depicting objects from the natural world; in this way, their correct and incorrect choices could be recorded and rewarded as appropriate. Using this apparatus, Herrnstein, Loveland, and Cable (1976) demonstrated that pigeons could correctly classify trees versus things that were not trees (e.g., a clump of celery) and could properly select a particular female human versus, for example, another female in the apartment of the subject female, or the subject's husband wearing her scarf. Thus, pigeons can apparently use some set of rules, applied to static pictures, to properly classify them, fitting some portion of a definition of cognition.

Several follow-up studies lend support to the foregoing conclusion. One study involved the classification of silhouettes of tree leaves (Cerella 1979). Pigeons, reared in indoor lofts where they could not have had prior experience with

FIGURE 11.17   Leaf patterns.
Leaf patterns used in classification testing with pigeons. The top row represents oak leaves and the bottom row, non-oak leaves.

leaves, were trained to respond positively (by pecking) to oak leaves and not to respond when other leaf forms were presented (figure 11.17). The pigeons readily learned this classification task, spontaneously generalizing from one oak leaf to others and also generalizing to be able to reject leaves that were not from oaks. Interestingly, the birds were not capable of discriminating among leaves from different oaks. Cerella feels that such discrimination of classes may be advantageous because, in nature, pigeons might learn to associate the visual form of oak leaves with finding acorns for food. A second study (Herrnstein and de Villiers 1980) involved discrimination of slides of underwater scenes according to whether there were fish present or not. Fish are not part of the natural world of pigeons, yet they represent a naturally occurring, open-ended category. Again, the pigeons performed at a level significantly above chance. The more clearly the exemplars were fishlike, the more likely the pigeons were to make positive identification.

Chimpanzees

Beginning more than 30 years ago (Gardner and Gardner 1969, 1971, 1985), various investigators have used several techniques to explore the language capacities and cognition of chimpanzees (*Pan troglodytes*) (Premack 1971; Fouts 1972; Rumbaugh 1977; Savage-Rumbaugh 1987; Rumbaugh et al. 1994; see also chapter 12). The techniques used have included American Sign Language, wherein chimpanzees are taught hand gestures for various words, the use of a computer console with symbols standing for words or actions on the keys, and a series of plastic shapes that are symbols for various words and actions. For all of these techniques, chimpanzees have been able to demonstrate that they can generally use symbols for words; classify objects by type, color, and other attributes; deal with the concept of same versus different; convey information about the future and about their intentions; and form what some are willing to agree are rudimentary sentences following some rules for grammar and syntax.

The ability of chimpanzees (and other animals) to acquire and use various language-related skills is significant,

both with regard to what this tells us about their cognitive capacities and because we can begin a rudimentary form of communication with them. Similar work has been carried out using three other great apes: the bonobo (*Pan paniscus*) (Savage-Rumbaugh et al. 1986; Rumbaugh and Savage-Rumbaugh 1994), the gorilla (*Gorilla gorilla*) (Patterson 1978), and the orangutan (*Pongo pygmaeus*) (Miles 1990). Related work has been done on bottlenosed dolphins (*Tursiops truncatus*) (Herman et al. 1984; Herman and Uyeyama 1999) and California sea lions (*Zalophus californianus*) (Schusterman and Krieger 1986; Schusterman and Kastak 1998; Schusterman et al. 2000). Ongoing work on marine mammals provides evidence of their ability to classify objects and referential symbols, and also their capacity for demonstrating equivalency. This latter ability involves using knowledge about relationships between A and B and between B and C to derive knowledge about a relationship between A and C.

Parrots

Pepperberg (1987a,b, 1990, 1994a, 2000; Pepperberg and Kozak 1986; Pepperberg and Brezinsky 1991) has worked for over a decade with an African grey parrot named Alex (*Psittacus erithacus;* figure 11.18) and more recently with two additional grey parrots (Pepperberg et al. 1998). In this case, the investigator is able to actually communicate with the subject using human vocal language. Alex is first trained to use a code, in the form of spoken English sounds, as referents for objects, actions, and attributes. We can then begin to explore his cognitive abilities. Pepperberg defines cognition as the processing stimuli and choosing to react in certain ways to those stimuli (Pepperberg 1998a). Processing stimuli involves two steps. The first step is the ability to use experience to solve a current problem. The second step involves the ability to select from among the previously acquired experiences to come up with the appropriate information to solve the problem. It is this latter ability that Pepperberg feels distinguishes the mental activity as cognition. Alex demonstrates his cognitive ability, for example, when he transfers information between modalities. The key here is not so much the language capacity of the parrot, but rather the utilization of the language channel as a means to explore the cognitive capacities of the parrot.

Alex has learned various attributes (e.g., color, shape, size, material) for more than 100 objects (Pepperberg 1998b, 2000). He can be asked relatively straightforward questions about the particular attributes of an object; in this sense, he is able to categorize the attributes of the exemplar using the word labels, for example, for color (green, red, yellow, blue, gray, orange, and purple). Or, of more interest here, he can be asked what is the same or what is different between two objects. Thus, if presented with two wooden cubes of the same color, but different sizes, the correct response could be to indicate that they differed with respect to the size attribute. He is also able to respond by indicating two objects are the same when their attributes match. Further, Alex demonstrates a degree of numerical competence. Once Alex had

FIGURE 11.18 **Alex, an African grey parrot.**
Alex has been used in a series of studies of cognition involving interspecific vocal communication with his trainers.
Source: Courtesy Irene M. Pepperberg.

been trained to indicate vocally the numbers of objects in groups of up to six homogenous objects presented simultaneously, he was then able to identify objects in heterogeneous groupings without further training. For example, when given a group of four items that varied with respect to two attributes (e.g., green and purple papers and chains of paper clips), Alex could then properly tell the trainer how many objects were defined by the unique conjunction of one color and one object category. Throughout most of the studies with Alex, his level of correct responses has varied around 80 percent, indicating a high degree of competence.

Vervet Monkeys

Another manifestation of cognition involves social relationships. This has been explored using vervet monkeys (*Cercopithecus aethiops*) (Cheney and Seyfarth 1985). Members of a social troop of this species recognize individual members of the group, dominance relationships, and matrilineal kin relationships. Both standard observational methods and playback experiments have been used to elucidate the recognition capacities of the vervets (Seyfarth et al. 1980; Cheney and Seyfarth 1982). The system apparently builds, beginning with recognition between parents and offspring. Developing young then extend their knowledge base to include other troop members, based on both kinship and, probably via observational learning as well as their own social interactions, dominance relationships among members of their troop (Cheney 1983). Gradually, as they mature, the scope of knowledge about relationships extends to members of other troops, particularly those in contiguous areas of habitat (Cheney and Seyfarth 1982, 1990). Using these data and similar evidence from other monkeys and apes, several primate researchers (Chance and Mead 1953; Jolly 1966; Humphrey 1976; Cheney, Seyfarth, and Smuts 1986; Byrne and Whiten 1988; Cheney and Seyfarth 1990) conclude that primate intelligence and the related cognitive processes evolved in a social context because of selection pressures rooted in greater reproductive success for monkeys that could learn, remember, and use information about social relationships. A similar line of reasoning has been used in recent work on spotted hyenas (*Crocuta crocuta*) in which recorded whoop calls from cubs were played back to mothers or to other breeding females (Holekamp et al. 1999). Responses were significantly greater to calls from the mothers' own cubs than to the calls from cubs that were not their own. In general, hyenas that were related to the cub whose call was played back responded more than individuals who were not related. Hyenas, however, differed from vervets used in similar tests in that the other breeding females did not look at the mother whose cub's call was played back; hyenas did not recognize a third-party relationship.

Corvids

The cache recovery capabilities of three species of corvids have been carefully examined (Balda and Kamil 1985; Balda and Kamil 1988, 1989; Marzluff and Balda 1992). Their basic hypotheses were that learning and spatial memory are important facets of the process of cache recovery and that these capacities might differ across the three species of corvids because of the differences in their habitat ecologies and the importance of storing and remembering cache locations. The experiments were conducted in a large room using captive adult birds taken from the wild. Birds from all three species were provided with individual sessions during which time they were permitted to obtain seeds from a central feeder and cache them in a series of holes in a raised floor in the test room. The test room also contained rocks, tree branches, and other landmarks. Birds were provided with habituation sessions so that they could become accustomed to caching in the holes in the floor rather than in the more customary crevices and holes in vertical surfaces; most birds learned this process without difficulty, though a few did not and were not tested further. After each caching session, the bird was removed from the room and returned to its home cage.

Seven days later, with one day of food deprivation, the birds were permitted back into the room with seeds placed in exactly the same holes where they had been cached by the bird in its previous session. The cache recovery efficiency of each bird was measured during the second test session. Cache recovery rates were higher than expected by chance for all three species, indicating that the birds do learn and retain characteristics of the cache sites where they put seeds. The recovery percentages were significantly higher for the nutcrackers and the pinyon jays than for the scrub jays. These differences in recovery rates may correlate with the natural history of the birds and with possible differences in spatial memory capabilities.

Balda and Kamil (1992) addressed the question of long-term spatial memory in Clark's nutcrackers using a seed-caching apparatus similar to that employed in their earlier work. All of the birds were placed individually in the test chamber and allowed to cache seeds in the floor, which contained 330 holes. The 25 test birds were then divided into four groups of 6 to 7 birds each. To determine retention of the spatial memory for where seeds were located, birds in each group were tested individually with seed caches arranged in the floor as they had left them. One group was tested at 11 days, and others at 82, 183, and 285 days after the initial caching experience. All of Clark's nutcrackers performed at levels significantly above random chance with respect to relocating seeds they had cached. There was, however, a higher error rate for the birds that were tested at 285 days. Overall, birds of this species appear well adapted in terms of a long-term spatial memory that permits relocating food resources that have been stored up to 9 months prior to their retrieval. This finding was supported for Clark's nutcrackers and extended to pinyon jays, scrub jays, and Mexican jays (*Aphelocoma coerulescens*), though the latter two species performed with only moderate accuracy regardless of the length of the delay between caching and recovery (Bednekoff et al. 1997).

In addition to the work on corvids, studies on spatial memory and food storing have been conducted on several other avian species. Black-capped chickadees (*Parus atricapillus*) and white-breasted nuthatches (*Sitta carolinensis*) can accurately recall the locations of caches (Petersen and Sherry 1996; Sherry 1998). Shettleworth (1995) summarized the available information on birds that store food and their ability to locate such caches. The hippocampus in the brain plays a significant role in spatial memory (Nadel 1991), and this certainly seems to be true for birds (Clayton and Lee 1998). Shettleworth and Hampton (1998) hypothesize that the size of the hippocampus relative to the telencephalon and to overall body size in birds is a function of the use of visual spatial memory. Thus, in birds that store food we would expect to find hippocampi that are relatively larger in size. The hippocampus and a related area called the parahippocampus are larger in both volume and in the number of neurons in food-storing birds than in those that do not store foods (Krebs et al. 1989; Sherry et al. 1989).

Further evidence for the relationship between hippocampal size and food storing comes from two studies. Lesions of the hippocampus in black-capped chickadees result in decrements in retrieval of storage caches (Sherry and Vaccarino 1989; Hampton and Shettleworth 1996). Experience with storing and retrieving food leads to increased number of neurons in and increased volume of the hippocampus in marsh tits (*Parus palustris*) that are given experience with storing and retrieving food (Clayton and Krebs 1994).

Song Learning

We have previously discussed the processes involved in song learning (chapter 10). Avian singing behavior is also an excellent system to explore aspects of animal cognition. There is wide variation in the underlying processes of song learning, significant variation in the roles of environmental input affecting song production, and different functions are served by songs in different social and ecological contexts

(Baptista et al. 1998; Todt and Hultsch 1998; Kroodsma and Byers 1998). Singing behavior provides us with a window on avian cognition.

Marsh wrens (*Cistothorus palustris*) have repertoires of 100–200 songs, which they use in lengthy singing bouts interacting with their neighbors (Verner 1976; Kroodsma and Verner 1997). Some of these interactions involve countersinging, in which two males alternate songs, the second bird matching the song of the first bird. In these interactions one bird typically is the leader or dominant and the second bird is the follower or subordinate. Additional work is needed to assess what behaviors are involved with these complex singing interactions and to determine how the singing relates to fitness. Several aspects of this singing process are indicative of the types of underlying cognitive processes used by these birds, including memorizing and remembering songs, categorizing songs, and discriminating which neighbor(s) are singing which songs.

Work on song sparrows (*Melospiza melodia*) has produced additional evidence concerning the cognitive processes that birds use in their singing and related behavior patterns (Stoddard et al. 1988, 1990, 1991; Podos et al. 1992; Beecher et al. 1996, 1998). Male song sparrows disperse within about one month of fledging. When they arrive at their new area, they memorize the songs of the males in that neighborhood. Their song-learning process is effectively delayed until after they have dispersed. The next breeding season they establish a territory in this new location, finalize their own song repertoire of 6–10 song types, and then generally remain there for the remainder of their life. Field studies have demonstrated some remarkable capabilities by these territorial male song sparrows. These males can classify songs either by the song type or according to the bird singing the song. In the latter instance, they also know the location of the singer—that is, they know who is in which territory (figure 11.19). When a song sparrow responds via song to a neighbor's song, he does so with a song type that is most like part of the repertoire of that particular neighbor, a phenomenon called repertoire matching. For song sparrows, females prefer males with larger song repertoires (Searcy 1984).

Another aspect of the study of avian cognition through song involves our extensive knowledge about the neural pathways that are involved (DeVoogd and Székely 1998). Investigations in this area have produced two types of results. First, there are detailed focal studies of particular birds and the brain regions involved in song production between sexes or for males during their annual cycle of reproduction. These studies provide evidence concerning the neural structures and specific pathways for both the input and storage of information from songs, and the sequence of steps involved in the motor control of song production. Second, comparative studies across different avian taxa can tell us a great deal about relationships involving similarities and divergences of neural anatomy and the varied functions of songs in different groups of birds. We are just at the beginning of a period where the integration of knowledge about the behavioral ecology (function and evolution) of bird songs can be related to the

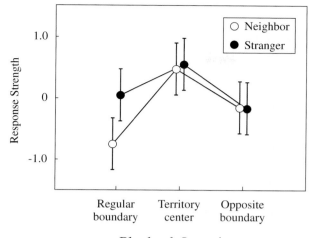

Playback Location

FIGURE 11.19   Song sparrow playback responses.
The playback technique was used with 14 male song sparrows to assess their responses when calls from either neighbors or strangers were played at one of three spatial locations. These locations consisted of (*a*) the regular boundary between the territories of the two males, where the song would normally be heard, (*b*) the center of the territory of the bird being tested with the playback song, and (*c*) a boundary opposite to that where the neighbor's song would normally be heard. The values shown are mean intensity responses with two standard errors above and below the mean; response intensity is a combination of the flights toward and the closeness of the approach to the playback speaker. The greatest intensity of responses occurs when the playbacks are at the center of the territory of the bird being tested. A strange bird's song at a normal location receives a greater response than the "normal" song of the neighbor at the customary boundary. Both the stranger and the neighbor receive responses when the songs are played at the opposite boundary.

Source: Figure from P. K. Stoddord et al. 1991. *Behavioral Ecology and Sociobiology* 29:211–15.

anatomy and physiology (mechanism and development); the next decade should be extremely exciting in this regard. In large measure, it has been the examination of animal cognition that has formed a basis for this bridging of ultimate and proximate causation of behavior.

Self-Recognition

Self-recognition has been considered a component of cognition in that it may indicate self-awareness. Gallup (1970, 1983) has explored this idea using chimpanzees. When chimpanzees are provided with a mirror located in close proximity to their cage, they initially engage in a number of social responses (e.g., head-bobbing, threats, etc.) toward the mirror. Eventually, they appear to be able to recognize the image as their own. One way to test this involves painting one eyebrow and the top of the opposite ear with a red dye. After putting on the dye, observations were first made of self-directed, particularly dye-directed, behavior without the mirror present. When the mirror was returned, the chimpanzees exhibited increased levels of activity directed to the dye-marked areas. Gallup tested several monkey species (*Macaca* spp.) and gorillas that did not exhibit any sort of self-awareness in the presence of a mirror.

Consciousness is also considered a part of cognition. Two books by Griffin (1981, 1984) have spurred considerable thinking about whether animals are, in fact, capable of self-awareness and about their thinking processes and consciousness. These books and related work have kindled an interest in a variety of studies, many of them under field conditions, which attempt to explore aspects of the possible mental processes in animals during the course of activities, such as foraging for food, social play, or avoiding predation.

Together, these various studies, from a number of different perspectives and involving an assortment of animal species and techniques, have begun to tell us a great deal about animal cognition. It seems likely, in light of the growing body of evidence, that Macphail's null hypothesis (1985) is not true—that is, there are species differences in intelligence and quite likely with regard to various learning processes. A commitment to studying learning phenomena in both laboratory and field settings will produce exciting new findings in the next decade. Moreover, with the progress being made in neuroscience, we can predict that there may be some genuine ties between what has been investigated from a learning and intelligence perspective and the actual neural and molecular processes that occur within the organism.

SUMMARY

Learning is the relatively permanent change or potential for change of behavior that results from experience. Habituation is the relatively persistent waning of a response that results from repeated stimulus presentations not followed by any form of reinforcement. In classical or Pavlovian conditioning, a neutral stimulus is paired with an unconditioned stimulus (US) to elicit an unconditioned response (UCR). After repeated pairings, the neutral stimulus becomes a conditioned stimulus (CS) and elicits the conditioned response (CR). In operant conditioning, sometimes called instrumental learning, the animal learns to associate a behavior (e.g., performing a particular action or task) with the consequences of that behavior. Modern conceptualization of classical conditioning

embodies new considerations concerning contiguity, information, inhibition, and salience. Among the many similarities in the processes of classical and operant conditioning and the procedures used to investigate them are acquisition, schedules of reinforcement, extinction, spontaneous recovery, and generalization. They differ in that classical conditioning involves a stimulus-stimulus pairing, and the animal does not actually control the sequence of events; whereas for operant conditioning, there is a stimulus-response pairing, and the sequence of events is contingent upon the animal's behavior and controlled by the investigator.

Other types of learning related to operant conditioning (instrumental learning) include latent learning, observational learning, and possibly imprinting. When an animal is presented with a series of similar problems and tested sequentially, the animal may develop a learning set.

A phylogenetic survey of the capacities of various groups of invertebrates to demonstrate various types of learning reveals some useful patterns. Protozoans and cnidarians exhibit habituation, but not classical or operant conditioning. Flatworms (*Platyhelminthes*) and segmented worms (*Annelida*) are capable of habituation responses and also exhibit both classical and operant conditioning in some test situations. Trails of chemical cues left by the organisms may be important for learning in these groups. Molluscs and arthropods are also capable of habituation, as well as classical and operant conditioning. Honeybees exhibit learning paradigms that are remarkably like those recorded for vertebrates. Very few species have been tested from each invertebrate phylum, and our knowledge is thus quite limited regarding the overall phylogeny of learning.

Vertebrates have been tested more extensively, but even within this phylum, there are large gaps in our knowledge; rats, pigeons, and monkeys have been used for a disproportionate amount of the research on vertebrate learning. There are no clear relationships between phylogenetic lineage and learning capacities among the vertebrate classes. Vertebrates, along with some invertebrate organisms, do exhibit rapid avoidance learning in feeding situations. One hypothesis concerning vertebrate intelligence, termed the null hypothesis, states there are no real differences, quantitatively or qualitatively, in terms of learning and related phenomena in vertebrates. Considerable data on comparative learning in vertebrates seem to run counter to this notion, though it has been useful for stimulating thinking and research. Exploratory behavior may form an important part of vertebrate learning phenomena.

Studies of comparative learning have been both quantitative and qualitative. Quantitative comparisons are impeded by individual differences in learning and motivation and by species differences in the form of biological constraints on learning. An animal presented with a particular learning situation may be prepared, unprepared, or contraprepared. Preparedness is defined in this context as the genetically based predisposition to exhibit certain behavior. Methods constraints in the study of learning make it impossible to equate the test conditions for each species. Also, we must be constantly aware of variations in sensory and perceptual worlds among different species.

Learning serves as the basis for adaptive behavior through which each animal is able to meet the challenges of everyday life and to cope with the problems of survival and reproduction. The types of learning involved in the natural lives of animals are exemplified by caching behavior in birds, acquiring food habits in rats, wild hunting dogs learning to kill prey, mice learning about refuges to avoid predation, learning species-specific communications signals for mate finding, and practicing to become a good parent.

Animal cognition, defined as the mental experiences and consciousness that characterize nonhuman animals, can be explained by examining the impact of evolution on these processes as well as the ontogeny and underlying neural bases. Pigeons exhibit the capacity to categorize objects from their natural world. Language-related phenomena in chimpanzees, other apes, some sea mammals, and African grey parrots all provide evidence of cognitive processes in vertebrates, particularly with regard to phenomena related to language. Data from several primate species indicate that the social context and social relationships may have been very important with regard to evolutionary selection pressures influencing the development of cognition among primates. Some species of corvids are capable of remembering the locations of seed caches over many months. Song learning in birds provides excellent examples of the integration of an understanding of the behavioral ecology and the underlying neural substrates for behavior via a perspective based on animal cognition. Finally, self-recognition has been used as a means of exploring self-awareness.

DISCUSSION QUESTIONS

1. Two important aspects of a young animal's life are its diet and its selection of a place to live. For each of the organisms listed below, indicate what type of learning processes are involved in acquisition of food habits and habitat selection: (a) minnow, (b) cardinal, (c) sea anemone, (d) chipmunk, (e) gypsy moth larva. How would you proceed to test your hypotheses with experiments?

2. In this chapter, we have examined how hypotheses about learning can be tested in applied or natural settings. Think of several ways you might test learning behavior in a natural setting using field methods for each of the following: (a) cockroach, (b) mountain goat, (c) crow, and (d) gorilla.

3. Imagine that we have just located a new species of butterfly. We are interested in obtaining an accurate assessment of the sensory/perceptual world of this organism and its learning capacities. Describe the procedures you would use to conduct the experiments necessary to answer this question.

4. One way to look at learning is to view it as a complex of adaptive behaviors that provide an organism with the flexibility and capacity to deal with the problems of survival and reproduction. How would you defend or contradict this statement: Invertebrates have less flexibility and a greater degree of preprogramming in their behavior than do vertebrates, and we would therefore expect invertebrates to exhibit fewer learning capacities than vertebrates?

5. The data in the following histograms are from the work of Galef and Clark (1971) regarding learning of dietary habits in rats. Depicted here are the feeding bouts and approaches by wild rat pups, starting at age 21 days, directed toward two diets, A and B, placed 5 cm apart. The parents of the test animals had been previously poisoned while feeding on diet B with the pups present. What conclusions can you draw concerning the dietary preferences of these pups? Of what adaptive value might their responses be in a natural setting where young rats are learning what to eat and what not to eat?

6. Draw together materials from several sections of this chapter that pertain to the importance of definitions, procedures, and constraints on learning and the study of learning. Make a list of these and come up with at least one additional example, beyond those in the text, for each of these potential pitfalls.

7. We have explored animal cognition in several ways in this chapter, using a variety of animal examples. With those studies in mind, what sorts of cognitive processes would you wish to explore in the following organisms? With these studies in mind, what sorts of cognitive processes would you wish to explore in the following organisms: (a) honeybees, (b) birds that make long-distance flights as in annual migrations, (c) decisions involved in foraging behavior in sharks? How would you proceed experimentally to test your ideas?

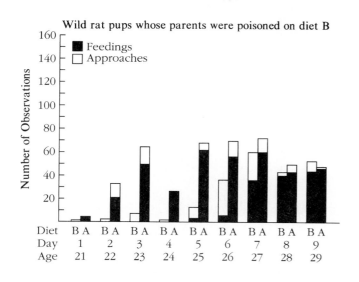

Wild rat pups whose parents were poisoned on diet B

SUGGESTED READINGS

Balda, R. P., I. M. Pepperberg, and A. C. Kamil, eds. 1998. *Animal Cognition in Nature.* San Diego: Academic Press.

Colgan, P. 1989. *Animal Motivation.* New York: Chapman and Hall.

Domjan, M. 1983. *The Principles of Learning and Behavior,* 4th ed. Pacific Grove, CA: Brooks/Cole Publishing Company.

Domjan, M. 1980. Ingestational aversion learning: Unique and general processes. *Adv. Stud. Behav.* 11:276–336.

Dukas, R., ed. 1998. *Cognitive Ecology.* Chicago: University of Chicago Press.

Galef, B. G. Jr. 1996. Social learning: Introduction. In C. M. Heyes and B. G. Galef Jr., eds. *Social Learning in Animals.* San Diego: Academic Press.

Hinde, R. A., and J. Stevenson-Hinde, eds. 1973. *Constraints on Learning.* New York: Academic Press.

Rescorla, R. A. 1988. Behavioral studies of Pavlovian conditioning. *Ann. Rev. Neurosci.* 11:329–52.

Roitblat, H. L. 1987. *Introduction to Comparative Cognition.* New York: W. H. Freeman.

Schwartz, B. 1989. *Psychology of Learning and Behavior.* New York: W. W. Norton.

Shettleworth, S. J. 1998. *Cognition, Evolution, and Behavior.* New York: Oxford University Press.

12

CHAPTER

Communication

 his final chapter on mechanisms of behavior sets the stage for the study of social interactions that comprise most of the chapters in part five. Animals convey information to members of their own species, and to other species as well, through an incredible diversity of sounds, colors, flashing lights, smells, and postures. The song of a male white-crowned sparrow (*Zonotrichia leucophrys*), for example, is specific not only to the species but also to an area and to the individual. By singing, the male may not only be warning neighboring males away, but may also be providing potential mates with information about his health, his social status, and even his place of birth.

We begin this chapter by defining communication and by examining how signals are coded to convey information. Next, we consider the kinds of information conveyed, the sensory channels through which the signals are received, and discuss the evolution of conspicuous body structures and the ritualization of behavior. Finally, we look at examples of symbolic "language" in nonhuman animals.

From extensive field observations, ethologists have found that often behavior occurs in regular patterns that recur nearly unchanged from event to event. These sequences are called **fixed action patterns (FAPs)**. The most striking FAPs are those that occur when animals interact with conspecifics in courtship, territory defense, or dominance encounters. Signaling is the primary function of these behaviors. Julian Huxley (1914) studied great crested grebes and called their strange postures **displays**. Moynihan (1956) defined display as any behavior pattern especially adapted in physical form or frequency to function as a social signal. Ethological interest focused on the ways in which displays evolve from other, noncommunicative, behavior patterns.

WHAT IS COMMUNICATION?

Wilson (1975) defined **biological communication** as an "action on the part of one organism (or cell) that alters the probability pattern of behavior in another organism (or cell) in a fashion adaptive to either one or both of the participants." The word **adaptive** implies that the signal or response is to some extent genetically controlled and under the influence of natural selection. But this definition presents some difficulty, as Marler (1967) pointed out: What about the mouse that rustles in the grass, making sound that enables the owl to catch it? This case fits Wilson's definition, but would we really say that the mouse is communicating with the owl? One way around this problem might be to add that the sender must *intend* to alter another's behavior; clearly, the mouse did not mean to attract the attention of the owl. But now we have another problem: How can we ever tell what an animal intends?

Who benefits from communication—the sender, the receiver, or both? Animal behaviorists have disagreed on this issue also. One classification scheme considers the value of any information communicated to both sender and receiver (Wiley 1983) (table 12.1).

1. Both sender and receiver benefit = true communication

2. Sender benefits, receiver is unaffected or harmed = manipulation or deceit

3. Sender is unaffected or is harmed, receiver benefits = eavesdropping or exploitation

4. Neither sender nor receiver benefits, both may be harmed = ignoring or spite

One view of communication has been that displays evolve in a way that maximizes the effectiveness of information transfer between sender and receiver to the benefit of both, referred to in table 12.1 as "true communication." Smith (1984) defined communication as "any sharing of information." Emphasis here is on the **coevolution** of signals among interacting organisms, Smith's definition is unsatisfactory under some circumstances, such as when sender and receiver do not both benefit as when a parent bird uses a broken wing distraction display to lead a predator away from her young. Thus, Slater (1983) defined communication as "the transmission of a signal from one animal to another such that the sender benefits, on average, from the response of the recipient." Thus, his definition includes both true communication and manipulation (or deceit) in table 12.1.

Still other views of communication stress that natural selection acts primarily at the level of the individual, and so both sender and receiver need not benefit. Often the sender manipulates others for his or her own benefit, as in advertising. The receiver may benefit or may be harmed. Displays evolve so that the sender can control the behavior of the receiver with a minimum of wasted energy (Dawkins and Krebs 1978, Krebs and Dawkins 1984). In this view, the purpose of the display is not so much to inform as to persuade. Exaggeration and redundancy are the rule; displays evolve to increase their persuasive power, not to maximize information transfer. Eavesdropping, where the receiver benefits but the sender does not, is also considered a type of communication, as is deceit. In this context, the sound of the mouse rustling in the grass would be a form of communication; in this instance, exploitation because the owl harms the mouse (table 12.1). These and other definitions of communication are reviewed by Hauser (1996), and Bradbury (1998).

Deceit, or sending signals that have a negative value for the receiver, might be expected both among and within different species. Interspecific deceit is rather common, particularly in predator-prey relationships, as we shall see later in this chapter and also in chapter 15. Intraspecific deceit seems to be much less common, possibly because the sender and receiver belong to a common gene pool (Dawkins and Krebs 1978). If a mutant "liar" with gene A appears in a population of nonliars (gene A), it may increase its fitness by deceiving conspecifics. As gene A spreads, the likelihood of deceiving another bearer of gene A increases. If bearers of gene A can recognize and lie only to bearers of gene a, gene A will become fixed in the population. The lying habit would eventually cease, however, because there would be no one left to deceive. Lying is an advantage only when most others tell the truth. Selection also would favor the ability to detect deceit by others and thus keep lying relatively rare.

TABLE 12.1   Terms used to describe interactions, depending on the value of the information to sender and receiver

Sender Value of Information	Receiver Value of Information	
	Positive	**Zero (or negative)**
Positive	True communication	Manipulation (deceit)
Zero (or negative)	Eavesdropping (exploitation)	Ignoring (spite)

Source: Data from R. H. Wiley, The evolution of Communication: Information and Manipulation, *Animal Behavior*, 2:156-89,1983.

One way to ensure that signals are honest is if they are costly to produce. If a signal costs more to produce than could be gained by sending phony information, then faking would not pay and signals would tend to be reliable. This notion forms the basis of the Handicap Principle, put forth by the Zahavi and Zahavi (1997). They argue that the cost of the signal, that is, the handicap borne by the sender, is what guarantees that the signal is reliable. When two individuals square off in a fight, threats usually precede any actual combat. The Zahavis argue that threats reliably indicate the sender's motivation to follow through. Often the threat consists of exposing vulnerable body parts, demonstrating that the threatening individual is confident of its ability to win and willing to risk injury. Thus, two dogs preparing to fight stand side by side, bodies stretched and hair raised. The traditional interpretation of such behavior is that it makes the animal appear bigger than it really is. But the Zahavis argue that the stretched body communicates confidence and willingness to invest in a fight, if necessary.

HOW DO SIGNALS CONVEY INFORMATION?

We saw in chapter 7 that information is coded in the nervous system by frequency modulation of electrical impulses. The information in displays is coded in a much more diverse fashion, as we shall see. A **signal** is the physical form in which a message is coded for transmission through the environment.

Discrete and Graded Signals

Some signals are **discrete** (digital or all or none) but others are **graded** (analog). For example, equids, such as zebras, communicate hostile behavior by flattening their ears and communicate friendliness by raising their ears (discrete signals) (figure 12.1). The intensity of either emotion is indicated by the degree to which the mouth opens (graded signal). The mouth-opening pattern is the same for both hostile and friendly behavior. Graded signals thus vary in intensity in proportion to the strength of the stimulus.

Distance and Duration

Some signals use the modality of smell. In some cases, the amount needed to trigger the receiver's response is no more than a few molecules, as is that of the sex attractants of many insects. Thus, the distance a signal travels may vary. A male several kilometers downwind can detect a small amount of material produced by a female. At the other extreme, visual displays usually operate over much shorter distances. The duration of a signal may also vary. Alarm signals, such as the chemicals produced by many invertebrates, may have a localized short-term effect and thus a rapid fade-out time. Bright plumage and other male adornments such as antlers may last the entire breeding season. In contrast, although the brightly colored epaulets on the red-winged blackbird are

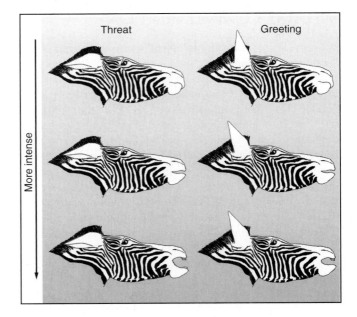

FIGURE 12.1   Composite facial signals in zebras.
Ears convey a discrete signal. They are either laid back as a threat or pointed upward as a greeting. The mouth conveys a graded signal and opens variably to indicate the degree of hostility or friendliness.
Source: After Trumler, E. 1959. Das "Rossigkeitsgesicht" und ähnliches Ausdrucksverhalten bei Einhufern. *Zeit. für Tierpsychol.* 16:478–88.

present during the entire breeding season, they are conspicuous only when the male exposes and erects these feathers during the song spread.

Composite Signals, Syntax, and Context

Although most species are limited to 20 to 40 different displays (Wilson 1975), signals can vary in several other ways that increase information content. Two or more signals can be combined to form a **composite** signal with a new meaning. In the zebra example, the meaning of the open mouth depends on whether the ears are forward (friendly) or backward (hostile) (see figure 12.1).

Animals can convey additional information with a limited number of displays by changing the **syntax,** or sequence of displays. For example, the two signals A and B would have different meanings depending on whether A or B came first. There is no evidence of natural syntax use in nonhuman animals; however, language learning in chimpanzees, in which chimps assemble words in novel ways as they communicate with humans, has been demonstrated (Rumbaugh and Gill 1976; Savage-Rumbaugh and Brakke 1990).

The same signals can have different meanings depending on the **context;** that is, depending on what other stimuli are impinging on the receiver. For example, the lion's roar can function as a spacing device for neighboring prides, as an aggressive display in fights between males, or as a means of maintaining contact among pride members. The song-spread

FIGURE 12.2 **Metacommunication in dogs.**
The play bow performed by the dog on the right communicates that behaviors that follow are play.
Source: Courtesy Marc Bekoff.

display of the male red-winged blackbird serves in courtship with females, as well as in conflict with other males.

Metacommunication

Increasing the information content of displays by **metacommunication,** or communication about communication, is theoretically possible: one display changes the meaning of those that follow. We can see good examples in play behavior: animals use aggressive, sexual, and other displays in play, but they precede such behavior by an act that communicates the message, "What follows is play, join in" (Bekoff 1977). Canids such as dogs and wolves precede play with the play bow (figure 12.2). Monkeys communicate play behavior through a relaxed, open-mouthed face.

We should be careful not to use the term metacommunication too loosely (Smith 1984). For example, male rhesus monkeys (*Macaca mulatta*) sometimes communicate dominance by carrying the tail elevated in an **S** shape over the back. While this posture conveys status and mood, it does not really change the meaning of behaviors that follow; rather, it communicates that aggressive behavior is likely to follow. Thus, this would not be a good example of metacommunication.

FUNCTIONS OF COMMUNICATION

Signals may be classified by their function. The ultimate function of any communication is increased fitness; this will be examined more closely in the section on the evolution of displays. Although many criteria can be used to classify proximate functions of communication, these classifications tend to be artificial and arbitrary and are made mainly for researchers' convenience to help keep things organized. The following functions of communication are modified from Wilson (1975), Smith (1984), and Bradbury and Vehrencamp (1998).

Group Spacing and Coordination

Group-living animals use a variety of signals that seem to keep members in touch. The highly arboreal *Cebus* monkeys of the South American rain forest forage in dense vegetation. A group of 15 may spread out over an area 100 meters in diameter as they search the treetops for fruits. In addition to the sound of moving branches, an observer hears a continual series of contact calls from the different members of the group. An individual that becomes isolated utters a "lost" call, which is much louder than the contact calls. Marler (1968) suggested that primates use the following types of spacing signals: (1) distance-increasing signals, such as branch shaking, which may result in another group's moving away; (2) distance-maintaining signals, such as the dawn chorus of howling monkeys (*Alouatta* spp.), which regulate the use of overlapping home ranges; (3) distance-reducing signals, such as the contact or lost calls of *Cebus* monkeys; and (4) proximity-maintaining signals, such as those that occur during social grooming within groups.

Color patterns of some coral reef fish are species-specific; these patterns attract conspecifics and hold them together in schools. Social insects, such as termites, ants, and bees, emit a variety of chemicals that result in assembly of conspecifics (Wilson 1971).

Recognition

Species Recognition

Species recognition before mating is assumed to be crucial to avoid infertile matings between members of closely related species. Note the striking differences in the courtship songs of three species of picture-winged Hawaiian *Drosophila* (figure 12.3). Their sound-producing mechanisms differ radically from those of mainland populations because they use their abdomens as well as their wings (Hoy et al. 1988). Differences among species do not necessarily mean that they evolved for

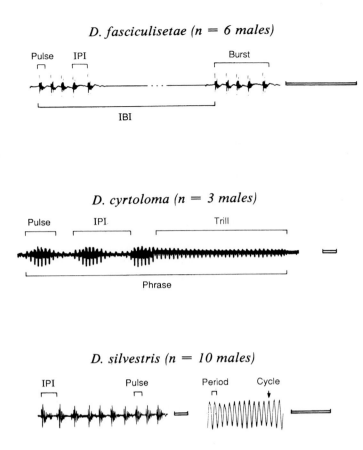

FIGURE 12.3 **Courtship songs of three species of picture-winged Hawaiian *Drosophila*.**
Oscillogram tracings of songs are shown. Note the large differences among species.
Source: R. R. Hoy, A. Hoikkala, and K. Kaneshiro, "Hawaiian Courtship Songs: Evolutionary Innovation in Communication Signals of *Drosophila*" *Science*, 240:218, copyright 1988 by the AAAS.

the purpose of species recognition, however. In this instance the songs may have evolved by means of sexual selection, where females choose males based on certain attributes of the song. In one of the Hawaiian species (*D. silvestris*), wing vibrations may serve as a sexual signal: the duration of wing vibration is positively associated with successful courtship (Boake and Hoikkala 1995). Females of other *Drosophila* species avoid mating with males whose wings have been clipped (Wilkinson 1987). In *D. silvestris,* however, clipping the male's wings does not seem to affect mating, and thoracic vibrations are thought to be important instead (Boake and Poulsen 1997). Thus, female choice for certain aspects of the male courtship may be responsible for generating song differences. The differences among species might secondarily come to serve as species-isolating mechanisms, reducing or eliminating matings among closely related species.

Zahavi and Zahavi (1997) also have argued that signals evolve through competition among individuals to demonstrate their quality, rather than because of an animal's need to identify other members of its own species. Smith's classic study (1966) demonstrated that species of Arctic gulls (genus *Larus*) use markings on the head to identify species. One

species has an eye ring, and when Smith painted some individuals' eye rings black to match the rest of the head, they were treated as members of a species that lacks an eye ring. However, current function says little about prior evolutionary history, and these species differences may have come about by means of sexual selection.

Deme Recognition

Local dialects in bird song have been demonstrated for a number of geographically separated populations of white-crowned sparrows. Females from one population in Colorado perform copulation-solicitation displays when they hear songs of males from their own population but rarely do so when they hear songs from another population (Baker 1983; Baker, in press) (figure 12.4). This mating preference may lead to mating with members of locally adapted demes. The question of why individuals should show a preference for mates from the same deme will be considered in chapter 14.

Class Recognition

Class recognition occurs mainly in social insect groups in which castes are treated differentially. In many species the nest queen receives preferential treatment (food and care) from workers because of pheromones she produces. Males are discriminated against as a group; they receive less food from the workers and, in times of food scarcity, are even driven from the colony (Wilson 1971).

Neighbor Recognition

Most animals stay in one area for much of their lives, and thus contact the same individuals repeatedly. It would be a waste of time to respond to the same known neighbors. The playback experiment has been a critical tool in the analysis of communication used in recognition, particularly for bird song. Researchers record vocalizations with a special directional microphone and a high-quality tape recorder, and loop the tape for repeated playbacks, placing the loudspeaker directly in the field. Observers score the responses of resident birds by the number of times they sing and approach the speaker before, during, and after playback.

To test whether white-throated sparrow males recognize neighbors individually or as a class, Falls and Brooks (1975) studied the effect of playback location on the resident male, moving the speaker in and around his territory. When they placed the speaker at the boundary between his territory and his neighbor's, his response to a stranger's song was stronger than his response to his neighbor's, as expected. But when they placed the speaker on the opposite boundary, he responded vigorously to the neighbor's song as if it were a stranger's. When they placed the speaker in the center of his territory, his response to the neighbor's song was intermediate. These results demonstrate that a territorial male can discriminate among the songs of his neighbors and can also associate each song with its singer's appropriate location. Similar results were reported for the European robin (*Eritha-*

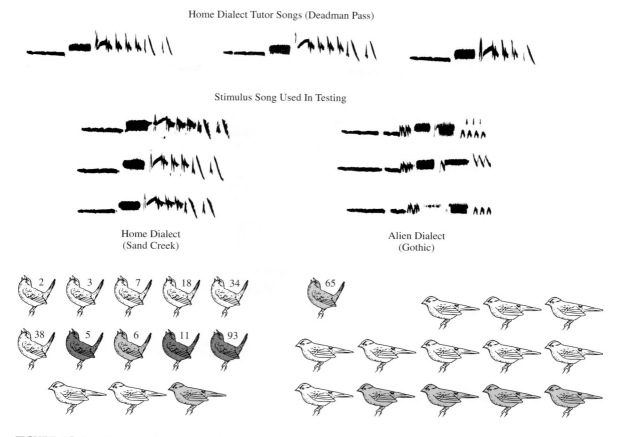

FIGURE 12.4   **The role of early experience in the response of female sparrows to male song.**
(a)Sound spectrograms of home-dialect tutor songs and stimulus songs in white-crowned sparrows. The spectrograms lie in the frequency range 3 to 6 kHz (vertical axis) and are about 2 seconds long (horizontal axis). (b)Postures showing the response of each female subject to the two stimulus song dialects. Juveniles are unshaded and adults are shaded. The lordotic posture is the copulation-solicitation display. Above each female is indicated the number of displays elicited from the subject during a 21-minute test. Juvenile and adult females responded more to the home dialect than to the alien dialect.

Reprinted with permission from M. C. Baker et al., "Early Experience Determines Song Dialect Responsiveness of Female Sparrows" in *Science*, 214: 819-21 (13 Nov 1981). Copyright 1981 by American Association for the Advancement of Science.

cus rubecula), a species with a much larger and more complex vocal repertoire (Brindley 1991).

Banner-tailed kangaroo rats (*Dipodomys spectabilis*) are solitary, nocturnal rodents that defend large dirt mounds in which they store seeds. One means of defending these mounds is by footdrumming, where they hit their hind feet on the ground and produce both airborne and seismic vibrations (Randall 1994). Footdrums occur in short bursts that vary in length and frequency, and individuals retain the same drumming pattern for life. Territory owners responded by footdrumming much more vigorously to the recorded sounds of strangers than they did to those of neighbors.

Kin Recognition

Communication may be involved in the differential responses of many organisms to their close relatives. For example, tadpoles of the American toad (*Bufo americanus*) prefer to associate with siblings over nonsiblings, even after being reared in isolation (Waldman 1982). Waldman hypothesized that some substance, contributed by the mother in the

egg jelly, was used as a cue. Other data from insects, amphibians, birds, and mammals demonstrate kin recognition even in the absence of interactions with kin early in life (Holmes and Sherman 1983). Chemical, auditory, and visual cues have all been implicated.

How can kin identify each other even if they have never interacted? One possibility is **phenotype matching,** where the individual "refers to" the kin whose phenotypes are learned by association—the referent. The referent is then compared with the stranger (Holmes 1986; Holmes and Sherman 1983). For example, female sweat bees (*Lasioglossum zephyrum*) guard the nest entrance and admit only nestmates (usually their sisters). In a laboratory experiment, Buckle and Greenberg (1981) reared guard bees either with groups of sisters or with groups of nonsisters; these bees thus became the referents of the guard bees. In later tests, the guard bees admitted strangers, but only if they were sisters of the referents (figure 12.5). A second possibility for kin recognition—one that does not require any previous experience—is the existence of "recognition genes" that enable the bearers to recognize the same genes in others (Hamilton

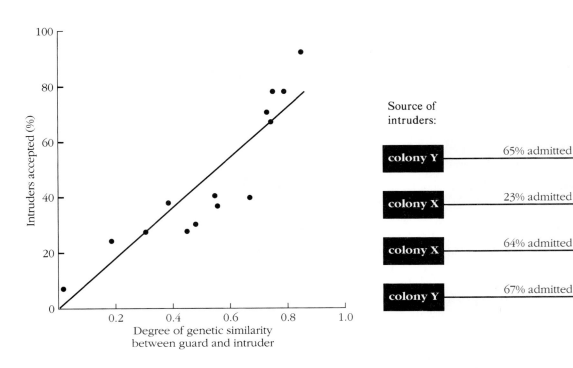

FIGURE 12.5   Nest guarding in sweat bees.
(a) Percentage of intruders accepted as a function of relatedness. (b)Female sweat bees were reared in nests of six bees composed either solely of sisters (colony X and colony Y) or of three sisters from one nest and three sisters from another (colony XY). Later, bees from X, Y, and XY colonies were tested as guards as shown; arrows represent unfamiliar intruders seeking entrance to the nest.
Source: (a) L. Greenberg, "Genetic Component of Bee Odor in Kin Recognition," *Science*, 206(1979):1095–97, copyright 1979 by the AAAS.

1964). This mechanism is called "**the green beard effect**" since such a trait could be selected for if genes controlling it confer not only the green beard, but also the preference for green beards on others.

Cues used to assess genetic relatedness may come from genes in the major histocompatibility complex (MHC), a cell recognition system that is used by the immune system to identify self and nonself. In rodents, and possibly other mammal species, genetic differences in this region produce urinary odor cues that can be used by whole organisms to recognize genetic similarities in other individuals (Brown and Eklund 1994; Yamazaki et al. 1976). House mice (*Mus domesticus*) and some other mammals, including humans, may use this information to choose genetically dissimilar individuals as mates. Such choices would lead to production of heterozygous offspring that would be more resistant to disease. Choosing genetically dissimilar mates would also result in less inbred offspring (Penn and Potts 1999). For a further discussion of kin recognition, see chapter 19.

Individual Recognition

Indigo buntings (*Passerina cyanea*) produce a complex song that is quite variable. Most of the phrases are paired (sweet-sweet, chew-chew, and so forth) (figure 12.6). Emlen (1972) analyzed tape recordings of the song and spliced pieces of the tapes together in varying orders. When he played them back in the field and observed the responses of territorial males, he discovered the significance of much of the song.

Part of the sequence was species-specific, communicating the message, "I am an indigo bunting," and part was variable from individual to individual. Changing the ordering of notes (syntax) or eliminating one member of each pair did not greatly affect the response of males. Changing the rhythm or the frequency of the notes themselves reduced the agonistic responses of resident birds, however, apparently because the sounds were no longer recognized as those of other indigo buntings.

When wild bottlenose dolphin (*Tursiops truncatus*) mothers were played the recorded signature whistles of their independent offspring, they responded more strongly than to whistles from a familiar, similar-aged offspring (Sayigh et al. 1998). Furthermore, the offspring responded more strongly to whistles from their mothers than to those from familiar, similar-aged females. These playback experiments suggest that the variation evident in these sounds functions in individual recognition.

Reproduction

Many of the most striking displays occur during courtship. Receptive females or males may advertise their condition, court a member of the opposite sex, form a bond, copulate, or perform postcopulatory displays. Although these behavior patterns may lead to correct species identification, assessment of a potential mate's condition and coordination of the neuroendocrine system between the sexes are other possible functions. Many closely related species that have overlap-

SONGS   RESPONSES

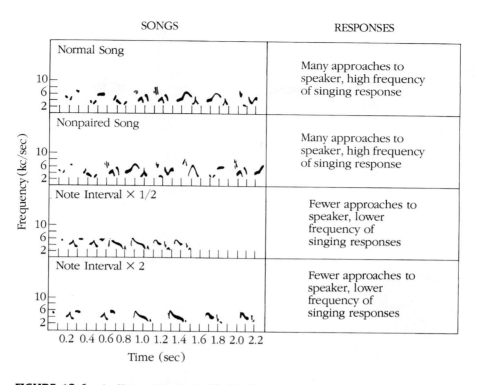

FIGURE 12.6   **Audiospectrograms of indigo bunting.**
Tapes of a normal song were cut and spliced together to test the function of various attributes of the song. The normal song and the nonpaired song elicited typical responses, but temporal changes, such as reducing the time between notes by one half or increasing it twofold, interfered with song recognition.

ping home ranges also have very distinct courtship patterns, as seen in the Hawaiian *Drosophila* (see figure 12.3).

We have also seen, below the species level, how female white-crowned sparrows preferentially solicit copulations from males of their own population (Baker 1983) (see figure 12.4). Another example is the different mating calls of two populations of cricket frogs (*Acris crepitans*) just 65 km apart in central Texas (Ryan and Wilczynski 1988). Females showed a strong preference for the calls from males of their own population. Electrophysiological studies also demonstrated that the basilar papillae of the inner ear are most responsive to sounds that match the local dialect. Thus, communication between sender and receiver has coevolved, and the two populations may be on their way to becoming reproductively isolated. The role of communication in reproduction is discussed more fully in chapters 17 and 18.

Agonism and Social Status

In social groups, when members of a species are in close proximity, it is sometimes beneficial for an individual to compete and fight with others for possession of a resource, be it food, space, or access to another individual. Physical combat is expensive in terms of energy; it increases the risk of death or injury, even for winners. Social species have evolved displays that communicate information about an individual's mood (recall the tail of the rhesus monkey) and the way that individual is likely to behave in the near future.

As a result of previous encounters, the animal may be dominant or submissive, and its behavior may thus be predictable. Presented with a limited resource, the submissive individual will yield to the dominant without an overt fight. In many cases, competition among members of a species is rare; thus, agonistic interactions make up only a small proportion of their behavioral repertoire. Dominant, aggressive displays tend to be opposites of submissive displays, exemplifying the principle of antithesis first mentioned by Darwin (1873) (figure 12.7).

Alarm

Animals use vocalizations and chemicals to alert group members to danger. Although male song varies greatly within and among closely related species, sympatric species (those with overlapping ranges) are likely to have simple, hard-to-locate alarm calls that differ little among species. Members of species that live together, and that are endangered by the same predators, benefit mutually by minimizing divergence in alarm vocalizations (Marler 1973). For example, Marler was unable to tell the difference between the "chirp" alarm calls of African blue monkeys (*Cercopithecus mitis*) and red-tailed monkeys (*C. ascanius*), whereas it was easy for him to differentiate the male songs of the two species—they differed greatly. Vervet monkeys (*Chlorocebus aethiops*) communicate **semantically** by using different signals to warn about different dangers in their environment.

FIGURE 12.7   Threatening and submissive postures in the dog.
Note the features that are opposites, such as ear and tail positions, shape of spine, and general posture.
Source: Charles Darwin, *On the Expression of the Emotions in Man and Animals,* copyright 1873, D. Appleton, New York.

FIGURE 12.8   Belding's ground squirrel giving alarm call.
This conspicuously calling squirrel, a lactating female, is more likely to be attacked by a predator than is a noncaller. Nearby squirrels benefit since they can remain hidden or take cover. Females with mothers, sisters, or offspring in the vicinity are most likely to call.
Source: Photo by George D. Lepp, courtesy of Paul Sherman.

Group members climb trees when they hear alarm calls given in response to leopards, they look up when they hear eagle alarms, and they look down when they hear snake alarms (Struhsaker 1967). Young vervets give alarm calls in response to a variety of animals, and their ability to classify predators and give appropriate alarm calls improves with age (Cheney and Seyfarth 1990; Seyfarth et al. 1980).

Both invertebrates and vertebrates produce chemical alarm substances. Sea urchins of the species *Diadema antillarium* move rapidly away from an area containing a crushed member of their own species (Snyder and Snyder 1970); earthworms (*Lumbricus terrestris*) also produce alarm pheromones (Ressler et al. 1968). Mice and rats excrete a substance in their urine when they are given electrical shocks, are beaten up by another animal, or otherwise stressed. This substance acts as an alarm and may cause others to avoid the area (Rottman and Snowden 1972).

Tracing the evolution of alarm calls presents a challenge to biologists because such signals seem unlikely to benefit the caller. Sherman (1977) studied individually marked Belding's ground squirrels (*Spermophilus beldingi*) (figure 12.8) and found that whenever a terrestrial predator (such as a weasel or a coyote) was spotted, the calling squirrel stared directly at the predator while sounding the alarm. Sherman suggested many hypotheses to explain this behavior, two of

which we will discuss. First, the predator may abandon the hunt once it is spotted by the potential prey (in this hypothesis, the caller is behaving selfishly). Second, others in the area may benefit from the warning, even though the caller may be harmed (in this hypothesis, the caller is behaving altruistically). Sherman demonstrated that callers attract predators and are more likely to be attacked after calling, and thus they are not behaving selfishly.

Because he kept records on mothers and offspring, Sherman knew that the males leave the area several months after birth and that the females are sedentary and breed near their birthplaces. He also found that adult and yearling females are much more likely to call than would be expected by chance, and that males are less likely to do so. Furthermore, females with female relatives in the area (such as mothers or sisters, but not necessarily with offspring) call more frequently in the presence of a predator than those with no female relatives living nearby. Sherman concluded that the most likely function of the alarm call is to warn family members. The behavior is phenotypically altruistic (reducing direct fitness) but genotypically selfish (increasing indirect fitness), and evolution by kin selection is indicated (see chapter 19).

Models have been useful in the study of the function of alarm calls, as seen in work on postures of brent geese (*Branta bernicla*) (Inglis and Isaacson 1978). When alarmed by sudden auditory or visual stimuli, geese adopt the extreme head-up posture (figure 12.9). When "flocks" of decoys were placed in grain fields where geese had been causing damage, real flocks' responses depended on the decoys' postures. In general, the real flocks avoided fields with a high proportion

Head Down: attractive

Head Up: aversive only in presence of extreme head-up shapes

Extreme Head Up: aversive

FIGURE 12.9   Models of brent goose postures.
When "flocks" of geese models are placed in grainfields, real geese avoid fields containing models in the extreme head-up (alarm) posture.
Source: Data from I. R. Inglis and A. J. Isaacson, "The Responses of Dark-Bellied Brent Geese to Models of Geese in Various Postures," *Animal Behavior* 26:953–58, 1978.

of extreme head-up decoys, while they were attracted to fields with mostly head-down decoys.

Finding Food

An important selective advantage of group living is increased efficiency in finding food. Many species of birds breed in large colonies or spend the night in communal roosts. One possible advantage of such behavior is that roosts serve as **information centers** (Ward and Zahavi 1973). A bird might learn the location of a rich food source by following other birds the next morning. Individuals going to rich food sources must be identified, however, and it is not clear how such information is communicated. Cliff swallows (*Hirundo pyrrhonota*) breed in colonies. Individuals that have been unsuccessful foraging for insects are more likely to follow successful foragers on subsequent trips (Brown 1986). In this instance the information may be gained by observing successful foragers feeding their offspring, a form of eavesdropping.

A number of species emit specific calls when they find food, especially primates and gallinaceous birds (Hauser 1996). In domestic chickens (*Gallus gallus*), roosters emit a characteristic pulsatile call when food is discovered, and they are more likely to call in the presence of hens (Evans and Marler 1994). If hens are in the vicinity, they come running, and may accept food from the male, who may then court them.

The African wild dog (*Lycaon pictus*) is a canid distantly related to domestic dogs and wolves. Just prior to hunting prey that often consists of large ungulates, members of the pack engage in an intense greeting ceremony, or "rally," consisting of a frenzy of nosing, lip licking, tail wagging, and circling (Creel and Creel 1995). Such behavior seems important in coordinating activities. Social carnivores may even communicate information about what type of prey they are about to hunt. Kruuk (1972) noted that hyenas hunting zebra on some occasions passed by prey they had hunted on other occasions.

Chimpanzees (*Pan troglodytes*) communicate the location of food and may actually lead others to it (Menzel 1971). Menzel removed captive chimps from their home pen. He then showed one chimp where food was hidden in the home pen and allowed that chimp to lead the others to it. No special signals were used; rather, the leader moved toward the hidden food purposively, looking back at the others periodically. The information transferred seemed to be, "something in those bushes ahead has aroused my expectations of edibles."

In an advancing wave of army ants, movements of the swarm are influenced by tactile stimulation with antennae and by the laying down of a pheromone trail. Having discovered a food source, a scouting forager dashes back and forth between the nearest raiding column and the food; in this way 50 to 100 ants are recruited to the food source within the first minute (Chadab and Rettenmeyer 1975; Franks 1989). Perhaps the most sophisticated communication system about food is that of honeybees, discussed later in this chapter. For further discussion of feeding behavior, see chapter 15.

Giving and Soliciting Care

A wide variety of signals is used between parent and offspring and among other relatives in the begging and offering of food. As Tinbergen (1951) demonstrated, the red spot on the herring gull's lower bill stimulates and directs a pecking response by the chick, and the chick's resultant pecking of the parent's beak stimulates the parent to regurgitate the food. Distress calls by the young are individually recognizable by the parent once the young are capable of leaving the birth site. When they are chilled, baby mice produce high-frequency sounds that are inaudible to humans, but audible to adult mice, who can then assist the young mice.

Soliciting Play

As we discussed earlier, play consists of behavioral patterns that may have many different functions in the adult: sex, aggression, exploration, and so forth. The play bow in canids (see figure 12.2) is communication about play and informs others that the motor patterns that follow are not the real thing. The function of play itself is a subject of debate. Observers usually agree that they can recognize play, but that they have had great difficulty ascribing definitions or functions to it. They most often suggest that the function of play is to help develop motor skills and behavior patterns used later in adult life (Bekoff and Byers 1998; Fagen 1981) (see also chapter 10).

Synchronization of Hatching

Precocial birds, such as pheasants and ducks, lay large clutches of eggs. Synchronous hatching is very important because to feed or to get into the water to escape predation by terrestrial vertebrates, the mother leaves the area with all her young following. Species that nest in tree holes leave the area permanently the day the chicks hatch. Late-hatching chicks are vulnerable to predation. A few days before hatching, the chicks begin to vocalize. This communication contributes to hatching synchrony, since similarly aged eggs incubated separately hatch over a period of several days (Vince 1969).

CHANNELS OF COMMUNICATION

The sensory channel is the physical form used to transmit the signal from sender to receiver. From an evolutionary perspective, there are costs and benefits associated with different channels, depending on the environment and the information being transmitted. Table 12.2 summarizes costs and benefits for four major channels.

Odor

From an evolutionary standpoint, the earliest type of communication was chemical, for odor is used throughout the animal kingdom, except by most bird species. Most pheromones are involved in mate identification and attraction, spacing mechanisms, or alarm. The greatest amount of research has been done on insects and mammals. There are several probable reasons why the widespread use of chemical signals has evolved: such signals can transmit information in the dark, can travel around solid objects, can last for hours or days, and are efficient in terms of the cost of production (Wilson 1975). However, because these pheromones must diffuse through air or water, they are slow to act and have a long "fade-out" time.

We have a relatively good understanding of insect pheromones. For example, the sex attractant bombykol, produced by the female silk moth, has been isolated; and we are reasonably familiar with the male's perceptual system (see chapter 7). A single molecule of bombykol triggers a nerve impulse in a receptor cell on a male's antenna. About 200 receptor-cell firings in one second lead to a behavioral response (Schneider 1974). The male responds by flying upwind, equalizing the pheromone concentration on both antennae, until he reaches the female. To control such pests as the gypsy moth in the northeastern United States, traps are baited with commercially synthesized pheromone to lure males. We mentioned earlier how social insects in the family Hymenoptera make extensive use of pheromones for class and kin recognition (Wilson 1971).

Mammals also use pheromones extensively; many of the recent studies have been done on rodents. Two general classes of substances that differ in effect have been identi-

fied: **priming pheromones** produce a generalized response, such as triggering estrogen and progesterone production that leads to estrus; and **signaling pheromones** produce an immediate motor response, such as the initiation of a mounting sequence. Bronson (1971) suggested that pheromones in mice can be classified by function:

- Priming pheromones
   Estrus inducer
   Estrus inhibitor
   Adrenocortical activator
- Signaling pheromones
   Fear substance
   Male sex attractant
   Female sex attractant
   Aggression inducer
   Aggression inhibitor

Examples of priming pheromones are substances excreted in male mouse urine that speed up maturation in young females (Lombardi and Vandenbergh 1977; Vandenbergh 1969). Other urinary products inhibit aggression, increase aggression, stimulate the adrenal cortex, block implantation of embryos, and so on. It is not clear how many different chemicals are involved; different functions may be served by the same pheromone (Drickamer 1989). The sources of these products include the sexual accessory glands and a number of specialized skin glands.

Many of the pheromones produced by mammals function as a means of staking out territories or home ranges, as does bird song. The advantage of pheromones, as mentioned previously, is that the odor may last for many days and nights; this advantage is significant because many mammals are nocturnal and visual signals are of little use. Since these substances are often associated with the urinary and digestive systems, eliminative behavior is often highly specialized. Hyena clans mark the boundaries of their territories by establishing latrine areas. Clan members defecate simultaneously in an area, then paw the ground. The feces turn white and become quite conspicuous. In the same or in another area, hyenas engage in "pasting." Both sexes have two anal glands that open into the rectum just

TABLE 12.2   General properties of the major sensory channels of communication

Signal Property	Sensory Channels			
	Olfactory	Auditory	Visual	Tactile
Range	long	long	medium	short
Transmission rate	slow	fast	fast	fast
Travel around objects?	yes	yes	no	no
Night use?	yes	yes	little	yes
Fade-out time	slow	fast	fast	fast
Locate sender	difficult	varies	easy	easy
Cost to send signal	low	high	medium	low

Source: Adapted from Alcock, *Animal Behavior: An Evolutionary Approach*, 4th edition, 1989, Sinauer Associates, Inc.

FIGURE 12.10   Pasting by spotted hyena.
Anal gland secretions are deposited on a grass stem.
Source: Photo by Kay Holekamp.

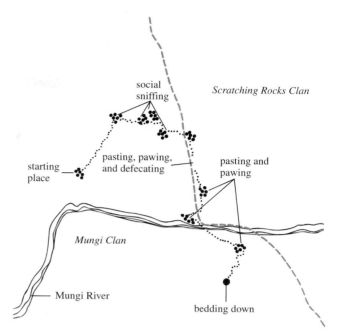

FIGURE 12.11   Activities of the Mungi hyena clan along its boundary.
On a particular evening six members of the clan left the starting place or "club" where they had been sleeping and moved toward the boundary with the Scratching Rocks clan, sniffing along the way. Once at the boundary they sought out tall grass stems for pasting and pawing. Two members of the clan also defecated at a site that many others had used previously. After more pasting and pawing, the clan bedded down, having covered about two kilometers in one hour's time. Members of one clan rarely trespass into the territory of another clan.

Source: Kruuk, H. 1972. *The Spotted Hyena: A Study of Predation and Social Behavior.* Chicago: University of Chicago Press.

inside the anal opening. When pasting, the hyena straddles long stalks of grass; as the stems pass underneath, the animal everts its rectum and deposits a strong-smelling whitish substance on the grass stems (Kruuk 1972) (figures 12.10 and 12.11).

Mule deer (*Odocoileus hemionus*) and other cervids produce pheromones from the tarsal and metatarsal glands on the hind legs, from the tail, and from urine. The leg is rubbed against the forehead; odors are then transmitted from the forehead to twigs (Muller-Schwarze 1971). It turns out, however, that the primary source of the tarsal gland scent is urine and that the gland has to be recharged about once a day with urine, a behavior pattern referred to as rub-urination (Sawyer et al. 1994). Apparently, fat-soluble pheromones present in the urine dissolve in sebum from the tarsal gland and convey information about individual identity and social status. Olfactory investigation in ungulates is characterized by **flehmen,** a retraction of the upper lip exhibited soon after sniffing the anogenital region of another or while investigating freshly voided urine (Estes 1972). This behavior pattern is especially common in males during the breeding season, but also occurs in females.

Sound

Information about immediate conditions can be transmitted faster by sound than by chemicals. A single organ can produce sound, it can travel around objects and through dense vegetation, and it can be used in the dark. Both frequency and amplitude modulation can convey information. The best frequency for an animal to use seems to depend on the environment. Brown and Waser (1984) demonstrated that there is a **sound window** for blue monkeys (*Cercopithecus mitis*) living in the forests of Kenya and Uganda (figure 12.12). At around 200 Hz, sounds are attenuated very little and are relatively unaffected by background noises. This frequency corresponds to that of the "whoop gobble" call produced by the adult male, who has a special vocal sac. Blue monkeys also seem to be unusually sensitive to sounds in this range compared to their more terrestrial cousins, the rhesus monkeys. Howling monkeys (*Alouatta* spp.) in the neotropical rain forest also signal to other groups with low-frequency calls. Animals with smaller home ranges, such as squirrel monkeys (*Saimiri sciureus*), use higher-frequency sounds, which dissipate rapidly. Such calls serve to maintain contact among group members.

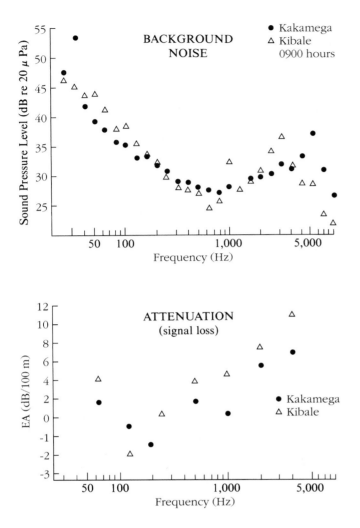

FIGURE 12.12   Sound window in the forest.
(a) Background noise levels measured in two East African forests at 9:00 A.M. (b) Excess attenuation (EA) as a function of signal frequency. Sound travels farthest in the 100–300 Hz range.

Source: Brown, C. H., and P. M. Waser. 1984. Hearing and communication in blue monkeys (*Cercopithecus mitis*). *Animal Behavior* 32:66–75.

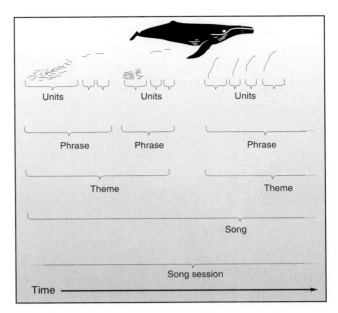

FIGURE 12.13   Song of humpback whale.
The song of a humpback whale can be broken up into units, phrases, themes, songs, and song sessions. Each whale sings its own variation of the song, which may last up to a half hour.

Source: Data from R. S. Payne and S. McVay, "Songs of Humpback Whales," *Science*, 173: 585–97 (13 August 1971). Copyright 1971 by the AAAS.

Ultrahigh-frequency sounds are used by a variety of animals, particularly by mammals. The distress calls of young rodents and some of the vocalizations of dogs and wolves are well above the range of human hearing, as are the echolocation sounds of bats. Although bat sounds are used mainly to locate food objects, communication also occurs between predator and prey. Noctuid moths, for example, possess tympanic membranes on each side of the body that receive sonar pulses from bats (Roeder and Treat 1961). Depending on the location and intensity of sound stimulation, the moth may fly away in the opposite direction, dive, or desynchronize its wingbeat to produce erratic flight.

Underwater sound has properties somewhat different from sound in air. Fish and invertebrates produce a variety of sounds, some of which have only recently been investigated. Marine mammals produce clicks, squeals, and longer, more complex sounds that incorporate many frequencies; the short-duration sounds are thought to function in echoloca-

tion. Baleen whales (suborder Mysticeti) produce lower and longer sounds than do the toothed whales, such as dolphins. Payne and McVay (1971) analyzed the sounds of the humpback whale (*Megaptera novaeangliae*), which vary individually, occur in sequences of 7 to 30 minutes' duration, and are then repeated (figure 12.13). The songs are associated with courtship, are sung by males, and apparently function much as does bird song (Tyack 1981). In spring, humpbacks congregate near Bermuda and sing basically the same song. Each year they tend to add new components and subtract old ones, so that the song changes gradually through the years (Payne and Payne 1985).

Low-Frequency Sound and Seismic Vibrations

Substrate-borne vibrations, referred to as seismic vibrations, may be used for communication by the white-lipped frogs (*Leptodactylus albilabris*) studied in the rain forests of Puerto Rico (Lewis and Narins 1985). Males calling from solid substances make thumps; females were found to be extremely sensitive to these vibrations.

Other species produce very low frequency sounds that may also have seismic properties. Desert rodents, such as kangaroo rats, have greatly inflated middle ears, which make them sensitive to low-frequency sounds and vibrations. As we saw previously, bannertail kangaroo rats (*Dipodomys spectabilis*) defend their territories by footdrumming (Randall 1984), producing sounds in the 200–2000 Hz range. Both Asian elephants (*Elephas maximus*) and African elephants (*Loxodonta africana*) use very low frequency rumbles in the range of 14–35 Hz to communicate over distances

LEARNING THE SECRETS OF INTERSPECIFIC COMMUNICATION

Up on a stage in front of a packed auditorium, Patricia McConnell conducts a simple demonstration. Before her is a wriggling young dog who periodically jumps up to try to lick her hands. First, Patricia shows the audience an ineffective way to teach a dog not to jump: she reaches toward the jumping dog with her hand to push it down. It's natural for primates to push, and it's a signal we readily understand. However, to a dog, outstretched front legs are an invitation to play. That's the message this dog is getting, and it wags its tail even harder and jumps again.

Next, Patricia demonstrates how to make this cross-species communication more effective. Dogs that want some personal space use a shoulder slam or body block to get it. This time when the dog jumps, Patricia swings her hips into it, keeping her hands at her side. The dog quickly gets the message, and soon sits back on its haunches when Patricia simply leans toward it. The dog rapidly learned what is wanted because the signal was naturally understandable.

Patricia McConnell, Ph.D., is a Certified Applied Animal Behaviorist and an Adjunct Assistant Professor at the University of Wisconsin, Madison. She runs a highly successful dog training business and hosts a nationally syndicated radio show on applied animal behavior. Her success comes from this key insight: it is vital to take into account the natural behavior of an animal in order to communicate with it effectively.

Auditory communication has been one of her main interests. For an undergraduate honors project, she analyzed the voice and whistle commands handlers give to sheepherding dogs to tell them to walk toward the sheep, circle, lie down, etc. Handlers had told her that it's not important which whistle is used for which behavior, as long as the sounds are easy to distinguish. However, when Patricia sorted the sonographs (pictures of the whistles) into piles in categories based on what the command meant, her eyes showed her patterns that her ears had not caught: certain types of commands shared the same structure. As she was thinking about her data while riding

horseback, she had another epiphany when she said "Whooooa" to the skitterish horse: long, continuous notes are used to soothe many animals, and short, repeated notes (like clicking to a horse) are used to make them speed up. She remembers the joy of the discovery: "The clouds parted and the angels sang!"

In her dissertation, Patricia looked across cultures. She recorded more than 105 animal handlers that spoke 19 different languages in communicating with rodeo and draft horses, obedience and sled dogs, camels, yaks, guard geese, and cats, among others. Across these languages she found the same patterns: short, repeated notes meant speed up ("Kittykittykitty!"). Slow, unmodulated notes were used to soothe, and single, modulated notes ("Whoa!") were used to stop animals that were already active. Could it be that mammals are predisposed to respond in particular ways to these sounds? Patricia tested this idea by training naive Border collie and beagle puppies to either four short tones with a rising frequency that meant "come," and one long tone with a descending frequency that meant "stay," or vice versa. As predicted, a command with four short tones was more effective than the long signal at eliciting approach and increasing motor activity.

Many people assume that they will automatically know the best way to communicate with their dog, but interspecific communication is more difficult than one might imagine. As Patricia says, "Imagine doing an ethogram on humans as an alien species, and figuring out what something as simple as a smile means: it could be joy, nervousness, tension, or something else." A dog faces a similar problem as it tries to understand its trainer.

The business Patricia started with $100 when she was fresh out of graduate school has grown into a success. She says it is more intellectually and emotionally challenging than she would ever have dreamed to apply her animal behavior degree in this way, and she is as proud of her abilities as an animal trainer as she is of her Ph.D. Days when she can solve a behavioral problem and prevent a dog from being put down are satisfying indeed.

---

of several kilometers (Payne et al. 1986; Poole et al. 1988). Such infrasonic sounds have very high pressure levels (over 100 decibels) and suffer little environmental attenuation. The main uses of these calls seem to be for long-distance coordination of group movements and location of mates.

Underwater vibrations are detected by some organisms. Male water mites (*Neumania papillator*) court females by vibrating their front legs while resting on vegetation (Proctor 1991). Females orient to these vibrations much as they would to those of their crustacean prey (figure 12.14). In this way the male takes advantage of the female's sensitivity to vibration in order to entice her to pick up spermatophores (packets of sperm) he has deposited on the substrate.

Ripple communication has been described in a number of aquatic and semiaquatic insect species (Wilcox 1995). Males of one species of water strider (*Gerris remigis*) send out ripples of a certain frequency, and receptive females respond by moving toward the source. When a female gets within a certain distance, the male switches to courtship (Wilcox 1972; 1979). Wilcox also observed that males generate high-frequency (HF) waves when they are close to another water strider. If return HF waves are not picked up from the second strider, the first attempts copulation. In order to demonstrate the function of HF signals, Wilcox used an ingenious playback method to pro-

gram females to send out HF signals. He glued a magnet to the female's foreleg and allowed her to move freely inside an electrical coil. When Wilcox played an electrical copy of the male HF signal through the coil, the magnet moved the female's leg, and she involuntarily sent out the HF signal. After Wilcox placed a rubber mask over a male's eyes to eliminate possible visual cues, that male always attempted to copulate when it approached a female not sending an HF signal and never attempted to do so when the female did send an HF signal. We might apply this type of playback experiment to studies of other types of substrate-transmitted signals, such as web communication among spiders.

Males of some orb-weaving species of spiders use vibrations to court females (Barth 1982). Once at the periphery of a female's web, the male attaches a special mating thread to it and vibrates the web to bring her out for mating. The male is at risk since the resident female is capable of attacking and eating him.

Touch

Many invertebrates, whose receptor-covered antennae are the first part of the body to contact other objects and organisms, use short-range communication in the form of physical

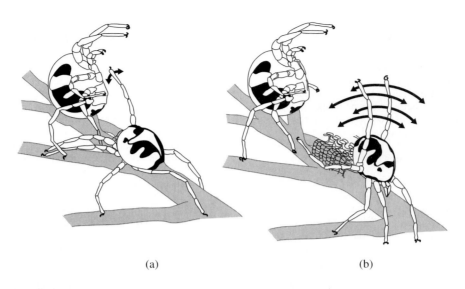

(a)   (b)

FIGURE 12.14   Underwater courtship by water mites (*Neumania papillator*).
(a) the male (right) vibrates his foreleg in front of the female, who is in the upright net stance assumed when catching food. (b) The male is now fanning with his legs over the spermatophores he has just deposited. This species does not copulate; sperm transfer occurs when the female picks up spermatophores.
Source: Redrawn from Proctor, H. C. 1991. Courtship in the water mite *Neumania papillator:* Males capitalize on female adaptations for predation. *Animal Behavior* 42:589–98.

contact. Antennae are used by subsocial insects, such as cockroaches, and by social insects, such as bees. The honeybee often performs the waggle dance in a dark hive. Therefore, much of the information about the type and location of food comes from tactile communication as the workers' antennae contact the dancer and, in the process, pick up taste cues.

Social insects, such as some bees, wasps, ants, and termites, have chemical and tactile signals that regulate feeding of larvae. The mutualistic relationship between some species of ants and aphids involves stereotypic tactile communication. The ants provide protection and the aphids repay them with secretions of honeydew (Wilson 1971, 1975.)

Perhaps the most widespread use of tactile stimuli occurs during copulation. In many rodents, stimulation of the back end of an estrous female produces concave arching of the back and immobility (lordosis). In some mammals, including cats, vaginal stimulation induces ovulation.

In most primates, grooming is an important social activity (figure 12.15) and seems to function not only in the removal of ectoparasites but also as a "social cement" in the reaffirmation of social bonds. Most grooming takes place between close relatives, but it occurs also between nonrelatives in long-term relationships (Sade 1965). In the large, multimale groups characteristic of macaques (*Macaca* spp.) and baboons (*Papio* spp.), grooming between the sexes is mostly confined to the mating season. South American titi monkeys (*Callicebus*), which live in monogamous groups, entwine their tails when resting. These signals may not be complex, but they are no less important than other signals.

FIGURE 12.15   Female rhesus monkey grooming offspring.
In addition to removing ectoparasites and other foreign matter from the skin and hair, grooming acts as "social cement," solidifying social bonds. Most grooming occurs between close relatives.
Source: Douglas B. Meikle.

Electric Field

Some sharks and electric fish have electroreceptors that they use both passively and actively in detecting objects and in communicating socially (Bradbury and Vehrencamp 1998). Sharks (*Scyliorhinus caniculus*) detect the electric field produced by prey flatfish that are buried in the sand (Kalmijn 1971). In contrast to strongly electric fish, such as electric eels that can produce **electric organ discharges (EODs)** of several hundred volts, weakly electric fish produce discharges of only a few volts. Freshwater fishes of the orders Mormyriformes

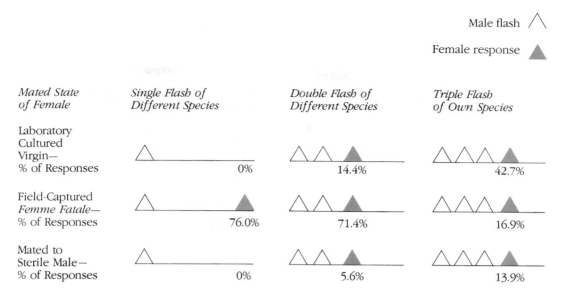

Male flash △

Female response ▲

Mated State of Female	Single Flash of Different Species	Double Flash of Different Species	Triple Flash of Own Species
Laboratory Cultured Virgin— % of Responses	0%	14.4%	42.7%
Field-Captured *Femme Fatale*— % of Responses	76.0%	71.4%	16.9%
Mated to Sterile Male— % of Responses	0%	5.6%	13.9%

FIGURE 12.16   **Responses of female *Photuris versicolor* to flashes of males.**
Unmated virgin fireflies of the same species (top row) usually answer only the triple flash of males of their own species. Mated females become *femmes fatales,* answering flashes of different species (middle row). Females mated to sterile males do not become *femmes fatales* (bottom row) and response significantly less.
Source: Data from J. E. Lloyd, "Aggressive Mimicry in *Photuris*: Firefly Femmes Fatales," *Science* 149:653–54, copyright 1965 by the AAAS.

from Africa and Gymnotiformes from South America communicate information about species identity (Hopkins and Bass 1981), individual identity, and sex by modulating the shape of the electric organ discharge. Möller (1976) demonstrated that members of this family also use electric organ discharges to maintain group coordination in schools. By altering either wavelengths or pulse duration, they can communicate threat, warning, submission, and so on (Bullock 1973). In several species of gymnotids, males modify their EOD to produce chirps that induce gravid females to spawn (Hagedorn and Heiligenberg 1985). Advantages of this sensory mode are that it is useful at night in murky waters, it can travel around and even through certain objects, and it provides precise information on location (see also chapter 7).

Vision

The need for a direct line of sight and ambient light limit their use, but within social groups, visual displays enable the receiver to locate the signaler precisely in space and time. Monkeys and apes are social, diurnal primates that, with a few exceptions, rely extensively on visual displays. Primates ourselves, we human observers have studied visual systems more than other communication systems.

If you live east of the Rocky Mountains, you have probably seen fields and lawns sparkle with flashes of fireflies. Beetles of the family Lampyridae that have specialized photogenic tissue in the abdomen emit these flashes. Such behavior is related in some way to mate attraction, and each species has its own flash code. The males' flashes vary in intensity, duration, and interval in a species-specific way, as do the females' responses (Carlson and Copeland 1978). Within a species, the

flash interval varies, depending on whether the male is searching for a female or courting one he has found. In one particular species (*Photuris versicolor*), the female, once she has mated, may mimic the flash response of females of closely related species, luring the males to her, and then devour them. Such females were aptly termed *femmes fatales* by Lloyd (1965) (figure 12.16). To make matters even more complicated, males of some species of the genus *Photuris* mimic males of other species in order to lure hunting femmes fatales of their own species into a second mating (Lloyd 1980). In other words, a mated female of species X who is mimicking the female of species Y in order to lure and eat a male of species Y is herself lured into another mating by a male of species X who is mimicking a male of species Y! Copeland (1983) argued that more data are needed before male mimicry can be assumed.

Fish and some invertebrates are able to change color within seconds by expanding and contracting chromatophores beneath the skin (figure 12.17). Most spectacular in this regard is the octopus; waves of color advance and recede according to the animal's mood. Because of the limitations of visual displays, these displays are usually coupled with other modes of communication, such as audition. For instance, in the song spread, the red-winged blackbird spreads its tail, lowers its wings, and raises its epaulets at the same time it renders the song.

EVOLUTION OF DISPLAYS

One of the questions asked early on by ethologists was how communicative displays evolved from noncommunicative behaviors. The evolutionary process is referred to as

a. Frightened or
   submissive fish

b. Fish in neutral state

c. Beginning appearance
   of yellow band used
   in courtship

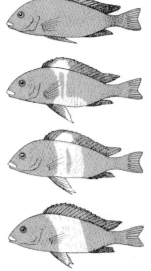

d. Increasing appearance
   of yellow band used
   in courtship

e. Maximum appearance
   of yellow band used
   in courtship

FIGURE 12.17   Graded visual signals in mouthbrooder cichlid fish.
An increasing expression of yellow band used in courtship shows in parts (c) through (e).
Source: Data from W. Wickler, "Zur Sociologie des Brabantbuntbarsches, *Tropheus moorei* (Pisces, Cichlidae)," *Zeitschrift für Tierpsychologie* 26:967–87, 1969.

ritualization. A behavior pattern may undergo the following changes during ritualization (Eibl-Eibesfeldt 1975):

1. Change in function
2. Change in motivation
3. Exaggeration of movements in frequency and amplitude, but concurrent simplification
4. "Freezing" of movements into postures
5. Stereotyping of the behavior, while keeping frequency and amplitude relatively constant even if motivation varies
6. Development of conspicuous body structures, such as ornamental feathers, enlarged claws, manes, sailfins

One of the behaviors thought to have given rise to displays is **intention movement** (low-intensity, incipient movement). More often, displays seem to have evolved from **displacement activities,** which sometimes occur in conflict situations when an animal is undecided as to the appropriate response to a stimulus. For instance, during courtship, male ducks sometimes preen their wings or just touch their feathers. Possibly this behavior is a displacement activity resulting from a conflict between sexual and agonistic behaviors. In mandarin ducks, this behavior has become a display involved in courtship. Several conspicuous feathers, or sails, have evolved and are exposed during this sham preening (figure 12.18). Other suggested origins of displays are food exchange, comfort movements, and thermoregulatory pat-

FIGURE 12.18   Male mandarin duck with modified primary feather or sail.
The mandarin duck provides an example of the evolution of conspicuous morphological traits in the ritualization of courtship display. The male points to the sail with its beak during courtship.
Source: Michael Hopiak/Courtesy of Cornell University Laboratory of Ornithology.

terns (displays that involve feather erection, the original function of which was the regulation of body temperature in birds) (Morris 1956).

Comparisons of closely related species can provide further information about the evolution of displays (see also chapter 6 for a discussion of the comparative method). A common feature of the courtship of pheasants and their relatives, including the domestic chicken, is food enticing (or calling). The male chicken (*Gallus gallus*) scratches several times with its feet and pecks at the ground while calling; if no food is present, he picks up stones as if they were food objects. The hen usually comes running, and the male can then attempt to copulate with her. The ring-necked pheasant (*Phasianus colchicus*) performs the same display. The male impeyan pheasant (*Lophorus impejanus*) bows low with a slightly spread tail and pecks the ground. When the hen approaches and searches for the food, he spreads his wings and tail feathers. The peacock pheasant (*Polyplectron bicalcaratum*) scratches the ground and then bows with wings and tail spread. If he is given food, he will offer it to the female. Finally, the peacock (*Pavo* sp.) male spreads his tail, shakes it, and moves back several steps, then points downward with his beak. The fanned tail arched over his head seems to focus the attention of the hen on the ground in front of him. Young male peacocks food-entice in the "original" form, with scratching and pecking, and develop the ritualized form as they mature (figure 12.19).

This series of behavior patterns shows that the more ancestral, or primitive, form of courtship involves the actual searching for food by the male and the offering of it to the female in order to attract her to him for possible copulation. The most advanced form, which is the farthest evolutionarily from the ancestral pattern, is shown by the peacock (see figure 12.19), whose search for food has become a highly ritualized display for mate attraction. The movements are highly exaggerated, and special plumage has evolved. See

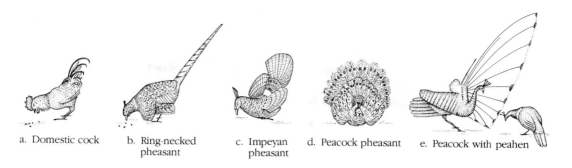

a. Domestic cock   b. Ring-necked   c. Impeyan   d. Peacock pheasant   e. Peacock with peahen
                       pheasant   pheasant

FIGURE 12.19   Food calling and courtship in phasianid birds.
The rooster shows food calling, or "tidbitting," the original or primitive behavior in (a). In parts (c) through (e), the food is no longer there, but the female is attracted to a spot on the ground by displays of the male. Such comparative series are taken as evidence for the evolution of displays.
Source: Data from R. Schenckel, "Zur Deutung der Balzleistungen einiger Phasianiden und Tetraoniden," *Ornithologische Beobachter* 53:182–201, 1956.

also chapter 5 for other examples of the evolution of ritualized displays.

The evolution of signaling systems can be influenced by **receiver-bias** if the perceptual organs of the receiver have undergone selection for other functions. For example, suppose that a hypothetical species of insect feeds preferentially on yellow flowers. Natural selection will likely have favored a perceptual system that is highly sensitive to the color yellow. Males with the color yellow might thus be especially attractive to females, assuming that the female's preference for yellow generalizes to mate choice. Males could thus be said to exploit this sensory bias to manipulate females to mate with them.

Female swordtail fish (*Xiphophorus helleri*) prefer males with swords, but so do female platyfish (*X. maculatus*), a species that lacks a sword (Basolo 1990, 1995). Phylogenetic analysis of this closely related group of fish species suggests that the ancestor of this lineage lacked a sword, and that the platyfish branched off before the sword appeared. Thus, the female preference for swords seems to have evolved before the sword itself, consistent with a pre-existing sensory bias. For other examples of the evolution of signaling systems, see chapter 6.

COMPLEX COMMUNICATION

We end this chapter by considering examples of complex communication from insects, birds, and mammals. The question motivating much of this research is whether or not humans are unique in their use of "true language."

Food Location in Honeybees

Hours or days may go by before a foraging honeybee (*Apis mellifera*) discovers a sugar or honey solution placed outdoors, but then new bees arrive within minutes. Aristotle thought that the other bees simply followed the forager to the food. However, von Frisch (1967) demonstrated that recruit-

ment takes place even when the forager is not allowed to return from the hive to the food source. He hypothesized that bees obtain information on food location through odor communicated in the hive by the forager's "dance." Hive members maintain antennal contact with the dancer's body and taste samples of regurgitated food. But von Frisch's observations later suggested to him that the bees were getting more specific information on location of food sources, and he went on to develop his famous dance-language hypothesis.

By placing food sources at varying distances and angles from enclosed observation hives, von Frisch found that foragers perform a dance on the comb. For food at short distances (from 20 to 200 meters, depending on the strain of bees used), returning foragers perform the round dance, a series of circles with reversals in direction every second or so. For food at greater distances, they perform the waggle dance, a figure eight with a straight portion in the middle of the figure (figure 12.20). The forager waggles its body and emits sound bursts during the straight, or wagging, run. The direction of the food relative to the sun is the same as the direction of the straight run relative to gravity. The duration of the straight run increases with distance at the rate of about one complete waggle per 30 meters. The area that the dance occupies on the comb, the duration of each complete figure-eight cycle, and the duration of sound bursts are all also correlated with distance to the food source. Von Frisch argued that this symbolic language is used to communicate information about the location of food.

Although these behaviors are highly correlated with food location, how do we prove that the bees actually use this information? Some have argued that since correlations exist, the dance must have a purpose, and that purpose must be communication. Such teleological reasoning has been unacceptable to others, however. First, many species of insects have the ability to transpose an angle flown or walked with respect to the sun into an angle with respect to gravity (Gould 1976). Correlations between waggling or buzzing and distance to food sources exist in species of insects besides honeybees, yet there is no indication that these correlations

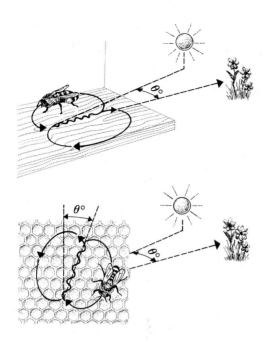

FIGURE 12.20   Waggle dance of the honeybee.
A foraging honeybee returns after discovering a food source. If the bee dances outside the hive, it waggles or vibrates its body as it passes through the straight run, which points directly toward the source. If it dances on an enclosed vertical comb, it orients itself by gravity and substitutes a point directly overhead for the sun. The angle θ between the sun and the food source is the same as that between a point directly overhead and the food source.

Source: Data from K. von Frisch, *Dance Language and Orientation of Bees*, copyright 1967 Harvard University Press, Cambridge, MA.

are part of a language. Second, von Frisch's studies had not eliminated the possibility that odor was solely responsible for the target accuracy of the recruits, since odors of specific locales could be transmitted during the dance. Wenner (1967) and others set out to test the olfactory hypothesis further. After many experiments, they concluded that the following foraging rule could account for what was observed: after the dance, recruits leave the hive, drop downwind, and pick up the odors to which they have been recruited. There were a number of differences between von Frisch and Wenner's experiments that could have accounted for the discrepancies. In general, Wenner used concentrated sucrose foods with strong odors at short distances from the hive. Von Frisch used low concentrations with weak odors at much greater distances. Dancing would be of less utility in the former situation than in the latter.

One way to demonstrate that the bees actually use the dance to find food is to design an experiment similar in principle to Wilcox's study of the role of high-frequency vibrations in water striders, which we discussed earlier. If the forager could be tricked into giving wrong information about food location during the dance but correct information about odor, we would be able to determine the method of communication by seeing where the recruits end up. Gould (1976) placed a light in the hive, thereby causing the foragers to orient their dance to it rather than to gravity. When the ocelli (the three sim-

ple eyes between the compound eyes) were painted over, the bees foraged and danced normally but were less sensitive to light. When they returned to the hive after discovering a food source, their waggle dance was oriented to gravity instead of the light in the hive, but the untreated recruits oriented to the light. If recruits used the dance information, they would be misinformed and would go to a place that corresponded to the angle of the dance and the light. If the food source were at an angle of 20 degrees to the right of the sun, the partially blind forager, using gravity, would dance at an angle of 20 degrees to the right of the vertical. But if a light were placed in the hive at an angle of 40 degrees to the right of the vertical, the untreated recruits should end up at a place 60 degrees to the right of the sun—off by 40 degrees. In fact, Gould's bees did just that, demonstrating that they used directional information from the dance (figure 12.21). In other experiments (reviewed in Gould 1976), researchers fed foragers a poison, which caused them to waggle at a rate slower than normal. Recruits wound up beyond the food supply by the predicted amount, demonstrating that bees also use distance information in the waggle dance.

More recently, European scientists constructed a "robot" bee of brass covered with beeswax that was controlled by a computer to dance, give out food samples, and make sounds like a real forager. The robot successfully communicated both distance and direction, although it was not as effective as live dancers (Michelsen et al. 1989). It was also possible to modify the dance components and separate the sound bursts from the wagging runs (Michelsen et al. 1992). These findings suggest that the main source of information is the direction and duration of the straight, or wagging, run. Both sound and wagging communicate information about distance and direction, but other components of the dance seem to be ignored. Odor cues are not necessary.

Language Acquisition

Communication using language has traditionally provided a clear-cut separation of humans from other animals. By true language we mean both the use of symbols for abstract ideas and the understanding of syntax, so that symbols convey different messages depending on their relative positions.

Great Apes

Numerous investigators have explored the potential of chimpanzees (*Pan troglodytes*) to learn language. An early subject was the Hayes' (1951) home-reared chimp that was taught to use human language. Its vocabulary consisted of only a few simple words, such as "cup" and "mama." Chimps seem to lack the motor ability to pronounce human sounds, but different approaches have demonstrated that they do not lack the ability to deal with the other aspects of learning complex language. The Gardners (1969, 1989) used the American Sign Language for the Deaf, and their chimp, Washoe, learned more than 100 words. Another chimp, Sarah, learned to use plastic pieces as word symbols and used them to communicate (Premak 1971).

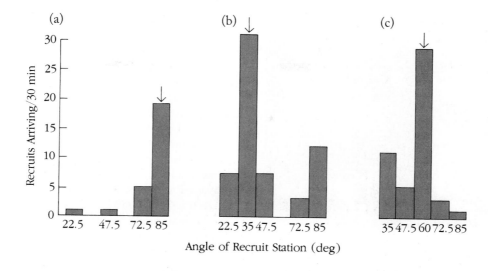

FIGURE 12.21   Effect of artificial light in the beehive on recruitment direction.
The presence of a light in a beehive causes the dancing foragers to use the light rather than the vertical axis as the reference point. When foragers are partially blinded so that they can see the sun but not the light in the hive, they use the vertical referent, opposite gravity, whereas recruits use the light. Bars denote the number of recruits to six stations within 30 minutes, and arrows denote the angle indicated by the dance in the three different experiments (a, b, and c). In all cases the food was located at 0 degrees. As the angle of the light was shifted, the dances indicated a new direction, and the distribution of recruits shifted accordingly. These misdirection experiments prove that bees use the direction information in the waggle dance.
Source: Gould J. L. 1976. The dance-language controversy. *Quart. Rev. Biol.* 51:211–44.

Using a computer, Rumbaugh and Gill (1976) taught their chimp, Lana, to press buttons with "Yerkish" symbols embossed on them in order to gain access to food and drink from vending machines, or to get human companionship. She had a vocabulary of several hundred words, used verbs and pronouns, and could assemble words in novel ways. Lana's conversations in Yerkish were pragmatic; once she obtained her immediate goal of food, drink, or companionship, the conversation ended. Her curiosity was related to her immediate needs, and she showed no interest in extending her knowledge of the world or how things in it work. This pragmatism contrasts with language development in humans, who use language at an early age to gain information about all aspects of the environment.

A controversy resulted when Terrace and his collaborators (Terrace et al. 1979) asserted that chimps could not really create sentences. They worked with their own chimp, Neam Chimpsky (Nim for short, named after the famous linguist Noam Chomsky), and also reanalyzed the videotapes and films made by other investigators. Nim mastered a respectable vocabulary of sign language words and, like other chimps, used them to convey information to another individual. Nim did use two-sign combinations that were syntactically consistent; he signed the correct "eat banana," for example, more often than "banana eat." But he was apparently imitating his teachers' previous utterances or responding to other cues, rather than creating sentences on his own. The Gardners said in response that Nim was trained in an environment unlikely to produce spontaneous behavior and that the film segments of Washoe that Terrace analyzed

were too short to demonstrate the complexity of communication (Marx 1980).

The Rumbaughs questioned whether Lana and her successors, Austin and Sherman, were using symbolic representation in the same way that humans do (Savage-Rumbaugh et al. 1980). Symbolization means the use of arbitrary symbols to refer to objects and events that are removed in time and space. Although chimps may learn to string words together in social interaction routines to attain goals, this accomplishment is not proof that they can do more than associate a word with an object—that is, their language learning does not demonstrate a referential relationship. However, in another paradigm, Sherman and Austin learned how to request tools from one another in order to obtain food that they then shared (Savage-Rumbaugh 1986). Thus, if Sherman was shown a container of food that required a wrench to open, he would punch the appropriate symbol on his keyboard. Austin then knew a wrench was needed, rather than some other tool, and would hand it to Sherman, who could then obtain the food. This example appears to be a case of symbolic communication between two nonhumans (figure 12.22).

What special linguistic abilities are needed for such communication? Epstein and colleagues (1980) set out to replicate some of Sherman and Austin's accomplishments using two pigeons named Jack and Jill. They were housed in adjacent chambers separated by a transparent partition. Using conventional procedures of shaping and reinforcement, Jack was trained to peck a "What Color" key. In response to this request, Jill was trained to look for a color hidden from Jack's view under a curtain, and then peck a key

FIGURE 12.22   Sherman and Austin at the computer keyboard.
Each key of the keyboard has a different symbol that represents a word. The positions of the keys are scrambled frequently to avoid the use of position cues. In order to obtain goals such as food, water, and companionship, chimps must press the keys in the correct order.
Source: Elizabeth Rubert/Courtesy of Yerkes Regional Primate Research Center.

corresponding to the color under the curtain. After seeing Jill perform this task, Jack then pressed a "Thank You" key, rewarding Jill with food. He then looked at the color selected by Jill and pecked the corresponding key in his own chamber, rewarding himself with food. Paraphrasing the words used by Savage-Rumbaugh and colleagues to describe the accomplishments of their chimps, Epstein and colleagues stated: "we have thus demonstrated that pigeons can learn to engage in a sustained and natural conversation without human intervention, and that one pigeon can transmit information to another entirely through the use of symbols."

Although it has been difficult to demonstrate that apes have language abilities that transcend those of other animals, a bonobo (*Pan paniscus*) named Kanzi has learned both English and Yerkish in a manner that is comparable to that of a 2-year-old child. Unlike the other apes, Kanzi was not formally trained to learn symbols. He learned meanings of symbols by observing and listening to his human caretakers, who spoke English, similar to the way a child learns language (Savage-Rumbaugh and Brakke 1990). He was able to carry out instructions such as "Get the telephone that is outdoors." He also understood novel sentences, and created novel combinations of words, such as "Car trailer" to convey that he wanted to travel in the car to the trailer, rather than to walk.

Parrots

We all know that parrots can talk, but we usually assume that they are simply mimicking what they hear without comprehension. An African gray parrot (*Psittacus erithacus*) named Alex, bought in a pet store in 1977, seems to demonstrate otherwise. As we discussed in chapter 11, he has learned, via spoken English, to identify more than eighty different objects; he can quantify collections of up to six objects; he can identify shapes and colors; and he understands concepts such as "same" and "different" (Pepperberg 1987a, b; Pepperberg 1994a, b). Although Pepperberg does not claim that Alex uses true language with its complexities of syntax and grammar, she does believe that he is using words to represent abstract concepts.

SUMMARY

Communication may be defined as an action on the part of one organism that alters the probability pattern of behavior in another organism in a fashion adaptive to either the sender or both the sender and the receiver. Behavior patterns that are specially adapted to serve as social signals are termed displays. Some definitions of communication imply that both sender and receiver must benefit and that there is evolution toward maximization of information transfer. An alternate, sociobiological view argues that communication is a means by which the sender manipulates the receiver, who may benefit or may be harmed: the purpose of a display is to persuade, not to inform.

Social signals, which vary in fade-out time, effective distance, and duration, convey information by being discrete or graded. They may be combined to form composite signals, and the order in which they appear may affect the information transmitted (syntax). Metacommunication, communication that alters the meaning of the message that is to follow, is seen among nonhuman animals mainly in conjunction with play behavior.

Communication functions in group spacing and coordination; individual, species, and class recognition; reproduction; agonism and social status; alarm; hunting for food; giving and soliciting care; soliciting play; and synchronization of hatching. Channels of communication include odor (mainly via pheromones), sound, touch, surface vibration, electric field, and vision.

Displays are thought to have evolved from such noncommunicative behaviors as intention movements and displacement acts. As behaviors change in function and motivation, natural selection produces exaggerated movements and postures, a process called ritualization. Conspicuous body structures, such as ornamental plumes on birds or claws on crabs, may have evolved to reduce ambiguity of the message and hence uncertainty of the response. By comparing differences in displays among closely related species, ethologists infer the evolutionary pathways of this process of ritualization.

Examples of complex communication exist in insects, birds, and mammals. By way of a dance on the comb, honeybees communicate information about the distance and direction of food to fellow workers. Odor cues also seem to be important, but several experiments have demonstrated that the dance information is of primary importance in locating the food. An example of complex visual communication is sign-language learning in great apes. Although they lack the motor ability to produce the sounds of human language, apes can acquire vocabularies of several hundred words by means of hand signs or substitute symbols. Apes can use nouns, pronouns, and verbs to converse about their immediate needs; they can be taught to communicate with each other, but their ability to create sentences and use true symbolism is still being debated. Studies of an African gray parrot suggest the ability to use simple words to identify objects and represent abstract concepts.

DISCUSSION QUESTIONS

1. Although we have discussed how communication operates in several discrete sensory channels, most displays involve more than one channel. Discuss examples in which two or more channels are used. Why have such multimedia displays evolved? What methods would you use to demonstrate the function of such displays?

2. Contrast the following two ideas about the evolution of communication systems in animals: (a) displays evolve so as to maximize information transfer from sender to receiver; (b) displays evolve so as to maximize manipulation of the receiver by the sender.

3. The table to the right shows the mean number of scent markings made during a 10-minute period by Maxwell's duikers—small, forest-dwelling antelopes—that were in three groups (I, II, and III). What do these data tell us about the function of marking and the social organization of this species? What can you say about the Type A and Type B females?

Group Membership	Marking Activity When With Own Group	Marking Activity After Presence of Additional	
		Male	Female
Males			
I	6.6	15.2	6.1
II	5.8	10.7	6.2
III	4.4	8.6	4.1
Type A Females			
I	3.5	3.7	18.6
II	3.4	3.1	12.2
III	1.5	0	1.7
Type B Females			
I	0.06	0	0.09
II	0	0.1	0
III	0.04	0	0.03

Source: K. Ralls, "Mammalian Scent Marking," *Science* 171 (1971) 443–49, copyright 1971 by the AAAS.

SUGGESTED READINGS

Bradbury, J. W., and S. L. Vehrencamp. 1998. *Principles of Animal Communication.* Sunderland, MA: Sinauer Associates.

Halliday, T. R., and P. J. B. Slater, eds. 1983. *Animal Behavior,* Vol. 2: *Communication.* New York: Freeman.

Hauser, M. D. 1996. *The Evolution of Communication.* Cambridge, MA: MIT Press.

Sebeok, T. A., ed. 1984. *How Animals Communicate.* 2d ed. Bloomington: Indiana University Press.

Smith, W. J. 1984. *Behavior of Communicating.* 2d ed. Cambridge: Harvard University Press.

Wilson, E. O. 1975. *Sociobiology: The New Synthesis.* Cambridge: Harvard University Press.

PART FOUR

Finding Food and Shelter

In part 3, we examined development and mechanisms of behavior, including sensory organs, endocrine regulation of behavior, biological rhythms, and learning. In part 4, we look at how those mechanisms help organisms interact with their abiotic and biotic environment. In particular, we now turn our attention to the factors that influence how animals find their way over both short and long distances (chapter 13), select a suitable place to live and reproduce (chapter 14), and find food and avoid being eaten by others (chapter 15).

© PhotoDisc

13 CHAPTER

Migration, Orientation, and Navigation

E ach year the arctic tern (*Sterna paradisaea*) flies from its breeding grounds in the tundra and fiords of the Arctic to the Antarctic ice pack, a flight of more than 20,000 km(Alerstam 1985, 1990). Some individuals fly across 3,000 km of open ocean, during which they do not feed or settle on the water. Many other species of birds— from tiny hummingbirds and warblers to cranes and hawks—travel spectacular distances, sometimes nonstop, over water and land from temperate and polar regions to warmer locations each winter; they then typically return to the same breeding area and even the same nesting site each spring. On a much smaller scale are the movements of the Eurasian milkweed bug (*Lygaeus equestris*), which travels from sites where they breed and feed on milkweed plants to sheltered overwintering sites, distances of only several hundred meters (Dingle 1996). Invertebrates sometimes move long distances as well, such as monarch butterflies (*Danaus plexippus*) that travel from the eastern United States to one of a few localized wintering sites in Mexico (Brower and Malcolm 1991). Many other animals disperse from natal sites, and most move about every day as they feed, nest, and search for mates. How do these animals find their way?

In the first part of the chapter, we look at examples of migration in different taxa, and its evolutionary history. In the second part, we examine the ways organisms orient and navigate in reference to cues in the environment, not only in long-distance migrations, but also within their home areas. In chapter 14 we further consider smaller-scale movements in habitat selection.

MIGRATION

Migration refers to persistent and straightened out movement effected by the animal's own locomotory exertions (Dingle 1996). It differs from routine daily movement in several ways. Migration involves special physiological changes, such as fat deposition, prior to the move. It is triggered by environmental signals, such as photoperiod, that lead to departure well in advance of deteriorating conditions. During migration the animal typically does not respond to the presence or absence of local resources until the trip is complete (Dingle 1996). Although we usually think of migration as a round-trip event, the term is also used to describe one-way movements, especially of insects and humans.

Birds

Best known among migrating forms are birds. We can ask several questions about their migratory flights: (1) What external cues and internal physiological events trigger migration? (2) What is the evolutionary history of migratory behavior? (3) What types of cues do birds use to guide them in their long-distance flights?

Techniques for Studying Migratory Behavior

Information about the species of birds that migrate, the locations of their summer and winter ranges, the routes they follow in flight, and the speed at which they travel over long distances is gathered through several techniques. A common method is banding, in which researchers fit small, colored or numbered cylinders of metal or plastic on birds' legs. Birds caught in mist nets or in traps (see chapter 3) are banded, and when the same bird is later caught, shot, or found dead in another location, the band number is sometimes reported. In the United States, the U.S. Fish and Wildlife Service is the clearinghouse for securing permission to band birds and for reporting all information pertaining to banding. Not surprisingly, only a small percentage of bands is ever recovered.

Examples of spectacular migration speeds have come from banding data (Griffin 1974). For instance, a sandpiper (*Actitis* spp.) flew 3,800 kilometers from Massachusetts to the Panama Canal Zone in 19 days (an average speed of 200 kilometers per day), and a 10-gram lesser yellowlegs (*Tringa flavipes*) traveled 3,100 miles from Massachusetts to the island of Martinique in the Caribbean in just 6 days (a rate of about 500 km per day).

Other techniques to study movement include observing individuals or flocks as they move past a fixed observation point, and even counting those that pass in front of the face of the moon. Radar is used to study flock migration to approximate the number of birds, their speed, their altitude, and their flight path. In addition, small radio transmitters can be attached to the backs of birds. Their signals can be used to track movement patterns and to ascertain flight speeds. For pelagic species that travel over large areas of uninhabited land or sea, the signals can be beamed up to satellites. In this way we have learned that wandering albatrosses (*Diomedea exulans*) breeding in the southwestern Indian Ocean covered 3,600–15,000 km on foraging trips lasting 3–33 days (Jouventin and Weimerskirch 1990). In a single day they were clocked at speeds of over 80 km/hr and traveled up to 900 kilometers.

More recently, a variety of intrinsic markers such as fatty acid profiles, molecular DNA analyses, and measurement of naturally occurring stable isotopes has been used to trace migration patterns (Hobson 1999). Isotope analysis relies on the fact that ecological communities have distinctive profiles that are reflected in the tissues of organisms that feed there. Individuals thus carry with them information about where they have fed previously.

Triggers

The timing of migration is under the control of endogenous biological clock mechanisms that are set by and adjusted to external stimuli (see chapter 9). Two stages are evident: a phase of preparation for migration and the actual initiation of migration. Of all the stimuli in the birds' environment, the most prevalent and consistent is daylength or photoperiod. The annual cycle of increasing daylength in the spring and declining daylength in the fall is consistent from year to year. Thus, photoperiod is the birds' best cue for initiating migration. Rowan (1925) was the first to show an association between daylength and migratory tendencies in birds when he exposed winter-caught juncos (*Junco hyemalis*) to artificially lengthened photoperiod. Experimentals moved north, while controls moved south. He also showed that this response was linked to sex hormones, since castrated males acted like controls, even when exposed to lengthened days.

The preparation phase is characterized by two major features: increasing fat deposition, and migratory restlessness, referred to as **Zugunruhe**. The metabolic system of most migratory birds goes through two cycles each year, one in the fall and another in spring. At these times, the birds add large amounts of fat, the food reserves that they need for energy during flight. For the bird species whose pre- and postmigratory body weights have been determined, the average weight loss during migration is 30 to 40 percent. Species that make long-distance flights over water or that fly nonstop from the summer residence to the winter resting ground store virtually all of the necessary energy for the flight before starting. Among the birds that employ this strategy are some warbler species (family Parulidae) that migrate over the Caribbean Sea nonstop or across the Gulf of Mexico (Gauthreaux 1971; Lowery 1946). Spring migrants that fly north across the Gulf from the Yucatan peninsula or farther south in Central America may have severely depleted fat reserves when they arrive on the Gulf Coast of the United States (Moore and Kerlinger 1987). Migrants with heavily depleted fat spend longer periods replenishing their food reserves (up to 7 days) than do birds that have made the journey with lower energy cost. Other species, like the white-crowned sparrow (*Zonotrichia leucophrys*) (figure 13.1), store some fat but make frequent stops to replenish their energy supply.

Control of metabolism is centered partly in the hypothalamus and pituitary gland. Hormones secreted by the pituitary gland affect the metabolism of foods and the accumulation of fat reserves (Meier 1973). After the fall migration, the pituitary apparently enters a refractory (unresponsive) phase but

FIGURE 13.1 White-crowned sparrow.
This long-distance migrant has been the focus of many studies. It stores some of its needed food reserves as fat before initiating its flight, but makes several stops along the route for additional fuel.
Source: © Michael Giannechini/Photo Researchers.

"reawakens" in the spring in time to trigger both the deposition of fat for migration and the production and secretion of sex hormones. The latter signal is to prepare for breeding after the birds return to their summer home. Initially, observers suggested that because changing photoperiod stimulated both migration and changes in sexual activity, these two activities were interrelated. Early experiments with golden-crowned sparrows (*Zonotrichia atricapilla*) (Morton and Mewaldt 1962) indicated that castrated birds still develop fat deposits; the two processes may involve some of the same hormones and triggers, but they are not directly linked.

A second characteristic of the preparation phase is migratory restlessness, or Zugunruhe. As birds approach the time of migration, they show increased activity levels. Automated perches placed in cages with captive birds measure the birds' movements, as shown in figure 13.2 (Farner 1955; Gwinner 1986). Restlessness is related to daylength and may also be affected by weather conditions, such as storms. Birds exhibit more restlessness when conditions are favorable for migratory flights.

Thus, some type of endogenous rhythm involving neuroendocrine mechanisms appears to regulate the processes of fat deposition and the onset of migratory restlessness (Gwinner 1986; Meier 1965). The initiation of migration is triggered by the accumulation of sufficient fat deposits, which is possibly monitored by neural and hormonal systems; and it is triggered by the appearance of favorable weather conditions. Once initiated, some migrations are all-or-none phenomena, in which there is no turning back; for example, the nonstop flights of golden plovers (*Pluvialis dominica*). For other species, like American robins (*Turdus migratorius*), red-winged blackbirds (*Agelaius phoeniceus*), and those that migrate primarily over land, the migrants may stop temporarily or even reverse direction if they encounter unfavorable weather, as they frequently do in the spring. Thus, there must be an internal clock regulating the annual pattern of

migratory activity that can be modified by external factors such as the weather or the energetic condition of the bird (Gwinner 1986, 1990).

Evolution of Migration

The selective factors leading to the evolution of migratory behavior are temporal and spatial variability in both climate and resource availability. Among all taxa, there is a general pattern of higher incidence of migration among species living in unstable or ephemeral habitats as compared to those in stable habitats (Dingle 1996). One obvious advantage for birds that migrate is that they can have an adequate food supply and favorable climatic conditions all year. By leaving the north temperate zones during winter, the birds avoid the costs of surviving in cold weather and the reduced availability of food. Some species are obligate insect feeders, and so must remain in relatively warm conditions. Others have adapted their diets and other behavior—for example, huddling for warmth at night—to permit them to remain at high latitudes throughout the year. Indeed, as more sources of artificial food (such as backyard bird feeders) have become available, some species have ceased to migrate. However, for many species, remaining in their summer locations throughout the year would mean that the demand for food would far outweigh available supply.

Other advantages of migration are related to reproduction. Although the breeding season is shorter at high latitude, daylength is increased; birds are able to concentrate their reproductive activities into fewer days, taking advantage of seasonally rich food supplies. In addition, the breeding season is a time when adults and their offspring (either eggs or pre-fledging young) are most vulnerable to predation. Because the breeding season is short, the probability of reproductive synchrony is enhanced. One advantage of reproductive synchrony is that predators are swamped by the presence of so many prey in one brief period.

Migration has several consequences for gene flow and evolution of birds. Geographic dispersal may be facilitated because of migration. As a result of migratory flights, individuals of a species may establish themselves in new areas, either by settling down in a new location along the migration route, or possibly by being forced off course by winds or inclement weather during migration. Because individuals of migratory species must reside in at least two different environments (and the interim environments during migration), they may be subjected to rigorous selection pressures. However, nonmigratory species are also subject to selection due to the harsh seasonal conditions they face by remaining in the same geographical area.

Cox (1985) developed a model for the evolution of migratory behavior based on two factors, time allocation and competition, to explain why some species migrate from a given temperate area and others do not. Among wood warbler species, clutch sizes of migrants and residents at high latitudes are larger than those of residents of the tropics. Fecundity (the number of young produced) is lower among

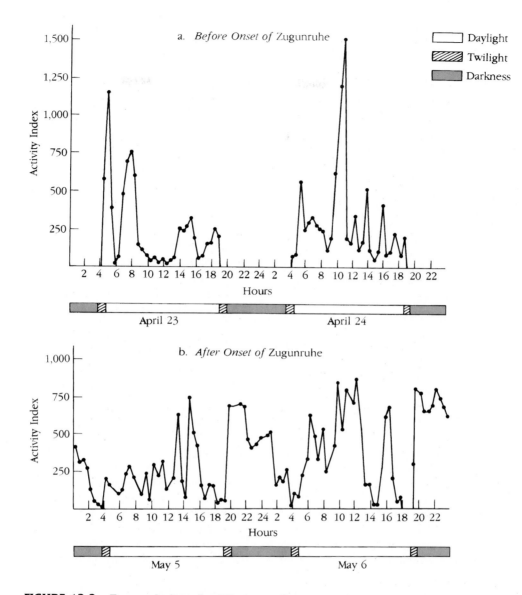

FIGURE 13.2   Zugunruhe in male white-crowned sparrow.
During the molting period, prior to the onset of Zugunruhe, or migratory restlessness, (a) the sparrow exhibits a different pattern of activity than it does after the onset of Zugunruhe (b). Lined bars represent the twilight, shaded bars indicate periods of darkness, and the open bars are daylight hours.
Source: Redrawn after D. S. Farner, "The Animal Stimulus for Migration: Experimental and Physiologic Aspects," in *Recent Studies in Avian Biology*, ed. by A. Wolson. Copyright © 1955 University of Illinois Press, Champaign, IL.

temperate-breeding migrants than it is among residents, however, because migrants have less time available for breeding. On the other hand, survival of juveniles in the tropics is low because of competition for food, while that of adults is high, compared to the temperate zone. Cox argues that it is the balance between competition, which is high in the tropics, and time available for breeding, which is short in the temperate zone, that determines whether or not migration will evolve in a particular species.

Long-distance migration covering thousands of kilometers may have evolved from species' making shorter trips in search for food. Temporal and spatial variability of resources requires that birds make relatively long trips in search of

food. Likely candidates are species that make foraging trips in search of ripe fruit, seeds, or nectar (Levey and Stiles 1992). Birds feeding on the reproductive parts of plants are likely to encounter more variability in supply than those feeding on insects that in turn feed on leaves, because most plants reproduce only during a small part of the year. Such trips require good spatial memory and navigational skills, likely preadaptations for long-distance migration. Indeed, bird species in Costa Rica that fed on fruits, seeds, or nectar were far more likely to be short-distance altitudinal migrants than were species feeding on insects or vertebrates. Furthermore, related species, in the same families as these short-distance migrants, are likely to be long-distance migrants.

Mammals

Bats

As the only mammals with true flight, bats might be expected to show migratory behavior comparable to birds. Such is not the case for most species, however. Bats have evolved relatively slow, maneuverable flight, necessary to catch insects while flying; the wings are even used to trap insects. Instead, many species of bats do not migrate but use hibernation, an energetically less costly way of dealing with cold temperatures and lack of food.

In spite of these flight constraints, many species of bats do migrate, most often to and from caves and other shelters used as hibernation sites (Griffin 1970). Information about bat migration comes from the seasonal appearances and disappearances at roosting sites coupled with recoveries of banded individuals (Fenton and Thomas 1985). Radiotelemetry has been used in a few instances, mostly to study foraging trips and roosting behavior (Fenton et al. 1993). In a study involving banding more than 73,000 little brown bats (*Myotis lucifugus*), individuals migrated more than 200 km from hibernation caves in southwestern Vermont. They generally moved southeast into Massachusetts and neighboring states for the summer (Davis and Hitchcock 1965). The endangered Indiana bat (*M. sodalis*) migrates from hibernation caves in Kentucky and southern Indiana as far north as Michigan (Barbour and Davis 1969).

Swifter-flying species, such as the hoary bat (*Lasiurus cinereus*) and free-tailed bat (*Tadarida brasiliensis*), move even longer distances. Hoary bats migrate from summer ranges in the Pacific Northwest as far north as Alaska, south into central California and Mexico for the winter. In winter they are not found above 37°N latitude, a limit probably set by the distribution of flying insects (Griffin 1970). Free-tailed bats seem to have both migratory and nonmigratory populations. Those in southern Oregon and northern California are year-round residents, but those in the southwestern United States migrate south into Mexico for the winter (Dingle 1980). For instance, a population in the Four Corners area (where Colorado, Utah, New Mexico, and Arizona meet) has a well-established flyway through the Mexican states of Sonora and Sinaloa west of the Sierra Madre Oriental Mountains. The routes of some southwestern U.S. populations have yet to be identified.

Long-distance migration is not restricted to bat species in temperate regions. Seasonal shifts in rain patterns trigger migration in some species of African bats (Fenton and Thomas 1985). For instance, West African fruit bats of the family Pteropodidae migrate distances of 1,500 kilometers each year, following rains into the Niger River basin (Thomas 1983).

Whales

Most species of baleen whales (suborder Mysticeti) spend summer months at high latitudes, feeding on plankton and small fish in the highly productive Antarctic and Arctic waters. As winter approaches in each area, whales migrate to warmer subtropical and tropical waters. Food supply does not drive this migration; tropical waters are relatively unproductive and, in fact, whales do not feed during migration or at their wintering grounds. Instead, they rely on fat deposits. The benefit of moving to warmer water is likely the energy savings from reduced heat loss, especially for calves. Calves are born in the tropical breeding grounds, and lactating females with their newborn calves move to feeding areas at higher latitudes as spring approaches. Breeding cycles of species that breed inshore, such as the humpback whale (*Megaptera novaeangliae*) and the California gray whale (*Eschrichtius robustus*), are fairly well known, but little is known about the breeding habits of the offshore species such as the blue whale (*Balaenoptera musculus*) and fin whale (*B. physalus*) (Dingle 1980).

California gray whales spend their summers feeding in the North Pacific and Arctic Oceans. In autumn they migrate south to subtropical breeding grounds off the coast of Baja California (Orr 1970). Humpback calves are born in September, and lactating females begin the trek back north with their calves in the spring, usually after males and newly pregnant females have already left (Dingle 1980).

Pinnipeds

Many species of seals and sea lions migrate thousands of kilometers from island breeding and molting areas to oceanic feeding areas. Island breeding sites are chosen because they are relatively free of predators. Northern elephant seals (*Mirounga angustirostris*) breed on island rookeries off California, migrate to foraging areas in the North Pacific and Gulf of Alaska, and later return to the islands to molt. By attaching recorders to the seals that recorded location, time, and depth, Stewart and DeLong (1995) found that seals travel linear distances of up to 21,000 kilometers during the 250 to 300 days they are at sea. Each individual thus makes two migrations per year, returning to the same foraging areas during both postbreeding and postmolt movements. Males migrate farther north than females, where they feed off the Alaskan coast.

Ungulates

Large ungulates migrate long distances as well. The best-studied northern species is the barren-ground caribou (*Rangifer tarandus*). Herds migrate north to calving grounds above the timberline in spring and return south in autumn, covering distances of more than 500 km (Orr 1970; figure 13.3). More recently, the movements of individuals have been monitored by satellite tracking (Craighead and Craighead 1987). Migrations seem to be made up of series of fairly straight movements that are little influenced by landmark features such as rivers (Bovet 1992; figure 13.4).

The mass migrations of wildebeest (*Connochaetes taurinus*) in East Africa are spectacular. The Serengeti population spends the wet season, usually December through April, in the southeastern Serengeti plains of Tanzania, where short grasses are lush and calving takes place. Large migratory herds form at the beginning of the dry season, in May and

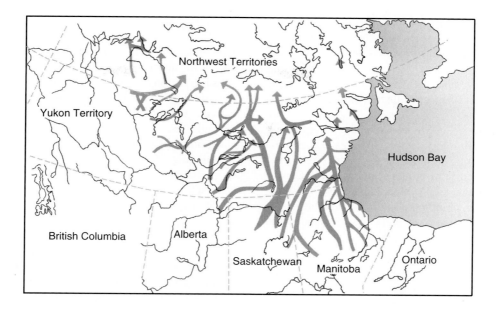

FIGURE 13.3   Migration routes of barren-ground caribou.
In spring, herds move north to calving grounds above timberline; in autumn, they return south
to the shelter of forests.
Source: Data from H.L. Gunderson, *Mammalogy*, 1976, McGraw-Hill, Inc., New York, NY.

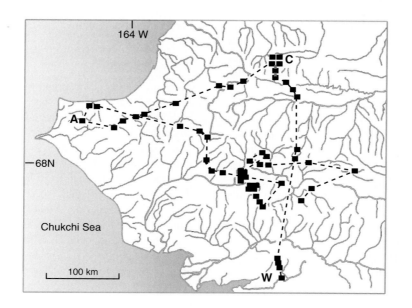

**FIGURE 13.4   Migration route of an adult, female caribou in north-
west Alaska in 1984.**
The dashed line connects successive satellite-fixes (filled squares). The caribou left the winter range (W) on 15 May and arrived on the calving ground
(C) on 30 May, where she calved on 5 June, and stayed until 16 June. She then
moved to the herd's aggregation area (A), where she stayed on 4 and 5 July.
She spent the summer traveling east, with occasional 1 to 2-week stays in
localized areas. The last fix was on 7 October, while she was moving toward
the winter range. Due to the hydrographic features of the area (thin lines =
rivers), the caribou's route was probably as often across, as it was parallel to,
valleys.

Source: Data after Craighead & Craighead, 1987, and Fairbanks of the World Map (1:2500000) of
the USSR Main Admin. of Geodesy and Cartography, Moscow, 1973; in J. Bovet, "Mammals" in
Animal Homing, (F. Papi, ed.), 1992, Chapman and Hall, New York.

June, as millions of animals move, sometimes single file, northwest toward Lake Victoria. In July and August, near the end of the dry season, herds move northeastward into the Masai-Mara of Kenya and return south to the breeding grounds between November and December. These patterns vary considerably, however, depending on the timing of rainfall (Dingle 1980).

Other ungulates engage in elevational migration; elk (*Cervus elaphus*) and mule deer (*Odocoileus hemionus*) move into high-elevation summer ranges that are relatively snow-free and then return to milder winter ranges at lower elevations (McCullough 1985). Mountain sheep follow the same routes each year, climbing up into isolated patches of lush grazing meadow as the snow melts in spring and return to lower elevations in autumn and winter.

Other Vertebrates

Although birds and mammals have received the most attention, a variety of fish, amphibian, and reptile species show migratory behavior (Dingle 1996). Pacific salmon (*Oncorhyncus* spp.) migrate as adults from the ocean to breed at the stream site where they were born years earlier (Hasler 1966). Certain eel species (*Anguilla* spp.) live much of their lives in fresh waters in North America and Europe and migrate to the Sargasso Sea region of the Atlantic where they breed and die (McCleave and Kleckner 1985). The larvae drift on ocean currents for a year or more, eventually entering rivers along the coasts of North America and Europe, where they take several more years to reach sexual maturity.

Most amphibian species live on land, but they must lay their eggs in water, and so migrate to ponds and streams to breed. For instance, wood frogs, *Rana sylvatica,* return repeatedly to the same breeding pond each year. Berven and Grudzien (1990) marked more than 11,000 adults, and every one returned to the pond in which it was marked. Among reptiles, sea turtles follow the opposite pattern, living in oceans but returning to land to lay eggs. Tagging studies show that, like the wood frogs, individuals usually return to the same sites, often on remote islands, each year.

Female green turtles (*Chelonia mydas*) deposit their eggs on sandy beaches on islands; they lay about a hundred eggs on each trip to the beach and make three to seven trips spaced about 12 days apart (Carr 1967). In nonreproductive periods, the turtles live in areas with abundant turtle grass along the coasts of continental landmasses. In one study (Carr 1967), turtles were banded at Ascension Island (figure 13.5). Young turtles that hatched on the island floated on currents that took them to the coast of South America. Adults had to swim against the same current to reach Ascension Island to reproduce. How did they find their way? The prevailing current may have certain features that allow the turtles to distinguish it from surrounding waters. This current could provide a general orientation, but not the precise information needed to target a small landmass in the midst of many square miles of ocean.

When the mysteries surrounding the movements of green sea turtles were first being unraveled, Carr and Coleman (1974) hypothesized that the turtles elected to return to the beaches of Ascension for breeding because those locations had once been much closer to South America; over millions of years, continental drift had moved the island farther from the continent. More recently, molecular techniques have been used to refute this notion (Bowen et al. 1989). A particular type of DNA in animals, mitochondrial DNA (mtDNA), is passed only from mothers to their offspring; this permits investigators to track female lineages. By examining the maps of mtDNA for differences, it is possible to estimate how long ago two populations became separated: the more divergence between their mtDNA makeup, the farther back in time they diverged. The mtDNA data do indicate some divergence among the three populations, confirming the notion that green sea turtles do return to their natal site for breeding. However, comparisons of mtDNA from turtles taken at three breeding locations—Ascension Island, Florida, and Venezuela—revealed that the population that uses Ascension Island for egg laying has a degree of difference small enough that any divergence between the groups of turtles took place on the order of 20,000–40,000 years ago, not 40 million years ago as postulated by the continental drift hypothesis. Thus, the patterns of migration cannot be explained by continental drift.

Invertebrates

Among the most widespread migrations are the daily vertical movements of many species of zooplankton living in both freshwater and marine environments (Huntley 1985). Distances moved vary from as little as two to several thousand meters. Most common is an ascent toward the surface in the evening twilight, but some species show the opposite pattern. Although changes in light intensity seem to trigger these movements in many species, there is uncertainty about their ultimate function. Most hypotheses relate to feeding and predator avoidance (Kerfoot 1985).

An intriguing pattern is that shown by Atlantic spiny lobsters (*Panulirus argus*) as they move from food-rich coastal waters to deep water with the approach of winter (Herrnkind 1985). Migrants walk continuously for several days in single-file queues of up to 50 individuals, covering 30–50 kilometers (figure 13.6).

Monarch butterflies (*Danaus plexippus*) of the eastern United States fly south to sites in Mexico for the winter, and monarchs from the West Coast move to locations along the central and southern California coast (figure 13.7). Unlike birds and mammals, monarchs do not restrict their breeding activities to their summer residences in northern locations. Adult butterflies migrate southward in the fall and overwinter in large aggregations in trees (Brower et al. 1977; Calvert and Brower 1981; Zahl 1963). Those that survive until the following spring breed at or near the overwintering site. Then, as the monarchs move northward, they complete

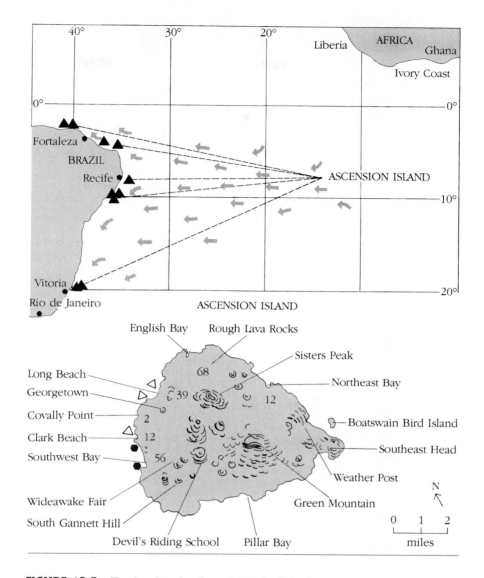

FIGURE 13.5 Turtle migration from Ascension Island.
The number of turtles tagged at each nesting beach location is shown by a figure. Solid triangles indicate mainland Brazil locations where researchers recovered turtles. Open triangles on nesting beach sites show turtles that returned after 3 years. Solid hexagons denote turtles that returned in the fourth year to nest, possibly having made two round trips to Brazil. Arrows indicate prevailing ocean currents.

additional generations. The summer residents of the northern United States and Canada are thus progeny of spring breeders in the southern United States (Brower and Malcolm 1991).

Some species of desert locusts (family Acrididae) produce phenotypes that exhibit migratory behavior as a result of their rearing environment. Hoppers (young locusts) that develop in low-density populations exhibit a low level of activity, and after metamorphosis into adult locusts they are rather solitary and do not engage in sustained flights. When young hoppers develop in dense populations, they tend to be more active and gregarious; when they become adults, they

remain more gregarious and engage in longer flights (Johnson 1969). The character of locust populations fluctuates between the sedentary and migratory phases, depending upon the interaction of social density with environmental triggers such as rainfall and food availability. When food supplies diminish, the physiology of the developing insects changes. This change (into migrators) combined with an increased population produces large swarms of migratory adult forms that may number in the millions. These swarms then set forth on long flights, and on the way, they strip any green vegetation in their path down to bare ground (Dingle 1996).

FIGURE 13.6   Migratory queue of spiny lobsters (*Panullirus argus*).
Up to 50 individuals march single file for several days, covering distances of 30–50 km.
Source: Photo from Herrnkind, W. F. 1985. Evolution and mechanisms of mass single-file migration in spiny lobster: Synopsis. pp. 197–211 in *Migration: Mechanisms and Adaptive Significance* (M. A. Rankin, ed.). Port Aransas, TX: Marine Science Institute.

FIGURE 13.7   Monarch butterflies (*Danaus plexippus*) at overwintering site in Mexico.
The monarch butterfly is a seasonal migrant, flying south from the eastern U.S. to Mexico for the winter. The butterflies that return north the following spring are the result of several generations of breeding en route.
Source: © Lincoln P. Brower.

ORIENTATION

Orientation refers to the way an organism positions itself in relation to external cues. One of the early investigators of animal orientation, Jacques Loeb (1918), theorized that asymmetrical stimulation of an animal's sensory organs results in differential contraction of muscles on the opposite side of the animal until the symmetry is restored for both sense organs and muscle actions. Other investigators concluded that not all animal orientation could be fitted to Loeb's sensory organ–muscle scheme, and they found that different organisms possessed different systems of orientation (Mast 1938).

Later, Fraenkel and Gunn (1940) summarized the work on animal orientation up to that date and defined general classes of orienting reactions. **Kineses** are random movement patterns in response to stimuli in which there is no orientation of the organism's body to the source of stimulation. The rate of movement increases with the intensity of the stimulus. **Taxes** (singular: **taxis**) are directed reactions involving (in a single-stimulus situation) an orientation of the long axis of the body in line with the stimulus source. Movements toward the stimulus are positive taxes; movements away from the source of stimulation are negative taxes. For instance, planaria (*Planaria* spp.) exhibit a negative **phototaxis**—movement away from a light source. Extensive studies have revealed other, often complex systems of larger-scale orientation and navigation in different organisms, a few of which we examine here.

Spiders constructing webs present an interesting opportunity to examine orientation in three-dimensional space (figure 13.8). Spiders that build vertical webs use at least three types of cues for orientation: light, gravity, and the threads or lines of the web they are constructing or moving about (Crawford 1984; Eberhard 1988). The spiders use gravity to determine their direction within the vertical plane, light to discriminate between the sides of the web, and a spatial memory of distances and directions on various lines involved in the web's construction.

NAVIGATION

Whether migrating thousands of kilometers or simply foraging within the home range, many species of animals return to a nest site, or den, a process called **homing**. Chitons are

FIGURE 13.8 **Orb-weaving spider and web.**
Some spiders, like the orb-weaving spider, *Araneus diadematus,* shown here, build vertical webs. These spiders apparently use light, gravity, and the threads of their webs for orientation.
Source: Courtesy Peter N. Witt.

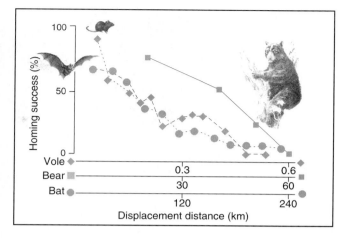

FIGURE 13.9 **Relationship between homing success and displacement distance.**
Meadow voles (*Microtus pennsylvanicus*), $n = 460$; black bears (*Ursus americanus*), $n = 112$; Indiana bats (*Myotis sodalis*), $n = 700$.
Source: Data after Robinson and Falls, 1965; MacArthur, 1981; Hassell, 1960; in J. Bovet, "Mammals" in *Animal Homing*, (F. Papi, ed.), 1992, Chapman and Hall, New York.

small molluscs that live in the intertidal zone. Members of one species, *Acanthopleura gemmata,* found in the Indian and Pacific Oceans, have custom-made scars on rocks where they hold fast during low tide. Having a tight fit is important to prevent dehydration, and also possibly to reduce predation. At high tide they forage on the rock surface several meters from home, returning home before getting dehydrated as the tide ebbs. Chitons use the most direct method of homing, leaving a chemical trail (like the bread crumbs in Hansel and Gretel) as they move, so they can retrace their outward path to get home. Disrupting the trail of an outward bound individual causes it to stop and search in the area of the trail interruption on the return trip (Chelazzi 1992). But other animals return home from places they have never been, and some can proceed directly home, rather than simply retracing their steps as do chitons.

Navigation is the process by which an animal uses various cues to determine its position in reference to a goal as it migrates or homes. Much of what we know about navigation comes from studies of homing pigeons, but enough work has been done with other taxa to suggest that they use many of the same mechanisms. One technique for studying homing is to displace the animal from its home range and record such variables as the direction it heads when released (the "vanishing bearing"), the time it takes to get home,

whether or not it makes it home at all, and the route it takes to get back home.

The roles of various senses can be studied via manipulations (e.g., use of blindfolds). In an early study in which deermice (*Peromyscus maniculatus*) were released at distances well beyond their home range, some individuals made it back home (Murie and Murie 1931). Three possible mechanisms were suggested: (1) the results were due to chance; (2) the mice were sufficiently familiar with the terrain; or (3) the mice had some sort of homing instinct, or sense of direction. These possibilities form the basis for much of the later work on homing mechanisms (Joslin 1977a). In at least some instances, alternatives (1) and (2) can be ruled out, and understanding (3) has been a fascinating challenge.

As might be expected, homing success decreases as a function of the distance the individual is displaced (Bovet 1992) (figure 13.9). Homing success also increases as a function of home-range size, even within the same species. For instance, house mice (*Mus domesticus*) with large home ranges homed over larger distances than did those with small home ranges (Anderson et al. 1977).

Several studies have added sensory deficits to displacement, with mixed results. Making white-footed mice (*Peromyscus leucopus*) **anosmic** (eliminating the sense of smell) had little effect on homing ability, whereas blinding them produced negative effects (Cooke and Terman 1977; Parsons and Terman 1978). Blindfolding bats produced little deficit over short distances (Mueller 1966), but at longer distances (>32 kilometers), vision appears to play a role in the orientation and navigation of bats (Williams and Williams 1967, 1970; Williams et al. 1966). As we consider the ways that animals navigate, keep in mind that more than one mechanism may be at work.

Path Integration

Aside from leaving a trail of bread crumbs, another way an animal might get back home is to keep track of all the turns, accelerations, and decelerations on the outward trip, integrate this information in the central nervous system, and use it to return, a process referred to variously as **path integration,** dead reckoning, or idiothetic. It involves internal, or egocentric, spatial localization in that the animal needs no external referent. This route-based orientation mechanism does not necessarily require retracing the outbound path. A desert ant (*Cataglyphis fortis*) follows a long, tortuous path out from its nest as it searches for food, but once it captures a food item, it proceeds straight back to the nest (figure 13.10). If the ant is moved just before it starts back home, it will take a path parallel to the one it would have taken, and at about the correct distance from home, it begins searching for its nest. Hamsters (*Mesocricetus auratus*) fed in the center of an arena are able to get back to their peripheral nestbox with food, even when tested in the dark and in the absence of other external cues (Etienne et al. 1988).

One problem with this type of navigation is that if the animal is accidentally shifted off course, as from wind or current, it may not be able to compensate and get back home without some external referent. Path integration also accumulates more and more error the greater the number of turns in the route. Thus, it is unlikely that such a mechanism would be reliable for long-distance homing.

Most mechanisms for navigation involve allocentric, or geocentric, cues that require an external frame of reference. Most can be placed into one of three categories: **piloting,** which refers to the ability to use fixed and familiar reference points or landmarks to orient; **compass orientation,** in which the animal uses external cues such as the sun to maintain a heading; and **true navigation,** in which the animal moves to a goal in unfamiliar territory in the absence of any sensory contact with that goal. True navigation requires both a compass and a map.

Piloting

Some animals can locate food or shelter based on their relationship to distant fixed cues, a mechanism sometimes

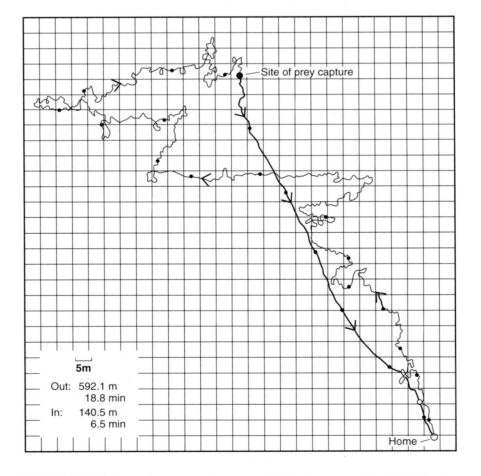

Site of prey capture

5m

Out: 592.1 m
       18.8 min
In: 140.5 m
       6.5 min

Home

FIGURE 13.10   **Foraging route and return path of a desert ant, *Cataglyphis fortis*.**
Note that the outward route is a meandering search, but once a prey item is captured, the ant follows a direct route home.
Source: Redrawn after Wehner, R. 1992. Arthropods. pp. 45–144 in *Animal Homing* (F. Papi, ed.). London: Chapman and Hall.

referred to as **piloting** (Bovet 1992). When a female digger wasp has dug a nest in the ground and laid eggs in it, she then leaves to capture prey to feed the young. But before doing so, she circles the nest site several times. When returning with food, she flies straight to the nest. Long ago, Niko Tinbergen (1951) hypothesized that the returning female uses local landmarks to identify the exact location of the nest. To test his idea, he moved the objects nearest the nest holes a short distance away when females were out foraging. Upon returning, they invariably searched for their nest hole by the displaced objects. Since Tinbergen's early work, the use of landmarks has been demonstrated in many species in both field and lab.

Landmarks would seem best for locating objects within the home range, such as when gray squirrels (*Sciurus carolinensis*) locate a food cache (McQuade et al. 1986). Piloting requires the existence of some sort of cognitive, or mental, map of the terrain, but it does not require a compass. Landmarks are assumed to be most useful within or close to the home range. In principle, however, they could be used on longer trips as well. An individual could head toward some distant landmark such as a mountain on the way out, and away from it on the way back. On the return trip, however, the animal would have to maintain a constant angle 180 degrees away from the landmark, that is, have a mental compass.

Birds congregate near and fly along coastlines or river valleys; and they may use landmarks for piloting, particularly in home areas at both ends of the migration, where they are more familiar with specific topographic features. However, the use of topography as a guidance system has inherent drawbacks. How do first-time migrants, who have never learned the landmarks, find their way? What if a storm blows birds off the normal route into areas they have never traversed? Also, visual landmarks alone cannot provide the proper orientation over long distances. Both landmarks and certain topographic features could be used in combination with some type of compass mechanism to provide sustained orientation during longer movements. In familiar terrain, landmarks may suffice without any compass (Able 1980).

Landmarks may play a critical role in finding particular stopover locations, as in the case of migrating waterfowl (Bellrose 1964, 1971). Evidence from radar studies of bird migration, however, has shown that nocturnal migrants in different parts of the country ignore most topographic features (Drury and Nisbet 1964; Gauthreaux 1971; Richardson 1972). Studies in which homing pigeons (*Columba livia*) were fitted with frosted lenses over their eyes (Schmidt-Koenig and Schlichte 1972; Schmidt-Koenig and Walcott 1973) revealed that upon their release from a new location, when they can locate the sun, even when the visibility is only 3 meters, they still can find their way home—or very close to home. When investigators used airplane tracking and ground radar to follow the flight paths of homing pigeons with frosted lenses, they found that the paths flown by the birds were generally in the direction of home. Thus, we can conclude that the pigeons' navigation system does not require detailed vision and can lead birds that have been released 13–20 kilometers from home to within a short distance of the loft.

Data from several species of mice indicate that when displaced, these animals can find their home range again, some from a considerable distance (up to 30 kilometers) and with speeds up to 300 meters per hour. In most studies, the rodents are usually displaced only up to 500 meters away and may take several days to return to the home range and reenter a trap (Joslin 1977a). Based on present data, the most important homing mechanism in rodents seems to be piloting—that is, location by using familiar landmarks or terrain—combined with random search when the displacement is for greater distances (Alyan and Jander 1994; Joslin 1977a,b). An animal learns about its habitat in several ways, including exploration of the home range. Some exploratory forays may take the rodent a distance beyond the home range. Rodents also learn about the habitat during dispersal when the young move away from the natal site to establish home ranges of their own.

Compass Orientation

Sun

One way to maintain direction in unfamiliar terrain is to use some celestial cue, such as the sun, as a compass and maintain a constant angle to it while traveling. Of course, the sun is not fixed in the sky, so an individual must have some sort of internal clock that enables it to compensate for the movement of the sun across the sky (about 15 degrees/hour). In a classic experiment with the ant *Lasius niger*, Santschi (1911) used a mirror to trick the ants into thinking the sun was coming from another direction. They changed their direction of travel so that they were still traveling at the same angle from the sun as before, but in relation to the "new" position of the sun. The assumption of a fixed position between the sun and a particular path is probably more common among animals that move short distances than those that travel long distances. For traveling longer distances, animals must be able to use the sun as a compass. Although the latitude at which an animal is positioned will determine the exact path the sun appears to follow, it always appears to rise in the east and set in the west. More importantly, the sun's rate of travel across the sky, on average, is roughly the same from season to season and year to year. Sun compass orientation has been found in a wide variety of animals, including crustaceans, insects such as ants and bees, and all the classes of vertebrates (Able 1980).

Kramer (1950, 1951) showed that European starlings (*Sturnus vulgaris*) placed in an outdoor cage with the sun visible, exhibited migratory restlessness in the appropriate direction in the spring and fall (figure 13.11*a,b*). When Kramer used mirrors to alter the apparent position of the sun, the pattern of the starlings' migratory restlessness shifted direction in a predictable manner (figure 13.11*c,d*).

Another way to test for the existence of a sun compass is to "clock-shift" the animal in the laboratory by delaying the onset of the light/dark cycle and then test it in the field.

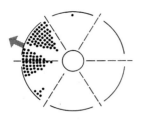

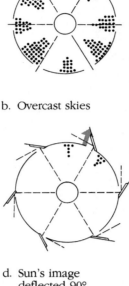

a. Clear skies

b. Overcast skies

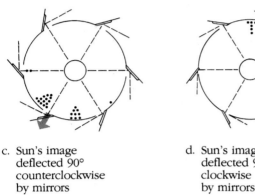

c. Sun's image
   deflected 90°
   counterclockwise
   by mirrors

d. Sun's image
   deflected 90°
   clockwise
   by mirrors

FIGURE 13.11 **Use of the sun as a compass by European starlings.**
Each circular diagram illustrates the orientation of diurnal sponta-neous migratory activity in a caged European starling. Experi-menters caged the bird in an outdoor pavilion with six windows and conducted tests during the migratory season (a) under clear skies, (b) under overcast skies, (c) with the sun's image deflected 90 degrees counterclockwise by mirrors, and (d) with the sun's image deflected 90 degrees clockwise by mirrors. Arrows denote mean direction of activity, and each dot within the circles represents 10 seconds of fluttering activity. The dotted lines indicate the direction of light coming from the sky.
Source: Redrawn after Kramer, G. 1951. Eine neue Methode zur Erforschung der Zugorientierung und die bisher damit erzielten Ergebnisse. *Proc. Xth Inter. Ornithol. Congr.,* Uppsala. 271–280.

For instance, if the test animal is shifted 6 hours in the lab-oratory, it should head 90 degrees off course when tested. As predicted, shifting the internal clocks of homing pigeons 6 hours fast (by housing them in an environmental chamber with a controlled light cycle) resulted in predictable 90-degree directional changes in the initial bearing of birds heading for the homesite. This finding is consistent with the notion that birds use the sun as a simple compass. As a fur-ther test, birds were clock-shifted 6 hours slow and released at a site 160 kilometers south of the home loft. If the birds were using true sun navigation, they should have decided that they were 6,400 kilometers east of home and headed directly west. However, the birds headed directly east, as would be predicted if they were using the sun only as a com-pass, and not as both a map and a compass (Keeton 1974).

Thirteen-lined ground squirrels (*Spermophilus tridecem-lineatus*) were tested in an outdoor arena 100 meters west of their home cages with only the sky visible (Haigh 1979). When released in the arena, they burrowed in the direction of

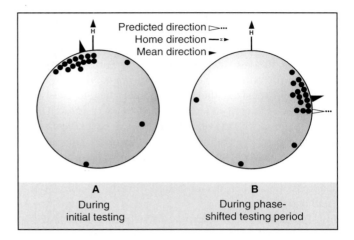

A
During
initial testing

B
During phase-
shifted testing period

FIGURE 13.12 **Use of the sun as a compass by thirteen-lined ground squirrels.**
Digging responses in outdoor enclosures (A) during initial testing, and (B) during the phase-shifted testing period.
Source: Data from G. R. Haigh, Sun-compass Orientation in the Thirteen-lined Ground Squirrel, *Spermophilus tridecemlineatus, in Journal of Mammalogy* 60; 629-32, 1979.

the home cages. These same individuals were then clock-shifted 6 hours in the laboratory. When retested, most shifted their burrowing direction 90 degrees in a clockwise direction, as predicted (figure 13.12). The ability to use the sun as a compass with time compensation has also been demonstrated in several other species of small rodents (Bovet 1992).

Nocturnal migrants may also use the sun by taking a bearing at sunset and using that bearing to fly at night (Able 1982). Big brown bats (*Eptesicus fuscus*) use the postsunset glow to travel from their roost to favorite foraging areas (Buchler and Childs 1982). They depart the roost in the evening at a colony-specific angle to the glow that is inde-pendent of landmarks. The mechanism by which birds use the sun near sunset as a bearing has been explored in some detail (Able 1980, 1989; Able and Able 1990b). The Ables have demonstrated that Savannah sparrows (*Passerculus sand-wichensis*) use polarization patterns of light from the sun as an indicator of direction and can take bearings from such cues. Furthermore, this process is learned in young birds. Groups of hand-raised young birds were provided with con-trolled experience of the daytime sky; conditions manipulated were sun azimuth, skylight polarization patterns, and mag-netic directions. In their first autumn, the birds were tested in Emlen funnels during the period between sunset and the first appearance of stars. In the Emlen funnel, the bird stands on an ink pad at the bottom of a funnel-shaped cage (figure 13.13). The funnel sides are covered with paper; the top is covered with a wire mesh screen that permits the bird to see the over-head sky. When the bird jumps up against the sides of the funnel in its restlessness, it leaves marks on the paper. Inves-tigators can turn the record of these marks into a vector dia-gram for analysis. The birds learned to perform migratory orientation based primarily on polarized light patterns.

After the initial discovery of sun-compass orientation in birds, theories were put forth to explain how birds use the sun

FIGURE 13.13   The Emlen funnel.
This circular cage enables researchers to record the direction of movements in migrating birds. A bird stands on the ink pad at the bottom of the cage, where it can view the stars overhead through a wire mesh top. The cage can be set up in an indoor planetarium, or outside. Each jump it makes as if to fly is recorded on paper on the sides of the funnel. (The cage is depicted here in cross section.)
Source: Redrawn after Emlen, S. T. 1967. Migratory orientation in the indigo bunting. I. Evidence for use of celestial cues. *Auk* 84:309–42.

cue (Mathews 1968). Some investigators theorized that birds use the sun only to gain a compass bearing to head in a particular direction. Others theorized that they use true navigation—the ability to orient toward a goal regardless of its direction and without the use of familiar landmarks (Griffin 1955).

Bees and some other insects, including ants, can receive and interpret information about polarized light (von Frisch 1967). The plane of polarization can provide an axis for orientation for these animals. It is also possible that the perception of polarized light may enable some animals to locate the position of the sun when the sky is partially overcast. The use of polarized light for orientation occurs in a variety of other vertebrates, including fish (Waterman and Hashimoto 1974) and salamanders (Adler and Taylor 1973).

Moon

The moon is apparently not as useful an aid in navigation as the sun for several reasons. It is visible on average only half of each night, and it moves, requiring compensation similar to the sun compass. Finally, the moon moves more slowly than

the sun (24.83 versus 24 hours/cycle) so that an animal that used both the moon and the sun would have to use two different clocks. However, the moon is a prominent feature of the night sky, and a variety of nocturnally active animals do have a moon compass (Brannon et al. 1981; Enright 1970; Moore and Kerlinger 1987; Papi and Pardi 1953; Scapini 1986).

Stars

Most long-distance migrants fly at night. When nocturnally migrating birds, such as blackcaps (*Sylvia atricapilla*) and lesser whitethroats (*Sylvia curruca*), both found in Europe, were placed in outdoor cages with a view of the clear night sky, they exhibited migratory restlessness in a direction appropriate to the seasonal migration (Kramer 1950; Sauer 1957). Sauer exposed birds to planetarium skies, which permitted him to manipulate star patterns experimentally. Not only were the results of outdoor studies confirmed, but when he shifted the planetarium's star patterns 180 degrees, the direction of the birds' activity also shifted. Under cloudy skies or with no stars visible, and with only diffuse, dim illumination in the planetarium, the birds exhibited random orientation. Sauer's results were confirmed and extended in both field and laboratory experiments by Emlen (1967a,b) in indigo buntings (*Passerina cyanea*).

In additional studies, Emlen (1970, 1975b) rotated the night sky in a planetarium. He conducted an experiment with three groups of young indigo buntings. Individuals in group 1 were raised in a windowless room with only diffuse light. Individuals in group 2 were allowed to see the normal night sky in the planetarium, with a normal rotation of the heavenly bodies around the pole star (the North Star or Polaris) once every other day. Birds in group 3 were raised the same way as those in group 2, except that the heavenly bodies were rotated around Betelgeuse, a bright star in the constellation Orion. The indigo buntings were later measured in the Emlen funnel for their migratory orientation under planetarium skies.

Two major conclusions resulted from this experiment. First, exposure to stellar sky patterns is necessary for normal southward migratory orientation in young buntings. Birds in group 1 exhibited random patterns of orientation when placed under the normal night sky. Second, birds in group 3 oriented 180 degrees away from Betelgeuse—as if headed south, using that star to define the southerly direction. Early experience thus plays a critical role in determining the migratory orientation of buntings, and they may use the sky pattern they learn at this time throughout life. These findings also help account for the evolution of stellar-cue orientation in spite of the change in the earth's magnetic poles that occurs about every 13,000 years that causes the positions of star patterns to change in the sky. Birds do not inherit a star map or knowledge of a specific star pattern. Rather, they inherit a predisposition to learn the sky pattern they see when they are very young, and then they use that sky pattern as the basis for orientation during migratory flights.

As was the case in our discussion of the compass/map role of the sun, we are again faced with the question of whether birds are using the stars merely as a compass or whether they are capable of true navigation using stellar configurations (figure 13.14). Use of the stars as a compass differs from use of the sun, because the night sky contains many more potential cues. Because the stars shift (and do so at varying speeds, depending on their position), birds must compensate, as was shown by Emlen (1967a). To use true star navigation, the birds would have to use several star patterns and would need several compensation rates for these different groupings. Emlen (1975a) and Able (1980, 1982, 1991) suggested that although no data as yet demonstrate star navigation convincingly, data do exist to support the hypothesis that birds use the stars as a directional or compass mechanism.

Olfactory Cues

Female Pacific salmon lay eggs in stream beds among the cracks and crevices in the rocks, where male salmon fertilize them by depositing sperm over the eggs. After hatching, the

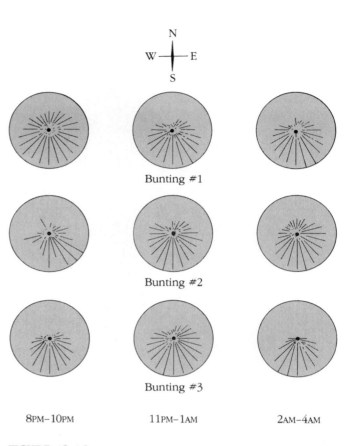

FIGURE 13.14   Response of indigo buntings to the night sky in a plantetarium setting.
Within each circle, the length of a vector (directional line) indicates the amount of migratory restlessness activity oriented in that particular compass direction. During three 2-hour time blocks at night, the orientation of indigo buntings does not shift. This indicates that the birds are capable of time compensation to account for the movement of the pattern of stars in the sky in an apparent circle.
Source: Redrawn after Emlen, S. T. 1967. Migratory orientation in the indigo bunting. *Passerina cyanea. Auk* 84:463–89.

fry develop in the home stream during the summer and then migrate to the ocean, where they mature. Some 3–5 years later (depending on the species), most of the surviving adults return to spawn and die in the stream where they hatched (figure 13.15).

How do the salmon find the stream where they hatched? The results of numerous experiments conducted over many years (Hasler 1960, 1966; Hasler et al. 1978) indicate that young salmon fry are imprinted with the odor of the natal stream. When ready to spawn, they apparently respond to subtle differences in the odor structure as they move from the ocean into their natal river mouth (Johnson 1987). They then follow the odor or pattern of odors from the home stream, making correct turns at each junction between streams until they arrive back at the natal site to spawn.

The importance of olfaction in pigeon homing has been much more controversial. Most early workers ignored the possible role of olfactory cues because birds had been widely assumed to have little or no sense of smell. A number of species of birds are now known to use odor to find

FIGURE 13.15   Pacific salmon migrating upstream.
After spending much of their adult life in the ocean, Pacific salmon migrate upstream to breed in the same location where they hatched. They respond to chemical cues in the water on which they imprinted as juveniles.
Source: Photo by Ronald Thompson/Frank W. Lane.

food and nest sites, however. Papi (1990, 1991) and Wallraff (1991) hypothesize that pigeons released in an unfamiliar location use an atmospheric odor profile to figure their position in space relative to home, that is, form a navigational map. For example, once a pigeon knows it is north of the loft, it then flies south, using the sun or stars as a compass.

The most direct test of such a hypothesis involves removing the olfactory capacity and testing for orientation and homing behavior. The olfactory capacity of animals has been removed by three different techniques: (1) nerve transection; (2) plugging the nostrils with plastic tubes that permit air passage, but preclude contact with the nasal mucosa; and (3) using local anesthesia (Wallraff 1986). When the olfactory sense is blocked or when nerves are transected, its influence on orientation can be determined. However, some behavioral effects can result that are not related to the homing ability being tested. Therefore, unequivocal evidence of the role of olfaction is difficult to show. Indeed, studies by Keeton and Brown (1976), Hartwick et al. (1978), and others found little or no effect on homing behavior when the nasal passages were blocked or a local anesthetic was used on the nasal epithelium. More recent studies, reviewed by Bingman (1998), tend support a role for olfaction in pigeon homing, however.

Magnetic Field

Evidence from several sources has accumulated that supports a role for geomagnetic cues in orientation and navigation in some birds. Early investigators failed to demonstrate that birds could sense the 0.05 gauss geomagnetic field of the earth (Emlen 1970; Kreithen and Keeton 1974a). Brookman (1978) conditioned pigeons to magnetic fields in a laboratory experiment. Measurement of Zugunruhe in European robins (*Erithacus rubecula*) provided a second line of evidence that animals use geomagnetic cues. Birds were placed in a cage surrounded by Helmholtz coils, which provide an artificial magnetic field. Wiltschko and Wiltschko (1972a,b) used this apparatus to demonstrate that the birds were not responding to the polarity of the horizontal component of the magnetic field; they did not shift the direction of their migratory restlessness when horizontal polarity was shifted. However, they did reverse direction when the vertical component was reversed; and they reversed the direction of orientation in a predictable manner. In an artificial magnetic field with a zero vertical component and strong horizontal component, the activity was random.

Other evidence for the role of magnetic fields in orientation comes from the work of Southern (1969, 1972) on ring-billed gulls (*Larus delawarensis*). Gulls in cages exhibited a strong tendency to walk in a southerly direction, except when storms produced temporary aberrations in the earth's magnetic field. Also, when gulls with small magnets affixed to their backs were taken some distance from home and then released, they displayed a random pattern of dispersion. Control gulls without the magnets exhibited normal and consistent directional headings to the south. Additional confirmation of the role of geomagnetic cues was reported by Keeton (1971) and Larkin and Keeton (1976), who used bar magnets attached to the wings of homing pigeons. More recent research (Phillips and Borland 1992; Semm and Demaine 1986; Wiltschko and Wiltschko 1988) suggests that the photopigment rhodopsin is capable of converting magnetic fields to nerve impulses. A hypothetical array of specialized photoreceptors in the eye may organize a directional compass that is calibrated through experience.

Investigators have used radar and airplane tracking to monitor the behavior of homing pigeons fitted with Helmholtz coils on their heads (Walcott 1972, 1977). Further, Walcott et al. (1979) reported the existence of an organ along the midline of the pigeon's brain that contains magnetite granules. This organ may prove to be a source of sensitivity to geomagnetism in the pigeon.

Able and Able (1990a) tested Savannah sparrows for geomagnetic orientation. Birds reared in outdoor cages exposed to the daytime or night sky oriented in the proper direction, whereas birds reared in the laboratory did not orient properly when tested in an Emlen funnel apparatus. If the direction of the apparent geomagnetic force was changed to east-southeast, the outdoor birds oriented in a northeast-southwest direction—significantly different from the control birds' orientation. Able and Able proposed

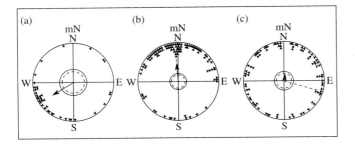

FIGURE 13.16 **Directional preferences of pied flycatchers. Birds were reared under natural sky in various magnetic fields and tested in the local geomagnetic field without visual cues.** (a)Controls (N = 41), reared in the natural local magnetic field. (b)Group WSW (N = 110), reared with magnetic north turned to 240° west-southwest. (c) Group ESE (N = 87), reared with magnetic north turned to 120° east-southeast. The headings of the individual test nights are marked at the periphery of the circle; the mean vector is shown as a solid arrow. In (b) and (c), the expected direction, based on the controls and the hypothesis that the course of the magnetic compass is set by celestial rotation, is given as a dashed radius. N: north; mN: magnetic north. The orientation of the WSW group was in accordance with the above hypothesis, but that of the ESE group was significantly more scattered. This may reflect a true asymmetry in the orientation system.

Arrows represent the mean vectors for nest location.

Source: Prinz K. and W. Wiltschko. 1992. Migratory orientation of pied flycatchers: interaction of stellar and magnetic information during ontogeny. Anim. Behav. 44:539-45. Copyright 1992 The Association for the Study of Animal Behavior. Used by permission of Academic Press Ltd. London.

that the geomagnetic compass of the birds is calibrated early in life and that celestial rotation provides the basis for these adjustments to the compass. Prinz and Wiltschko (1992) have shown that in some experimental conditions, celestial cues override geomagnetic information during the premigratory period (figure 13.16).

Much less clear is whether mammals can obtain directional information from geomagnetic cues. Frequent reports of otherwise healthy whales that strand themselves on beaches suggest that these whales have made navigational errors in areas where magnetic minima intersect the coast (Klinowska 1985). When aerial sightings of fin whales off the northeast U.S. coast were plotted on maps, no association could be drawn between location and measurements such as bottom depth or slope. Sightings of migrating, but not feeding, animals, however, *were* associated with areas of low geomagnetic field intensity and gradient in autumn and winter (Walker et al. 1992). On the other hand, sightings of common dolphins (*Delphinus delphis*) off the coast of southern California were related to bottom topography but not to magnetic patterns (Hui 1994).

When white-footed mice were displaced 40 meters from their home areas in woods and released in a circular arena in adjacent fields, exploratory and escape behavior was concentrated in the homeward direction. A second group of mice was treated exactly the same, except that a magnetic field opposite to that of the earth's geomagnetic field was established in the transport tube. Those mice concentrated their activity in the opposite direction from home, suggesting that they had a magnetic sense and used geomagnetic fields as a

compass cue (August et al. 1989). Similar results have been found in several other species of rodents (Bovet 1992) and apparently even in humans (Baker 1987).

Although a wide range of other animals can sense the earth's magnetic field, relatively little is known about the physiological mechanisms involved in this sensory ability (Lohmann and Johnsen 2000).

If an animal is going to get home in unfamiliar terrain without the use of landmarks, it needs a map as well as a compass to know its location in relation to home. For instance, it might be possible to use the magnetic isoclines to get information about longitude. It would, however, need another gradient along a second axis in order to get information about latitude, to fix itself in two-dimensional space. Such a grid-based bicoordinate navigation system has not yet been demonstrated conclusively in any animal.

Meteorological Cues

If we watch migrating birds in the spring or fall, we find that migratory activity tends to be concentrated into a few days or nights. Birds respond to favorable weather conditions (Drury and Keith 1962; Kreithen and Keeton 1974b). Birds often fly downwind on their migratory flights (Bellrose 1967, Bruderer and Steidinger 1972, Richardson 1971). Do birds select favorable winds after they have oriented themselves using other types of cues? Or, as some (Able 1974) have suggested, might the birds be using the wind itself as a directional cue?

Further experiments (Able 1982; Able et al. 1982) involved monitoring the natural nocturnal migratory flights of individuals of several *Passerina* species (buntings) under various weather and wind conditions. When migrants could observe the sun near sunset, or the stars, they flew in the appropriate migratory direction regardless of wind direction. If observations were made under totally overcast skies with no view of sun or stars possible, birds often flew downwind, sometimes in an inappropriate direction. When migrant white-crowned sparrows (*Zonotrichia leucophrys*) were fitted with frosted lenses and released from balloons aloft, they headed downwind, even though that direction was sometimes seasonally inappropriate. These results indicate the visual cues' importance in determining appropriate migratory direction; when deprived of visual cues, some species can use wind direction for orientation, although sometimes inappropriately.

Neural Mechanisms

Much progress was made in the 1970s in new field and laboratory techniques for studying bird orientation and also in overall knowledge of what cues may be important (Able 1978; Emlen 1975a). In the 1980s and 1990s the focus shifted toward understanding the neural mechanisms involved. It is clear from much of the foregoing discussion that many bird species possess multiple systems for orientation; the use of available cues apparently follows some type of hierarchical scheme. The mechanism employed depends on the bird's preferences for cues and the prevailing weather conditions at its location.

An important part of the brain for spatial learning in both mammals and birds is the hippocampus, part of the telencephalon or forebrain. In young homing pigeons (*Columbia livia*), lesions in this area strongly impair the ability to learn a spatial map (Bingman and Jones 1994). In adult, experienced birds, such lesions do not impair homing ability, but if these birds are taken to a new home loft and forced to learn a new map, hippocampal lesions do impair homing performance. Thus, the hippocampus is important for learning the navigational map, but not necessarily in its retention and use (Bingman et al. 1998). The hippocampus is also important for navigation in the vicinity of the loft, where familiar landmarks are used; young, inexperienced pigeons with hippocampal lesions do poorly. The ability to actually use the map seems to reside, in part, in another area of the telencephalon called the piriform cortex. This region is closely connected to the olfactory bulb and lesions to it impair the ability of young homing pigeons to learn a navigational map based on olfactory cues. Several other areas have been implicated in the learning and retention of navigational maps, and

the picture is far from complete. Still less is known about the areas of the brain involved in the compass. Sensory processing seems to take place in the optic tectum, but the more complex learning processes may take place at higher levels such as the hippocampus (Bingman et al. 1998).

Species of North American birds that store food, such as red-breasted nuthatches (*Sitta canadensis*), blackcapped chickadees (*Parus atricapillus*), and scrub jays (*Aphelocoma coerulescens*), have larger hippocampi than do closely related species that do not store food (Sherry et al. 1989). Among species of European titmice (genus *Parus*), the hippocampus grows at about the same rate as the rest of the telencephalon until the birds begin to feed on their own. At that point, hippocampi of food-storing species continue to grow while those of nonstoring species do not (Healy et al. 1994). Lesions of the hippocampus impair the ability of blackcapped chickadees to recover stored food (Sherry and Vaccarino 1989). It has long been known that the hippocampus is involved in spatial learning in mammals (O'Keefe and Nadel 1978). Although rats with lesions in this region can learn single associations in order to recognize a goal, they are unable to form relational representations involving multiple stimuli (Eichenbaum et al. 1990).

SUMMARY

Birds demonstrate the most spectacular feats of migration, with some species traveling tens of thousands of kilometers each year from wintering to breeding sites. Much of what we know about their travels is learned from banding individuals and the use of radar, but radio transmitters are now commonly used. The timing for migration is under the control of endogenous biological clocks that are adjusted by external stimuli, primarily changes in photoperiod. The hypothalamus and pituitary gland are involved in regulating physiological changes prior to migration. These changes include fat deposition and migratory restlessness, or Zugunruhe, and the initiation of migration itself. Migratory behavior has evolved most often among species living in unstable habitats, those with temporal and spatial variability in climate and resource availability. Long-distance migration may have evolved in species that make relatively long trips in search for ephemeral food sources such as ripe seeds, fruit, or nectar.

Among mammals, long-distance migrants include bats, whales, pinnipeds, and ungulates. Most spectacular among fish are the upstream movements of salmon, but eels move long distances from oceans to coastal areas to breed as well. Sea turtles travel thousands of kilometers to lay eggs on the same beaches where they were born. Best known among the invertebrates are the long-distance movements of monarch butterflies.

Whether migrating or simply returning home after a foraging bout, species ranging from chitons to caribou need some means of finding their way. The simplest way is to leave a trail, which works when trips are short and of brief duration. Some species use path integration, in which they somehow keep track of all the turns and accelerations during the outward trip with no external referent. For long-distance migration, when months may pass before returning home, piloting, which involves the use of landmarks, is used when in familiar locations. Many species can maintain a constant bearing when traveling in unfamiliar territory by using the sun or stars as a compass. However, they need other information to figure out which bearing to maintain when in unfamiliar ground; that is, they need a map as well as a compass. Both olfactory and geomagnetic cues have been tentatively proposed to provide the information for a map.

Recent research has been directed at understanding the brain mechanisms involved in migration and homing. The hippocampus is important in spatial learning; birds with lesions in this area have problems learning the positions of landmarks necessary for piloting. Similarly, food-storing species of birds have relatively large hippocampi, and lesions in this area reduce their ability to recover stored food. Other areas of the brain are also involved in the retention and use of information for navigation.

DISCUSSION QUESTIONS

1. We saw that use of molecular techniques has helped answer questions about the evolution of homing behavior of green sea turtles. How might these techniques be used to help understand the evolution of migration in birds?

2. Before the advent of global positioning satellites, how did sailors determine their location on the high seas? Could these techniques be used by nonhuman animals?

3. African mole rats (*Cryptomys hottentotus*) are colonial, subterranean rodents that dig the longest underground burrow systems of any mammal. Burrow systems are linearly arranged, with the main tunnel often more than 200 m long. The main tunnel is usually oriented in a north/south direction. Burda and colleagues (1990) explored the possible role of the geomagnetic field as a cue for underground orientation, considering

that these animals are virtually blind. Family groups were provided with nesting material in a light-proof arena. They were then exposed to the local geomagnetic field (control) or to experimental fields produced by Helmholtz coils surrounding the arena. The symbols in the figure denote the location of nests along the walls of the arena. What conclusions can you draw from these results? What additional experiments might be necessary?

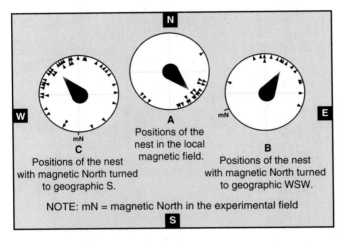

Positions of the nest in the local magnetic field.

A

C
Positions of the nest with magnetic North turned to geographic S.

B
Positions of the nest with magnetic North turned to geographic WSW.

NOTE: mN = magnetic North in the experimental field

Arrows represent the mean vectors for nest location.

Source: Data from H. Burda et al., "Magnetic Compass Orientation in the Subterranean Rodent *Cryptomys hottentotus* (Bathyergidae)" in *Experientia*, 46:528–530, 1990.

4. Recent studies support the hypothesis that birds can use geomagnetic cues for orientation. Data from Baker (1980) suggest that the same may be true for humans. The following data shown here are taken from a hypothetical study similar to a study conducted by Baker. Individuals were blindfolded, driven 10 km from home, and released one by one (just like homing pigeons). The initial orientation bearing for these individuals at two different sites is recorded here (each dot represents the initial bearing for one individual). Fifteen people were tested at each site. What conclusions can you draw about geomagnetic cues and orientation by humans? What methods problems might you encounter in conducting such an experiment?

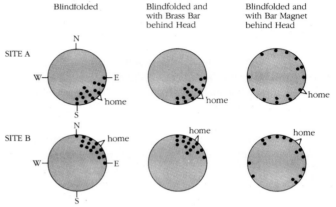

SUGGESTED READINGS

Dingle, H. 1996. *Migration: The Biology of Life on the Move.* New York: Oxford University Press.

Papi, F., ed. 1992. *Animal Homing.* London: Chapman & Hall.

Rankin, M. A., ed. 1985. *Migration: Mechanisms and Adaptive Significance.* Port Aransas, TX: Marine Science Institute.

CHAPTER 14

Habitat Selection

The earliest humans no doubt were aware that different kinds of plants and animals live in different habitats. The surface of our planet consists of a mosaic, or patchwork, of habitat types, and the distribution of organisms is neither uniform nor random. What causes these differences?

Plants are generally dependent on natural agents for dispersal, such as water or air currents or other organisms. The result is an essentially random dissemination of plant individuals; few ever reach environments conducive to survival and reproduction. As we saw in chapter 13, animals have well-developed locomotive abilities, at least during some point in the life cycle, and they play more active roles in finding places to live. **Habitat selection** can be defined as the choosing of a place in which to live (Rosenzweig 1990). This definition does not imply that the choice is necessarily a conscious one or that individuals make a critical evaluation of the entire constellation of factors confronting them. More often the choice is an "automatic" reaction to certain key aspects of the environment.

This chapter concerns the distribution of species (the presence or absence of a species in a particular habitat) and the dispersal of the individuals of a species within the habitat. We first consider how dispersal ability, other organisms, and physical and chemical factors restrict habitat use. Next, we examine the choice of breeding sites and why so many animals disperse from their place of birth. We then look at the role of proximate factors in habitat selection—the environmental and social cues that influence the choice of a place to live. Finally, the roles of genetics, early experience, learning, and tradition in the development of habitat preferences are examined.

When examining the distribution and abundance of most species, we often find areas of high population density near the geographic center of the species' range; abundance

Presence or Absence of Species: Factors Restricting Habitat Use
Dispersal Ability
Behavior
Interactions with Other Organisms
Physical and Chemical Factors

Dispersal from the Place of Birth
Inbreeding Avoidance Hypothesis
Intraspecific Competition Hypothesis
Examples of Dispersal
Inbreeding Versus Outbreeding

Habitat Choice and Reproductive Success
Proximate Factors: Environmental Cues
Determinants of Habitat Preference
Heredity
Learning and Early Experience
Tradition

Theory of Habitat Selection

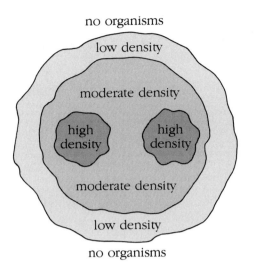

no organisms

low density

moderate density

high density

high density

moderate density

low density

no organisms

FIGURE 14.1 **Typical relationship between distribution in space and density for a species.**
Areas of optimal habitat support the highest densities and are surrounded by suboptimal habitats with lower densities. This distribution pattern contrasts with that shown in figure 14.3.

decreases outwardly (figure 14.1). The **fundamental niche** (the multidimensional space that a species occupies under ideal conditions with no competition) is often much larger than the **realized niche** (the space occupied under real-world conditions that involve competitors, predators, and disease).

PRESENCE OR ABSENCE OF SPECIES: FACTORS RESTRICTING HABITAT USE

If a species occupies an area and reproduces there, we know that all its needs are met and that it can compete with other species successfully. A useful way to identify the factors affecting the distribution of a species is to determine why it is absent from a place. We can examine possible reasons one by one, by using transplant experiments, by giving individuals a choice of artificial habitats in the laboratory, or by building enclosures in the field that encompass different habitats.

In a transplant experiment, organisms are moved to a new environment, and their survival and reproduction are monitored. Because organisms sometimes survive but fail to reproduce, a long-term project covering several generations is necessary. In a study of the distribution of heathland ants in England, Elmes (1971) dug up colonies and moved them to sites that differed in temperature and in moisture content. He moved 18 colonies of one species *(Lasius niger)* that normally inhabits low, wet heathland to higher, drier sites, and moved 6 colonies to other low, wet areas as controls. Elmes monitored their survival over a 5-year period and found that the higher the colony was moved, the less likely it was to survive (table 14.1). Those at the 13-m level did as well as the controls at 11 m. One of the main factors that affected survival was competition with other species that normally inhabit the higher, drier part of the heath. This experiment

demonstrates the need for controls in research and the need for long-term monitoring of transplants.

If a transplant is successful, two possible factors may explain why a species is not found naturally in the transplant area: (1) the area is inaccessible because the dispersal ability of the organism is limited, or (2) the organism fails to recognize the area as a suitable habitat (referred to by Lack [1933] as the "psychological factor"). If a transplant fails, the causal factors may be the presence of other species (competitors, parasites, pathogens) or physical and chemical factors (temperature, pH, and so on). When trying to understand why a species is absent from an area, we can thus proceed through a series of steps (Krebs 1985), as shown in figure 14.2.

Dispersal Ability

Is a species absent from a place because it cannot get there? Many cases of successful introduction by humans have demonstrated that locomotive abilities adequately explain a species' absence. The European starling *(Sturnus vulgaris)* originally was found in most of Europe and Asia. After several unsuccessful introductions of small numbers of starlings into the United States, 80 pairs were released in New York City's Central Park in 1890 by Eugene Scheifflin of the Acclimatization Society. The "goal" of this group was to familiarize Americans with all the birds in Shakespeare's plays (Miller 1975). Within 50 years, starlings had reached the West Coast, and today they are probably the most numerous bird species in the country. Starlings nest in tree cavities and are more aggressive than most native species; they will even evict larger species such as flickers *(Colaptes auratus)* from nest holes. The eastern bluebird *(Sialia sialis)* is one native species that now occupies only a fraction of its former range partly because of its unsuccessful competition with the starling. Although they are insectivorous during the summer, starlings are generalists and switch to seeds in the winter; many native insectivorous species must migrate south to find insects. Other aspects of the starlings' habits have led to their success in modern industrialized countries: a tolerance of loud noises and air pollution; a willingness to roost and perch in various places, from trees to bridge supports; and a preference for feeding in grassy areas.

Ecologists sometimes characterize species as being r- or K-selected; r refers to the rate of population increase, and K refers to the number in the population at the upper limit, or carrying capacity, of the environment. Species that are r-selected have high reproductive rates, rapid development, and great powers of dispersal (MacArthur and Wilson 1967); they tend to live in ephemeral environments in which recolonization of areas is necessary. Diamond (1974) studied the birds of New Guinea and nearby islands and found that certain species were always the first to recolonize islands that had lost their fauna due to volcanic explosions or tidal waves; he referred to these r-selected species as "supertramps." K-selected species have low reproductive rates, slow development, and limited powers of dispersal; characteristically inhabitants of stable environments, these species may fail to colonize new areas separated by relatively small

TABLE 14.1 Transplants of heathland ant colonies in England

Ant colonies of *Lasius niger,* which naturally occur in low, wet areas, were moved either to a similar habitat as a control or to higher, drier sites. Each transplanted colony was observed for 5 years.

Location of Transplant	\multicolumn{6}{c}{Years Survived}	Total Number of Colonies Transplanted
	0	1	2	3	4	5+	
Control							
Lasius niger zone 11-m level	1		1			4	6
Dry heath							
13-m level			1		1	3	5
17–20-m level	6	3				1	10
25-m level	1	1	1				3

Source: Data from G. W. Elmes, "An Experimental Study on the Distribution of Heathland Ants," *Journal of Animal Ecology,* 40:495–99, 1971.

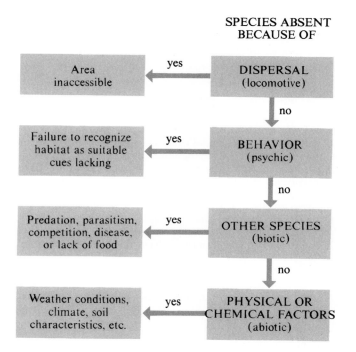

FIGURE 14.2 **Methodological approach for studying the geographical distribution of a species.**
A useful way to proceed in the analysis of a species' distribution is to determine why that species is *not* in a particular place. Four basic factors are listed; behavior may be involved in each and the factors may interact.
Source: Data from C. J. Krebs, *Ecology: The Experimental Analysis of Distribution and Abundance,* 3d ed., copyright 1985 Harper & Row Publishers, Inc., New York.

barriers. One island Diamond studied was separated from New Guinea by only 10 m of water, yet it had only half the New Guinea species expected on the basis of the availability of comparable habitats in the two areas.

Behavior

Sometimes behavior patterns keep species from occupying apparently suitable habitats. Two closely related species of British birds, the tree pipit (*Anthus trivialis*) and the meadow pipit *(A. pratensis)*, are both ground nesters and eat similar types of food, but the tree pipit is absent from many treeless areas that the meadow pipit inhabits. The reason has to do with behavior associated with song: the tree pipit ends its aerial song on a perch, such as a tree or pole; the meadow pipit ends its aerial song on the ground. Thus, the tree pipit is excluded from areas that it could otherwise occupy because of a specific behavior pattern (Lack 1933).

Female mosquitoes of the genus *Anopheles* (many species of which transmit such diseases as malaria) are very particular about where they lay their eggs. In the southern part of India, *Anopheles culifacies* eggs and larvae are found only in new rice fields, where the plants are less than 12 inches high; however, eggs transplanted to old fields have yielded normal numbers of larvae and adults, and other species of *Anopheles* lay their eggs in these mature fields. Russell and Rao (1942) demonstrated that the mechanical obstruction of the rice plants inhibited *A. culifacies* females from laying eggs. Glass rods or bamboo strips "planted" in the water had the same effect; shade was not a factor. *A. culifacies* females oviposit while they are on the wing, performing a hovering dance 2–4 inches above the water; possibly the obstructions interfered with this oviposition dance. However, in one experiment researchers partially submerged a box with no lid in the shallow water; although the box did not directly interfere with oviposition, few eggs were laid.

Why should a species not take advantage of a suitable habitat? One possibility is that the habitat is not actually suitable, perhaps because of competition, predation, or other factors the scientist may have failed to detect. Or such habitats may not have been suitable in the past: If organisms responding to certain environmental cues in previous optimal habitats left more offspring, their genetically influenced behaviors would become widespread and persist. New environments, although suitable, may not contain those cues and therefore are not utilized.

Interactions with Other Organisms

Do other organisms keep a particular species out? Even if an individual can and "wants" to get to a place, other factors

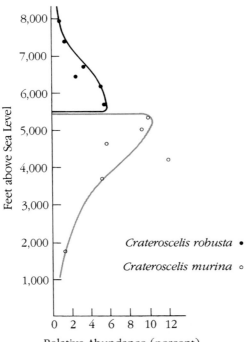

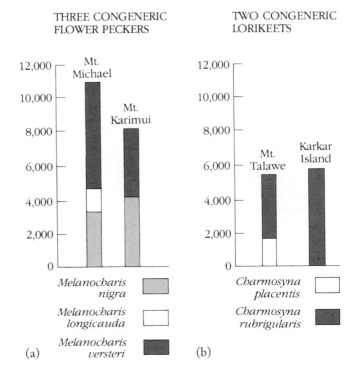

FIGURE 14.3 **Influence of other species on distribution of warblers.**

Abundance is measured as a percentage of all bird individuals observed. As one sample from the side of a mountain in New Guinea shows, *Crateroscelis murina* reaches maximum abundance at 5,400 feet and is abruptly replaced by *C. robusta*. This distribution can be compared with that in figure 14.1.

Reprinted with permission from J. M. Diamond, "Distributional Ecology of New Guinea Birds," in *Science*, 179(1973):759–69. Copyright 1973 American Association for the Advancement of Science.

FIGURE 14.4 **Altitudinal ranges of species of birds on New Guinea mountains and surrounding islands.**

In (a), three similar congeneric flower peckers, *Melanocharis nigra, M. longicauda,* and *M. versteri,* occupy nonoverlapping areas up to about 11,000 feet on Mt. Michael. On Mt. Karimui, which is smaller and more isolated than Mt. Michael, *M. longicauda* is absent. In (b), two congeneric lorikeets, *Charmosyna placentis* and *C. rubrigularis,* occupy an area on Mt. Talawe, but only *C. rubrigularis* colonized Karkar Island. Altitudinal ranges in all cases are nonoverlapping and are larger in the absence of other species.

may prevent its becoming established. These factors could be predators, parasites, disease agents, allelopathic agents (poisons or antibiotics), or competitors. Demonstrating conclusively that one species prevents an area from being colonized by another is difficult, but it has been indicated with experimental and observational data. For instance, an elaborate series of experiments by Kitching and Ebling (1967) showed how mussels *(Mytilus edulis)* were kept out of protected bays along the coast of Ireland by three species of crabs and one species of starfish. Where the coast was unprotected, crabs were restricted by wave action and small mussels could survive. In sheltered waters, mussels survived only in areas such as steep rock faces that the predators could not reach. Kitching and Ebling proposed that four criteria must be met before we can conclude that a predator restricts the habitat of its prey: (1) if they are protected from predators, prey will survive when transplanted to a site where they normally do not occur; (2) if the distributions of prey and predator are negatively correlated; (3) if the predator is observed eating the prey; and (4) if the predator can be shown to destroy prey in transplant experiments.

The more similar two species are, the more likely they are to compete intensely and thereby restrict each other's distribu-

tion, but once again positive proof is difficult to obtain. Most often we rely on presumptive evidence from comparisons of the closely related species' distributions in several habitats. Although a species is usually less numerous at the edge of its distribution than at the center (as shown in figure 14.1), it is sometimes abruptly replaced by a close relative, with both species at maximum density at the interface (figure 14.3). If interspecific competition restricts distribution, we would expect one species to extend its range in the absence of the other. This phenomenon was demonstrated by Diamond (1978) with bird distributions on mountain tops in New Guinea, where a second, closely related species may be absent due to its inability to disperse (figure 14.4). In these cases, the single species occupied a much larger altitude range than it did on islands where both were present. Competitive exclusion was actually witnessed by Orians and Collier (1963) when colonial tricolored blackbirds *(Agelaius tricolor)* moved into a marsh already occupied by red-winged blackbirds *(Agelaius phoeniceus)*. After the invasion, the red-wing territories were restricted to the periphery.

The studies mentioned above, however suggestive, do not clearly show that competition by one species excludes the other. The responses are correlational, and experiments

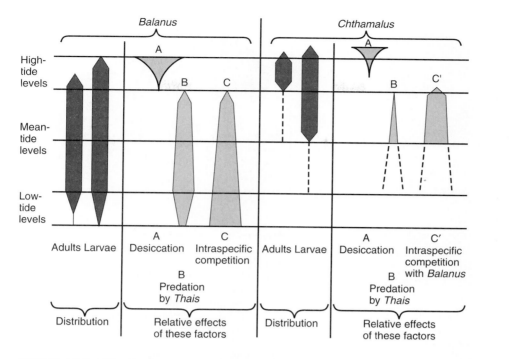

FIGURE 14.5 **The distribution of adult and newly settled larvae of two species of barnacles.**
The upper limit to the distribution is set by desiccation, while the lower limit is set by a combination of competition for space and predation by a species of snail.
Source: Redrawn from Connell, J. H. 1961. The influence of interspecific competition and other factors on the distribution of the barnacle *Chthamalus stellatus*. *Ecology* 42:710–23.

are needed to demonstrate cause and effect. One such experiment is to remove one species and note changes in nearby competitors, or to introduce a closely related species and monitor the success of each, as Vaughan and Hansen (1964) did with two species of pocket gophers *(Thomomys bottae* and *Thomomys talpoides)*. Slight differences between species in dispersal powers and environmental tolerances led to one or the other species' winning out.

Physical and Chemical Factors

If a transplant experiment fails, and no evidence exists that biotic factors have eliminated the species, some combination of physical and chemical factors may be involved. Each organism has a range of tolerances for these factors, and much of its behavior is directed toward staying within these limits. Temperature and moisture are the main factors that limit the distribution of life on earth, but physical factors (such as light, soil structure, and fire) and chemical factors (such as oxygen, soil nutrients, salts, and pH) are important as well. Most research on the effects of physical and chemical factors on organisms is done by physiological ecologists and will not be discussed further here. The effect of these factors on reproduction is treated in chapters 8 and 17.

The classic studies of Connell (1961) show that a combination of factors influences the distribution of organisms. He studied two species of barnacles *(Balanus balanoides* and *Chthamalus stellataus)* that live in the intertidal zone of rocky coastlines of Britain. Physical factors, especially tolerance to

desiccation and high temperatures, predation by snails (genus *Thais*), and interspecific competition for space all interacted to determine distribution and abundance (figure 14.5).

DISPERSAL FROM THE PLACE OF BIRTH

The discussion thus far has dealt with the problem of habitat selection at the species distribution level. Animals may also make decisions about whether to remain at (or return to) the natal site or to disperse to other breeding locations. Natal dispersal means leaving the site of birth or social group (emigration), traversing unfamiliar habitat, and settling into a new area or social group (immigration). Moving away from familiar ground is risky because the individual is unfamiliar with the location of food and shelter and is no longer in the presence of familiar neighbors and relatives.

In most species of birds and mammals, members of one sex tend to disperse, while members of the other sex are **philopatric,** breeding near the place where they were born. Among mammals, it is usually the males that disperse, while among most species of birds (perching birds of the order Passeriformes), the opposite is true (Greenwood 1980). The reason for this difference may be that most bird species are monogamous, and the male defends a territory that contains resources vital to him and his mate. It is probably easier for a male to establish such a territory and attract a mate in or near his natal site, where he is familiar with the location of resources

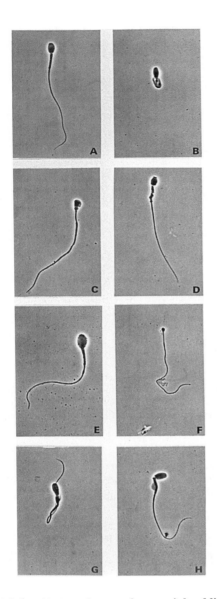

FIGURE 14.6 **Abnormal sperm from an inbred lion population in Africa.**
(a) Normal; (b) tightly coiled flagellum; (c) missing mitochondrial sheath; (d) abnormal acrosome and deranged midpiece; (e) macrocephalic with abnormal acrosome; (f) microcephalic with missing mitochondrial sheath; (g) bent flagellum; (h) bent neck with residual cytoplasmic droplet.
Source: Dave Wildt.

and/or predators. On the other hand, many species of mammals are polygynous. The females form the stable nucleus, and the males attempt to maximize their access to them, frequently moving from one group to another (Greenwood 1980).

The causes of dispersal can be understood at several different levels. At the proximate level we wish to know the immediate reasons why an individual leaves the natal area. For instance, it might be forced out by its parents or other residents, or it might respond involuntarily to increases in testosterone levels associated with sexual maturation. At the ultimate level we wish to know the long-term, evolutionary causes of dispersal. For instance, individuals that fail to disperse may have lower reproductive success because their off-

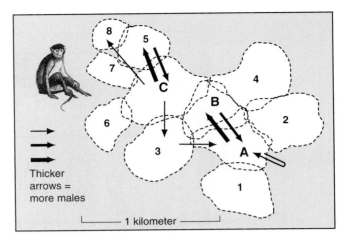

FIGURE 14.7 **Group transfer by natal and young adult male vervet monkeys from the three study groups between March 1977 and July 1982.**
Arrows indicate direction of movement; each line indicates one male. Letters indicate ranges of the study groups; numbers indicate ranges of regularly censused groups. Only groups with ranges adjacent to the study group are shown. Males usually transferred to neighboring groups with brothers or age peers.
Source: Data from D.L. Cheney and R.M. Seyfarth, "Nonrandom Dispersal in Free-ranging Vervet Monkeys: Social and Genetic Consequences," *The American Naturalist*, 122:392-412, 1983.

spring are inbred and therefore less viable. Natural selection would then favor dispersers.

Inbreeding Avoidance Hypothesis

The ultimate cause of dispersal from the natal site has been argued by many to be inbreeding avoidance. The costs of inbreeding, referred to as **inbreeding depression,** have been documented in many laboratory and zoo populations (Ralls et al. 1979), but only more recently have they been studied in natural populations. Inbreeding depression manifests itself through reduced reproductive success and survival of offspring from closely related parents compared to offspring of unrelated parents. It is caused by increased homozygosity of the inbred offspring and the resulting expression of deleterious recessive alleles. For example, in African lions (*Panthera leo*), males from a small, inbred population showed lower testosterone levels and more abnormal sperm than did males from a large, outbred population (Wildt et al. 1987) (figure 14.6). When both inbred and outbred white-footed mice (*Peromyscus leucopus*) were released back into natural habitat, the inbred stock survived less well than the outbred stock, although differences between the two stocks were not great in the laboratory environment (Jiménez et al. 1994).

If one or the other sex disperses, there will be less chance of matings between related individuals. Among black-tailed prairie dogs (*Cynomys ludovicianus*), young males leave the family group before breeding; females remain. Also, adult males usually leave groups before their daughters mature (Hoogland 1982). Among primates such as vervet monkeys (*Chlorocebus aethiops*) and baboons (*Papio anubis*), males usually leave the natal group at sexual maturation or shortly after. They usually transfer to a neighboring group with age

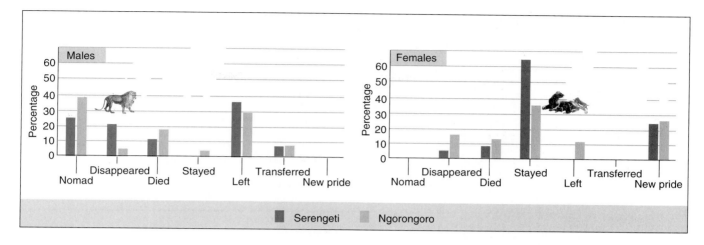

FIGURE 14.8   **The fate of subadult lions by 4 years of age at two African sites.**
Note the differences between the sexes in dispersal pattern.
Source: Data from Anne E. Pusey and C. Parker, "The Evolution of Sex-biased Dispersal in Lions" *Behavior,* 101:275-310, 1987, E.J. Brill Publishers, Leiden.

peers or brothers (figure 14.7). Several years later they may again transfer alone to a third group. Cheney and Seyfarth (1983) argued that this pattern of nonrandom movement followed by random movement minimizes the chances of mating with close kin. Packer (1979) reported that a male baboon, after failing to disperse at sexual maturity and after mating with relatives, sired offspring with low survival rates compared to the offspring of outbred males. Thus, in this case there seems to be a real cost associated with inbreeding.

Intraspecific Competition Hypothesis

A second cause of dispersal from the natal site may be competition with conspecifics (Dobson 1982; Moore and Ali 1984). Most species of mammals are polygynous, with males mating with more than one female. Males may be forced to disperse as they compete for access to females. Although the inbreeding avoidance hypothesis predicts that one sex should disperse, it doesn't predict which sex should disperse; the competition hypothesis predicts that males should be the dispersing sex in polygynous species. According to Greenwood (1980), in such systems the reproductive success of males is limited by the number of females with which they can mate, and males are likely to range farther than females as they search for mates. Females, on the other hand, are limited by resources (food and nesting sites) that can best be obtained and defended by staying at home. Among group-living mammals, females typically form the stable nucleus, and the males attempt to maximize their access to them, frequently moving from one group to another.

Hamilton and May (1977) proposed a somewhat different competition model in which animals disperse so as to avoid local resource competition with close relatives and thus avoid lowering their indirect fitness. In a new habitat they are likely to be competing with nonrelatives and therefore would suffer no such cost. These models are not contradictory, but more research is needed on both the proximate and ultimate causes of dispersal.

Examples of Dispersal

Lions

Dispersal patterns in lions follow the typical mammalian pattern: females usually remain in or near their natal pride, whereas males always leave, usually before 4 years of age, to become nomads and/or take over new prides (figure 14.8) (Pusey and Packer 1987). Competition with other males seems to be an important factor, because departures most often occur when a new coalition of males takes over the pride. However, some males appear to leave voluntarily, either in search of mating opportunities or to avoid breeding with kin. Once a coalition takes over a new pride, it is usually ousted by another coalition within a few years, or it leaves to take over another pride. In all cases, males leave the pride before their daughters start mating. An additional factor is that new coalitions of males kill the young cubs in the new pride (see chapter 16). Thus, breeding males must remain in a new pride long enough to ensure the survival of their cubs. Pusey and Packer conclude that male-male competition, mate acquisition, protection of young cubs, and inbreeding avoidance all play roles in the evolution of dispersal patterns of lions.

Belding's Ground Squirrels

The question of why individuals disperse has been addressed at several different levels of analysis by Holecamp and Sherman (1989). Belding's ground squirrels (*Spermophilus beldingi*) follow the typical mammalian pattern: females remain in the natal area for life, while males disperse. The proximate causes of male dispersal seem to be the effects of prenatal exposure to testosterone (organizational effects, chapter 8) and the attainment of a critical body weight. Effects of testosterone later in life (activational effects) seem less important, since castration of males just prior to natal dispersal did not prevent dispersal. Holecamp and Sherman were not able to test the inbreeding avoidance and competition

TABLE 14.2   **Why juvenile male Belding's ground squirrels disperse**

Answers have been found at each of four levels of analysis.

Levels of analysis	Summary of findings
Physiological mechanisms	Dispersal by juvenile males is apparently caused by organizational effects of male gonadal steroid hormones. As a result, juvenile males are more curious, less fearful, and more active than juvenile females.
Ontogenetic processes	Dispersal is triggered by attainment of a particular body mass (or amount of stored fat). Attainment of this mass or composition apparently also initiates a suite of locomotory and investigative behaviors among males.
Effects on fitness	Juvenile males probably disperse to reduce chances of nuclear family incest.
Evolutionary origins	Strong male biases in natal dispersal characterize all ground squirrel species, other ground-dwelling sciurid rodents, and mammals in general. The consistency and ubiquity of the behavior suggest that it has been selected for directly across mammalian lineages.

From "Why Male Ground Squirrels Disperse," by Kay E. Holecamp and Paul W. Sherman, *American Scientist*, 77:232–39. Copyright © 1989. Reprinted by permission of *American Scientist*, journal of Sigma Xi, The Scientific Research Society.

hypotheses directly, but concluded that inbreeding avoidance was the more likely means by which dispersal increased fitness (table 14.2).

Inbreeding Versus Outbreeding

If inbreeding depression were the only factor involved, we might expect individuals to disperse as far as possible from relatives. However, such is not usually the case. According to Shields (1982), most species that have been adequately studied are philopatric, remaining close to the place of birth. He cites apparent cases of **outbreeding depression,** in which matings between members of different populations within a species yield less fit offspring. Members of a population may possess adaptations to local conditions that are lost with outbreeding—two areas might differ slightly in temperature, humidity, or in the types of food available. If each population is genetically adapted to these conditions, they would be better off mating with individuals with those same adaptations. Perhaps that is why white-crowned sparrow (*Zonotrichia leucophrys*) females respond sexually to the songs of males from their natal area but not to males from other areas (Baker 1983; see also chapter 12).

A certain degree of inbreeding may be advantageous for other reasons. In sexually reproducing organisms, the loss of genes in the offspring can be reduced by one-half if the parents are related. Furthermore, according to Shields (1982), gene complexes are less likely to be disrupted in matings between relatives. Finally, kin selection (chapter 4), which results in the evolution of cooperative behaviors, can operate only when relatives are in proximity. We predict more cooperative behavior within philopatric species and within the philopatric sex. Indeed, female Belding's ground squirrels are philopatric and engage in altruistic alarm calling, while males, who disperse away from relatives, do not alarm call (Sherman 1981; see also chapter 12).

From the preceding arguments we might predict some sort of optimal inbreeding strategy, in which matings between very close relatives (siblings, or parents and offspring) are avoided but matings with more distant relatives

are favored. Bateson (1982) found that female Japanese quail (*Coturnix coturnix*) spend more time in proximity to first cousins than in proximity to siblings or more distant relatives. If these tendencies reflect later sexual preferences, an optimal level of inbreeding might be among first cousins.

Similar results were reported for white-footed mice (*Peromyscus leucopus*). In laboratory tests, estrous females preferred males who were first cousins over nonrelatives or siblings (Keane 1990). Heavier pups and larger litters resulted from matings between first cousins than for other degrees of relationship. However, it is not yet clear how much inbreeding goes on in natural populations of this species; Wolff et al. (1988) used biochemical tests to determine maternity and paternity and concluded that little if any mating between close relatives seems to occur. Cues used to assess genetic relatedness may come from genes in the major histocompatibility complex (MHC), a cell recognition system that is used by the immune system to identify self and nonself (Penn and Potts 1999). See chapter 12 for a further discussion of kin recognition.

HABITAT CHOICE AND REPRODUCTIVE SUCCESS

Although it is assumed that an individual that makes a "correct" choice of habitat has higher reproductive success than one that makes an "incorrect" choice, few studies have actually measured this relationship. Witham (1980) studied the life history of the aphid *Phemphigus betae,* a plant parasite about 0.6 mm long that feeds on leaves of the cottonwood tree. In the spring, after hatching from eggs laid the previous fall in the bark of the tree, females (called stem mothers) move up the trunk and select a leaf on which to feed. This activity triggers the formation of a hollow gall on the leaf, within which the female produces offspring parthenogenetically.

Witham found that females settling on large leaves have higher reproductive success than females on small leaves. Not surprisingly, aphids select the largest leaves, leaving small ones vacant. However, latecomers may have to choose whether to take an already occupied large leaf or an unoccupied small one. If she takes an occupied leaf, she will have to settle farther from the base, where there is less food. Stem mothers farther from the base were found to be smaller in size, and they produced fewer young than those closer to the base. Stem mothers may engage in shoving and kicking contests that last for days, with the largest aphid usually getting the basal position. Choice of habitat in these insects is nonrandom and results in higher average fitness than would random leaf selection.

Up to this point, we have assumed that habitat selection operates to increase the reproductive success of the individuals involved. Are there cases where members of one species manipulate the habitat selection of another species to their own benefit? It has long been known that parasites can modify the behavior of their intermediate hosts to increase the parasites' chances of being transmitted to their final hosts. The parasitic wasp *Aphidius nigripes* is an endoparasitoid of the potato aphid *Macrosiphum euphorbiae* (Brodeur and McNeil 1989). The wasp completes pupal development within the aphid, which becomes mummified as a result. Nonparasitized aphids are usually found on the undersurface of potato leaves, their favored feeding area, while aphids parasitized by nondiapausing larvae (those that will not overwinter as pupae) usually move to the upper surface shortly before being mummified (figure 14.9) (Brodeur and McNeil 1989). Aphids parasitized by diapausing larvae (those that will overwinter as pupae) tended to leave the host plant and mummify in concealed sites. Therefore, such pupae would be protected over winter and would less likely be eaten by predators or attacked by hyperparasitoids (i.e., parasites of the parasites). Thus, in this example, we see that an animal may select a habitat that is optimal not for itself, but for one of its parasites.

PROXIMATE FACTORS: ENVIRONMENTAL CUES

Organisms may respond directly to abiotic environmental factors. Speed or frequency of locomotion may be dependent on the intensity of stimulation. Isopods, such as *Porcellio,* are found in moist environments; when placed in a humidity gradient, they move faster in drier air and eventually wind up at the moist end of the gradient (Fraenkel and Gunn 1940). Likewise, the turning rates of protozoans are dependent on intensity of stimulation. (Types of orientation behavior involved in habitat selection were discussed in chapter 13.)

Organisms may integrate more than one environmental variable: for instance, temperature and humidity. Several species of fruit flies *(Drosophila)* prefer warm temperatures in a laboratory gradient apparatus (Prince and Parsons 1977), but only if the humidity is high. At low humidity they move to the cooler area, thereby reducing water loss and increasing the probability of survival. This adaptive pattern is a response not only to external cues but also to the individual's physiological state.

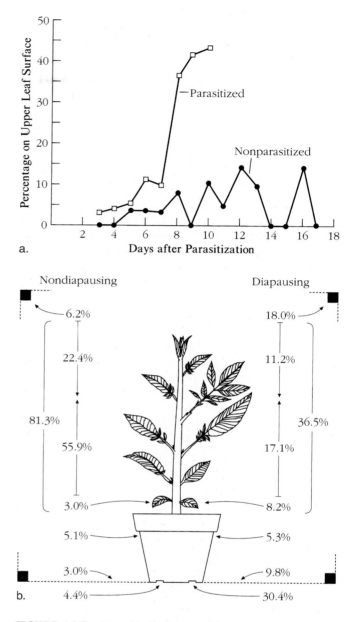

FIGURE 14.9   The effects of a parasitic wasp on the behavior of the potato aphid.
(a) The effect of parasitism by wasp larvae on the distribution of aphid adults. Squares denote parasitized aphids, circles denote nonparasitized aphids. (b) The distribution of aphid mummies containing diapausing and nondiapausing individuals of the parasitoid wasp.

Reprinted with permission from J. Brodeur and J. N. McNeil, "Seasonal Microhabitat Selection by an Endoparasitoid through Adaptive Modification of Host Behavior," in *Science,* 244:226–28. Copyright 1989 American Association for the Advancement of Science.

Sale (1970) hypothesized the existence of a simple mechanism of habitat selection in fish that is based on levels of exploratory behavior. Sense organs monitor specific stimuli in the environment and send a summation of pertinent stimuli back to central nervous system centers, which regulate the amount of exploration. As the constellation of cues approaches some optimum level, exploratory behavior ceases and the animal stays where it is.

a. TEST ENVIRONMENTS

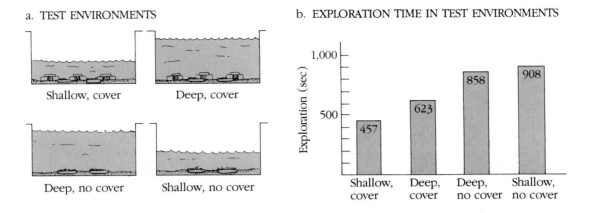

Shallow, cover   Deep, cover

Deep, no cover   Shallow, no cover

b. EXPLORATION TIME IN TEST ENVIRONMENTS

FIGURE 14.10   **Laboratory environments for testing habitat selection in fish.**
(a) In all four test environments small, flat rocks were present bearing a film of algae as a food source. Variables were water level (deep or shallow) and covers under which the fish could hide (present or absent). Exploration time, as shown in (b), was least in the environment with shallow water and cover present. This habitat is the one most similar to the preferred habitat in the field.

An alternative hypothesis is that an animal has a cognitive map of the ideal habitat and that its behavior is goal directed. Working with a species of surgeonfish, the Hawaiian manini *(Acanthurus triostegus)*, Sale (1970) tested juveniles in laboratory tanks with various water depths and bottom covers (figure 14.10). Exploration time was least in the tank with shallow water and bottom cover and highest in the tank with shallow water and no bottom cover. In choice tests and field observations, most fish preferred shallow areas with bottom cover. Thus, there is no need to suggest the inheritance of complex cognitive maps and goal-directed behaviors; rather, the animal simply moves more in an unsuitable habitat and less in a suitable one.

Sale's model still does not explain how the animal "knows" what is suitable and what is not, or how stimuli from multiple cues are integrated. Nor does it explain the role of photoperiod in the response of dark-eyed juncos *(Junco hyemalis)* to photographs of their natural habitat. Wild-caught birds were presented a choice of viewing one of two 35-mm color slides showing different habitats. Birds kept in the lab under a winter photoperiod of 9L:15D (9 hours of light and 15 hours of darkness) preferred (spent more time in front of) slides of their southern winter habitat (pine and hardwood forest). After daylength was increased to 15L:9D, the birds' viewing preferences shifted to the northern summer habitat (grassland and conifer forest) (Roberts and Weigl 1984).

Social cues may also affect choice of habitat. Large-sized juncos (usually males) dominate smaller individuals (usually females and juveniles) in wintering flocks. Ketterson (1979) explained the finding that females usually migrate farther south than males by hypothesizing that subordinate birds are forced to migrate farther to avoid competing with dominants. In their lab study, Roberts and Weigl (1984) found that during the short days (simulating winter),

small subordinate juncos showed the strongest preference for winter scenes. Social cues may also be important among colonially nesting birds. Laughing gulls *(Larus atricilla)* were observed nesting on only one of two similar islands. When a storm destroyed the nesting island, they moved over to the other one. Thus a "suitable" habitat cannot be defined by its structural features alone; social features also play a role (Klopfer and Hailman 1965).

Risk of predation and competition are other factors that may affect habitat use. Hairy-footed gerbils *(Gerbillurus tytonis)* live in vegetated islands in a sea of sand in the Namib Desert of southern Africa (Hughes et al. 1994). Habitat use was determined by tracks in the sand and by how quickly they gave up feeding at stations containing seeds mixed with sand. Gerbils preferred sites around bushes or grass clumps over open areas and were more active on new moon nights than on full moon nights. They also gave up feeding at seed trays sooner in open areas and on full moon nights (figure 14.11). These differences were likely caused by greater risk of predation in open areas and when the moon was full. When striped mice *(Rhabdomys pumilio)*, a close competitor of the gerbil, were removed, gerbils increased foraging activity, especially in the grass clumps.

The proximate cues to which animals respond when selecting a habitat may not be the same as the ultimate factors that have brought about the evolution of the response (Hildén 1965). The ultimate reason a mouse chooses to live in a wooded area rather than a field might be an inability to survive the higher temperatures in the field. But rather than responding directly to temperature, it might cue on the geometric shapes of trees.

The blue tit *(Parus caeruleus)*, a European relative of the chickadee, lives in oak woodlands where most of its preferred food is found (Partridge 1978). But the blue tit establishes its breeding territory each year before leaves and

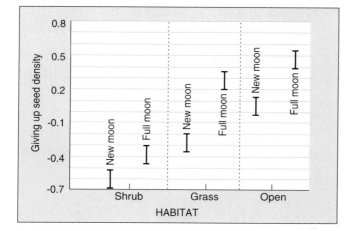

FIGURE 14.11   **The effect of predation risk on habitat use in hairy-footed gerbils.**
Values along the y axis are giving up seed densities (GUDs) ($\log_{10}$ scale) in food trays containing seeds mixed with sand that were placed in different habitats. High values mean that gerbils left the trays when many seeds were still present. N = new moon, and F = full moon; bars denote 95 percent confidence intervals. Gerbils gave up at higher seed densities in open areas and on full moon nights.

caterpillars (the staple food) have even appeared, so it must be using other features, such as shape of the trees, as cues to the habitat. Although birds have been studied intensively, we know little about the signals they respond to when choosing their habitat. Possible cues are stimuli from landscape; sites for nesting, singing, or feeding; food itself; or other animals. In migratory species, it is not even clear when in the life cycle a choice is made. Breeding sites may be selected in late summer or fall before migration, rather than in the spring, as is usually assumed (Brewer and Harrison 1975).

Wood frogs *(Rana sylvatica)* lay their eggs during a brief period each spring in temporary ponds that dry up in the summer. All the egg masses are deposited in one place in the pond. The physical features of this location seem less important than does the presence of an egg mass, which triggers other females to lay their eggs there (Howard 1980). By placing an egg mass in one part of the pond before any other eggs had been laid, Howard was able to induce all the females to lay their eggs there. Eggs in the center of such a mass may be protected from predators and from temperature fluctuations (Berven 1981).

Previous breeding success affects the habitat choice of many bird species; typically, individuals disperse to a new nest site if their previous attempt fails. For example, mountain bluebirds *(Sialia currucoides)* were presented with two types of nest boxes. When breeding for the first time, most birds selected the same type of nest as the one in which they had been raised, even though they were breeding in a new territory. If they had an unsuccessful breeding attempt in one year, the next year they shifted nest sites and nest box type (Herlugson 1981). Collared flycatchers *(Ficedula albicollis)* may select breeding habitat based on the reproductive suc-

cess of other flycatchers in the vicinity (Doligez et al. 1999). Females tended to remain in patches with high average reproductive success, while juvenile males and unsuccessful adult males tended to disperse from such sites, presumably because competition in good patches was intense.

In trying to develop realistic models of habitat selection, it has become clear that rarely, if ever, is a single factor sufficient to explain the choice. More recent studies attempt to incorporate several factors simultaneously to more closely approximate the real world. Thus, a small, nocturnal species of gecko from Australia *(Oedura lesueurii)* is faced with a number of trade-offs when selecting a daytime refuge (Downes and Shine 1998). As an ectotherm, it selects habitats, in part, based on their thermal regime. Ideally, it should find a warm place so it can digest the previous night's food, avoid other larger and dominant geckos, and certainly avoid a place that has the scent of a broadheaded snake *(Hoplocephalus bungaroides)*, a predator of geckos. In laboratory tests using males, avoidance of predators was a higher priority than thermoregulation; lizards would avoid a warmer retreat-site with predator scent in favor of a cooler, unscented site. Dominance interacted with thermal preferences in determining refuge selection, with smaller males forced to use cooler retreat-sites when larger males were present. Finally, smaller males were forced to use either predator-scented retreat-sites or no retreat-site when larger males were present.

There also may be a temporal component to the decisions an animal makes in choosing a habitat. Among species that migrate long distances between breeding and wintering sites, habitat choice may involve a sequence of decisions that is arranged hierarchically (Klopfer and Ganzhorn 1985). Thus, in spring a bird migrating to its nesting grounds may cue on large-scale features of the habitat to get to the general vicinity and shift to small-scale features, such as presence of certain species of plants, to find the exact nest site. In autumn the bird initially makes decisions, possibly genetically programmed, at a large spatial scale, for example, to head south (Hutto 1985). Once in the vicinity of an overwintering area it makes a series of choices at progressively smaller scales, possibly based on the costs and benefits associated with each habitat (figure 14.12). Presumably, food availability is an important factor at these smaller scales.

Understanding the process of habitat selection has become a major goal of conservation biologists in their efforts to save endangered and threatened species. For instance, habitat choice of California spotted owls *(Strix occidentalis occidentalis)* was examined at three spatial scales: landscape, habitat patch, and microsite (Moen and Gutierrez 1997). Roost and nest sites of owls contained fewer habitat patches, that is, the forest was more continuous than surrounding areas not used by owls. Ninety-seven percent of the habitat patches in which owls roosted contained one or more large trees (> 1 m in diameter). These sites also contained high structural diversity compared to unoccupied sites.

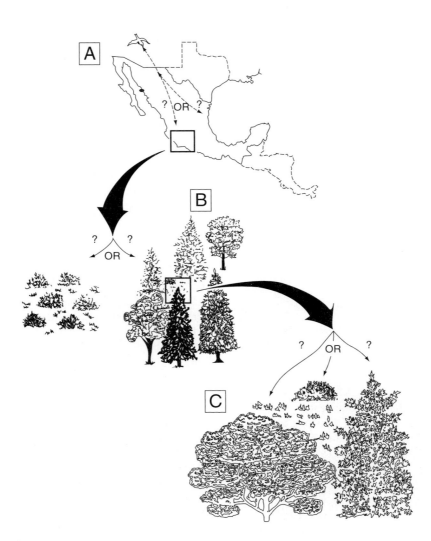

FIGURE 14.12   The hierarchical decision-making process as a migratory bird leaves the breeding area and selects an overwintering site.
Level A represents a large-scale decision as the bird leaves the breeding area and heads generally south for the winter. Once in the vicinity of the wintering grounds, level B, the bird chooses a general habitat type (e.g., forest versus grassland). At a still smaller scale, level C, it chooses a specific vegetation type (e.g., pine trees versus deciduous trees).

Source: Redrawn from Hutto, R.L. 1985. Habitat selection by nonbreeding, migratory land birds. Pp. 455–76 in *Habitat Selection in Birds* (M. L. Cody, ed.). Academic Press, New York.

DETERMINANTS OF HABITAT PREFERENCE

Heredity

How can we sort out the roles that genetics, imprinting, and learning play in habitat choice? If two animals reared from birth in identical environments are found to differ in habitat preference when they are tested as adults, we can conclude that those differences must be due to hereditary factors. When coal tits *(Parus ater)* and blue tits were reared in aviaries with no vegetation and then presented with a choice between oak and pine branches, coal tits preferred pine and blue tits preferred oak (figure 14.13). The

differences correspond to the distribution of these birds in nature and to the response of wild titmice in aviaries (Partridge 1974, 1978). It is perhaps not surprising that blue tits feed more efficiently in oak than do coal tits (Partridge 1976).

A genetic basis for habitat selection has been shown for a wide variety of invertebrates, including insects, molluscs, and crustaceans. Included are preferences of adults for general features of the environment (e.g., meadow versus dense woods in different species of *Drosophila*), background matching in moths and butterflies, where to lay eggs (e.g., different species of plants in butterflies), and preferences for different habitats by larvae (e.g., *Drosophila*) (reviewed in Jaenike and Holt 1991).

TABLE 14.3   Vegetation preference of chipping sparrows

One side of the test chamber contained oak branches; the other; pine. Findings indicate that the preference for pine could be modified to some extent by early experience.

Chipping Sparrows	Percentage of Time Spent in Pine	Percentage of Time Spent in Oak
Wild-caught adults	71	29
Laboratory-reared, no foliage exposure	67	33
Laboratory-reared, oak foliage exposure only	46	54

Source: Data from P. H. Klopfer, "Behavioral Aspects of Habitat Selection: The Role of Early Experience," *Wilson Bulletin,* 75:15–24, 1963.

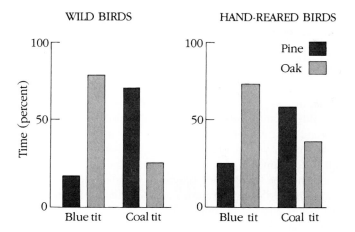

FIGURE 14.13   **Time tits spent in oak and pine habitats in the laboratory.**
Both wild and hand-reared blue tits preferred the oak habitat to the pine habitat. Wild and hand-reared coal tits, however, preferred the pine habitat. These preferences reflect those of each species in the wild.

Learning and Early Experience

Some experimenters have modified the environments of young birds to test whether their genetic predisposition to respond to certain stimuli can be altered by early experience, or **habitat imprinting.** For example, when Klopfer (1963) placed wild-caught chipping sparrows (*Spizella passerina*) in a room containing both pine and oak branches, they preferred the pine, as they usually do in nature (table 14.3). Klopfer reared a group of eight nestlings in a covered cage that prevented their seeing any foliage. When tested at 2 months of age, these birds, like their wild-caught counterparts, preferred pine to oak. Klopfer reared a third group of ten birds in the presence of oak leaves to see if their preference could be shifted. When tested at 2 months, they were ambivalent, spending about equal time in each habitat. This result suggests that their inherited predisposition for pine could be affected to some extent by early experience.

After these tests, the oak-reared birds were removed to a large aviary located in a pine-hardwood forest. Within the aviary itself, however, were only some small broadleaf shrubs. When Klopfer retested the birds, their preference for

pine had increased. After 8 months in an aviary containing pine, the birds' preference for pine was even stronger. These results suggest that the birds retain the ability to shift to the naturally preferred vegetation type.

One of the classic studies on habitat selection in mammals is Wecker's on deermice (*Peromyscus maniculatus*) (Wecker 1963). This species, one of the more common North American rodents, is divided into many geographically variable subspecies of two general types: the long-eared, long-tailed forest form and the smaller, short-eared, short-tailed grassland form. In the laboratory, the grassland form (*Peromyscus maniculatus bairdii*) does well in forest conditions, where its food preference and temperature tolerance are similar to those of the forest subspecies; thus, experimenters assumed that the avoidance of forests in the grassland deermouse is a behavioral response (Harris 1952).

Wecker's objective was to assess the genetic basis of this behavior and to test the idea that habitat imprinting (Thorpe 1945) is important. He constructed an enclosure halfway in a forest and halfway in a grassland, released the mice in the middle, and recorded their locations. The animals he tested were of the grassland subspecies (*bairdii*) and were of three basic types: (1) wild-caught in grassland; (2) offspring, reared in laboratory, of wild-caught; (3) reared in laboratory for 20 generations. Both wild-caught mice and their offspring selected the grassland half of the enclosure, regardless of previous experience. Laboratory stock and their offspring showed no preference, whether or not they had been raised in forest conditions. However, laboratory stock reared in a grassland enclosure until after weaning showed a strong preference for the grassland when tested later (table 14.4). Wecker reached the following conclusions:

1. The choice of grassland environment by grassland deermice is predetermined genetically.

2. Early grassland experience can reinforce this innate preference but is not a necessary prerequisite for subsequent habitat selection.

3. Early experience in forest or laboratory is not sufficient to reverse the affinity of this subspecies for the grassland habitat.

4. Confinement of these deermice in the laboratory for 12 to 20 generations results in a reduction of the hereditary control over the habitat selection response.

5. Laboratory stock retain the capacity to imprint on early grassland experience but not on forest.

TABLE 14.4   **Habitat selection by deermice *(Peromyscus maniculatus bairdii)* as a function of hereditary background and early experience**

The outdoor test enclosure was forest on one side and grassland on the other. Preference was measured by the percentage of time, amount of activity, and depth of penetration by mice in each side of the enclosure.

Number of Mice Tested	Hereditary Background	Early Experience	Habitat Preference
12	Grassland	Grassland	Grassland
13	Laboratory	Grassland	Grassland
12	Grassland	Laboratory	Grassland
7	Grassland	Forest	Grassland
13	Laboratory	Laboratory	None
9	Laboratory	Forest	None

Source: Data from S. C. Wecker, "The Role of Early Experience in Habitat Selection by the Prairie Deer Mouse, *Peromyscus maniculatus bairdii*," *Ecological Monographs* 33:307–25, copyright 1963 by Ecological Society of America, Tempe, AZ.

Wecker also suggested that learned responses, such as habitat imprinting, are the original basis for the restriction of this subspecies to grassland environments; genetic control of this preference is secondary.

In a series of laboratory experiments designed to explore the importance of early experience on bedding preference in inbred mice *(Mus domesticus)*, Anderson (1973) raised animals either on cedar shavings or on a commercial cellulose material. When he tested them later, he found that the mice preferred the bedding on which they had been raised, although females raised on cellulose drifted toward cedar shavings in subsequent tests. Naive mice preferred cedar shavings (figure 14.14).

Habitat imprinting occurs in migratory vertebrates, which commonly tend to return to the vicinity of their birth to breed. In the case of **iteroparous** organisms (those that have their young at intervals) such as birds, the adults of most species return to the same area to nest each year. **Semelparous** breeders (those that have their young all at once) such as salmon breed only once. After feeding in the open ocean for several years, salmon return to the same upstream spawning bed where they hatched. The salmon's olfactory system is programmed to respond to unique odors of the home stream during a critical period in the first few weeks of life (Hasler et al. 1978).

Tradition

Inherited tendencies and imprinting may be involved in restricting habitat choice to a small part of the potential range, but **tradition**—behavior passed from one generation to the next through the process of learning—may also be an important factor. For instance, many species of waterfowl have staging areas, where they rest and feed during migration from breeding to wintering grounds. The same areas tend to be used year after year.

Mountain sheep *(Ovis canadensis)* live in unisexual groups; females are likely to stay in the natal group but may switch to another female group when they are between one and two years of age (Geist 1971). Young rams desert the natal group after the second year of life and join all-ram bands. Mothers do not tend to chase their young away at weaning, as most other mammals do. Females follow an older, lamb-leading female; males follow the largest-horned ram in the band. When rams mature, they are followed by younger rams and pass on their habitat preferences to them.

Until the last century, mountain sheep occupied a much larger range in North America and Asia than they do today. Measures enacted to protect sheep, mainly hunting regulations, have done little to increase their numbers. A formerly inhabited part of the range that appears intact is not colonized, and transplants to suitable areas are often unsuccessful. In contrast, deer *(Odocoileus* spp.) and moose *(Alces alces)* have recolonized areas rapidly and have reached population densities higher than ever (Geist 1971).

Why have sheep failed to extend their range, while moose and deer have done so? Geist pointed out that deer and moose, which are relatively solitary beasts, establish ranges by individual exploration after being driven out of the mother's range; sheep, in contrast, transmit home-range knowledge from generation to generation and often associate with group members for life.

Understanding the niches of these species can help explain the sheep system, which may seem a rather poor adaptation to the environment. The moose is associated over much of its range with early successional plant communities that follow in the wake of forest fires. Moose habitats are subject to rapid expansion after fires, but are relatively short-lived, and moose must continually colonize new habitats. Each spring when her new calf is born, the cow drives away her yearling, which may wander some distance before establishing its new range.

Mountain sheep habitats are formed by stable, long-lasting climax grass communities, which exist in small patches. Geist argued that, given the distance between patches and the ease with which wolves can pick off sheep, the best strategy for the sheep is to stay on familiar ground. Sheep also have at least two and as many as seven seasonal home ranges that may be separated by 20 miles or more. These areas are visited regularly by the same sheep year after year at the same time; knowledge of the location of these ranges and the best

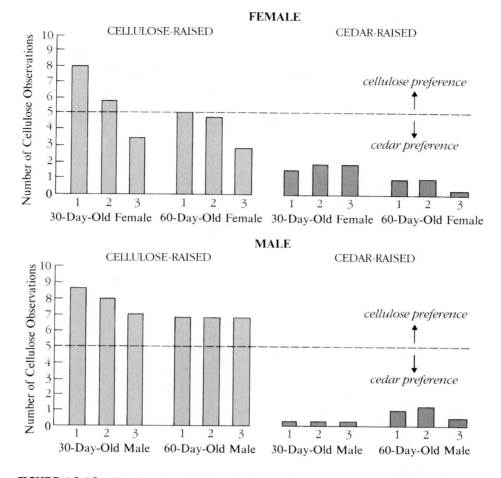

FIGURE 14.14 **Sleeping site selection of mice as a function of prior bedding experience and test day.**

Mice were born and raised on either cellulose or cedar bedding and were then given a choice of bedding types at either 30 or 60 days of age. Mice generally preferred the bedding type on which they had been raised. There was, however, a "drift" toward cedar preference across test days in females. Older females raised on cellulose initially showed no preference, but by test day 3 they chose cedar.

Source: Data from L.T. Anderson, "An Analysis of Habitat Preferences in Mice as a Function of Prior Experience," *Behaviour,* 47:302–39, 1973.

times to visit them is transmitted from one generation to the next. Because new habitats rarely become available, there is no advantage to an individual's dispersing and attempting to colonize other areas.

THEORY OF HABITAT SELECTION

Theoretical approaches to the problem of habitat selection are in a rather early stage of development, and no single general theory is currently accepted (Rosenzweig 1985). Several different approaches have been used. One is to think of habitats as patches, or areas of suitable habitat interspersed among areas of unsuitable habitat, and to apply optimal foraging theory, first developed by MacArthur and Pianka (1966) (see chapter 15). This theory enables us to predict which habitat patches an animal should select and when it should leave one habitat and move to another so as to get the greatest benefit for the least cost. This economic model incorporates such factors as the availability of resources in various patches and the costs of getting from one patch to another. Although the resource is usually assumed to be food (i.e., energy), nest sites or mates are other possibilities. For a review of optimal foraging theory, see Stephens and Krebs (1986).

A second approach is the **ideal free distribution,** which predicts how individuals distribute themselves so as to have the highest possible fitness (Fretwell and Lucas 1970). It assumes that animals have complete and accurate knowledge about the distribution of resources (ideal) and that they are passive toward one another and can go to the best possible site (free). Individuals settle in habitats so that the first arrivals get the best resources. As density increases, less-desirable areas are occupied, and animals spread themselves out so that all have the same fitness in the absence of intraspecific competition. One obvious result of such a distribution is that rich habitats will have more individuals than poor ones. Figure 14.15 illustrates how this might work for three habitats that differ in quality. Note that the lines slope

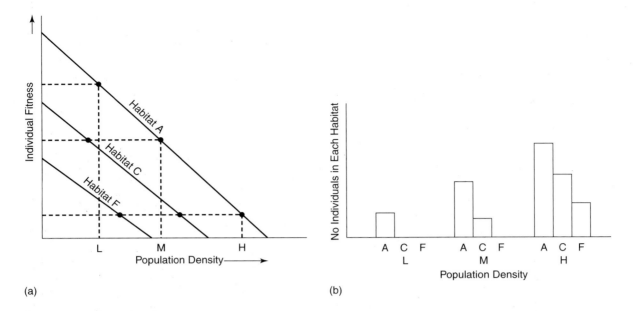

(a)

(b)

FIGURE 14.15   Model of habitat selection based on the ideal free distribution.
Three habitats are shown, with A = good, C = fair, and F = poor. (a) Along the y axis is plotted the fitness of individuals and along the x axis is population density, with L = low, M = medium, and H = high. In all cases fitness declines with population density, as would be expected as intraspecific competition for resources increases. At the lowest densities, individuals will do best in A, and should settle there first. But as A fills up, above density L, some individuals would do better in C. As A and C continue to fill up, eventually individuals will settle in F. At the highest densities individuals will have to choose among good but crowded habitat or poor but less crowded habitat. If the distribution is ideal and free, fitness will be the same in all three habitats at density H. (b) The number of individuals expected in each habitat at low, medium, and high densities.
Source: Redrawn from Fretwell, S. D. 1972. *Populations in a Seasonal Environment.* Princeton University Press, Princeton, NJ.

downward, because individual fitness declines as population density increases. At low densities new arrivals will settle in the best habitat, A. As A fills up, later arrivals settle in A and the next best habitat, C. Last to be settled is F. At high densities (H), most individuals are in the best habitat, but all should have the same average fitness.

If intraspecific competition occurs via dominance or territory (chapter 16), a despotic distribution develops, with some individuals monopolizing the best resources (Fretwell 1972). Another variation on the ideal free distribution is the ideal preemptive distribution (Pulliam and Danielson 1991). Potential breeding sites differ in quality, that is, the expected reproductive success of their occupants, and individuals choose the best unoccupied sites. These best sites are thus preempted and are no longer available to others, but their occupancy does not influence the expected reproductive success of occupants of other sites. Several studies have shown

that individuals in the preferred habitat have higher fitness, as measured by reproductive success and survival, than those in less preferred habitat. For instance, Grant (1975) found that meadow voles *(Microtus pennsylvanicus)* in the preferred grassland habitat had higher survival and reproductive success than in the less preferred woodland.

To conclude this chapter, we should point out that habitat selection may have an impact above the level of the individual, affecting population structure and even the formation of new species (chapter 4). Although the usual models of speciation involve the formation of geographic barriers to gene flow, reproductive isolation could evolve if members of two parts of a population came to prefer different microhabitats in the same region. If the preferences were heritable and individuals mated assortatively, a barrier to gene flow would be created that could eventually lead to the formation of new species (Rice 1987). However, such a mechanism has yet to be demonstrated.

SUMMARY

Habitat selection refers to the choice of a place to live. Factors affecting this choice can be best understood by a stepwise approach to why a species is absent from a particular place. The transplant experiment, in which organisms are relocated outside their natural ranges and their survival and reproduction are monitored, is an important tool. First, if the transplant is a success, we can assume that the organism's dispersal powers could not overcome geographical barriers. The success of introductions of foreign species by humans and

data on colonization of islands suggest that this inability to disperse often explains a species' absence. Second, in a more complex case, organisms may fail to colonize an otherwise suitable area because of behavioral responses to specific features of the habitat. Third, the presence of such other factors as competitors, parasites, predators, or diseases may also explain a species' absence. Evidence for exclusion due to competition comes in large part from range expansion by one species in the absence of another. Experimental demonstration of exclusion by predators requires the species to survive in the absence of the predator. Fourth, physical and chemical factors that are beyond the range of tolerance of organisms can restrict a species' distribution. Temperature and moisture are very important and may interact with other factors.

Behavior patterns may restrict organisms to a fraction of the habitat that they seem to us to be equipped to occupy. However, apparently suitable areas may in fact not be so in the long run. Behavior patterns that restrict a species' distribution may have evolved because of negative factors associated with other habitats in the evolutionary past.

Within a species' range individuals may disperse or may remain in the natal area to breed. In mammals, males typically disperse, while females are philopatric. In birds, the opposite is generally true. By dispersing, the chances of matings between close relatives are reduced, minimizing inbreeding. Animals may also disperse to reduce competition. Although there are demonstrated costs to extreme inbreeding, outbreeding may also incur genetic costs. Several recent studies suggest that in some species there may be an optimal degree of inbreeding where first cousins are favored as mates.

A few studies have measured the impact of habitat selection on reproductive success. Among insects such as aphids, information about habitat quality and the density of competitors is used to make choices that will maximize reproductive success. However, parasites may alter the habitat selection of their hosts to maximize their own success rather than that of their host.

The proximate cues animals use in habitat selection include direct locomotive responses to such environmental variables as temperature and humidity. Little is known about the cues used by vertebrates; fish may use olfaction, and birds may respond to vegetation type. The cues may be only indirectly related to the ultimate factors that determine survival; for example, birds select breeding habitats before the leaves and staple insect foods have emerged. Competition and risk of predation also influence habitat selection.

Several experiments with birds and mammals have been conducted to determine the roles of genes and experience in habitat selection. Selection of the "correct" habitat is under some degree of genetic control, as has been demonstrated by studies in which animals have been reared in isolation and later tested in various habitats. However, early experience can modify later choices.

Tradition, the transmission of knowledge of habitats from one generation to the next, is thought to be important in mountain sheep. These animals live in close social groups in stable habitats but seasonally occupy two or more ranges. Other ungulates—for example, moose—often live in unstable habitats and gain knowledge of suitable places to live through individual exploration.

Theoretical approaches to habitat selection include optimal foraging models, in which individuals choose patches (habitats) and stay in them in order to maximize the gain of some resource. The ideal free distribution assumes that, in the absence of competition, individuals settle in the best habitats first, with later arrivals moving into suboptimal sites.

DISCUSSION QUESTIONS

1. Review the life history of an organism of your choice. Try to explain the dispersal patterns of each sex in light of our discussion of the costs and benefits of inbreeding and outbreeding. Include ecological as well as genetic factors in your answer.

2. Males are typically the dispersing sex among mammals, but there are exceptions, including the chimpanzee *(Pan troglodytes)* and humans. Can you think of both proximate and ultimate causes of this reversed pattern?

3. Pick a species that is on the endangered species list and illustrate why an understanding of habitat selection is necessary to develop a plan to protect it.

4. Although tradition is claimed to play a role in the habitat preference of some animals, firm data are lacking. Design an experiment to demonstrate the role of tradition in habitat choice.

5. The figure shows the distribution of two species of chipmunk, *Eutamias dorsalis* and *Eutamias umbrinus,* in the Great Basin area of Nevada. What conclusions about animal distribution can be drawn from these data?

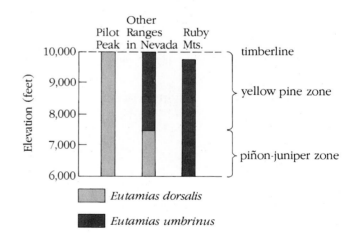

Source: Data from E.R. Hall, *Mammals of Nevada,* copyright 1948 University of California Press, Berkeley, CA.

SUGGESTED READINGS

Chepko-Sade, B. D., and Z. T. Halpin, eds. 1987. *Mammalian Dispersal Patterns: The Effects of Social Structure on Population Genetics.* Chicago: University of Chicago Press.

Cody, M. L., ed. 1985. *Habitat Selection in Birds.* New York: Academic Press, Inc.

Krebs, C. J. 1994. Habitat Selection, chapter 5 in *Ecology: The Experimental Analysis of Distribution and Abundance,* 4th ed. New York: Harper & Row.

Orians, G. H., et al. 1991. Habitat selection, a symposium organized by Gordon H. Orians. *Am. Nat.* 137:S1–S130.

Shields, W. M. 1982. *Philopatry, Inbreeding and the Evolution of Sex.* Albany: State University of New York Press.

Foraging Behavior

ll animals, by definition, must acquire food for energy. In this chapter, we first consider the decisions faced by an animal that is foraging, or searching for food. We then look at the behavioral techniques that animals use to acquire food, and the relationship between social organization and foraging. Finally, we examine predation from the prey's perspective with a discussion of antipredator techniques.

OPTIMALITY THEORY

Because foraging is so clearly related to an animal's fitness, animals are likely to be under natural selection to be effective foragers. A tool of behavioral ecologists called optimality modeling has been frequently used in the study of foraging behavior. Optimality models predict which decisions an animal should make in order to maximize its inclusive fitness under a given set of conditions hypothesized to drive the behavior. Comparisons of actual and predicted behavior shape our understanding of the behavior's function. Optimality models have three parts (Stephens and Krebs 1986):

1. A set of **decisions,** or strategies, that are available to the animal. For example, a foraging bird may choose to eat a particular piece of food, or search for another one instead; a spider may stay where it is, or move its web to a new place. By the words "decision" and "strategy" we do not imply that animals are consciously aware, or that they undergo a decision-making process such as we do. All that is meant is that an animal performs one action out of a variety of alternatives available to it.

2. The **currency,** or the criterion used to compare the value of different decisions. In order to decide among the decisions available to the animal, we must be able to compare them with a common measure. For example, many foraging models use the rate of energy intake as the currency. The best strategy would be the one that maximizes this rate. In other models, animals might minimize the time they spend foraging. Later,

we'll discuss how the choice of currency may influence the outcome of a model.

3. The **constraints,** or limits on the animal. Constraints can be internal, or intrinsic to the animal (such as particular nutritional needs, or the ability to see only certain colors) or external (such as temperature or light levels that affect an animal's ability to forage effectively). An animal can optimize its behavior only within the range of its capabilities and needs, and constraints define this range.

Optimality models may also include other variables besides these three basic parts, as we will see in some of the examples below. In addition, optimality models can be used to describe other aspects of behavior, such as mate acquisition (chapters 17 and 18) and habitat selection (chapter 14); the basic ideas are the same.

Of course, as discussed in chapter 5, we can't expect animal behavior to be perfectly optimized. Many forces, such as genetic drift, or rapid environmental change coupled with evolutionary lag, may prevent animals from being optimal. Of what use, then, is an optimality modeling approach? First, modeling helps us clarify our thinking. When constructing a model we must clearly lay out all the elements that we think might be important in determining an animal's behavior. Second, the results of a model can provide a quantitative prediction that can then be tested. These predictions are often more precise than those one can make without the aid of a model. For example, it might be logical to predict that a bird eating berries on a bush should not necessarily search for every last berry, no matter how long it takes, before leaving for another bush. This still leaves many questions open, however: When should it leave? How does the availability of other bushes affect its decision? By modeling this problem, we can generate more exact predictions. Finally, a model that successfully predicts the decisions of animals gives us more confidence that we really understand the factors that affect behavior.

FORAGING MODELS

A lion encounters a wildebeest. Should it try to capture it or not? A foraging shorebird finds a small worm. Should it eat it or keep searching? Very different species of animals face very similar questions. Modelers have paid particular attention to two types of models: **diet selection models,** or prey models and **patch models.** The former deal with the types of prey a forager should eat, while the latter deal with how long a forager should stay in a food-containing patch.

Choice of Food Items: The Prey Model

Most species of animals are surrounded by all manner of things that they might consider eating. The barn owl *(Tyto alba)* is a nocturnal predator of small mammals. In southwestern New Jersey, these owls roost in tree cavities or silos and forage in fields over a radius of several kilometers. Colvin (1984) found that more than 90 percent of the available small mammals are white-footed mice *(Peromyscus leucopus)* and house mice *(Mus domesticus)*, and less than 5 percent are meadow voles *(Microtus pennsylvanicus)*. However, 70 percent of the owls' diets were meadow voles (figure 15.1). Clearly, owls are not simply taking prey species in proportion to their abundance in the habitat. How do we model a problem of prey choice such as this?

This problem can be modeled relatively simply; variations of this model were derived independently by a number of different authors (references in Stephens and Krebs 1986).

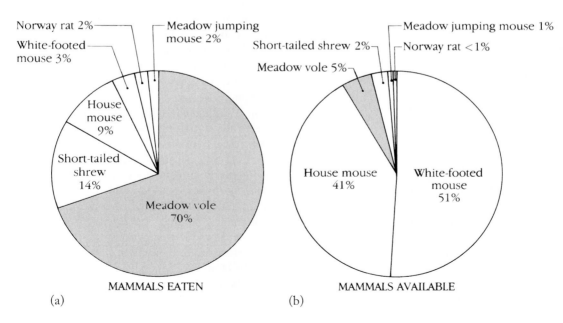

FIGURE 15.1 **The percentage of small mammals eaten by barn owls (a), compared to the percentage available based on trapping (b).**
In this study from southwestern New Jersey, barn owls are meadow vole specialists.

The scenario is that a forager is searching for food, and it finds one prey at a time. Let's consider the three parts of this model. The decision variable is whether the forager should eat the prey it finds, or whether it should continue searching for another type of prey. (For simplicity's sake, modelers refer to all types of food as "prey," whether it is another animal or a piece of plant material.) The currency is the rate of energy (caloric) intake: we assume that the animal benefits by maximizing this rate, and we will measure the relative value of the available decisions in this currency. Finally, there are various constraints that we can include. Prey, once found, need to be processed in some way: nuts may need to be cracked open, live prey may need to be subdued. This takes a certain amount of time, and this is called **handling time.** Different types of prey may have different handling times. We will also assume that foragers cannot handle prey and search for it at the same time. (If they could, their problems would be solved!)

The easiest way to put all these variables together is to use a mathematical expression. Let's define some variables to represent different numbers in the model. First, let's simplify matters by considering only two different types of available prey, types 1 and 2. Each of these might provide a different amount of energy for the animal, which we define as E_i, where the subscript i represents different prey types. So, for example, the number of calories in an item of prey type 1 would be represented as E_1, and that in type 2 as E_2. Similarly, each prey type can have its own handling time, usually measured in seconds, which we will represent as h_i.

Because we are interested in maximizing the rate of energy intake, we need to figure out what this rate would be for each prey item. We need a new term: **profitability,** or the ratio of energy gained to the handling time of each type of prey.

$$\text{Profitability of prey } i = \frac{E_i}{h_i} \qquad \text{(eq. 15.1)}$$

To simplify matters further, let's define prey type 1 as the more profitable prey type, so that

$$\frac{E_1}{h_1} > \frac{E_2}{h_2}. \qquad \text{(eq. 15.2)}$$

Finally, each prey type can have its own search time, which we will represent as S_i. Search time is the amount of time it takes for an animal to locate an item of a particular type. As the density of a prey item increases in the environment, search time for that prey item goes down. There is one more very important assumption: the foragers know the values of the variables we have defined. They know, for example, the probability of finding a prey of a particular type, much as a good gambler knows the odds of drawing a particular set of cards.

Imagine a searching animal has found a prey item. We can now use the variables we have defined to ask the question: should the animal eat the prey item or continue searching to find a new one? If the prey item is type 1, the more profitable prey, the choice is obvious: the animal should eat it, because it will never find anything better. So we have the first prediction of the model: always eat the most profitable prey.

The question becomes more challenging if the animal finds prey of type 2. Should the animal eat the less profitable prey or continue searching until the better prey is found? We must compare the rate of energetic intake for these two choices:

Gain from eating prey type 2, once it is found: $\dfrac{E_2}{h_2}$ (eq. 15.3)

Gain from searching for and eating prey type 1: $\dfrac{E_1}{S_1 + h_1}$ (eq. 15.4)

The denominator of equation 15.4 contains both the search time and the handling time, because the animal must first find prey type 1, and then consume it. Note that both of these are rates of energy intake (calories per second). This means that we now can compare the decisions available to the animal in the currency of the model by comparing these values. So, a forager should eat prey of type 2 when

$$\frac{E_2}{h_2} > \frac{E_1}{S_1 + h_1} \qquad \text{(eq. 15.5)}$$

Now we have predictions that can be tested. First, the model says that the decision to eat prey type 2 should be partially based on the search time for prey type 1. If prey of type 1 are very abundant, S_1 will be low, and the value on the right side of the inequality will be large. A second prediction is that search time for prey of type 2 should not be important in the forager's decision to eat it. An animal could be knee-deep in the lower-quality prey and would still be predicted to ignore them, as long as the inequality is met. Third, the model predicts that a forager should instantly switch back and forth between eating both kinds of prey to eating only the higher-quality prey, depending on whether or not this inequality is met. This is known as the "zero-one" rule: an animal should eat the less profitable prey either none of the time (i.e., with a probability of zero) or all the time (i.e., with a probability of one).

This model can be easily expanded to include a series of different prey items, not just two. The model then predicts that prey types should be added to the diet in order of their profitability, and, as in the two-prey case, the inclusion of a particular type does not depend on its own encounter rate (Stephens and Krebs 1986).

How do these predictions hold up to empirical work? Krebs et al. (1977) conducted an elegant test of this model with birds called great tits (*Parus major*). They set up a little conveyor belt that carried food past the birds. The prey were either large (eight segment) or small (four segment) chunks of mealworms. By varying the number of worms placed on the belt, the researchers could regulate the birds' search time for different prey items. They directly measured the handling times of both prey types. Several predictions of the model held up well. The birds selected prey on the basis of profitability. Selectivity depended on the encounter rate with profitable prey, rather than the encounter rate with unprofitable prey. However, the zero-one rule did not bear out. As the researchers changed the frequency at which the birds encountered the different prey, they failed to switch cleanly from one to two prey types, but instead exhibited partial preferences: they ate prey with a probability somewhere between zero and one. Partial preferences have been found in many other species as well. Why might this be so? Let's examine further our assumptions and possible constraints and how they affect the outcome of the model.

The Effect of Other Currencies and Constraints on the Model

The prey model we are discussing is very simple indeed, and we can envision many additions or changes to the model that might make it reflect reality more accurately. One assumption that is not likely to be met is that animals are equipped with perfect knowledge of the variables in the model. They may have to sample their environments and learn about the new conditions. This is one possible explanation for the failure of most tested animals to follow the zero-one rule: animals may have to sample their environment and gather information before making a good decision. This can be explicitly incorporated into foraging models (e.g., Hirvonen et al. 1999).

Another question is whether the proper currency has been chosen. We have assumed that animals are attempting to maximize the rate of energy intake, but instead they may try to minimize time spent searching (e.g., Schoener 1971) or perhaps may pay attention to both calories and some other aspect of food. Sometimes we can experimentally test which variable is being chosen by foragers. For example, overwintering wild white-tailed deer (Odocoileus virginianus) were given a choice of four artificial foods that were either high or low in protein, and high or low in energy. They chose diets high in calories but low in protein (Berteaux et al. 1998). In another example, the golden-winged sunbird (Nectarinia reichenowi), an African nectar-feeder that resembles a hummingbird, defends a feeding territory even in the nonbreeding season. Pyke (1979a), using data collected by Gill and Wolf (1975), tested the notion that these birds were maximizing the rate of energy intake. The birds spent far less time foraging and defending and more time just sitting than was predicted. Instead of maximizing energy intake, what they seemed to be doing was minimizing costs—eating and defending just enough to maintain themselves. However, these data were obtained during the nonbreeding season. Presumably, foraging rates and territory defense would increase to meet the demands of reproduction in the breeding season. Thus, the season during which an animal is observed may influence which variation of the model is most appropriate.

In other cases, animals may need to reduce their intake of toxins in certain foods. Mammalian herbivores, for example, eat only a limited quantity of plants that produce high concentrations of toxins (McArthur et al. 1991). Some animals, such as pikas (Ochotona princeps) get around this constraint by gathering food and storing it until the toxins degrade (Dearing 1997).

These models also assume that the probability that an animal finds a particular prey item depends only on its abundance in the environment. However, not all prey types may be equally easy for foragers to find. In fact, animals sometimes get faster and faster at finding a particular prey type when they have had experience with it. This has been called a **search image.** Pietrewicz and Kamil (1979, 1981) tested whether blue jays (Cyanocitta cristata), after previous encounters with cryptic moths (Catocala spp.), improve their ability to detect that species and thus show a specific search image (figure 15.2). The jays viewed photographic slides of either one of the two cryptic moth species on a matching substrate, or of the substrate with no moth. When the experimenters projected a picture of a moth, the jays could get a food reward by pecking a key ten times. When no moth was present, pecking the food key not only resulted in no reward, but in a delay of the next slide as well; the correct response was to peck the advance key that started the next slide. When jays saw a series of slides of the same moth species, they significantly improved their ability to detect it. When slides of both species were mixed together, they did not improve. Thus, their search image seemed to be restricted to one species at a time.

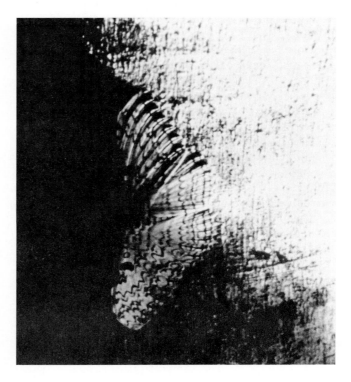

FIGURE 15.2   Moth on tree trunk.
Blue jays improve in their ability to detect these cryptic prey by forming a search image.
Source: Steve Pollick.

Even animals that depend on senses other than vision can form search images. Striped skunks can form olfactory search images (Nams 1997). When naive skunks were given repeated experiences with a certain type of food, over time they could detect it both faster and from greater distances.

How Long to Stay in a Patch

Whether or not to eat a particular food item is not the only problem facing foraging animals. Many types of food are not distributed randomly in the environment, but instead occur in patches; imagine scattered trees that are bearing fruit, or tide pools holding tasty snails. Foragers seeking these foods must decide how long to stay in a particular patch, or when to leave and find a new patch. If you have picked berries, you have faced the same problem: should you search a bush so well so you get every last berry, or should you move on to another bush?

Several factors should play into your decision. The average richness of the patches should be important: if berries are in short supply, you might do better to stay and search for every last one. If other bushes have many berries, you will probably do better to move on. In addition, the distance between the patches should be important: if you can reach another bush by taking a few steps, you will probably make a different decision about leaving your patch than if you have to walk half a mile.

We can look at these factors together with a graphical model called the **marginal value theorem** (Charnov 1976). In this model, we again assume that the forager maximizes the rate of energy intake; this is the currency. The decision variable is to stay in a patch, or leave to seek a new patch. Constraints include: searching for patches and foraging within them are mutually exclusive, patches are found one at a time, and the forager knows the parameters in the model and does not have to learn them. In addition, we assume that the cumulative rate of food intake within each patch can be characterized by a gain curve (figure 15.3a). When the forager first enters a patch, it has gained no energy. As it forages within a patch, its net energy gain increases as it finds food. However, as the food within the patch is depleted, the gain curve flattens out: it becomes harder and harder to find each food item. It is important to remember that the plot describes the net, or cumulative, energy gain. In this model, all patches have the same gain curve.

We can add to this graph the average travel time to a new patch (figure 15.3b). Food is found only in patches, so before a forager finds a new patch, its net energy gain is zero. Now we have all the information needed to calculate when a forager should leave a patch. If it is maximizing its net energy gain, it should leave when its expected net gain from staying declines to its expected net gain from traveling to and foraging in a new patch (Charnov 1976). In other words, it should stay until it can do better elsewhere, travel costs included. This can be determined by drawing a line, tangent to the gain curve, to the point on the x axis that represents the travel time. The slope of this line is in the units of the currency

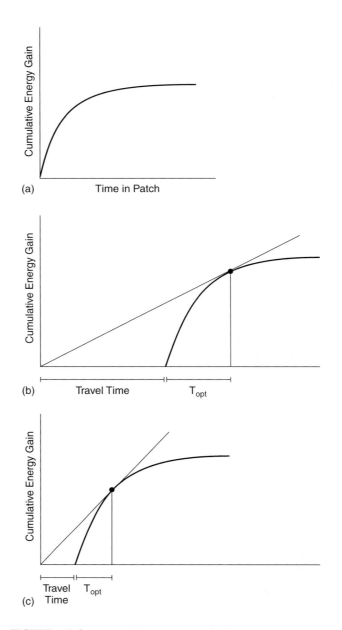

FIGURE 15.3   Predicting when to leave a patch: The marginal value theorem.
(a) A forager's cumulative energy gain in a patch is illustrated by the gain curve. The gain curve rises rapidly as the forager first enters a patch, but then levels out as food resources are depleted. (b) Now travel time between patches is added. The animal does not gain energy while it is traveling. The optimal time to stay in a patch (T_{opt}) is determined by drawing a tangent from the origin to the gain curve. (c) Here, travel time is shortened compared to (b), and the optimal time in a patch is also shorter.

(energy gain per unit time). Of all the lines that could be drawn between the origin and the gain curve, the tangent is the one with the highest possible slope, or the greatest energy gain per unit time.

Now we can modify some of the variables and see how the predictions change. For example, we can make the travel time between patches exceptionally short (figure 15.3c). The optimal time spent in the patch is now shortened.

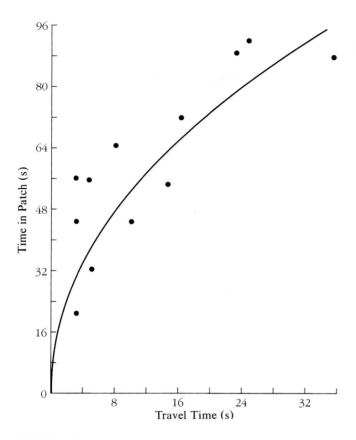

FIGURE 15.4   Test of the marginal value theorem in great tits.
Time in patch as a function of travel time between patches. Line is the predicted curve based on the theory, adjusted for energy expenditure.

FIGURE 15.5   Bumblebee foraging on sedum flowers.
Behavioral ecologists have studied these central-place foragers in order to understand currencies, decisions, and evolutionary constraints in tests of optimality models.
Source: Jerome Wexler/Photo Researchers, Inc.

Despite the simplicity of this model, studies of a variety of insects, birds, and mammals generally support its predictions (McNair 1982). One test of the marginal value theorem was done in an aviary using great tits *(Parus major)* (Cowie 1977). Each "patch" consisted of a plastic cup filled with sawdust in which six pieces of worm were hidden. Six cups were on each of five artificial trees. Cardboard lids were placed on the cups in two ways: hard-to-remove lids fit tightly inside the cups and had to be pried off by the birds, whereas easy-to-remove lids simply lay on top of the cup and could be flipped off. Birds were tested in two situations: either all hard or all easy. Hard-to-remove lids corresponded to long travel times between patches, easy lids to short times. As can be seen in figure 15.4, time spent in each patch increased as travel time between patches increased, and the fit to the predicted curve was quite close.

As with the diet selection model, all the assumptions of the model are not necessarily met in real life. For example, animals may not know in advance what the rate of energetic return will be in each patch. Foragers faced with two patches of unknown prey density might be expected to sample both (Shettleworth 1984). In a laboratory study that again used great tits, the birds were given a choice of two feeders, one at each end of an aviary. Each feeder required the bird to hop on a perch a certain number of times to get a mealworm, and one feeder required more hops than the other. Initially, the birds fed at both feeders, but they gradually shifted to the one that required fewer hops (Krebs et al. 1978). Similarly, foraging bluegill sunfish feed on prey that vary both spatially and temporally, so the assumption that they have complete knowledge of all variables in their environment is unjustified. They seem to be able to incorporate information into their assessment of patches as they forage (Wildhaber et al. 1994).

Central-Place Foraging

An extension of these models has been developed for **central-place foragers,** animals that carry food back to a central location for storage or for feeding to offspring (Orians and Pearson 1979; Schoener 1979). For example, nesting birds and insects that live in hives are central-place foragers (figure 15.5). The problem here is not only when to leave the patch, but also which and how much food to collect before returning to home base. Time in the patch, load size, and selectivity are predicted to increase as the travel distance increases. This prediction has been upheld in a number of studies. For example, as travel distance increases, pine squirrels gather cones with a larger number of seeds per cone (Elliot 1988). Another interesting example that combines central-place foraging and diet selection is that of a marine diving bird, the rhinoceros auklet *(Cerorhinca monocerata)*. Davoren and Burger (1999) found that adults collected different prey items when feeding young than when feeding themselves. Most meals brought back to the nest for chicks comprised one or two large fish, which are more efficient to deliver, whereas adults generally selected smaller prey when foraging for themselves.

Risk-Sensitive Foraging

The models we have discussed so far concern a forager's response to the mean (average) amount of prey available.

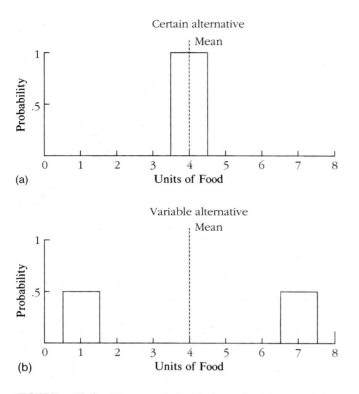

(a)

(b)

FIGURE 15.6   The standard design of risk sensitivity experiments.
The forager is offered a choice between (a) a certain alternative and (b) a probabilistically determined alternative with the same mean as the certain alternative.

However, mean is not the only characteristic that may vary across patches. Suppose that a small bird, a junco *(Junco hyemalis)*, is given a choice between two trays of food. When it selects one tray, the experimenter removes the other. One tray always has four seeds in it. The other tray varies: on half the trials, it has one seed, and on the other half of the trials, it has seven seeds (figure 15.6; Stephens and Krebs 1986). Thus, these two patches have the same mean amount of seeds, but the second patch has a higher variance around that mean. We say the second patch has higher risk. (Note that "risk" in this context does not refer to danger from predators; it is analogous to the risk a gambler takes.) If an animal can distinguish between these two patches, we say they are **risk-sensitive.** If a forager prefers the variable patch, it is **risk-prone;** if it prefers the stable patch, it is **risk-averse** (Stephens and Krebs 1986).

Why would juncos, or any animal, ever be risk-sensitive? One way to approach this problem is to consider **utility,** or the value that a resource has to an animal. Often this value is measured in units of fitness. The relationship between each unit of a resource and its utility is called the utility function, and it may have different shapes. For example, in figure 15.7*a* is a straight-line utility function: for each seed that a bird finds, its fitness increases by the same amount. This is unlikely to be very common. More common is the utility function pictured in figure 15.7*b,* where the utility function

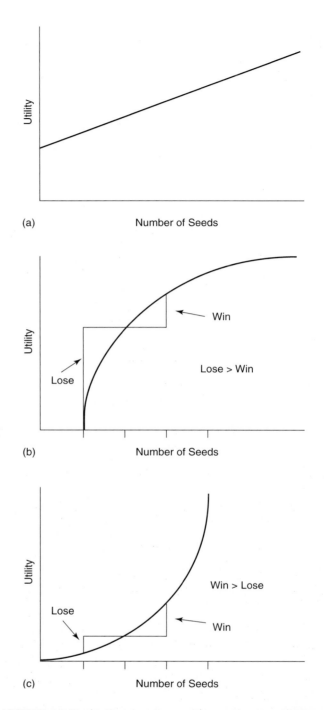

(a)

(b)

(c)

FIGURE 15.7   Utility functions and an animal's willingness to accept risk.
Utility is the value that a resource has to an animal. (a) In a linear utility function, every unit of resource has the same value to the animal. (b) In a concave-down utility function, the resource becomes less valuable to the animal as it gains more of it. Imagine an animal can gamble and either win or lose one unit of resource. It stands to lose more utility than it can win, and so it should avoid gambling and be risk-averse. (c) In a concave-up utility function, the resource becomes more valuable to the animal as it gains more of it. Here, a gamble may pay off in a large win, or a relatively small loss. This animal should be risk-prone, and take the gamble.

flattens out with increasing number of seeds. For example, this might occur if an animal can consume only a certain number of seeds, and seeds found after that are useless. The shape of this function is described as concave-down. In other cases, the utility function is concave-up (figure 15.7c): each unit brings more fitness than the previous one. Imagine, for example, that an animal is 20 steps from a water hole: taking 20 steps instead of 19 steps will make a bigger difference in its fitness than taking 19 steps instead of 18 steps (Stephens and Krebs 1986). We can imagine that at least part of many utility functions may be concave-up. Clearly, utility functions can take on a variety of shapes.

The shape of the utility function is tied to risk. Imagine that an animal is at a point on the concave-down utility function in figure 15.7b, and has a choice about whether to stay where it is, or to gamble. If it gambles, it will either get one more unit of resource, or lose one unit. Because of the shape of the curve, it stands to lose more than it wins. Therefore, it should be risk-averse, and choose not to gamble at all. In contrast, if the curve is concave-up, it stands to win more than it loses, and the best strategy is to gamble (figure 15.7c).

Which option did the foraging juncos choose? It depended on their energy budgets (Caraco et al. 1980). If the amount of food provided in the no-risk patch was enough to meet their energetic needs and sustain their body weights, they were risk-averse. Their utility function was concave-down: as the number of seeds increased, each seed became less and less valuable. However, if the juncos were kept at cold temperatures and were energetically stressed, each additional seed meant a lot. These juncos had a concave-up utility function, and they chose the risky patch. Another intuitive way to think of this is that if an animal cannot meet its energy needs with the low-risk option (e.g., the known reward of four seeds is not enough to survive on), its best choice is to gamble and take the risky option. The animal then has at least a chance of getting enough resources to survive.

In empirical studies, animals generally respond to variance in food availability (Stephens and Krebs 1986, reviewed in Bateson and Kaulnik 1997). An example comes from two closely related spider species (Uetz 1996). Each spider builds an orb web in which to capture prey (see a description of orb webs in the section on trap building). When orb webs are placed near to one other, the variance in prey intake is reduced. The primary reason for this is the "ricochet effect" (Uetz 1989): prey that are missed by one spider often bounce into the web of a neighbor. *Metepeira atascadero* lives in desert grassland, where the average biomass of the available prey is less than or equal to individual needs. Here, most spiders forage solitarily because it is better to take the high-risk option of foraging alone. In contrast, *M. incrassata* lives in the moist tropical mountains where prey are abundant, two to three times the number needed to meet the spiders' needs. Hundreds of individual spiders join together in large groups, where the variance in their prey capture rate is lowered.

An interesting problem to consider with models of risk is the time period over which energy gain is measured. We might make different predictions if we assume an animal is maximizing long-term gain versus short-term gain (Stephens and Krebs 1986). If an animal maximizes long-term gain, it may be willing to endure periods of no food in order to get periods of high food, whereas if it maximizes short-term gain, it will not be willing to do so. Different assumptions may be appropriate for different species.

Effects of Competitors on Foraging Behavior

So far we have discussed foragers as if they were unaffected by other animals. However, the presence of competitors and predators may also be important in determining foraging behavior. For example, competition by members of the same or different species may force them to forage in suboptimal habitats or include food items they would not otherwise consume. For instance, stickleback fish that had parasite infestations fed on smaller-sized daphnia rather than compete with uninfested sticklebacks for larger, more profitable prey (Milinski 1984).

American crows (*Corvus brachyrhynchos*) change their behavior in the presence of conspecific competitors (Cristol and Switzer 1999). Crows fly up and drop walnuts onto the ground in order to break them open. When other crows are present, they do not drop them from as great a height, presumably so they have a better chance of swooping down to pick up the nut before competitors snatch it.

One way to increase net energy gain is by defending a territory against potential competitors. (For a more detailed discussion of territoriality see chapter 16.) Since an animal must spend energy advertising its presence and chasing out intruders, defending a territory will be economical only under certain circumstances. Pied wagtails (*Motacilla alba*) are insectivorous birds. Some of them defend winter feeding territories along riverbanks, feeding on insects that are washed up on shore. The others feed in flocks in nearby pools. Territory holders, usually males, follow a circuit up one bank and down the other, a pattern that maximizes food intake as new insects are washed ashore (Davies and Houston 1983, 1984); intruding wagtails are chased away. When food in the territories was very scarce, the owners fed elsewhere in flocks but kept returning to the territory to evict intruders. When food was abundant, owners often shared the territory with a satellite (usually a juvenile or a female) that walked about one-half a circuit behind the owner (figure 15.8). This meant that the food available to the owner was reduced by one-half, but the owner sometimes gained because the satellite helped chase away intruders. If food declined, the owner chased the satellite away. If food became extremely abundant, owners made no effort to defend their territories. Sometimes territory holders fed on the territory even when they could have done better in the flock. Thus, they were not maximizing energy intake in the short run. This system differs from that of the nectar-feeders such as the golden-winged sunbirds (Gill and Wolf

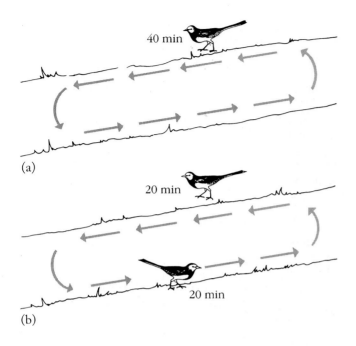

FIGURE 15.8   Pied wagtails feeding along a riverbank.
(a) Pied wagtails exploit their territories systematically. The circuit of the riverbank takes, on average, 40 minutes to complete. (b) When a territory is shared between two birds, each walks, on average, half a circuit behind the other and so crops only 20 minutes worth of food renewal.

1975) whose territory sizes vary as a function of food availability. The wagtails keep the same size area but vary their responses to intruders as food supply fluctuates. Davies and Houston suggest that territory maintenance is a long-term investment to protect a food source that is more reliable than the pools where the flocks feed.

Effects of Predators on Foraging Behavior

Many foraging animals are also potential prey themselves, and must balance the risk of predation with the benefits of foraging. Often, an animal is less vigilant to the presence of danger when it is attacking or handling prey. In addition, animals that rely on crypsis (blending in with the background) to avoid detection by predators may "blow their cover" when attacking prey, and thus become more vulnerable to attack (reviewed in Houtman and Dill 1998). The effect of the presence of a predator on diet selectivity has been directly measured in a handful of studies, and it varies across situations. The rule seems to be that as it becomes more dangerous to capture more profitable prey, selectivity decreases; when it is more dangerous to capture less profitable prey, selectivity increases (Houtman and Dill 1998).

Risk of predation may also affect the choice of patch (reviewed in Lima and Dill 1990). Often there is a trade-off: high-quality patches may sometimes pose the greatest risk of predation. There are many examples across a variety of taxa of foragers that choose less-valuable food resources if they

are closer to cover. For example, desert baboons (*Papio cynocephalus ursinus*) spent more time foraging in a low-risk, but food-poor habitat (Cowlishaw 1997). Similarly, juvenile hoary marmots foraged close to their burrows rather than risk predation by eagles or coyotes in more distant but better-quality patches (Holmes 1984). Namib Desert gerbils foraging for seeds left patches sooner under the full moon, when the risk of predation was high, than under the darker skies of the new moon (Hughes et al. 1994).

Researchers have manipulated the relative costs and benefits of different patches in order to see exactly how animals balance them. Arenz and Leger (1999) manipulated the cost of vigilance in thirteen-lined ground squirrels (*Spermophilus tridecemlineatus*) in a clever experiment. They constructed Plexiglas boxes of varying lengths, open on one end. The boxes were opaque on all sides but had clear tops so the squirrels did not feel that they were in a safe burrow. Squirrels were attracted into the boxes by peanut butter bait, but once inside their vision was obstructed by the sides of the box. In order to scan for potential predators, they had to withdraw their heads from the boxes. Longer boxes thus had a higher cost of vigilance. As box length increased, the cost of vigilance increased, and squirrels decreased the time they were vigilant.

State-Sensitive Models of Foraging

As we saw in the example with the risk-sensitive juncos, all members of a population may not make the same foraging decision. In fact, the same individual may make different foraging decisions at different times. For example, a very hungry animal may choose to go to a risky patch of food if the quantity of food is higher, whereas a well-fed animal may give up food in favor of avoiding predators. For example, hungry scorpions (*Buthus occitanus israelis*) are willing to forage under bright moonlight when they are in danger from predators, but well-fed scorpions hide when the moon is bright (Skutelsky 1996). We can describe changeable characteristics of individuals, such as hunger level, age, or body size, as states. We can include measures of these states, or state variables, into models in order to more accurately predict animal behavior.

As you might imagine, adding these extra elements can substantially increase the difficulty of creating a model. Help has come in the form of computers and a form of programming called dynamic programming. This tool has enabled researchers to calculate the optimal decisions of animals with different states and of the same animals at different times in their lives. It is also easy to incorporate other variables that may affect fitness, such as the risk of predation. For more details on this type of modeling, see Clark and Mangel (2000).

TECHNIQUES FOR ACQUIRING FOOD

Modifying Food Supply

Some animals modify their food supply so that it increases. For example, grazing animals in the grasslands stimulate the growth of some species of grass and prevent succession of

FIGURE 15.9   Leaf-cutter ants, *Atta cephalotes,* carrying leaf fragments to nest in lowland tropical rain forest in Costa Rica.

Source: © Gregory G. Dimijian/Photo Researchers, Inc.

grassland communities to other community types, such as forests. Some species of intertidal limpets increase their food supply with the mucous trail they secrete during locomotion (Connor and Quinn 1984). Two species of solitary limpets (*Lottia gigantea* and *Collistella scabra*) routinely return home to a depression on a rock, depositing a mucous trail that acts as an adhesive trap for algae. The mucus also stimulates algal growth along the trail, and the limpets feed on this algae as they retrace their path home.

The ultimate in nonhuman manipulation of producers in an ecosystem may be the "agriculture" practiced by certain fungus-growing ants. Their activities have been described in Hölldobler and Wilson's (1990) Pulitzer-prize-winning treatise on ants. Leaf-cutter ants (cover photo, figure 15.9) are easy to spot in the New World tropics as they walk in long queues, carrying leaf sections back to the nest. The reason for this was initially very puzzling: were they thatching their roof? using the leaves for thermoregulation? Over many years, observers put together the following story. Once inside the nest, ants cut the leaf sections into smaller pieces, and work the edges with their mandibles to make the pieces wet and pulpy. They may deposit an anal droplet on the leaf before inserting the leaf into the garden. These processed leaves form the basis for their garden. The ants then plant their crop: they place bits of fungal mycelium on the leaves, and the fungus rapidly grows (figure 15.10). The ants feed on the tips of the fungus. This arrangement is mutualistic. By

FIGURE 15.10   Fungus garden of leaf-cutter ants.

Ants of the species *Acromyrmex octospinosus* cultivate fungus as a food source. This garden, measuring approximately 31 by 11 centimeters, was found under a log in Trinidad.

Source: Courtesy Neal A. Weber.

converting the indigestible cellulose of the leaves into digestible sugars, the fungus makes the vast energy supply contained in the forest leaves available to the ants. The ants spread the fungus and care for it by weeding out alien fungi, fertilizing it, and producing antibiotics that act against competing fungi and microorganisms.

Other ant species act as livestock farmers, and raise aphids or other plant-feeding insects in or near their nests, protect them, and feed on their "honeydew" excretions from their digestive tract (Hölldobler and Wilson 1990). The ants and the aphids appear to communicate with each other. Whereas unattended aphids merely eject the honeydew excretions from their anus, aphids that are caressed by an ant's antennae and forelegs will excrete the droplet slowly and hold it on the tip of the abdomen while the ant consumes it. Again, as with the leaf-cutter ants, both partners benefit from this relationship.

Trap Building

Some invertebrates, such as ant lion (genus *Myrmeleon,* in the order Neuroptera) larvae, make traps to capture food. They make funnel-shaped pits in the sand and bury themselves just below the pit. When ants tumble in, the ant lions knock them down to the bottom of the pit with grains of sand that they hurl at them with tosses of their head. Other traps include the underwater nets of caddis fly larvae (order Trichoptera).

The traps of spiders are described in detail in Foelix (1996). Orb webs are the most well known, and have several different elements. Radial threads are similar to the spokes of a wheel, radiating out from the hub of the web. A fine thread, covered with glue, spirals outward from the hub; it is this sticky spiral that traps insects. Most orb webs are flat, and the spider sits either at the hub or on a connecting thread and monitors the vibration of prey hitting the web. Some species, however, modify the orb web further: *Theridiosoma* attaches a lateral tension thread to the hub and tightens it, turning the web into an inside-out umbrella. The spider pulls on the radii near the hub with its hind legs, and pulls on the tension thread with its front legs, and sits motionless waiting for prey. When an insect flies near the web, the spider lets go with its front legs so that the web snaps back into a two-dimensional shape, and traps the insect.

Still other spiders build sheet webs, some with funnels attached in which to hide. These webs are not glue covered, and the spider must rely on speed to snatch up the prey before it disentangles itself. The bolas spider uses a bizarre method of prey capture: it uses a short thread with a large drop of glue on the end and throws it after insects that fly nearby. Still other variations on spider webs are illustrated in figure 15.11.

Electromagnetic Fields

Unusual channels of communication may be used to capture food. For example, Kalmijn (1971) demonstrated that sharks can detect the weak electric fields of fish even when the latter are buried in sand (figure 15.12). Electric fish (families Gymnotidae, Mormyridae, and Gymnarchidae) generate their own electric fields and can identify the presence of potential prey by the alternation of this field by the prey (Lissmann 1958). They can even judge the distance of the prey by comparing the amplitude and gradient of the voltage distribution (von der Emde 1999). The platypus, an

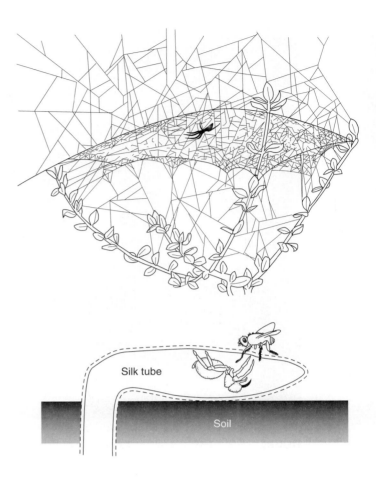

FIGURE 15.11   Spider webs.
(a) *Linyphia triangularis* hangs upside down from a dome-shaped web. (b) The purse-web spider *Atypus* lives in a silken tube, and bites prey through the walls of the tube and pulls it inside.

egg-laying monotreme mammal with webbed feet, also can use electroreception to find prey: they will even attack batteries hidden underwater (reviewed in Pettigrew 1999).

Aggressive Mimicry

A taxonomically widespread strategy for capturing prey is the use of lures, referred to as **aggressive mimicry.** Fish of the order Lophiiformes have a modified first dorsal fin spine on the tip of the snout. At the end of the modified spine may be a fleshy appendage, a tuft of filaments, or, in deep-sea forms, an organ containing light-emitting bacteria. The bait often resembles worms or crustacea, and the rest of the fish resembles an unmoving object, such as an algae-encrusted rock or a sponge. The fish wriggles the bait and keeps the rest of its body still. See figure 15.13 for a case in which the lure is a nearly exact replica of a small fish (Pietsch and Grobecker 1978).

Siphonophores of the phylum Cnidaria capture prey using tentacles armed with stinging cells called nematocysts. Some species have tentacles with branches and clustered nematocysts that look like copepods or fish larvae.

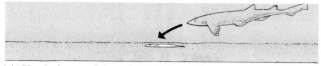

(a) Shark detects fish under sand.

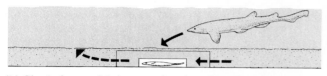

(b) Shark detects fish in agar chamber. Chamber doesn't block electric field.

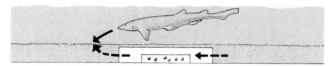

(c) Shark is unable to detect pieces of fish in agar chamber.

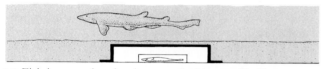

(d) Fish in agar chamber is covered with metallic film.

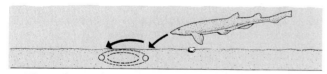

(e) Electrodes produce dipole field that shark detects.

FIGURE 15.12   Feeding responses of sharks to objects buried in sand.
Solid arrows denote responses of the shark; dashed arrows denote the flow of seawater through the agar chamber. The shark responds to the magnetic field of intact fish (a) and (b). Pieces of fish in (c) produce no magnetic field, and the shark responds to odor carried outside the agar chamber by the current. Metallic film in (d) blocks the magnetic field of the intact fish, and the shark fails to detect the fish. In (e), the shark responds to an artificially produced magnetic field, ignoring a piece of fish.
Source: Data from A. J. Kalmijn, "The Electric Sense of Sharks and Rays," *Journal of Experimental Biology,* 55:371–83, 1971.

When the siphonophore moves its tentacles, it lures zooplankton predators into the web of nematocysts, and the predators become food for the siphonophore instead (Purcell 1980).

Aggressive mimicry can be purely behavioral as well as morphological. In chapter 12 we saw an example of sexual enticement by fireflies, in which a female mimics the mating signal of another species and eats the male that

FIGURE 15.13   Aggressive mimicry by fish.
Angler fish (*Antennarius* spp.) and its lure are shown. In the bottom photo, a 2-second time exposure shows the pattern of movement of the luring apparatus. Note the lure resembles a small fish.
Source: Courtesy David B. Grobecker.

responds. The jumping spider *Portia fimbriata* displays a very impressive repertoire of behaviors. This spider enters the webs of other spiders, and vibrates the web in the same way that a struggling prey would. Web-building spiders do not have good vision, so they are attracted to the "prey," but are instead attacked by *Portia. Portia* seems to adjust the signals it makes according to the response of its target: it begins by giving a variety of vibrational signals in different patterns, but then focuses on producing the signals that provoke a reaction from the target spider (Jackson and Wilcox 1993).

Tools

Although the use of tools was once considered an exclusively human trait, it has evolved independently in several different lineages. In most cases the tool is an unmodified inanimate object. The sea otter *(Enhydra lutris),* for example, holds a rock on its chest and cracks shellfish, such as mussels, against it. Similarly, the Egyptian vulture *(Neophron percnopterus)*

FIGURE 15.14   Chimpanzee using a tool to obtain ants.
Chimps select sticks, modify them, and insert them into termite or ant mounds to extract the clinging insects. The infant observes the process from its mother's abdomen.
Source: © James Moore/Anthro-Photo.

picks up rocks and drops them on ostrich eggs. The chimpanzee *(Pan troglodytes)*, however, sometimes makes a modified tool by stripping leaves from a twig, which it then inserts into an ant or termite nest (figure 15.14). The insects cling to the stick, and the chimp eats those that hang on after it removes the stick. The woodpecker finch *(Cactospiza pallida)* of the Galápagos Islands uses sticks in a similar way to extract larvae from dead wood and may modify the stick by shortening it (reviewed in Lawick-Goodall 1970; Beck 1980).

FORAGING AND SOCIAL BEHAVIOR

When food is spread widely and irregularly over the environment and cannot be defended, individuals of a species may group into flocks or herds and may associate with other species. Foragers may directly benefit from neighbors in several ways.

Sharing Information

Foragers may get information about food sources from one another, especially if food occurs in dense, rare patches which are unpredictable in time or space. Animals may monitor the area around them for other foragers that have found food (Thorpe 1956). Some species produce distinctive sounds when they discover food that give information to conspecifics. For example, when chickens *(Gallus gallus)* hear food calls given by conspecifics, they look at the ground as if searching for food (Evans and Evans 1999). Groups may also act as "information centers" (Ward and Zahavi 1973). In this case, an unsuccessful forager follows previously successful group members back to a food source. An example is ravens, who depend on finding large carcasses to survive cold winters. Ravens share roost sites at night, and often the entire group is seen departing in one direction in the morning. Marzluff et al. (1996) captured ravens and held them in captivity so they were naive about the location of carcasses that the researchers put out. Some naive ravens were released into a roost with others that knew the location of the carcasses. Of these naive ravens, 14 of 14 found the carcass upon which the rest of the group was feeding. Only 4 of 15 naive birds that were kept away from knowledgeable conspecifics found food.

Animals feeding in groups may forage more efficiently because each individual spends less time scanning the environment for predators. For example, downy woodpeckers *(Picoides pubescens)* wintering in mixed-species flocks scanned less and fed more than did solitary individuals (Sullivan 1984). This phenomenon is described in more detail in the section on defense against predators.

Some individuals in social groups may specialize in stealing food from others, or in joining others that have already located food. These asymmetrical roles have been called "producing" (generating food) and "scrounging" (stealing food) (Barnard and Sibly 1981). This creates a problem for the producers, which must decide whether to stay with the scroungers or leave for another foraging group.

Cooperative Hunting

Some animals can take down larger or more dangerous prey when they hunt as a group. For example, some tropical spider species are social, and cooperatively construct large communal webs, which may be several meters across (figure 15.15). Together, the spiders attack trapped prey items, drag them back to a central retreat, and feed on them communally (Buskirk 1981).

Perhaps the most spectacular food-catching enterprise is the march of an army ant colony (figure 15.16) (Franks 1989). T. C. Schneirla spent most of his life trying to understand the complex life cycle of several New World species.

FIGURE 15.15   Communal spider web.
Social spiders *(Anelosimus eximius)* from southeast Peru capturing a deerfly.
Source: Photo by Karen R. Cangialosi.

FIGURE 15.16   Emigrating army ants.
Army ants *(Eciton burchelli)* moving along a woody vine carrying white larvae. Along both sides of the column there are rows of "guard workers" including a few soldiers with large white heads. These protect the emigration column from being disturbed.
Source: Carl Rettenmeyer, Connecticut Museum of Natural History.

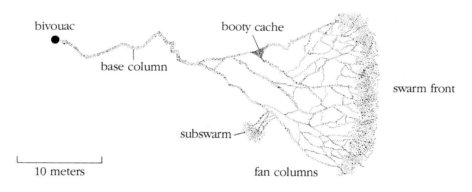

FIGURE 15.17   Pattern of raiding employed by army ants.
In this swarm raid of army ants *(Eciton burchelli)*, which can be found on Barro Colorado Island in the Panama Canal Zone, the advancing front is made up of a large mass of workers. The swarm flushes a variety of prey, mostly invertebrates, but also snakes, lizards, and birds. The queen and immature forms remain at the bivouac site.
Source: Data from C. W. Rettenmeyer, "Behavioral Studies of Army Ants," *Kansas University Science Bulletin*, 44:281–465, 1963.

Each night the colony, which consists of a queen, workers, larvae, pupae, and eggs, forms a bivouac (Schneirla and Piel 1948). Workers, upwards of 500,000 strong, form a protective net (the bivouac) by hooking their legs and bodies together. Each morning the bivouac dissolves and the ants begin moving outward. A column emerges along the path of least resistance and heads away from the bivouac site. There are no true leaders; workers in the lead turn back into the swarm behind them every few centimeters. The ants lay down pheromone trails that guide those that follow, and the column may branch into a fan-shaped affair (figure 15.17).

Virtually all animals in the path are stung, cut into pieces, and transported to the rear. Army ants are able to attack and consume prey items as large as snakes, lizards, and small birds that they would not be able to handle as individuals. Arthropod life in these areas is temporarily depleted. When larvae are developing, the bivouac location changes each night (nomadic phase); during the egg-laying and egg-development phases of the reproductive cycle, the bivouac remains in the same place each night (stationary phase).

Perhaps the most familiar and most intensively studied cooperative hunters are the mammalian carnivores. One member of the family Felidae and several members of the family Canidae have evolved complex social behavior related to the cooperative capture of prey.

The lion *(Panthera leo)* lives in closed social units and is most abundant in the grasslands and open woodlands of Africa. They use a **stalk-and-rush** method of hunting. They rely on getting very close to the prey, and then surprising it with a sudden burst of speed. By hunting in groups, they add to their diet two species that an individual lion could never attack alone: buffalo *(Syncerus caffer)* and giraffe *(Giraffa camelopardalis)*. They are also at least twice as successful as lions that hunt alone. Groups can also drive other predators and scavengers from the food. Schaller (1972) found that plains-dwelling prides kill less than half their food, relying on other predators, such as hyenas *(Crocuta crocuta)*, to make kills for them (figure 15.18).

Lions have been used as a model system to examine the question of optimal group size. Caraco and Wolf (1975) used Schaller's data, and estimated that individuals hunting in groups of two get the highest rate of food intake (figure 15.19). However, lions typically forage in larger groups. Numerous explanations for this observation have been proposed (see references in Packer et al. 1990). For example, lions may forage in larger groups because group members are kin, and they are more likely to aid relatives (see chapter 19). In addition, the variance in prey intake may also be important in determining foraging decisions, as described earlier in this chapter. Next, group size may not be stable. If individuals cannot be prevented from joining a group, group size may be pushed above the optimum size. This is understandable from the perspective of a solitary individual that has the choice between hunting alone or joining a group: it should join the group if this provides a bigger payback than hunting alone. Finally, lions face other demands besides foraging success. They must protect their young and maintain a long-term territory. Packer et al. (1990) point out that for lions, all these factors must be included in order to understand the size of hunting groups.

FIGURE 15.18   Male lions with zebra.
Male lions do relatively little killing of prey. Male on the right drags a zebra carcass that he probably scavenged from hyenas.
Source: © George Schaller/Bruce Coleman, Inc.

The African wild dog *(Lycaon pictus)* may seem an unlikely big game predator, as it weighs only about 18 kg (figure 15.20). However, these dogs typically catch prey weighing as much as 250 kg (Schaller 1972). Rather than using the stalk-and-rush tactic of lions, these canids are **coursers,** typically pursuing prey for many kilometers. The dogs live in mixed-sex packs that average ten adults. Of particular interest is the prehunt "ceremony," in which dogs draw back their lips to expose their teeth, nibble and lick each other's mouths, and run whining from pack member to pack member. The greeting seems to represent ritualized food begging. Setting out on a hunt, the dogs travel at a trot and fan out loosely over the terrain. When a prey is encountered (such as Thomson's gazelles, *Gazella thomsoni,* or wildebeest, *Connochaetes taurinus;* Carbone et al. 1997) they cooperatively work to cut off its escape route and bring it down. Wild dogs face competition from the spotted hyena, *Crocuta crocuta* (Gorman et al. 1998). Hyenas act as kleptoparasites and steal food from dogs. Hunting is very energetically costly for wild dogs, so even a small amount of kleptoparasitism can have grave effects. In fact, hyenas may be a contributing factor to the decline of wild dogs; there are only about 5,000 individuals remaining in the wild (Gorman et al. 1998). Lions are another problem for wild dogs; they are responsible for 39 percent of natural pup deaths and 43 percent of natural adult deaths. In fact, dogs are at their lowest density where food is most abundant because they seem to be avoiding areas used by lions (Mills and Gorman 1997). The conservation of wild dogs presents a challenging problem.

Although hyenas do scavenge food from other carnivores, they get more of their food from hunting (Kruuk 1972). The basic social unit is the clan, which consists of 10–60 hyenas of both sexes and all ages. There is little exchange between clans; each defends a territory with a centrally located den. Because the prey, mainly wildebeests and zebras, migrate to the woodlands in the dry season, most hyena clans break up at that time, and some individuals become nomadic and follow the prey. Thus, food availability has a profound effect on social organization in hyenas. In the Masa Marai reserve in Kenya, three-quarters of the hunts observed over 7 years were made by solitary individuals (Holekamp et al. 1997). Holekamp et al. (1997) described the chases: hyenas typically rushed a group of animals, paused briefly to watch, selected an individual, and chased it down over distances ranging from 75 m to 4 km. Prey were then grabbed and disembowelled, taking 0.5–13 minutes to die. Other hyenas arrived at the kill site and competed for the prey.

The social carnivore of northern temperate and arctic regions is the wolf *(Canis lupus)*, which typically preys on deer *(Odocoileus* spp.), moose *(Alces americana)*, buffalo *(Bison bison)*, sheep *(Ovis* spp.), caribou *(Rangifer tarandus)*, and elk *(Cervus canadensis)*. Pack size varies greatly, but most packs have fewer than eight members (Mech 1970). Most often they use scent to detect prey; the lead animals stop when they detect the odor of prey. They may also use chance encounter and tracking. If they detect the prey at

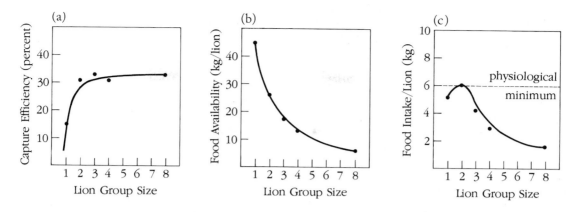

FIGURE 15.19   Capture efficiency, food availability, and estimated food intake as functions of lion group size for Thomson's gazelle prey.
Capture efficiency (a) increases as the lion group size increases, showing the benefits of cooperative hunting. Little increase occurs beyond hunting parties of two, however, since Thomson's gazelle is a small prey-type for lions. In (b), as the lion group size increases, the food availability declines, since there are more mouths to feed. Given the relationships in (a) and (b), the projected food intake per lion as shown in (c) is maximized at a group size of two. If Thomson's gazelle were the only prey taken, two would be the only group size in which lions would be able to obtain enough food to survive.

some distance, they proceed slowly and stalk sometimes to within 10 meters of the prey. Once the prey is in flight, the wolves rush. Wolves must get close to their prey during the stalk-and-rush or they will quickly give up. If they cannot make an attack, they may chase the prey, usually for less than half a mile. Some earlier studies had described wolves as coursers, but Mech concluded that they are mainly stalkers and rushers (figure 15.21).

DEFENSE AGAINST PREDATORS

Now we turn to the prey's point of view. Most animals, except for a few upper trophic-level carnivores, are potential prey. Many animals have behavioral adaptations to reduce the chances of being eaten. First we'll examine strategies undertaken by individuals, and then we will turn to antipredator strategies used by group-living animals.

Individual Strategies

Escaping and Freezing

Many animals keep close to a nest, burrow, or other refuge near the center of their home range. For example, white-footed mice (*Peromyscus leucopus*) usually nest in hollow trees or logs, and the probability of finding the mouse at a particular spot declines rapidly with distance from the nest, so they are rarely in unfamiliar space or far from shelter (Vessey 1987). Familiarity with escape routes makes it more likely that animals can avoid potential predators (reviewed in Stamps 1995).

Another predator response strategy is to freeze. The presence of protective or cryptic coloration is often associated with this behavior, as in the spotted white-tailed deer fawn (*Odocoileus virginianus*). Some animals carry this

FIGURE 15.20   African wild dogs attacking zebra.
Extensive cooperation is necessary among small social carnivores, such as African wild dogs, to enable them to capture much larger prey, such as zebra.
Source: © George Schaller/Bruce Coleman, Inc.

strategy further and feign death. For example, the opossum (*Didelphis virginiana*), if harassed by a predator such as a dog, remains motionless on its back. Hog-nosed snakes (*Heterodon platyrhinos*) exhibit similar behavior, although they may precede it by threat behavior in which the hog-nosed snake resembles a cobra. Freezing and feigning death may work because many predators seem to respond only to moving prey. For example, wolves generally will not attack prey that stand motionless (Mech 1970).

Deception

Another strategy used by vulnerable organisms is deception. When threatened, larvae of the moth *Hemeroplanes* sp. inflates its anterior end to form an excellent representation of

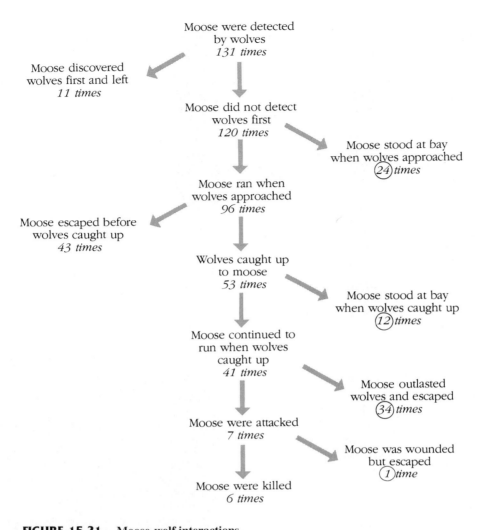

FIGURE 15.21   Moose-wolf interactions.
Observations at Lake Superior's Isle Royale National Park show the results of 131 separate moose-wolf interactions. Circled numbers indicate moose actually encountered by wolves. Only 6 of the 131 moose detected by wolves were killed.

a snake's head, which it then waves (figure 15.22a). Other lepidopteran larvae resemble inanimate objects in the environment, such as twigs, leaves, bark, or even bird droppings.

Many animals have behaviors or markings that may serve to misdirect or surprise predators. Some reef fish have conspicuous eyespots on their tail end that are much larger than the real eyes; these could frighten away potential predators or misdirect their attack to less vulnerable parts of the body. *Pieris rapae* butterflies were artificially given wing markings, including eyespots, that resembled a false head on their posterior wings. Marked individuals were more likely to escape from blue jays (*Cyanocitta cristata*) than were controls (Wourms and Wasserman 1985). Some moths (figure 15.22b) have similar spots. The moths are cryptically colored when at rest on vegetation, but if startled, they spread their forewings and reveal large, brightly colored eyespots on the hindwings.

Disruptive color patterns consist of high-contrast markings that are thought to break up the outline of an organism. Merilaita (1998) analysed the spots on the marine isopod *Idotea baltica,* and found that they did not directly match the background pattern (indicating that the spot pattern does not serve as crypsis) and were in contact with the edge of the isopod more often than expected by chance (suggesting that they function to break up the outline). Further indirect evidence that patterns provide protection against predation comes from the finding that chemically protected or unpalatable species usually lack the markings found in palatable species. However, few experimental field tests of this question have been attempted. Silberglied et al. (1980) obliterated the wing stripes on a species of tropical butterfly (*Anartia fatima*) with a felt-tipped pen and compared their survival with that of unaltered controls (figure 15.23). The lack of a difference between experimentals and controls in survival rate or frequency of wing damage (an indicator of predatory attack) suggests that the color patterns have some other function.

(a)   (b)

FIGURE 15.22   **Insect defense displays.**
The anterior end of an alarmed caterpillar (a) resembles the head of a snake and frightens predators. A moth *(Antharea polyphemus)* (b) exposes large eye spots on its underwing in order to frighten away or deflect attacks by predators.

Source: Photos (a) © Lincoln P. Brower, (b) © Thomas Eisner, Cornell University.

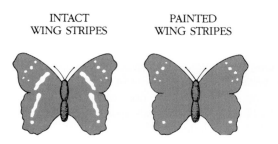

INTACT   PAINTED
WING STRIPES   WING STRIPES

FIGURE 15.23   **Field experiment to test function of butterfly markings.**
Butterflies *(Anartia fatima)* with wing stripes intact, left, and covered with ink, right. If stripes function to reduce predation, butterflies with stripes obliterated should suffer higher mortality. Results of experiments, however, indicated no such function for the stripes.

Source: Data from R. E. Silberglied, et al., "Disruptive Coloration in Butterflies: Lack of Support in *Anartia fatima," Science,* 209:617-19, 1980.

FIGURE 15.24   **Bombardier beetle spraying.**
Quinones and hydrogen peroxide are mixed, heat to the boiling point and are sprayed on an attacker.

Source: © Thomas Eisner and Daniel Aneshansley, Cornell University.

Toxicity

Many organisms are harmful to predators. Plants may contain toxins such as alkaloids; animals may have armor plates (as in the armadillo) or spines (as in the porcupine, hedgehog, and three-spined stickleback). Many invertebrates inject venom into attackers. Eisner (1966, 1970) discovered a large number of arthropod defense mechanisms. For instance, the bombardier beetle *(Brachinus* spp.) has a plumbing system that resembles a liquid-fueled rocket. It stores quinones and hydrogen peroxide in separate reservoirs. When the beetle is disturbed, these two liquids are mixed in an outer vestibule, and in the presence of enzymes, the solution heats to the boiling point and is discharged as a noxious spray (figure 15.24).

Noxious animals tend to be conspicuously colored, presumably so that predators can easily recognize and thus avoid them. Such **warning coloration,** or **aposematism,**

occurs, for example, in skunks *(Mephitis mephitis)* and coral snakes *(Micrurus fulvius)* (Wickler 1968). Many animals sequester poisons for defense. Orange and black monarch butterflies *(Danaus plexippus)* are toxic because of poisons (cardiac glycosides) they obtain from the milkweed plants they feed on as larvae. Adult females lay eggs on the toxic species of milkweed; the larvae have evolved a resistance to the poisons and are able to store them. Milkweed species vary in the amount of toxins they have, so monarchs also vary in their toxicity.

How could warning coloration evolve? Any butterfly with a new mutation enabling it to store the toxins through metamorphosis would probably not survive, since "uneducated" predators would be common. One possibility is that the predator need not consume the entire butterfly to reject it,

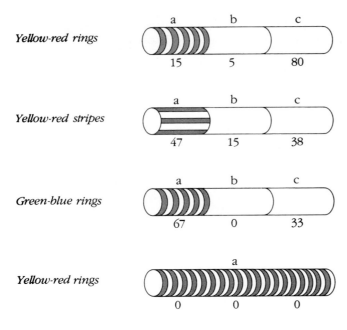

Yellow-red rings

a   b   c
15   5   80

Yellow-red stripes

a   b   c
47   15   38

Green-blue rings

a   b   c
67   0   33

Yellow-red rings

a
0   0   0

FIGURE 15.25   Wooden snake models are used to test hand-reared motmots.
Area (a) of each model was painted as indicated at left; areas (b) and (c) were plain wood. Numbers beneath each model show the percentage of pecks by the motmots. Motmots avoided those models or portions of models painted with yellow-red rings, the same colored rings that the coral snake has.

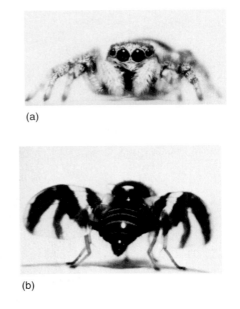

(a)

(b)

FIGURE 15.26   Prey mimics predator.
(a) Front view of zebra spider and (b) posterior view of snowberry fly. Note similarity between spider legs and fly wing markings.
Source: Courtesy of B. D. Roitberg, R. LaBlonde, Photographer.

and so the toxic butterfly could live to reproduce and pass on its ability to store toxins to its progeny; in fact, many bad-tasting prey are released before they are seriously harmed (Wiklund and Jarvi 1982). But when illness is delayed, the prey may not survive. Since the predator, who is now "educated," will probably never again eat another butterfly resembling that one, we seem to have a case of altruism on the part of the consumed butterfly. A second possibility is that through the process of kin selection, the toxic adult protects relatives that live in the area and share its genes—by sacrificing its life to educate the predator (Hamilton 1963).

Avoidance of dangerous or unpalatable prey may occur also in the absence of opportunities for learning. For example, even laboratory-reared motmots (*Eumomota superciliosa*)—lizard- and snake-eating birds—avoid models painted with the color patterns of the coral snake (Smith 1975) (figure 15.25). Naive domestic chicks were less likely to eat mealworms that were painted black and yellow than those painted green or olive (Schuler and Hesse 1985, but see Sklorz and Volz 1990).

Mimicry

Among arthropods such as butterflies, unpalatable species tend to resemble each other and are referred to as Müllerian mimics—for example, the monarch and queen (*Danaus gilippus*) butterflies are usually both toxic because of the plants on which they feed. Müllerian mimics seem to have evolved because by looking alike and acting similarly, they provide fewer different prey types for predators to learn to avoid and thus are less likely to be eaten by mistake. Some palatable

species, called Batesian mimics, may evolve morphologies and behaviors similar to those of unpalatable species. Batesian mimics enjoy protection because of the predators' learned avoidance of the toxic species. A well-studied example of mimicry is that of monarch butterflies. Brower et al. (1968) demonstrated "one-trial conditioning" of blue jays (*Cyanocitta cristata*) fed monarch butterflies: consumption of a single monarch butterfly often led to vomiting by the jay and avoidance of other monarchs or their mimics.

An unusual case of mimicry is that of a sheep in wolf's clothing: a fly mimics its predator, a spider (Mather and Roitberg 1987). Snowberry flies (*Rhagoletis zephria*) are among the prey of solitary zebra jumping spiders (*Salticus scenius*). The flies have a wing banding pattern that resembles the legs of the jumping spider (figure 15.26). Flies display with their wings, mimicking a walking spider. Other spiders flee from displaying flies, apparently mistaking them for an aggressive conspecific. Nondisplaying flies, or those with their bands covered, are more likely to be eaten by the spiders.

Distraction Displays

Individuals may use distraction displays—for example, the broken-wing displays of avocets (*Recurvirostra americana*)—to attract the attention of a predator and draw it away from the nest (figure 15.27) (Sordahl 1981). Many mammals have white rump or tail patches; in the presence of predators the hairs are erected and the tail waved, making a conspicuous display. Observers have suggested several functions for this **flagging** behavior, such as distracting the predator from other members of the group, warning other group members, confusing the predator when many group members are displaying, signaling the predator that it has been detected, and eliciting premature pursuit (figure 15.28) (Smythe 1977;

FIGURE 15.27 American avocet performing broken-wing display.
This behavior pattern presumably functions to distract predators from the nest or young. Although considered by some to be altruistic behavior, it really is most easily understood as a form of parental investment.
Source: © Leonard Lee Rue III/Animals/Animals/Earth Scenes.

Guthrie 1971). These hypotheses are difficult to tease apart, as both the predator and conspecifics are likely to be in a position to receive a signal from a flagging animal. In white-tailed deer, deer that flag generally run faster than deer that are not flagging, suggesting that flagging is meant to deter the predator from pursuing; the hypothesis that flagging warns conspecifics was not supported (Caro et al. 1995).

Social Strategies

Experimental evidence from a wide range of taxa demonstrates that animals can reduce their risk of predation by associating with conspecifics in groups (reviewed in Lima and Dill 1990). There are a number of hypotheses concerning why this might be so, and they can often be teased apart experimentally. These are discussed further in chapter 19; see especially figure 19.1.

The Encounter Effect

Groups of prey may be less likely to be encountered by searching predators than are solitary prey. The detection of three-dimensional groups does not increase proportionally with group size (i.e., a predator is not five times as likely to find a group with ten members as with two members) (Inman and Krebs 1987).

"Forget these guys."

FIGURE 15.28 Prey might benefit by advertising their alertness and condition to predators.

The Dilution Effect

Whereas the encounter effect acts when predators are searching for prey, **the dilution effect** acts after a predator has found a group. Simply stated, this is "safety in numbers." If a predator that attacks a group can eat only one prey, it is better to be in a larger group than a smaller group from a purely probabilistic perspective: the chances of being the individual that is selected by the predator is lower in a large group (Foster and Terherne 1981). The encounter and the dilution effects may operate simultaneously (see chapter 19).

The Selfish Herd

Some early ethologists felt that natural selection at the population level was necessary to explain why animals form groups; that is, the behavior had to evolve "for the good of the group" rather than for the good of the individual. Several authors, notably W. D. Hamilton, described how selfish behavior can lead to aggregation. Imagine, said Hamilton (1971), frogs surrounding the rim of a pond that are trying to evade a water snake. The snake will snatch the nearest frog. Thus, it is in the best interest of each frog to attempt to get between two other frogs. This rule alone leads to the aggregation of frogs into heaps.

There is much empirical evidence in support of the idea that the edges of groups are more susceptible to predation.

Many flocking or schooling animals cluster in the presence of a predator, as Hamilton points out. Both tadpole groups (Watt et al. 1997) and fish schools (Krause 1993) become more compact in the presence of chemical cues indicating that a predator is near. Direct evidence of differences in predation risk in different areas of a group comes from studies of a Mexican spider, *Metepeira incrassata,* which we met earlier in this chapter in our discussion of risk-sensitive foraging. Each spider has its own orb web, but hundreds of individuals are clustered together in a colony supported by shared frame lines. Predation attempts by wasps, hummingbirds, and other predators were directly observed by Rayor and Uetz (1993), and risk of predation was highest on the edge, especially for large spiders, which are favored by predators. Animals may also be protected from parasites in the same way: ungulates in the center of groups are bitten less often than those on the periphery (Mooring and Hart 1992). However, not every predator preferentially attacks the edge of a group: black-crowned night herons (*Nycticorax nycticorax*) are more likely to attack nests in the center of a least-tern colony (*Sterna antillarum*) (Brunton 1997).

Increased Detection of Predators

The more animals in a group, the more likely it is that one of them will detect an approaching predator. This has been called the "many eyes" hypothesis.

Some animals give alarm calls when they detect a predator. When alarm calling is not particularly dangerous to the caller, it is easy to see how it might evolve. In other species, such as Belding's ground squirrels, callers put themselves at risk of predation. As we saw in chapter 12, Sherman (1977, 1981) showed that close relatives of the caller were likely to be close by and thus could benefit from this seemingly altruistic behavior.

Alarm calls may be specific to the type of predator. Vervet monkeys give different calls to leopards, eagles, and snakes. Monkeys that hear playbacks of these calls respond appropriately: they run into trees when they hear the leopard alarm, look up or run into bushes with the eagle alarm, and stand bipedally and look down when they hear the snake alarm. Interestingly, infants gave alarm calls to animals that were not dangerous, but were likely to give leopard calls to other terrestrial animals, and snake calls to long, snakelike objects (Cheney and Seyfarth 1990). Chickens also produce qualitatively different calls in response to terrestrial and aerial predators. Hens crouch when they hear an aerial call, and stand erect and vigilant when they hear a terrestrial call (Evans et al. 1993).

Because of the added protection offered by groupmates, animals may be able to devote more time to other activities besides watching for predators. This has received a great deal of experimental attention, especially in species that have special, easy-to-recognize behaviors when they are searching for predators. A classic study species, especially for animal-behavior laboratory courses, are foraging geese that take a heads-up vigilance posture that is easily distinguishable from their heads-down feeding behavior. Individual geese in large groups generally spend less time with their heads down than do geese in smaller groups. This effect has been found in other species

as well. This shift in behavior has often been interpreted as a result of the "many eyes" effect, though Roberts (1996) notes that if individual risk is reduced in groups through any mechanism, one would expect vigilance to decrease.

Underlying the "many eyes" hypothesis is the assumption that all members of a group are alerted to a threat as long as one member senses it, or "collective detection." Lima (1995) tested mixed flocks of juncos (species) and tree sparrows (species) foraging for corn meal. In order to carry out this test, he needed to selectively target only one bird. He rolled a rubber ball down a high-walled ramp. Only the bird directly in front of the ramp could see the alarming approach of the ball, which was stopped by a pad before it left the ramp. He monitored the behavior of other birds near the targeted birds. Most birds (38 of 64) did not flush when the targeted bird flushed; 18 partially flushed; and 8 completely flushed. Thus, in these species, there is only weak support for collective detection. However, this mechanism may function in other species, especially those that give alarm calls when startled.

The Confusion Effect

A variety of predators exhibit a lower prey capture rate and a longer period of hesitation when they attack grouped prey than when they attack solitary prey (Miller 1922). Factors that enhance this **confusion effect** are the number and density of prey swarms and the uniformity of appearance of swarm members (Milinski 1979). The predator tries to isolate an individual from the group, and any individual that strays from the group or is in any way different from the rest is likely to be attacked. For example, when experimenters presented hawks with sets of ten mice, the hawks were more likely to catch the oddly colored mouse (Mueller 1971). Predators can improve with experience: in laboratory experiments with three-spined stickleback fish (*Gasterosteus aculeatus*) feeding on water fleas (*Daphnia*), fish with experience in feeding on dense swarms fed more efficiently on them than did fish with experience only with low-density populations. (Milinski 1979).

Group Defense: Mobbing

Some prey species are able to turn the tables and attack the predator. Mobbing of predators by many members of the prey species is common. Many species from a variety of taxa mob snakes, including Formosan squirrels (*Callosciurus erythraeus thaiwanensis*) (Tamura 1989), scrub jays (*Aphelocoma c. coerulescens*) (Francis et al. 1989), and cotton-top tamarins (*Saguinus oedipus*) (Hayes and Snowden 1990). Raptors are also often mobbed, often by mixed groups of birds from several species. For example, the mobbing call of the blackcapped chickadee (*Parus atricapillus*) is recognized by at least ten other bird species that then join in mobbing (Hurd 1996). Mobbing can be very effective. Captive owls (*Athene noctua* and *Strix aluco*) showed great distress when mobbed by blackbirds and moved their perch sites (*Turdus merula*). Powerful owls (*Ninox strenua*) abandoned their daytime roosts during 20 percent of mobbing bouts and responded by calling or actively monitoring mobbers during 54 percent of bouts (Pavey and Smyth 1998).

SUMMARY

This chapter deals with foraging and fitness from the viewpoints of both the prey and predator. Successful foraging is generally tightly linked with an animal's survival and reproductive success. Optimality modeling can help us predict the behavior of animals under particular sets of circumstances. These models have three parts: the decision variable (the choices available to the animal), the currency in which the optimal choice is measured, and constraints that limit the animal's behavior. Two common models that are useful for understanding foraging behavior are the diet model (about the decision whether to eat a particular food) and the patch model (about whether to leave a patch of food and seek a new one). Modelers have also examined the effect of variance in food availability on behavior. We can modify the simplest forms of the model to make them more realistic.

Animals employ a range of interesting feeding behaviors. Some modify their own food supply, such as leaf-cutter ants that plant fungal gardens. Spiders and ant lions build traps to capture insect prey. Some aquatic animals use electric fields to capture food. Others use morphological and behavioral adaptations to mimic prey and lure predators to their deaths. Some foragers use tools to capture or manipulate prey.

Social organization and food resources are often closely related. Foragers may share information about the location of food, or may cooperate in hunting. Army ants and social carnivores provide interesting examples of cooperative hunting.

Adaptations that provide defense against predation are just as elaborate as those that increase feeding efficiency. Individual tactics include escaping, having cryptic coloration and behavior, feigning death, practicing deception, and having markings or behaviors that surprise the predator or that misdirect its attack. Many organisms produce or store toxic substances to gain protection, and nontoxic species may mimic the species with these toxic substances. Living in groups may provide protection through the encounter effect, the dilution effect, increased detection of predators, the confusion effect, and mobbing.

DISCUSSION QUESTIONS

1. Which assumption of the diet model do you think is most restrictive, that is, unlikely to apply to most species? Defend your answer. Do you think it would be possible to modify the basic model to take this into account?

2. Let's examine the marginal value theorem in more detail.

   a. Imagine that a particular type of food is very low in energy. How would this change the shape of the gain curve in figure 15.3? How would the optimal time in a patch compare with an environment where the food was high in energetic value?

   b. The gain curve describes the net energy gain. What would a plot of the instantaneous energy gain look like?

3. Animals might maximize long-term or short-term energy gain. Which species do you think might maximize each of these? Why?

4. Discuss the similarities and differences between the evolution of alarm calls and the evolution of the ability to store substances toxic to predators.

5. Imagine you are studying a species of fish that shows schooling behavior. You hypothesize that schooling functions as a defense against predators. Design an experiment, or a series of experiments, that will enable you to tell whether the fish benefit from (a) the dilution effect and (b) the confusion effect.

6. Normal butterflies (*Nymphais io*) with large eyespots, and altered butterflies with eyespots blotted out were presented equal numbers of times to six yellow buntings (*Emberiza citrinella*), predatory birds. Data on the buntings' escape responses are presented here (Blest 1957). Discuss these results.

Number of fright responses by birds (yellow buntings) when presented with butterflies

Yellow Bunting Individuals	Normal Butterfly	Eyespots Removed
1	56	16
2	11	5
3	8	4
4	18	1
5	18	3
6	17	2

Source: Data from A. D. Blest. "The function of eyespot patterns in the Lepidoptera." *Behaviour* 11:209–56, 1957.

SUGGESTED READINGS

Hamilton, W. D. 1995. *Narrow Roads of Gene Land.* Vol. 1: *Evolution of Social Behaviour.* New York: W. H. Freeman and Company.

Preston-Mafham, R., and K. Preston-Mafham. 1993. *The Encyclopedia of Land Invertebrate Behavior.* Cambridge: MIT Press.

Stephens, D. W., and J. R. Krebs. 1986. *Foraging Theory.* Princeton: Princeton University Press.

Wickler, W. 1978. *Mimicry in Plants and Animals.* New York: McGraw-Hill.

PART FIVE

Social Organization and Mating Systems

The common thread in the final four chapters is interaction among members of the same species, first in terms of competition and conflict (chapter 16), then mating and rearing offspring (chapters 17 and 18), culminating in the evolution of cooperation and complex social systems (chapter 19).

16 CHAPTER

Conflict

Konrad Lorenz (1966) defined aggression as "the fighting instinct in beast and man which is directed against members of the same species." His use of the word *instinct* points to one facet of the old nature-nurture controversy that has direct bearing on human behavior. Is aggression a universal property of social animals, including humans, and is it an inborn trait whose expression is inevitable? How can we explain the widespread tendency of animals in a fight to use restraint and not to fight to the death, even when the most aggressive individuals control resources such as food or mates?

In this chapter, we define aggression, agonistic behavior, and competition; consider forms of aggression; and then consider ways in which aggression is expressed. We look briefly at the internal causes of aggressive behavior— genetic, neural, and hormonal—and at external causes (i.e., the role of environment). Next, we discuss control of aggression, explore the use of game theory models of conflict, and conclude with implications of the study of animal aggression for understanding human behavior.

AGGRESSION, AGONISTIC BEHAVIOR, AND COMPETITION

Definitions

Aggression is a complex phenomenon that has been defined in many ways. Some definitions include predatory behavior, in which the animal being attacked is eaten in the process. Psychologists have defined aggression as behavior that appears to be intended to inflict noxious stimulation or destruction on another organism (Moyer 1976). The notion of intent is necessary to exclude such destructive behaviors as a male elephant seal's trampling a pup while trying to copulate with the mother. Use of the word *aggression* emphasizes offensive behavior. Behavioral ecologists take a more functional approach and consider aggression as a form of resource competition, in which an animal actively excludes rivals from some resource such as food, shelter, or mates (Archer 1988).

A term with a more precise definition is **agonistic behavior** (not to be confused with agnostic behavior), which is a system of behavior patterns with the common function of adjustment to situations of conflict among conspecifics. The term includes all aspects of conflict, such as threats, submissions, chases, and physical combat, but it specifically excludes predatory aggression, since, according to Scott (1972); ingestive behavior is part of a separate behavioral system with a different function.

We can list forms of aggressive behavior as follows (Moyer 1976; Wilson 1975):

- Territorial—exclusion of others from some physical space

- Dominance—control of the behavior of a conspecific as a result of a previous encounter

- Sexual—use of threats and physical punishment, usually by males, to obtain and retain mates

- Parental—attacks on intruders when young are present

- Parent-offspring—disciplinary action by parent against offspring (mostly in mammals, usually associated with weaning)

- Predatory—act of predation, possibly including cannibalism

- Antipredatory—defensive attack by prey on predator, such as mobbing

With the exception of predation, these involve conflict among conspecifics and so would also be included under agonistic behavior. These forms of aggression serve very different functions within and between species, and independent regulatory centers in the brain may even have evolved for them. Therefore, the term aggression is not a unitary concept. Most of our discussion here concerns agonistic behavior among conspecifics; we considered interactions between different species in chapter 15.

Competition for Resources

Most agonistic behavior involves competition for some limited resource (food; water; access to a member of the opposite sex; or space for nesting, wintering, or safety from predators). Such behavior is distributed throughout the animal kingdom. Even sessile organisms such as sea anemones (phylum Cnidaria) have special weapons they use against members of their own species (Francis 1973, 1988). Along the rocky shores of the West Coast of the United States *Anthopleura elegantissima* actively compete for space. If two polyps from genetically distinct clones come into contact, one or both will retract their feeding tentacles and inflate a special set of tentacles called acrorhagi that are laden with nematocysts (stinging cells) (figure 16.1). Once the tips of the acrorhagi touch the other anemone, they detach and the nematocysts are discharged into the recipi-

ent. This process may be reciprocated, until one of the contestants withdraws its tentacles and retreats (Ayre and Grosberg 1995). If retreat is impossible, the loser may die from its injuries.

Exploitation and the Ideal Free Distribution

Competition can be divided into two forms: **exploitation,** in which organisms passively use up resources, and **interference,** in which organisms interact so as to reduce one another's access to, or use of, resources. Suppose that there is a group of 12 ducks on a pond in a city park. Two people, one at each end of the pond, are feeding bread to the ducks. One is feeding the ducks twice as fast as the other. How should the ducks distribute themselves so that each gets the most food? Common sense tells us that if the birds don't guard the access to the bread, and if they have all the information they need about the feeding rates, an average of four should be at the end with the low feeding rate and eight at the end with the high feeding rate. This theoretical pattern is called an **ideal free distribution—ideal,** because it assumes that animals have accurate and complete information about the distribution of resources, and **free,** because individuals are passive toward each other and free to go wherever they can exploit the most resources (Fretwell 1972; Fretwell and Lucas 1970) (see also figure 14.15). When just such an experiment was conducted using 33 mallard ducks on a lake in England (Harper 1982), the ducks distributed themselves as predicted (figure 16.2).

Interference Competition and Resource Defense

More often it is the case that the individuals are not passive. Some individuals may establish territories and defend resources, or some may be dominant and control the access of others to resources. In the duck experiment by Harper (1982), it turned out that even though the results fit an ideal free distribution, some ducks got more than their share of food and became **despots.** In species of birds, the best habitat quickly gets taken over; younger or less aggressive individuals are excluded and forced to breed in less suitable habitats. For instance, prime breeding habitat for great tits (*Parus major*) is oak woodland. Excluded individuals attempt to breed in hedgerows, where they have lower reproductive success than those in prime habitat. Should a breeder in oak woodland die, it is quickly replaced by a hedgerow resident (Krebs 1971).

CONFLICT ABOUT PHYSICAL SPACE

An area occupied more or less exclusively by an animal or group and defended by overt aggression or advertisement is a **territory.** The defense of space against members of the

(2)

(3)

(4)

(5)

(6)

(7)

FIGURE 16.1   Agonistic behavior in sea anemones.

2. The anemone is shown at rest.

3. Inflation: the white tips of the inflating acrorhagi have become visible; the column has become more elongated.

4. Movement of application: the fully inflated acrorhagi are drawn upward and back and are just about to begin moving downward toward the victim.

5. Movement of application: a second movement of application is underway and more fully inflated acrorhagi have been recruited (*cf.* fig. 4); transparent areas are obvious at the tips of several acrorhagi where acrorhagial ectoderm was released from the acrorhagi and applied to the victim during the first movement of application.

6. Movement of application: the acrorhagi have reached the bottom of their downward sweep and are being wiped against the body of the victim; swelling of the fosse area between the inflated acrorhagi and the tentacles is apparent here.

7. Movement of application: seen from overhead as the acrorhagi reach the level of the victim, elongation of the oral disc in the area proximal to the inflated acrorhagi is apparent; scraps of applied acrorhagial ectoderm are visible on the body of the victim (→).

Source: Francis, L. 1973. Intraspecific aggression and its effect on the distribution of *Anthopleura elegantissima* and some related sea anemones. *Biol. Bull.* 144:73–92.

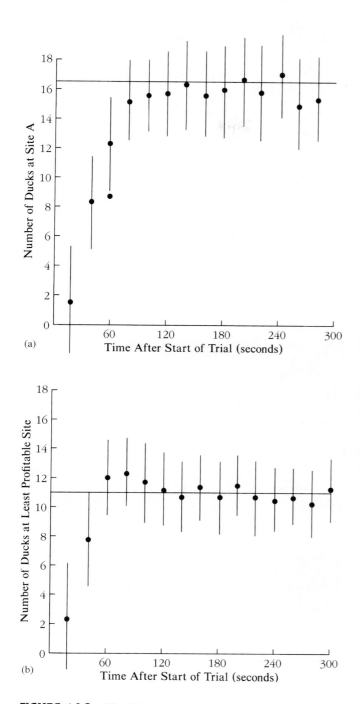

FIGURE 16.2 Ideal free ducks.
(a) The mean number of ducks at site A plotted against time since start of trial, when the patch profitability ratio was unity. The horizontal line is the ideal free prediction: half the 33 ducks at site A. (b)The mean number of ducks at the least profitable site plotted against time since start of trial, when the patch profitability was 2:1. The horizontal line is the ideal free prediction: one third of the 33 ducks at the least productive site.

same species accounts for much of the agonistic behavior in a host of animals, ranging from limpets to long-billed marsh wrens. This is not to say that all animals defend their turf, as might be assumed from reading such books as Robert Ardrey's *The Territorial Imperative* (Ardrey 1966). In fact, the majority of species probably do not.

The area used habitually by an animal or group, in which the animal spends most of its time, is its **home range.** Most organisms spend their lives in a relatively restricted part of the available habitat and learn the locations of food, water, and shelter in this area. The area of heaviest use within the home range is the **core area.** This location may contain a nest, sleeping trees, water source, or a feeding tree. As with home range, the designation of a core area is somewhat arbitrary, but useful in understanding the behavior and ecology of different species, or the same species in different habitats or at different population densities. Figure 16.3 illustrates home ranges and core areas of baboons in Africa.

The minimum distance that an animal normally keeps between itself and other members of the same species is its **individual distance.** In birds, it is frequently the distance that one individual can reach to peck another; we see this spacing in starlings lined up on a telephone wire, in sea gull nests in a colony, or in green heron (*Butorides virescens*) fledglings on a tree branch (figure 16.4). Individual distance, or personal space, exists for humans as well. For example, when talking to another person, most of us become uncomfortable and back off if the other person comes too close. The amount of personal space required varies from culture to culture and from situation to situation (Hall 1966).

The size of the territory or home range depends on the size of the animal as well as on the particular resource the animal is defending. The type of territory or home range most common in small mammals and insect-eating birds is relatively large and includes the food supply, courtship, and nesting areas. Its size is an approximate function of the animal's weight and metabolic rate (McNab 1963). For mammals, $Area = 6.76\ W^{0.63}$, where *Area* equals the expected home range or territory in acres, and *W* equals the body weight in kilograms. Thus, a 20-gram mouse should have an area of 0.57 acres; this figure is not far off the mark. The productivity of the habitat is important, and white-footed mice range farther in less productive habitats. If the area is defended against conspecifics, we would expect a smaller value. This relationship between home range or territory size and body weight suggests that ultimately what most vertebrates defend is the food source.

Territory

The observation that male birds defend space dates back to classical Greek civilization, but the first scientific study of the phenomenon is usually attributed to Bernard Altum, in a book written in 1868 (Wilson 1975). Altum noted that birds tended to adjust the size of their territories to meet their ecological requirements, such that territories were smaller in areas of good habitat. Howard (1920) presented the first

FIGURE 16.3 **Home ranges and core areas of groups of baboons in Nairobi Park, Kenya.**
Although home ranges overlap extensively among groups of baboons (*Papio anubis*), core areas overlap little. Shading indicates core areas.

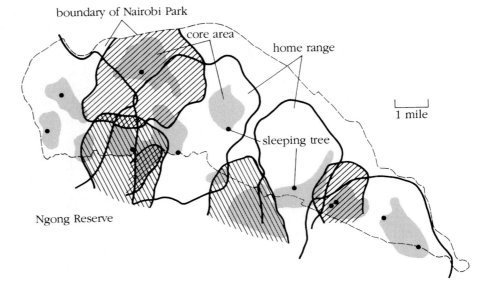

FIGURE 16.4 **Individual distance in young green herons.**
Each bird maintains a uniform distance from its nearest neighbor.
Source: Photo by Roy Lowe/U.S. Fish and Wildlife Service.

modern study of territoriality. Working mainly with aquatic birds, he described the defense of an area by a mated pair, pointed out the birds' need to defend a resource in order to breed, and suggested a possible role of territory in population regulation.

To demonstrate territoriality, we must show that an individual, mated pair, or group has exclusive use of some space, and we must also document their defense of that area. Defense is easily observed in birds such as great tits (*Parus major*), which sing from perches at the borders of their territories. If territorial holders are removed, new birds move in or neighbors expand their territories, often within hours (Krebs 1971) (figure 16.5). Replacement birds typically come from less suitable habitats in the vicinity, suggesting competition for sites.

Although territorial behavior is easy to document in conspicuous, diurnal species such as sea gulls (figure 16.6)

or lizards, nocturnal, cryptic species are another story. For example, salamanders typically live under rocks and logs and in burrows in the forest. Early anecdotal accounts of their behavior suggested that agonistic behavior was rare or nonexistent. Careful observations and lab experiments have shown that many species are territorial, marking the boundaries of their space with scents, and actively repelling intruders when necessary (Jaeger 1986).

One of the best-studied salamander species is the redbacked salamander, *Plethodon cinereus,* which inhabits the forest floor of eastern North America. Both males and females tend to be found singly under rocks and logs, especially during dry spells (Jaeger 1984). Experiments in the field demonstrate that when residents are removed from these cover objects, they are quickly replaced by smaller individuals, consistent with the notion that salamanders compete for and defend such sites (Mathis 1990).

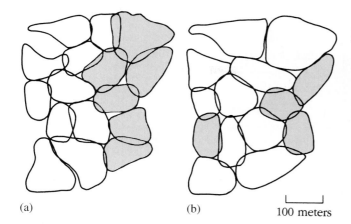

FIGURE 16.5   Territories in breeding birds.
Six pairs of great tits, which were defending territories as indicated by shaded areas on map (a), were removed. After three days, four new pairs had moved in from less suitable habitats, as indicated by shaded areas on map (b). Some of the remaining residents also expanded their territories.
Source: Krebs, J. R. 1971. Territory and breeding density in the great tit, *Parus major* L. *Ecology* 52:2–22.

FIGURE 16.6   California gull defending territory.
The bird in the left foreground attacks an intruder. Both parents defend the nest site in this monogamous species.
Source: Courtesy Bruce Pugesek.

Individuals mark the substrate with pheromones that are recognized individually (Jaeger 1986). When they encounter scent marks of intruders, they "overmark" with their own scent. If scent marking fails to repel intruders, agonistic displays follow. In the laboratory, territorial residents employed a series of threat postures from lower to higher intensity: most intense was a form of "all trunk raised" with both body and tail arched. Intruders were more likely to show submissive postures, called "move away" and "flat," in response to the more intense threats (Jaeger and Schwarz 1991).

If displays fail to repel intruders, physical combat, with biting, may follow. Some salamander species, such as *Aneides flavipunctatus,* found in the western United States, engage in vigorous fights in which the aggressor bites the other and pins it to the substrate with its trunk and tail (Staub 1993) (figure 16.7). Although fighting rarely has been observed in the field, scars attributable to bite wounds frequently are found on animals collected in the wild (Staub 1993).

In salamanders, as with most other territorial species, the "home field" advantage is powerful, and residents usually win regardless of their fighting ability (Mathis et al. 2000). We will consider the effects of residency on the outcome of agonistic encounters again, at the end of this chapter.

Territorial animals spend much time patrolling the boundaries of their space, singing, visiting scent posts, and making other displays. Such behavior takes more time and energy than does simple exploitative competition. However, these displays often have evolved so as to require relatively little energy, and once the territory has been established, the neighbors have been conditioned and need only occasional reminders to keep out. The tendency to respond less aggressively to neighbors than to strangers has been called the "dear enemy" phenomenon (Wilson 1975). It has been

FIGURE 16.7   Fighting in salamanders.
Bite-hold behavior in *Aneides flavipunctatus.* In this aggressive species about half the males found in the wild had scars from fighting.
Source: Staub, N. L., 1993. Intraspecific agonistic behavior of the salamander *Aneides flavipunctatus* (Amphibia: Plethodontidae) with comparisons to other plethodontid species. *Herpetologica* 49:271–82.

reported widely in birds, such as white-throated sparrows (*Zonotrichia albicollis*) (Falls and Brooks 1975). Returning to our example of red-backed salamanders, in laboratory tests, the resident performed more intense agonistic displays toward and more often bit strangers than familiar individuals (Jaeger 1981).

Possible fitness costs of territoriality include decreased survivorship associated with lower feeding rates, increased chances of injury from fighting, and increased predation due to conspicuousness during displays. Reproductive output could also be reduced because of less time and energy available for parental care. In laboratory tests with red-backed

FIGURE 16.8 **Costs of territoriality in lizards.**
Frequencies of various behavior patterns in testosterone-implanted male lizards (*Sceloporus jarrovi*).

Source: Marler, C. A., and M. C. Moore. 1989. Time and energy costs of aggression in testosterone-implanted free-living male mountain spiny lizards (*Sceloporus jarrovi*). *Physiol. Zool.* 62:1334–50.

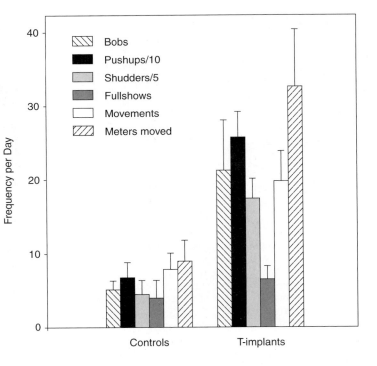

salamanders, the more intense the threat, the more time the resident spent in territorial defense and the less time spent it foraging (Jaeger et al. 1983).

One way to identify these costs in the field is to manipulate the behavior of individuals by means of hormone injections. When male mountain spiny lizards (*Sceloporous jarrovi*) were given testosterone implants in the non-breeding season, they became territorial, exhibiting head bobs, push-ups, and other social displays normally seen only during the breeding season (Marler and Moore 1989) (figure 16.8). They also foraged less than control males and had lower survival rates than control males. Territorial males increased their expenditure of energy by 31 percent over controls, perhaps explaining why their survival rates were lower (Marler et al. 1995).

When should an animal establish a territory? The key seems to be **economic defendability** (Brown 1964), such that the costs (energy expenditure, risk of injury, etc.) are outweighed by the benefits (access to the resource). Important are such things as the distribution of the limited resource in space and whether or not the availability of the resource fluctuates seasonally. A limited resource (food) that is uniformly distributed in time and space is most efficiently used if members of the population spread themselves out through the habitat, possibly defending areas. A resource that is spaced in clumps and that is unpredictable might favor colonial living or possibly nomadism.

Territorial Model

It's easy enough to say that we should compare the costs and benefits of holding a territory, but where should we begin?

One way is to construct a mathematical model that takes into account all the variables we think are important.

Carpenter and MacMillen (1976) studied a nectar-feeding bird, the Hawaiian honeycreeper *Vertiaria coccinea,* in the non-breeding season and constructed a model to predict when territorial behavior should and should not occur. Let E be the energy an animal needs to sustain itself, that is, the basic cost of living. Let P be the productivity of the environment and a be the fraction of P that a nonterritorial animal can obtain. Thus,

$$E \leq aP \qquad \text{(eq. 10.1)}$$

or the animal will starve. If the animal is to reproduce, however, then the inequality must be strict, that is, $E < aP$ or the animal will have no extra energy for production of progeny.

Next, let T be the energetic cost of defending a territory. Territoriality is expected to occur only if the benefit, in units of energy, exceeds the cost T of being territorial. So let b be the additional fraction of P that can be obtained by defending a territory. For it to be worthwhile for an animal to defend a territory,

$$T < bP \qquad \text{(eq. 10.2)}$$

The law of inequalities allows us to combine inequality (10.2) with the strict version of inequality (10.1) to give:

$$E + T < aP + bP$$

Because $aP + bP = P(a + b)$,

$$E + T < P(a + b) \qquad \text{(eq.10.3)}$$

the model presented by Carpenter and MacMillan (1976). The premise of their model is that the energetic cost of living plus the additional cost of being territorial, or $E + T$, must be less than the sum of energetic inputs in the absence of territoriality and the energy gained by being territorial, or $P(a + b)$, if territoriality is to occur.

Further development of the model leads to the prediction that at very high levels of food productivity, the birds can get enough food without excluding other birds, so territoriality disappears. Below a certain level of food availability, territorial behavior also should disappear since the resource is no longer worth defending. The birds then should switch to other foods or leave the area. Carpenter and MacMillen were able to estimate the parameters needed to test their model, and the honeycreepers behaved as predicted.

One way to examine the strength of a model is to test it under a variety of different circumstances. Carpenter and MacMillan tested their model during the non-breeding season. How well do the predictions of this model hold up during the breeding season and in other systems? The relationship between nectar availability and territorial aggressiveness was explored in males of several species of honeyeaters (*Phylidonyris* spp.) on the east coast of Australia. In general, aggressiveness toward conspecifics declined as nectar availability increased (Armstrong 1991), as would be predicted by the Carpenter and MacMillan model. However, when Armstrong increased food by adding feeding stations in neutral sites, the birds continued to defend their territories aggressively, suggesting that other factors were responsible for the changes in aggressiveness. Similarly, male blue-throated hummingbirds (*Lampornis clemenciae*), found in mountain canyons of Mexico, seldom chase other species of hummingbirds from their territory when food is abundant, but continue to chase members of their own species, suggesting that territorial behavior has functions other than protecting food resources (Powers and McKee 1994). Thus, the Carpenter and MacMillan model works only under some conditions.

As we have just seen, food is most often assumed to be the resource that is defended, but other resources may be critical, and thus be the focus of competition (Maher and Lott 2000). For example, male bullfrogs (*Rana catesbeiana*) defend the sites that are most suitable for development of the larvae (Howard 1978) (figure 16.9). Juvenile lizards defend space that provides a refuge from predators (Stamps 1983). Colonial species may exhibit some territoriality; for example, the tricolor blackbird (*Agelaius tricolor*) defends only a small area around the nest (Orians 1961). A mating system involving a peculiar type of territory is the **lek** (see also chapter 18). In this case, the only resource that the organism defends is the space where mating takes place. Feeding and nesting occur away from the site. Lek, or arena, systems are characterized by promiscuous, communal mating; the males are likely to have evolved elaborate ornamental plumages, as exemplified by the sage grouse (*Centrocerus urophasianus*) studied by Wiley (1973) and Gibson and Bradbury (1985).

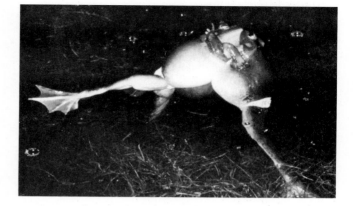

FIGURE 16.9   Male bullfrogs fighting.
Males establish territories through contests, with larger males usually winning and controlling the better sites. Females preferentially mate with these males and lay eggs in their territories. Thus the outcomes of male aggression are access to more females and higher reproductive success.
Source: Courtesy Richard D. Howard.

DOMINANCE

When two adult male laboratory mice of an aggressive inbred strain are socially isolated for a few weeks and then put together, they begin to fight. One or both may assume a hunched posture and advance with mincing steps; they may also vibrate their tails to produce a rattling sound, and one may roughly groom the other on the lower back. They may then start directing bites at each other's face, shoulders, and lower back, wrestling vigorously. Before long, one begins to get the best of things, and the other may try to escape. Once cornered, the newly submissive mouse will rear up, squeak, and box with its forepaws. If the two are separated for a day and then reintroduced, the previous day's victor will establish his superiority more quickly, and the submissive mouse will retire sooner. By including other mice, we can demonstrate that mice individually recognize each other; they remember which mice beat them up previously and which ones they were able to dominate. Among vertebrates, the establishment of dominance hierarchies typically involves both individual recognition and learning and does not necessarily involve defense of space.

Although the displays used differ, one can observe a similar series of events when crayfish are introduced into an observation tank. Fighting is most intense initially, with threat postures, grabbing the opponent with the large claws, and attempting to tear off the appendages of the opponent. The intensity of encounters diminishes as a dominance hierarchy develops (Goessmann et al., in press).

Dominance Hierarchies

We can say that one animal is dominant to another if it controls the behavior of that animal (Scott 1966). In another sense, a prediction is being made about the outcome of future competitive interactions (Rowell 1974). If four or five mice that are strangers to each other are put together, several outcomes are possible. Most likely a despot (A) will take over, and all the subordinates (B, C, and D) will be more or less equal:

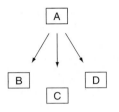

Another possibility, common in crayfish, chickens, and monkeys, is a linear hierarchy, where A dominates B, B dominates C, and so on. Such relationships are called transitive because, for instance, if A dominates B and B dominates C, then A dominates C (Chase 1982).

Sometimes triangular, intransitive relationships form, where:

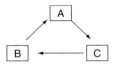

Such nonlinear arrangements tend to be temporary, and are relatively uncommon (Chase 1982).

Among nonhuman primates third parties may intervene, sometimes supporting the aggressor and sometimes the recipient of the aggression. The term **alliance** is used when two individuals repeatedly form coalitions. Alliances have been documented not only in nonhuman primates but also in bottlenose dolphins (*Tursiops truncatus*) (Connor et al. 1992):

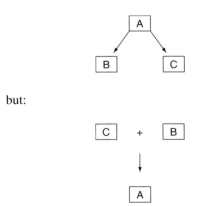

but:

The idea of dominance was first put forth by Schjelderup-Ebbe (1922), who worked out the pecking order in domestic chickens. Dominance hierarchies occur most often in arthropods and vertebrates living in permanent or semipermanent social groups, but they have been reported even in nonsocial

sea anemones (Brace et al. 1979). There is much variation in the intensity of dominance and in the frequency of reversals. Peck-right hierarchies are formed when *all* the aggression goes from dominant to subordinate (table 16.1). In peck-dominance hierarchies, only a *majority* of agonistic acts go from dominant to subordinate. Among closely related species, there may be a clear-cut hierarchy in one species and not in the other. For example, African green monkeys (*Chlorocebus sabaeus*) commonly kept in zoos show little or no dominance hierarchy, even when access to some highly prized food is limited; the vervet (*Chlorocebus aethiops*), a close relative, has a pronounced linear hierarchy. Species that are territorial in the wild may change over to a dominance hierarchy when crowded in captivity. Sometimes the dominance rank order is a function of the resource for which the animals are competing; for example, one individual may have first access to water, another to a favored breeding site.

Rowell (1974) and others argued that the whole concept of dominance should be reassessed because it is largely a product of stress induced by captivity. The phenomenon of social control is widespread among natural groups, however, particularly in primates. We saw in chapter 4 that differences in competitive ability provide the means by which natural selection acts. Dominance hierarchies seem to be a common result when socially living organisms engage in competition, and the concept has proved a useful one for researchers (Bernstein 1981).

TABLE 16.1 **Method for ranking individuals based on outcomes of dyadic (two-way) interactions**

In a hypothetical group of five individuals identified as *A, B, C, D, E,* (a) Matrix of wins and losses in which individuals are arbitrarily arranged alphabetically. For instance, *B* was the winner and *A* was the loser in eight contests; *A* never defeated *B*. (b) The order has been arranged to maximize the values in the upper right half of the matrix. Thus, the correct order is *C, D, B, E, A.* Some reversals did occur, as seen by numbers in the lower left half of the matrix. Thus, *E* is dominant to *A* because *E* defeated *A* four times, and *A* defeated *E* only once. This type of rearrangement works only if the hierarchy is linear.

(a)		Loser				
		A	*B*	*C*	*D*	*E*
	A		0	0	0	1
	B	8		0	1	4
Winner	*C*	7	7		6	9
	D	12	8	0		5
	E	4	0	0	1	

(b)		Loser				
		C	*D*	*B*	*E*	*A*
	C		6	7	9	7
	D	0		8	5	12
	B	0	1		4	8
Winner	*E*	0	1	0		4
	A	0	0	0	1	

Dominance Hierarchy in the Rhesus Monkey

To understand the nuances of social control, let us examine in some detail the dominance relationships in a relatively aggressive primate with a well-developed linear hierarchy. An important determinant of social interactions among rhesus monkeys is the degree of relatedness. Beginning at birth, monkeys associate more closely with maternal relatives, and have more positive interactions with them (Berman 1982). By the time infant rhesus monkeys (*Macaca mulatta*) are 3–4 months old, they have developed most of the threat and submission postures and vocalizations of the adult (figure 16.10).

(a)

(b)

FIGURE 16.10   Threat and submission in rhesus monkeys.
The open-mouthed threat (a) by an adult female rhesus monkey is directed at a monkey out of view to the right. The submissive grimace (b) by the young male on the left is in response to the alert posture of the dominant male on the right.
Source: (a) Douglas B. Meikle and (b) Stephen H. Vessey.

Although fights are not common among infants, by the end of the first year the infants in a group have established a linear pecking order (Vessey and Meikle 1984).

Dominance in Females

Position in the hierarchy is determined by the mother's rank (Marsden 1968; Sade 1967). Adult females are linearly ranked, with sons and daughters generally ranked just below their mothers. The infants get support from mothers and siblings; offspring of lower-ranking mothers soon learn to submit. Sometimes a 1-year-old infant chases a 2- or 3-year-old juvenile, but only if the latter comes from a lower-ranking family. Hierarchies among families are very stable; those observed at La Parguera rhesus colony in Puerto Rico remained stable for more than 20 years. Because female rhesus monkeys never change groups, extensive genealogies develop over the years. As the heads of genealogies die off, the highest-ranking daughters take over. At about puberty (3 years), each female becomes dominant over all her older sisters for reasons that are not clear; perhaps the greater support received from her mother gives the young female an advantage in agonistic encounters. The ultimate reason for this phenomenon might be that a female has her highest reproductive value at puberty, with all of her reproductive years ahead of her. If the mother should die, the new dominant female could be as young as 3 years old.

Dominance in these families manifests itself subtly but constantly at a food source, water source, or resting spot, with dominants supplanting subordinates. Fights break out sporadically in the group for no apparent reason, and family members are quick to support each other. In some groups, the highest-ranking female and some of her daughters collaborate with high-ranking males to break up fights among other families and to oppose threats from outside the group.

Dominance in Males

Dominance among males is more complex because males leave the natal group shortly after puberty (3½ years). While in the natal group, they rank below their mothers. In a new group, of course, family membership is meaningless, and the males must form new alliances. As males move into new groups, dominance relationships are quickly formed. Surprisingly, size and previous fighting experiences do not seem to influence rank directly; as long as they stay in the new group, their rank is positively correlated with seniority: those in the group the longest rank the highest (Drickamer and Vessey 1973). Establishing ties with the central high-ranking females of the group may be of importance in this respect. Males generally move into a group that contains an older brother (half-sib), and brothers may collaborate to stay in the group and attain high rank, potentially increasing their inclusive fitness (Meikle and Vessey 1981).

Early monkey watchers were much impressed with dominant (alpha) males and wrote about their confident walk, upright tail carriage, and sexual exploits; about how they led the group from place to place; and about how they

TABLE 16.2   Change in spatial relations and agonistic behavior of group's second-ranking male after removal of alpha male

Days After Removal of Alpha Male	Hours Observed	Percentage of Time Seen at			Fights Broken Up/Hr	Attacks on Females/Hr
		Edge	Inside	Midst		
−30 to 0	4.2	90	10	0	0.5	0.2
0 to 8	2.2	12	38	50	0.3	2.7
8 to 34	6.0	13	9	78	1.3	1.0
34 to 90	5.8	5	10	85	0.9	0.3

Source: Stephen H. Vessey, "Free-Ranging Rhesus Monkeys: Behavioural Effects of Removal, Separation, and Reintroduction of Group Members," in *Behaviour* 40 (1971) pp. 216–27, E.J. Brill, Leiden.

determined the rank of their group in the intergroup dominance hierarchy. The usual method of collecting field data was ad libitum note-taking: the observer simply wrote down behaviors that caught his or her attention. This technique led to biased conclusions because the most conspicuous individuals were the large high-ranking males, and the most conspicuous behavior was aggression. More recent field studies use bias-free techniques such as focal animal observations, in which the observer singles out and observes certain individuals exclusively for a fixed time interval (Altmann 1974).

Much of the mystique of the dominant male evaporated when observers noted that these males are usually the last to get involved in a skirmish, situate themselves safely in the center of the group, and let the lower-ranking peripheral males do most of the fighting (Vessey 1968). The alpha male has little to do with the group's rank, he does not actually lead the group, and he may not be the most sexually active. The ascription to alpha males of the control role—breaking up fights by charging one or both of the combatants, and protecting the group against serious extragroup threats—does seem to have substance (Bernstein 1966) (table 16.2).

Costs and Benefits of Dominance

Studies of group-living birds and mammals indicate that the dominant animals are well fed and healthy. Subordinates may be malnourished or diseased and thus suffer higher mortality. Christian and Davis (1964) reviewed data for mammals showing that low-ranking individuals (frequent losers in fights) have higher levels of adrenal cortical hormones than do dominants. These hormones elevate blood sugar and prepare the animal for "fight or flight." The cost is a reduction in antigen-antibody and inflammatory responses—the body's defense mechanisms—and a reduction in levels of reproductive hormones.

There should be some selective advantage to high rank. Furthermore, in times of food shortage when access to food is restricted, high-ranking individuals have a distinct advantage. All group members, even the lowest ranking, probably benefit from a dominance hierarchy because there is less energy spent in scrambling for resources. Also, as we shall see later, the advantages of group living, even as a subordinate, may more than compensate for the stress endured.

Baboons (*Papio anubis*) in the Masai Mara National Reserve in Kenya have a social organization similar to that

of rhesus monkeys. By briefly anesthetizing males with a syringe fired from a blowgun and obtaining blood samples, Sapolsky (1990, 1991) studied relationships among dominance, adrenal cortical hormones, and sex hormones. When dominance hierarchies were stable, with few rank reversals, dominant individuals had lower concentrations of cortisol than did subordinates, as one would expect (Sapolsky 1990). However, when ranks were unstable, absolute rank was not a good predictor of stress levels; males being challenged from below in the hierarchy showed higher cortisol levels than did those that were challenging males above them (Sapolsky 1992) (figure 16.11). This result makes sense in that individuals rising in rank would be under less stress than those falling in rank. Even when ranks are stable, not all subordinate males had high levels of cortisol. Some subordinants were active sexually and showed patterns similar to dominants (Virgin and Sapolsky 1997).

Testosterone levels also respond to stressors. Males living in groups with unstable hierarchies or under drought conditions had lower testosterone titers than did those males living in groups under nonstressful conditions. Within socially stable groups high- and low-ranking males had similar basal concentrations of circulating testosterone. However, under the stress of being darted and held in captivity for several hours, low-ranking males showed a rapid decline in testosterone, whereas high-ranking males actually showed an increase during the first hour (Sapolsky 1991) (figure 16.12). Thus, subordinate baboons likely suffer a greater impact on their reproductive function when exposed to stressful situations than do dominant individuals.

The relationship between dominance and social stress is further complicated by studies in other species showing that subordinates may have lower, not higher, levels of adrenal cortical hormones. In both African wild dogs (*Lycaon pictus*) and dwarf mongooses (*Helogale parvula*), the dominant, highly aggressive individuals seem to be under more stress (Creel et al. 1996). Among birds, nonbreeding male scrub jays (*Aphelocoma coerulescens*) had similar levels of adrenal cortical hormones as did breeders, while in females the nonbreeders actually had lower levels (Schoech et al. 1991).

Dominance is often directly correlated with reproductive success, as in northern elephant seals, in whom the highest ranking males do virtually all of the mating (Haley et al. 1994). But in some systems the highest-ranking male may be so busy displaying that lower-ranking males copulate with

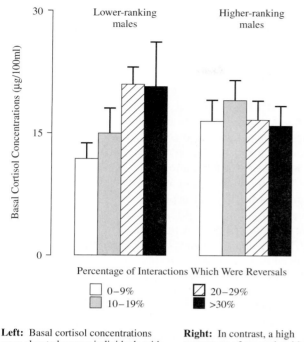

Left: Basal cortisol concentrations were elevated among individuals with a high percentage of interactions that were reversals of the direction of dominance with the three males immediately lower ranking in the hierarchy.

Right: In contrast, a high percentage of reversals with the three males immediately higher ranking in the hierarchy was not associated with elevated cortisol concentrations.

FIGURE 16.11 Stress hormones (cortisol) in response to dominance-rank challenges by lower- versus higher-ranking male baboons.

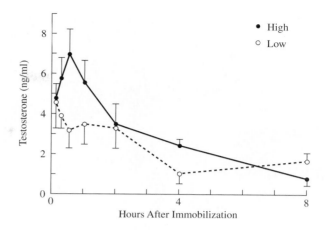

FIGURE 16.12 Differences in testosterone levels between high- and low-ranking male baboons in response to the stress of immobilization.

Low-ranking males (open circles) showed continuous declines in testosterone concentrations throughout the period. High-ranking males (closed circles) showed a transient elevation of testosterone concentrations during the first post-stress hour.

the females. In many species the male dominance hierarchy is age-graded, with younger, lower-ranking males working their way up the hierarchy as older males die off or leave the group. Thus, low rank does not necessarily mean low lifetime reproductive success. Much-needed studies on lifetime reproductive success began to be published only in the last two decades (Clutton-Brock 1988).

Among primates, dominant males usually have priority of access to estrus females, but such access does not necessarily translate into paternity. Surprisingly, in captive groups the use of molecular techniques to actually assign parentage reveals no consistent relationship between male dominance and reproductive success, with some studies showing a positive correlation and others no correlation (de Ruiter et al. 1992; Smith 1981). Among wild groups, however, the expected pattern emerges. Long-tailed macaques (*Macaca fascicularis*) live in mixed sex social groups in the rain forests of Indonesia. Using molecular techniques to analyze DNA taken from blood samples, alpha males sired between 50 and 83 percent of the offspring in each of three groups (de Ruiter et al. 1992). Beta males accounted for about half of the remaining offspring, with lower-ranking males siring the rest (de Ruiter et al. 1992).

Dominance does not necessarily translate into higher reproductive success for female primates either. For exam-

ple, female olive baboons (*Papio cynocephalus anubis*) show a linear dominance hierarchy, and dominant females have priority of access to scarce resources. Although high-ranking females have shorter inter-birth intervals and higher infant survival rates than low-ranking females, they also suffer more miscarriages and long-term infertility (Packer et al. 1995). These results suggest that qualities essential to achieving high rank may also carry reproductive costs.

Many popular treatments have emphasized the importance of dominance and aggression in "keeping the species fit," since the survivors of fierce battles are the most vigorous and therefore "improve" the species by passing on those traits. This interpretation is no longer accepted, however; the simplest and most likely explanation of aggression and dominance is that individuals benefit. As with other traits, there should be an optimum level of aggression that depends on the individual's particular social and physical environment. Individuals that are too aggressive would be selected against, as would those that are too passive. Costs of aggression include the risk of injury and the risk of predation. But there are also the energetic costs associated with fighting. Thus shore crabs (*Carcinus maenas*) in staged agonistic encounters in the lab accumulated metabolites such as L-lactate in their hemolymph and tissues, and their glycogen stores were depleted (Sneddon et al. 1999). These costs were especially apparent when crabs fought in water with low oxygen levels, as occurs naturally in tide pools where these animals live.

Reconciliation

When two chimpanzees have a fight, often one will approach the other with outstretched arm and open palm, followed by mouth-to-mouth kissing (figure 16.13). Such behavior is

FIGURE 16.13   Reconciliation in chimpanzees.
An adult male (left) and female chimpanzee engage in a mouth-to-mouth kiss after an aggressive incident between them.
Source: Photograph by Frans de Waal.

termed **reconciliation** by de Waal (1993). Agonistic interactions between members of territorial species tend to space individuals out, but among those species that live in permanent social groups, the participants are bound together by strong bonds. In such groups the costs of aggression could persist. In primates, especially great apes, there is an increased probability of contact between individuals following agonistic interactions, and special reassuring and appeasing behavior patterns are observed. The presumed function of reconciliation is to repair disturbed social relationships (de Waal 1993).

INFANTICIDE AND SIBLICIDE

The Indian langur monkey (*Semnopithecus entellus*) lives in a variety of habitats. In some areas, social groups of langurs contain only one adult male, five to ten females, and their young. Several observers have reported instances in which a new male came into the group, chased out the old male, and killed some or all of the infants (figure 16.14). One interpretation of infanticidal behavior is that the group is socially disorganized by the change in males, and the usual restraints on overt aggression are absent. Once the new male has established himself, he ceases his attack on the young. From the standpoint of the species or group, such behavior is maladaptive and should be rare, which in fact it is. A second interpretation of these events is that infanticide is adaptive for the new male, since he removes the offspring of the presumably unrelated male and causes the females to come into estrus sooner to bear his own offspring (Hrdy 1977a,b). Hrdy's view is consistent with the notion that social behavior results from the action of natural selection on the individual.

When new male lions first enter a pride in an attempted takeover, they kill the smaller cubs and evict the older cubs and subadults (Pusey and Packer 1987). Infanticide by incoming males accounts for about one-fourth of cub mortality (Packer et al. 1990); females therefore group their young into crèches to protect them from nomadic males. Males benefit by killing cubs because they eliminate the offspring of competing males and bring the females into estrus sooner. Infanticide is also performed by a variety of rodent species. Under crowded conditions rodents often eat their young, sometimes those already dead.

The killing and *eating* of conspecifics is referred to as **cannibalism.** Cannibalism occurs in a wide variety of organisms, from protozoa to primates, and its costs and benefits have been the subject of much interest (Elgar and Crespi 1992). As with infanticide, we could interpret such behavior as social pathology, which results from crowded conditions and leads to maladaptive behavior. We should note, however, that under crowded conditions or when food is scarce, the young would probably not survive anyway, so the parents save their reproductive investment for more propitious times by consuming their young. Under adverse conditions rodents and rabbits resorb developing embryos in the uterus, with the same effect as cannibalism. In kangaroos most of the development of the young takes place in the pouch, and females can terminate "pregnancy" by simply throwing the young out of the pouch. Cannibalism is widespread among amphibians, especially between larval forms. Some species of salamanders and toads have enlarged mouths and strong jaws adapted for a carnivorous diet that includes members of their own species (Crump 1992). Larvae of some species of amphibians, such as spadefoot toads (*Scaphiopus bombifrons*), are polymorphic: one morph is herbivorous and the other carnivorous. The carnivorous morph tends to nip at conspecifics and eat them if they are not relatives (Pfennig et al. 1993).

An important benefit of cannibalism seems to be nutritional, since individuals are more likely to eat conspecifics when they are hungry, and cannibals tend to grow faster than noncannibals. Therefore cannibalism perhaps should be considered a form of feeding behavior rather than agonistic behavior. Cannibalism also increases with population density in many species, however, suggesting competition as a driving force.

Older and larger nest mates sometimes kill their younger and smaller siblings, referred to as **siblicide.** (Mock et al. 1990; Mock and Parker 1997)(see chapter 17). Among spotted hyenas, the young fight vigorously while in the den, forming dominance relationships within hours. One sibling often kills the other, especially when the two are of the same sex (Frank et al. 1991). When food supplies are adequate, however, both cubs typically survive and become close allies (Smale et al. 1995). For additional papers on infanticide, see the books edited by Hausfater and Hrdy (1984), Elgar and Crespi (1992), and Parmagiano and vom Saal (1994).

FIGURE 16.14 Infanticide in langur monkeys.
The male langur monkey has stolen the infant of a third female (not visible) and is being attacked by them. Males frequently wound or kill infants sired by other males.
Source: Sarah Blaffer-Hrdy/Anthro-Photo.

INTERNAL FACTORS IN AGGRESSION

Hormones

The physiological bases for agonistic behavior are usually considered from the perspective of the nervous system (chapter 7) and the endocrine system (chapter 8). The two systems are closely linked, however, since hormones can be produced by nerve cells as well as endocrine glands. The effect of castration on sexual and aggressive behavior in males has been known since the time of Aristotle. Castration reduces aggression, and replacement with testosterone restores aggression, in a wide variety of fish, reptiles, birds, and mammals (reviewed by Huntingford and Turner 1987). Some of the hormone changes are rapid; male rhesus monkeys that have been defeated in fights show declines in circulating testosterone within hours (Rose et al. 1975).

In many species of mammals, exposure to sex steroids during critical periods in development is a primary factor in the appearance of aggressive behavior in adulthood; thus, they have organizational effects as well as activational effects (chapter 8). Female rhesus monkeys that received testosterone before birth and that were tested as juveniles threatened more and played more roughly than did untreated females (Phoenix 1974). In species of mammals where multiple embryos are present in the uterus, hormones produced by one sex may affect the other (vom Saal 1983). (See figure 8.8.) When tested as adults, males that were between males in utero were more aggressive than males that were between females. Similarly, females that were between males as embryos fought more than females that were between other females. Exposure to androgens in utero causes females to have a larger distance between the anus and genitalia (AGD), making them look more like males (see also chapter 8). Such females' tails rattled more in the presence of intruders, an aggressive response, than did females with short AGDs (Palanza et al. 1995).

The notion that testosterone is *the* universal substance causing aggression in all vertebrates is much too simplistic, however. There are some species in which castration and replacement with testosterone has little effect on agonistic behavior. In Siamese fighting fish (*Betta splendens*), gonadectomy had no effect on levels of aggression (Weis and Coughlin 1979). Nor does castration reduce aggression in adult male prairie voles (*Microtus ochrogaster*) when paired with intact intruders (Demas et al. 1999).

Luteinizing hormone (LH), from the anterior pituitary, increases aggression in some species of birds (Davis 1963). Castrated zebra finches (*Poephilia guttata*) receiving estrogens become just as aggressive as those receiving testosterone. Once injected, hormones undergo chemical changes as they are broken down by the body; it appears that the metabolites of these hormones are the substances actually causing the increase in aggression (Harding 1983). In fishes, there is evidence that thyroid hormone is involved in control of aggression (Villars 1983), and the adrenal glands have been implicated in dominance-related color changes in reptiles (Greenberg and Crews 1983). In humans, Reinisch (1981) showed that pregnant women who were treated with synthetic progesterone produced offspring who were more aggressive when tested as adolescents than similar adolescents not treated prenatally.

Invertebrate hormones are much different from those of vertebrates, and much less is known about how they work (Breed and Bell 1983). Most of our information comes from crustaceans and insects. Males of many species of crustaceans have enlarged claws used for display and fighting. Development of these claws seems to be under the influence of a hormone produced by andronergic glands near the testes. Larger glands produce more hormones, which leads to larger claws and greater fighting efficiency (Nagamine and Knight 1980). In juvenile lobsters (*Homaraus americanus*), levels of aggression are related to the molt cycle. When levels of the molting hormone ecdysterone are high, aggression is low. This makes sense because after a molt the exoskeleton is soft and the animals are vulnerable. Once the shell hardens, and the animal competes for a shelter for the next molt, aggression is high (Tamm and Cobb 1978).

Among insects, the paired corpora allata, neurosecretory glands in the head, produce substances called juvenile hormones (see figure 8.4). In grasshoppers, females shift from responding aggressively to males to becoming sexually receptive under the influence of juvenile hormone. If the corpora allata are removed, females continue to reject males. In male field crickets (*Gryllus* spp.), however, removal of the corpora allata had no affect on agonistic behavior (Adamo et al. 1994). Roseler et al. (1986) implicated both juvenile hormone and ecdysteroids in the establishment of dominance hierarchies of female paper wasps (*Polistes gallicus*), as they initiate nests in the spring. Subordinate females have lower levels of juvenile hormone and reduced gonad development.

Neural Mechanisms

The vertebrate brain structures most involved in aggression are part of the limbic system, which includes structures such as the amygdala and the hypothalamus. The hypothalamus is involved in defense and escape behavior in a wide variety of vertebrates, ranging from fish to primates (Huntingford and Turner 1987). Using lesions, electrical stimulation, and single neuron recordings from specific areas of the brain, researchers have found that different brain sites are responsible for different types of aggression. For example, in cats, electrical stimulation of the ventromedial nucleus of the hypothalamus produces growling, hissing, and attacking with claws (defensive attack) (Flynn 1967). Stimulation of the lateral hypothalamic area produces a biting attack with no defensive elements. Thus, the notion that all types of agonistic behavior involve a single neural system may be an oversimplification.

Areas of the brain that are involved in predatory aggression are the amygdala of the forebrain and the central gray of the midbrain. These areas are connected by nerve pathways, and they interact. For example, electrical stimulation of certain areas in the thalamus causes cats to attack rats (Bandler and Flynn 1974). Using special staining techniques, axons in the thalamus were traced to areas in the central gray. Stimulation of those latter sites elicited similar attacks.

The use of radio transmitters permitted electrical stimulation of specific brain areas in seminatural social groups (Delgado 1967; Herndon et al. 1979). Monkeys with electrodes implanted in certain parts of the thalamus, hypothalamus, or central gray, become aggressive when the electrodes are activated (figure 16.15). In some cases, when stimulation is applied to the hypothalamus, lower-ranking monkeys become dominant as a result (Robinson et al. 1969).

FIGURE 16.15   Radio-stimulated attack in rhesus monkey.
Social aggression in rhesus monkeys can be elicited by telestimulation of hypothalamic structures.
Source: Courtesy Yerkes Regional Primate Research Center.

Immunological techniques can be used to trace the paths of nerve axons to different brain areas. For instance, in one part of the amygdala of the rat brain, called the medial amygdalar nucleus, nerves from the dorsal region seem to be associated with circuits controlling reproductive behavior, while those in the ventral region are associated with agonistic behavior (Canteras et al. 1995).

Further evidence that mating and agonistic behavior are controlled in part by separate areas of the brain comes from studies of expression of Fos protein in Syrian hamster brains (Kollack-Walker and Newman 1995). Adult males were allowed to interact either with a sexually receptive female or an intruder male. They were then sacrificed and their brains sectioned and stained immunologically for Fos protein. In this way, active sites in the brain could be seen microscopically. The results show that while some areas of the limbic system were activated by both mating and agonistic behavior, others were selectively activated.

Chemical messengers, or neurotransmitters, are involved in the transmission of nerve impulses across synapses (see figure 7.4) or across nerve-muscle junctions (see figure 7.5). These can produce excitatory or inhibitory effects, depending on the nature of the postsynaptic receptors. Best known perhaps is acetylcholine, which is involved in the neuromuscular junction and in the autonomic nervous system. Norepinephrine (noradrenaline) is widespread in the pons and medulla, with fibers projecting anteriorly to the mid- and forebrain. Dopamine and serotonin are found in the midbrain, with fibers projecting anteriorly to the hypothalamus, amygdala, and striatum. These and others, such as substance P (Shaikh and Siegel 1997), have been specifically implicated in the control of aggression.

Considerable evidence links serotonin to aggression: a rise in brain serotonin levels accompanies lowered aggression in a variety of vertebrates, from rainbow trout (*Oncorhynchus mykiss*) (Winberg and Lepage 1998) to lab rats (Olivier et al. 1995), while administration of drugs that

FIGURE 16.16   **Experimental setup for delivering substances into freely moving crayfish.**
(a) A cannula connected to a syringe pump is implanted in the pericardial sinus of the crayfish on the right.

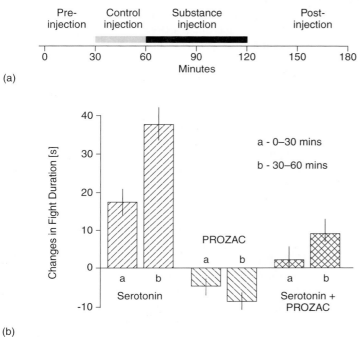

Pre-injection   Control injection   Substance injection   Post-injection

0   30   60   90   120   150   180
Minutes

(a)

a - 0–30 mins

b - 30–60 mins

Changes in Fight Duration [s]

PROZAC

a   b   Serotonin

a   b

a   b   Serotonin + PROZAC

(b)

Changes in fight duration with serotonin and Prozac treatments.
(b) Each pair of bars represents the mean fight duration during the first and second 30 minutes of treatment with the agent.

Source: Huber, R., K. Smith, A. Delago, K. Isaksson, and E. A. Kravitz. 1997. Serotonin and aggressive motivation in crustaceans: Altering the decision to retreat. *Proc. Natl. Acad. Sci.* 94:5939–42.

lower brain serotonin is accompanied by increased aggression. Among invertebrates such as crustaceans, however, the effect is the opposite: serotonin typically increases aggression (Huber et al. 1997). When subordinate crayfish (*Astacus astacus*) were given injections of serotonin into their hemolymph, they escalated contests, and fights were more intense and lasted longer than prior to administration (Huber and Delago 1998). When given Prozac, a selective serotonin reuptake inhibitor, along with the serotonin injections, there was no increase in fighting (figure 16.16).

Returning to rodents, when two strange adult male hamsters are placed together in a neutral area, after a brief period of fighting a dominant/subordinate relationship emerges. During subsequent encounters, the animals will flank mark (see figure 8.9) rather than fight, with the dominant animal doing most of the marking (Johnston 1975). Although the production of pheromones by the flank glands depends on androgens, marking behavior is also under the influence of neurochemicals such as arginine vasopressin (AVP) (Ferris and Delville 1994). Flank marking can be "turned on" by microinjections of AVP into the anterior hypothalamus and

"turned off" by microinjections of AVP-receptor antagonist (a substance that blocks the uptake of AVP) in the same area. Similarly, overt aggression such as biting is turned on and off by these same substances.

Although microinjection of AVP into the ventrolateral hypothalamus increases aggression in male hamsters, the response is androgen dependent and thus does not occur in castrated animals. Other neurotransmitters, especially serotonin, interact with AVP in mediating aggression. When males were given a serotonin-reuptake inhibitor (similar to Prozac), which increases levels of serotonin in the brain, the AVP-induced aggression was blocked (Delville et al. 1996).

Genetics

The synthesis of nerve structures, neurosecretions, and other compounds is under genetic control, and agonistic behavior is shaped by natural selection, as is any other behavior. Agonistic behavior has long been known to have a heritable basis (Maxson 1981): artificial selection can lead to significant changes in levels of aggression within just a few generations.

For example, Ebert and Hyde (1976) tested wild female house mice for aggressiveness, and by selecting high- and low-scoring mice, they produced two lines: one with highly aggressive females and one with passive females. The unselected control lines were, as expected, intermediate. Domestic laboratory strains of mice differ widely in aggressiveness (Southwick and Clark 1968), as do dog breeds (Scott and Fuller 1965). Siamese fighting fish (*Betta splendens*), fighting cocks, and even crickets have been artificially selected over the years for performance in contests with large sums of money riding on the outcome.

More recent work has attempted to identify specific genes and evaluate their role in behavior (Nelson and Young 1998). One technique is targeted disruption of single genes, referred to as knockout mutants. (see chapter 5). By comparing the behavior of wild-type animals with knockouts, the function of the gene can be assessed. Several of these mutations produce effects on aggression in lab stains of mice. For example, knocking out a gene for serotonin receptors leads to a marked increase in aggression (Saudou et al. 1994), as does one that encodes neuronal nitric oxide synthase (Nelson et al. 1995). On the other hand, mice with targeted disruption of the gene for estrogen receptors are less aggressive than the wild-type controls (Nelson and Young 1998).

Genes controlling agonistic behavior may be localized on certain chromosomes. Offensive behavior in male mice, which involves bite-and-kick attacks on the flanks and rump of the opponent, is influenced by a region of the Y chromosome. Genes in this area are hypothesized to affect synthesis of testosterone-dependent pheromones, as well as the perception of olfactory stimuli triggering attacks (Maxson 1996; Monahan and Maxson 1998)

A genetic link to aggression in humans was postulated to explain the fact that males with two Y chromosomes (XYY males) were found more frequently than expected in prison populations, given their frequency in the population at large. However, it turned out that XYY males also tended to be of below-average intelligence, and thus may have been more likely to get caught after committing a crime. Also, there was no indication that they were more aggressive than were normal males (Mazur 1983), although they did have slightly elevated levels of testosterone (Schiavi et al. 1984). In sum, effects of an extra Y chromosome on aggression in human males are slight or nonexistent.

EXTERNAL FACTORS IN AGGRESSION

Learning and Experience

Although genes control production of the hormones, neurosecretions, and neurons that produce aggressive behavior, researchers generally believe that some factor outside the animal triggers the response. Previous experience can produce semipermanent changes in the expression of agonistic behavior, and animals can be conditioned to win or lose. If we give a naive male laboratory mouse that has been socially isolated for a few weeks a few easy victories over less aggressive mice, he will turn into a fighter mouse and will quickly attack any other male in sight. Likewise, we can "create" a loser by repeatedly exposing a naive male laboratory mouse to highly aggressive mice (Scott and Fredericson 1951).

Nor is this phenomenon limited to rodents. The intertidal owl limpet *Lottia gigantea* shows behavior ranging from forceful aggression to evasion, in response to intraspecific contact in the field. Laboratory experiments demonstrated that limpets given an experience of consecutive agonistic victories responded aggressively to subsequent contact. Limpets that were given an experience of consecutive agonistic defeats responded evasively to subsequent contact (Wright and Shanks 1993).

In nonhuman primates interest has centered on the importance of early experience on later levels of aggression. For instance, we know that monkey societies differ widely in the frequency and intensity of aggression. Bonnet macaques (*Macaca radiata*) have a rather loose dominance hierarchy, with little fighting and much friendly interaction. Pigtail macaques (*M. nemestrina*) have a more rigid hierarchy, with more fighting and a low tolerance of strangers. Treatment of infants also differs between the two species. Pigtail mothers are very restrictive and do not let others handle or carry the infant as much as bonnet mothers do. The latter tolerate "aunting" by young females. Bonnet infants are cared for by others if the mother is removed; in contrast, pigtail infants become very depressed and huddle in the corner (Rosenblum and Kauffman 1968). Rhesus macaques are similar to pigtail macaques in terms of having a rigid dominance hierarchy and higher levels of aggression. In a comparison with bonnet mothers, rhesus mothers were more restrictive of their infants' social interactions and were more aggressive toward them (Mason et al. 1993). Whether this difference in early experience is responsible for the different social structures that characterize the adults of the two species remains to be demonstrated.

Scott (1975, 1976) emphasized the importance of early experience and learning in the development of aggression and suggested that one of the most important controls of agonistic behavior is passive inhibition. Animals form the habit of not fighting, particularly during critical periods of development. Some evidence in support of this view comes from studies of rats and mice (Scott 1966), and monkeys (Rosenblum and Kauffman 1968). Although it is clear that animals can be trained in the laboratory to be passive and nonaggressive, researchers have not clarified the role of early experience in the restraint of aggression in natural populations.

Pain and Frustration

More direct causes of aggression are pain and frustration. Researchers have shown that noxious stimuli, such as loud noises, foot shocks, tail pinches, and intense heat, cause different lab species to attack a wide variety of objects, from conspecific cage mates to tennis balls (Hutchinson 1983; Ulrich 1966; Ulrich et al. 1965). Although referred to as attack behavior, the response often seems to be defensive in nature; for instance, shocked rats assume the upright "boxing" posture that rats losing a fight normally exhibit. These defensive attacks occur in a laboratory situation in which the victim is prevented from escaping (i.e., the victim is cornered). The full range of attack behaviors seen in the wild is not elicited by painful stimuli.

Researchers have explored the frustration-aggression connection in humans. In some early experiments, researchers allowed children to view a room full of toys but denied them access to it; when they finally let them into the room, the children frequently smashed toys and fought more than did a control group that was allowed immediate access to the room (Berkowitz 1969; Johnson 1972). Frustration is now considered to be only one of several causes of aggression, however (Feshbach 1997).

Social Factors

Another strong stimulus for aggression is the presence of strangers. In most species, the introduction of a strange animal into an established social group produces the violent reaction of xenophobia, usually in members of the same sex as the stranger (Southwick et al. 1974). The amount of fighting tends to decrease with time, but acceptance of strangers is usually a slow process. A related phenomenon is isolation-induced aggression, in which socially isolated animals become hyperactive and prone to attack other animals (Scott 1966).

Crowding per se tends to increase interaction rates but does not always increase aggression. Overall group size, presence of strangers, or restricted access to resources has a much more powerful effect on aggression than does reducing the space available to an already established social unit.

Sex and Hunger

The relationship between sex hormones, particularly testosterone, and aggression is striking in seasonally breeding species, as we have already seen. As the gonads increase in size in response to environmental changes in photoperiod, rainfall, vegetation, and so forth, fighting and wounding increase as well. Most of this increase is related to competition for breeding territories, social rank, or access to females. In some monogamous species (e.g., sea gulls), fighting occurs during the early phases of pair formation before a sexual bond develops. In others, courtship and copulation seem to have aggressive components, as evidenced by the appalling racket mating cats make, which makes the listener wonder whether it is a fight to the death or courtship. Other carnivores also have violent mating behavior. When polecats mate, the male grabs the female by the scruff of the neck and drags her back to his nest. Some of this aggression at the time of mating probably stems from the fact that males tend to court females rather indiscriminately, often making advances when the females are not sexually receptive. Some mammals, mainly carnivores and lagomorphs, are induced ovula-

tors, and vigorous courtship plus copulation may be needed to trigger the release of an egg into the oviduct as copulation occurs. Although sex and aggression frequently seem linked, recall our earlier discussion indicating that the neural circuitry is distinct.

Some evidence has shown that in groups of fishes, the initiation of feeding increases intragroup agonistic behavior. Albrecht (1966) proposed that predatory (feeding) and agonistic behaviors, although functionally distinct, are motivationally linked. However, Poulsen and Chiszar (1975) found that receipt of aggression inhibited feeding in submissive bluegill sunfish (*Lepomis macrochirus*). Feeding was unaffected by levels of aggression in dominant bluegills. They concluded that feeding and aggression are independently motivated.

EVOLUTION OF RESTRAINT AND APPEASEMENT

Given the obvious benefits gained by the winners of competitive interactions, one might expect that natural selection would lead to the evolution of highly aggressive individuals with sophisticated armamentaria. However, most individuals show considerable restraint in their use of force. Why is it that animals tend to use restraint?

Displays, discussed in chapter 12, are behaviors that evolved to communicate information. Many displays communicate agonistic behavior, such as a threat or a submission. For instance, figure 12.7 shows a threatening dog and a submissive dog; the postures are in many ways opposites of one another (Darwin's theory of antithesis). Displays that communicate an animal's submissive intentions may inhibit attack by another. Lorenz (1966), Scott (1966), and others argued that displays evolved for the benefit of the species in order to reduce wasteful large-scale wounding and killing. However, this is unsatisfactory; as we have seen (chapter 4), natural selection does not act "for the good of the species," but will favor a behavior if it benefits the individual, even if the species as a whole is harmed. Tinbergen (1951), pointed out that ritualized contests (displays) would reduce the risk of injury to the aggressor, and that threats take less energy than physical combat and are of advantage to individuals. More recently, behavioral ecologists have applied game theory, a branch of applied mathematics, to better understand the evolution of fighting strategies in animals.

Game Theory Models

Maynard Smith and Price (1973) independently developed a model of how displays, rather than all-out fighting, could be selected for at the level of the individual. During agonistic encounters, the decision about whether to attack, display, or flee depends on what the other participant does. As in other models we have discussed (chapter 12), these alternative courses of action are called **strategies.** The participants' choice of strategy is assumed to be heritable. The outcome of each game is determined by a **payoff matrix,** or a predetermined set of outcomes, the costs and benefits that result from different strategies. These costs and benefits are given in terms of expected Darwinian fitness measured by reproductive success (the number of offspring produced), which is the currency of the model. An **Evolutionarily Stable Strategy (ESS)** is a strategy, which, if adopted by most members of the population, cannot be successfully invaded by any rare alternative strategy (Maynard Smith 1974).

One version is the Hawk-Dove game (Krebs and Davies 1987). There are just two strategies: Hawk, attack immediately, flee only if injured, and Dove, display to another Dove, but flee immediately without getting injured if attacked by a Hawk. If Hawk encounters Dove at a resource, Hawk always wins and Dove always loses. If the resource has value V, Hawk has no cost and obtains $+V$. Dove, which flees immediately, does not gain or lose anything. If Hawk encounters Hawk at a resource, and we assume both Hawks are identical, each has a 50 percent chance of getting the resource and a 50 percent chance of being injured. So, each Hawk gets on average $(1/2\ V - 1/2\ I)$, or $(V - I)/2$. Finally, what happens if two Doves meet? Assuming that they are identical and that both display, each has a 50 percent chance of getting the resource. Half the time, a Dove will win $V - D$, and half the time, it will lose and get $-D$. Thus, on average, each Dove will get $1/2\ V - D$. The payoff matrix for this game would be as follows:

	Against:	
Payoff to:	**Hawk**	**Dove**
Hawk	$(V - I)/2$	V
Dove	0	$V/2 - D$

If we substitute some hypothetical numbers for these variables, we can compute the ESS.

Let the resource value, or V, = 50;

Cost of injury, I, = 100;

Cost of displaying, D = 10.

Plugging these values into the matrix above, we obtain:

	Against:	
Payoff to:	**Hawk**	**Dove**
Hawk	-25	$+50$
Dove	0	$+15$

If all individuals in the population played Dove, the average payoff would be $+15$. But what if a mutant that plays Hawk enters the population? It would do well against Dove, gaining $+50$ fitness units. So Hawk mutants would

spread. Clearly, under these conditions pure Dove is not an ESS because it is invaded by another strategy. But as Hawk increases, the chances of Hawk meeting Hawk increase, as does the cost of injury, and so the payoff approaches -25, pure Hawk. At this point Dove could invade, since Dove gets 0 against Hawk, not very high, but better than -25. So pure Hawk is not an ESS either. The ESS in this case is some mixture of Hawks and Doves.

We can calculate the stable mixture as follows: Let p equal the proportion of Hawks in the population, so $1 - p$ would be the proportion of Doves. The average payoff for each strategy is the payoff for interacting with each type of opponent times the probability of meeting that type. For Hawk, this is

$$p(V - I)/2 + (1 - p)V.$$

For Dove, this is

$$p(0) + (1 - p)(V/2 - D).$$

At equilibrium the average payoff for Hawk equals that for Dove. So, setting these two equations equal to each other and solving for p, we get $p = 35/60$, or 58 percent Hawk for the ESS, and 42 percent Dove.

In general, if $V > I$, pure Hawk is the ESS. If $V < I$, a mixed ESS results. Pure Dove is never an ESS. One conclusion we may draw from this model is that, depending on the various costs and benefits, noncombative individuals that display and retreat may be following an adaptive strategy; they are not simply the losers or less fit victims of the aggressive dominants.

More realistic (and complex) models have been developed that include situations where the contestants differ in some quality. Three kinds of asymmetry have been considered: (1) cases where one individual already owns the resource or occupies the area, which leads to home-field advantage; (2) cases where contestants differ in resource holding power (RHP), as when one is larger or stronger than the other; (3) cases where the resource has different values for each of the contestants.

Maynard Smith (1976) added a conditional strategy to the Hawk-Dove model called "bourgeois" to deal with case (1). These individuals play a Hawk strategy if they are on home ground (or already own the resource in dispute) but play as a Dove if it is someone else's territory. Assuming that there are no other asymmetries, bourgeois turns out to be a pure ESS. Male speckled butterflies (*Pararge aegeria*) occupy mating territories on the forest floor; the resident male always wins when intruders enter, and the outcomes reverse if ownership changes (Davies 1978). As predicted by the model, both contestants behave like hawks if they can be tricked into ownership of the same territory.

Asymmetries in RHP and resource value have been manipulated in laboratory studies of spiders. Working with female funnel-web spiders (*Agelenopsis aperta*) (See also page 67), Riechert (1984) found that residency was usually unimportant in determining the outcome and that assessment of relative weights occurred early in the contest. Subsequent behavior depended on the results of this assessment. If a spider had a large weight advantage, it immediately escalated the fight, whereas smaller spiders tended to retreat. If they were the same size, then the resident won. Austad (1983) studied contests between male bowl and doily spiders (*Frontinella pyramitella*) over access to a female. He tested his data against a model referred to as the "war of attrition" (Maynard Smith 1974). In this game, two individuals display or fight continuously until one wins; costs are assumed to increase linearly with persistence time. Austad varied resource value by introducing an intruder male at various times while the resident male mated with the female. As more and more of her eggs were fertilized over time, her value to the resident male declined, and he was less inclined to fight. Males varied in RHP in terms of size: larger males usually won. As predicted by the model, fights were longest when males were of similar weight, and also when the female was of greatest value to the resident. For a review of game theory and animal contests, see Riechert (1998).

RELEVANCE FOR HUMANS

Whatever its evolutionary origins, the ritualization of agonistic displays into contests instead of struggles has occurred in practically all species with aggressive behavior. As we have seen, killing and wounding do occur; however, given the weapons available and the competition for resources that probably occurs at some point in the lives of all organisms, serious injury is much rarer than is potentially possible.

Control of aggression in nonhuman animals may have special relevance for us. Lorenz (1966) argued that humans are aggressive for the same reasons as are other animals. However, humans have only recently acquired the means of inflicting serious injury through the use of weapons; we have not evolved the ritualized displays typical of species with such natural weapons as canine teeth or horns. Thus, aggression in humans escalates with violent consequences.

Humans also become violent, according to Lorenz, because we have few harmless outlets for aggression—a particular problem in modern society, where so little of our time and energy is expended in subsistence activities. Lorenz, in his hydraulic model of behavior, argued that action-specific energy (in this case for aggression) builds up until it is released in some way. Although other research discredits the idea that specific "energies" are stored in the brain, the discovery that different chemicals are involved in transmitting information to specific parts of the brain could give credence to his theory. If aggression is inevitable, Lorenz argued, then we must find harmless ways to vent it. This cathartic approach suggested to him the importance of ritualized tournaments in which participants and spectators alike can work out their hostilities; for example, sports events. The majority of studies show that the opposite is true: people engaged in aggressive behavior, including contact sports, tend to become more, not less aggressive (Geen and Quantry 1977)

The idea that we are more aggressive if we are deprived of aggression is not supported by experiments on laboratory animals either. For instance, rats trained to kill mice showed no tendency to increase their rate of killing after being deprived of mice for a few days (Van Hemel and Myer 1970). A number of studies on humans have tried to evaluate the effect of the observation of aggression in movies and television on subsequent aggressive behavior. Possible changes, if any, after viewing violence would be either a decrease in aggressiveness through catharsis or an increase in aggressiveness through some type of learning, through a generalized arousal, or through a reduction in inhibitions. A 1982 report by the National Institutes of Mental Health concluded that violence on television does lead to aggressive behavior in children and teenagers who watch the programs. Not surprisingly, television networks challenged the findings (Walsh 1983). Controlled laboratory experiments indicate that various visual stimuli (for example, comic book violence) do increase the subjects' aggressive behavior (Berkowitz 1969, 1993).

What about the long-term effects of viewing violence? In a study of factors in childhood that predict later criminal behavior, a sample of 220 children in Finland was followed from age 7–9 into adulthood (Viemerö 1996). For males, the best predictor of physical aggression in adolescence was previously behaving aggressively as a child. For females, the best predictor was viewing violence on television as a child. When predicting the number of arrests as adults, viewing violence on television was important in both sexes.

Conditioning to situational stimuli may be important in humans, as Berkowitz and LePage (1967) showed. They instructed a subject to play the role of an experimenter trying to teach others a task. When the "subjects" made a mistake, the "experimenter" (i.e., the real subject) was supposed to administer a punishment in the form of an electric shock. The researchers could then measure the aggressiveness of the real subject by the number of shocks he or she delivered. A second variable was the presence or absence of a gun on a table near the "experimenter." The real subject administered more shocks when the weapon was present than when it was absent. Presumably the gun facilitated aggression because of its association with violence (Berkowitz 1993).

Social Control and Social Disorganization

Observations of cichlid fish (*Cichlasoma biosallatum*) (aptly named the Jack Dempsey fish, after the heavyweight boxer) show that when strange fish of the same species are continually introduced into the group, fighting remains at a high level; if group membership is allowed to stabilize, fighting declines to a relatively low level (Scott 1975). Thus, one cause of aggression seems to be social disorganization, as seen in newly formed groups or groups that have been disturbed. When a high-ranking male or female rhesus monkey dies or leaves the group, aggression within the group may increase. New males moving into the central hierarchy attack females and are sometimes mobbed by them in return. The absence of a high-ranking control animal contributes to the increase of aggression. As new relationships are established within the group, and as control animals emerge and begin to break up fights, group aggression decreases (Vessey 1971). The following tactics, modified from Marler (1976), can be used by any social animal to avoid or deescalate aggression:

- Keep away
- Evoke behavior that requires proximity, is physically incompatible with aggression, and reduces arousal (e.g., grooming)
- Avoid provoking extreme arousal of frustration
- Use submissive displays (antithesis of aggressive displays)
- Behave predictably
- Divert attack elsewhere

Aggression in human societies can be traced, in part, to social disorganization and xenophobia. Sociologists use the term *community* to refer to a social system with stable membership whose members interact extensively among themselves. When group membership is unstable and when relationships and roles are uncertain, alienation and aggressive interactions are frequent (Scott 1976). The high rate of violent crime in inner cities may be spurred in part by unstable family structure.

There is consensus that human aggression, just like nonhuman aggression, is not a unitary phenomenon dependent on one major cause (Feshbach 1997). Nor is aggression an abnormal form of behavior; rather, it is part of an adaptive response to the environment shared by most animals. Biological factors, such as hormones and neurotransmitters, play a role in human aggression, but reducing aggression in human societies will mainly require sociological, economic, and political solutions (Huntingford and Turner 1987).

SUMMARY

Agonistic behavior, defined as social fighting among conspecifics, includes all aspects of conflict, such as threat, submission, chasing, and physical combat, but excludes predation. Aggression emphasizes overt acts intended to inflict damage on another, and may include predation, defensive attacks on predators by prey, and attacks on inanimate objects. Many species of invertebrates and nearly all vertebrates engage in some type of agonistic behavior, which may include the formation of a dominance hierarchy, territoriality, combat for mates, or parent-offspring conflict.

Much conflict behavior involves the social use of space. Home range is the area used habitually by an individual or group; core area is the zone of heaviest use within the home range. Many social animals maintain a personal space around themselves, also referred to as individual distance. Territory is the space that is used exclusively by an individual or group and that is defended. Individual territories occur most often when a needed resource is predictable and evenly distributed in space. Some territories include food and water supply, nest site, and mates; other territories contain only one resource that is defended—for example, a place where only mating takes place, as in the lek.

Species that live in more or less permanent groups usually develop dominance hierarchies, in which individuals control the behavior of conspecifics on the basis of the results of previous encounters. Dominance hierarchies are not always determined by fighting ability; age, seniority, maternal lineage, and formation of alliances have been shown to be important in some mammals.

Agonistic behavior is usually highly ritualized and communicative in nature; killing or wounding is infrequent. Recent observations interpret overt violence as an expression of genetically selfish behavior on the part of an individual rather than as a maladaptive response to abnormal conditions.

Researchers have studied the causes of aggression from both an internal and external perspective. Internal factors include the subcortical limbic system, in particular the thalamus, hypothalamus, amygdala, and central gray. Various neurotransmitters and hormones are linked with the exhibition of different forms of aggression. For example, testosterone is most often associated with aggression, but other hormones are important in producing aggression in some species. Artificial selection in the laboratory for high and low aggression demonstrates a genetic component of aggression upon which natural selection can act. In the laboratory, animals can be conditioned to be dominant or subordinate, which leads us to conclude that previous experience and learning affect later aggressive behavior. External factors that influence aggression include early learning and experience, pain and frustration due to restricted access to resources, xenophobia, and crowding.

Models based on game theory demonstrate that natural selection acting at the level of the individual can produce an evolutionary stable strategy (ESS), in which a certain proportion of individuals in a population are aggressive and a certain proportion are passive. The behavioral strategy used depends on the relative benefits and costs of winning and losing, and on what the other members of the population are doing.

Control of aggression is of prime interest because it relates to noninstitutionalized violence in humans. Some researchers have emphasized the inevitability of aggression in humans and have sought ways to divert it into harmless channels. Others have suggested that the modification of the social environment through early experience and avoidance of social disorganization can help to control aggression. Encouraging noncompetitive habits during early experience and creating stable communities may help reduce violent agonistic behavior.

DISCUSSION QUESTIONS

1. Distinguish between agonistic behavior and aggression.

2. Discuss ways aggression is restrained in animals. Evaluate the arguments put forth to explain the evolution of such restraint.

3. In the Hawk-Dove game discussed in this chapter, a third strategy, "bourgeois" was described: play Hawk if you own the resource in dispute, play Dove if you are an interloper. Prepare the payoff-matrix for this three-strategy game.

4. The data shown here present ranks and maternal lineages among rhesus monkeys at La Parguera, Puerto Rico in June 1969. Part of A Group, which was formed in 1962 when unrelated adult females were placed together in a large cage, is shown. Females 34, 227, 184, and 236 established a linear dominance hierarchy in the order listed; they were then released. The offspring of those females are shown connected to the females with a line; their ages are given in parentheses. Most male offspring left the group at 3 or 4 years of age and are, therefore, not shown. The bottom row of numbers and letters shows the monkeys ranked on the basis of agonistic encounters, with those having won the greatest number of encounters on the left, and the least on the right. Males are in squares. What kinds of rules seem to apply in determining dominance orders in rhesus monkeys? Can you explain how or why such rules come about?

Ranks and maternal lineages among rhesus monkeys

34(13+)				227(13+)				184(13+)			236(13+)			
313(6)	A9(4)	G4(3)	I3(2)	268(6)	A2(4)	E9(3)	I2(2)	B2(4)	D0(3)	F9(2)	269(6)	300(5)	D7(4)	I9(2)
313–68(1)												K4(1)		

34, G4, A9, 313, 313–68, I3, 227, E9, A2, 268, I2, 184, B2, D0, F9, 236, D7, 300, K4, 269, I9

SUGGESTED READINGS

Archer, J. 1988. *The Behavioural Biology of Aggression.* New York: Cambridge University Press.

Feshbach, S., and J. Zagrodzka. 1997. *Aggression: Biological, Developmental, and Social Perspectives.* New York: Plenum Press.

Huntingford, F., and A. Turner. 1987. *Animal Conflict.* London: Chapman and Hall.

Mason, W.A., and S. P. Mendoza, eds. 1993. *Primate Social Conflict.* Albany, NY: SUNY Press.

Svare, B. B., ed. 1983. *Hormones and Aggressive Behavior.* New York: Plenum Press.

CHAPTER

Sexual Reproduction and Sexual Selection

After reaching maturity at sea, coho salmon (*Oncorhynchus kisutch*) find the entrance to their natal river, work their way up rapids to reach their birthplace, expend their remaining energy by shedding masses of eggs or sperm, and then die. We can ask a number of "why" questions about the evolutionary and long-term ecological determinants of salmon reproductive behavior. For example, why does a salmon return to its birthplace to reproduce and not to some other stream? Why does it put all its effort into a single explosive reproductive episode rather than produce smaller numbers of young at intervals? For that matter, why does it reproduce sexually at all? Male salmon actually change morphology as they come into breeding condition: some develop bright red colors and enlarged, hook-shaped jaws with large teeth, while other males resemble females (Gross 1985, 1991). Why have such differences evolved?

In the first part of this chapter, we consider the forces that may determine the existence of sex in the first place. We then explore the differences between the sexes and the evolution of sexually selected traits. Chapter 18 deals with mating systems and parental care. In both these chapters we emphasize the ultimate factors that have shaped the evolution of reproductive behavior. For a discussion of proximate mechanisms in reproduction, see chapters 7 and 8.

COSTS AND BENEFITS OF SEX

Many unicellular organisms, plants, and invertebrates reproduce asexually during one part of their life cycle and reproduce sexually at other times. During the sexual phase, the organisms will often produce seeds or eggs that are resistant to environmental stress and that are capable of being dispersed. Some sort of genetic recombination occurs in virtu-

a.

b.

FIGURE 17.1   Sequentially hermaphroditic coral reef fish.
One male in the species *Anthias squamipinnis* defends a territory containing several females. The dominant female in the group (a) changes into a male (b) upon the death or disappearance of the male. Protogyny is common in reef fish.
Source: Courtesy Douglas Shapiro.

ally all organisms, with the exception of a large group of rotifers (phylum Rotifera) in which no sexual reproduction or genetic exchange seems to have occurred for millions of years (Welch and Meselson 2000). Some organisms—for example, earthworms, snails, and certain species of fish—are hermaphrodites: they possess functional sexual organs of both males and females. Usually hermaphrodites do not self-fertilize, possibly because the new genetic combinations created would not be novel enough to outweigh the costs of breaking up previously successful combinations. Some species of fish and some invertebrates are sequentially hermaphroditic; they change from female to male (**protogyny**), or from male to female (**protandry**) usually once in their lives (figure 17.1).

Costs of Sex

Why is sexual reproduction so widespread? By reproducing asexually, an organism maximizes its genetic contribution to offspring by creating carbon copies of itself. By reproducing sexually, genotypes are broken up at meiosis (chapter 4), and the genes combine with those of another individual during fertilization. The result is a 50 percent reduction in transmis-

sion of genes to the next generation, which is referred to as the **cost of meiosis** (Williams 1975). The energetic cost of raising young is about the same whether the offspring are produced sexually or asexually, so a female would have to produce two sexual offspring for each asexual offspring in order to pass the same number of genes to the next generation. If a female can get a male to share in raising the offspring, the cost of meiosis is reduced. If they share equally, a female can rear two sexual offspring for the same cost as a single asexual offspring, and the relative cost of meiosis might be reduced to zero (Wittenberger 1981). A further cost of meiosis, however, is that the unique combination of traits in the parent is broken up in sexually produced offspring. This aspect of the cost of meiosis can be reduced if individuals mate with relatives, since they are likely to share some of the same combinations of genes (Shields 1982).

A second cost of sex is sometimes referred to as the cost of producing males. A single male produces enough sperm to fertilize the eggs of many females. However, most sperm are wasted and many, if not most, males never even fertilize one egg (Clutton-Brock 1988, 1991); from this perspective males waste resources that could be used for reproduction in asexual form. A third cost of sexual reproduction is the cost of courtship and mating, since it takes energy to secure a partner, and there is often an increased risk of being injured or killed by a competitor or a predator (Daly et al. 1978). A related cost of mating is the possibility of acquiring sexually transmitted diseases and parasites from the partner.

Benefits of Sex

Given these costs, one might wonder how sexual reproduction got started in the first place and how it is maintained in a population. We might expect parthenogenetic (asexual) individuals to "infect" sexual populations and replace the sexually reproducing organisms. What are the possible benefits of sex and how can it be maintained in a population? Most often mentioned is that of faster evolution. Fisher (1958) first pointed out that sexually reproducing populations can evolve faster in a changing environment. He stated that the rate of evolutionary change in a population is a function of the degree of genetic variation available. Through sexual recombination, mutations occurring at different loci in different individuals can appear in novel combinations in the offspring. A second, related advantage was noted by Muller (1964) and is referred to as **Muller's ratchet.** Suppose that a deleterious mutation occurs in an individual member of an asexual population. The gene will be passed on to all its offspring. The only way an individual that is free of this mutation can arise is by back (or reverse) mutation, which for some loci is an unlikely event. Thus, the ratchet turns one notch each time a deleterious mutation occurs, and mutations accumulate in the population. However, if sexual reproduction occurs, recombination between two (diploid) individuals with different mutations could produce offspring with neither mutation. In this way, harmful mutations can be edited out of a population.

However, the ability of sexual reproduction to edit mutations out of a population depends on the chromosomal locations of alleles that affect a particular trait. Sexual reproduction can either enhance or impede the short-term response of quantitative characters to selection (Lynch and Deng 1994). In addition, most mutations appear to arise in the male germ line, and the ratio of mutations in males compared to females is quite high (Redfield 1994). Computer simulations suggest that the cost of male mutations can easily exceed the benefits of recombination in terms of reducing the impact of deleterious mutations (Redfield 1994).

The preceding arguments have emphasized the advantages of sex for the population, or group: faster evolution and the elimination of mutations from the gene pool. Williams (1975) championed benefits that might accrue to individuals, rather than populations, since natural selection is thought to act primarily at the level of the individual (chapter 4). He noted that in species with both sexual and asexual reproduction, the sexual phase usually occurs before unpredictable changes in environmental conditions begin. Under this argument, if parents produce genetically variable offspring, the likelihood increases that a few will be adapted to one of the new environments. Williams used a raffle analogy: if there is only one survivor (winner) in each different habitat, reproducing asexually would be like having tickets that all have the same number printed on them. The chances of picking a winner are much less than if all the tickets have different numbers—the case with sexually produced offspring. From this we might predict that species living in highly variable and unpredictable environments would be more likely to have sexual reproduction; those in stable and predictable environments would be more likely to have asexual reproduction. The emphasis here is on variation of the physical environment.

Others have argued that biotic fluctuations make it even more necessary to produce genetically variable progeny (Bell 1982; Van Valen 1973). For instance, as predators are selected to become more efficient at catching their prey, the prey must continually evolve new ways to avoid being eaten. Such evolutionary "races" take place between competing species and between pathogens and hosts as well. Because of their short life spans, bacteria, viruses, and parasites rapidly evolve into new strains. The host produces variable offspring "in the hopes" that some will be resistant. This explanation for the maintenance of sex has been called the **Red Queen hypothesis** because of Alice's encounter with the Red Queen in *Through the Looking Glass.* After a wild run they ended up where they started, and the Queen told Alice: "Now here, you see, it takes all the running you can do, to keep in the same place."

One prediction of the raffle hypothesis is that sexual species should predominate in environments that are highly and unpredictably variable, while asexual species should predominate in habitats that are more stable. In fact, just the opposite is usually the case, and the Red Queen hypothesis seems the more likely explanation. In stable environments, biotic interactions are likely to be more intense, and animals have more predators, parasites, and pathogens from which to escape. However, in species that can alternate their mode of reproduction during the year, there is a tendency for reproduction to be asexual in spring and summer and sexual in the fall before harsh winter conditions set in. Often the zygote is highly resistant to extreme temperature and moisture conditions. These cases seem to support the raffle hypothesis.

Further support for the Red Queen hypothesis comes from a species of snail (*Potamopyrgus antipodarum*) from New Zealand. Those living in lakes, where rates of parasitism are high, are more likely to reproduce sexually than are those from streams, where parasitism is lower (Lively 1987). Also, the percentage of males in different populations (an indicator of the extent of sexual reproduction) increases with the degree of parasitism. The advantage of sexual reproduction here may be to "keep up" defensively with the rapidly evolving parasites.

ANISOGAMY AND THE BATEMAN GRADIENT

In most animal species, there are only two sexes that are anatomically different and that are produced in about equal numbers. The possible genetic combinations for two sexes are astronomical—three or more sexes would add little except confusion in mate selection. In most microorganisms, fungi, and algae, the sexes are anatomically similar, and the gametes are all of the same size (**isogamy**). However, division of labor is the rule in most plants and animals: females produce few large, sessile, energetically expensive eggs, and males produce many small, motile, energetically cheap sperm. This difference in gamete size, called **anisogamy,** may have resulted from disruptive selection (Parker et al. 1972) and has set the stage for many differences in the reproductive behavior of males and females (Trivers 1972). Since females by definition produce a relatively small number of the expensive gametes, they are often a limited resource for which males compete. A male's reproductive success is usually a function of how many different females he can inseminate, while a female's is a function of how many eggs she can produce.

One of the results of this difference between the sexes is that the reproductive success of males is likely to vary more than that of females. In laboratory populations of fruit flies (*Drosophila melanogaster*), Bateman (1948) demonstrated that nearly all females mated; however, many males failed to mate, and some males mated several times. In other words, the variance in copulatory success was higher for males than for females. The flies had chromosomal markers so that parents of particular offspring could be identified. Males that copulated most also sired the most offspring, referred to as the **Bateman gradient** (Andersson 1994). Females, on the other hand, needed only one mate to produce the maximum number of offspring (figure 17.2).

When reproductive success is determined over the entire lifetime, the Bateman gradient becomes evident, especially in species such as lions, where male-male competition for

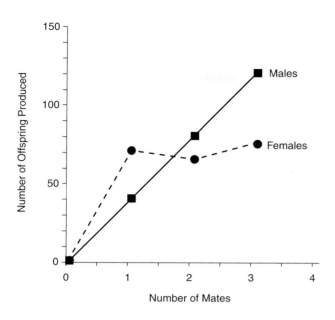

FIGURE 17.2   Reproductive success in fruit flies (*Drosophila*) as a function of the number of mates.

Note that female reproductive success does not increase with number of mates, while that of males does.

Source: Data from A. J. Bateman, Intra-sexual selection in *Drosophila*, in *Heredity*, 2:349–68, 1948.

access to pride females is intense (figure 17.3). However, in species such as scrub jays, where the sexes form stable pair bonds, no such effect is present.

SEX DETERMINATION AND THE SEX RATIO

We are accustomed to thinking that sex is determined by sex chromosomes, as it is in birds and mammals. However, things are a bit more complicated in other organisms. Some animals have physiological and behavioral control over sex determination. In social insects such as bees and ants (Hymenoptera), males are derived from unfertilized (haploid) eggs and females from fertilized (diploid) eggs. Such a mode of sex determination is called **haplodiploidy.** In many species, early in the season, after the queen is fertilized, she does not release stored sperm when she lays her eggs, and male offspring result. Later in the season she releases sperm when she lays her eggs. All-female broods that become workers result. In many species of turtles, sex is determined by the incubation temperature: those incubated at a low temperature become males, while those incubated at a higher temperature become females (Bull 1980).

Among organisms that undergo a sex change are protogynous fishes such as the sea bass (*Anthias squamipinnis*) studied by Shapiro (1979) on Aldabra Island off the east coast of Africa. The mechanism that controls the fish's change from female to male is not completely understood. It is triggered, however, by the death or departure of the (usually) single resident territorial male. Shapiro observed that

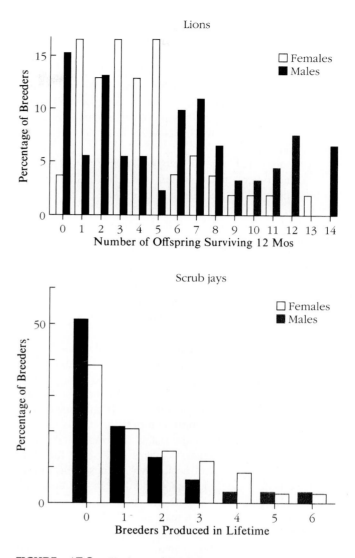

FIGURE 17.3   Variance in lifetime reproductive success (LRS).

(a) LRS for African lions. Most females have from 1 to 5 surviving offspring, whereas male reproductive success varies widely. (b) LRS for Florida scrub jays. Note that the variances are similar for both sexes.

upon the permanent absence of the male, the largest female undergoes a marked change in behavior, engaging in more aggressive displays, as she assumes the male color pattern and reproductive role (see figure 17.1). In another species, the saddleback wrasse (*Thalassoma duperrey*), sex change is a function of the relative size of neighbors. When few larger fish (usually males) are present, females change into males (Ross et al. 1983).

The sex ratios of populations of most species tend to be about 1:1 at birth or hatching, but may deviate significantly from equality among adults. The **operational sex ratio** considers only the reproductively active members of the population. Deviations from a 1:1 ratio can have a large impact on the mating system, since members of the abundant sex will compete for access to the scarcer sex.

Why does the sex ratio tend to be 1:1 at birth? This problem was solved by Fisher (1958) as follows: Consider a population containing more males than females. Some males will be unable to find mates. Females that tend to produce more females than males will be favored through natural selection because all of their female offspring will probably find mates, and the sex ratio will converge toward 1:1. The symmetrical argument also holds when the ratio is skewed in favor of females.

The argument was developed further by Hamilton (1967) and others, who pointed out that equality of the investment by parents in offspring of each sex is the important factor in the ratio. If twice as much effort by the parents is required to raise male young to maturity, we would expect a 1:2 male to female sex ratio in young that reach maturity. In species of wasps of the genus *Symmorphus,* the sex ratios of offspring were inversely proportional to the amount of food provided by the mother to the developing larvae (Trivers and Hare 1976). In those species where more food was provided to female larvae, the ratio was biased toward males, and vice versa.

Deviations from 1:1 also occur in populations that are spatially divided into smaller units. The parasitoid wasp *Nasonia vitripennis* lays eggs directly in the larvae of their host—a blowfly. Developing wasps feed on the host, and the flightless males mate on or near the host. If only one female lays eggs in a host, **local mate competition** occurs as male sibs try to mate with the same sisters (Werren 1983). From an evolutionary perspective, it makes little sense for close relatives to compete with each other because this lowers inclusive fitness (see chapters 4 and 19); in addition, one male can fertilize many females. In such situations, the primary sex ratio is strongly female biased. What if a female lays eggs in a host that has already been parasitized? In these cases, the female somehow can assess the proportion of the total eggs in the host that are hers and adjusts the sex of her eggs accordingly: the fewer that are hers, the more male-biased the sex ratio (Werren 1983).

According to a model based on maternal condition proposed by Trivers and Willard (1973), the Bateman gradient sets the stage for another possible deviation from a 1:1 sex ratio. The model assumes that mothers in the best physical condition produce healthier offspring that are better able to compete for mates or other resources. Because males are more likely than females to compete for additional mates, the reproductive success of males is highly variable, and males should require greater maternal investment. For example, male Antarctic fur seal pups (*Arctocephalus gazella*) are heavier at birth, grow faster, and weigh more at 60 days than their sisters (Goldsworthy 1995). Trivers and Willard (1973) argue that reproductive success of sons should be high if their mothers are in good condition, but low, perhaps zero, if their mothers are in poor shape. Female offspring are likely to breed anyway, regardless of their mother's condition. Citing data from mammals, such as mink, deer, seals, sheep, and pigs, these investigators predicted that a female would produce male offspring if she were in good condition and female offspring if she were in poor condition.

A number of studies on mammals lend support to Trivers and Willard's model. Dominant red deer (*Cervus elaphus*) hinds (females) have access to the best feeding sites and are able to invest more in their offspring via lactation (Clutton-Brock et al. 1984). Male offspring in this species grow faster than do females and would seem to benefit more from this greater investment. As adults, males compete intensely to control harems. A female that produces a successful son can achieve more than twice the reproductive success of a female producing a daughter. As predicted by Trivers and Willard (1973), there was a positive correlation between dominance rank of hinds and the percentage of sons produced (figure 17.4). Similarly, in a field study of opossums (*Didelphis marsupialis*) in Venezuela, females receiving supplemental food produced more sons than daughters (Austad and Sunquist 1987). Furthermore, in a sample of old females in poor condition the sex ratio was skewed toward female offspring, as predicted.

Models different from that of Trivers and Willard have been proposed to explain biased sex ratios in other mammals. In studies of a South African prosimian primate, the bush baby (*Otolemur crassicaudatus*), Clark (1978) noted that the sex ratio was male-biased. Female offspring tended to remain near the mother's home range, but males dispersed, the usual patterns for mammals (see chapter 14). Daughters therefore compete with their mothers and sisters for food. By producing fewer daughters, Clark (1978) argues, local resource competition is reduced. A troop of baboons (*Papio anubis*) living in Amboseli National Park, Kenya, has been studied for many years by Altmann et al. (1988). Dominant females produce more daughters than sons, whereas subordinate females produce more sons than daughters, exactly the opposite result predicted by Trivers and Willard. Altmann et al. (1988) point out that daughters, who remain in their

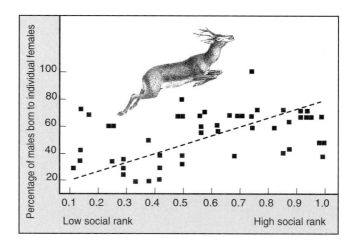

FIGURE 17.4   Red deer birth sex ratios.
Birth sex ratios of individual red deer females differing in social rank over their life spans. Measures of maternal rank were based on the ratio of animals the subject threatened or displaced to animals that threatened or displaced it. High-ranking females tended to have sons.
Source: Data from T. D. Clutton-Brock, S. D. Albon, and F. E. Guinness. "Maternal Dominance, Breeding Success and Birth Sex Ratios in Red Deer," in *Nature* 308:358–360, 1984.

natal group for life, share the social rank of their mothers and can benefit from their mother's high rank. Presumably high-ranking families get larger and even more powerful as more and more daughters are born. On the other hand, as with bush babies, sons usually leave the natal troop and thus do not benefit by their mother's rank. In these baboons, the best strategy for dominant females is to produce daughters, whereas for subordinate females it is to produce sons.

Offspring sex ratio data for primates are somewhat confusing since about equal numbers of studies show high-ranking mothers producing a greater proportion of sons or a greater proportion of daughters than low-ranking females. In addition, a number of studies show no relationship between maternal dominance rank and offspring sex ratio. Van Schaik and Hrdy (1991) argued that when primate population growth rates are high and resources abundant, high-ranking mothers are expected to respond by producing a higher proportion of sons than daughters. Under such conditions, sons of high-ranking mothers may have much greater success at intrasexual competition and, therefore, disproportionately high reproductive success. However, when resources are limited and resource competition is more intense, high-ranking females are expected to produce a greater proportion of daughters than low-ranking females, since daughters of high-ranking females would have an advantage in competition for resources. The relationship between population growth rate and proportion of males born to high- and low-ranking females supports the van Schaik and Hrdy model (figure 17.5).

It is not clear how such adjustments of sex ratios occur at the proximate level. Some evidence points to the timing of insemination within the estrous cycle, such that more of one sex is conceived early and more of the other sex later in the cycle, as in white-tailed deer (*Odocoileus virginiana;* Verme and Ozoga 1981), Norway rats (*Rattus norvegicus*)

(Hedricks and McClintock 1990), hamsters (*Mesocricetus auratus*) (Huck et al. 1990), and humans. A number of studies, including those in humans, suggest that levels of sex hormones present at the time of conception affect offspring sex ratios (James 1996). In many species of mammals, the sex ratio at conception is male-biased. Intrauterine mortality rates are typically higher for male than for female embryos, and these rates are probably highest among mothers in poor condition. In this way the maternal condition hypothesis could be explained, as females in poor condition would lose their male embryos and possibly come into estrus again. After birth, other mechanisms may operate to select for one sex or the other.

SEXUAL SELECTION

The success of an individual is measured not only by the number of offspring it leaves, but also by the quality or probable reproductive success of those offspring. Thus, it becomes important who its mate will be. Darwin (1871) introduced the concept of **sexual selection,** a special process that produces anatomical and behavioral traits that affect an individual's ability to acquire mates. Sexual selection can be divided into two types: **intersexual selection,** in which members of one sex choose certain mates of the other sex and **intrasexual selection,** in which individuals of one sex compete among themselves for access to the other sex. A result of either type of selection is the evolution of sex differences—that is, the sexes become **dimorphic.**

Darwin was not able to explain why it was usually males that competed for access to females in order to mate with them, while females usually seemed to be choosing among males as mates. As we saw previously, Bateman (1948) argued that this general difference between the sexes arose from differences in the factors that limited their reproduction. He proposed that the fertility of females is limited by egg production while in males, fertility is limited more often by the number of mates. Trivers (1972) generalized Bateman's concept and argued that the breeding system (e.g., monogamy) as well as the adult sex ratio are functions of a single variable—parental investment—that controls sexual selection. Trivers (1972) defined **parental investment** as "any investment by the parent in an individual offspring that increases the offspring's chance of surviving (and hence reproductive success) at the cost of the parent's ability to invest in other offspring." Trivers argued that the reproductive success of the sex that invests less in offspring (usually males) is limited by their ability to mate with the sex that invests more (usually females). On the other hand, the reproductive success of the sex that invests more is limited by resources necessary for investment, not by access to the other sex. This led Trivers to propose that the larger the difference in the potential maximum net reproductive successes of the sexes (the ratio of L/M from figure 17.6), the greater the potential for sexual selection in that species.

When females invest far more than males, there will be strong selection among males for traits that allow them to

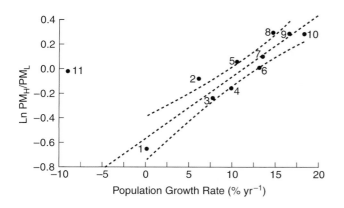

FIGURE 17.5 **The relationship between population growth rate and the proportion of sons born to high- and low-ranking females in populations of primates.**
Each data point is the long-term sex ratio of births from one colony. PM = proportion of males born to (H) high-ranking and (L) low-ranking females. Populations with high growth rates tended to have male-biased sex ratios among high-ranking females.
Source: C. van Schaik and S. B. Hardy, "Intensity of Local Competition Shapes the Relationship Between Maternal Rank and Sex Ratios at Birth to Cercopithecine Primates," *American Naturalist* 138:1555–62, 1991.

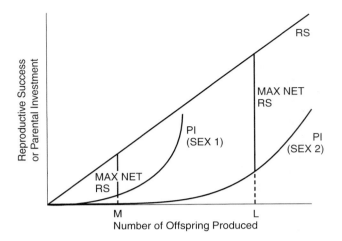

FIGURE 17.6   Relationship between reproductive success and parental investment.
Theoretical reproductive success (RS) and decrease in future reproductive success resulting from parental investment (PI) are graphed as functions of the number of offspring produced by individuals of the two sexes. At M and L, the net RS reaches a maximum for sex 1 and 2 respectively. The RS of sex 2 is limited by access to sex 1.
From R.L. Trivers, "Parental Investment and Sexual Selection," in *Sexual Selection and the Descent of Man,* edited by Bernard Campbell. Copyright © 1972 by Aldine de Gruyter, Hawthorne, NY. Reprinted with permission.

FIGURE 17.7   Satin bowerbird at bower.
The male satin bowerbird (*Ptilonorhynchus violaceus*) constructs the bower as part of his courtship ritual to attract a female. Males that build structures with more sticks and brightly colored objects are more successful in attracting mates.
Source: Photo by Richard A. Forster, courtesy of the Massachusetts Audubon Society.

somehow gain access to females. Investment by females is critical to the reproductive success of males in this situation. On the other hand, selection should act on females to result in the evolution of choosiness for mates that will invest in offspring and/or will contribute good genetic material to offspring. There would not be a strong selective force favoring females that competed to mate with males, since access to their sperm does not limit the reproductive success of females. As we will discuss, it is not necessarily males that compete for females and females that choose; in some species, the opposite occurs. What appears to be important in determining the evolution of those traits is the relative amount of parental investment made by the sexes (Trivers 1972).

Intersexual Selection

In intersexual selection, individuals of one sex (usually the males) "advertise" that they are worthy of an investment; then members of the other sex (usually the females) choose among them. Most naturalists after Darwin discounted the importance of mate choice in evolution, but in the last two decades it has become widely accepted as an important evolutionary force.

Bowerbirds

As an example of intersexual selection in action, consider the bowerbirds of Southeast Asia. Bowerbirds are unique in that they build a structure whose only function seems to be to attract mates. The male satin bowerbird (*Ptilonorhynchus violaceus*) of Australia and New Guinea stations itself alone in the forest, clears a space, weaves two freestanding walls from sticks, and decorates this avenue bower with brightly

colored objects, including snail shells, flowers, leaves, and even plastic (figure 17.7). A visiting female moves into the avenue between the walls, while the male goes through an elaborate display sequence. If the female is receptive, she crouches and the male enters the avenue and mounts her for a 3-second copulation, after which she shakes her feathers and leaves. Although she may visit several bowers, she usually mates with only one male, and less than 10 percent of courtships lead to copulation (Borgia 1995). Gilliard (1963, 1969) reasoned that the bower functions in place of showy plumage and noted that those species with the fanciest bowers had the drabbest plumage. He argued that such behavior arose to synchronize mating between males and females. Borgia (1995) hypothesizes that bowers probably first evolved to provide the female with an escape route from unwanted matings; subsequently, they have taken on other functions involving female choice of male quality.

Borgia (1985) assessed the relative contribution of the male satin bowerbird's decorations to his mating success. Human observers and cameras controlled by infrared detection devices monitored activities at the bower. Removing decorations reduced mating success. Aspects of the bower that were positively related to mating success were the number of blue feathers, snail shells, and yellow leaves. The general construction of the bower and the density of sticks in the walls were also important. Borgia suggested that bowers in this species function in part as "markers," conveying information to females about the male's genetic quality. Dominant males have high-quality bowers that attract the most females.

It is important to keep in mind that although intersexual choice is most often observed among females, it is not exercised exclusively by them. For example, male water striders (*Gerris lacustris*) prefer to mate with larger, long-winged females than with short-winged females. This preference is not influenced by whether the male is long- or short-winged,

FIGURE 17.8   Peacock in display.
In this extreme example of sexual selection the courting male spreads his tail feathers and shakes them in front of the female. Such plumage would seem to confer little advantage to male survival, but it may indicate to the female the male's genetic superiority.
Source: © Terence A. Gili/Animals Animals/Earth Scenes.

and it appears to be adaptive, since larger, long-winged females lay more eggs per day (Batorczak et al. 1994). In addition, male bush crickets will readily mate with unparasitized females but will be choosy if females are parasitized (Simmons 1994).

Runaway Selection

A number of hypotheses have been proposed to explain the evolution of sexually dimorphic traits. Fisher (1958) proposed a model for sexual selection, and used birds as an example. Suppose a plumage characteristic in males, such as a longer tail, is attractive, for some reason, to females. Further imagine that tail length is heritable (sons have tails like their fathers) and preference for the tail length is heritable (daughters prefer the same tail lengths their mothers do). Further development of the trait will proceed in males, as will the preference for that trait in the females, resulting in a self-reinforcing, **runaway selection** process. An extreme result of this process is seen in the peacock (figure 17.8). Sexually selected traits may become so exaggerated that survival of the males that bear them may be reduced. Counterselection in favor of less ornamented males will then occur—due, for example, to greater predation on the more ornamented males—and sexual selection will be brought to a steady state. (See also chapter 5.)

Female preference for male ornamentation was demonstrated experimentally in a small African species of widow bird (*Euplectes progne*). Females preferred males with long tails. When tails were artificially shortened or lengthened,

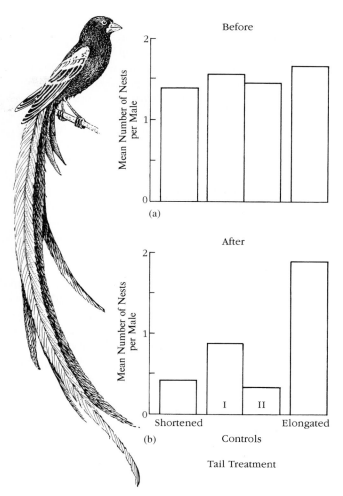

FIGURE 17.9   Sexual selection for tail length in long-tailed widow birds.
(a) There was no difference among the four groups before the tails were altered. (b) After the tails were shortened or lengthened, the mating success went down and up respectively. The two kinds of control birds were (I) unmanipulated, and (II) cut and glued back without altering length. Mating success is measured as the number of active nests in each male's territory (Andersson 1982).

females still preferred those with the longest tails, showing that they were responding to tail length itself, and not some other correlated trait (Andersson 1982) (figure 17.9).

Laboratory studies of mate choice reveal that subtle and arbitrary differences can indeed influence mate choice, differences that seem to have no relation to the quality of the male. Burley et al. (1982) were able to influence choice of mates in a colony of zebra finches simply by using different colored leg bands. Males with red bands and females with black bands were preferred as mates and had higher reproductive success than did males with green and females with blue.

Selection for Good Genes

As alternatives to Fisher's runaway selection hypothesis are several **"good genes,"** or **indicator,** models, which assume that the trait favored by females in some way indicates male fitness. One is the **handicap hypothesis** proposed by Zahavi

(1975). Agreeing that sexual selection can produce traits that are detrimental to survival, Zahavi adds that they must be both costly to produce and linked to superior qualities in the males. Thus, a bird that can survive in spite of conspicuous and costly-to-produce tail feathers, or a stag that can afford to spend energy on a huge bony growth of antlers, must be genetically superior and a worthy mate. Important to this hypothesis is the notion of "truth in advertising"; the male's handicap must be linked to overall genetic fitness. Only in this way will females benefit by picking a male with this handicap. One problem with this model is that when a male with a handicap is picked by a female, not only are his favorable traits passed on to his offspring, but also the genes for the costly handicap, which should be selected against. However, if the handicap trait of males is facultative, that is, it is displayed only when the male can afford to do so, the model is not so problematic.

There is direct evidence that some aspect of courtship enables females to choose males with good genes. Male butterflies of the genus *Colias* vary in enzymes involved in glycolysis, the process that provides energy to the flight muscles. Certain genotypes are superior in terms of flight capacity and longevity. In tests of mate choice, the superior male genotypes are favored, especially by older, more discriminating females who had already mated once (Watt et al. 1986).

Although Darwin, Fisher, and Zahavi thought of sexual selection as a process distinct from natural selection, some biologists argue that the two are inseparable. Kodric-Brown and Brown (1984) claim that most sexually selected traits are aids to survival, rather than handicaps. Thus, a stag with a big set of antlers may be dominant over other males and may have better access to a food supply. They further argue that the sexually selected trait must be a reliable index of the male's condition. Therefore, the trait must not be genetically fixed, but be influenced by environmental conditions.

The Role of Parasites and Disease   Evidence for an indicator role for a sexually selected trait comes from a hypothesis put forth by Hamilton and Zuk (1982). They argued that sexually selected traits have evolved to indicate an animal's state of health, specifically whether or not it is free from disease or parasitism. They predicted that species in which the males have the showiest plumage would be the most prone to infestation with blood parasites. Comparisons of plumage brightness among museum bird specimens confirmed their prediction. If brightness is linked to overall condition, the brighter males within a species should be relatively disease-free and thus preferred as mates. Studies of red jungle fowl (Zuk et al. 1990) and barn swallows (Møller 1990) provide experimental support as well. Support for the Hamilton and Zuk model has been found for a number of taxa in addition to birds, including fishes (Houde and Torio 1992) and insects (Simmons 1994), but other studies have failed to support it (Møller et al. 1999; Read and Harvey 1989).

Hamilton and Zuk (1982) proposed that only males in prime condition can afford to express the most costly secondary sexual characters, and that they would be the ones mostly likely to carry genes for resistance to disease. Several other hypotheses have been proposed to explain how parasites could influence sexual selection, however. One is the parasite-avoidance hypothesis: Females mate with the brightest males not to get the male's good genes into their offspring, but simply to avoid exposure to parasites from the male during mating. Borgia and Collis (1989) found that among satin bowerbirds, male mating success is inversely related to numbers of lice on the males, as predicted by Hamilton and Zuk. However, there was no correlation between a male's condition, as evidenced by weight and the brightness of his plumage, and the number of parasites. Lice are most often found on the head near the eyes. They are white and easily seen against the dark blue plumage of the male, revealing his condition. During courtship a female can inspect males, avoiding those with many lice. In this way a female could gain direct benefits of choosing a parasite-free male, rather than the indirect benefits of mating with a genetically superior male.

Møller et al. (1999) reviewed studies in a variety of species (mostly birds) on the effects of parasites and immune function on the expression of secondary sexual characteristics of their hosts. In a majority of studies there were negative relationships between both parasite burden and immune function and the expression of sexually selected traits within species.

Although we usually think of sexually selected traits as morphological features, such as combs, antlers, or feather plumes, the traits may be auditory or even olfactory. For instance, the song repertoire, that is, the number of different song types sung by a male, is positively correlated with reproductive success in several species of birds because females prefer such males (Møller et al. 2000). When comparing among different species, there is a positive relationship between repertoire size and the size of the spleen and other organs involved in the immune system's response to parasites and disease. These results further suggest that host-parasite interactions have played a role in sexual selection.

So far there have been relatively few tests of role of disease in sexual selection in mammals. In humans, Low (1990) found no direct evidence that sexual selection was linked to pathogen stress, although societies with high levels of pathogen stress tended to be polygynous, with some males having more than one mate. In house mice (*Mus musculus*), females prefer males who differ from them at the MHC complex, a large chromosomal region that contains genes involved in the immune response to disease (Penn and Potts 1999). Offspring of such matings would be more heterozygous and thus able to respond to rapidly evolving parasites (the Red Queen hypothesis).

Fluctuating Asymmetry   In addition to the size or color of sexually selected traits, the symmetry of paired traits may also indicate fitness. **Fluctuating asymmetry** refers to random deviations from bilateral symmetry in paired traits (Andersson 1994). This means, for example, that when a paired trait, such as horns or canines in mammals, is measured for length, thickness, or some other attribute, the right and left sides may differ. These deviations are thought to reflect the inability of the organism to maintain developmental homeostasis (i.e., symmetry) in the presence of environmental variation and stress. Greater asymmetry is associated with low food quality and quantity; pollution; disease; and genetic factors, such as inbreeding, hybridization, and mutation. Symmetrical individuals are likely to be dominant and to be preferred as mates. Thus, male fallow deer (*Dama dama*) with symmetrical antlers were dominant over those with asymmetrical antlers (Maylon and Healy 1994), and male oribi (*Ourebia ourebi*) with symmetrical horns had larger harems than did asymmetrical males (Arcese 1994).

Fluctuating asymmetry may indicate fitness in both sexes. Among gemsbok (*Oryx gazella*), both males and females with asymmetrical horns were in poorer condition and lost more aggressive encounters than did those with symmetrical horns (Møller et al. 1996). Furthermore, symmetrical males were more often territorial breeders than were asymmetrical males, and symmetrical females more often had calves than did asymmetrical females.

Barn swallows (*Hirundo rustica*) have two long outer tail feathers resembling streamers. They vary both in overall length and symmetry. Møller (1992) manipulated both the length and the symmetry of these feathers in males in the field, and then followed their reproductive success. He shortened feathers by simply cutting off the distal 20 mm, or lengthened them by gluing on an additional 20 mm of feather. Symmetry was altered by performing treatments to either one or both tail feathers. Control groups either had their feathers cut and glued back on, or were unmanipulated. Males with shorter and less symmetric tail feathers took longer to acquire mates, their mates took longer to lay eggs, and they produced

fewer fledglings than controls or males with longer and more symmetrical feathers (figure 17.10). The manipulations did not seem to affect the ability to fly or competition with other males. The differences appeared to be due to preferences of females for males with long, symmetrical tails.

Fluctuating asymmetry also is associated with mating success among some insects. Male Japanese scorpion flies (*Panorpa japonica*) with asymmetrical forewings tended to lose fights with symmetrical males over nuptial gifts (insect food items offered to females during mating). Females generally avoided mating with males that had no gifts (Thornhill 1992b). Moreover, in lab tests, females preferred the sex pheromone from symmetrical males over that from asymmetrical males (Thornhill 1992a). Thus, asymmetrical males tend to lose both in competition with other males and in their ability to attract females (figure 17.11).

The vertical bars on male swordtail fish (*Xiphophorus cortezi*) function both in deterring rival males and in attracting females. These bars can be experimentally removed by freeze branding. When females were given a choice of males with similar or different numbers of bars on each side, they preferred the males with similar numbers (Morris and Casey 1998).

Fluctuating asymmetry is most pronounced in sexually selected traits. For example, canines are used as weapons in male-male competition in a variety of primates. In general, males (but not females) from species subject to the strongest sexual selection (as indicated by size dimorphism and male-male competition) showed the highest asymmetry in canines (Manning and Chamberlain 1993). There was no relationship between canine asymmetry and body mass or diet type. The connection between fluctuating asymmetry and environmental stress is suggested by the marked increase in asymmetry of canines (but not of premolars) of lowland gorillas (*Gorilla gorilla*) during the twentieth century (Manning and Chamberlain 1994). Such an increase is consistent with environmental degradation associated with increasing rates of deforestation during this century.

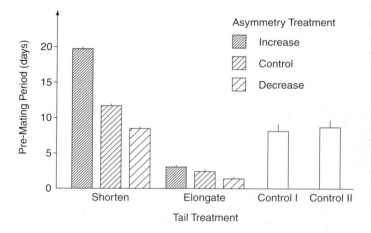

FIGURE 17.10   The effects of male tail length and symmetry of males on the time it took to acquire a mate in barn swallows (*Hirundo rustica*).
In these field experiments, tails were altered both in length and symmetry. Males with long, symmetrical tails mated the fastest, their mates laid eggs soonest, and the pair fledged more offspring than those with short, asymmetrical tails.
Source: Møller, A. P. 1992. Female swallow preference for symmetrical male sexual ornaments. *Nature* 357:238–40.

FIGURE 17.11   Male scorpion fly (Panorpa sp.) In the pheromone-emitting posture.
Source: © Hans Pfletschinger/Peter Arnold.

Cryptic Female Choice   We usually assume that females choose among males prior to copulation, but it is possible for females to choose among the sperm of different males after copulation. In most animals the male does not place his sperm directly on the eggs, but places them in some sort of receptacle. Sperm transport and storage is under control of the female, and when the female mates with more than one male, she has the potential to control which male's sperm fertilizes her eggs (Eberhard 1996, 1997). The term cryptic refers to the fact that the choice would not be apparent by simply determining which males copulate.

In many species of insects, the male continues to perform courtship displays during and after copulation. Presumably, these displays function to improve the chances that the female will use the sperm from him rather than from some other male. In one species of fly, for example, the number of genital taps by the male during copulation was positively correlated with the proportion of eggs that he fertilized (Otronen and Siva-Jothy 1991). Females discard droplets of sperm immediately before laying eggs: fewer taps means more of that male's sperm were discarded.

Another line of evidence for cryptic female choice comes from the presence of extremely long, thin, and tortuous spermathecal ducts, through which sperm must pass before reaching fertilization sites. Eberhard argues that one function of such complexity is to allow females to discriminate among the sperm of different males. Other interpretations are possible, however, and it remains to be seen how widespread this phenomenon is (Birkhead 1998).

Intrasexual Selection

Competition Before Mating

Intrasexual selection involves competition within one sex (usually males), with the winner gaining access to the opposite sex. Competition may take place before mating, as it does with ungulates such as deer (family Cervidae) and antelope (family Bovidae). Typically, males live most of the year in all-male herds; as the breeding season approaches, males engage in highly ritualized battles with their antlers or horns. The winners of these battles gain dominance and so most of the mating (figure 17.12). Antlers are better developed in those cervid species where males compete strongly for large groups of females (Clutton-Brock et al. 1982).

It is often difficult to determine which type of sexual selection is operating to produce an observed effect, since other members of both sexes may be present during courtship. The red belly of the male stickleback both attracts females and intimidates other males. Similarly, the chirp of male field crickets (family Gryllidae) is used to exclude other males from an area and to attract females (Alexander 1961). As a final example of this problem, you may recall that we have already discussed how antlers of deer demonstrate the effects of both female choice and male-male competition. Females may incite competition among males and thus maintain some control over the choice of mate. For example, female elephant seals (*Mirounga angustirostris*) vocalize loudly whenever a

FIGURE 17.12   Use of antlers in combat between male Pere David deer.
Winners of contests become dominant and control access to resources such as mates.
Source: ©Wildlife Conservation Society/Bronx Zoo.

male attempts to copulate. This behavior attracts other males and tests the dominance of the male attempting to copulate (Cox and Le Boeuf 1977). In response to the female's sounds, the dominant harem master will drive off low-ranking, potentially inferior mating partners.

Males may fight directly at the time of copulation, as exemplified by the male of the yellow dung fly (*Scatophaga stercoraria*) (Parker 1970a). A male will sometimes attack a mated pair, displace the male, and mate with the female (figure 17.13).

Competition After Mating

Nor does competition among males to sire offspring cease with the act of copulation. Females of many species store sperm and may remate before sperm from the previous mating are used up, creating the possibility of **sperm competition** (Parker 1970b). Sperm competition does not involve individual sperm actually fighting it out to gain access to eggs; rather, it involves a selection pressure that has led to two opposing types of adaptation in males: those that reduce the chances that a second male's sperm will be used (first-male advantage) versus those that reduce the chances that the previous male's sperm will be used (second-male advantage) (Gromko et al. 1984).

First-male adaptations include mate-guarding behavior and the deposition of copulatory plugs, both of which reduce the chance of sperm displacement by a second male. Among mammals, copulatory plugs occur in rodents (Dewsbury 1988), bats (Fenton 1984), and some primates (Strier 1992). Although the plugs in guinea pigs (*Cavia porcellus*) appear to block subsequent inseminations (Martan and Shepherd 1976), those in deermice (*Peromyscus maniculatus*) are ineffective (Dewsbury 1988). In the latter species the plugs probably function to retain the sperm within the female's reproductive tract. Dewsbury (1984) found that in the muroid rodents he tested, the last male to mate or the male ejaculating most often sired most of the offspring.

FIGURE 17.13   Male dung flies fighting over female.
The attacking male (on the left) is attempting to push the paired
male (on the right) away from the female (only her wings are visible). Sometimes the attacking male succeeds in taking over the
female and mating with her; she then continues to oviposit eggs fertilized by the second attacking male.
Source: Courtesy G. A. Parker.

FIGURE 17.14   Sperm competition in sheep.
Dominant rams copulate at high frequency immediately after a
"sneak" copulation by a subordinate ram.
Source: Irene Vandermolen/Visuals Unlimited.

In fruit flies (*Drosophila melanogaster*), first males may
also transfer to females antiaphrodisiac substances that
inhibit courtship by other males (Jallon et al. 1981). Female
spiders store sperm for long periods and often mate with several males. In laboratory studies of the bowl and doily spider
(*Frontinella pyramitela*), Austad (1982) found that the first
male's sperm had priority in fertilizing eggs over sperm of
subsequent males, the usual case in spiders. Recall from our
discussion of cryptic female choice that female anatomy may
"set the stage" for sperm competition though the evolution of
complex sperm transport and storage organs.

Among insects, the advantage usually accrues to the
sperm of the second (or subsequent) male, as in the yellow
dung flies studied by Parker (1970b). The last male to mate
sired 80 percent of the offspring subsequently produced. The
ultimate second-male adaptation may be the penis of the
damselfly (*Calopteryx maculata*) which has a dual function:
it removes sperm deposited in the female by a previous male
via a special "sperm scoop" and then replaces it with its own
(Waage 1979).

In mammals, adaptations of second males are probably
restricted to dilution of the first male's sperm by frequent
ejaculation of large amounts of sperm from a second male.
Female Rocky Mountain bighorn sheep (*Ovis canadensis*)
usually mate with more than one male during estrus. Dominant rams guard estrous females from forced copulations by
subordinate males but are not always successful. If a subordinate male does achieve a copulation, the dominant male
immediately copulates with that female himself (figure
17.14), probably reducing the chances that the subordinate
male's sperm will fertilize the egg (Hogg 1988).

Sperm competition has been suggested as the cause of
peculiar behavior in the dunnock (*Prunella modularis*), a
small, European sparrowlike bird. When a male's mate is

likely to have mated with another male, the first male repeatedly pecks at the cloaca of his mate until she everts it, sometimes ejecting a sperm bundle. He then reinseminates her
(Davies 1983, 1992) (figure 17.15).

Some species of moths and butterflies (Lepidoptera)
produce two kinds of sperm. One kind is the usual type
(eupyrene) that fertilizes the eggs. The second type (apyrene)
contains no nuclear material but may comprise more than
half the sperm complement. Why should males waste energy
on these dud sperm? Silberglied et al. (1984) suggested that
apyrene sperm are the result of sperm competition, possibly
displacing eupyrene sperm from first males or delaying
remating by the female.

A somewhat similar suggestion has been made to
explain the relatively large number of deformed sperm in
mammals, up to 40 percent in humans, for example (Baker
and Bellis 1988). Baker and Bellis argue that these deformed
sperm play a "kamikaze" role, staying behind to form a plug
to inhibit passage of sperm from a second male. Meanwhile,
a small number of "egg-getter" sperm proceed to the oviduct
to attempt fertilization. The kamikaze sperm hypothesis has
been criticized on several grounds: selection should favor
use of seminal fluids rather than sperm to form the plug, and
there should be more deformed sperm in ejaculates from
species where the female is likely to mate with several males.
However, no such relationship seems to exist (Harcourt
1989). Clearly, more data are needed to adequately test this
intriguing hypothesis.

In primates, the amount and quality of sperm that the
male produces are related to the type of mating system. In
gorillas (*Gorilla gorilla*) and orangutans (*Pongo pygmaeus*),
the winners of male-male competition have relatively free
access to females. In chimpanzees (*Pan troglodytes*), however, several males may attempt to mate with a female in
estrous. Møller (1988) argued that in this case competition

a.

b.

FIGURE 17.15   Cloaca-pecking in dunnocks.
(a) Prior to copulating with a female, the male repeatedly pecks her cloaca, sometimes causing her to eject sperm from a previous copulation. (b) During the very brief copulation, the male appears to jump over the female, and cloacal contact lasts for a fraction of a second.

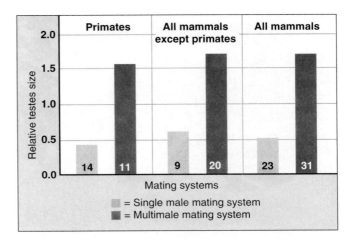

FIGURE 17.16   Mean relative testes size of mammals in relation to mating system.
Sample sizes are at the bottom of each column. Species in which more than one male mates with a female have relatively large testes.
Source: Data from G. J.Kenagy and S. C. Trombulak, "Size and Function of Mammalian Testes in Relation to Body Size" in *Journal of Mammalogy*, 67:1–22, 1986.

takes place in the female's fallopian tubes and that the male with the most and best sperm fertilizes the egg. In fact, chimps have larger testes than other apes and produce a high-quality ejaculate—greater numbers of sperm and more motile sperm.

Among woolly spider monkeys (*Brachyteles arachnoides*), males sometimes form aggregations, taking turns copulating with an estrous female at intervals of about 20 minutes (Milton 1985; Strier 1992). Copious amounts of ejaculate are produced, and a plug forms in the female's vagina. This plug is rather easily removed by subsequent males, so it may not function as a deterrent. As might be predicted from the previous discussion, these monkeys have extraordinarily large testes. Rather than competing aggressively for access to females, these males appear to engage in sperm competition (Milton 1985).

Among mammals in general, a similar relationship exists: species in which only one male has access to one or more females have smaller testes than do species in which more than one male has access to females (Kenagy and Trombulak 1986) (figure 17.16). Some notable exceptions do occur, however. The supposedly monogamous western grasshopper mouse, *Onychomys torridus,* has extremely large testes (nearly five times the predicted size), in spite of

the presumed lack of sperm competition. Lions, on the other hand, have rather small testes, in spite of the observation that one male may copulate more than 50 times in 24 hours with the same female, and a female often mates with more than one pride male (Kenagy and Trombulak 1986).

Brown and colleagues (1995) have suggested that factors other than sperm competition among males may explain the differences in relative testes mass among primates. They argue that the size and length of the female's vagina may determine how much sperm are needed to effect fertilization, a larger or longer vagina requires more ejaculate. They note that the vagina of a female gorilla is approximately 10 cm long, while that of a chimpanzee is about 17 cm. The latter is made even longer by the genital swelling during estrus. The factors that affect vagina size and shape are yet to be explored, but could possibly be a form of cryptic female choice.

Following conception, male-male competition may take a different form. In some species of mice the **Bruce effect** operates early in pregnancy: a strange male (or his odor) causes the female to abort and become receptive (Bruce 1966). Among langur monkeys (*Semnopithecus entellus*) strange males may take over a group, driving out the resident male (see figure 16.14). The new male may then kill the young sired by the previous male (Hrdy 1977a,b). Females who have lost their young soon become sexually receptive, and the new resident male can inseminate them. Infanticide by adult males thus may be viewed as a second-male adaptation. Similar findings have been made for lions—coalitions of males kill cubs on taking over a pride (Packer 1986).

SUMMARY

Asexual reproduction, by which an organism produces exact copies of itself in the next generation, might seem the most likely outcome of natural selection since it maximizes passage of one's genes into the next generation. Most organisms, however, reproduce sexually at some point in their life cycles, incurring the costs of meiosis and of finding a mate. The benefits of being able to mask harmful recessive genes and of increased variability in an ever-changing and unpredictable environment apparently outweigh the costs. Many invertebrates have both sexual and asexual reproduction. Some animals have functional male and female sex organs; and some fish and invertebrates change sexes during adult life.

Females produce large, sessile, and energetically expensive eggs, while males produce small, motile, and energetically cheap sperm. This sets the stage for the reproductive success of males to be limited by access to females, while that of females is limited by lack of access to resources.

Natural selection tends to maintain a 1:1 ratio of the two sexes. Individuals of some species are hermaphoditic, while others can change sex. In both vertebrates and invertebrates, females may adaptively adjust the sex ratio of their offspring in order to produce the sex that will have the highest reproductive success.

Sexual selection affects anatomy and behavior at the time of mating. It may lead to the evolution of elaborate secondary sexual characteristics, particularly in males. Intersexual selection involves choices made between males and females, with the females usually choosing the males. The evolution of elaborate traits may come about either by (1) a process of runaway selection, in which some arbitrary trait evolves in males because females prefer it and so produce sons with the trait and daughters that prefer the trait. The trait thus continues to evolve until countered by natural selection, or (2) a linkage of the trait with the health or genetic makeup of the male, so that females selecting males with the trait will obtain more fit mates. In intrasexual selection, competition takes place between members of one sex (usually the male), with the winner gaining access to the opposite sex. The traits evolve as aids to competition, such as for fighting. Intrasexual competition may take place before copulation, or after copulation in the form of sperm competition, blocked implantation, or infanticide.

DISCUSSION QUESTIONS

1. Describe several different hypotheses for why sexual reproduction evolved. How do you think some groups of rotifers have managed to exist for millions of years without sex?

2. When we see differences between the sexes, we usually assume that they have come about by sexual selection. Can you think of selection pressures other than obtaining mates that might produce such differences?

3. In many species, females mate with more than one male, setting the stage for cryptic female choice or sperm competition, as discussed in this chapter. Can you think of observations or experiments that would distinguish between these two possibilities? For a further discussion on criteria for demonstrating cryptic female choice, see the following references: Birkhead 1998, and Eberhard 2000.

4. Using the figure shown here, describe the general relationship between testes weight and body weight for different primate genera, and then explain why some genera are above the line and some below it. What can you surmise about human mating systems from this graph?

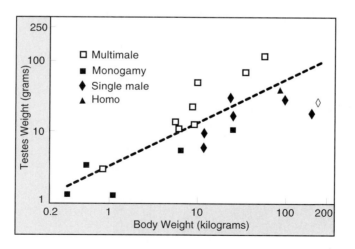

Paired testes weight (g) versus body weight (kg) for different primate genera.

Source: Data from A.H. Harcourt et al., "Testis Weight, Body Weight and Breeding System in Primates" in *Nature,* 293:55–57, 1981.

SUGGESTED READINGS

Andersson, M. 1994. *Sexual Selection.* Princeton: Princeton University Press.

Choe, J. C., and B. J. Crespi, eds. 1997. *The Evolution of Mating Systems in Insects and Arachnids.* Cambridge, U.K: Cambridge University Press.

Michod, R. E., and B. R. Levin. 1988. *The Evolution of Sex.* Oxford, England: Blackwell Scientific Publications, Ltd.

Trivers, R. L. 1972. Parental investment and sexual selection. Pp. 136–79 in *Sexual Selection and the Descent of Man, 1871–1971* (B. Campbell, ed.). Chicago: Aldine.

n chapter 17, we explored the evolution and maintenance of sex and the evolution of sexually selected traits. We saw that the sexes often differ greatly in how much investment they make in each offspring and that these differences greatly influence sexual selection. In this chapter, we first discuss the different types of mating systems seen in animals and then relate them to types of parental care.

MATING SYSTEMS

As we saw in chapter 17, anisogamy prevails in nearly all animals, with females investing more into each egg than males invest in each sperm. According to Trivers (1972), this difference sets the stage for male-male competition for access to females, and for attempts by males to mate with more than one female, a condition referred to as **polygyny.** Polygyny results in greater variation in the reproductive success of males than of females. For each male that fertilizes the eggs from a second female, another male is likely to fertilize none, as we saw earlier for lions (see figure 17.3). We have also seen that sexual selection tends to act more strongly on males than on females. However, not all species are polygynous.

In trying to evaluate the adaptive significance of differences in mating systems, we must look at ecological factors as well as historical ones. For example, group size may be related to predator pressure and food distribution. In the open plains, where large predators are present and food is widely distributed, omnivorous primates and grazing mammals such as ungulates live in large groups, in which mating with several members of the opposite sex is likely for both males and females. In densely forested areas, where communication over long distances is difficult, small family units and

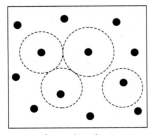

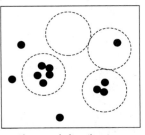

Uniform distribution.
Little polygamy potential

Clumped distribution.
High polygamy potential

FIGURE 18.1   The influence of the spatial distribution of resources (food, nest sites) or mates on the ability of individuals to monopolize those resources.
Dots are resources and circles are defended areas. Uniform distribution of resources on the left offers little opportunity for monopolization. Monogamy is the likely mating system here.

monogamy prevail. Figure 18.1 illustrates how the spatial distribution of resources (food, nest sites, predators, or mates) might influence the type of mating system.

Most classifications of mating systems are based on the extent to which males and females associate (bond) during mating. **Monogamy** refers to association between one male and one female at a time. **Polygyny** refers to association between one male and two or more females at a time. **Polyandry** refers to association between one female and two or more males at a time. **Promiscuity** refers to the absence of any prolonged association and to multiple mating by at least one sex. The term **polygamy** is more general, and incorporates all multiple-mating and nonmonogamous-mating systems. One problem with classification systems based on bonding is that a judgment is needed to determine what constitutes an association. Does it include parental care? What about species, such as many nonhuman primates, that live in year-round social groups? A prolonged association exists, but during breeding females mate with several males, and males may mate with many females. Is this polygyny, polyandry, or promiscuity?

Emlen and Oring (1977) developed an ecological classification of mating systems that may eliminate the need to make these judgments because it is based on the ability of one sex to monopolize or accumulate mates, and it emphasizes the ecological and behavioral potential for monopolization. Although developed primarily for birds, the classification of mating systems seems generally applicable to most vertebrate and insect taxa.

- **Monogamy.** Neither sex is able to monopolize more than one member of the opposite sex.
- **Polygyny.** Males control access to more than one female.

    Resource-defense polygyny. Males control access to females indirectly by monopolizing critical resources.

    Female-defense polygyny. Males control access to females directly, usually because females are grouped for other reasons.

    Male-dominance polygyny. Mates or resources are not monopolizable; females select mates from aggregations of

FIGURE 18.2   Monogamous pair of California gulls with young.
There is little sexual dimorphism in these male and female gulls (*Larus californicus*) and both sexes care for the young.
Source: Courtesy Bruce Pugesek.

males, as in leks, based on the quality of the male's display or his territory.

Scramble polygyny. Males actively search for mates without overt competition.

- **Polyandry.** Females control access to more than one male.

    Resource-defense polyandry. Females control access to males indirectly by monopolizing critical resources.

    Female-access polyandry. Females do not defend resources essential to males, but they interact among themselves to limit access to males.

Monogamy

In monogamous systems, neither sex is able to monopolize more than one member of the opposite sex. When the habitat contains scattered renewable resources or scarce nest sites, monogamy is the most likely strategy. If there is no opportunity to monopolize mates, an individual will benefit from remaining with its initial mate and helping to raise the offspring. The formation of long-term pair bonds also seems advantageous because less time need be spent finding a mate during each reproductive cycle. Long-lived birds such as sea gulls that breed with former mates have higher reproductive success, probably because of less aggression between mates and greater synchronization of sexual behaviors (Coulson 1966). About 90 percent of all bird species are more or less monogamous (figure 18.2). Many species are opportunistically polygamous. Another factor promoting monogamy could be predation risk. Some species live in small social units and behave secretively in order to reduce the chances of being eaten.

Polygyny

In polygynous systems, individual males have access to more than one female. In **resource-defense polygyny,** males defend areas containing the feeding or nesting sites critical for reproduction, and a female's choice of a mate is influ-

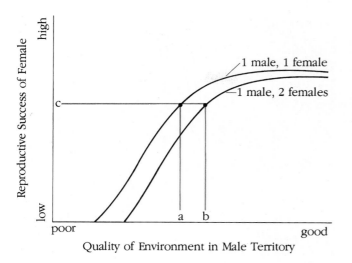

FIGURE 18.3   Reproductive success of female as a function of quality of male territory.
One curve is for a monogamous pair of birds and another for a female who has to share the territory with another female. As territorial quality improves, so does the female's reproductive success. If territories vary enough from one male to the next, a female may have greater reproductive success by joining an already mated male with a good territory, a process that leads to polygyny. Here, a female mating with a male that is defending a territory of quality (a) would have the same reproductive success (c) as a female joining an already mated male in a superior territory of quality (b). For a test of this and several other models of polygyny in redwings, see Lenington (1980).

Source: Data from G.H. Orians, "On the Evolution of Mating Systems in Birds and Mammals," *American Naturalist,* 103:589–603, 1969.

enced by the quality of the male and of his territory. For example, male walnut flies defend sites that are used by females to lay eggs (Papaj 1994). Territories that vary sufficiently in quality may attain the **polygyny threshold,** the point at which a female may do better to join an already mated male possessing a good territory than an unmated male with a poor territory (Orians 1969). Thus, some males may get two or more mates while others get none (figure 18.3). Typically, males do not provide parental care and usually delay breeding until later in life, when their ability to obtain a good territory has increased.

Many species are probably facultatively (i.e., optionally) polygynous. In habitats where feeding or nesting resources cannot be monopolized by males, monogamy is likely, while habitats with defensible clumped resources would favor polygyny. Thus, we may expect to find variation in the mating system of a single species.

Female-defense polygyny may occur when females are gregarious for reasons unrelated to reproduction. Some males monopolize females and exclude other males from their harems. In many species of seals, the females haul out on land to give birth, and they mate soon after. The females are gregarious because there are a limited number of suitable sites, and the males monopolize the females for breeding. Intense competition among males results in marked sexual dimorphism and a large variance in male reproductive success.

FIGURE 18.4   Two male sage grouse displaying on lek.
Males occupy small territories in one of the mating centers of the lek, as females congregate there. Males with more peripheral territories of the lek, away from the mating center, seldom have the opportunity to mate.

Source: Photo courtesy R. Haven Wiley.

If males are not involved in parental care and have little opportunity to control resources or mates, **male dominance polygyny** may occur. If female movements or their areas of concentration are predictable, the males may concentrate in these areas and pool their advertising and courtship signals. Females then select a mate from the group of males. These areas where males congregate and defend small territories in order to attract and court females are called **leks** (figure 18.4). Lek, or arena, systems are characterized by promiscuous, communal mating; the males are likely to have evolved elaborate ornaments, as exemplified by the sage grouse (*Centrocerus urophasianus*) studied by Wiley (1973) and Gibson and Bradbury (1985). The same area is typically used year after year. Females select a mate, copulate, then leave the area and rear their young on their own. Females therefore get no direct benefits from males in lekking species.

Most studies show that mating success on leks is highly skewed in favor of a small number of males. Male mating success may depend on the dominance of the male, female preferences for certain territories, or phenotypic traits of the males. Older, more dominant males may occupy the preferred territories and/or have the most attractive displays, and thus do most of the copulating, but often the reasons for the skew are not clear. One factor that contributes to reproductive skew in some species is the tendency for females to copy other females in making choices of which male to breed with. Copying has been demonstrated experimentally in guppies by Dugatkin (1992).

Male hammer-headed bats (*Hypsignathus monstrosus*) from central and west Africa display at traditional sites along riverbanks (Bradbury 1977). Each territory is about 10 m apart, and the males emit a loud clanking noise to attract females. Once chosen, a male copulates with the female and resumes calling immediately. Some males, for whatever reason, have much higher success than others at attracting females, and in one year 6 percent of the males achieved 79 percent of the copulations. As might be expected, this species shows extreme sexual dimorphism, with males nearly twice the size of females. Each male has a huge muzzle that ends in flaring lip flaps and a large larynx associated with the clanking sound they make.

Leks also are seen in several species of ungulates, such as Uganda kob (*Kobus kob*), (Buechner and Roth 1974) and topi (*Damaliscus korrigum*), (Gosling and Petrie 1990). In topi, the largest and reproductively most successful males defend single territories, while the smaller males cluster in leks. In Uganda kob, a dual male strategy also exists, except that males on single territories are less successful than those in leks (Balmford 1991). In another study of kob, there was no relationship between lek size and mating success (Deutsch 1994). The reproductive success of lekking male fallow deer (*Dama dama*) varies widely, with a few males having spectacular success while most males do no breeding at all (Appolonio et al. 1992). Competitively inferior males follow a low-risk strategy and defend single territories; there they can count on getting a few copulations but spend less time and energy fighting and displaying.

Lek behavior occurs in a wide variety of insects, and is of two general types (Shelly and Whittier 1997). In substrate-based mating aggregations, males spend most of their time perched at particular stations that they aggressively defend against other males. For instance, male botflies of the genus *Cuterebra* may collide violently in midair chases and incur permanent damage to their wings (Alcock and Schaefer 1983). Mating may occur on the substrate or in the air. In many species, the sites are at the tops of hills or ridges, and the behavior is referred to as hilltopping. Other insects show aerial aggregations or swarms. In these, the males usually do not defend specific sites or engage in aggression with other males. Rather, the males move chaotically in the swarm, searching for incoming females. Although male courtship is often present in substrate-based systems, it is lacking in aerial-based systems (Shelly and Whittier 1997).

In the absence of territory or dominance, in some species **scramble polygyny** takes place as the males try to mate. Female wood frogs (*Rana sylvatica*) congregate in small temporary ponds, often during a single night in early spring. Large numbers of males rush about attempting to mate with fecund females (Berven 1981), sometimes dislodging smaller, already-mating males (Howard 1988). A somewhat similar situation occurs in horseshoe crabs (*Limulus polyphenus*). Females come ashore to lay eggs on certain beaches on just a few nights of the year. Males gather offshore and intercept females as they come in; once a male attaches to a female with his special claws, he is difficult to dislodge by other males, and rides on the back of the female up on the beach, fertilizing eggs as they are laid (Brockman 1990). Under almost opposite conditions, where females are widely dispersed, the same thing may happen. For instance, male thirteen-lined ground squirrels (*Spermophilus tridecemlineatus*) actively search out estrous females, showing little evidence of dominance or territoriality (Schwagmeyer 1988).

Polyandry

In polyandrous systems, females control access to more than one male. Because female investment in eggs exceeds

FIGURE 18.5   Male red-necked phalarope with eggs.
In this polyandrous species, the male cares for the eggs and young while the female seeks additional mates.
Source: © Eric Hosking/Photo Researchers, Inc.

that of males in sperm, polyandry is rare. In most cases, females provide parental care while males seek new mates. If food availability at the time of breeding is highly variable, or if breeding success is very low due to high predation on the young or the eggs, females may have to produce many offspring. In birds, male incubation is common; a few cases of polyandry in which males do all the incubating and females lay multiple clutches have been documented (figure 18.5).

Breeding sites of the American jacana (*Jacana spinosa*), a large wading bird found in Central and South America, are limited and are divided into small territories by males (Jenni 1974). Female jacanas control superterritories that may encompass the nesting areas of several males. Frequently, several males incubate the clutches of one female, and she provides replacement clutches for them if nests are lost through predation, as often happens. Breeding females are 50 percent larger than the males, who they dominate; and these females provide little parental care. In this reversal of polygyny, the females specialize only in egg production.

Females of the migratory spotted sandpiper (*Actitis macularia*) compete for control of breeding territories, and males provide most of the offspring care. (Oring and Lank 1982). Females arrive on the breeding grounds before the males and are **philopatric** (they return to the place where they were born). However, the vast majority of birds are either monogamous or polygynous, and the reverse is generally true: males arrive first to establish territories and are more likely than females to return to the natal site to breed (Greenwood 1980).

Females in many species of mammals mate with more than one male, and sometimes littermates have more than one father, for example, in deermice (Birdsall and Nash 1973). Such systems, however, are usually referred to as promiscuous rather than polyandrous, because males also mate with multiple females, males provide little or no offspring care, and there is no lasting bond between the partners. Several species of larger canids show most of the features of true polyandry. Although mainly monogamous, the African wild dog (*Lycaon pictus*) sometimes exhibits polyandry; females are occasionally mated by several males, males provide extensive care of pups, and females are the dispersing sex (Moehlman 1986).

Ecology and Mating Systems

A classic example of the way in which mating systems are related to resource distribution is provided by Orians' comparative study of blackbird social systems (Orians 1961). The red-winged blackbird (*Agelaius phoeniceus*) is usually polygynous; a male defends a territory containing two or three females. The male arrives three to four weeks before the females (frequently at the same site as the previous year), mates, and then expends a considerable amount of energy defending his territory until the young are fledged. Males rarely help to raise the young; females do all of the nest building and incubating and nearly all of the feeding of the young, as is typical of polygynous species.

In parts of northern California, a very closely related species, the tricolored blackbird (*Agelaius tricolor*), is **sympatric** (coexists) with the redwing (Orians 1961); its mating system, however, is quite different from that of the redwing. Male and female tricolored blackbirds pair off in a nomadic colony of anywhere from 100 to 200,000 birds. They establish territories, find mates, build nests, and lay eggs—all within one week. Activities are highly synchronous within each colony. As we can see from figure 18.6, both sexes of the tricolored blackbird make a large investment—but in a shorter time frame than that of the redwings.

Why do two very closely related species do things so differently in the same place and at the same time? The answer seems to be related to their energy source. Redwings have a relatively stable diet of seeds and insects, which are available for several months. A male redwing defends the same territory several years in a row, and each territory contains most of the food needed to support the females and the young. In contrast, tricolors go out from the colony on mass feeding flights, possibly to assess concentrated food sources; and they attack rice and other grain fields at the time of seed maturity. Thus, the tricolor mating system seems designed to take advantage of an ephemeral, but rich and concentrated, food source. When tricolors locate such an energy supply, they move in, establish a new colony, and reproduce before the source is gone. A similar strategy is used by African weaver finches (*Quelea* spp.), which move all over the continent, attack rice and wheat fields, breed quickly, and then move to another ripening area

---

Playing Lizard Games

The realization came over both of them simultaneously and Barry Sinervo doesn't remember whether he or his collaborator, Curt Lively, said it first: "They're playing the rock-paper-scissors game!" This game is usually played by a pair of children, who simultaneously put out their hands in a symbol for a rock, paper, or scissors. No single strategy is invincible: rock breaks scissors, scissors cut paper, and paper wraps around rock. The players of this round, however, were sided-blotched lizards, *Uta stansburiana*.

These lizards appear in three color morphs, each with their own behavioral strategies. Orange-throated males are aggressive, mate with many females, and have large territories that they can readily defend from either of the other morphs. Blue-throated males are milder in temperament, more monogamous, defend smaller territories, and lose in conflicts with orange males. However, blue males win against males with yellow-striped throats. Yellow males are the least aggressive, and do not hold territories. They resemble females in both color and behavior, and sneak about, attempting to mate with females on the sly. They are most successful in fooling orange males; blue males are more likely to evict yellow males. Thus, the dynamics are those of the rock-paper-scissors game: orange beats blue, blue beats yellow, yellow beats orange.

How can we understand this "game"? Although this textbook describes different proximate and ultimate factors that influence behavior, in reality many may come into play simultaneously. Good study systems allow the opportunity to make sense of complexity, and Barry recognized that this species had a lot going for it. On rocky outcrops in the coast range of California, he can see 300 lizards in a 5-minute walk. Like many other lizards, *Uta* communicates by visual displays that are obvious to humans, such as doing push-ups. Their territorial nature means researchers can find the same individuals repeatedly. Movement among outcrops is limited, so the ratios of different morphs can be manipulated. Their genetics can be worked out because *Uta* can be reared easily in the lab. Surgical manipulation, done out on the rocks, provides a simple way to manipulate hormone levels: the lizards quickly recover from the anesthetic and resume fighting and mating.

Barry has found that these morphs have a genetic basis: color segregates in the pattern of a Mendelian single-gene locus with three alleles. Male hybrids between the morphs have physical characteristics of both and are intermediate in behavior. In the lab, day-old hatchlings behave according to type. Hormones interact physiologically with both gene expression and behavior; for example, feisty orange males have more testosterone, and testosterone affects throat color, which in turn elicits different behavior from conspecifics. Testosterone also shortens life. These proximate mechanisms interact to produce the phenotype upon which natural selection acts. Because of the nature of the rock-paper-scissors game, the population cycles predictable through about 5 years from a high frequency of blue morphs, which is replaced by a high frequency of orange morphs, then by a high frequency of yellow. Thus we can see frequency-dependent selection in action.

Barry's advice to budding behaviorists is to be persistent, put in time in the field, and take techniques from the lab back to the field whenever possible. He can follow his own advice easily now, as he's a professor at the University of California at Santa Cruz, the university nearest the study site he discovered in graduate school.

Barry Sinervo will take you on a manic tour of Lizards Land at this website, http://bionet.ucsc.edu/people/barrylab/public_html/index.html.

FIGURE 18.6 Time expenditure for pair of red-winged blackbirds and tricolored blackbirds during breeding season. The reproductive effort is spread out over more than 4 months in the redwing (a), but it is completed in about half that time by the tricolor (b). Tricolored blackbirds engage in mass feeding flights to concentrated food sources away from the breeding colony.

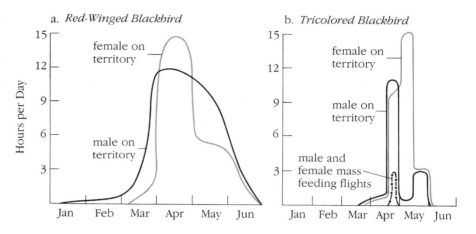

a. *Red-Winged Blackbird*

b. *Tricolored Blackbird*

(Ward 1971). Because of their nomadism, these serious economic pests have been very difficult to control.

Alternative Reproductive Tactics

Thus far we have emphasized differences between species and between the sexes, showing that males and females often have different interests in terms of how much to invest in individual offspring. But behaviorists have long known of instances where members of the same sex use different tactics. Most often observed are species in which some males fight for territories or social status, while other males sneak copulations by being inconspicuous or by mimicking females. An alternative tactic seen in bullfrogs (*Rana catesbeiana*) is the nonaggressive parasitic, or satellite, male that lurks at the edge of a dominant male's territory and attempts to copulate with females as they respond to the calls of the territorial male (Howard 1978) (figure 18.7). These different tactics are related to age in some species, with younger, smaller males being lower ranking or nonterritorial, and thus having to sneak copulations. But in other species, such as bullfrogs, age is not a factor. It was long thought that such subordinate males were "making the best of a bad job," but more recent work suggests that these opposing tactics are often conditional strategies, meaning that the best tactic for an individual depends on its status (Gross 1996).

Recall the example of the coho salmon that led off chapter 17. Breeding males are of two types, called hooknose and jack (Gross 1985, 1991). Hooknose males fight to gain access to females that are about to lay eggs. Their bright red color; enlarged, hook-shaped mandibles with large teeth; and protective cartilage deposits on their back are all equipment for fighting. Jacks, at the other extreme, are small and have none of those features, and are thus ill-equipped to fight. One might assume that jacks have low reproductive success. Under certain conditions, however, when refuges are available, the jacks can get close enough to spawning females to fertilize their eggs. Hooknose males typically take a year longer to mature than jacks, accounting for their being larger when they migrate upstream to spawn (figure 18.8). It is not clear what environmental or genetic factors predispose some males to become hooknoses and others jacks, but fish that are larger as larvae tend to become jacks. The path to becoming a

FIGURE 18.7 Territorial male bullfrog with parasitic male. The parasitic male is the smaller of the two and is facing forward. Females are attracted to the call of the territorial male; when a female enters the territory, the parasitic male attempts to mate with her before the territorial male has a chance.
Source: Courtesy Richard D. Howard.

hooknose or a jack is thus irreversible. The decision to sneak or fight is conditional, however, because an individual adjusts its behavioral tactic to its morphology, using the tactic that gives it the highest reproductive success (Gross 1996).

The number of tactics males can use is not limited to two. Inside sponges in the Gulf of California lives a species of marine isopod (*Paracerceis sculpta*), a crustacean relative of the sowbug. Although females are all more or less the same size, males come in three sizes and configurations (figure 18.9). The big alpha males defend harems inside the sponges. The medium-size beta males mimic female behavior and morphology so well that alpha males actually try to mate with them. The tiny gamma males are highly mobile, invading harems and hiding from the alpha males (Shuster 1992). In a 2-year field study, the reproductive success of all three male morphs was about the same (Shuster and Wade 1991). In this species the morphs are under genetic control (Shuster and Sassaman 1997). The alleles that are thought to control the expression of these morphs are in Hardy-Weinberg equilibrium, suggesting that there is no differential selection favoring one male morph over another (Shuster and Wade 1991).

MALE SALMON STRATEGY

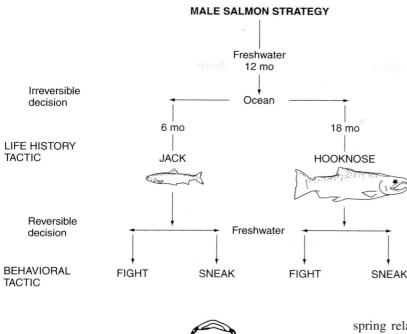

FIGURE 18.8   **Alternative life histories for male coho salmon.**
Males have two irreversible life history tactics, hooknose or jack, and two reversible behavioral tactics, fight or sneak. For hooknose, fight is the predominant tactic, while for jack, sneak predominates.
Source: Data from Gross, M. R. 1991. Salmon breeding behavior and life history evolution in changing environments. *Ecology* 72:1180–86.

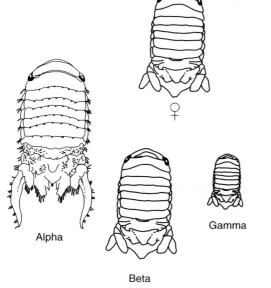

FIGURE 18.9   **Three male morphs in the marine isopod,** *Paracerceis sculpta.*
Females are all about the same size. Note that the beta males resemble females in both size and shape. Although the different morphs use different tactics to obtain copulations, the reproductive success for all three is about the same.
Source: Shuster, S. M. 1992. The reproductive behavior of alpha male, beta male, and gamma male morphs in *Paracerceis sculpta,* a marine isopod crustacean. *Behaviour* 121:231–58.

PARENTAL CARE

The extent of parental care varies tremendously in the animal kingdom. The amount of care given generally increases as the complexity of the organism increases. While many aquatic invertebrates simply shed eggs and sperm into the water, primate care may last for many years, amounting to 25 percent of the offspring's life span. The kinds of parent-off-spring relationships are varied and not simple to describe, however. For instance, birds display a wide range of care techniques—from burying their eggs in rotting vegetation that provides heat for incubation (thus freeing the parents), to sitting on the eggs without feeding until the eggs hatch.

Ecological Factors

The term **reproductive effort** is used to denote both the energy expended and the risk taken for breeding, which reduce reproductive success in the future. Finding mates and caring for young take extra energy, and parents may run a greater risk of predation. Individuals are faced with the decision—conscious or otherwise—of whether to breed now or wait until later. If the choice is to breed now, should some effort be spared for another attempt later?

What environmental factors influence the investment that parents make in their young after birth? Ecologists have attempted to relate environmental conditions to parental care. For example, species adapted to stable environments have a tendency toward larger body size, slower development, longer life span, and having young at intervals (**iteroparity**) rather than all at once (**semelparity**). Typically, individuals of such species will occupy a home range or territory (see chapter 16). Stable conditions favor production of small numbers of young that receive extensive care and thus have a low mortality rate. Such species are said to be **K-selected,** in reference to the fact that populations are usually at or near K, the carrying capacity of the environment. Intraspecific competition is likely to be intense, and the emphasis is on producing high-quality offspring, rather than high quantity.

Species that are adapted to fluctuating environments have high reproductive rates, rapid development, small body size—and need little parental care. Their populations tend to be controlled by physical factors, and their mortality rate is high. Such species are said to be **r-selected,** where r refers to the reproductive rate of the population. Coho salmon, which reproduce far upstream from feeding areas, must expend a great deal of energy before they can breed. Once they incur the great cost of migrating upstream, they breed explosively and die.

Other species, which do not have such a high initial cost before breeding, may defer reproduction or spread it out over time. In environments where survival of offspring is low and unpredictable, parents may "hedge their bets" and put in a small reproductive effort each season. The California gull (*Larus californicus*) uses such tactics; the birds live 15 years or more but rear only one or two chicks per year (Pugesek 1981, 1983). As the parents age, they increase their effort, laying more eggs, feeding the chicks more food, and defending them more vigorously, possibly because the parents' chances of surviving another year become smaller.

Note that the predictions of "bet-hedging" contradict those of r and K selection. In unstable, unpredictable environments, r selection, which favors high reproductive rates, should be important. However, bet-hedging theory suggests low reproductive rates and spreading reproductive effort across many breeding seasons. More data are needed from natural populations before we can resolve this apparent contradiction.

Prolonged dependency and extensive parental care are also favored when a species—for example, large mammalian carnivores such as the felids and canids—depends on food that is scarce and difficult to obtain. Much effort is spent searching for prey, and in some species, cooperation is needed for the kill. During the prolonged developmental period, the young benefit from a considerable amount of learning through observation of parents and through play.

The Old World monkeys and the great apes have the longest periods of dependency. Typical of these species is an infancy of 18–40 months and a juvenile phase of 6–7 years, making up nearly one-third of the total life span. The reason for this prolonged dependency may be related to their flexible behavior, which is shaped largely by learning. The complexity of monkey and ape social systems and the importance of kinship depend on a knowledge of individuals and an extensive behavioral repertoire.

Which Sex Should Invest?

Because of anisogamy, an egg requires a greater investment of energy than does a sperm; therefore, male and female strategies are expected to differ not just with regard to obtaining mates, but also to how much to invest in the care of offspring. Since eggs are likely to be limited in number, we expect that males will compete for the opportunity to fertilize them and thus will be subject to sexual selection. A female is likely to mate, but given her already large investment, her ability to invest further may be limited; thus, she will be particular about which male fertilizes her precious eggs. Males, on the other hand, will try to inseminate as many females as possible, and be unlikely to provide care of offspring (see figure 17.6) (Trivers 1972).

However, just because only one sperm fertilizes an egg, millions are generally required in each ejaculation to ensure fertilization—even of a single egg (Dewsbury 1982). Also, there is a limit on the number of times most males can ejaculate within a certain period. Remember also the evidence that sperm competition in males of some species leads to selection for increased sperm production. Therefore, a male's investment in sperm is not necessarily trivial, and he too can be expected to be somewhat choosy about his mate. In some species of insects, the male contributes nutritive substances in addition to sperm and is the choosier sex (Gwynne 1981). Nevertheless, it is generally assumed that in most species, the female's investment in gametes is greater than the male's.

Other factors besides anisogamy seem to affect the contribution of each sex to parental care. Certain taxonomic groups of animals seem to be predisposed to a particular pattern. Thus, birds tend to be monogamous, and both parents care for young. Mammals are generally polygynous, and the male contributes little to raising the offspring. Most species of fish do not care for their young at all; however, among those species that do, it is usually only the male.

What factors could be responsible for these taxonomic differences? Trivers reasoned that confidence of parentage might explain the results (Trivers 1972). In species with internal fertilization and where sperm competition could take place, the male might be inclined to desert and seek additional mates. The female is confident of her genetic relationship to the offspring, so she invests further in it. Another possibility, suggested by Williams, is that parental care evolved in the sex that is most closely associated with the embryos (Williams 1975). Where eggs are fertilized internally, the evolution of embryo retention and live birth might be more likely, followed by further care of the young by the mother. Where eggs are laid and fertilized externally in a male's territory, the male becomes most closely associated

TABLE 18.1 Parental care and fertilization mode in teleost fishes

The table shows number of families; a single family may appear in more than one category but is not listed under 'no parental care' unless care is completely unknown in the family.

	Number of Families When Fertilization Mode	
Parental Care by	Internal	External
Male	2	61
Female	14	24
Neither	5	100

Source: Data from Gross and Shine.

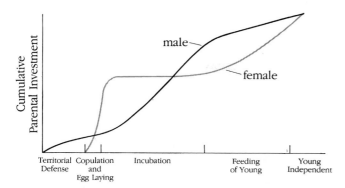

FIGURE 18.10 **Parental investment in offspring by male and female.** In this hypothetical example we find that the male establishes and defends a territory, so his cost is initially higher; but the female invests heavily in eggs, and her cost soon exceeds that of the male. The male then incubates the eggs, and his cumulative investment exceeds hers until termination of parental care. Differences in investment may affect such behaviors as mate desertion.

Adapted with permission from R. L. Trivers, "Parental Investment and *Sexual Selection*," in Sexual Selection and the Descent of Man, edited by Bernard Campbell. Copyright © 1972 by Aldine de Gruyter, Hawthorne, NY.

with them and additional paternal care is likely. Table 18.1 illustrates that internal fertilization favors female care, and external fertilization, male care, as both these theories predict (Gross and Shine 1981).

Still other factors may predispose a taxon to a particular pattern. In birds, males can incubate eggs and feed young—in the case of pigeons, they can even provide crop milk. Thus monogamy prevails in birds. In mammals, however, gestation and milk production are restricted to the female, and there is relatively little the male can do to provide direct care for the young. In mammals whose young are relatively advanced at birth (precocial), the opportunities for male investment are even lower, and males compete for multiple mates more than they do in species whose young are immature at birth (altricial), and whose males and females can more equally invest (Zeveloff and Boyce 1980). Thus, polygyny prevails in mammals.

Among monogamous species, the investment by each parent is not equal at all times (figure 18.10). It may be advantageous for one partner to desert the other and find a new mate if the remaining partner's investment is greater and the offspring are likely to survive anyway. Because of its large investment, the remaining partner has the burden of caring for the young (Trivers 1972).

But should a parent continue to invest just because it has made a previous commitment in terms of time and energy? According to Dawkins and Carlisle (1976), such an individual would be committing the "Concorde fallacy," named in honor of the supersonic transport that was completed even though a profitable return was unlikely, because of a large previous financial investment. It has been argued that organisms should behave so as to increase future reproductive success regardless of prior investments in offspring. However, Trivers' concept of parental investment actually does consider future ability to invest. If one parent has already made a large commitment, that parent may be physically unable to begin another breeding cycle; its reproductive success is thus increased if it remains with its young.

Examples of Parental Investment

Mating of northern elephant seals, *Mirounga angustirostris*, takes place in colonies; males establish dominance hierarchies, and only the high-ranking males breed (Haley et al. 1994; Le Boeuf 1974; Le Boeuf and Reiter 1988). Male-male competition is intense: less than one-third of the males copulate at all, and the top five males do at least 50 percent of the copulating (figures 18.11 and 18.12). The males make no investment in offspring beyond the sperm, as evidenced by the fact that males may trample pups as they strive to inseminate females. The males probably have no way of knowing which young are their own because pups are born a year after copulation.

Although less common than in birds, in some mammal species the sexes share more or less equally in care of young, as in canids such as silver-backed jackals (Moehlman 1983). Both parents defend the territory and hunt cooperatively. The males play a crucial role, because in cases in which the father disappears, the female and her offspring dies (Moehlman 1986). Among tamarins and marmosets (family Callitrichidae), females produce twins. Adult males typically show strong interest in the infants soon after birth and carry them as much or more than the female does (Vogt 1984).

In a few species of vertebrates, the situation is completely reversed, and the male does most of the caring for young. Since the males' investment in offspring becomes larger than the females', we would expect females to compete for and try to attract males, rather than vice versa. Indeed, this reversal of the typical roles occurs in fish such as pipefishes and sea horses; in polyandrous birds such as phalaropes, jacanas, tinamous, and sandpipers;

As with vertebrates, one also finds a complete range of investment patterns among insects, from none to extensive care by the male (figure 18.13). In some orthopteran insects, more than 25 percent of the male's weight may be transferred to the female in a spermatophore (figure 18.14). This protein-rich meal increases the number and size of eggs produced by the female (Gwynne 1984). In cases where the male's investment in the spermatophore exceeds that of the female in eggs, males should become the choosy sex and females should compete for access to them. Gwynne (1981) found that male Mormon crickets (*Anabrus simplex*) mated more often with heavy females that contained more eggs. Females, on the other hand, competed aggressively for access to singing males.

Some species of insects also provide extensive care after the young hatch, such as the social insects we will discuss further in the next chapter. Even the male sometimes helps out, as in some species of burying beetles (genus *Nicrophorus*) (Scott 1990). When these insects encounter a dead animal, such as a mouse, they excavate an underground chamber containing the mouse, and the female lays eggs near it. Upon hatching, the larvae are fed the partially digested carcass by both parents, for which the young "beg." Scott demonstrated that the main reason the male helps, rather than deserting and seeking additional mates, is to guard the crypt and brood from other, infanticidal, burying beetles.

FIGURE 18.11   Sexual dimorphism in elephant seals. Among a herd of females, two males fight to establish dominance. Males differ strikingly from females, are about three times larger, possess an enlarged snout, or proboscis, and have cornified skin around the neck. In this highly polygynous species males invest nothing in their offspring other than DNA from sperm.

Source: Data from M. P. Haley, C. J. Deutsch, and B. J. Le Boeuf, "Size Dominance, and Copulatory Success in Male Northern Elephant Seals, *Mirounga Angustirostris,* in *Animal Behaviour,* 48:1249–1260, 1994

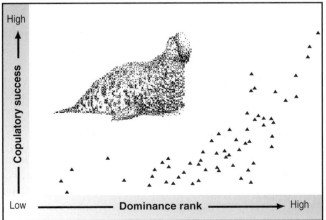

FIGURE 18.12   Dominance and copulatory success in elephant seals.

Source: Data from Haley, M. P., C. J. Deutsch, and B. J. Le Boeuf. 1994. Size, dominance and copulatory success in male northern elephant seals, *Mirounga angustirostris. Anim. Behav. 48:1249–60.*

FIGURE 18.13   Egg brooding by male water bug (*Abedus herberti*). The female lays the eggs directly on the back of the male.

Source: John Cancalosi.

a.

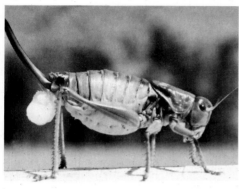

b.

c.

FIGURE 18.14   Reversed sex roles in Mormon crickets.
A female Mormon cricket is shown (a) mounting a male, (b) with attached spermatophore from male, and (c) consuming the spermatophore.

Source: Gwynne, D.T., 1981, Sexual Differences Theory: Mormon crickets show role reversal in mate choice, *Science* Vol. 213, 14 August, pp.779–80, Fig. 1.

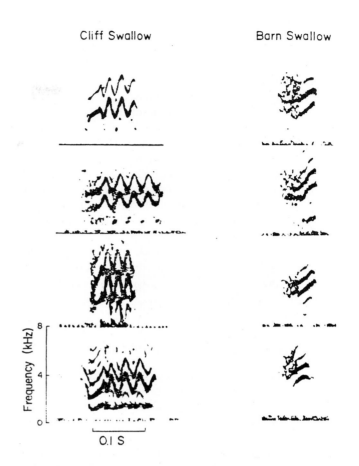

FIGURE 18.15   Calls of young cliff and barn swallows.
Note the greater complexity of the cliff swallow calls. Cliff swallows nest in large colonies, where the chance of a parent's misdirecting parental care is high.

Source: Michael D. Beecher, et al., "Acoustic Adaptations for Parent-Offspring Recognition," *Experimental Biology,* 45:179–93. Copyright © 1986 Springer-Verlag, Heidelberg.

PARENT-OFFSPRING RECOGNITION

One practical problem faced by parents that provide care for their offspring is being able to recognize them. Mechanisms of recognition should evolve when there is a risk of misdirecting parental care toward nonrelatives. Among gulls and terns that nest on the ground in dense colonies, the precocial chicks begin running around several days after hatching. At about this time, the parents learn to recognize their own chicks (Tinbergen 1960). Thus, ring-billed gull (*Larus delawarensis*) parents will accept chicks from other parents until the chicks are about 5 days old; after this time, the acceptance rate declines sharply (Miller and Emlen 1975). Another species of gull, the kittiwake (*Rissa tridactyla*), nests on tiny ledges on steep cliffs. For obvious reasons, chicks don't leave the nest until they are several weeks old and can fly. Mechanisms of recognizing chicks have not evolved in this species: parents will adopt older chicks, even those of other species (Cullen 1957).

The cues used to recognize offspring (or other kin) vary with the taxonomic group (see chapter 12 and 19 for more on kin recognition). Vocalizations are thought to be important in birds, as parents learn to recognize the calls of their own young. Some species of swallows produce distinctive calls, and each young has its own "signature" (Beecher 1982; Beecher et al. 1986; Medvin et al. 1993). Other species do not have such signatures (figure 18.15). The two species' social organization differs: northern rough-winged (*Stelgidopteryx serripennis*) and barn (*Hirundo rustica*) swallows nest by themselves or in small groups, and it is unlikely that a parent would confuse its young with those of other parents. Bank (*Riparia riparia*) and cliff (*Hirundo pyrrhonota*) swallows live in large colonies, where breeding is highly synchronized, and there is a good chance that parents could misdirect parental behavior, as young are all of similar age and often return to the wrong burrow entrance. Information analysis confirmed that the calls of the colonial species contained more information than those of the solitary species (Beecher 1982).

Dolphins produce whistles that are unique to each individual. These sounds develop during the first few months of life, as young mimic the sounds of others and form an individual signature that remains stable for life (Tyack 1997). Signature whistles are used primarily when individuals are out of sight of one another. They seem to function in mother-offspring recognition, but also may be used to recognize other group members (Sayigh et al. 1998). Other group members also may match the sounds of different individuals, suggesting that they are addressing each other individually (Janik 2000).

PARENT-OFFSPRING CONFLICT

Anyone who has raised children or grown up with a sibling has observed frequent disagreement between parent and child. It is not unusual to see a mother rhesus monkey (*Macaca mulatta*) bat her 10-month-old infant or raise her hand over her head, thereby pulling her nipple from its mouth. Frequently, the infant responds by throwing a "temper tantrum." We can view much of the conflict as a disagreement over the amount of time, attention, or energy the mother should give to the offspring (i.e., the infant wants more than the parent wants to give). We often interpret such conflict in humans as being maladaptive and related to psychological problems of parent or child, or to some negative cultural influence.

Another hypothesis, based on the coefficient of relationship (chapter 4) and on parental investment, is that the conflict arises because natural selection operates differently on the two generations (Trivers 1974). From the mother's standpoint, she should invest a certain amount of time and energy in her offspring and then wean the young and invest in new young. When the cost to the mother in fitness exceeds the benefit, she should reject the young. From the offspring's standpoint, however, the offspring will profit from continued care until the cost to its mother is twice the benefit (since the offspring shares half of its mother's genes). When the cost is twice the benefit to the mother, then the offspring's own inclusive fitness will start to decline. At this point, the offspring should leave the mother and allow her to increase the offspring's inclusive fitness by having more offspring (figure 18.16).

Alexander (1974) challenged this idea; he argued that selection will work against behaviors of the offspring that allow them to cheat and to receive more than their share of care from the parent. Such offspring will pass those traits on to their own offspring, who will cheat them in turn and lower their fitness as parents. Thus, parents will win in cases of parent-offspring conflict. Alexander coined the term **parental**

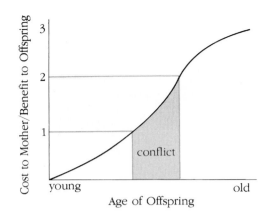

FIGURE 18.16   Parent-offspring conflict: Ratio of cost to mother and benefit to offspring as a function of age of offspring. When the offspring is young, the ratio is less than 1, and both the mother and offspring gain in terms of fitness from the relationship. Above 1, the fitness of the mother begins to decline, and so she should try to cease caring for this offspring and invest in new offspring. Between 1 and 2 is a zone of conflict (shaded area) because the fitness of the offspring is still increasing. Above 2, the fitness of the offspring also declines, and the offspring willingly becomes independent

Source: Data from R.L. Trivers, "Parental Investment and Sexual Selection," in *Sexual Selection and the Descent of Man*, ed. by Bernard Campbell, copyright 1972 by Aldine de Gruyter, Hawthorne, N.Y.

manipulation to refer to cases where a parent selectively provides care to certain offspring at the expense of others. In many species, the finite nature of parental care is apparent, because time spent with one offspring reduces the amount of care that a parent can provide for other offspring (Johnson 1986). Such trade-offs can even result in higher mortality for some offspring (figure 18.17).

SIBLING RIVALRY

Conflict is not always limited to parents and offspring. In species where young share a confined space, such as a nest, and must share limited resources, the conflict may be among the siblings themselves (Mock and Parker 1997). Among birds, siblings may engage in deadly combat, a process studied by Mock and colleagues (Forbes 1994; Mock 1984; Mock and Ploger 1987) in great egrets (*Casmerodius albus*) and cattle egrets (*Bubulcus ibis*). **Siblicide** is when one young kills the other outright or forces another out of the nest (figure 18.18). The parents rarely intervene in such encounters. One cause of it seems to be the size of the food objects brought by the parents. Small fish are monopolizable by one chick and promote aggression and dominance. In contrast to great egrets, great blue herons (*Ardea herodius*) feed their chicks large pieces that cannot be monopolized; siblicide is infrequent in this species (Mock 1984).

A second factor that increases the chances that young will dominate and even kill each other is hatching asynchrony. When incubation is begun after the first egg is laid, chicks hatch at different times. The chick from the first egg laid hatches first and gets a head start. That chick is likely to

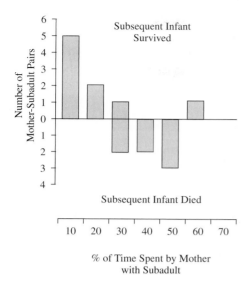

FIGURE 18.17 **The cost of reproduction in wallabies.**
The survival of newborn red-necked wallabies in relation to the frequency of association between mothers and their previous subadult offspring during the 6 months following their weaning.
Source: C. N. Johnson, "Philopatry Reproductive Success of Females and Maternal Investment in Red-necked Wallaby," *Behavioral Ecology and Sociobiology,* 18:143–60. Copyright 1986 Springer-Verlag, Heidelberg, Germany.

FIGURE 18.18 **Cattle egret chicks fighting to the death.**
Due to asynchronous hatching of eggs, the chicks differ in age and size. Here the oldest chick is delivering a blow to a younger sibling. Siblicide is common to this species, and the parents do not intervene.
Source: Doug Mock.

kill younger siblings and get more than its share of food. Clearly, it is not in the chicks' best interests to run the risk of being bludgeoned or thrown out of the nest by their older siblings. Why does this behavior persist? Mock and Ploger (1987) experimented with cattle egrets by moving newly hatched chicks from one nest to another in order to create artificial synchronous and asynchronous broods. They found that chicks in synchronous broods fought more, demanded more

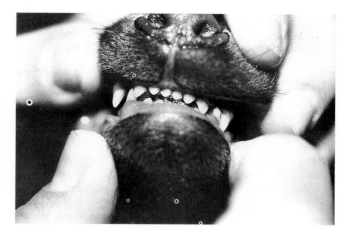

FIGURE 18.19 **Teeth of spotted hyena on day of birth.**
Source: Photo courtesy of Laurence Frank, U. C. Berkeley.

food, and survived less well than did the asynchronous broods. Thus, although hatching asynchrony seems to work against the well-being of all but the oldest chicks, it is in the parents' interest to create inequalities in chick size. In this form of parental manipulation of offspring, brood size adjusts to the optimum number of chicks for the food available.

Competition among siblings can be intense among some species of mammals as well. Pigs (*Sus scrofa*) are born with razor-sharp teeth that they use in fights with siblings to gain access to favored teats for nursing (Fraser 1990; Fraser and Thompson 1991). Among spotted hyenas (*Crocuta crocuta*), litters are usually twins, and pups are also born with fully erupted canines and incisors (figure 18.19). Fighting begins immediately after birth, and if littermates are of the same sex, one is often killed (Frank et al. 1991). Spotted hyenas are unusual in that young are precocial and both sexes have high levels of circulating androgens at birth. In trying to understand why siblings should compete, keep in mind that they share only one quarter to one half their genes, and thus their interests are not identical.

In chapter 16 we noted examples of cannibalism among aquatic larval salamanders and toads. Many of these species breed in temporary ponds and larvae are often consumed by carnivorous siblings. Although a number of studies show that larvae recognize kin and avoid eating siblings, a few show the opposite trend. When larval marbled salamanders of different sizes were paired, the cannibalistic ones ate their siblings preferentially (Walls and Blaustein 1995). One possible explanation for these results is that the smaller larvae had such a small probability of surviving to reproduce that the nutritional benefit to their older siblings would more than make up for any reduction of indirect fitness. But this doesn't explain why they should prefer to eat siblings over nonsiblings.

Among invertebrates, especially marine species, eggs and sperm are shed directly into the water and become pelagic. There is no nursery or other opportunity for sibs to interact, so there is little competition among siblings. Among insects, however, some species have broods with extensive parental care, such as the burying beetles discussed earlier. In these systems there is the potential for competition among siblings. Most parasitoid wasps lay their eggs on or in insect hosts, and in some species the female lays more than one egg. The parasitoid larvae then consume the host from the inside out. Larvae of solitary parasitoids usually posses enlarged mandibles for fighting with other larvae, and only one wasp ultimately emerges from the insect host (Fisher 1961). In chapter 19 we will consider more fully the circumstances in which organisms can be expected to compete or cooperate.

SUMMARY

Anisogamy sets the stage for competition among males for access to females, and for attempts by males to mate with more than one female, a condition referred to as polygyny. Ecological factors, however, such as food distribution and predator pressure, affect the distribution of both males and females, and therefore the type of mating system. Monogamy, in which one male and one female form a pair bond for one or more breeding seasons, occurs when neither sex is able to monopolize more than one member of the opposite sex. In resource-defense polygyny, males defend areas containing feeding or nesting sites critical for reproduction and thus gain access to more than one female. If females are gregarious, as in female-defense polygyny, males may form harems. In male-dominance polygyny, males may concentrate in an area and display to attract females for mating. Occasionally females monopolize males, as in resource-defense polyandry or female-access polyandry, but such cases of polyandry are restricted to relatively few species.

Members of one sex may use more than one tactic to obtain fertilizations. Often some males in the population act as nonaggressive satellites and attempt to steal copulations from the dominant or territorial males. In other species males are of two different sizes, as with salmon, one adapted for fighting, the other for sneaking. Some crustaceans have males of three sizes and shapes, each adapted for a different tactic, and the differences are genetically controlled.

The more complex the organism, the greater the amount of parental care given to the young. Among species adapted to stable environments, body size tends to be larger, their life span longer, and they tend to have young at intervals (iteroparity), rather than all at once (semelparity). These conditions favor production of small numbers of young, which receive extensive care and thus are subject to a reduced mortality rate. An organism's dependence on food that is scarce and difficult to obtain also favors prolonged dependence of the young and extensive parental care. When survival of offspring is low and unpredictable, parents may bet-hedge, producing few offspring each year.

In species with internal fertilization, the male's confidence of paternity may be low, forcing him to seek additional mates rather than risk caring for another male's offspring. Where there is external fertilization, and where the female lays eggs in a male's territory, the male tends to provide the care. In birds, both sexes tend to care for young, while in mammals, females gestate and lactate, and thus provide most of the care.

In species with parental care, mechanisms have evolved to reduce the chances of misdirecting parental investment to unrelated offspring. Such mechanisms are particularly well developed in species with highly mobile offspring and in species that breed in large groups.

Parent-offspring conflict has traditionally been viewed as maladaptive or as a necessary part of the weaning process. More recent ideas point out that natural selection operates differently on the two generations, so it is in the offspring's interest to receive more care than the parent is willing to give. Parents usually win such conflicts and manipulate offspring to maximize their own reproductive success. In species where young are raised in a confined space such as a nest, sibling rivalry may occur in which siblings compete and may even kill each other.

DISCUSSION QUESTIONS

1. Why is polyandry uncommon? What sorts of ecological and phylogenetic circumstances might favor polyandry in animals?

2. The graph to the right shows the relationship between harem size, or the number of females that a male monopolizes, and the ratio of male-to-female body length, a measure of sexual dimorphism, in a variety of species of seals. (a) What is the most likely cause of this relationship? (b) What prediction could you make about the breeding habitat occupied by a species at the lower left of the curve versus one at the upper right?

3. Males typically compete for access to females and often obtain more than one mate, while females are particular about their choice of mates. In some species, however, the male supplies the female with large "gifts" in the form of captured prey or large, protein-rich sperm packets. What differences might you expect to see in the reproductive behavior of such species when compared with the typical pattern?

4. When egret chicks in the nest fight to the death, why don't the parents intervene?

5. Describe the possible relevance to humans of Trivers' concept of parent-offspring conflict.

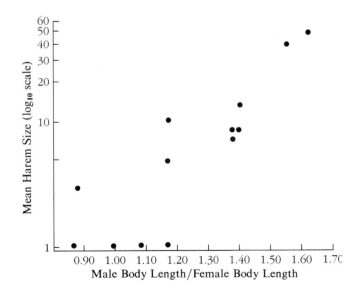

Source: From R. D. Alexander, et al., "Sexual Dimorphisms and Breeding Systems in Pinnipeds, Ungulates, Primates and Humans," in *Evolutionary Biology and Human Social Behavior*, ed. by N. A. Chagnon and W. Irons. Copyright © 1979 Duxbury Press. Reprinted by permission of the editor.

SUGGESTED READINGS

Choe, J. C., and B. J. Crespi. 1997. *The Evolution of Mating Systems in Insects and Arachnids.* Cambridge, U.K: Cambridge University Press.

Clutton-Brock, T. H. 1991. *The Evolution of Parental Care.* Princeton, NJ: Princeton University Press.

Mock, D. W., and G. W. Parker. 1997. *The Evolution of Sibling Rivalry.* New York: Oxford University Press.

19 CHAPTER

Social Behavior

 dozen spiders bite a locust struggling in their communal web. A vampire bat regurgitates warm blood to another. A mass of 300 ladybugs overwinter together in a cavity in a dead tree. A ground squirrel gives an alarm call and its sister dodges into a burrow.

Each of these animals lives at least part of its life in a group, but the details of their behavior vary tremendously. Some species barely tolerate conspecifics, and get together only briefly for mating. Others live in groups but show no cooperation. Other species spend every moment of their lives immersed in interactions with conspecifics. The term group, or aggregation, does not imply any particular form of special behavior or cooperation; some groups of animals are together simply because they are attracted to the same resource. Many species, however, do change their behavior when in the presence of conspecifics. We can define a **society** as a group of individuals of the same species that is organized in a cooperative manner extending beyond sexual and parental behavior, and we call their behavior social behavior. An extreme version of social behavior is altruism, in which one animal helps another to the decrement of its own direct fitness. We'll look at this entire range of behavior in this chapter.

The main ideas driving the study of social behavior have changed over the years, and we'll begin with a historical overview. Next, we'll discuss the selection pressures that favor grouping, as well as those that act against it. We will next examine cases of altruism, where animals help others in spite of costs to their own fitness. Finally, we'll examine some examples of social systems in greater detail.

CHANGING PERSPECTIVES IN THE STUDY OF SOCIAL BEHAVIOR

More so than most other aspects of animal behavior, how researchers approach the study of social behavior has changed fundamentally over the course of the last century.

Early Studies

W. C. Allee, in *Cooperation Among Animals* (1951), synthesized much of the thinking in the first half of this century about animal societies. Allee recognized the importance of natural selection at the level of the individual and worked extensively on behaviors such as competition and dominance (Allee 1951). However, he felt that altruistic or cooperative forces were somehow stronger, and he devoted much study to what he called "natural cooperation," a force supposedly separate from natural selection. Allee identified several grades of social behavior. At the lower end, he included the coloniality of invertebrates, such as Cnidarians; he believed that their aggregation was involuntary and somehow less advanced. He also included passive aggregations of animals carried by the tides or winds, those resulting from orienting toward or moving to favorable stimuli such as light or food, and sleeping groups in animals that moved about independently during the day. At the top of the list were complex societies.

Although Allee appreciated the negative effects of crowding, he also pointed out the unfavorable consequences of too few conspecifics in an area (or "undercrowding"). Typical of his experiments was the finding that goldfish (*Cyprinus* spp.) reared in a toxic colloidal suspension of silver survived longer in groups than they did alone; the larger amount of slime produced by groups of fish precipitated the silver on the bottom of the tank, thus protecting them. Thus, groups generate epiphenomena above and beyond the characteristics of the individuals.

Animal behaviorists debated the importance of natural selection acting at the level of the group versus that of the individual. They were spurred on by the book *Animal Dispersion in Relation to Social Behavior,* in which Wynne-Edwards suggested that social behavior evolves, in part, for the function of population regulation (Wynne-Edwards 1962), and animals might forego reproduction for the greater good. In this view, differential survival and reproduction of groups, rather than individuals, is emphasized.

Later Studies

There are problems, however, with the idea that animals will behave for the good of the species, as we discussed in chapter 4. The publication in 1966 of *Adaptation and Natural Selection* by G. C. Williams led to a rapid shift in thinking among researchers working on social behavior. Williams argued that when considering any adaptation, we should assume that natural selection operates at that level necessary to explain the facts, and no higher—usually at the level of parents and their young. For example, a group of imaginary antelope has limited resources, and their population will go extinct if reproduction is not curtailed. Imagine that some individuals do not reproduce, whereas others have many offspring. The genes of the reproducers will increase in the next generation; this is selection at the level of the individual. William's argument says that the population may be doomed, but individual selection does not act to preserve it. His book influenced decades of researchers, and many puzzling examples of social behavior have been carefully examined and found to be understandable in light of selection at the level of the individual or gene.

In recent years, as we discussed in chapter 4, biologists have revived, reassessed, and reframed the idea of group selection, and generated some clear examples of group selection in action. In addition, some of Allee's ideas, after a period of being ignored, have also received recent attention (reviewed in Stephens and Sutherland 1999). Of particular interest is Allee's observation that many species suffer a decrease in their per capita rate of increase when their populations are at low densities (the Allee effect; reviewed in Courchamp et al. 1999). This is particularly relevant for rare or endangered species, as it means that these small populations are more likely to go extinct. For example, a New Zealand flightless parrot, the kakapo (*Strigops habroptilus*) was reduced to less than 60 individuals. A breeding program resulted in nine chicks, but only two were female: this sort of stochasticity in sex ratio would be unimportant in a large population, but may be a contributing factor to extinction in a small population. The risks of stochastic fluctuation may be especially high depending on the mating system. For example, monogamous species with biparental care might be exceptionally sensitive. Other species depend more directly on having a minimum group size. Hunting groups of African wild dogs need to be large enough to be energetically efficient: when population sizes drop too low, they will be more likely to disappear (reviewed in Courchamp et al. 1999).

Studies of group selection and the Allee effect will continue to prove interesting. However, most of the rest of this chapter will concern selection at the level of the individual or the genes.

WHY LIVE IN GROUPS?

At first glance, you may assume that living in groups is somehow superior to living a more solitary life. Yet, like nearly all the behaviors we study, there are costs and benefits associated with both. A great deal of research has focused on figuring out the balance between these factors in particular species. In this section, we'll briefly discuss each of these selective forces, returning to some in more detail later in the chapter.

Benefits of Group Living

Here is a list of possible benefits of group living. Keep in mind that, even if we see that an advantage is currently operating, it may not have led to the evolution of sociality in any particular case; rather, it may be a secondary benefit appearing after sociality had already evolved. Also remember that these are not mutually exclusive.

1. **Protection from physical factors.** Abiotic factors, especially cold, rainy, or snowy weather, can promote grouping. For example, bobwhite quail (*Colinus virginianus*) survive low temperatures better when grouped than when isolated (Gerstell 1939). Gregarious butterfly larvae (*Aglais urticae* and *Inachis io*) experience a less variable range of body temperatures than do solitary larvae (*Polygonia calbum* and *Vanessa atalanta*, Bryant et al. 2000). This benefit would result in aggregations, but not necessarily organized social groups.

2. **Protection against predators.** One of the most often documented selective advantages of living in a group is protection against predators. As discussed in chapter 15, there are several mechanisms by which animals can avoid predation by being in a group. To illustrate and reinforce these points, we'll look at a nicely worked-out example on a spider we've met before, *Metepeira incrassata,* studied by Uetz and his colleagues (reviewed in Uetz and Hieber 1997). This is a colonial spider that has a shared frame web to which all spiders contribute silk. Within that frame web, each individual spider builds its own orb web, which it defends from others (figure 19.1*a*). Spiders are attacked by wasps and hummingbirds. Being in a colony, as opposed to being alone, benefits an individual in several ways. First is the encounter effect, which affects the probability that a predator locates a colony. This relationship is somewhat complex: as colony size increases, the rate that colonies encounter predators also increases. However, the rate does not increase as quickly as expected (figure 19.1*b*), probably because of the apparency of the colonies to predators. A colony of 1,000 spiders is not twice as visible to a potential predator as is a colony with 500 spiders, because some of the spiders are hidden behind others.

   The second factor that reduces predation risk for these spiders is the dilution effect, which reduces individual risk once a colony is encountered. This, as we discussed in chapter 15, is simply safety in numbers: if a predator is going to eat only one prey, it is better to be in a large group than alone. In *Metepeira,* wasps sometimes attack more than one spider in a colony, which reduces the dilution effect. On the other hand, spiders are informed about the presence of a predator through vibrations transmitted through the web, and this early warning effect reduces the capture success of wasps. Overall, there is a decrease in risk with increasing group size that is even steeper than predicted by numerical dilution (figure 19.1*c*).

   Position effects within a colony are also important. Recall Hamilton's (1971) selfish herd (chapter 15): when frogs aggregate to avoid a water snake, it is in each frog's best interest to be in the center of the group. Similarly, spiders on the inside of a colony have reduced risk of predation because they are surrounded by conspecifics, but the trade-off is a reduction in food supply as insect prey are also unlikely to make it to the core (Rayor and Uetz 1990, 1993).

   Young are often especially vulnerable to predators, and in many species group living appears to be an adaptation to protect them. For example, a female "parent bug," a species of Hemipteran (*Elasmucha grisea*), defends her clutch of eggs and developing nymphs by covering them with her body and showing aggressive behavior toward disturbances. Sometimes two parent bugs guard their eggs side by side on a leaf. Mappes et

al. (1995) studied whether pairs of females were more effective at guarding their offspring by experimentally constructing pairs of females with clutches by cutting off pieces of leaves with eggs attached and placing them near existing clutches. (By experimentally creating pairs instead of using natural pairs, the researchers could control for any differences between solitary and paired bugs in their guarding efficiency.) Jointly guarding females lost fewer eggs to predators than did single females.

   Predator protection in groups has been shown for herds of ungulates, flocks of birds, colonial mammals, and schools of fish, among others. However, while the mechanisms described here might explain some aggregations, more is required to explain systems where individuals cooperate.

3. **Assembly of sexual species for finding mates.** Solitary sexual species may expend considerable time and energy in simply locating a potential mate. Group-living animals can often find a mate more readily than can solitary species. Some species group together specifically for reproduction; for example, mating swarms are common in insects and in some vertebrates (see chapter 18).

4. **Locating and procuring food.** As we discussed in chapter 15, animals in groups may have better foraging success than solitary animals. We discussed information sharing, where animals learn about the location of food sources. A related advantage is created by tradition; knowledge about resource location can be transmitted to subsequent generations, as it is in sheep (*Ovis canadensis*) (Geist 1971) (see also chapter 14). Animals also participate in cooperative hunting, resulting in increased capture rates and capture of larger prey. For example, Harris's hawks (*Parabuteo unicinctus*) form hunting parties of two to six in the non-breeding season and cooperate to capture rabbits several times larger than themselves (figure 19.2) (Bednarz 1988).

5. **Resource defense against conspecifics or competing species.** Many examples of group territoriality are included here. Among invertebrates, large colonies may have a competitive advantage over smaller groups. The bryozoan *Bugula turrita* occurs in dense colonies on pilings and rocks in shallow water along the coasts of North America. Another species, *Schizoporella errata,* frequently overgrows them, unless the *B. turrita* larvae group together. The resulting dense colonies are less likely to be overgrown by *S. errata* (Buss 1981).

6. **Division of labor among specialists.** In some societies, different individuals have different tasks. This is especially pronounced in the castes of insects such as ants, wasps, bees, and termites. We will discuss this in more detail in the section "Eusociality" later in the chapter.

7. **Richer learning environment for young.** Some species, especially many mammals (primates in particular) depend on learning. Learning provides great plasticity but requires a long period of physiological and psychological dependence on others. Consider hyenas. As we saw in chapter 15, hyenas are highly social species that cooperatively attack prey. In fact, they can learn about which prey are palatable from one another. Yoerg (1991) trained hyenas to avoid a particular type of food (corned beef hash or one of two flavors of cat food) by lacing it with lithium chloride. When trained hyenas were allowed the chance to join with others in eating the same type of food, they overcame their aversion and joined in, demonstrating the importance of social interactions in diet choice for this species.

8. **Aiding (or receiving aid from) offspring or other relatives.** As we will see in more detail later in the chapter, living with relatives gives animals the opportunity to pass on their genes either by improving the chances of success of their offspring or by helping other relatives.

9. **Modifying their environment.** Many species build structures, such as nests, burrow systems, dams, etc. In many cases,

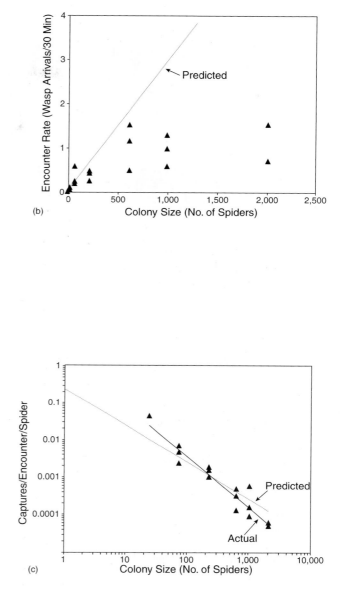

a.

FIGURE 19.1   The benefits of group-living in colonial web-building spiders.
(a) *Metepeira incrassata* spiders share a communal frame web, but build and defend individual orbs within that frame web. (b) The encounter effect acts in these colonies: encounter rates with wasp predators at the colony level increase with group size, but at a rate lower than predicted (the line indicates expected encounter rates if wasp attacks increase in proportion to colony size). (c) The dilution effect also comes into play after wasps encounter the colonies. The number of captures per encounter per spider decrease with colony size even more rapidly than predicted (dashed line is predicted line).

Fig. 1a. Source: Courtesy G. W. Vetz.

Fig. 1b–c. Modified from Vetz, G. W. and C. S. Hieber. 1997. Colonial web-building spiders: balancing the costs and benefits of group-living. In *The Evolution of Social Behavior in Insects and Arachnids,* ed. J. C. Choe and B. J. Crespi, 458–475. Cambridge: Cambridge University Press.

a group can accomplish what a single individual may not. For example, honeybees not only build elaborate nests together, but can actively cool them by fanning, or warm them with microvibrations of their powerful wing muscles (Seeley 1993).

Costs of Group Living

1. **Increased competition for resources.** Conspecifics are likely to have the same resource requirements, so competition for food, shelter, and mates is likely. For example, in prairie dog colonies (*Cynomys* spp.), the amount of agonistic behavior per individual increases as a function of group size. Black-tailed prairie dogs are more highly social and have higher rates of aggression than do the less social white-tails (Hoogland 1979a).

Burying beetles (*Nicrophorus* sp.) compete for an unusual resource: carcasses. Adult beetles prepare a carcass by removing its hair and rolling it into a ball as they bury it, and depositing anal secretions upon it that affect its decomposition. Females lay eggs near a prepared carcass and the larvae make their way to it, where they feed upon it as well as on liquefied carrion regurgitated by adults. Adults sometimes share carcasses and rear their offspring together. In four different species of burying beetles, the total number of larvae per female declines when a carcass is shared (reviewed in Trumbo and Fiore 1994), demonstrating the effect of competition. So why do beetles share? In some species (such as *N. defodiens*), the probability that the nest fails altogether is quite high, so the first adult to arrive at a carcass might be better off to tolerate a newcomer rather than fighting for a resource that is of low value (Trumbo

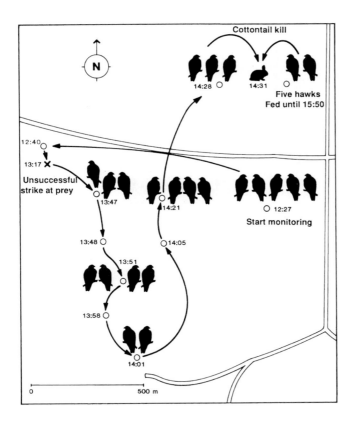

FIGURE 19.2   Cooperative hunting among hawks.
Sequence of movements of a Harris' hawk implanted with a radio transmitter.

Reprinted with permission from J. C. Bednarz, "Cooperative Hunting Among Hawks," in *Science*, 239:1526, 1988. Copyright 1988 American Association for the Advancement of Science.

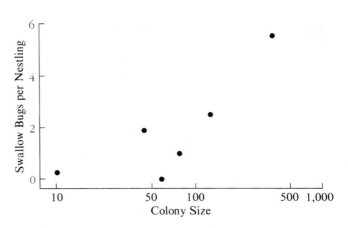

FIGURE 19.3   A cost of sociality.
The degree of parasitism by bugs as a function of swallow colony size. As colony size increases, so does the extent of bug infestation among 10-day-old nestlings.

Source: C. R. Brown and M. B. Brown, "Ectoparasitism as a Cost of Coloniality in Cliff Swallows (*Hirunda pyrrhonata*)," in *Ecology*, 67:1206–18. Copyright © 1967 by the Ecological Society of America, Tempe, AZ.

1995; but see Robertson et al. 1998). In another species (*N. tomentosus*), beetles commonly compete with fly larvae, and three or more adult beetles are better able to rid the carcass of fly eggs and larvae than are smaller numbers (Scott 1997).

Costs of competition may vary across group members. For example, the long-legged web-building spider *Holocnemus pluchei* is facultatively group living; sometimes it lives alone and sometimes it shares a web with conspecifics. When an insect hits the web, spiders charge toward it and will fight vigorously for it, sometimes to the death. The largest spider in the web nearly always wins, so the effect of competition is much more pronounced for small individuals. Nonetheless, many small spiders live in groups, probably in part to save the energetic costs of building their own webs (Jakob 1991).

2. **Increased chance of spread of diseases and parasites.** The correlation between the risk of infection and living in groups has now been well documented. Ectoparasites such as fleas and lice are more numerous in larger and denser prairie dog (*Cynomys* spp.) colonies than in smaller ones (Hoogland 1979a). This is not trivial, as fleas transmit bubonic plague; plague epidemics periodically decimate prairie dog colonies, so members of dense colonies are at risk. Similarly, as cliff swallow (*Hirundo pyrrhonota*) nesting colony size increases, the number of blood-sucking swallow bugs per nest also increases. Cliff swallow nestlings in parasitized nests lose weight and suffer higher mortality than do those in parasite-free nests (Brown and Brown 1986) (figure 19.3).

Sometimes animals battle both internal and ectoparasites. Rubenstein and Hohmann (1989) studied feral horses and parasitic infections on the barrier island of Shackleford Banks,

North Carolina. Horses in larger groups have a higher incidence of infection with internal parasites, as indicated by the number of parasite eggs in their feces. As in the prairie dogs and swallows, infection has serious implications: the intensity of infection at the end of one breeding season correlates with body condition at the beginning of the next. Horses are also plagued with ectoparasitic flies, sometimes up to 200 at a time. However, in contrast with endoparasites, group size negatively correlates with the number of flies per horse. Rubenstein and Hohmann suggest that endoparasites play a larger role in structuring horse society than ectoparasites, because even when flies are exceptionally active, females do not leave small harems to join larger ones.

It seems that it would be adaptive if an individual could avoid infected group mates, thereby reducing its own risk of infection. Bullfrog tadpoles (*Rana catesbeiana*) can in fact do this: they can detect and avoid chemical cues from conspecifics that are infected with a pathogen (*Candida humicola*) (Kiesecker et al. 1999).

Although it makes sense that risk of disease should increase with group size, this is not always true. Bluegill sunfish that nest in colonies are less likely to lose their eggs to fungus than are solitary nesters; one likely explanation is that solitary males spend more time chasing predators and less time fanning their eggs, which reduces disease (Côte and Gross 1993).

3. **Interference with reproduction, such as killing of young by nonparents.** For example, when male lions take over a pride, they almost invariably kill small cubs (Pusey and Packer 1987) (see chapter 14). Mexican jays often interfere in the nest of another, stealing nest lining or even destroying the nest (Brown and Brown 1990). Proximity to conspecifics can also increase the risk of cuckoldry, brood parasitism, or sexual harassment.

THE EVOLUTION OF COOPERATION AND ALTRUISM

Group living often has direct fitness benefits for each individual, as we have seen; for example, an animal might be more likely to survive winter weather if it is in an aggregation rather than alone. In cases like these, the evolution of

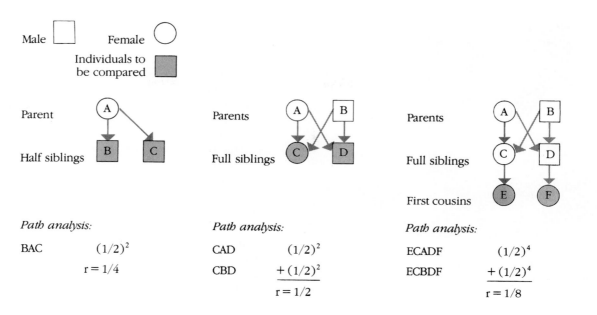

FIGURE 19.4   **Coefficients of relatedness.**
To compute the coefficient of relatedness between two organisms, begin by counting the number of parent-offspring relationships that connect them. For example, half siblings are connected by two, because they share one parent. Next, raise ½ to this number. If there is more than one common ancestor, such as in full siblings or first cousins, compute each path separately and sum them.

grouping behavior is not difficult to understand. What is more puzzling are cases where one animal behaves in such a way as to increase the fitness of another at the expense of its own fitness, such as by putting themselves in danger by giving warning calls about predators or by foregoing reproduction to help raise another individual's offspring. Why might animals behave altruistically?

Kin Selection

To explain the evolution of altruistic behavior, Hamilton (1963, 1964) developed the theory of kin selection, which we introduced in chapter 4. This idea relies on the fact that animals do not pass on genes to the next generation only through their own offspring, or descendant kin. Genes also may be passed on through the offspring of relatives, or nondescendant kin. Because of this, a mutant gene that causes its bearer to behave altruistically toward kin may spread through a population.

In order to understand this concept, we must first understand how to calculate the **coefficient of relatedness** between two organisms. The coefficent of relatedness is the probability that two individuals possess the same allele because they inherited it from a common ancestor. Between nonrelatives, the coefficient of relatedness is 0. When a diploid parent produces a sperm or an egg, there is a 50 percent chance that any particular allele the parent has is donated to its offspring. Thus, parent and offspring have a coefficient of relatedness of 0.5. For more complex relationships, we can calculate the coefficient of relatedness by counting up the number of parent-offspring connections between them (in other words, the number of meiotic, or gamete-forming, events). The complete procedure is illus-

TABLE 19.1	Common coefficients of relatedness in diploid species		
Relationship	**r**	**Relationship**	**r**
Parent and offspring	$\frac{1}{2}$	Cousins	$\frac{1}{8}$
Full siblings	$\frac{1}{2}$	Aunt (or uncle) and niece (or nephew)	$\frac{1}{4}$
Half siblings	$\frac{1}{4}$	Grandparent and grandchild	$\frac{1}{4}$

trated in figure 19.4, and table 19.1 shows some of the most common coefficients of relatedness.

How are coefficients of relatedness used to understand altruism? Hamilton's rule says that whether an altruistic trait will spread through a population depends on three elements. A trait will spread if

$$\frac{b}{c} > \frac{1}{r}$$

where

b = benefit to recipient of an altruist's help, in terms of offspring produced by the recipient that it would not otherwise have;

c = cost to altruist, in terms of offspring *not* produced by the altruist that it would have had if it had not helped;

r = coefficient of relationship, or the proportion of genes shared by two individuals because of descent from a common ancestor.

Hamilton's ideas about the evolution of cooperative behavior through kin selection has stimulated vast amounts of research. Although Allee and others had observed that

FIGURE 19.5   Female white-footed mouse nursing young of two litters (note different sizes of young).
The mother of the older litter is often the sister of the nursing mother.

Source: Photo by K. Fletcher.

most complex social groups are made up of relatives, kin-selection theory makes specific predictions that can be tested: animals should be more likely to direct helping behavior toward kin, especially close kin. One of the most well-known tests of kin selection is Sherman's study of alarm calls in ground squirrels, which was described in chapter 12 (Sherman 1977). To recap, individuals gave alarm calls such that those most likely to benefit were close kin. Males, who disperse as subadults and thus are unlikely to have relatives nearby, usually do not give alarm calls.

For another example of kin selection, recall the rhesus monkeys described in chapter 16. Monkeys aid each other in fights in proportion to their degree of relatedness. When males leave the natal group, they usually join groups containing older brothers. In contrast to their behavior with non-brothers, they associate with brothers in the new group, aid each other in fights, and avoid disrupting each other's sexual relationships (Meikle and Vessey 1981). Kin selection may also explain the nursing of nonoffspring by female white-footed mice (figure 19.5). In those cases where young of both litters had been previously marked, the mother of the older pups was the sister of the nursing female (Jacquot and Vessey 1994). We will discuss more examples of kin selection later in the chapter.

How do animals recognize kin in order to correctly distribute their altruistic acts? We touched on this earlier in the chapter on communication. There are four major ways. (1) Sometimes animals seem to use a simple rule of thumb: be altruistic toward animals that are in a particular location (such as nestmates). This rule works when there is little chance of a nonrelative showing up in the wrong place. It can backfire rather spectacularly: cuckoos exploit other species of birds by laying eggs in their nests, and parent birds will treat the baby cuckoo as its own young, even though it may be markedly larger. (2) Other animals follow

another simple rule, which is to treat familiar animals differently than unfamiliar ones. For example, like many species of fish, guppies (*Poecilia reticulata*) prefer to school with familiar individuals. However, they do not distinguish unfamiliar kin from nonkin (Griffiths and Magurran 1999).

(3) Animals might use phenotype matching, in which they use some feature of conspecifics as a cue that they are kin. This feature might be genetically or environmentally based (such as odors from foods that relatives are more likely to eat). Beavers (*Castor canadensis*) can identify kin because of the chemical profile of their anal gland secretions (Sun and Muller 1998a,b). We already discussed (chapter 7) the ability of golden hamsters to learn their own odor and match it with the odor of strangers they meet to judge relatedness (Mateo and Johnston 2000). (4) Finally, animals may inherit "recognition alleles" that allow them to recognize others with the same alleles without having to learn them. A good example is the colonial marine invertebrate called the sea squirt, whose planktonic larvae settle more closely to larvae that share the same allele in a particular region of DNA (Grosberg and Quinn 1986). See chapter 12 for a further discussion of kin recognition mechanisms.

Reciprocity

Why might unrelated animals cooperate? Axelrod and colleagues (Axelrod and Hamilton 1981; Axelrod and Dion 1988) developed what is now the classic mathematical model for the evolution of cooperation. They used the type of modeling called game theory, which we have already seen in chapter 16.

The basis for their analysis was a game called prisoner's dilemma. Imagine a scenario in which two criminals are caught and jailed. The police have good evidence that the men committed a crime, but they are also suspected of a more serious crime. They are questioned separately in different rooms. Each player has the choice to either cooperate with the other by denying all knowledge of the major crime, or they may defect (or squeal), and accuse the other of the major crime. As a reward for defecting, the defector is forgiven the minor crime. As in other game theory models, the payoff of each strategy depends on the behavior of the opponent. If both players cooperate (neither accuses the other), they each get a reward, R, and pay the penalty only for the minor crime. If both defect (squeal on each other), they each get a punishment, P, and have to serve the time for the major crime. If one cooperates and one defects, the defector is set free and gets the best possible reward, T (the temptation to defect). The player that cooperates when the other defects gets the sucker's payoff of S, because he pays the price for both crimes. The exact value of the payoffs don't matter: as long as $T > R > P > S$, as illustrated in the payoff matrix in table 19.2, the conditions for the prisoner's dilemma are met.

What should a prisoner do? Look at the payoff matrix again. In a single round of the game, Player A's best strategy

TABLE 19.2   **The payoff matrix for player A in the prisoner's dilemma**

The numbers shown are the years saved off of a maximum prison sentence of 12 years (2 years for the minor crime and 10 years for the major crime). The exact values are not important, as long as T > R > P > S.

Player A	Player B Cooperation	Defection
Cooperation	R = 10	S = 0
Defection	T = 12	P = 2

FIGURE 19.6   **Vampire bat *(Desmodus rotundus).***
Blood is the sole food source for this species.
Source: Photo by William A. Wimsatt, courtesy of the Mammal Slide Library at Cornell University.

is to defect regardless of what player B does: if player B cooperates, T > R, and if player B defects, P > S. The same logic applies to player B. However, players will do worse if they both defect rather than if they both cooperate (R > P); hence the dilemma. Players are predicted to defect when they play a single round even though payoffs are lower than if they cooperated.

What happens when individuals play multiple rounds of the game (in a series of unknown length) with the same opponents? After conducting an international computer tournament for the best solution, the highest score was obtained by the simple strategy "tit-for-tat," in which one player cooperates on the first move and then does whatever the other player did on the preceding move. This strategy is thus nice (it begins by cooperating), it is quick to retaliate (it immediately responds to defection by defecting on the next round), and it is forgiving (it immediately cooperates once the opponent cooperates). Axelrod and Hamilton argue that this strategy is evolutionarily stable and that it shows how cooperation based on reciprocity could get started in an asocial group.

The prisoner's dilemma and the tit-for-tat strategy have generated an enormous amount of literature, especially more permutations of the model (reviewed in Dugatkin 1998). One recent twist is especially interesting. Instead of only two strategies, cooperate or defect, now add a series of intermediate strategies ("cooperate a little"). This may be more typical of animal behavior: for example, animals may groom each other for 5 minutes or an hour or any length in between. Under these conditions, cooperation is even more likely to evolve (Roberts and Sherratt 1998).

Do any species play tit-for-tat? Guppies have been suggested as an example. Instead of fleeing from predatory fish, they often move in and inspect it. A cooperating fish moves toward the predator, and a defecting fish moves away. In a series of papers, Milinski, Dugatkin, and their colleagues argue that predator inspection fulfills the requirements for tit-for-tat: the response of guppies depends on what their partner did in the previous interaction (reviewed in Dugatkin 1998; also see references cited therein about other possible explanations). Other behaviors are also candidates for tit-for-tat. For example, impala groom each other and remove ticks, and there is almost a perfect match between the number of grooming bouts an individual delivers and the number it receives (Hart and Hart 1992). However, it seems unlikely that we will find a perfect example of the classic tit-for-tat strategy being played in nature without any other variables clouding the issue.

As the prisoner's dilemma model shows, individuals may cooperate and behave altruistically if there is a chance that they will be the recipients of such acts at a later time. Think about the conditions under which this is likely. One would not expect reciprocal altruism to evolve in mating swarms of insects, for example, because it is unlikely that two animals will meet each other again. Reciprocal altruism is most likely to evolve if (1) animals are in contact long enough to permit reciprocation, suggesting that longer-lived organisms are better candidates, (2) the benefit to the receiver exceeds the cost to the donor, and (3) donors should be able to recognize cheaters (those that don't reciprocate) and refrain from helping them.

A good candidate for reciprocal altruism comes from an unlikely source. Vampire bats *(Desmodus rotundus)* live in the New World tropics. At night these bats feed on blood from cattle and horses (and possibly unwary sleeping ecologists), then they return to a hollow tree to roost during the day (figure 19.6). Wilkinson (1984) marked nearly 200 bats that roosted in 14 trees and spent 400 hours observing them in their roosts. He recorded 110 cases of blood sharing, where one bat regurgitated blood and fed it to another bat. Not surprisingly, most of these exchanges were between mothers and offspring. In most of the other feedings, he was able to determine both the coefficient of relationship between the pair and an index of association based on how often the pair had been together in the past. Wilkinson concluded that both relatedness and association contributed significantly to the pattern of exchange (figure 19.7); it is important to note that these are not mutually exclusive. Vampire bats may be able to assess each other's ability to give blood via grooming, as their stomachs get hugely full when they have had a good meal (Wilkinson 1986).

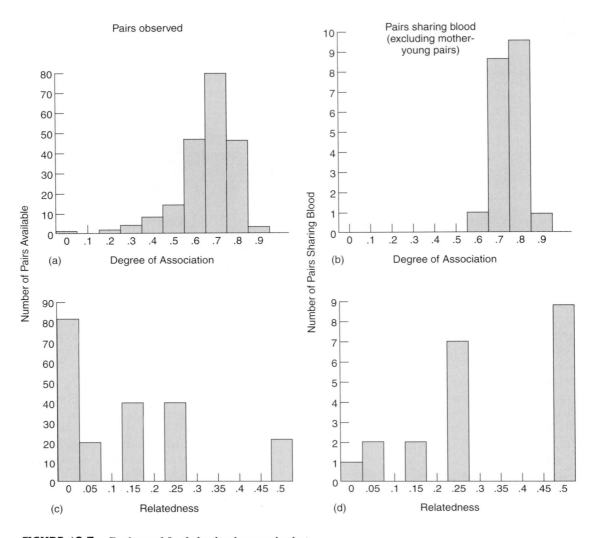

FIGURE 19.7   Reciprocal food sharing in vampire bats.
Figures a and b show that bats most likely to share food are more likely to associate together, suggesting that recipro-cal altruism may function. In addition, figures c and d show that relatedness also significantly affects whether two indi-viduals will share food.

Mutualism and the Appearance of Altruism

In some cases, what appears to be altruistic helping behavior may be of direct benefit to the donor as well as the recipient. In some cases benefits to the helper are obvious: a bird that joins in a mob that attacks a hawk benefits directly from the hawk's departure. In other cases, however, the benefits to the donor are subtle. For example, long-tailed manakins (*Chi-roxiphia linearis*) in the forests of Costa Rica have an unusual lek-mating system in which two males cooperate in a complex display in order to attract a receptive female. The display consists of a duet of song ("To-lay-do") and dual-male leapfrog hops and "butterfly" flights (figure 19.8). The benefit to the dominant (alpha) males is clear, as females will not copulate with them if they are alone. However, alpha males get virtually all of the copulations (259 of 263), yet subordinate (beta) males do much of the work displaying

(McDonald and Potts 1994). Why should beta males cooper-ate and give up reproduction? Indirect fitness benefits do not seem to be a factor, since DNA fingerprinting revealed alpha and beta males to be no more closely related than randomly selected pairs. Alpha males never seem to reciprocate in later years, so reciprocal altruism is not a likely explanation either. It turns out that the subordinate males increase their direct fitness, but the payoff is delayed from 5 to 13 years! Beta males eventually ascend to alpha status, and females tend to remain faithful to a particular lek from one year to the next. The best explanation of this pattern is one of mutualism, where both alpha and beta males increase their direct fitness by cooperating to attract females. Although the beta males appear to be gaining little, they may eventually become alpha males and "inherit" the territory.

Another example is meerkats (*Suricata suricatta*), desert mongooses that engage in sentinel behavior: they stand up their hind legs and watch alertly for predators, giv-

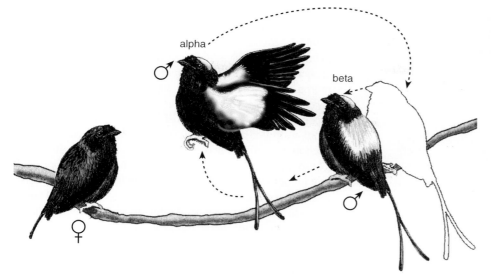

FIGURE 19.8   Dual courtship of the lek-breeding, long-tailed manakin. Dominant and subordinate males cooperate to attract females, but only the dominant male gets to copulate.

alpha

beta

ing alarm calls when one approaches (figure 19.9). This seems like it could be altruistic behavior. However, mongoose guards are no more likely to be preyed upon than are others, indicating that they are not paying an exceptionally high cost. In addition, they spent 30 percent more time guarding when researchers supplemented their diet with hard-boiled eggs. Thus, guarding might be an individual's optimal activity when its stomach is full and no other animal is on guard. It is not necessary to invoke reciprocity or kin selection to explain this behavior, although other aspects of their behavior may require these explanations (Clutton-Brock et al. 1999). These examples demonstrate the difficulty of demonstrating unequivocally the selection pressures underlying a seemingly altruistic trait.

EXAMPLES OF SOCIAL SYSTEMS

A Taxonomic Overview

We'll begin this section with a brief taxonomic tour, and then look at some well-studied examples in more detail. By necessity, we are painting a picture with a broad brush, and many exceptions exist.

We don't normally think of single-celled organisms as having complex social behavior, but there are some remarkable exceptions. Slime molds (especially *Dictyostelium discoideum*) have a unique developmental cycle, described in Hickman et al. (1984). They normally live as individual amoebas in forest detritus, and reproduce by binary fission. However, when food runs short, the amoebas aggregate to form a large slug (figure 19.10). This slug crawls to a favorable location and develops a stalk with a fruiting body on top. The stalk cells do not reproduce, but the fruiting body cells do. This poses an intriguing puzzle for biologists that study social behavior: if individual cells are genetically distinct, why would some forego reproduction? It is especially interesting

FIGURE 19.9   Meerkats in guarding position.

FIGURE 19.10   The slug phase of the slime mold *Dictyostelium discoideum*, which is formed by the aggregation of individual amoebas.

that amoebas from different clones will combine and some clones "cheat" by not contributing a proportional share of cells to the sterile stalk (Strassman et al. 2000).

Many multicellular invertebrates have highly developed social systems. In fact, it is sometimes difficult to decide what constitutes an individual and what constitutes a society. Hydrozoans of the phylum Cnidaria are a striking example of

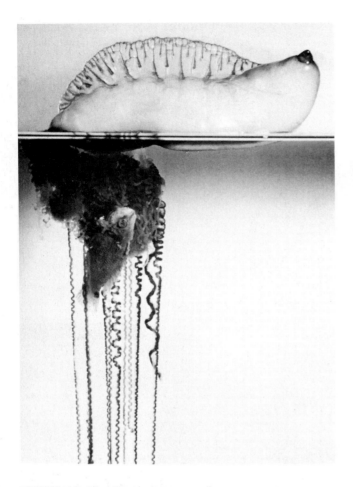

FIGURE 19.11   Portuguese man-of-war.
This colonial invertebrate of the genus *Physalia* is actually a group
of individuals. Four types of polyps (zooids) make up the colony,
with each type specializing in either flotation, feeding, defense, or
reproduction.
Source: © D. P. Wilson/Frank Lane Picture Agency.

a colonial system. Individuals are physically united in colonies
and specialize as reproductive or nonreproductive types. The
basic unit is the polyp, which consists of a tubular body formed
from two cell layers, with a central mouth surrounded by soft
tentacles. Although one genus (*Hydra*) represents organisms
that are single, free-living polyps, most of the genera are colonial, and the polyps (also called zooids) show a division of
labor. For example, some polyps of the Portuguese man-of-war
(*Physalia*) (figure 19.11), produce a gas-filled float, while other
polyps have stinging cells. In deciding whether or not an organism like *Physalia* is a society or a single individual, consider
that the entire colony comes from a single zygote. The different zooids form by budding off asexually. Since all the zooids
are genetically identical, the man-of-war may be considered a
single individual with several organ systems. However, in other
species the zooids forming the colony are unrelated (Wilson
1975), and they are capable of distinguishing among relatives
with great precision. For example, *Hydractinia symiolongicar-
pus* colonizes gastropod shells inhabited by hermit crabs. As
the colonies grow asexually, they compete for space. If they
meet another colony that is a clone, they fuse with it; if it is not

related, they produce specialized stinging cells and attack. Full
sibs will fuse in about 30 percent of pairings (Grosberg et al.
1996). Recognition is based on polymorphic genes that allow
them to distinguish self from conspecifics, and competition
appears to be important in promoting the evolution of this polymorphism (Grosberg and Hart 2000).

Social insects, such as bees and wasps, have probably
received the most attention of all invertebrates. However,
other arthropods also exhibit some degree of social behavior,
such as the spiders and beetles we have already discussed.
These species can give us insight into the selection pressures
that favor the initial evolution of group living. Entomologists
have coined a number of terms to describe arthropod social
behavior. The term subsocial refers to parent/offspring associations, and parasocial refers to cooperative associations
between two individuals of the same generation. Social
behavior evolved numerous times in arthropods, sometimes
through a subsocial route and sometimes a parasocial route.

Social behavior also varies across vertebrate classes. In
fish, schooling is perhaps the most conspicuous form of social
behavior, and it often seems to be a result of predator pressure.
For example, when fish detect a predator, they often bunch
closer together. Other advantages to schooling are increased
ability to locate patchy food resources, conservation of energy
because heat is generated, creation of currents that facilitate
locomotion, and increased likelihood of finding mates.

While amphibians and reptiles do not show evidence of
large, complex societies with division of labor, they do
demonstrate many complex social interactions related to territory defense and obtaining mates; a few species engage in
parental care. Large groupings are uncommon, but occur in
hibernacula (overwintering aggregations), in mating frogs
and toads, in schooling tadpoles, and at egg-laying sites. The
latter makes some species particularly vulnerable to human
intrusion; the same beaches favored by sea turtles are popular with developers as well. Some reptiles are social during
only part of the year. For example, grass snakes (*Natrix
natrix*) have complex social behavior during the mating season. When male snakes compete for access to females, they
roll together into balls and engage in tail wrestling (Luiselli
1996). Parental care is generally not extensive among reptiles. Exceptions are in the crocodilians: females make and
then defend large nests until the young hatch (Greer 1971).
Upon hatching, the young are led or actually carried by the
parents to the water's edge.

A majority of birds are socially monogamous territorial
breeders. However, many species aggregate in feeding,
migratory, and roosting flocks, and well-organized breeding
colonies are common. Perhaps the most complex avian
social systems are those involving cooperative breeding, in
which nonparents share in the rearing of young. We'll discuss these in more detail later.

In contrast to birds, mammals tend to be either solitary,
where the most complex social unit is a mother with her
young, or polygynous, where males monopolize several
females each and harem formation is common (Eisenberg
1981). However, complex social organization has evolved in
some species in nearly all mammalian orders, especially in

marsupials, carnivores, ungulates, and primates. Most of these systems are organized matrilineally; mothers and daughters may stay together, and groups are thus composed of mothers, daughters, sisters, aunts, and nieces. Because of the prevalence of polygyny and the associated tendency of males to disperse as they reach sexual maturity, adult males are usually unrelated to the other adults in the group.

Some of the most complex mammalian societies have evolved in relatively open habitats. Grouping seems to function largely as defense against predators and protection of resources through group territory. Although cooperative rearing of young—nonmothers nursing or providing food for young—is not common, it does occur in social carnivores and a handful of other species. For example, lionesses share the nursing of cubs in the pride, and subordinate wolves regurgitate food for the alpha female and her litter.

Within primates, we see variation in sociality. For example, rhesus monkeys have large troops with closed membership, whereas chimpanzees have loose communities. Social system structure is influenced by kinship, coalitions and friendship among nonrelatives, the distribution of resources (especially food), and learning and early experience. Of particular interest is the effect that a large brain and a good memory have on social interactions. For example, capuchins sometimes engage in cooperative hunting, where several individuals pursue prey but only one makes a capture. In a clever laboratory experiment, pairs of capuchins were tested to see if they were more likely to share food that they had "captured" cooperatively versus food that they got through solo efforts (de Waal and Berger 2000). Monkeys were in adjoining cages, and had an opportunity to pull a tray with transparent food bowls toward them (figure 19.12a). The tray could be weighted so that only one monkey or both of the monkeys were needed to pull it in. In "solo trials," one monkey had access to a cup of food and a pull bar, and the tray was weighted so it could pull it alone. In "cooperation trials," only one food bowl was baited, but both monkeys had to pull to move the tray. Monkeys could share food through "facilitated taking": they dropped food near the divider between the cages so the other could grab it. Monkeys were significantly more likely to share food after successful cooperation trials rather than after solo trials (figure 19.12b).

Communal Breeding and Helpers at the Nest

Animals that are rearing young often interact with conspecifics. This can range from simple proximity of nests (e.g., colonies of breeding birds), to sharing of the same nest (either aggressively or amicably; an example would be the burying beetles we met earlier whose "nest" is a carcass), to individuals that help to rear young that are not their own. This range of associations is found in both vertebrates and invertebrates, although historically the arthropod and vertebrate literatures have been rather separate and have developed different vocabularies (Brockmann 1997). Across taxa, however, communal associations are predicted to be more likely to arise when ben-

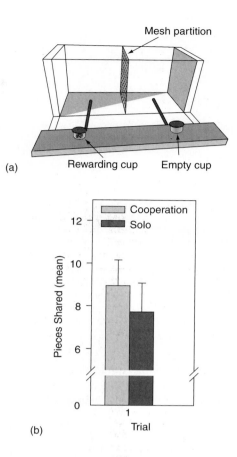

FIGURE 19.12   Cooperation in monkeys.
(a) Monkeys were in cages side by side. Each had a handle with which to pull in a food tray. The tray could be weighted so that it was harder to pull in some trials than in others. In cooperation trials, the strength of both monkeys was needed to pull the tray in, but only one food bowl was filled. In solo trials, only one monkey was needed to pull the tray in, and again only one food bowl was filled. (b) After cooperation trials, monkeys that received the food were more likely to drop it within reach of their partner than after solo trials.

efits are high, additional nesting sites are rare, females have limited clutch sizes, and dominant females are able to skew reproduction in their favor (Robertson et al. 1998).

An example of the diversity in social nesting comes from jays, New World members of the family Corvidae. In California scrub jays (*Aphelocoma coerulescens*), a pair of birds defends a territory and rears young without help from others. Steller's jays (*Cyanocitta cristata*) overwinter as pairs with extensively overlapping home ranges. Mexican jays (*Aphelocoma ultramarina*), in contrast, share the same winter home ranges with no separated core areas, and in fact defend these same areas jointly during the breeding season (Brown and Brown 1990). Several pairs in a territory may breed at any one time, but they do not cooperate with one another, and in fact will destroy each other's nests. Nonbreeders generally help feed the young. In Florida scrub jays (figure 19.13, *Aphelocoma coerulescens*), young from previous years help their

FIGURE 19.13 **Scrub jays in nest with helper nearby.**
Helpers are usually older siblings of the nestlings; they help feed young and defend the nest against predators such as snakes. (Note that three adults are present.)
Source: Courtesy Glen E. Woolfenden.

parents, feeding offspring and helping to protect them from predators.

Explaining the behavior of **helpers at the nest** in these cooperatively breeding species has been a challenge. Do helpers really have a meaningful impact on the reproductive success of those they help? Why don't they find their own mate and breed on their own? And if they can't breed, why do they bother helping instead of spending time foraging or preparing themselves for breeding in some other way?

Emlen (1991) reviews the research on these questions. In a number of species, such as dwarf mongooses, hyenas, and red-cockaded woodpeckers, the presence of helpers correlates with increased breeding success of the animals that are assisted. Correlational studies can be tricky to interpret, however; for example, what if animals that are being helped also have the best territory? In the woodpeckers, the effect of helpers disappeared when Walters (1990) compared pairs breeding on the same territory with and without helpers in different years. In other cases, experimenters have removed helpers and demonstrated that nest success is lower than in control groups. Florida scrub jays, for example, lose significantly more offspring to predators when they do not have helpers (table 19.3). Brown and colleagues (1982) removed helpers from breeding groups of grey-crowned babblers, an Australian species (*Pomatostomus temporalis*). Intact control groups with a full complement of helpers produced an average of 2.4 young, while the depleted groups produced an average of 0.8 young (figure 19.14). Thus, helpers of many species really do help.

Why don't helpers just breed on their own? There are a number of reasons that are related to ecological constraints. First, in saturated habitats, it might be impossible for an animal to find a territory or nest site (Emlen 1982). This seems to be the case for the Florida scrub jay (Woolfenden and Fitzpatrick 1984), which is particular about its habitat requirements. When Walters provided red-cockaded woodpeckers, which normally spend at least 10 months constructing nesting

cavities, with artificial nests, helpers gave up helping and moved into them (reviewed in Emlen 1991). Second, even when other territories are available, they might not be of high quality, and it might be better for helpers to wait to inherit a good territory or to compete for a territory near their natal area (Stacey and Ligon 1987). Third, in some species, habitat modifications made by the group may constrain striking out on one's own. Acorn woodpeckers, for example, store acorns in shared granaries, and it would not be feasible for an individual to leave the group (Koenig and Mumme 1987). Fourth, animals may have difficulty in finding a mate. And fifth, dispersing away from the natal area to a breeding site may be very risky. Finally, life history characteristics also appear to play a role in whether an animal breeds on its own or stays to cooperate, as a comparative analysis by Arnold and Owens (1999) suggests: cooperative breeding is restricted to certain families, and is associated with small clutch size and low adult mortality. Life history traits and ecological restraints are likely to act in concert to promote the evolution of delayed dispersal (Hatchwell and Komdeur 2000).

Even if a helper is unable to breed independently, why should it help? In many species of birds, unsuccessful males simply become floaters, and wait for breeding opportunities, so it's not obvious why cooperative breeders are different. There may be multiple adaptive reasons, as demonstrated by pied kingfishers (Reyer 1990). These fish-eating birds from Asia and Africa fit the model described above: females are a limited resource and many males, especially young, inexperienced males, do not attract a mate. Instead, males delay reproduction, and may help breeding pairs. There are two categories of helpers in this species. Primary helpers associate with only one mated pair, beginning during incubation. They feed the breeding adults as well as the offspring, and help to defend the nest. Secondary helpers begin helping only after the nestlings have hatched. Primary helpers offer more help than do secondary helpers: they spend more time nest guarding and bring more fish to the young. In fact, primary helpers provide as much parental care as the breeders (figure 19.15).

Why are there two types of helpers in pied kingfishers? Primary helpers are generally the sons of at least one of the pair that they help; their average coefficient of relatedness to the nestlings is 0.32. Thus, by helping to raise their siblings, they improve their indirect fitness. More young survive than would be possible without help, and the parents themselves are more likely to survive. Secondary helpers, in contrast, are not especially closely related to the offspring they help ($r \leq 0.05$). However, they improve their chances of nesting in the following season. When a male breeder dies, it is usually his former secondary helper who takes over the female. Delayers, who neither breed nor help in their first year, have lower fitness overall. The fitness benefits for the three strategies are compared in table 19.4.

In general, a helper may benefit from helping in four ways (Emlen 1991). First, its inclusive fitness may be increased by producing nondescendant kin, as we saw in the primary helpers in the kingfishers. Second, it may enhance its likelihood of breeding in the future, as do kingfisher secondary helpers. Third, its own survivorship may be increased. It

TABLE 19.3   **Effect of helpers on reproductive success in Florida scrub jays**

Year	Number of Pairs		Fledglings per Pair ($\overline{x}$)		Independent Young per Pair ($\overline{x}$)	
	Without Helper	With Helper	Without Helper	With Helper	Without Helper	With Helper
1969	2	5	0	2.6	0	2.0
1970	8	8	2.0	3.5	1.3	1.8
1971	6	19	1.3	2.1	1.0	1.2
1972	13	17	0.3	1.6	0.1	1.1
1973	18	10	1.3	1.8	0.4	1.1

From G. E. Woolfenden, "Florida Scrub Jay Helpers at the Nest," in *Auk*, 92:1–15, copyright © 1975 by the American Ornithologists' Union.

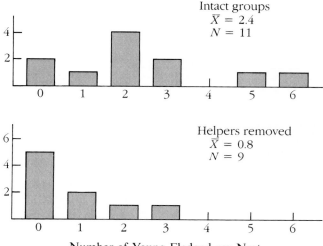

FIGURE 19.14   The effect of artificial removal of helpers upon reproductive success in groups of grey-crowned babblers.
Source: Based on data from J. L. Brown and Brown, "Helpers: Effects of Experimental Removal on Reproductive Success," *Science*, 215:421–22, 22 January 1982, copyright 1981 by the AAAS.

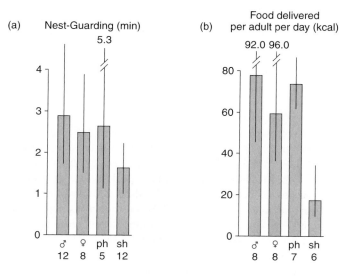

FIGURE 19.15   Two kinds of helpers in pied kingfishers.
Primary helpers (ph) spend as much time nest-guarding and deliver as much food as do either the mother or father. Secondary helpers (sh) do far less.

may gain access to resources, it may benefit through augmenting group size (recall the Allee effect), or by decreasing predation risk. Fourth, it might gain useful experience at parenting, or may receive help in the future. For example, unrelated dwarf mongooses guard and feed young of the dominant pair. Rood (1983) observed one case where a female was later assisted by the very young she had helped rear.

Finally, parents may manipulate their offspring in order to get them to help. For example, in Seychelles warblers (*Acrocephalus sechellensis*), the presence of helpers in high-quality territories improved the reproductive success of their parents, but the presence of helpers on low-quality territories decreased future reproductive output of their parents because of competition for food and breeding (Komdeur 1994). Female offspring remain on their parents' territories longer than male offspring, so it is better for parents on high-quality territories to produce daughters and parents on low-quality territories to produce sons. This is exactly the case: breeding parents with no helpers on low-quality territories produce 77 percent sons, while those

on high-quality territories produce only 13 percent sons. When Komdeur removed helpers from high-quality territories, parents produced more daughters (Komdeur 1996, 1997).

Another example of parental manipulation comes from white-fronted bee-eaters (*Merops albicollis*), cooperatively breeding birds in Africa. Here there is an unusual pattern: male parents actively disrupt the breeding attempts of their sons, although they leave other birds alone. The result is that the sons often give up their own breeding attempts and return to assist their parents (Emlen and Wrege 1992).

Eusociality

At first glance, the ultimate in unselfishness appears to be the worker honeybee. She sacrifices her life when stinging intruders in the vicinity of the colony, and she sacrifices her own reproductive success for that of the colony: she produces no eggs and raises the young of others (except under unusual circumstances), and her direct fitness is zero. Honeybees are

TABLE 19.4   Gains in direct, indirect, and inclusive fitness values for primary and secondary helpers, and delayers that do not help at all, in pied kingfishers during their first 2 years of life

	Year	Direct	Indirect	Inclusive
First-year breeder	1	0.96	0	0.96
	2	0.80	0	0.80
Total		1.76	0	**1.76**
Primary helper	1	0	0.45	0.45
	2	0.42	0.20	0.62
Total		0.42	0.65	**1.09**
Secondary helper	1	0	0.04	0.04
	2	0.87	0.01	0.87
Total		0.87	0.05	**0.92**
Delayer	1	0	0	0
	2	0.30	0	0.3
Total		0.30	0	**0.3**

Source: Reyer 1990.

termed eusocial ("truly social"). Wilson (1971) described three traits that separate eusocial insects from others: (1) cooperation in the care for young; (2) reproductive castes cared for by nonreproductive castes; and (3) overlap between generations such that offspring assist parents in raising siblings. In addition, many eusocial insects have a suite of other complex behaviors, such as trophyllaxis (oral and anal exchange of food), recruitment to food sources, and an alarm pheromone to rally defenses against intruders.

Most eusocial insects, including ants, bees, and wasps, are in the order Hymenoptera (although the converse—that most Hymenoptera are eusocial—is not true). In fact, eusociality evolved nearly a dozen times in this group. This observation led to an interesting hypothesis based on the fact that Hymenopterans have an unusual method of sex determination. They are haplodiploid: males are haploid (1n), containing only one set of chromosomes that come from an unfertilized egg; females are diploid (2n), containing two sets of chromosomes from a fertilized egg. In this system, all the sperm from one male are genetically identical. If only one male mates with one queen to form the colony, the daughters will share identical genes from their father and will share half the genes from their mother. Among the queen's daughters, then, the coefficient of relationship (r) will be 0.75 instead of the usual 0.5 (table 19.5). Sisters are more closely related to each other than they would be to their own daughters, where r = 0.5.

By now you will be able to see where this argument leads us. Hamilton (1964) argued that this peculiarity predisposes members of this order toward greater cooperation through kin selection. Daughters should increase their inclusive fitness more by rearing siblings than by rearing their own offspring.

Support for this idea came from Trivers and Hare (1976) in a study of investment of workers into their siblings.

TABLE 19.5   Coefficients of relatedness in haplodiploid species. Compare to diploid species in Table 19.1.

Relationship	r
Mother and daughter	$\frac{1}{2}$
Full sisters	$\frac{3}{4}$
Mother and son	$\frac{1}{2}$

Because female workers are related to sisters by r = 0.75 and to brothers by r = 0.25, workers should prefer to raise three times as many sisters than brothers. From the queen's point of view, both males and females are equally valuable, so there is a conflict of interest between queens and workers; however, workers have many opportunities to manipulate the sex ratio as they are the ones who care for the larvae. In twenty species of ants, the investment into reproductive females (measured as dry weight) is about three times larger than into reproductive males. In contrast, in slave-making ant species, the queen's brood is reared not by her daughters but by slaves, workers of other species stolen from their own nests. Since the slaves are unrelated to the brood they rear, they should not favor one sex over another. Trivers and Hare found that the investment ratio in the slave species was 1:1, as predicted.

Problems remain, however. These calculations are based on the assumption that only one male fertilizes the queen, but we now know that in many species of eusocial insects females mate multiple times. This reduces the relatedness among daughters and will reduce the bias in the predicted investment. Other factors may come into play as well: for example, Alexander and Sherman (1977) argue that female-biased sex ratios result if sons compete with one another for matings;

FIGURE 19.16   Queen termites (*Mucrotermes*) surrounded by workers.
The reproductive male is in the upper left. Termites are eusocial insects in the order Isoptera.
Source: © Edward S. Ross.

there is no point in making "extra" sons if they will not all be able to mate under conditions of local mate competition. In addition, there are eusocial species that are founded by multiple queens, and others that are diploid. Where does this leave us? Haplodiploidy is neither necessary (there are eusocial species that are not haplodiploid) nor sufficient (there are haplodipoid species that are not eusocial) for eusociality. The validity of the haplodiploidy hypothesis is uncertain at present, pending more data (Bourke and Franks 1995).

Other factors are surely important in the evolution of eusociality in the Hymenoptera. Among them are reliance on a jointly constructed nest, extensive maternal care of larvae, possession of a sting, and the ability of females to reproduce without males (reviewed in Bourke and Franks 1995). Additionally, Queller (1989) suggested that a female that leaves her natal nest to found her own nest will face a longer start-up time. It will take longer for her colony to grow and produce offspring, and it might be better to stay home and produce more, albeit less closely related, offspring. In addition, if a female that is alone in a nest dies, all her offspring will die too, but in a nest with others, her offspring will still be reared (Gadagkar 1990). Finally, we should remember that kin selection is still certain to be important in Hymenoptera, even if the increased relatedness associated with haplodiploidy is not as important as was once hypothesized.

Let's take a look at non-Hymenoptera eusocial species. Termites (figure 19.16) are diploid, eusocial insects. Over the years, a number of explanations have been proposed for the evolution of eusociality. For example, termites are dependent on symbiotic protozoans or other mutualists that digest cellulose in the termite gut, and they get these through contact with group members. This, however, does not necessitate the evolution of sterile castes. Probably the best explanation for termite eusociality involves the same sort of selection pressures that have led to sociality in other taxa, including family associations in food-rich habitats, slow development, high-risk dispersal, opportunities for nest inheritance, and advantages of group defense (Thorne 1997).

Jarvis (1981) demonstrated the existence of mammalian eusociality in the naked mole rat (*Heterocephalus glaber*)

FIGURE 19.17   The naked mole rat (*Heterocephalus glaber*, Bathyergidae) of East Africa.
These rodents are eusocial.
Source: Photo by Graham C. Hickman, courtesy of the Mammal Slide Library.

(figure 19.17). She captured 40 members of a colony from their burrow system in Kenya and studied them for 6 years in an artificial burrow in the laboratory. Only one female in the colony ever had young; mother and young were fed by male and female adults of the worker caste; members of this caste were not seen breeding. Another caste of nonworkers assisted in keeping the young warm; males of this caste bred with the reproductive female. This species meets our criteria for eusociality. There is cooperative care of young; there is a reproductive caste, with nonreproductives caring for reproductives; and there are offspring assisting parents in the rearing of siblings (Sherman et al. 1991; Honeycutt 1992). We now know of two eusocial species of naked mole rats. Both occupy arid regions of Africa where rainfall is unpredictable and prolonged droughts are the rule. Foods are underground tubers that are rich but patchily distributed resources. Most of the time, the soil is too dry to dig burrow systems. Jarvis et al. (1994) argue that these conditions favor coloniality

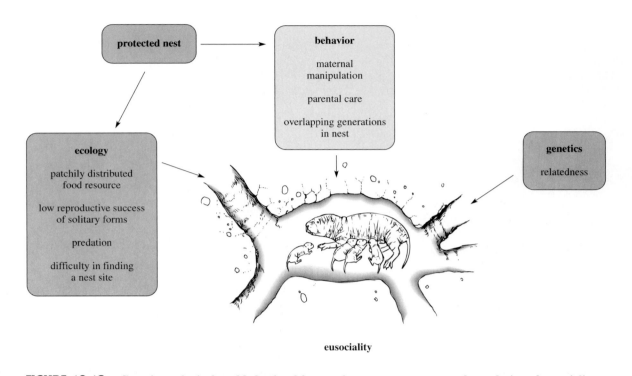

FIGURE 19.18   Genetic, ecological, and behavioral factors that appear to promote the evolution of eusociality.

because many individuals are needed to dig, on those rare occasions when it rains enough to make digging possible. The presence of a protected nest and genetic relatedness among colony members may then promote the evolution of eusociality (Honeycutt 1992) (figure 19.18).

Snapping shrimps (*Synalpheus regalis*) live inside sponges on coral reefs. There may be more than 300 individuals in a group, but only one breeds (Duffy 1996). The selective pressures favoring eusociality include a high degree of relatedness (most appear to be offspring of the queen and possibly a single male) as well as pressure from competitors. Food is not limiting, but empty sponges are. Nonbreeding shrimps vigorously attack intruders from other colonies, snapping at them with their powerful claws.

Humans

Do any of the preceding discussions have relevance to the evolution of human social systems? Some argue that human behavior is too heavily shaped by culture and that our behavior is so far removed from genetic control that natural selection can have little to do with our present societies, and that evolutionarily based explanations of human behavior can lead to politically dangerous conclusions. Others argue that an evolutionary approach can give us insights both into the particulars of societal structure, as well as to the psychological mechanisms that underlie decision making (e.g., Daly and Wilson 1999). Here we will consider a few examples of evidence for kin selection in humans.

Kin selection has been offered as an explanation of inheritance patterns in humans. Inheritance can be viewed as a form of investment. In societies in which the certainty of paternity is low (males are generally not confident which offspring are

theirs), inheritance follows the matriline (investment is in one's sister's children rather than one's own children) rather than the patriline (investment is in one's own children). However, as pointed out in Borgerhoff Mulder's (1991) review, this interpretation is still problematic; for example, we are not really sure about of the reliability of the certainty of paternity estimates.

The Yanomamo, an Amerindian tribe in southern Venezuela, live in villages of several hundred individuals. Chagnon and Bugos (1979) recorded a crisis situation—an axe fight—precipitated when members of a recent splinter group visited their original village. One of the male visitors (Mohesiwa) was insulted by a resident woman, and he beat her. When she told her story, her kinsmen were angered, and the dispute escalated into an axe fight with several injuries. Chagnon and Bugos analyzed the fight in detail and compared the coefficients of relatedness. The supporters of Mohesiwa were related to him on the average 7.8 times more than they were to the principal opponent, who was a half-brother of the beaten woman. Chagnon and Bugos argued that kinship behavior in this society is consistent with predictions based on Hamilton's theory.

Closer to home, Daly and Wilson (1981, 1982, 1988, 1996) studied violence against stepchildren. Out of 98 murders among cohabitants in Detroit, 76 (77 percent) were by nonrelatives living in the home, usually the spouse. When corrected for their representation in the family, relatives were 10 times less likely to kill each other than were nonrelatives. Kinship is also important in predicting patterns of child abuse. Parents are far more likely to abuse their stepchildren than their genetic offspring (Daly and Wilson 1981). An even stronger relationship is evident when comparing cases of child homicide, where stepparents are greatly overrepresented.

Other approaches besides analyses of kinship have been used in the studies of humans. For example, researchers have long used game theory to help us understand human behavior, including the analysis of cold war politics. A recent example is Wilson (1998), who discusses the variation across humans in their tendency to manipulate others for personal gain in terms of game-theoretic strategies.

Our ability to apply evolutionarily based explanations to human behavior is limited. It is impossible to conduct controlled experiments in humans, and other explanations (such as familiarity) can be invoked to explain many of the preceding observations. In addition, although some social systems can be understood as partly the result of kin selection, humans engage in reciprocity, which may transcend kinship as an explanation for much of human behavior. The constraints of rigid kin selection have been broken by our use of written and spoken language to fashion long-remembered agreements upon which civilizations can be built.

SUMMARY

A society may be defined as a group of individuals of the same species, organized in a cooperative manner that extends beyond sexual behavior. We began with a historical overview of the discipline in the last century, with its initial emphasis on characteristics of groups that were above and beyond the characteristics of individuals. Later, researchers favored selection at the level of the individual or the gene as the most parsimonious explanation of grouping behavior, although in recent years some of the earlier ideas have once again been given attention.

Group living has both costs and benefits, and any number of these may operate simultaneously. The potential benefits of social behavior include protection from physical elements, detection of and defense against predators, assembly of sexual species for finding mates, locating and procuring food, resource defense against conspecifics or competing species, division of labor among specialists, a richer learning environment for the young, aid (or receiving aid from) offspring or other relatives, and modifying the environment. Costs include increased competition, spread of contagious diseases and parasites, and interference in reproduction.

Evolutionary biologists have been puzzled by animals that aid one another to the detriment of their own fitness. One explanation for these altruistic acts is based on the fact that animals share genes with nondescendant kin as well as their own kin. By helping to produce nondescendant kin, animals may increase their inclusive fitness. Hamilton's rule predicts when an altruistic act will spread through the population.

While examples of kin selection abound in nature, there are fewer examples of cooperation among nonrelatives. Cooperation may also evolve through reciprocity. The prisoner's dilemma, a game theory model, has been a provocative model of the conditions under which cooperation may become established. We end this section with some examples that appear, at first glance, to be altruistic, but in fact lead to an increase in the actor's direct fitness.

The rule in social organizations across taxa is that there are exceptions to every rule. For example, birds are generally socially monogamous and territorial, yet many cooperatively breed. Individuals forego breeding to help other birds, often (but not always) their parents. The key to understanding cooperative breeding is to keep in mind both immediate and long-term, and both direct and indirect, fitness benefits. The same can be said of eusociality, where the reproductive division of labor is even more striking.

Finally, we briefly discuss the sociobiology of human behavior. Intriguing patterns in such phenomena as inheritance and assistance of kin during aggressive interactions are in line with evolutionary interpretations of human behavior. However, human behavior is particularly difficult to study and interpret, so caution is warranted.

DISCUSSION QUESTIONS

1. A bird can help its sister raise five additional offspring, or it can raise three offspring of its own. What should it do? What additional information would you like to have before you feel confident in your answer? (Hint: think about long-term gains.)

2. Is kin recognition necessary for kin selection? Justify your answer.

3. Imagine you are initiating a study of a little-known cooperatively breeding bird. Outline a series of tests to determine why helpers help. Which explanations are mutually exclusive? Which are not?

4. We know that much of human behavior has a genetic basis, but also a great deal of flexibility. In other species, we can undertake manipulative experiments to investigate the costs and benefits of social behavior, but these techniques are not available to us for work on humans. What do you think about how (or if) we should approach the issue of the study of the evolution of human social behavior?

SUGGESTED READINGS

Choe, J., and B. Crespi, eds. 1997. *The Evolution of Social Behavior in Insects and Arachnids.* Cambridge, England: Cambridge University Press.

Rubenstein, D. I., and R. W. Wrangham, eds. 1986. *Ecological Aspects of Social Evolution. Birds and Mammals.* Princeton: Princeton University Press.

Stacey, P., and W. Koening. 1990. *Cooperative Breeding in Birds: Long-Term Studies of Ecology and Behavior.* Cambridge, England: Cambridge University Press.

Wilson, E. O. 1975. *Sociobiology: The New Synthesis.* Cambridge, MA: Harvard University Press.

REFERENCES

Able, K. P. 1974. Environmental influences on the orientation of nocturnal bird migrants. *Anim. Behav.* 22:225–39.

———. 1978. Field studies of the orientation cue hierarchy of nocturnal songbird migrants. In *Animal Migration, Navigation, and Homing,* ed. Schmidt-Koenig and W. T. Keeton. New York: Springer-Verlag.

———. 1980. Mechanisms of orientation, navigation, and homing. In *Animal Migration, Orientation, and Navigation,* ed. S. A. Gauthreaux. New York: Academic Press.

———. 1982. Field studies of avian nocturnal migration orientation. I. Interaction of sun, wind and stars as directional cues. *Anim. Behav.* 30:761–67.

———. 1989. Skylight polarization patterns and the orientation of migratory birds. *J. Exp. Biol.* 141:241–56.

———. 1991. Common themes and variations in animal orientation systems. *Amer. Zool.* 31:157–67.

Able, K. P., and M. A. Able. 1990a. Ontogeny of migratory orientation in the Savannah sparrow (*Passerculus sandwichensis*): Mechanisms of sunset orientation. *Anim. Behav.* 39:905–13.

———. 1990b. Ontogeny of migratory orientation in the Savannah sparrow (*Passerculus sandwichensis*): Calibration of the magnetic compass. *Anim. Behav.* 39:1189–98.

Able, K. P., V. P. Bingman, P. Kerlinger, and W. Gergits. 1982. Field studies of avian nocturnal migratory orientation. II. Experimental manipulation in white-throated sparrows (*Zonotrichia albicollis*) released aloft. *Anim. Behav.* 30:768–73.

Abramson, C. I. 1986. Aversive conditioning in honeybees (*Apis mellifera*). *J. Comp. Psychol.* 100:108–16.

Abu-Gideiri, Y. B. 1966. The behaviour and neuroanatomy of some developing teleost fishes. *J. Zool.* 149:215–41.

Adamo, S. A., K. Schildberger, and W. Loher. 1994. The role of the corpora allata in the adult male cricket (*Gryllus campestris* and *Gryllus bimaculatus*) in the development and expression of its agonistic behaviour. *J. Insect Phys.* 40:439–46.

Ader, R., and P. M. Conklin. 1963. Handling of pregnant rats: Effects on emotionality of their offspring. *Science* 142:411–12.

Adkins-Regan, E. 1981. Early organizational effects of hormones. In *Neuroendocrinology of Reproduction,* ed. N. Adler. New York: Plenum.

———. 1987. Sexual differentiation in birds. *Trends. Neurosci.* 10:517–22.

Adkins-Regan, E., and M. Ascenzi. 1987. Social and sexual behaviour of male and female zebra finches treated with oestradiol during the nestling period. *Anim. Behav.* 35:1100–12.

Adler, K., and D. H. Taylor. 1973. Extraocular perception of polarized light by orienting salamanders. *J. Comp. Physiol.* 87:203–12.

Albrecht, H. 1966. Zur Stammesgeschichte einiger Bewegungsweisen bei Fischen untersucht am Verhalten von *Haplochromis* (Pisces, Cichlidae). *Zeit. Tierpsychol.* 23:270–301.

Alcock, J., and J. E. Schaefer. 1983. Hilltop territoriality in a Sonoran desert bot fly (Diptera: Cuterebridae). *Anim. Behav.* 31:518–25.

Alerstam, T. 1985. Strategies of migration flight, illustrated by arctic and common terns, *Sterna paradisaea* and *Sterna hirundo.* Pp. 580–603 in *Migration: Mechanisms and Adaptive Significance* (M. A. Rankin, ed.). Port Aransas, TX: Marine Science Institute.

Alerstam, T. 1990. *Bird Migration.* Cambridge: Cambridge University Press.

Alexander, R. D. 1961. Aggressiveness, territoriality, and sexual behavior in field crickets (Orthoptera: Gryllidae). *Behaviour* 17:130–223.

———. 1974. The evolution of social behavior. *Ann. Rev. Ecol. Syst.* 5:325–83.

Alexander, R. D. 1974. The evolution of social behavior. *Ann. Rev. Ecol. Syst.* 5:325–83.

Alexander, R., and P. Sherman. 1977. Local mate competition and parental investment in social insects. *Science* 196:494–500.

Alexander, R. D., J. L. Hoogland, R. D. Howard, K. M. Noonan, and P. W. Sherman. 1979. Sexual dimorphisms and breeding systems in pinnipeds, ungulates, primates and humans. In *Evolutionary Biology and Human Social Behavior: An Anthropological Perspective,* ed. N. A. Chagnon and W. Irons. N. Scituate, MA: Duxbury.

Alexander, R. D., and K. M. Noonan. 1979. Concealment of ovulation, parental care, and human social evolution. In *Evolutionary Biology and Human Social Behavior: An Anthropological Perspective,* ed. N. A. Chagnon and W. Irons. N. Scituate, MA: Duxbury.

Alkon, D. L. 1980. Cellular analysis of a gastropod (*Hermissenda crassicornis*) model of associative learning. *Biol. Bull.* 159:505–60.

———. 1983. Learning in a marine snail. *Sci. Amer.* 249(July):70–85.

Allee, W. C. 1951. *Cooperation among Animals.* Chicago: University of Chicago Press.

Altmann, J. 1974. Observational study of behavior: Sampling methods. *Behaviour* 49:227–67.

Altmann, J., G. Hausfater, and S. A. Altmann. 1988. Determinants of reproductive success in savannah baboons, *Papio cynocephalus.* In *Reproductive Success,* ed. T. H. Clutton-Brock. Chicago: University of Chicago Press.

Alyan, S., and R. Jander. 1994. Short-range homing in the house mouse, *Mus musculus:* Stages in the learning of directions. *Anim. Behav.* 48:285–98.

Ambrose, H. W. III. 1972. Effect of habitat familiarity and toe-clipping on rate of owl predation in *Microtus pennsylvanicus.* *J. Mammal.* 53:909–12.

American Ornithologists' Union. 1988. Report of the committee on use of wild birds in research. *Auk,* 105(Suppl.): 1A–41A.

American Society of Mammalogists. 1987. Acceptable held methods in mammalogy: Preliminary guidelines approved by the American Society of Mammalogists. *J. Mammal.* 68(Suppl.): 1–18.

Anderson, B., and S. M. McCann. 1955. A further study of polydipsia evoked by hypothalamic stimulation in the goat. *Acta Physiol. Scand.* 33:333–46.

———. 1956. The effect of hypothalamic lesions on water intake of the dog. *Acta Physiol. Scand.* 35:312–20.

Anderson, L. T. 1973. An analysis of habitat preference in mice as a function of prior experience. *Behaviour* 47:302–39.

Anderson, P.K., G.E. Heinsohn, P.H. Whitney, and J.P. Huang. 1977. *Mus musculus* and *Peromyscus maniculatus*: Homing ability in relation to habitat utilization. *Can. J. Zool.* 55:169–82.

Andersson, M. 1982. Female choice selects for extreme tail length in a widowbird. *Nature* 299:818–20.

Andersson, M. 1994. *Sexual Selection.* Princeton: Princeton University Press.

Animal Behaviour Society and Association for the Study of Animal Behaviour. 1986. Guides for the use of animals in research. *Anim. Behav.* 34:315–18.

Appolonio, M., M. Festa-Bianchet, F. Mari, S. Mattioli, and B. Sarno. 1992. To lek or not to lek: Mating strategies of male fallow deer. *Behav. Ecol.* 3:25–31.

Arcese, P. 1994. Harem size and horn symmetry in oribi. *Anim. Behav.* 48:1485–88.

Archer, J. 1973. Tests for emotionality in rats and mice: A review. *Anim. Behav.* 21:205–35.

———. 1988. *The Behavioural Biology of Aggression.* New York: Cambridge University Press.

Ardrey, R. 1966. *The Territorial Imperative.* New York: Atheneum.

Arendash, G. W., and R. A. Gorski. 1982. Enhancement of sexual behavior in female rats by neonatal transplantation of brain tissue from males. *Science* 217:1276–78.

Arendt, J., M. Aldhous, and V. Marks. 1986. Alleviation of jet lag by melatonin: Preliminary results of controlled double blind trial. *Brit. Med. J.* 292:1170.

Arenz, C. L., and D. W. Leger. 1999. Thirteen-lined ground squirrel (Sciuridae: *Spermophilus tridecemlineatus*) antipredator vigilance decreases as vigilance cost increases. *Anim. Behav.* 75:97–103.

Arey, L. B. 1954. *Developmental Anatomy.* Philadelphia: W. B. Saunders.

Arling, G. L., and H. F. Harlow. 1967. Effects of social deprivation on maternal behavior of rhesus monkeys. *J. Comp. Physiol. Psychol.* 64:371–77.

Armstrong, D. P. 1991. Aggressiveness of breeding territorial honeyeaters corresponds to seasonal changes in nectar availability. *Behav. Ecol. Sociobiol.* 29:103–11.

Arnold, A. P., and R. A. Gorski. 1984. Gonadal steroid induction of structural sex differences in the central nervous system. *Ann. Rev. of Neurosci.* 7:413–42.

Arnold, K., and I. Owens. 1999. Cooperative breeding in birds: The role of ecology. *Behav. Ecol.* 10:465–71.

Aronson, L. R., and M. L. Cooper. 1966. Seasonal variation in mating behavior in cats after desensitization of the glans penis. *Science* 152:226–30.

Aschoff, J. 1960. Exogenous and endogenous components in circadian rhythms. *Cold Spr. Harb. Symp. Quant. Biol.* 25:11–28.

———. 1963. Comparative physiology: Diurnal rhythms. *Ann. Rev. Physiol.* 25:581–600.

———. 1979. Circadian rhythms: Influences of internal and external factors on the period measured in constant conditions. *Zeit. Tierpsychol.* 49:225–49.

———. 1981. *Handbook of Behavioral Neurobiology.* Vol. 4: *Biological Rhythms.* New York: Plenum.

Askenmo, C. E. H. 1984. Polygyny and nest selection in the pied flycatcher. *Anim. Behav.* 32:972–80.

Attneave, F. 1959. *Applications of Information Theory to Psychology.* New York: Holt, Rinehart and Winston.

August, P. V., S. G. Ayvazian, and J. G. T. Anderson. 1989. Magnetic orientation in a small mammal, *Peromyscus leucopus. J. Mamm.* 70:1–9.

Austad, S. N. 1982. First male sperm priority in the bowl and doily spider, *Frontinella pyramitela* (Walckenaer). *Evolution* 36:777–85.

———. 1983. A game theoretical interpretation of male combat in the bowl and doily spider (*Frontinella pyramitela*). *Anim. Behav.* 31:59–73.

Austad, S. N., and M. E. Sunquist. 1987. Sex ratio manipulation in the common opossum. *Nature* 324:58–60.

Ayre, D. J., and R. K. Grosberg. 1995. Aggression, habituation, and clonal coexistence in the sea anemone *Anthopleura elegantissima. Am. Nat.* 146:427–53.

Axelrod, R., and D. Dion. 1988. The further evolution of cooperation. *Science* 242:1385–90.

Axelrod, R., and W. D. Hamilton. 1981. The evolution of cooperation. *Science* 211:1390–96.

Bailey, E. D. 1966. Social interaction as a population regulating mechanism in mice. *Can. J. Zool.* 44:1007–12.

Baker, L. M., and J. C. Smith. 1974. A comparison of taste aversions induced by radiation and lithium chloride in CS-US and US-CS paradigms. *J. Comp. Physiol. Psychol.* 87:644–54.

Baker, M. C. 1983. The behavioral response of female Nuttall's white-crowned sparrows to male song of natal and alien dialects. *Behav. Ecol. Sociobiol.* 12:309–15.

Baker, Myron C. in press. *Bird Song Research: The Past One Hundred Years.*

Baker, R. 1980. Goal orientation by blindfolded humans after long-distance displacement: Possible involvement of a magnetic sense. *Science* 210:555–57.

Baker, R. R. 1987. Human navigation and magnetoreception: The Manchester experiments do replicate. *Anim. Behav.* 35:691–704.

Baker, R. R., and M. A. Bellis. 1988. 'Kamikaze' sperm in mammals? *Anim. Behav.* 36:936–38.

Balda, R. P., I. M. Pepperberg, and A. C. Kamil (eds.). 1998. *Animal Cognition in Nature.* San Diego, CA: Academic Press.

Balda, R., and A. Kamil. 1998. The ecology and evolution of spatial memory in corvids of the southwestern USA: The perplexing pinyon jay. In R. P. Balda, I. M. Pepperberg, and A. C. Kamil (eds.), *Animal Cognition in Nature.* New York: Academic Press.

Balda, R. P. 1980. Recovery of cached seeds by a captive *Nucifraga caryocatactes. Z. Tierpsychol.* 52:331–46.

———. 1987. Avian impacts on pinyon-juniper woodlands. 525–33, In *Proceedings: Pinyon-Juniper Conference,* ed. R. L. Everett. U.S. Forest Service Technical Report INT -215.

Balda, R. P., and A. C. Kamil. 1988. The spatial memory of Clark's nutcrackers (*Nucifraga columbiana*) in an analogue of the radial arm maze. *Anim. Learn. Behav.* 16:116–22.

———. 1989. A comparative study of cache recovery by three corvid species. *Anim. Behav.* 38:486–95.

———. 1992. Long-term memory in Clark's nutcracker, *Nucifraga columbiana. Anim. Behav.* 44:761–69.

Balmford, A. P. 1991. Mate choice on leks. *Trends Ecol. Evol.* 6:87–92.

Balthazart, J. A. Foidart, C. Surelmont, and N. Harada. 1990. Neuroanatomical specificity in the co-localization of aromatase and estrogen receptors. *J. Neurobiol.* 22:143–57.

Balzer, I., and R. Hardeland. 1992. Multiple ultradian frequencies in dark motility of *Euglena. J. Interdiscipl. Cycle Res.* 23:47–55.

Bandler, R. J., Jr., and J. P. Flynn. 1974. Nerve pathways from the thalamus associated with regulation of aggressive behavior. *Science* 183:96–99.

Banfield, A. W. 1954. Preliminary investigation of the barren ground caribou. II. Life history, ecology, and utilization. *Can. Wild. Serv. Wild. Mgmt. Bull,* No. 10B.

Baptista, L. F., and L. Petrinovich. 1986. Song development in the white-crowned sparrow: Social factors and sex differences. *Anim. Behav.* 34:1359–71.

Baptista, L. F., D. A. Nelson, and S. L. L. Gaunt. 1998. Cognitive processes in avian vocal acquisition. In R. P. Balda, I. M. Pepperberg, and A. C. Kamil (eds.), *Animal Cognition in Nature.* San Diego: Academic Press.

Barash, D. P. 1973a. The social biology of the Olympic marmot. *Anim. Behav. Monogr.* 6:171–249.

———. 1973b. Social variety in the yellow-bellied marmot (*Marmota flaviventris*). *Anim. Behav.* 21:579–84.

———. 1974a. The evolution of marmot societies: A general theory. *Science* 185:415–20.

———. 1974b. Social behavior of the hoary marmot (*Marmota caligata*). *Anim. Behav.* 22:257–62.

Barbour, R. W., W. H. Davis, and M. D. Hassell. 1966. The need of vision in homing by *Myotis sodalis. J. Mammal.* 47:356–57.

Barbour, R. W., and W. H. Davis. 1969. *Bats of America.* Lexington: University of Kentucky Press.

Barfield, R. J. 1971. Activation of sexual and aggressive behavior activated by androgen implanted into the male ring dove brain. *Endocrinology* 89:1470–76.

Barfield, R. J., D. E. Busch, and K. Wallen. 1972. Gonadal influence on agonistic behavior in the male domestic rat. *Horm. Behav.* 3:247–59.

Barinaga, M. 1994. A new tool for examining multigenic traits. *Science* 264:1691.

Barlow, G. W. 1968. Ethological units of behavior. In *The Central Nervous System and Fish Behavior,* ed. D. Ingle. Chicago: University of Chicago Press.

Barnard, C. J., and R. M. Sibly. 1981. Producers and scroungers: A general model and its application to captive flocks of house sparrows (*Passer domesticus*). *Anim. Behav.* 29:543–50.

Barnes, B. M. 1989. Freeze avoidance in a mammal: Body temperatures below 0°C in an arctic hibernator. *Science* 244:1593–95.

Barnett, S. A. 1956. Behavior components in the feeding of wild and laboratory rats. *Behaviour* 9:24–43.

———. 1963. *The Rat: A Study in Behavior.*Chicago: Aldine.

Barnwell, F. H. 1966. Daily and tidal patterns of activity in individual fiddler crabs (genus *Uca*) from the Woods Hole region. *Biol. Bull.* 130:1–17.

Barth, F. G. 1982. Spiders and vibratory signals: Sensory reception and behavioral significance. In *Spider Communication: Mechanisms and Ecological Significance,* eds. P. N. Witt and J. S. Rovner. Princeton, NJ: Princeton Univ. Press.

Barth, R. H. 1965. Insect mating behavior: Endocrine control of a chemical communication system. *Science* 149:882–83.

———. 1968. The comparative physiology of reproductive processes in cockroaches. I. Mating behavior and its endocrine control. *Adv. Reprod. Physiol.* 3:167–201.

Basmajian, J. V., and D. A. Ranney. 1961. Chemomyelotomy: Substitute for general anesthesia in experimental surgery. *J. Appl. Physiol.* 16:386.

Basolo, A. L. 1990. Female preference predates the evolution of the sword in swordtail fish. *Science* 250:808–10.

Basolo, A. L. 1995. A further examination of a pre-existing bias favoring a sword in the genus *Xiphophorus. Anim. Behav.* 50:365–375.

Bateman, A. J. 1948. Intra-sexual selection in *Drosophila. Heredity* 2:349–68.

Bateson, M., and A. Kacelnik. 1997. Starlings' preferences for predictable and unpredictable delays to food. *Anim. Behav.* 53:1129–42.

Bateson, P. P. G. 1978. Early experience and sexual preferences. In *Biological Determinants of Sexual Behavior,* ed. J. B. Hutchinson. New York: Wiley.

———. 1982. Preference for cousins in Japanese quail. *Nature* 295:236–37.

Batorczak, A., P. Jablonski, and A. Rowinski. 1994. Mate choice by male water striders (*Gerris lacustris*): Expression of a wing morph preference depends on a size difference between females. *Behav. Ecol.* 5:17–20.

Baum, M. J., M. S. Erskine, E. Kornberg, and C. E. Weaver. 1990. Prenatal and neonatal testosterone exposure interact to affect differentiation of sexual behavior and partner preference in female ferrets. *Behav. Neurosci.* 104:183–98.

Beach, F. A. 1950. The snark was a boojum. *Amer. Psychol.* 5:115–24.

———. 1960. Experimental investigations of species-specific behavior. *Amer. Psychol.* 15:1–18.

———. 1976. Sexual attractivity, proceptivity, and receptivity in female mammals. *Horm. Behav.* 7:105–38.

Beach, F. A., and A. M. Holz-Tucker. 1949. Effects of different concentrations of androgen upon sexual behavior in castrated male rats. *J. Comp. Physiol. Psychol.* 42:433–53.

Beck, B. B. 1980. *Animal Tool Behavior: The Use and Manufacture of Tools by Animals.* New York: Garland STPM.

Beck, S. D. 1968. *Insect Photoperiodism.* New York: Academic Press.

Becker, J. B., S. Marc Breedlove, and D. Crews, eds. *Behavioral Endocrinology.* Cambridge, MA: MIT Press.

Bednarz, J. C. 1988. Cooperative hunting in Harris' hawks (*Parabuteo unicinctus*). *Science* 239:1525–27.

Bednekoff, P. A., R. P. Balda, A. C. Kamil, and A. G. Hile. 1997. Long-term spatial memory in four seed-caching corvid species. *Anim. Behav.* 53:335–41.

Beecher, M. D. 1982. Signature systems and kin recognition. *Amer. Zool.* 22: 477–90.

Beecher, M. D., M. B. Medvin, P. K. Stoddard, and P. Loesch. 1986. Acoustic adaptations for parent-offspring recognition in swallows. *Exp. Biol.* 45:179–93.

Beecher, M. D., P. K. Stoddard, S. E. Campbell, and C. L. Horning. 1996. Repertoire matching between neighbouring songbirds. *Anim. Behav.* 51:917–23.

Beecher, M. D., S. E. Campbell, and J. C. Nordby. 1998. The cognitive ecology of song communication and song learning in the song sparrow. In R. Dukas (ed.), *Cognitive Ecology.* Chicago: University of Chicago Press.

Begun, D., J. L. Kubie, M. P. O'Keefe, and H. Halpern. 1988. Conditioned discrimination of airborne odorants by garter snakes (*Thamnophis radix* and *T. sirtalis sirtalis*). *J. Comp. Psychol.* 102:35–43.

Behrend, E. R., and M. E. Bitterman. 1963. Sidman avoidance in the fish. *J. Exp. Anal. Behav.* 6:47–52.

Bekoff, M. 1974a. Social play in coyotes, wolves, and dogs. *Bioscience* 24:225–30.

———. 1974b. Social play in mammals. *Amer. Zool.* 14:265–436.

———. 1974c. Social play and play-soliciting by infant canids. *Amer. Zool.* 14:323–40.

———. 1977. Social communication in canids: Evidence for the evolution of a stereotyped mammalian display. *Science* 197:1097–99.

———. 1978. Social play: Structure, function and the evolution of a cooperative social behavior. In *Development of Behavior,* ed. G. Burghardt and M. Bekoff. New York: Garland STPM Press.

———. 1992. Time, energy, and play. *Anim. Behav.* 44:981–82.

Bekoff, M., and J. A. Byers. 1981. A critical reanalysis of the ontogeny and phylogeny of mammalian social and locomotor play: An ethological hornet's nest. In *Behavioral Development,* ed. K. Immelmann et al. New York: Cambridge University Press.

Bekoff, M., and J. A. Byers. 1992. Time, energy and play. *Anim. Behav.* 44:981–82.

Bekoff, M., and J. A. Byers. 1998. *Animal Play: Evolutionary, Comparative, and Ecological Perspectives.* New York: Cambridge University Press.

Beletsky, L. D., and G. H. Orians. 1987. Territoriality among red-winged black birds. I. Site fidelity and movement patterns. *Behav. Ecol. Sociobiol.* 20:21–34.

Beletsky, L. D., and G. H. Orians. 1989. Territoriality among red-winged black birds. III. Testing hypotheses of territorial dominance. *Behav. Ecol. Sociobiol.* 24:333–39.

Beletsky, L. D., G. H. Orians, and J. C. Wingfield. 1989. Relationships of steroid hormones and polygyny to territorial status, breeding experience, and reproductive success in male red-winged blackbirds. *Auk* 106:107–17.

———. 1990a. Effects of exogenous androgen and antiandrogen on territorial and nonterritorial red-winged blackbirds (Aves: Icterinae). *Ethology* 85:58–72.

———. 1990b. Steroid hormones in relation to territoriality, breeding density, and parental behavior in male yellow-headed blackbirds. *Auk* 107:60–68.

Beletsky, L. D., and G. H. Orians. 1991. Effects of breeding experience and familiarity on site fidelity in female red-winged blackbirds. *Ecology* 72:787–96.

Beletsky, L. D., and G. H. Orians. 1991. Effects of breeding experience and familiarity on site fidelity in female red-winged blackbirds. *Ecology* 72:787–96.

———. 1992. Year-to-year patterns of circulating levels of testosterone and corticosterone in relation to breeding density, experience, and reproductive success of the polygynous red-winged blackbird. *Horm. Behav.* 26:420–32.

Bell, G. 1982. *The Masterpiece of Nature: The Evolution and Genetics of Sexuality.* Berkeley: University of California Press.

Bell, W. J. 1991. *Searching Behaviour: The Behavioural Ecology of Finding Resources.* London: Chapman and Hall.

Bellrose, F. C. 1964. Radar studies of waterfowl migration. *Trans. N. Amer. Wild. Nat. Conf.* 29:128–43.

———. 1967. Orientation in waterfowl migration. *Proc. Ann. Biol. Colloq., Oreg. State Univ.* 27:73–99.

———. 1971. The distribution of nocturnal migrants in the air space. *Auk* 88:397–424.

Belvin, M. P., and J. C. P. Yin. 1997. Drosophila learning and memory: Recent progress and new approaches. *BioEssays* 19:1083–89.

Bennett, M. A. 1940. The social hierarchy in ring doves. II. The effect of treatment with testosterone propionate. *Ecology* 21:148–65.

Bentley, D., and R. R. Hoy. 1970. Postembryonic development of adult motor patterns in crickets: A neural analysis. *Science* 170:1409–11.

———. 1974. The neurobiology of cricket song. *Science* 231:34–44.

Benzer, S. 1973. Genetic dissection of behavior. *Sci. Amer.* 229:24–37.

Bercovitch, F. B. 1986. Male rank and reproductive activity in savanna baboons. *Int. J. Primatol.* 7:533–50.

Berkowitz, L. 1969. *The Roots of Aggression.* New York: Atherton.

Berkowitz, L. 1993. *Aggression: Its Causes, Consequences, and Control.* New York: McGraw-Hill.

Berkowitz, L., and A. LePage. 1967. Weapons as aggression-eliciting stimuli. *J. Per. Soc. Psychol.* 7:202–7.

Berman, C. M. 1980. Mother-infant relationships among free-ranging rhesus monkeys on Cayo Santiago: A comparison with captive pairs. *Anim. Behav.* 28:860–73.

Berman, C. M. 1982. The ontogeny of social relationships with group companions among free-ranging infant rhesus monkeys. I. Social networks and differentiation. *Anim. Behav.* 30:149–62.

Bernstein, I. S. 1966. Analysis of a key role in a capuchin (*Cebus albifrons*) group. *Tulane Stud. Zool.* 13:49–54.

Bernstein, I. S. 1981. Dominance: The baby and the bathwater. *Behav. and Brain Sci.* 4:419–57.

Beroza, M., and E. F. Knipling. 1972. Gypsy moth control with the sex attractant pheromone. *Science* 177:19–27.

Berteaux, D., M. Crete, J. Huot, J. Maltais, and J. P. Ouellet. 1998. Food choice by white-tailed deer in relation to protein and energy content of the diet: A field experiment. *Oecologia* 115:84–92.

Berven, K. A. 1981. Mate choice in the wood frog, *Rana sylvatica. Evolution* 35:707–22.

Berven, K. A., and T. A. Grudzien. 1990. Dispersal in the wood frog (*Rana sylvatica*): Implications for genetic population structure. *Evol.* 44:2047–56.

Beugnon, G. 1986. Learned orientation in landward swimming in the cricket *Pteronemobius lineolatus. Behav. Proc.* 12:215–26.

Bingman, V. P. 1998. Spatial representations and homing pigeon navigation. Pp. 69–85 in *Spatial Representation in Animals* (S. Healy, ed.). Oxford: Oxford University Press.

Bingman, V. P., and T.-J. Jones. 1994. Sun compass-based spatial learning impaired in homing pigeons with hippocampal lesions. *J. Neurosci.* 14:6687–94.

Bingman, V. P., L. V. Riters, R. Strasser, and A. Gagliardo. 1998. Neuroethology of avian navigation. Pp. 201–26 in *Animal Cognition in Nature* (R. P. Balda, I. M. Pepperberg, and A. C. Kamil, eds.). New York: Academic Press.

Birdsall, D. A., and D. Nash. 1973. Occurrence of successful multiple insemination of females in natural populations of deer mice (*Peromyscus maniculatus*). *Evolution* 27:106–10.

Birkhead, T. R. 1998. Cryptic female choice: Criteria for establishing female sperm choice. *Evolution* 52:1212–18.

Bitterman, M. E. 1960. Toward a comparative psychology of learning. *Amer. Psychol.* 15:704–12.

———. 1965a. The evolution of intelligence. *Sci. Amer.* 212:92–100.

———. 1965b. Phyletic differences in learning. *Amer. Psychol.* 20:396–410.

———. 1975. The comparative analysis of learning. *Science* 188:699–709.

———. 1988. Vertebrate-invertebrate comparisons. In *Intelligence and Evolutionary Biology,* ed. H. J. Jerison and I. Jerison. New York: Springer-Verlag.

Blaich, C. F., and D. B. Miller. 1986. Call responsivity of mallard ducklings (*Anas platyrhynchos*). IV. Effects of social experience. *J. Comp. Psychol.* 100:401–5.

———. 1974. Character displacement in frogs. *Amer. Zool.* 14:1119–25.

Blest, A. D. 1957. The function of eyespot patterns in the Lepidoptera. *Behaviour* 11:209–56.

Block, G. D., and T. L. Page. 1978. Circadian pacemakers in the nervous system. *Ann. Rev. Neurosci.* 1:19–34.

Block, R. A., and J. V. McConnell. 1967. Classically conditioned discrimination in the planarian, *Dugesia dorotocephala. Nature* 215:1465–66.

———. 1975. The comparative analysis of learning. *Science* 188:699–709.

Boag, P. T., and P. R. Grant. 1981. Intense natural selection in a population of Darwin's finches (Geospizinae) in the Galápagos. *Science* 214:82–85.

Boake, C. R. B. (ed.). 1994. *Quantitative Genetic Studies of Behavioral Evolution.* Chicago: University of Chicago Press.

Boake, C. R. B., and A. Hoikkala. 1995. Courtship behaviour and mating success of wild-caught *Drosophila silvestris* males. *Anim. Behav.* 49:1303–13.

Boake, C. R. B., and T. Poulsen. 1997. Correlates versus predictors of courtship success: Courtship song in *Drosophila silvestris* and *D. heteroneura. Anim. Behav.* 54:699–704.

Bock, J., and K. Braun. 1999. Blockade of N-methyl-D-aspartate receptor activation suppresses learning-induced synaptic elimination. *Proc. Nat. Acad. Sci. USA* 96:2485–90.

Boice, R. 1972. Some behavioral tests of domestication in Norway rats. *Behaviour* 42:198–231.

Boice, R., and M. R. Denny. 1965. The conditioned licking response in rats as a function of the CS-UCS interval. *Psychonom. Sci.* 3:93–94.

Bolles, R. 1970. Species-specific defense reactions and avoidance learning. *Psychol. Rev.* 77:32–48.

Bolles, R. C. 1985. The slaying of Goliath: What happened to reinforcement theory? In *Issues in the Ecological Study of Learning,* ed. T. D. Johnston and A. T. Pietrewicz. Hillsdale, NJ: L. Erlbaum Assoc.

Borg, K. E., K. L. Eisenshade, B. H. Johnson, D. D. Lunstra, and J. J. Ford. 1992. Effects of sexual experience, season, and mating stimuli on endocrine concentrations in the adult ram. *Hormones and Behavior* 26:87–109.

Borgerhoff Mulder, M. 1991. Human behavioural ecology. Pp. 69–104 in *Behavioural Ecology: An Evolutionary Approach,* 3rd edition. (J. Krebs and N. Davies, eds.). Oxford: Blackwell Scientific.

Borgia, G. 1985. Bower quality, number of decorations and mating success of male satin bowerbirds (*Ptilonorhynchus violaceus*): An experimental analysis. *Anim. Behav.* 33:266–71.

Borgia, G. 1995. Why do bowerbirds build bowers? *Amer. Sci.* 83:542–47.

Borgia, G., and K. Collis. 1989. Female choice for parasite-free male satin bowerbirds and the evolution of bright male plumage. *Behav. Ecol. Sociobiol.* 25:445–54.

Bottjer, S. W., S. L. Glaessner, and A. P. Arnold. 1985. Ontogeny of brain nuclei controlling song learning and behavior in zebra finches. *J. Neurosci.* 5:1556–62.

Bouchard, T. J., Jr. 1994. Genes, environment, and personality. *Science* 264:1700–1.

Bouissou, M. F. 1978. Effect of injections of testosterone propionate on dominance relationships in a group of cows. *Horm. Behav.* 11:388–400.

Bouissou, M. F., and V. Gaudioso. 1982. Effect of early androgen treatment on subsequent social behavior in heifers. *Horm. Behav.* 16:132–46.

Bourke, A., and N. Franks. 1995. *Social Evolution in Ants.* Princeton: Princeton University Press.

Bovet, J. 1992. Mammals. Pp. 321–61 in *Animal Homing* (F. Papi, ed.). New York: Chapman and Hall.

Bowen, B. W., A. B. Meylan, and J. C. Avise. 1989. An odyssey of the green sea turtle: Ascension Island revisited. *Proc. Nat. Acad. Sci. U.S.A.* 86:573–76.

Boycott, B. B. 1965. Learning in the octopus. *Sci. Amer.* 212(3):42–50.

Brace, R.C., J. Pavey, and D.L. Quicke.1979. Intraspecific aggression in the colour morphs of the anemone *Actinia equina*: the 'convention' governing dominance ranking. *Anim. Behav.* 27:553–61.

Bradbury, J. W., and S. L. Vehrencamp. 1998. *Principles of Animal Communication.* Sunderland, Mass: Sinauer Associates.

Bradbury, J. W. 1977. Lek mating behavior in the hammer-headed bat. *Zeit. Tierpsychol.* 45:225–55.

Brady, J. 1967a. Control of the circadian rhythm of activity in the cockroach. I. The role of the corpora cardiaca, brain, and stress. *J. Exp. Biol.* 47:153–63.

———. 1967b. Control of the circadian rhythm of activity in the cockroach. II. The role of the subesophageal ganglion and ventral nerve cord. *J. Exp. Biol.* 47:165–78.

———. 1969. How are insect circadian rhythms controlled? *Nature* 223:781–84.

———. 1979. *Biological Clocks.* Baltimore: University Press.

———. 1988. The circadian organization of behavior: Timekeeping in the tsetse fly, a model system. *Adv. Stud. Behav.* 18:153–91.

Brannon, E. L., T. P. Quinn, G. L. Lucchetti, and B. D. Ross. 1981. Compass orientation of sockeye salmon fry from a complex river system. *Can. J. Zool.* 59:1548–53.

Braude, S., Z. Tang-Martinez, and G. T. Taylor. 1999. Stress, testosterone, and the immunodepression hypothesis. *Behav. Ecol.* 10:345–50.

Braun, K., J. Bock, M. Metzger, S. Jiang, and R. Schnabel. 1999. The dorsocaudal neostriatum of the domestic chick: A structure serving higher associative functions. *Behav. Brain Res.* 98:211–18.

Bredenkoetter, M., and K. Braun. 1996. Changes in neuronal responsiveness in the mediostriatal neostriatum/hyperstriatum after auditory filial imprinting in the domestic chick. *Neurosci.* 76:355–65.

Breed, M. D., 1983. Nestmate recognition in honey bees. *Anim. Behav.* 31:86–91.

Breed, M. D. and W. J. Bell. 1983. Hormonal influences on invertebrate aggressive behavior. In *Hormones and Aggressive Behavior,* ed. B. B. Svare, 577–90. New York: Plenum.

Brewer, R., and K. G. Harrison. 1975. The time of habitat selection in birds. *Ibis* 117:521–22.

Brindley, E. L. 1991. Response of European robins to playback of song: Neighbour recognition and overlapping. *Anim. Behav.* 41:503–12.

Broadhurst, P. L. 1963. The Choice of Animal for Behaviour Studies. *Laboratory Animals Centre Collected Papers* 12:65–80.

Brockman, H. 1997. Cooperative breeding in wasps and vertebrates: The role of ecological constraints. Pp. 347–71 in *Social Behavior in Insects and Arachnids,* (J. Choe and B. Crespi, eds.) Cambridge: Cambridge University Press.

Brockman, H. J. 1990. Mating behavior of the horseshoe crabs, *Limulus polyphemus. Behaviour* 114:206–20.

Brodeur, J., and J. N. McNeil. 1989. Seasonal microhabitat selection by an endoparasitoid through adaptive modification of host behavior. *Science* 244:226–28.

Bronson, F. 1971. Rodent pheromones. *Biol. Reprod.* 4:344–57.

Bronson, F. H. 1979. The reproductive ecology of the house mouse. *Quart. Rev. Biol.* 54:265–99.

———. 1989. *Mammalian Reproductive Biology.* Chicago: Chicago University Press.

Brookman, M. A. 1978. Sensitivity of the homing pigeon to an earth-strength magnetic field. In *Animal Migration, Navigation, and Homing,* ed. K. Schmidt-Koenig and W. T. Keeton. New York: Springer-Verlag.

Brooks, D. R., and D. A. McLennan. 1991. *Phylogeny, ecology, and behavior.* Chicago: University of Chicago Press.

Brookshire, K. H. 1970. Comparative psychology of learning. In *Learning: Interactions,* ed. M. H. Marx. New York: Macmillan.

Brower, L. P., and J. V. Z. Brower. 1962. Investigations into mimicry. *Nat. Hist.* 71:8–19.

Brower, L. P., and S. B. Malcolm. 1991. Animal migrations: Endangered phenomena. *Am. Zool.* 31:265–76.

Brower, L. P. 1985. New perspectives on the migration biology of the monarch butterfly, *Danaus plexippus* L. *Contrib. Mar. Sci Suppl.* 27: 748–86.

Brower, L. P., W. H. Calvert, L. E. Hedrick, and J. Christian. 1977. Biological observations on an overwintering colony of monarch butterflies (*Danaus plexippus Danaidae*) in Mexico. *J. Lepid. Soc.* 31:232–42.

Brower, L. P., W. N. Ryerson, L. L. Coppinger, and S. C. Glazier. 1968. Ecological chemistry and the palatability spectrum. *Science* 161:1349–51.

Brown, C. H., and P. M. Waser. 1984. Hearing and communication in blue monkeys (*Cercopithecus mitis*). *Anim. Behav.* 32:66–75.

Brown, C. R. 1986. Cliff swallow colonies as information centers. *Science* 234:83–85.

Brown, C. R., and M. B. Brown. 1986. Ectoparasitism as a cost of coloniality in cliff swallows (*Hirundo pyrrhonota*). *Ecology* 67:1206–18.

Brown, J., and E. Brown. 1990. Mexican jays: Uncooperative breeding. Pp. 269–88 in *Cooperative Breeding in Birds: Long-Term Studies of Ecology and Behavior* (P. Stacey and W. Koening eds.). Cambridge: Cambridge University Press.

Brown, J. L. 1964. The evolution of diversity in avian territorial systems. *Wilson Bull.* 76:160–69.

Brown, J. L. 1997. A theory of mate choice based on heterozygosity. *Behav. Ecol.* 8:60–65.

Brown, J. L., and A. Eklund. 1994. Kin recognition and the major histocompatibility complex: An integrative review. *Amer. Nat.* 143:436–61.

Brown, J. L., E. R. Brown, S. D. Brown, and D. D. Dow. 1982. Helpers: Effects of experimental removal on reproductive success. *Science* 215:421–22.

Brown, J. R., H. Ye, R. T. Bronson, P. Kiddes, and M. E. Greenberg. 1996. A defect in nurturing in mice lacking the immediate early gene fosB. *Cell* 86:297–309.

Brown, L., R. W. Shumaker, and J. F. Downhower. 1995. Do primates experience sperm competition? *Am. Nat.* 146:302–06.

Brown, R. E. 1986. Social and hormonal factors influencing infanticide and its suppression in adult male Long-Evans rats (*Rattus norvegicus*). *J. Comp. Psychol.* 100:155–61.

Brown, S. D., R. J. Dooling, and K. O'Grady. 1988. Peripheral organization of acoustic stimuli by budgerigars (*Melopsittacus undulatus*). III. Contact calls. *J. Comp. Psychol.* 102:236–47.

Brown, W. L., and E. O. Wilson. 1956. Character displacement. *Syst. Zool.* 5:49–64.

Bruce, H. M. 1961. Observations on the suckling stimulus and lactation in the rat. *J. Reprod. Fertil.* 2:17–34.

———. 1966. Smell as an exteroceptive factor. *J. Anim. Sci.,* Suppl. 25:83–89.

Bruderer, B., and P. Steidinger. 1972. Methods of quantitative and qualitative analysis of bird migration with a tracking radar. *NASA Spec. Publ.* NASA SP-262:151–67.

Bruner, J. S., A. Jolly, and K. Sylva, eds. 1976. *Play: Its Role in Development and Evolution.* New York: Basic Books.

Brunton, D. 1997. Impacts of predators: Center nests are less successful than edge nests in a large nesting colony of Least Terns. *Condor* 99:372–80.

Bryant, S., C. Thomas, and J. Bale. 2000. Thermal ecology of gregarious and solitary nettle-feeding nymphalid butterfly larvae. *Oecologia* 122(1):1–10.

Buchler, E. R., and S. B. Childs. 1982. Use of post-sunset glow as an orientation cue by the big brown bat (*Eptesicus fuscus*). *J. Mamm.* 63:243–47.

Buchsbaum, R. 1938. *Animals without Backbones.* Chicago: University of Chicago Press.

Buckle, G. R., and L. Greenberg. 1981. Nestmate recognition in sweat bees (*Lasioglossum zephyrum*): Does an individual recognize its own odour or only odours of its nestmates? *Anim. Behav.* 29:802–9.

Buechner, H. K., and H. D. Roth. 1974. The lek system in Uganda kob. *Am. Zool.* 14:145–62.

Bull, J. J. 1980. Sex determination in reptiles. *Quart. Rev. Biol.* 55:3–21.

Bull, J. J., W. H. N. Gutzke, and D. Crews. 1988. Sex reversal by estradiol in three reptilian orders. *Gen. Comp. Endocrinol.* 70:425–28.

Bullock, T. H. 1973. Seeing the world through a new sense: Electroreception in fish. *Amer. Scientist* 61:316–25.

Burda, H., S. Marhold, T. Westenberger, R. Wiltschko, and W. Wiltschko. 1990. Magnetic compass orientation in the subterranean rodent *Cryptomys hottentotus* (Bathyergidae). *Experientia* 46:528–30.

Burghardt, G. M. 1982. Comparisons matters: Curiosity, bears, surplus energy, and why reptiles do not play. *Behav. Brain. Sci.* 5:159–60.

Burghardt, G. M. 1998a. Play. In G. Greenberg and M. Haraway (eds.), *Comparative Psychology: A Handbook.* New York: Plenum.

Burghardt, G. M. 1998b. The evolutionary origins of play revisited. In M. Bekoff and J. A. Byers (eds.), *Animal Play.* New York: Cambridge University Press.

Burghardt, G. M. 1967. The primacy of the first feeding experience in the snapping turtle. *Psychonomic Science* 7:383–84.

Burghardt, G. M., and E. H. Hess. 1966. Food imprinting in the snapping turtle, *Chelydra serpentia. Science* 151:108–9.

Burghardt, G. M., and M. A. Krause. 1999. Plasticity of foraging behavior in garter snakes (*Thamnophis sirtalis*) reared on different diets. *J. Comp. Psychol.* 113:269–76.

Burley, N. 1979. The evolution of concealed ovulation. *Amer. Nat.* 114:835–58.

Burley, N., G. Krantzberg, and P. Radman. 1982. Influence of colour-banding on the conspecific preferences of zebra finches. *Anim. Behav.* 27:686–98.

Burnet, B., and K. Connolly. 1974. Activity and sexual behavior in *Drosophila melanogaster.* In *Genetics of Behaviour,* ed. J. H. F. van Abeelen. Amsterdam: North-Holland.

Burnet, B., K. Connolly, M. Kearney, and R. Cook. 1973. Effects of male paragonial gland secretion on sexual receptivity and courtship behaviour of female *Drosophila melanogaster. J. Insect. Physiol.* 19:2421–31.

Burtt, E. T. 1974. *The Senses of Animals.* London: Wykeham.

Buskirk, R. E. 1981. Sociality in Arachnida. In *Social Insects,* ed. H. R. Hermann. New York: Academic Press.

Buss, L. W. 1981. Group living, competition, and the evolution of cooperation in a sessile invertebrate. *Science* 213:1012–14.

Butler, R. A. 1953. Discrimination learning by rhesus monkeys to visual-exploration motivation. *J. Comp. Physiol. Psychol.* 46:95–98.

———. 1954. Incentive conditions which influence visual exploration. *J. Comp. Physiol. Psychol.* 48:19–23.

Byers, J. A. 1998. Biological effects of locomotor play: Getting into shape, or something more specific? In Bekoff, M. and J. A. Byers (eds.). *Animal Play.* New York: Cambridge University Press.

Byers, J. A. 1997. *American Pronghorn.* Chicago: University of Chicago Press.

Byrne, R. W., and A. Whiten. 1988. *Machiavellian Intelligence.* Oxford: Clarendon Press.

Caldero, J., D. Prevette, and X. Mun. 1998. Peripheral target regulation of the development and survival of spinal sensory and motor neurons in the chick embryo. *Journal of Neuroscience* 18:356–70.

Calhoun, J. B. 1962. Population density and social pathology. *Sci. 'ner.* 206:139–48.

———. 1973. Death squared: The explosive growth and demise of a mouse population. *Proc. Roy. Soc. Med.* 66:80–88.

Calvert, W. H., and L. P. Brower. 1981. The importance of forest cover for the survival of overwintering monarch butterflies (*Danaus plexippus Danaidae*). *J. Lepid. Soc.* 35:216–25.

Calvert, W. H., and L. P. Brower, 1986. The location of the monarch butterfly (*Danaus plexippus* L.) overwintering sites in relation to topography and climate. *J. Lepid. Soc.* 40:164–87.

Canady, R. A., D. E. Kroodsma, and F. Nottebohm. 1984. Population differences in complexity of a learned skill are correlated with the brain space involved. *Proc. Natl. Acad. Sci.* 81:6232–34.

Canteras, N. S., R. B. Simerly, and L. W. Swanson. 1995. Organization of projections from the medial nucleus of the amygdala: A PHAL study in the rat. *J. Comp. Neurol.* 360:213–45.

Capaldi, E., G. Robinson, and S. Fahrbach. 1999. Neuroethology of spatial learning: The birds and the bees. *Ann. Rev. Psych.* 50:651–82.

Capranica, R. R. 1976a. Morphology and physiology of the auditory system. In *Handbook of Frog Neurobiology,* ed. R. Llinas and W. Precht, 551–75. Berlin: Springer-Verlag.

———. 1976b. The auditory system of anurans. In *Physiology of Amphibia,* Vol. 3, ed. B. Lofts, 443–66. New York: Academic Press.

Capranica, R. R., and A. J. M. Moffat. 1977. Place mechanism underlying frequency analysis in the toad's inner ear. *J. Acoust. Soc. Amer.* 62 (Suppl.):S36.

———. 1980. Nonlinear properties of the peripheral auditory system of anurans. In *Comparative Studies of Auditory Processing in Vertebrates,* ed. A. Popper and R. Fay, 139–65. Berlin: Springer-Verlag.

———. 1983. Neurobehavioral correlates of sound communication in anurans. In *Advances in Vertebrate Neuroethology,* ed. J. P. Ewert, R. R. Capranica, and D. J. Ingle, 701–30. New York: Plenum Publishing.

Caraco, T., and L. L. Wolf. 1975. Ecological determinants of group sizes of foraging lions. *Amer. Nat.* 109:343–52.

Caraco, T., S. Martindale, and T. S. Whitham. 1980. An empirical demonstration of risk-sensitive foraging preferences. *Anim. Behav.* 28:820–30.

Carbone, C., J. T. Du Toit, and I. J. Gordon. 1997. Feeding success in African wild dogs: Does kleptoparasitism by spotted hyenas influence hunting group size? *J. Anim. Ecol.* 66:318–26.

Carlisle, D. B., and P. E. Ellis. 1959. La persistance des glandes ventrales céphaliques chez les criquets solitaires. *Comptes Rendu* 249:1059–60.

Carlson, A. D., and J. Copeland. 1978. Behavioral plasticity in the flash communication systems of fireflies. *Amer. Scientist* 66:340–46.

Caro, T. M., L. Lombardo, A. W. Goldizen, M. Kelly. 1995. Tail-flagging and other antipredator signals in white-tailed deer: New data and synthesis. *Behav. Ecol.* 6(4):442–50.

Carpenter, F. L., and R. E. MacMillen. 1976. Threshold model of feeding territoriality and test with a Hawaiian honeycreeper. *Science* 194:639–42.

Carr, A. 1965. The navigation of the green turtle. *Sci. Amer.* 212:78–86.

———. 1967. Adaptive aspects of the scheduled travel of Chelonia. *Proc. Ann. Biol. Colloq.,* Oreg. State Univ. 27:35–36.

Carr, A. J., and P. J. Coleman. 1974. Seafloor spreading theory and the odyssey of the green turtle. *Nature* 249:128–30.

Cerella, J. 1979. Visual cues and natural categories in the pigeon. *J. Exp. Psychol.: Human Percept. Perf.* 5:68–77.

Chadab, R., and C. W. Rettenmeyer. 1975. Mass recruitment by army ants. *Science* 188:1124–25.

Chadwick, C. S. 1951. Further observations on the water drive in *Triturus viridescens.* II. Induction of the water drive with the lactogenic hormone. *J. Exp. Zool.* 86:175–87.

Chagnon, N. A., and P. E. Bugos, Jr. 1979. Kin selection and conflict: An analysis of a Yanomamo ax fight. In *Evolutionary Biology and Human Social Behavior: An Anthropological Perspective,* ed. N. A. Chagnon and W. Irons. N. Scituate, MA: Duxbury Press.

Chance, M. R. A., and A. P. Mead. 1953. Social behaviour and primate evolution. *Symp. Soc. Exp. Biol. VII (Evolution):* 395–439.

Charniaux-Cotton, H., and L. H. Kleinholz. 1964. Hormones in invertebrates other than insects. In *The Hormones,* vol. 4, ed. G. Pincus, K. V. Thimann, and E. B. Astwood. New York: Academic Press.

Charnov, E. L. 1976. Optimal foraging: The marginal value theorem. *Theor. Pop. Biol.* 9:129–36.

Charnov, E., and J. Finerty. 1980. Vole population cycles: A case for kin-selection? *Oecologia* 45:1–2.

Chase, J., and R. A. Suthers. 1969. Visual obstacle avoidance by echolocating bats. *Anim. Behav.* 17:201–7.

Chase, I. D. 1982. Dynamics of hierarchy formation: The sequential development of dominance relationships. *Behaviour* 80:218–40.

Chelazzi, G. 1992. Invertebrates (excluding Arthropods). Pp. 19–43 in *Animal Homing* (F. Papi, ed.). London: Chapman & Hall.

Cheney, D. L. 1983. Extrafamilial alliances among vervet monkeys. In *Primate Social Relationships,* ed. R. A. Hinde. Sunderland, MA: Sinauer.

Cheney, D. L., and R. M. Seyfarth. 1982. Recognition of individuals within and between groups of free-ranging vervet monkeys. *Amer. Zool.* 22:519–29.

———. 1983. Nonrandom dispersal in free-ranging vervet monkeys: Social and genetic consequences. *Amer. Nat.* 122:392–412.

———. 1985. Social and non-social knowledge in vervet monkeys. *Phil. Trans. Royal Soc. Lond.* B. 308:187–201.

———. 1990. *How Monkeys See the World.* Chicago: University of Chicago Press.

Cheney, D. L., R. M. Seyfarth, and B. Smuts. 1986. Social relationships and social cognition in nonhuman primates. *Science* 234:1361–66.

Cheng, M. F. 1979. Progress and prospects in ring dove research: A personal view. *Adv. Stud. Behav.* 9:97–130.

Cheng, M. F. 1992. For whom does the female dove coo? A case for the role of vocal self-stimulation. *Anim. Behav.* 43:1035–44.

Chevalier-Skolnikoff, S. 1974. Male-female, female-female, and male-male sexual behavior in the stumptail monkey, with special attention to the female orgasm. *Arch. Sex. Behav.* 3:95–116.

Chitty, D. 1960. Population processes in the vole and their relevance to general theory. *Can. J. Zool.* 38:99–113.

———. 1967. The natural selection of self-regulatory behavior in animal populations. *Proceedings of the Ecological Society of Australia* 2:51–78.

Christenson, T. E. 1984. Behaviour of colonial and solitary spiders of the Theridid species *Anelosimus eximus. Anim. Behav.* 32:725–34.

Christian, J. J. 1950. The adreno-pituitary system and population cycles in mammals. *J. Mammal.* 31:247–59.

———. 1970. Social subordination, population density, and mammalian evolution. *Science* 168:84–90.

———. 1978. Neurobehavioral endocrine regulation of small mammal populations. In *Populations of Small Mammals under Natural Conditions,* ed. D. P. Snyder. The Pymatuning Symposia in Ecology 5:143–58.

Christian, J. J., and D. E. Davis. 1964. Endocrines, behavior, and population. *Science* 146:1550–60.

Christiansen, K., and R. Knussman. 1987. Androgen levels and components of aggressive behavior in men. *Horm. Behav.* 21:170–80.

Clark, A. B. 1978. Sex ratio and local resource competition in a prosimian primate. *Science* 201:163–65.

Clark, C. W., and M. Mangel. 2000. *Dynamic State Variable Models in Ecology.* New York: Oxford University Press.

Clarke, C. A., and P. M. Sheppard. 1960. Supergenes and mimicry. *Heredity* 14:175–85.

Clarke, J. D., and G. J. Coleman. 1986. Persistent meal-associated rhythms in SCN-lesioned rats. *Physiol. Behav.* 36:105–13.

Clayton, N. S., and D. W. Lee. 1998. Memory and the hippocampus in food-storing birds. In R. P. Balda, I. M. Pepperberg, and A. C. Kamil (eds.), *Animal Cognition in Nature.* San Diego: Academic Press.

Clayton, N. S., and J. R. Krebs. 1994. Hippocampal growth and attrition in birds affected by experience. *Proc. Nat. Acad. Sci. USA* 91:7410–14.

Clemens, L. G. 1974. Neurohormonal control of male sexual behavior. In *Reproductive Behavior,* ed. W. Montagna and S. Sadler. New York: Plenum.

Clemens, L. G., B. A. Gladue, and L. P. Coniglio. 1978. Prenatal endogenous androgenic influences on masculine sexual behavior and genital morphology in male and female rats. *Horm. Behav.* 10:40–53.

Clemmons, J. R., and R. Buchholz. 1997. *Behavioral Approaches to Conservation in the Wild.* New York: Cambridge University Press.

Cloudsley-Thompson, J. L. 1952. Studies in diurnal rhythms. II. Changes in the physiological responses of the woodlouse *Oniscus aspellus* L to environmental stimuli. *J. Exp. Biol.* 29:295–303.

———. 1960. Adaptive functions of circadian rhythms. *Cold. Spr. Harb. Symp. Quant. Biol.* 24:361–67.

Clutton-Brock, J. 1995. Origins of the dog: Domestication and early history. In J. Serpell (ed.), *The Domestic Dog: Its Evolution, Behaviour, and Interactions with People.* Cambridge: Cambridge University Press.

Clutton-Brock, T. H., ed. 1988. *Reproductive Success.* Chicago: University of Chicago Press.

Clutton-Brock, T. H. 1991. *The Evolution of Parental Care.* Princeton: Princeton University Press.

Clutton-Brock, T. H., S. D. Albon, and F. E. Guinness. 1984. Maternal dominance, breeding success and birth sex ratios in red deer. *Nature* 308:358–60.

Clutton-Brock, T. H., F. E. Guinness, and S. D. Albon. 1982. *Red Deer: Behavior and Ecology of Two Sexes.* Chicago: University of Chicago Press.

Clutton-Brock, T., M. O'Riain, P. Brotherton, D. Gyanor, R. Kansky, A. Griffin, and M. Manser. 1999. Selfish sentinels in cooperative mammals. *Science* 284:1640–44.

Colby, D. R., and J. G. Vandenbergh. 1974. Regulatory effects of urinary pheromones on puberty in the mouse. *Biol. Reprod.* 11:268–79.

Cole, J. E., and J. A. Ward. 1970. An analysis of parental recognition by the young of the cichlid fish, *Etroplus maculatus* (Bloch). *Z. Tierpsychol.* 27:156–276.

Cole, L. J. 1916. Twinning in cattle with special reference to the free-martin. *Science* 43:177.

Collier, G., and C. R. Rovee-Collier. 1982. A comparative analysis of optimal foraging behavior: Laboratory simulations. In *Foraging Behavior: Ecological, Ethological, and Psychological Approaches,* ed. A. C. Kamil and T. D. Sargent, 39–76. New York: Garland STPM Press.

Colvin, B. A. 1984. Barn owl foraging behavior and secondary poisoning hazard from rodenticide use on farms. Ph.D. dissertation, Bowling Green State University.

Connaughton, M. A., and M. H. Taylor. 1995. Seasonal and daily cycles in sound production associated with spawning in the weakfish (*Cynoscion regalis*). *Environ. Biol. Fishes* 42:233–40.

Connell, J. H. 1961. The influence of interspecific competition and other factors on the distribution of the barnacle *Chthamalus stellatus*. *Ecology* 42:710–23.

Connor, R. C., R. A. Smolker, and A. F. Richards. 1992. Dolphin alliances and coalitions. Pp. 415–43 in *Coalitions and Alliances in Humans and Other Animals* (A. H. Harcourt and F. B. M. de Waal, eds.). New York: Oxford University Press.

Connor, V. M., and J. F. Quinn. 1984. Stimulation of food species growth by limpet mucus. *Science* 225:843–44.

Cook, A. 1971. Habituation in a freshwater snail (*Limnaea stagnalis*). *Anim. Behav.* 17:679–82.

Cook, L. M., G. S. Mani, and M. E. Varley. 1986. Postindustrial melanism in the peppered moth. *Science* 231:611–13.

Cook, L. M., R. L. H. Dennis, G. S. Mani. 1999. *Proc. Roy. Soc. Lond. Ser. B Biol. Sci.* 266:293–97.

Cooke, J. A., and C. R. Terman. 1977. Influence of displacement distance and vision on homing behavior of the white-footed mouse (*Peromyscus leucopus noveboracensis*). *J. Mamm.* 58:58–66.

Cooper, R., and J. Zubek. 1958. Effects of enriched and restricted early environments on the learning ability of bright and dull rats. *Can. J. Psychol.* 12:159–64.

Copeland, J. 1983. Male firefly mimicry. *Science* 221:484–85.

Coppola, D. M., and J. G. Vandenbergh. 1987. Induction of a puberty-regulating chemosignal in wild mouse populations. *J. Mammal.* 68:86–91.

———. 1960. Patterns of circadian rhythms in insects. *Cold Spr. Harb. Symp. Quant. Biol.* 25:357–60.

Corning, W. C., and R. von Burg. 1973. Protozoa. In *Invertebrate Learning,* Vol. 1, ed. W. C. Corning, J. A. Dyal, and A. O. D. Willows. New York: Plenum.

Corning, W. C., and S. Kelly. 1973. Platyhelminthes: The turbellarians. In *Invertebrate Learning,* Vol. 1, ed. W. C. Corning, J. A. Dyal, and A. O. D. Willows. New York: Plenum.

Corrent, G., and A. Eskin. 1982. Transmitterlike action of serotonin in phase shifting a rhythm from the *Aplysia* eye. *Amer. J. Physiol.* 242:R333–38.

Côte, I., and M. Gross. 1993. Reduced disease in offspring: A benefit of coloniality in sunfish. *Behav. Ecol. Sociobiol.* 33:269–74.

Coulson, J. C. 1966. The influence of the pair bond and age on the breeding biology of the kittiwake gull, *Rissa tridactyla*. *J. Animal. Ecol.* 35:269–79.

Courchamp, F., T. Clutton-Brock, and B. Grenfell. 1999. Inverse density dependence and the Allee effect. *TREE* 14:405–10.

Couvillon, P. A., and M. E. Bitterman. 1980. Some phenomena of associative learning in honeybees. *J. Comp. Physiol. Psychol.* 94:878–85.

———. 1982. Compound conditioning in honeybees. *J. Comp. Physiol. Psychol.* 96:192–99.

———. 1985. Analysis of choice in honeybees. *Anim. Learn. Behav.* 13:246–52.

———. 1986. Performance of honeybees in reversal and ambiguous-cue problems: Tests of a choice model. *Anim. Learn. Behav.* 14:225–31.

Cowie, R. J. 1977. Optimal foraging in great tits (*Parus major*). *Nature* 268:137–39.

Cowlishaw, G. 1997. Trade-offs between foraging and predation risk determine habitat use in a desert baboon population. *Anim. Behav.* 53:667–86.

Cox, C. R., and B. J. Le Boeuf. 1977. Female incitation of male competition: A mechanism in sexual selection. *Amer. Nat.* 111:317–35.

Cox, G. W. 1985. The evolution of avian migration systems between temperate and tropical regions of the New World. *Am. Nat.* 126:451–74.

Craig, W. 1914. Male doves reared in isolation. *J. Anim. Behav.* 4:121–33.

Craighead, D. J., and J. J. Craighead. 1987. Tracking caribou using satellite telemetry. *Nat. Geog. Res.* 3:462–79.

Crawford, J. D. 1984. Orientation in a vertical plane: The use of light cues by an orb-weaving spider, *Araneus diadematus* Clerk. *Anim. Behav.* 32:162–71.

Creel, S., and N. M. Creel. 1995. Communal hunting and pack size in African wild dogs, *Lycaon pictus*. *Anim. Behav.* 50:1325–39.

Creel, S., N. M. Creel, and S. L. Monfort. 1996. Social stress and dominance. *Nature* 379:212.

Crews, D. 1974. Castration and androgen replacement on male facilitation of ovarian activity in the lizard, *Anolis carolinensis*. *J. Comp. Physiol. Psychol.* 87:963–72.

Crews, D. 1975. Psychobiology of reptilian reproduction. *Science* 189:1059–65.

———. 1977. The annotated *Anole*: Studies on the control of lizard reproduction. *Amer. Scientist* 65:428–34.

———. 1979. Neuroendocrinology of lizard reproduction. *Biol. Reprod.* 20:51–73.

———. 1980. Interrelationships among ecological, behavioral, and neuroendocrine processes in the reproductive cycle of *Anolis carolinensis* and other reptiles. *Adv. Stud. Behav.* 11:1–75.

———. 1983. Alternative reproductive tactics in reptiles. *BioScience* 33:562–66.

Crews, D. 1998. On the organization of individual differences in sexual behavior. *Amer. Zool.* 38:118–32.

Crews, D., and N. Greenberg. 1981. Function and causation of social signals in lizards. *Amer. Zool.* 21:273–94.

Crews, D., J. S. Rosenblatt, and D. S. Lehrman. 1974. Effects of unseasonal environmental regime, group presence, group composition, and males' physiological state on ovarian recrudescence in the lizard *Anolis carolinensis*. *Endocrinology* 95:102–6.

Cristol, D. A., and P. V. Switzer. 1999. Avian prey-dropping behavior. II. American crows and walnuts. *Behav. Ecol.* 10(3):220–26.

Cross, B. A., and R. G. Dyer. 1972. Ovarian modulation of unit activity in the anterior hypothalamus of the cyclic rat. *J. Physiol.* (Lond.) 222:25P.

Crowcroft, P. 1973. *Mice All Over.* Brookfield, IL: Chicago Zoological Park.

Crowcroft, P., and F. Rowe. 1957. Social organization and territorial behaviour in the wild house mouse (*Mus musculus* L.). *Proc. Zool. Soc. Lond.* 140:517–31.

Crump, A. J., and J. Brady. 1979. Circadian activity patterns in three species of tsetse fly: *Glossina palpalis, austeni,* and *morsitans*. *Physiol. Entomol.* 4:311–18.

Crump, M. 1992. Cannibalism in amphibians. Pp. 256–76 in *Cannibalism: Ecology and Evolution Among Diverse Taxa*

(M. A. Elgar and B. J. Crespi, ed.). New York: Oxford University Press.

Csányi, V. 1986. Ethological analysis of predator avoidance by the paradise fish (*Macropodus opercularis* L.) II. Key stimuli in avoidance learning. *Anim. Learn. Behav.* 14:101–9.

Cullen, E. 1957. Adaptations in the kittiwake to cliff nesting. *Ibis* 99:275–302.

———. 1960. Experiment on the effect of social isolation on reproductive behavior in the three-spined stickleback. *Anim. Behav.* 8:235.

Czeisler, C. A., R. E. Kronauer, J. S. Allan, J. F. Duffy, M. E. Jewett, E. N. Brown, and J. M. Ronda. 1989. Bright light induction of strong (Type 0) resetting of the human circadian pacemaker. *Science* 244:1328–33.

Daan, S., and J. Aschoff. 1982. Circadian contributions to survival. In *Vertebrate Circadian Systems,* ed. J. Aschoff, S. Daan, and G. A. Groos. Berlin: Springer-Verlag.

Daan, S., and A. J. Lewy. 1984. Scheduled exposure to daylight: A potential strategy to reduce jet-lag following transmeridian flight. *Psychopharmacol. Bull.* 20:566–68.

Daly, M. 1978. The cost of mating. *Amer. Nat.* 112:771–74.

Daly, M., and M. Wilson. 1996. Violence against stepchildren. *Curr. Dir. Psych. Sci.* 5:77–81.

Daly, M., and M. Wilson. 1999. Human evolutionary psychology and animal behaviour. *Anim. Behav.* 57:509–19.

Daly, M., M. I. Wilson, and S. F. Faux. 1978. Seasonally variable effects of conspecific odors upon capture of deermice (*Peromyscus maniculatus gambelli*). *Behav. Biol.* 23:254–59.

Daly, M., and M. I. Wilson. 1981. Abuse and neglect of children in evolutionary perspective. In *Natural Selection and Social Behavior,* ed. R. D. Alexander and D. W. Tinkle. New York: Chiron Press.

———. 1982. Homicide and kinship. *Amer. Anthrop.* 84:372–78.

———. 1988. Evolutionary social psychology and family homicide. *Science* 242:519–24.

Darchen, R., and B. Delage. 1970. Facteur déterminant les castes chez les Trigones. *Compt. Ren. Acad. Sci.* (Paris) 270:1372–73.

Darwin, C. 1845. *The Voyage of the* Beagle. London: Dent.

———. 1859. *The Origin of Species.* London: Dent.

———. 1871. *The Descent of Man, and Selection in Relation to Sex.* London: John Murray.

———. 1873. *On the Expression of the Emotions in Man and Animals.* New York: D. Appleton.

Datta, L. G., S. Milstein, and M. E. Bitterman. 1960. Habitat reversal in the crab. *J. Comp. Physiol. Psychol.* 53:275–78.

Davey, G. 1989. *Ecological Learning Theory.* New York: Routledge.

Davidson, J. M. 1966a. Characteristics of sex behavior in male rats following castration. *Anim. Behav.* 14:266–72.

———. 1966b. Activation of the male rat's sexual behavior by intracerebral implantation of androgen. *Endocrinology* 79:783–94.

Davies, N. B. 1978. Territorial defence in the speckled wood butterfly (*Pararge aegeria*): the resident always wins. *Anim. Behav.* 26:138–47.

———. 1983. Polyandry, cloaca-pecking, and sperm competition in dunnocks. *Nature* 302:334–36.

Davies, N. B., and A. I. Houston. 1983. Time allocation between territories and flocks and owner-satellite conflict in foraging pied wagtails, *Motacilla alba. J. Anim. Ecol.* 52:621–34.

———. 1984. Territory economics. In *Behavioural Ecology: An Evolutionary Approach,* 2d ed., ed. J. R. Krebs and N. B. Davies. Oxford, England: Blackwell Scientific Publications, Ltd.

Davies, N. B. 1992. *Dunnock Behaviour and Social Evolution.* Oxford: Oxford University Press.

Davis, D. E. 1951. The relation between level of population and pregnancy of Norway rats. *Ecology* 32:459–61.

———. 1963. The physiological analysis of aggressive behavior. In *Social Behavior and Organization among Vertebrates,* ed. W. Etkin. Chicago: University of Chicago Press.

Davis, W. H., and H. B. Hitchcock. 1965. Biology and migration of the bat, *Myotis lucifungus,* in New England. *J. Mammal.* 46:296–313.

Davoren, G. K., and A. E. Burger. 1999. Differences in prey selection and behaviour during self-feeding and chick provisioning in rhinoceros auklets. *Anim. Behav.* 58:853–63.

Dawe, A. R., and W. A. Spurrier. 1972. The blood-borne "trigger" for natural mammalian hibernation in the 13-lined ground squirrel and the woodchuck. *Cryobiology* 9:163–72.

Dawkins, R. 1976. *The Selfish Gene.* New York: Oxford University Press.

———. 1982. *The Extended Phenotype.* Oxford: Oxford University Press.

Dawkins, R. 1982. *The Extended Phenotype.* San Francisco: W. H. Freeman.

Dawkins, R., and T. R. Carlisle. 1976. Parental investment, mate desertion, and a fallacy. *Nature* 262:131–33.

Dawkins, R., and J. R. Krebs. 1978. Animal signals: Information or manipulation? In *Behavioural Ecology: An Evolutionary Approach,* eds. J. R. Krebs and N. B. Davies. Oxford, England: Blackwell Scientific Publications, Ltd.

de Beer, G. 1958. *Embryos and Ancestors.* 3d ed. London: Oxford University Press.

De Long, K. T. 1967. Population ecology of feral house mice. *Ecology* 48:611–34.

De Loof, A., and R. Huybrechts. 1998. "Insects do not have sex hormones": A myth? *Gen. Comp. Endocrinol.* 111:245–60.

De Loof, A., R. Huybrechts, and T. Briers. 1981. Do insects have steroid sex hormones and do compounds with juvenile hormone activity occur in vertebrates? *Ann. Soc. R. Zool. Belg.* 110:179–84.

de Ruiter, J. R., W. Scheffrahn, G. J. J. M. Trommelen, A. G. Uitterlinden, R. D. Martin, and J. A. R. A. M. van Hooff. 1992. Male social rank and reproductive success in wild long-tailed macaques. Pp. 175–91 in *Paternity in Primates: Genetic Tests and Theories* (R. D. Martin, A. F. Dixson, and E. J. Wickings, eds.). Karger: Basel.

de Souza, H. M. L., A. B. da Cunha, and E. P. dos Santos. 1970. Adaptive polymorphism of behavior evolved in laboratory populations of *Drosophila willistoni. Amer. Nat.* 104:175–89.

de Waal, F., and M. Berger. 2000. Payment for labour in monkeys. *Nature* 404:563.

de Waal, F. B. M. 1993. Reconciliation among primates: A review of empirical evidence and unresolved issues. Pp. 111–44 in *Primate Social Conflict* (W. A. Mason and S. P. Mendoza, eds.). Albany, NY: SUNY Press.

deWilde, J. 1975. An endocrine view of metamorphosis, polymorphism and diapause in insects. *Amer. Zool.* (sup. 1): 13–28.

Dearing, M. Denise. 1997. The manipulation of plant toxins by a food-hoarding herbivore, *Ochotona princeps. Ecology* 78(3):774–81.

DeCoursey, P. J. 1960. Phase control of activity in a rodent. *Cold. Spring Harb. Symp. Quant. Biol.* 25:49–56.

———. 1961. Effect of light on the circadian activity rhythm of the flying squirrel, *Glaucomys volans. Zeit. Physiol.* 44:331–54.

———. 1983. Biological timekeeping. *Biology of Crustacea* 7:107–62.

———. 1986. Circadian photoentrainment: Parameters of phase delaying. *J. Biol. Rhythms* 1:171–86.

Delcomyn, F. 1998. *Foundations of Neurobiology.* New York: W. H. Freeman and Company.

Delgado, J. M. R. 1963. Cerebral heterostimulation in a monkey colony. *Science* 141:161–63.

———. 1966. Aggressive behavior evoked by radio-stimulation in monkey colonies. *Amer. Zool.* 6:669–81.

———. 1967. Social rank and radio stimulated aggressiveness in monkeys. *J. Nerv. Ment. Disease* 144:383–90.

Delville, Y., K. M. Mansour, and C. F. Ferris. 1996. Serotonin blocks vasopressin-facilitated offensive aggression: Interactions within the ventrolateral hypothalamus of golden hamsters. *Physiol. & Behav.* 59:813–16.

Demas, G. E., C. A. Moffatt, D. L. Drazen, and R. J. Nelson. 1999. Castration does not inhibit aggressive behavior in adult male prairie voles (*Microtus ochrogaster*). *Physiol. & Behav.* 66:59–62.

Denenberg, V. H., and K. M. Rosenberg. 1967. Nongenetic transmission of information. *Nature* 216:549–50.

Denenberg, V. H., and A. E. Whimbey. 1963. Behavior of adult rats is modified by the experiences their mothers had as infants. *Science* 142:1192–93.

Derscheid, J. M. 1947. Strange parrots. *Avic. Mag.* 53:44–49.

Dethier, V. G. 1962. *To Know a Fly.* Englewood Cliffs, NJ: Prentice-Hall.

———. 1976. *The Hungry Fly.* Cambridge: Harvard University Press.

Dethier, V. G., and D. Bodenstein. 1958. Hunger in the blowfly. *Zeit. Tierpsychol.* 15:129–40.

Dethier, V. G., and A. Gelperin. 1967. Hyperphagia in the blowfly. *J. Exp. Biol.* 47:191–200.

Dethier, V. G., and E. Stellar. 1961. *Animal Behavior: Its Evolutionary and Neurological Basis.* Englewood Cliffs, NJ: Prentice-Hall.

Deutsch, J. A. 1960. *Structural Basis of Behavior.* Chicago: University of Chicago Press.

———. 1978. *Comparative Animal Behavior.* New York: McGraw-Hill.

Deutsch, J. C. 1994. Uganda kob mating success does not increase on larger leks. *Behav. Ecol. Sociobiol.* 34:451–59.

Deverill, J. I., C. E. Adams, and C. W. Bean. 1999. Prior residence, aggression and territory acquisition in hatchery-reared and wild brown trout. *Journal of Fish Biology* 55: 868–75.

Devlin, B., M. Daniels, and K. Roeder. 1997. The heritability of IQ. *Nature* 388:468–71.

DeVoogd, T., and F. Nottebohm. 1981. Gonadal hormones influence dendritic growth in the adult avian brain. *Science* 214:202–4.

DeVoogd, T. J., and T. Székely. 1998. Causes of avian song: Using neurobiology to integrate approximate and ultimate levels of analysis. In R. P. Balda, I. M. Pepperberg, and A. C. Kamil (eds.), *Animal Cognition in Nature.* San Diego: Academic Press.

Dewsbury, D. A. 1972. Patterns of copulatory behavior in male mammals. *Quart. Rev. Biol.* 47:1–33.

———. 1975. Diversity and adaptation in rodent copulatory behavior. *Science* 190:947–54.

———. 1978. *Comparative Animal Behavior.* New York: McGraw-Hill.

———. 1982. Ejaculate cost and male choice. *Amer. Nat.* 119:601–10.

———. 1984. *Comparative Psychology in the Twentieth Century.* Stroudsburg, PA: Hutchinson Ross.

Dewsbury, D. A. 1984. Sperm competition in muroid rodents. Pp. 547–71 in *Sperm Competition and the Evolution of Animal Mating Systems* (R. L. Smith, ed.). New York: Academic Press.

———. 1985. *Leaders in the Study of Animal Behavior.* Lewisburg, PA: Bucknell University Press.

Dewsbury, D. A. 1988. A test of the role of copulatory plugs in sperm competition in deer mice (*Peromyscus maniculatus*). *J. Mamm.* 69:854–57.

Dewsbury, D. A. 1999. The proximate and ultimate: Past, present, and future. *Behav. Proc.* 46:189–99.

Diamond, J. M. 1974. Colonization of exploded volcanic islands by birds: The supertramp strategy. *Science* 184:803–6.

———. 1978. Niche shifts and the rediscovery of interspecific competition. *Amer. Scientist* 66:322–31.

Dickemann, M. 1975. Demographic consequences of infanticide in man. *Ann. Rev. Ecol. Syst.* 6:107–37.

Dill, P. A. 1977. Development of behaviour in alevins of Atlantic salmon, *Salmo salar,* and rainbow trout, *S. Gairdneri. Anim. Behav.* 25:116–21.

Dillon, L. S. 1978. *Evolution: Concepts and Consequences.* 2d ed. St. Louis: Mosby.

Dingle, H. 1980. Ecology and evolution of migration. In *Animal Migration, Orientation and Navigation,* ed. S. A. Gauthreaux. New York: Academic Press.

Dingle, H. 1996. *Migration: The Biology of Life on the Move.* New York: Oxford University Press.

Dobson, F. S. 1982. Competition for mates and predominant juvenile male dispersal in mammals. *Anim. Behav.* 30:1183–92.

Dobzhansky, T. 1937. *Genetics and the Origin of Species,* 3rd edition. New York: Columbia University Press.

Doligez, B., E. Danchin, J. Clobert, and L. Gustafsson. 1999. The use of conspecific reproductive success for breeding habitat selection in a noncolonial, hole-nesting species, the collared flycatcher. *J. Anim. Ecol.* 68:1193–1206.

Domjan, M. 1980. Ingestional aversion learning: Unique and general processes. *Adv. Stud. Behav.* 11:276–337.

———. 1983. Biological constraints on instrumental and classical conditioning: Implications for general process theory. In *The Psychology of Learning and Motivation,* vol. 17, ed. G. H. Bower. New York: Academic Press.

Dörner, G., and M. Kawakami, eds. 1978. *Hormones and Brain Development.* New York: Elsevier North-Holland.

Doty, R. L. 1981. Olfactory communication in humans. *Chemical Senses* 6:351–76.

Douglas, J. 2001. Migrational orientation variation between different subpopulations of the monarch butterfly (*Danaus plexippus* L.). INTEL project report. Psyche (in press).

Douglas, M. M. 1986. *The Lives of Butterflies.* Ann Arbor: University of Michigan Press. 241 pages.

Downes, S., and R. Shine. 1998. Heat, safety or solitude? Using habitat selection experiments to identify a lizard's priorities. *Anim. Behav.* 55:1387–96.

Drickamer, L. C. 1970. Seed preferences in wild caught *Peromyscus maniculatus bairdi* and *P. Leucopus noveboracensis. J. Mammal.* 51:191–94.

———. 1972. Experience and selection behavior as factors in the food habits of *Peromyscus:* Use of olfaction. *Behaviour* 41:269–87.

———. 1974a. Sexual maturation of female house mice: Social inhibition. *Develop. Psychobiol.* 7:257–65.

———. 1974b. A ten-year summary of population and reproduction data for free-ranging *Macaca mulatta* at La Parguera, Puerto Rico. *Folia Primatologica* 21:61–80.

———. 1975. Daylength and sexual maturation of female house mice. *Develop. Psychobiol.* 8:561–70.

———. 1979. Acceleration and delay of first vaginal estrus in wild *Mus musculus. J. Mammal.* 60:215–16.

———. 1981. Selection for age of sexual maturation in mice and the consequences for population regulation. *Behav. Neur. Biol.* 31:82–89.

———. 1982a. Delay and acceleration of puberty in female mice by urinary chemosignals from other females. *Develop. Psychobiol.* 15:433–42.

———. 1982b. Acceleration and delay of sexual maturation in female house mice by urinary cues: Dose levels and mixing urine from different sources. *Anim. Behav.* 30:456–60.

———. 1983. Male acceleration of puberty in female mice. *J. Comp. Psychol.* 97:191–200.

———. 1989. Pheromones: Behavioral and biochemical aspects. In *Advances in Comparative and Environmental Physiology,* ed. J. Balthazart, 269–348. Berlin: Springer-Verlag.

Drickamer, L. C. 1986. Puberty-influencing chemosignals in mice: Ecological and evolutionary considerations. In D. Duvall, D. Muller-Schwarze, and R. M. Silverstein (eds.), *Chemical Signals in Vertebrates* IV:441–55.

———. 1992. Behavioral selection of odor cues by young female mice affects age of puberty. *Develop. Psychobiol.* 25:461–70.

Drickamer, L. C. 1988. Acceleration and delay of sexual maturation in female house mice (*Mus domesticus*) by urinary chemosignals: Mixing urine sources in unequal proportions. *J. Comp. Psychol.* 102:215–21.

Drickamer, L. C., ed. 1998. Animal behavior: Integration of proximate and ultimate causation. *Amer. Zool.* 38:39–259.

Drickamer, L. C., and P. L. Brown. 1998. Age-related changes in odor preferences by house mice living in seminatural enclosures. *J. Chem. Ecol.* 24:1745–56.

Drickamer, L. C., and J. E. Hoover. 1979. Effects of urine from pregnant and lactating female house mice on sexual maturation of juvenile females. *Develop. Psychobiol.* 12:545–51.

Drickamer, L. C., and R. X. Murphy. 1978. Female mouse maturation: Effects of excreted and bladder urine from juvenile and adult males. *Develop. Psychobiol.* 11:63–72.

———. 1986. Puberty-influencing chemosignals in mice: Ecological and evolutionary considerations. In *Chemical Signals in Vertebrates*, vol. 4, ed. D. Duvall, D. Muller-Schwarze, and R. M. Silverstein, 441–55. New York: Plenum.

———. 1988. Acceleration and delay of sexual maturation in female house mice (*Mus domesticus*) by urinary chemosignals: Mixing urine sources in unequal proportions. *J. Comp. Psychol.* 102:215–21.

Drickamer, L. C., and J. Stuart. 1984. Peromyscus: Snow tracking and possible cues used for navigation. *Amer. Mid. Nat.* 111:202–4.

Drickamer, L. C., and J. G. Vandenbergh. 1973. Predictors of dominance in the female golden hamster (*Mesocricetus auratus*). *Anim. Behav.* 21:564–70.

Drickamer, L. C., J. G. Vanderbergh, and D. R. Colby. 1973. Predictors of dominance in the male golden hamster (*Mesocricetus auratus*). *Anim. Behav.* 21:557–63.

Drickamer, L. C., and S. H. Vessey. 1973. Group changing in male free-ranging rhesus monkeys. *Primates* 14:359–68.

Drummond, H. 1981. The nature and description of behavior patterns. In *Perspectives in Ethology*, vol. 4, eds. P. P. G. Bateson and P. H. Klopfer. New York: Plenum.

Drury, W. H., and J. A. Keith. 1962. Radar studies of songbird migration in coastal New England. *Ibis* 104:449–89.

Drury, W. H., and I. C. T. Nisbet. 1964. Radar studies of orientation of songbird migrants in southeastern New England. *Bird Banding* 35:69–119.

Dubnau, J., and T. Tully. 1998. Gene discovery in *Drosophila*: New insights for learning and memory. *Ann. Rev. Neurosci.* 21:407–44.

D'Udine, D., and E. Alleva. 1983. Early experience and sexual preferences in rodents. In *Mate Choice*, ed. P. P. G. Bateson, 311–27. New York: Cambridge University Press.

Duffield, G. E., and F. J. P. Ebling. 1998. Maternal entrainment of the developing circadian system in the Siberian hamster. *J. Biol. Rhythms* 13:315–29.

Duffy, J. 1996. Eusociality in a coral-reef shrimp. *Nature* 381:512–14.

Dufty, A. M. 1989. Testosterone and survival: A cost of aggressiveness? *Horm. Behav.* 23:185–93.

Dugatkin, L. A. 1992. Sexual selection and imitation: Females copy the mate choice of others. *Am. Nat.* 139:1384–89.

Dugatkin, L. 1998. Game theory and cooperation. In L. Dugatkin, and H. Reeve, eds. *Game Theory and Animal Behavior,* Oxford: Oxford University Press.

Dukas, R. 1999. Ecological relevance of associative learning in fruitfly larvae. *Behav. Ecol. Sociobiol.* 45:195–200.

Dukas, R., ed. 1998. *Cognitive Ecology.* Chicago: University of Chicago Press.

Dunbar, R. I. M. 1976. Some aspects of research design and their implications in the observational study of behavior. *Behaviour* 58:78–98.

Duncan, J. R., and D. M. Bird. 1989. The influence of relatedness and display effort on the mate choice of captive female American kestrels. *Anim. Behav.* 37:112–17.

Dunlap, J. C. 1990. Closely watched clocks: Molecular analysis of circadian rhythms in *Neurospora* and *Drosophila*. *Trends Genet.* 6:159–65.

Dunlap, K. D., and H. H. Zakon. 1998. Behavioral actions of androgens and androgen receptor expression in the electrocommunication system of an electric fish, *Eigenmannia virescens*. Horm. Behav. 34:30–38.

———. 1993. Genetic analysis of circadian clocks. *Ann. Rev. Physiol.* 55:683–728.

Dunlap, K. D., M. L. McAnnelly, and H. H. Zakon. 1998. Diversity of sexual dimorphism in electrocommunication signals and its androgen regulation in a genus of electric fish, *Apteronotus. J. Comp. Physiol. A* 183:77–86.

Dyal, J. A., and W. C. Corning. 1973. Invertebrates learning and behavior taxonomies. In *Invertebrate Learning*, vol. 1., ed. W. C. Corning, J. A. Dyal, and A. O. D. Willows. New York: Plenum.

Ealey, E. H. M. 1963. The ecological significance of delayed implantation in a population of the hill kangaroo (*Macropus robustus*). In *Delayed Implantation*, ed. A. C. Enders. Chicago: University of Chicago Press.

Eberhard, W. G. 1988. Memory of distances and directions moved as cues during temporary spiral construction in the spider *Leucauge mariana* (Araneae: Araneidae). *J. Insect. Behav.* 1:51–66.

Eberhard, W. G. 1996. *Female Control: Sexual Selection by Cryptic Female Choice.* Princeton, NJ: Princeton University Press.

Eberhard, W. G. 1997. Sexual selection by cryptic female choice in insects and arachnids. Pp. 32–57 in *The Evolution of Mating Systems in Insects and Arachnids* (J. C. Choe and B. J. Crespi, eds.). Cambridge: Cambridge University Press.

Eberhard, W. G. 2000. Criteria for demonstrating postcopulatory female choice. *Evolution* 54:1047–50.

Ebert, P. D., and J. S. Hyde. 1976. Selection for agonistic behavior in wild female *Mus musculus*. *Behav. Genet.* 6:291–304.

Edwards, D. A. 1968. Mice: Fighting by neonatally androgenized females. *Science* 161:1027–28.

Ehrenfeld, D. W., and A. Carr. 1967. The role of vision in the sea-finding orientation of the green turtle (*Chelonia mydas*). *Anim. Behav.* 15:25–36.

Ehrman, L., and P. A. Parsons. 1981. *Behavior Genetics and Evolution.* New York: McGraw-Hill.

Eibl-Eibesfeldt, I. 1975. *Ethology: The Biology of Behavior.* 2d ed. New York: Holt, Rinehart and Winston.

Eichenbaum, H., C. Stewart, and R. G. M. Morris. 1990. Hippocampal representation in place learning. *Neurosci.* 10:3531–42.

Eisenberg, J. 1981. *The Mammalian Radiations.* Chicago: University of Chicago Press.

Eisner, T. E. 1966. Beetle spray discourages predators. *Natur. Hist.* 75:42–47.

———. 1970. Chemical defenses against predators in Arthropods. In *Chemical Ecology,* ed. E. Sondheimer and J. B. Simeone. New York: Academic Press.

Elgar, M. A., and B. J. Crespi. 1992. *Cannibalism: Ecology and Evolution Among Diverse Taxa.* New York: Oxford University Press.

Elias, M. 1981. Serum cortisol, testosterone, and testosterone-binding globulin responses to competitive fighting in human males. *Aggressive Behavior* 7:215–24.

Elliot, P. F. 1988. Foraging behavior of a central place forager: Field tests of theoretical predictions. *Am. Nat.* 131:159–74.

Elmes, G. W. 1971. An experimental study on the distribution of heathland ants. *J. Anim. Ecol.* 40:495–99.

Elner, R. W., and R. N. Hughes. 1978. Energy maximization in the diet of the shore crab, *Carcinus maenas. J. Anim. Ecol.* 47:103–16.

Eltringham, S. K. 1978. Methods of capturing wild animals for marking purposes. In *Animal Marking,* ed. B. Stonehouse, 13–23. Baltimore: University Park Press.

Emlen, S. 1991. Evolution of cooperative breeding in birds and mammals. In *Behavioural Ecology: An Evolutionary Approach,* 3rd edition (J. Krebs and N. Davies, eds.). Oxford: Blackwell Scientific.

Emlen, S., and P. Wrege. 1992. Parent-offspring conflict and the recruitment of helpers among bee-eaters. *Nature* 356:331–33.

Emlen, J. T. 1952a. Social behavior in nesting cliff swallows. *Condor* 54:177–99.

———. 1952b. Flocking behavior in birds. *Auk* 69:160–70.

Emlen, S. T. 1967a. Migratory orientation in the indigo bunting. *Passerina cyanea. Auk* 84:463–89.

———. 1967b. Migratory orientation in the indigo bunting. I. Evidence for use of celestial cues. *Auk* 84:309–42.

———. 1970. Celestial rotation: Its importance in the development of migratory orientation. *Science* 170:1198–1201.

———. 1972. An experimental analysis of the parameters of bird song eliciting species recognition. *Behaviour* 41:130–71.

———. 1975a. Migration, orientation, and navigation. In *Avian Biology,* vol. 5, ed. D. S. Farner and J. R. King. New York: Academic Press.

———. 1975b. The stellar-orientation system of a migratory bird. *Sci. Amer.* 233(2):102–11.

———. 1978. The evolution of cooperative breeding in birds. In *Behavioural Ecology: An Evolutionary Approach,* eds. J. R. Krebs and N. B. Davies. Oxford, England: Blackwell Scientific Publications, Ltd.

———. 1982. The evolution of helping. I. An ecological constraints model. *Amer. Nat.* 119:29–39.

———. 1984. Cooperative breeding in birds and mammals. In *Behavioural Ecology: An Evolutionary Approach,* 2d ed., ed. J. R. Krebs and N. B. Davies. Oxford, England: Blackwell Scientific Publications, Ltd.

Emlen, S. T., and L. W. Oring. 1977. Ecology, sexual selection, and the evolution of mating systems. *Science* 197:215–23.

Emlen, S. T., and P. H. Wrege. 1988. The role of kinship in helping decisions among white-fronted bee-eaters. *Behav. Ecol. Sociobiol.* 23:305–15.

Enright, J. T. 1970. Ecological aspects of endogenous rhythmicity. *Ann. Rev. Ecol. Syst.* 1:221–38.

Epstein, R., R. P. Lanza, and B. F. Skinner. 1980. Symbolic communication between two pigeons (*Columbia livia domestica*). *Science* 207:543–45.

Erickson, C. J., and D. S. Lehrman. 1964. Effect of castration of male ring doves on ovarian activity of females. *J. Comp. Physiol. Psychol.* 58:164–66.

Erwin, J., G. Mitchell, and T. Maple. 1973. Abnormal behavior in non-isolate-reared rhesus monkeys. *Psychol. Reports* 33:515–23.

Eskin, A., G. Corrent, C.-Y. Lin, and D. J. McAdoo. 1982. Mechanism of shifting the phase of a circadian oscillator by serotonin: Involvement of cAMP. *Proc. Nat. Acad. Sci. USA* 79:660–64.

Eskin, A., and J. S. Takahashi. 1983. Adenylate cyclase activation shifts the phase of a circadian pacemaker. *Science* 220:82–84.

Esser, A. H. 1971. *Behavior and Environment: The Use of Space by Animals and Men.* New York: Plenum.

Estes, R. D. 1972. The role of the vomeronasal organ in mammalian reproduction. *Mammalia* 36:315–41.

Etheredge, J., S. Perez, O. R. Taylor, and R. Jander. 1999. Monarch butterflies (*Danaus plexippus* L.) use a magnetic compass for navigation. *PNAS* (in press).

Etienne, A. S., R. Maurer, and F. Saucy. 1988. Limitations in the assessment of path dependent information. *Behaviour* 106:81–111.

Evans, C. S., and L. Evans. 1999. Chicken food calls are functionally referential. *Anim. Behav.* 58:307–19.

Evans, C. S., and P. Marler. 1994. Food calling and audience effects in male chickens, *Gallus gallus:* Their relationships to food availability, courtship and social facilitation. *Anim. Behav.* 47:1159–70.

Evans, C. S., L. S. Evans, and P. Marler. 1993. On the meaning of alarm calls: Functional reference in an avian vocal system. *Anim. Behav.* 46:23–38.

Ewert, J., H. Buxbaum-Conradi, M. Glagow, A. Röttgen, E. Schürg-Pfeiffer, and W. Schwippert. 1999. Forebrain and midbrain structures involved in prey-catching behaviour of toads: Stimulus-response mediating circuits and their modulating loops. *Eur. J. Morph.* 37:111–15.

Ewert, J. P. 1980. *Neuroethology.* Berlin: Springer-Verlag.

———. 1984. Tectal functions that underlie prey-catching and predator avoidance behaviors in toads. In *Comparative Neurology of the Optic Tectum,* ed. H. Vanegas, 247–416. New York: Plenum Publishing.

———. 1985. Concepts in vertebrate neuroethology. *Anim. Behav.* 33:1–29.

Fabricius, E. 1964. Crucial periods in the development of the following response in young nidifugous birds. *Zeit. Tierpsychol.* 21:326–37.

Fagen, R. 1981. *Animal Play Behavior.* New York: Oxford University Press.

Falconer, D. S., and T. F. C. MacKay. 1996. *Introduction to Quantitative Genetics,* 4th edition. Boston: Addison-Wesley Publishing Company.

Falls, J. B., and R. J. Brooks. 1975. Individual recognition by song in white-throated sparrows. II. Effects of location. *Can. J. Zool.* 53:1412–20.

Farner, D. S. 1955. The annual stimulus for migration: Experimental and physiologic aspects. In *Recent Studies in Avian Biology,* ed. A. Wolfson. Urbana: University of Illinois Press.

———. 1964. The photoperiodic control of reproductive cycles in birds. *Amer. Scientist* 52:137–56.

Feder, H. 1981. Experimental analysis of hormone actions on the hypothalamus, anterior pituitary, and ovary. In *Neuroendocrinology of Reproduction,* ed. N. Adler. New York: Plenum.

Feldman, J. F., and M. Hoyle. 1976. Isolation of circadian clock mutants of *Neurospora crassa. Genetics* 75:605–13.

Feng, A. S., P. M. Narins, and R. R. Capranica. 1975. Three populations of primary auditory fibers in the bullfrog (*Rana catesbiana*): Their peripheral origins and frequency sensitivities. *J. Comp. Physiol.* 100:221–29.

Fenton, M. B. 1984. Sperm competition? The case of vespertilionid and ronolophid bats. Pp. 573–87 in *Sperm Competition and the Evolution of Animal Mating Systems* (R. L. Smith, ed.). New York: Academic Press.

Fenton, M. B., and D. W. Thomas. 1985. Migrations and dispersal of bats (Chiroptera). Pp. 409–24 in *Migration: Mechanisms and Adaptive Significance* (M. A. Rankin, ed.). Port Aransas, TX: Marine Science Institute.

Fenton, M. B., D. Audet, D. C. Dunning, J. Long, C. B. Merriman, D. Pearl, D. M. Syme, B. Adkins, S. Pedersen, and T. Wohlgenant. 1993. Activity patterns and roost selection by *Noctilio albiventris* (Chiroptera: Noctilionidae) in Costa Rica. *J. Mamm.* 74:607–13.

Ferguson, D. E., H. F. Landreth, and J. P. McKeown. 1967. Sun compass orientation of the northern cricket frog, *Acris crepitans. Anim. Behav.* 15:45–53.

Ferguson, D. E., H. F. Landreth, and M. R. Turnipseed. 1965. Astronomical orientation of the southern cricket frog, *Acris gryllus. Copeia* (1965):58–66.

Ferreira, A. J. 1965. Emotional factors in the prenatal environment. *J. Nerv. Ment. Disorders* 141:108–18.

Ferris, C. F., and Y. Delville. 1994. Vasopressin and serotonin interactions in the control of agonistic behavior. *Psychoneuroendocrinology* 19:593–601.

Ferris, C. F., T. Stolberg, and Y. Delville. 1999. Serotonin regulation of aggressive behavior in male golden hamsters (*Mesocricetus auratus*). *Behav. Neurosci.* 113:804–15.

Feshbach, S., and R. D. Singer. 1971. *Television and Aggression.* San Francisco: Jossey-Bass.

Feshbach, S. 1997. The psychology of aggression: Insights and issues. Pp. 213–35 in *Aggression: Biological, Developmental, and Social Perspectives* (S. Feshbach and J. Zagrodzka, eds.). New York: Plenum Press.

Fisher, R. A. 1930. *The Genetical Theory of Natural Selection.* Oxford: Clarendon Press.

Fisher, R. A. 1958. *Genetical Theory of Natural Selection.* New York: Dover.

Fisher, R. C. 1961. A study in insect multiparasitism: II. The mechanisms and control of competition for the host. *J. Exp. Biol.* 38:605–28.

Fletcher, H. J. 1965. The delayed-response problem. In *Behavior of Nonhuman Primates,* vol. 1, ed. A. M. Schrier, H. F. Harlow, and F. Stollnitz. New York: Academic Press.

Fletcher, T. J. 1975. The environmental and hormonal control of reproduction in male and female red deer (*Cervus elaphus*). Ph.D. Thesis. University of Cambridge, England.

———. 1978. The induction of male sexual behavior in red deer (*Cervus elaphus*) by the administration of testosterone to hinds and estradiol 17B to stags. *Horm. Behav.* 11:74–88.

Flynn, J. P. 1967. The neural basis of aggression in cats. In *Neurophysiology and Emotion,* ed. D. C. Glass. New York: Rockefeller University Press and Russell Sage Foundation.

Foelix, R. 1996. *Biology of Spiders,* 2nd edition. New York: Oxford University Press.

Forbes, L. S. 1994. Avian brood reduction and parent-offspring "conflict." *Amer. Nat.* 142:82–117.

Foster, J. B. 1966. The giraffe of Nairobi National Park: Home range, sex ratios, the herd, and food. *E. Afr. Wildl. J.* 4:139–48.

Foster, S. A., and J. A. Endler. 1999. *Geographic Variation in Behavior: Perspective on Evolutionary Mechanisms.* New York: Oxford University Press.

Foster, W. A., and J. E. Treherne. 1981. Evidence for the dilution effect in the selfish herd from fish predation on a marine insect. *Nature* 293:466–67.

Fouad, K., F. Libersat, and W. Rathmayer. 1996. Neuromodulation of the escape behavior of the cockroach *Periplaneta americana* by the venom of the parasitic wasp *Ampulex compressa. J. Comp. Physiol. A* 178:91–100.

Fouts, R. S. 1972. Use of guidance in teaching sign language to a chimpanzee. *J. Physiol. Comp. Psychol.* 80:515–22.

———. 1973. Acquisition and testing of gestural signs in four young chimpanzees. *Science* 180:978–80.

Fox, C. A., A. A. Ismail, D. N. Love, H. E. Kirkham, and J. A. Loraine. 1972. Studies on the relationship between plasma testosterone levels and human sexual activity. *J. Endocrinol.* 52:51–58.

Fox, M. W. 1975. The behavior of cats. In *Behavior of Domestic Animals,* ed. E. S. E. Hafez. Baltimore: Williams & Wilkins.

Fox, M. W., and A. L. Clark. 1971. The development and temporal sequencing of agonistic behavior in the coyote (*Canis latrans*). *Z. Tierpsychol.* 28:262–78.

Fraenkel, G. 1975. Interactions between ecdysone, burisicon, and other endocrines during puparium formation and adult emergence in flies. *Amer. Zool.* 15 (sup. 1):29–41.

Fraenkel, G., and D. L. Gunn. 1940. *Orientation of Animals: Kineses, Taxes, and Compass Reactions.* New York: Oxford University Press.

Franchina, C. R., and P. K. Stoddard. 1998. Plasticity of the electric organ discharge waveform of the electric fish *Brachyhypopomus pinnicaudatus.* I. Quantification of day-night changes. *J. Comp. Physiol. A.* 183:759–68.

Francis, A. M., J. P. Hailman, G. E. Woolfenden. 1989. Mobbing by Florida scrub jays: Behavior, sexual asymmetry, role of helpers and ontogeny. *Anim. Behav.* 38:795–816.

Francis, L. 1973. Intraspecific aggression and its effect on the distribution of *Anthopleura elegantissima* and some related sea anemones. *Biol. Bull.* 144:73–92.

Francis, L. 1988. Cloning and aggression among sea anemones (Coelenterata: Actiniaria) of the rocky shore. *Biol. Bull.* 174:241–53.

Frank, L. G., S. E. Glickman, and P. Licht. 1991. Fatal sibling aggression, precocial development, and androgens in neonatal spotted hyenas. *Science* 252:702–4.

Frank, S. 1998. *Foundations of Social Evolution.* Princeton: Princeton University Press.

Franks, N. R. 1989. Army ants: A collective intelligence. *Amer. Scientist* 77:138–45.

Fraser, D. F. 1976. Coexistence of salamanders in the genus *Plethodon:* A variation on the Santa Rosalia theme. *Ecology* 57:238–51.

Fraser, D. 1990. Behavioral perspectives on piglet survival. *J. Reprod. Fertil.* Suppl. 40:355–70.

Fraser, D., and B. K. Thompson. 1991. Armed sibling rivalry by domestic piglets. *Behav. Ecol. Sociobiol.* 29:1–15.

Free, J. B. 1965. The allocation of duties among worker honeybees. *Symp. Zool. Soc. Lond.* 14:39–59.

Freedle, R., and M. Lewis. 1977. Prelinguistic conversations. In *Interaction, Conversation, and the Development of Language,* ed. M. Lewis and L. A. Rosenblum. New York: Wiley.

Freedman, J. L. 1980. Human reactions to population density. In *Biosocial Mechanisms of Population Regulation,* ed. M. N. Cohen, R. S. Malpass, and H. G. Klein. New Haven: Yale University Press.

Freeman, L. M., T. Arora, and E. F. Rissman. 1998. Neonatal androgen affects copulatory behavior in the female musk shrew. *Horm. Behav.* 34:231–38.

Fretwell, S. D. 1972. *Populations in a Seasonal Environment.* Princeton, NJ: Princeton University Press.

Fretwell, S. D., and H. L. Lucas. 1970. On territorial behaviour and other factors influencing habitat distribution in birds. I. Theoretical development. *Acta Biotheoretica* 19:16–36.

Frye, M. A., and R. M. Olberg. 1995. Visual receptive field properties of feature detecting neurons in the dragonfly. *J. Comp. Physiol. A.* 177:569–76.

Fugger, H. N., S. C. Cunningham, E. F. Rissman, and T. C. Foster. 1998. Sex differences in the activational effect of Era on spatial learning. *Horm. Behav.* 34:163–70.

Fullard, J. H. 1994. Jamming bat echolocation: The dogbane tiger moth *Cycnia tenera* times its clicks to the terminal attack calls of the big brown bat *Eptesicus fuscus. J. Exp. Biol.* 194: 285–98.

Fuller, C. A., R. Lydic, F. M. Sulzman, H. E. Albers, B. Tepper, and M. C. Moore-Ede. 1981. Circadian rhythm of body temperature persists after suprachiasmatic lesions in the squirrel monkey. *Am. J. Physiol.* 241:R385–91.

Fuller, J. L., and E. C. Simmel, eds. 1983. *Behavior Genetics: Principles and Applications.* Hillsdale, NJ: Lawrence Erlbaum Associates.

Furisch, F. T., and D. Jablonski. 1984. Late Triassic naticid drill holes: Carnivorous gastropods gain a major adaptation but fail to radiate. *Science* 224:78–80.

Futuyma, D. J. 1987. *Evolutionary Biology.* 2d ed. Sunderland, MA: Sinauer Assoc.

Futuyma, D. J. 1998. *Evolutionary Biology,* 3rd edition. Sunderland, MA: Sinauer.

Gadagkar, R. 1990. Evolution of eusociality: The advantage of assured fitness returns. *Phil. Trans. Roy. Soc. Lond. Ser. B.* 329:17–25.

Galef, B. G. 1971. Social factors in the poison avoidance and feeding behavior of wild and domesticated rat pups. *J. Comp. Physiol. Psychol.* 75:341–57.

———. 1976. Social transmission of acquired behavior: A discussion of tradition and social learning in vertebrates. *Adv. Stud. Behav.* 6:77–100.

———. 1977. Mechanisms for the social transmission of food preferences from adult to weanling rats. In *Learning and Mechanisms in Food Selection,* ed. L. M. Barker, M. Best, and M. Domjan. Waco: Baylor University Press.

———. 1981. Development of olfactory control of feeding-site selection in rat pups. *J. Comp. Physiol. Psychol.* 95:615–22.

———. 1990. Tradition in animals: Field observations and laboratory analyses. In *Interpretation and explanation in the study of animal behavior,* ed. M. Bekoff and D. Jamieson. San Francisco: Westview Press.

Galef, B. G., and M. M. Clark. 1971. Social factors in the poison avoidance and feeding behavior of wild and domesticated rat pups. *J. Comp. Physiol. Psych.,* 75:341–57.

Galef, B. G., and P. W. Henderson. 1972. Mother's milk: A determinant of the feeding preferences of weaning rat pups. *J. Comp. Physiol. Psychol.* 78:213–19.

Galef, B. G., and L. Heiber. 1976. The role of residual olfactory cues in the determination of feeding site selection and exploration patterns of domestic rats. *J. Comp. Physiol. Psychol.* 90:727–39.

Galef, B. G., D. J. Kennett, and S. W. Wigmore. 1984. Transfer of information concerning distant foods in rats: A robust phenomenon. *Anim. Learn. Behav.* 12:292–96.

Galef, B. G., D. J. Kennett, and M. Stein. 1985. Demonstrator influence on observer diet preference: Effects of simple exposure and the presence of a demonstrator. *Anim. Learn. Behav.* 13:25–30.

Galef, B. G., Jr. 1986. Social transmission of acquired behavior: A discussion of tradition and social learning in vertebrates. *Adv. Stud. Behav.* 6:77–100.

Galef, B. G., A. Mischinger, and S. A. Malenfant. 1987. Hungry rats' following of conspecifics to food depends on the diets eaten by potential leaders. *Anim. Behav.* 35:1234–39.

Galef, B. G., J. R. Mason, G. Preti, and N. J. Bean. 1988. Carbon disulfide: A semiochemical mediating socially-induced diet choice in rats. *Physiol. Behav.* 42:119–24.

Galef, B. G., Jr. 1996. Social learning: Introduction. In C. M. Heyes and B. G. Galef, Jr., eds., *Social Learning in Animals.* San Diego: Academic Press.

Gallistel, C. R., C. T. Piner, T. O. Allen, N. T. Adler, E. Yadin, and M. Negin. 1982. Computer assisted analysis of 2-DG autoradiographs. *Neurosci. Biobehav. Rev.* 6:409–20.

Gallup, G. G., Jr. 1970. Chimpanzee: Self-recognition. *Science* 167:86–87.

———. 1983. Toward a comparative psychology of mind. In *Animal Cognition and Behavior.* ed. R. L. Mellgren. New York: North-Holland Publishing Co.

Gandelman, R., F. S. vom Saal, and J. M. Reinisch. 1977. Contiguity to male fetuses affects morphology and behavior of female mice. *Nature* 266:722–24.

Ganzhorn, J. U. 1986. Feeding behavior of *Lemur catta* and *Lemur fulvus. Int. J. Primatol.* 7:17–30.

Garcia, J., and R. A. Koelling. 1966. Relation of cue to consequence in avoidance learning. *Psychonom. Sci.* 4:123–24.

Garcia, J., B. K. McGowan, and K. F. Green. 1972. Biological constraints on conditioning. In *Classical Conditioning II: Current Research and Theory,* ed. A. H. Black and W. F. Prokasy. New York: Appleton-Century-Crofts.

Garcia, J., W. G. Hankins, and K. W. Rusiniak. 1976. Flavor aversion studies. *Science* 192:265–66.

Gardner, B. T., and R. A. Gardner. 1971. Two-way communication with an infant chimpanzee. In *Behavior of Nonhuman Primates,* eds. A. M. Schrier and F. Stollnitz. New York: Academic Press.

———. 1985. Signs of intelligence in cross-fostered chimpanzees. *Phil. Trans. Royal Soc. Lond.* B. 308:159–76.

Gardner, R. A., and B. T. Gardner. 1969. Teaching sign language to a chimpanzee. *Science* 165:664–72.

Gardner, R. A., B. T. Gardner, and T. E. Van Cantfort. 1989. *Teaching sign language to chimpanzees.* Albany, NY: SUNY Press.

Gass, C. L., and R. D. Montogomerie. 1981. Hummingbird foraging behavior. Decision making and energy regulation. In *Foraging Behavior: Ecological, Ethological, and Psychological Approaches,* ed. A. C. Kamil and T. D. Sargent. New York: Garland Press.

Gaston, S. 1971. The influence of the pineal organ on the circadian activity rhythm in birds. In *Biochronometry,* ed. M. Menaker. Washington, DC: National Academy of Sciences.

Gauthreaux, S. A., Jr. 1971. A radar and direct visual study of passerine spring migration in southern Louisiana. *Auk* 88:343–65.

Geen, R. G., and N. B. Quantry. 1977. The catharsis of aggression: An evaluation of an hypothesis. *Adv. Exp. Soc. Psychol.* 10:1–37.

Geist, V. 1971. *Mountain Sheep: A Study in Behavior and Evolution.* Chicago: University of Chicago Press.

Gelber, B. 1965. Studies of the behavior of *Paramecium aurelia. Anim. Behav. Suppl.* 1:21–29.

Gerhardt, C. 1979. Vocalizations of some hybrid treefrogs: Acoustic and behavioral analyses. *Behaviour* 49:130–51.

Gerhardt, H. C. 1974a. Significance of some spectral features in mating call recognition in the green treefrog (*Hyla cinerea*). *J. Exp. Biol.* 61:229–41.

———. 1974b. Vocalizations of some hybrid treefrogs: Acoustic and behavioral analyses. *Behaviour* 49:130–51.

———. 1975. Sound pressures levels and radiation patterns of the vocalizations of some North American tree frogs and toads. *J. Comp. Physiol.* 102:1–12.

———. 1982. Sound pattern recognition in some North American treefrogs (Anura: Hylidae): Implications for mate choice. *Amer. Zool.* 2:581–95.

Gerlai, R. 1993. Can paradise fish (*Macropodus opercularis*) recognize its natural predator? An ethological analysis. *Ethology* 94:127–36.

Gerstell, R. 1939. Certain mechanisms of quail losses revealed by laboratory experimentation. Trans. *Fourth N. Amer. Wildl. Inst.* 462–67.

Gibbon, J., M. D. Baldock, C. Locurto, L. Gold, and H. S. Terrace. 1977. Trial and intertrial durations in autoshaping. *J. Exp. Psychol.: Anim. Behav. Proc.* 3:264–84.

Gibbs, H. L., and P. R. Grant. 1987. Oscillating selection on Darwin's finches. *Nature* 327:511–13.

Gibson, R. M., and J. W. Bradbury. 1985. Sexual selection in lekking sage grouse: Phenotypic correlates of male mating success. *Behav. Ecol. Sociobiol.* 18:117–23.

Gilbert, L. I. 1974. Endocrine action during insect growth. *Rec. Prog. Horm. Res.* 30:347–84.

Gilhousen, H. C. 1927. The use of the vision and of the antennae in the learning of crayfish. *Univ. Calif. Publ. Physiol.* 7:73–89.

———. 1977. Nonrandom foraging by sunbirds in a patchy environment. *Ecology* 58:1284–96.

Gill, F. B., and L. L. Wolf. 1975. Economics of feeding territoriality in the golden-winged sunbird. *Ecology* 56:333–45.

Gilliard, E. T. 1963. The evolution of bower birds. *Sci. Amer.* 209:38–46.

———. 1969. *Birds of Paradise and Bower Birds.* Garden City, NY: Natural History Press.

Gilraldeau, L. A., and D. L. Kramer. 1982. The marginal value theorem: A quantitative test using load size variation in a central place forager, the eastern chipmunk *Tamias striatus. Anim. Behav.* 30:1036–42.

Girardie, J. 1995. Molecular approaches to study invertebrates hormones, with particular reference to insects/Netherlands *J. Zool.* 45:10–14.

Glendinning, J. I., M. Tarre, and K. Asaoka. 1999. Contribution of different bitter-sensitive taste cells to feeding inhibition in a caterpillar (*Manduca sexta*). *Behav. Neurosci.*113:840–54.

Glossup, N. R. J., L. C. Lyons, and P. E. Hardin. 1999. Interlocked feedback loops within the *Drosophila* circadian oscillator. *Science* 286:766–68.

Gladue, B., M. Boechler, and K. D. McCaul. 1989. Hormonal responses to competition in human males. *Aggressive Behav.* 15:409–22.

Gochee, C., W. Rasband, and L. Sokoloff. 1980. Computerized densitometry and color coding of [^{14}C]-deoxyglucose autoradiographs. *Ann. Neurol.* 7:359–70.

Goessmann, C., H. C., and R. Huber. In press. The formation and maintenance of crayfish hierarchies: Behavioral and self-structuring properties. *Behav. Ecol. Sociobiol.* 48:418–28.

Goldfoot, D. A., S. M. Essock-Vitale, C. S. Asa, J. E. Thornton, and A. I. Leshner. 1978. Anosmia in male rhesus monkeys does not alter copulatory activity with cycling females. *Science* 199:1095–96.

Goldfoot, D. A., H. Loon, W. Groeneveld, and A. K. Slob. 1980. Behavioral and physiological evidence of sexual climax in the female stumptailed macaque (*Macaca arctoides*). *Science* 208:1477–79.

Golding, D. W. 1972. Studies in the comparative neuroendocrinology of polychaete reproduction. *Gen. Comp. Endocr.* (sup. 3): 580–90.

Goldsmith, T., J. Collins, and D. Perlman. 1981. A wavelength discrimination function for the hummingbird *Archilochus alexandri. J. Comp. Physiol.* 143:103–10.

Goldsworthy, S. D. 1995. Differential expenditure of maternal resources in Antarctic fur seals, *Arctocephalus gazella,* at Heard Island, southern Indian Ocean. *Behav. Ecol.* 6:218–28.

Gorman, M. L., M. G. Mills, J. P. Raath, and J. R. Speakman. 1998. High hunting costs make African wild dogs vulnerable to kleptoparasitism by hyenas. *Nature* 391:479–81.

Gosling, L. M., and M. Petrie. 1990. Lekking in topi: A consequence of satellite behaviour by small males at hotspots. *Anim. Behav.* 40:272–87.

Gottlieb, G. 1963. A naturalistic study of imprinting in wood ducklings (*Aix sponsa*). *J. Comp. Physiol. Psychol.* 56:86–91.

———. 1968. Prenatal behavior of birds. *Quart. Rev. Biol.* 43:148–74.

———. 1971. *Development of Species Identification in Birds.* Chicago: University of Chicago Press.

Gottlieb, G., and J. G. Vandenbergh. 1968. Ontogeny of vocalization in duck and chick embryos. *J. Exp. Zool.* 168:307–26.

Gould, J. L. 1975. Honeybee recruitment: The dance-language controversy. *Science* 189:685–93.

———. 1976. The dance-language controversy. *Quart. Rev. Biol.* 51:211–44.

Gould, S. J. 1977. *Ontogeny and Phylogeny.* Cambridge, MA: Harvard University Press.

Gowaty, P. A., and S. J. Wagner. 1988. Breeding season aggression of female and male eastern bluebirds (*Sialia sialis*) to models of potential conspecific and interspecific egg dumpers. *Ethology* 78:238–50.

Gowaty, P. A. 1995. Battles of the sexes and origins of monogamy. Pp. 21–52 in J. L. Black, ed., *Partnerships in Birds.* Oxford: Oxford University Press.

Goy, R. W. 1970. Experimental control of psycho-sexuality. *Phil. Trans. Roy. Soc. Lond.,* ser. B. 259:149–62.

Goy, R. W., F. B. Bercovitch, and M. C. McBrair. 1988. Behavioral masculinization independent of genital masculinization in prenatally androgenized female rhesus macaques. *Horm. Behav.* 22:552–70.

Graber, R. R. 1968. Nocturnal migration in Illinois: Different points of view. *Wilson Bull.* 80:36–71.

Grant, E. C., and J. H. MacKintosh. 1963. A comparison of the social postures of some common laboratory rodents. *Behaviour* 21:246–59.

Grant, P. R. 1975. Population performance of *Microtus pennsylvanicus* confined to woodland habitat and a model of habitat occupancy. *Can. J. Zool.* 53:1447–65.

Grant, P. R. 1981. Speciation and the adaptive radiation of Darwin's finches. *Amer. Sci.* 69:653–63.

———. 1986. *Ecology and evolution of Darwin's finches.* Princeton, NJ: Princeton University Press.

Grant, P. R., and I. Abbott. 1980. Interspecific competition, null hypotheses and island biogeography. *Evolution* 34:332–41.

Grant, P. R., and B. R. Grant. 1980. The breeding and feeding characteristics of Darwin's finches on Isla Genovesa, Galápagos. *Ecol. Monogr.* 50:381–410.

Green, R. G., C. L. Larson, and J. F. Bell. 1939. Shock disease as the cause of the periodic decimation of the snowshoe hare. *Amer. J. Hyg.* 30:83–102.

Green, S. 1975. Dialects in Japanese monkeys. *Zeitschrift fur Tierpsychologie.* 38:305–314.

Greenberg, N., and D. Crews. 1983. Physiological ethology of aggression in amphibians and reptiles. In *Hormones and Aggressive Behavior,* ed. B. B. Svare, 469–506. New York: Plenum.

Greenwood, P. J. 1980. Mating systems, philopatry, and dispersal in birds and mammals. *Anim. Behav.* 28:1140–62.

Greer, A. E. Jr. 1971. Crocodilian nesting habits and evolution. *Fauna* 2:20–28.

Greer, M. A. 1955. Suggestive evidence of a primary "drinking center" in the hypothalamus of the rat. *Proc. Soc. Exp. Biol.* 89:59–62.

Griffin, D. A. 1955. Bird navigation. In *Recent Studies in Avian Biology,* ed. A. Wolfson. Urbana: University of Illinois Press.

Griffin, D. R. 1958. *Listening in the Dark.* New Haven: Yale University Press.

———. 1970. Migrations and homing of bats. In *Biology of Bats,* vol. 1, ed. W. A. Wimsatt. New York: Academic Press.

———. 1974. *Bird Migration.* New York: Dover.

———. 1981. *The Question of Animal Awareness.* New York: Rockefeller University Press.

———. 1984. *Animal Thinking.* Cambridge, MA: Harvard University Press.

Griffin, D. A., and R. Galambos. 1941. The sensory basis of obstacle avoidance by flying bats. *J. Exp. Zool.* 86:481–506.

Griffiths, S., and A. Magurran. 1999. Schooling decisions in guppies (*Poecilia reticulata*) are based on familiarity rather than kin. *Behav. Ecol. Sociobiol.* 45: 437–43.

Gromko, M. H., D. G. Gilbert, and R. C. Richmond. 1984. Sperm transfer and use in the multiple mating system of *Drosophila.* In *Sperm Competition and the Evolution of Animal Mating Systems,* ed. R. L. Smith. New York: Academic Press.

Grosberg, R., and J. Quinn. 1986. The genetic control and consequences of kin recognition by the larvae of a colonial marine invertebrate. *Nature* 322:456–59.

Grosberg, R., D. Levitan, and B. Cameron. 1996. Evolutionary genetics of allorecognition in the colonial hydroid *Hydractinia symbiolongicarpus. Evolution* 50:2221–24.

Grosberg, R., and M. Hart. 2000. Mate selection and the evolution of highly polymorphic self/nonself recognition genes. *Science* 289:2111–14.

Gross, M. R. 1985. Disruptive selection for alternative life histories in salmon. *Nature* 313:47–48.

Gross, M. R., and R. Shine. 1981. Parental care and mode of fertilization in ectothermic vertebrates. *Evolution* 35:775–93.

Gross, M. R. 1991. Salmon breeding behavior and life history evolution in changing environments. *Ecology* 72:1180–86.

Gross, M. R. 1996. Alternative reproductive strategies and tactics: Diversity within sexes. *Trends. Ecol. Evol.* 11:92–98.

Grossfield, J., and B. Sakri. 1972. Divergence in the neural control of oviposition in *Drosophila. J. Insect Physiol.* 18:237–41.

Grota, L. J., and K. B. Eik-Nes. 1967. Plasma progesterone concentrations during pregnancy and lactation in the rat. *J. Reprod. Fertil.* 13:83–91.

Groves, P. M., and G. V. Rebec. 1988. *Introduction to Biological Psychology,* 3d ed. Dubuque, IA: W. C. Brown.

Gruss, M., and K. Braun. 1996. Stimulus-evoked increase of glutamate in the mediorostral neostriatum/hyperstriatum ventrale of the domestic chick after auditory filial imprinting: An in vivo microdialysis study. *J. Neurochem.* 66:1167–73.

Guhl, A. M. 1961. Gonadal hormones and social behavior. In *Sex and Internal Secretions,* ed. W. C. Young. Baltimore: Williams & Wilkins.

Gulick, W. L., and H. Zwick. 1966. Auditory sensitivity of the turtle. *Psychol. Rec.* 16:47–53.

Guthrie, R. D. 1971. A new theory of mammalian rump patch evolution. *Behaviour* 38:132–45.

Gwinner, E. 1986a. *Circannian Rhythms.* Berlin: Springer-Verlag.

———. 1986b. Circannual rhythms in the control of avian migration. *Adv. Stud. Anim. Behav.* 16:191–228.

Gwinner, E. 1990. Circannual rhythms in bird migration: Control of temporal patterns and interactions with photoperiod. Pp. 257–68 in *Bird Migration: Physiology and Ecophysiology* (E. Gwinner, ed.). Springer-Verlag, Berlin.

Gwinner, E., and M. Hau. 1996. Food as a circadian zeitgeber for house sparrows: The effect of different food access durations. *J. Bio. Rhythms* 11:196–207.

Gwynne, D. T. 1981. Sexual difference theory: Mormon crickets show role reversal in mate choice. *Science* 213:779–80.

———. 1984. Courtship feeding increases female reproductive success in bushcrickets. *Nature* 307:361–63.

Haartman, L. V. 1956. Territory in the pied flycatcher *Muscicapa hypoleuca. Ibis* 98:460–75.

Hagedorn, M., and W. Heiligenberg. 1985. Court and spark: Electric signals in the courtship and mating of gymnotoid fish. *Anim. Behav.* 33:254–65.

Haigh, G. R. 1979. Sun-compass orientation in the thirteen-lined ground squirrel, *Spermophilus tridecemlineatus. J. Mamm.* 60:629–32.

Hailman, J. P. 1967. The ontogeny of an instinct: The pecking response in chicks of the laughing gull (*Larus atricilla* L.) and related species. *Behavior,* Suppl. No. 15:1–159.

———. 1969. How an instinct is learned. *Sci. Amer.* 221:98–106.

———. 1973. Anatomical and physiological basis of embryonic motility in birds and mammals. In *Studies on the Development of Behavior and the Nervous System,* ed. G. Gottlieb. New York: Academic Press.

Hale, E. B., W. M. Schleidt, and M. W. Schein. 1969. The behavior of turkeys. In *Behavior of Domestic Animals,* ed. E. S. E. Hafez. Baltimore: Williams & Wilkins.

Hall, E. T. 1966. *The Hidden Dimension.* Garden City, NY: Doubleday.

Hall, J. C. 1994. The mating of a fly. *Science* 264:1700–14.

Hall, J. C., and C. P. Kyriacou. 1990a. Genetics of biological rhythms in *Drosophila. Science* 232:494–97.

Hall, W. G., and R. W. Oppenheim. 1987. Developmental psychobiology: Prenatal, perinatal, and early postnatal aspects of behavioral development. *Ann. Rev. Psychol.* 38:91–128.

Haley, M. P., C. J. Deutsch, and B. J. Le Boeuf. 1994. Size, dominance and copulatory success in male northern elephant seals, *Mirounga angustirostris. Anim. Behav.* 48:1249–60.

Hall, J. C. 1994. Central processing of communication sounds in the anuran auditory system. *Amer. Zool.* 34:670–84.

Hall, S. 1998. Object play by adult animals. In *Animal Play* (M. Bekoff and J. A. Byers, eds.). New York: Cambridge University Press.

Hall, J. C., and C. P. Kyriacou. 1990b. Genetics of biological rhythms in *Drosophila. Adv. Insect Physiol.* 22:221–98.

Halliday, T. R. 1983. Motivation. 1. Causes and Effects. In *Animal Behaviour,* ed. T. R. Halliday and P. J. B. Slater, 100–33. New York: W. H. Freeman.

Hamburger, V. 1963. Some aspects of the embryology of behavior. *Quart. Rev. Biol.* 38:342–65.

Hamburger, V. 1973. Anatomical and physiological basis of embryonic motility in birds and mammals. In G. Gottlieb (eds.), *The Evolutionary Synthesis.* Cambridge, MA: Harvard University Press.

Hamilton, W. D. 1963. The evolution of altruistic behavior. *Amer. Nat.* 97:354–56.

———. 1964. The genetical evolution of social behaviour I, II. *J. Theoret. Biol.* 7:1–52.

———. 1967. Extraordinary sex ratios. *Science* 156:477–88.

———. 1971. Geometry for the selfish herd. *J. Theoret. Biol.* 31:295–311.

Hamilton, W. D., and R. M. May. 1977. Dispersal in stable habitats. *Nature* 269:578–81.

Hamilton, W. D., and M. Zuk. 1982. Heritable true fitness and bright birds: A role for parasites? *Science* 218:384–87.

Hamilton, W. D., and M. Zuk. 1984. Heritable true fitness and bright birds: A role for parasites? *Science* 218:384–87.

Hampton, R. R., and S. J. Shettleworth. 1996. Hippocampal lesions impair memory for location but not color in passerine birds. *Behav. Neurosci.* 110:946–64.

Hansell, M. H. 1984. *Animal architecture and building behavior.* New York: Longman.

Hansen, E. W. 1966. The development of maternal and infant behavior in the rhesus monkey. *Behaviour* 27:107–49.

Hansen, L. P., and G. O. Batzli. 1978. The influence of food availability on the white-footed mouse: Populations in isolated woodlots. *Can. J. Zool.* 56:2530–41.

Harcourt, A. H. 1989. Deformed sperm are probably not adaptive. *Anim. Behav.* 37:863–64.

Hardin, G. 1968. The tragedy of the commons. *Science* 162:1243–48.

Harding, C. F. 1983. Hormonal influences on avian aggressive behavior. In *Hormones and Aggressive Behavior,* ed. B. B. Svare. New York: Plenum.

———. 1965. Sexual behavior in the rhesus monkey. In *Sex and Behavior,* ed. F. Beach. New York: Wiley.

Harker, J. E. 1960. Endocrine and nervous factors in insect circadian rhythms. *Cold Spr. Harb. Symp. Quant. Biol.* 25:279–87.

———. 1964. The physiology of diurnal rhythms. *Camb. Monogr. Exp. Biol.* 13:1–114.

Harlow, H. F. 1949. The formation of learning sets. *Psychol. Rev.* 56:51–65.

———. 1950. Learning and satiation of response in intrinsically motivated complex puzzle performance by monkeys. *J. Comp. Physiol. Psychol.* 43:289–94.

———. 1951. Primate learning. In *Comparative Psychology,* ed. C. P. Stone. Englewood Cliffs, NJ: Prentice-Hall.

———. 1962. Development of affection in primates. In *Roots of Behavior,* ed. E. L. Bliss. New York: Harper & Row.

Harlow, H. F., and M. K. Harlow. 1962a. Social deprivation in monkeys. *Sci. Amer.* 207:136–46.

———. 1962b. The effect of rearing conditions on behavior. *Bull. Menninger Clinic* 26:213–24.

———. 1969. Age-mate or peer affectional system. *Adv. Study Behav.* 2:333–83.

Harlow, H. F., M. K. Harlow, and D. R. Meyer. 1950. Learning motivated by a manipulation drive. *J. Exp. Psychol.* 40:228–34.

Harlow, H. F., M. K. Harlow, and S. J. Suomi. 1971. From thought to therapy: Lessons from a primate laboratory. *Amer. Scientist* 59:538–49.

Harlow, H. F., and S. J. Suomi. 1971. Social recovery by isolate-reared monkeys. *Proc. Nat. Acad. Sci. USA* 68:1534–38.

Harlow, H. F., and R. R. Zimmerman. 1959. Affectional responses in the infant monkey. *Science* 130:421–32.

Harnly, M. H. 1941. Flight capacity in relation to phenotypic and genotypic variations in the wings of *Drosophila melanogaster. J. Exp. Zool.* 88:263–73.

Harper, D. G. C. 1982. Competitive foraging in mallards: 'ideal free' ducks. *Anim. Behav.* 30:575–84.

Harris, V. T. 1952. An experimental study of habitat selection by prairie and forest races of the deer mouse, *Peromyscus maniculatus. Contrib. Lab. Vert. Biol., Univ. Mich.* 56:1–53.

Harrison, J. L. 1952. Moonlight and pregnancy of Malayan forest rats. *Nature* 170:73–74.

Hart, B., and L. Hart. 1992. Reciprocal allogrooming in impala, *Aepyceros melampus. Anim. Behav.* 44:1073–83.

Hartl, D. L. 1988. *A Primer of Population Genetics.* Sunderland, MA: Sinauer Assoc. 305 pp.

Hartl, D. L., and A. G. Clark. 1997. *Principles of Population Genetics,* 3rd edition. Sunderland, MA: Sinauer Associates.

Hartwick, P., J. Kiepenheuer, and K. Schmidt-Koenig. 1978. Further experiments on the olfactory hypothesis of pigeon homing. In *Animal Migration, Navigation, and Homing,* ed. K. Schmidt-Koenig and W. T. Keeton. New York: Springer-Verlag.

Harvey, L. A., and C. R. Propper. 1997. Effects of androgens on male sexual behavior and secondary sex characters in the explosively breeding spadefoot toad, *Scaphiopus couchii. Horm. Behav.* 31:89–96.

Haskins, K. E., A. Sih, and J. J. Krupa. 1997. Predation risk and social interference as factors influencing habitat selection in two species of stream-dwelling waterstriders. *Behav. Ecol.* 8:351–63.

Hasler, A. D. 1960. Guideposts of migrating fishes. *Science* 132:785–92.

———. 1966. *Underwater Guideposts: Homing of Salmon.* Madison: University of Wisconsin Press.

Hasler, A. D., A. T. Scholtz, and R. M. Horrall. 1978. Olfactory imprinting and homing in salmon. *Am. Sci.* 66:347–55.

Hatchwell, B., and J. Komdeur. 2000. Ecological constraints, life history traits and the evolution of cooperative breeding. *Anim. Behav.* 59:1079–86.

Hausfater, G., and S. B. Hrdy, eds. 1984. *Infanticide: Comparative and Evolutionary Perspectives.* New York: Aldine.

Hauser, M. 1996. *The Evolution of Communication.* Cambridge: MIT Press.

Hawkins, R. D., T. W. Abrams, T. J. Carew, and E. R. Kandel. 1983. A cellular mechanism of classical conditioning in *Aplysia:* Activity-dependent amplification of presynaptic facilitation. *Science* 219:400–5.

Hayes, J. H. 1951. *The Ape in Our House.* New York: Harper Brothers.

Hayes, S. L., and C. T. Snowdon. 1990. Predator recogntion in cotton-top tamarins (*Saguinus oedipus*). *Amer. J. Primatol.* 20:283–92.

Hazlett, B. A., and W. H. Bossert. 1965. A statistical analysis of the aggressive communications systems of some hermit crabs. *Anim. Behav.* 13:357–73.

Healy, S. D., N. S. Clayton, and J. R. Krebs. 1994. Development of hippocampal specialization in two species of tit (*Parus* spp.). *Behavioural Brain Res.* 61:23–28.

Hedrick, A. V. 1986. Female preferences for male calling bout duration in a field cricket. *Behav. Ecol. and Sociobio.* 19:73–77.

Hedrick, A. V. 1988. Female choice and the heritability of attractive male traits: An empirical study. *Am. Nat.* 132:267–76.

Hedrick, A. V. 1994. The heritability of mate-attractive traits: A case study on field crickets. Pp. 228–50 in Boake (ed.), *Quantitative Genetic Studies of Behavioral Evolution.* Chicago: University of Chicago Press.

Hedrick, A. V., and L. M. Dill. 1993. Mate choice by female crickets is influenced by predation risk. *Anim. Behav.* 46:193–96.

Hedricks, C., and M. K. McClintock. 1990. Timing of insemination is correlated with the secondary sex ratio of Norway rats. *Physiol. Behav.* 48:625–32.

Hedwig, B. 2000. Control of cricket stridulation by a command neuron: Efficacy depends on behavioral state. *J. Neurophysiol.* 83(2):712–22.

Hegner, R. E., and J. C. Wingfield. 1986. Social modulation of gonadal development and circulating hormone levels during autumn and winter. In *Behavioural Rhythms,* ed. Yvon Queinec and N. Delvolve, 109–17. Toulouse, France: Privat.

———. 1987a. Effects of experimental manipulation of testosterone levels on parental investment and breeding success in male house sparrows. *Auk* 104:462–69.

———. 1987b. Effects of brood-size manipulation on parental investment, breeding success, and reproductive endocrinology of house sparrows. *Auk* 104:470–80.

Heinrich, B. 1983. Do bumblebees forage optimally, and does it matter? *Amer. Zool.* 23:273–81.

Helfrich-Förster, C. 1995. The period clock gene is expressed in central nervous system neurons which also produce a neuropeptide that reveals the projections of circadian pacemaker cells within the brain of *Drosophila melanogaster. Proc. Nat. Acad. Sci. USA* 92:612–16.

Hendrickson, A. E., N. Wagoner, and W. M. Cowan. 1972. Autoradiographic and electron microscopic study of retinohypothalamic connections. *Zeit. Zellforsch.* 125:1–26.

Herlugson, C. J. 1981. Nest site selection in mountain bluebirds. *Condor* 83:252–55.

Herman L., and R. Uyeyama. 1999. The dolphin's grammatical competency: Comments on Kako. *Anim. Learn. and Behav.* 27:18–23.

Herman, L. M., J. P. Wolz, and D. G. Richards. 1984. Comprehension of sentences by bottlenosed dolphins. *Cognition* 16:1–90.

Herndon, J. G., A. A. Perachio, and M. McCoy. 1979. Orthogonal relationship between electrically elicited social aggression and self-stimulation from the same brain sites. *Brain Research* 171:374–80.

Herrnkind, W. F. 1985. Evolution and mechanisms of mass single-file migration in spiny lobster: Synopsis. Pp. 197–211 in *Migration: Mechanisms and Adaptive Significance* (M. A. Rankin, ed.). Port Aransas, TX: Marine Science Institute.

Hernstein, R. J., D. H. Loveland, and C. Cable. 1976. Natural concepts in pigeons. *J. Exp. Psychol.: Anim. Behav. Proc.* 2:285–302.

Hernstein, R. J., and P. A. de Villiers. 1980. Fish as a natural category for people and pigeons. *Psychol. Learn. Motiv.* 14:59–95.

Hess, E. 1959. Imprinting. *Science* 30:133–41.

Hews, D. K., and M. C. Moore. 1995. Influence of androgens on differentiation of secondary sex characters in tree lizards, *Urosaurus ornatus. Gen. Comp. Endocrinol.* 97:86–102.

Heyes, C. M. 1996. Introduction: Identifying and defining imitation. In *Social Learning in Animals.* (C. M. Heyes and B. G. Galef, Jr., eds.). San Diego: Academic Press.

Hickman, C., L. Roberts, and A. Larson. 1998. Dubuque: WCB/McGraw-Hill.

Hickman, C., L. Roberts, and F. Hickman. 1984. *Integrated Principles of Zoology,* 7th edition. St. Louis: Times Mirror/Mosby College.

Hildén, O. 1965. Habitat selection in birds. A review. *Ann. Zool. Fenn.* 2:43–75.

Hilgard, E. R., and G. H. Bower. 1975. *Theories of Learning,* 4th ed. Englewood Cliffs, NJ: Prentice-Hall.

Hill, P. S. M. 1998. Environmental and social influences on calling effort in the prairie mole cricket (*Gryllotalpa major*). *Behav. Ecol.* 9:101–8.

Hinde, R. A. 1965. Interaction of internal and external factors in integration of canary reproduction. In *Sex and Behavior,* ed. F. A. Beach. New York: Wiley.

———. 1970. *Anim. Behav.* 2d ed. New York: McGraw-Hill.

———. 1973. Constraints on learning—An introduction to the problems. In *Constraints on Learning: Limitations and*

Predispositions, ed. R. A. Hinde and J. Stevenson-Hinde. New York: Academic Press.

―――. 1973. On the design of check-sheets. *Primates* 14:393–406.

Hinde, R. A., and L. Davies. 1972. Removing infant rhesus from mother for 13 days compared with removing mother from infant. *J. Child. Psychol. Psychiat.* 19:199–211.

Hinde, R. A., M. E. Leighton-Shapiro, and L. McGinnis. 1978. Effects of various types of separation experience on rhesus monkeys five months later. *J. Child. Psychol. Psychiat.* 19:199–211.

Hinde, R. A., and L. McGinnis. 1977. Some factors influencing the effects of temporary mother-infant separation—Some experiments with rhesus monkeys. *Psychol. Med.* 7:197–212.

Hinde, R. A., and Y. Spencer-Booth. 1967. The behaviour of socially living rhesus monkeys in their first two-and-a-half years. *Anim. Behav.* 15:169–96.

―――. 1970. Individual differences in the responses of rhesus monkeys to a period of separation from their mothers. *J. Child. Psychol. Psychiat.* 11:159–76.

―――. 1971. Effect of brief separation from mother on rhesus monkeys. *Science* 173:111–18.

Hinde, R. A., and J. Stevenson-Hinde, eds. 1973. *Constraints on Learning: Limitations and Predispositions.* New York: Academic Press.

Hirsch, J. 1967. Behavior-genetic analysis. In *Behavior-Genetic Analysis,* ed. J. Hirsch. New York: McGraw-Hill.

―――. ed. 1967. *Behavior-Genetic Analysis.* New York: McGraw-Hill.

―――. 1975. Jensenism: The bankruptcy of "science" without scholarship. *Educ. Theory* 25:3–27.

―――. 1981. To "unfrock the charlatans." *Sage Race Rel. Abstr.* 6:1–67.

Hirsch, J., T. R. McGuire, and A. Vetta. 1980. Concepts of behavior genetics and misapplications to humans. In *Evolution of Human Social Behavior,* ed. J. S. Lockard. New York: Elsevier.

Hirvonen, H., E. Ranta, H. Rita, and N. Peuhkuri. 1999. Significance of memory properties in prey choice decisions. *Ecological Modelling* 115:177–89.

Hobson, K. A. 1999. Tracing origins and migration of wildlife using stable isotopes: a review. *Oecologia* 120:314–26.

Hodgkin, A. L. 1971. *Conduction of the Nervous Impulse.* Springfield, IL: Charles C. Thomas.

Hodos, W., and C. B. G. Campbell. 1969. Scala naturae: Why is there no theory in comparative psychology? *Psychol. Rev.* 76:337–50.

Hoffman, K. 1959. Die Aktivitätsperiodik von im 18- und 36-Stunden-tag erbrüteten Eideschsen. *Zeit. Vergleich. Physiol.* 42:422–32.

Hogan, J. A., A. J. Hogan-Warburg, L. Panning, and C. A. Moffatt. 1998. Causal factors controlling the brooding cycle of broody junglefowl hens with chicks. *Behaviour* 135:957–80.

Hogg, J. T. 1988. Copulatory tactics in relation to sperm competition in Rocky Mountain bighorn sheep. *Behav. Ecol. Sociobiol.* 22:49–59.

Holecamp, K. E., and P. W. Sherman. 1989. Why male ground squirrels disperse. *Amer. Scientist* 77:232–39.

Holekamp, K. E., L. Smale, R. Berg, and S. M. Cooper. 1997. Hunting rates and hunting success in the spotted hyena (*Crocuta crocuta*). *J. Zool., Lond.* 242:1–15.

Holekamp, K. E., and L. Smale. 1998. Dispersal status influences hormones and behavior in the male spotted hyena. *Horm. Behav.* 33:205–16.

Holekamp, K. E., E. E. Boydston, M. Szykman, I. Graham, K. J. Nutt, S. Birch, A. Piskiel, and M. Singh. 1999. Vocal recognition in the spotted hyena and its possible implications regarding the evolution of intelligence. *Anim. Behav.* 58:383–95.

Hölldobler, B., and C. D. Michener. 1980. Mechanisms of identification and discrimination in social hymenoptera. In *Evolution of Social Behavior: Hypotheses and Empirical Tests,* ed. H. Markl. Deerfield Beach, FL: Verlag Chemie.

Hölldobler, B., and E. O. Wilson. 1990. *The Ants.* Boston: Harvard University Press.

Holling, C. S. 1959. The components of predation as revealed by a study of small mammal predation of the European pine sawfly. *Can. Entomol.* 91:293–320.

Holmes, W. G. 1984. Predation risk and foraging behavior of the hoary marmot in Alaska. *Behav. Ecol. Sociobiol.* 15:293–302.

―――. 1988. Kinship and development of social preferences. In *Developmental Psychobiology and Behavioral Ecology,* ed. E. M. Blass, 389–413. New York: Plenum Press.

Holmes, W. G. 1986. Identification of paternal half-siblings by captive Belding's ground squirrels. *Anim. Behav.* 34:321–27.

Holmes, W. G., and P. W. Sherman. 1983. Kin recognition in animals. *Amer. Scientist* 71:46–55.

Honeycutt, R. L. 1992. Naked mole-rats. *Amer. Scient.* 80:43–53.

Hoogland, J. L. 1979. Aggression, ectoparasitism, and other possible costs of prairie dog (Sciuridae: *Cynomys* spp.) coloniality. *Behaviour* 69:1–35.

―――. 1982. Prairie dogs avoid extreme inbreeding. *Science* 215:1639–41.

Hopf, S., M. Herzog, and D. Ploog. 1985. Development of attachment and exploratory behavior in infant squirrel monkeys under controlled rearing conditions. *Int. J. Behav. Develop.* 8:55–74.

Hopkins, C. D. 1974. Electric communication in fish. *Amer. Scientist* 62:426–37.

Hopkins, C. D., and A. H. Bass. 1981. Temporal coding of species recognition signals in an electric fish. *Science* 212:85–87.

Horseman, G., and F. Huber. 1994. Sound localisation in crickets: II. Modelling the role of a simple neural network in the prothoracic ganglion. *J. Comp. Physiol. A.* 175:399–413.

Hotta, Y., and S. Benzer. 1972. The mapping of behavior in *Drosophila* mosaics. *Nature* 240:527–35.

―――. 1973. Courtship in *Drosophila* mosaics: Sex-specific foci for sequential action patterns. *Proc. Nat. Acad. Sci. USA* 73:4154–58.

Houde, A. E., and A. J. Torio. 1992. Effect of parasitic infection on male color pattern and female choice in guppies. *Behav. Ecol.* 3:346–56.

Houtman, R., and L. M. Dill. 1998. The influence of predation risk on diet selectivity: A theoretical analysis. *Evol. Ecol.* 12:251–62.

Howard, H. E. 1920. *Territory in Bird Life.* New York: Atheneum.

Howard, R. D. 1978. The evolution of mating strategies in bullfrogs, *Rana catesbeiana. Evolution* 32:850–71.

―――. 1980. Mating behaviour and mating success in wood frogs. *Anim. Behav.* 28:705–16.

―――. 1988. Reproductive success in two species of anurans. In *Reproductive success,* ed. T. H. Clutton-Brock. Chicago: University of Chicago Press.

Hoy, R. R., A. Hoikkala, and K. Kaneshiro. 1988. Hawaiian courtship songs: Evolutionary innovation in communication signals of *Drosophila. Science* 240:217–19.

Hrdy, S. B. 1977a. Infanticide as a primate reproductive strategy. *Amer. Scientist* 65:40–49.

―――. 1977b. *Langurs of Abu: Female and Male Strategies of Reproduction.* Cambridge: Harvard University Press.

Hubel, D. H., and T. N. Wiesel. 1965. Receptive fields and functional architecture in two non-striate visual areas of the cat. *J. Neurophysiol.* 28:229–89.

―――. 1976. Spatial vision in anurans. In *The Amphibian Visual System: A Multidisciplinary Approach,* ed. K. V. Fite. New York: Academic Press.

Huber, F. 1978. The insect nervous system and insect behaviour. *Anim. Behav.* 26:969–81.

―――. 1983a. Implications of insect neuroethology for studies on vertebrates. In *Advances in Neuroethology,* ed. J. P. Ewert, R. R. Capranica, and D. J. Ingle, 91–138. New York: Plenum Press.

———. 1983b. Neural correlates of orthopteran and cicada phonotaxis. In *Neuroethology and Behavioral Physiology,* ed. F. Huber and H. Markl, 108–35. Berlin: Springer-Verlag.

Huber, R., and A. Delago. 1998. Serotonin alters decisions to withdraw in fighting crayfish, *Astacus astacus:* The motivational concept revisted. *J. Comp. Physiol. A.* 182:573–83.

Huber, R., K. Smith, A. Delago, K. Isaksson, and E. A. Kravitz. 1997. Serotonin and aggressive motivation in crustaceans: Altering the decision to retreat. *Proc. Natl. Acad. Sci.* 94:5939–42.

Huck, U. W., J. Seger, and R. D. Lisk. 1990. Litter sex ratios in the golden hamster vary with time of mating and litter size and are not binomially distributed. *Behav. Ecol. Sociobiol.* 26:99–109.

Huck, U. W., and E. M. Banks. 1980. The effects of cross-fostering on the behaviour of two species of North American lemmings, *Dicrostonyx groenlandicus* and *Lemmus trimucronatus. Anim. Behav.* 28:1046–52.

Hudson, D. J., and M. E. Lickey. 1977. Weak negative coupling between the circadian pacemakers of the eyes of *Aplysia. Neurosci. Abst.* 3:179.

Hughes, J. J., D. Ward, and M. R. Perrin. 1994. Predation risk and competition affect habitat selection and activity of Namib Desert gerbils. *Ecology-Tempe.* 75:1397–1405.

Hui, C. A. 1994. Lack of association between magnetic patterns and the distribution of free-ranging dolphins. *J. Mamm.* 75:399–405.

Humphrey, N. K. 1976. The social function of intellect. In *Growing Points of Ethology,* ed. P. P. G. Bateson and R. A. Hinde. Cambridge: Cambridge University Press.

Hunt, G. J., R. E. Page, Jr., M. K. Fondrk, and C. J. Dullum. 1995. Major quantitative trait loci affecting honey bee foraging behavior. *Genetics* 141:1537–45.

Hunt, R. M., X. Xiang-Xu, and J. Kaufman. 1983. Miocene burrows of extinct bear dogs: Indication of early denning behavior of large mammalian carnivores. *Science* 221:364–66.

Huntingford, F. A. 1986. Development of behaviour in fish. In *The Behavior of Teleost Fishes,* ed. T. J. Pitcher, 47–68. Baltimore: Johns Hopkins University Press.

Huntingford, F., and A. Turner. 1987. *Animal Conflict.* London: Chapman and Hall.

Huntley, M. 1985. Experimental approaches to the study of vertical migration of zooplankton. Pp. 71–90 in *Migration: Mechanisms and Adaptive Significance* (R. M. A., ed.). Port Aransas, TX: Marine Science Institute.

Hurd, C. R. 1996. Interspecific attraction to the mobbing calls of black-capped chickadees *(Parus atricapillus). Behav. Ecol. Sociobiol.* 38:287–92.

Hutchinson, J. B. 1978. Hypothalamic regulation of male sexual responsiveness to androgen. In *Biological Determinants of Sexual Behavior,* ed. J. B. Hutchinson. New York: Wiley.

Hutchinson, R. R. 1983. The pain-aggression relationship and its expression in naturalistic settings. *Aggressive Behav.* 9:229–42.

Hutt, C. 1966. Exploration and play in children. *Symp. Zool. Soc. Lond.* 18:23–44.

———. 1967a. Temporal effects on response decrement and stimulus satiation in exploration. *Brit. J. Psychol.* 58:365–73.

———. 1967b. Effects of stimulus novelty on manipulatory exploration in an infant. *J. Child Psychol. Psychiat.* 8:241–47.

———. 1970a. Specific and diversive exploration. *Adv. Child Dev. Behav.* 5:119–80.

———. 1970b. Curiosity in young children. *Science J.* 6:68–71.

Hutt, C., and R. Bhanvani. 1972. Predictions from play. *Nature* 237:171–72.

Hutto, R. L. 1985. Habitat selection by nonbreeding, migratory land birds. Pp. 455–76 in *Habitat Selection in Birds* (M. L. Cody, ed.). New York: Academic Press.

Huxley, J. S. 1914. The courtship-habits of the great-crested grebe *(Podiceps cristatus);* with an addition to the theory of sexual selection. *Proc. Zool. Soc. Lond.* 35:491–562.

Immelmann, K. 1965. Objektfixierung geschlechtlicher Triebhandlung bei Prachtfinken. *Naturwissenschaften* 52:169–70.

———. 1972. Sexual and other long-term aspects of imprinting in birds and other species. *Adv. Stud. Behav.* 4:147–74.

Immelmann, K., and C. Beer, 1989. A *Dictionary of Ethology.* Cambridge, MA: Harvard University Press.

Inglis, L. R., and A. J. Isaacson. 1978. The responses of dark-bellied brent geese to models of geese in various postures. *Anim. Behav.* 26:953–58.

Ingram, J. C. 1978. Primate markings. In *Animal Marking,* ed. B. Stonehouse, 169–74. Baltimore: University Park Press.

Irwin, R. E. 1988. The evolutionary importance of behavioural development: The ontogeny and phylogeny of bird song. *Anim. Behav.* 36:814–24.

Inman, A. J., and Krebs. 1987. Predation and group living. *Trends Ecol. Evol.* 2:31–32.

Irwin, R. E. 1988. The evolutionary importance of behavioural development: The ontogeny and phylogeny of bird song. *Anim. Behav.* 36:814–24.

Itani, J. 1958. On the acquisition and propagation of a new food habit in the troop of Japanese monkeys at Takasakiyama. *Primates* 1:131–48.

Jacklet, J. W. 1969. Circadian rhythm of optic nerve impulses recorded in darkness from isolated eye of *Aplysia. Science* 164:562–63.

———. 1973. Neuronal population interactions in a circadian rhythm in *Aplysia.* In *Neurobiology of Invertebrates,* ed. J. Salanki. Budapest: Akademiai Kiado.

———. 1976. Circadian rhythms in the nervous system of a marine gastropod, *Aplysia.* In *Biological Rhythms in the Marine Environment,* ed. P. J. DeCoursey, 17–31. Columbia, SC: University of South Carolina Press.

Jacklet, J. W., and J. Geronimo. 1971. Circadian rhythm: Populations of interacting neurons. *Science* 174:299–302.

Jackson, J. R. 1963. Nesting of keas. *Notornis* 10:319–26.

Jackson, R. R., and R. S. Wilcox. 1993. Spider flexibly chooses aggressive mimicry signals for different prey by trial and error. *Behaviour* 127:21–36.

Jacobson, M. 1974. A plentitude of neurons. In *Aspects of Neurogenesis,* vol. 2, ed. G. Gottlieb. New York: Academic Press.

———. 1978. *Developmental Neurobiology,* 2d ed. New York: Plenum.

Jacobssen, S., and T. Jarvik. 1976. Anti-predator behaviour of two-year-old hatchery-reared Atlantic salmon *(Salmo salar)* and a description of the predatory behaviour of burbot *(Lota lota). Zool. Rev.* 38:57–70.

Jacquot, J. J., and S. H. Vessey. 1994. Non-offspring nursing in the white-footed mouse, *Peromyscus leucopus. Anim. Behav.* 48:1238–40.

Jaeger, R. G. 1981. Dear enemy recognition and the costs of aggression between salamanders. *Am. Nat.* 117:962–74.

Jaeger, R. G. 1984. Agonistic behavior of the red-backed salamander. *Copeia* 1984:309–14.

Jaeger, R. G. 1986. Pheromonal markers as territorial advertisement by terrestrial salamanders. Pp. 191–203 in *Chemical Signals in Vertebrates* (D. Duvall, D. Müller-Schwarze, and R. M. Silverstein, eds.). New York: Plenum.

Jaeger, R. G., and J. K. Schwarz. 1991. Gradational threat postures by the red-backed salamander. *J. Herpetol.* 25:112–14.

Jaeger, R. G., K. C. B. Nishikawa, and D. E. Bernard. 1983. Foraging tactics of a terrestrial salamander: Costs of territorial defense. *Anim. Behav.* 31:191–98.

Jaenike, J., and R. D. Holt. 1991. Genetic variation for habitat preference: Evidence and explanations. *Amer. Nat.* 137:S67–S90.

Jakob, E. M. 1991. Costs and benefits of group living for pholcid spiders: Losing food, saving silk. *Animal Behaviour* 41:711–22.

Jallon, J. M., C. Antony, and O. Benamar. 1981. Un antiaphrodisiaque produit par les males de *Drosophila melanogaster* et transfere aux femelles los de la copulation. *C. R. Acad. Sci. Paris* 292:1147–49.

Janowitz, H. D., and M. I. Grossman. 1949. Some factors affecting the food intake of normal dogs and dogs with esophagotomy and gastric fistules. *Amer. J. Physiol.* 159:143–48.

James, W. H. 1996. Evidence that mammalian sex ratios at birth are partially controlled by parental hormone levels at the time of conception. *J. Theor. Biol.* 180:271–86.

Janik, V. M. 2000. Whistle matching in wild bottlenose dolphins (*Tursiops truncatus*). *Science* 289:1355–57.

Jarvis, J. U. M. 1981. Eusociality in a mammal: Cooperative breeding in naked mole rat colonies. *Science* 212:571–73.

Jarvis, J. U. M., M. J. O'Riain, N. C. Bennett, and P. W. Sherman. 1994. Mammalian eusociality: A family affair. *Trends Ecol. Evol.* 9:47–51.

Jenni, D. A. 1974. Evolution of polyandry in birds. *Amer. Zool.* 14:129–44.

Jensen, A. R. 1967. Estimation of the limits of heritability of traits by comparison of monozygotic and dizygotic twins. *Proc. Nat. Acad. Sci.* (USA) 58:149–56.

———. 1973. *Educability and Group Differences.* New York: Harper & Row.

Jensen, D. D. 1965. Paramecia, planaria, and pseudo-learning. *Anim. Behav. Suppl.* 1:9–20.

Jerison, H. J. 1973. *Evolution of the Brain and Intelligence.* New York: Academic Press.

Jiménez, J. A., K. A. Hughes, G. Alaks, L. Graham, and R. C. Lacy. 1994. An experimental study of inbreeding depression in a natural habitat. *Science* 266:271–73.

Joerman, G., U. Schmidt, and C. Schmidt. 1988. The mode of orientation during flight and approach to landing in two phyllostomid bats. *Ethology* 78:332–40.

Johansson, B. 1959. Brown fat: A review. *Metabolism* 8:221–40.

John, E. R. 1972. Switchboard versus statistical theories of learning and memory. *Science* 177:850–64.

Johnson, C. G. 1969. *Migration and Dispersal of Insects by Flight.* London: Methuen.

Johnson, C. N. 1986. Philopatry, reproductive success of females and maternal investment in the red-necked wallaby. *Behav. Ecol. Sociobiol.* 19:143–50.

Johnson, D. F., and G. Collier. 1987. Caloric regulation and patterns of food choice in a patchy environment: The value and cost of alternative foods. *Physiol. Behav.* 39:351–59.

———. 1989. Patch choice and meal size of foraging rats as a function of the profitability of food. *Anim. Behav.* 38:285–97.

Johnson, P. B. 1987. New directions in fish orientation studies. In *Sign Posts in the Sea* (W. F. Herrnkind and A. B. Thistle, eds.). Tallahassee: Florida State University Press.

Johnston, R. E. 1975. Scent marking by male golden hamsters (*Mesocricetus auratus*): I. Effects of odors and social encounters. *Zeit. Tierpsychol.* 37:75–98.

Johnson, R. N. 1972. *Aggression in Man and Animals.* Philadelphia: Saunders.

Johnston, T. D., and G. Gottlieb. 1985. Effects of social experience on visually imprinted maternal preferences in Peking ducklings. *Devel. Psychobiol.* 18:261–71.

Jolly, A. 1966. Lemur social behavior and primate intelligence. *Science* 153:501–6.

Joslin, J. K. 1977a. Rodent long distance orientation ("homing"). *Adv. Ecol. Res.* 10:63–90.

———. 1977b. Visual cues used in orientation in white-footed mice, *Peromyscus leucopus:* A laboratory study. *Amer. Mid. Nat.* 98:303–18.

Jouventin, P., and H. Weimerskirch. 1990. Satellite tracking of wandering albatrosses. *Nature* 343:746–48.

Jung-Hoffman, I. 1966. Die Determination von Königin und Arbeiterin der Honigbee. *Z. Bienenforschung* 8:296–322.

Jussiaux, M., and C. Trillaud. 1979. Comparaison entres des techniques de vasectomie et d'androgenisation pour la détection des chaleurs. *Cereopa, Journee d'Etude,* March 1979.

Kalmijn, A. J. 1971. The electric sense of sharks and rays. *J. Exp. Biol.* 55:371–83.

Kamil, A. C. 1978. Systematic foraging by a nectar-feeding bird, the amakihi (*Loxops virens*). *J. Comp. Physiol. Psychol.* 92:388–96.

Kamil, A. C. 1988. A synthetic approach to the study of animal intelligence. In D. W. Leger (ed.), *Comparative Perspectives in Modern Psychology:* University of Nebraska *Symposium on Motivation,* 35:230–57.

Kamil, A. C. 1998. On the proper definition of cognitive ethology. In *Animal Cognition in Nature* (R. P. Balda, I. M. Pepperberg, and A. C. Kamil eds.). San Diego: Academic Press.

Kamil, A. C., and R. P. Balda. 1985. Cache recovery and spatial memory in Clark's nutcrackers (*Nucifraga columbiana*). *J. Exp. Psychol.: Anim. Behav. Proc.* 11:95–111.

Kamil, A. C., R. P. Balda, and K. Grim. 1986. Revisits to emptied cache sites by Clark's nutcrackers (*Nucifraga columbiana*). *Anim. Behav.* 34:1289–98.

Kamil, A. C., R. P. Balda, D. J. Olson, and S. Good. 1993. Returns to emptied cache sites by Clark's nutcrackers, *Nucifraga columbiana:* a puzzle revisited. *Anim. Behav.* 45:241–52.

Kamil, A. C., and T. D. Sargent, eds. 1981. *Behavior: Ecological, Ethological, and Psychological Approaches.* New York: Garland Publishing.

Kaufman, G. A. 1989. Use of fluorescent pigments to study social interactions in a small nocturnal rodent, *Peromyscus maniculatus. J. Mammal.* 70:171–74.

Kawai, M. 1965. Newly acquired precultural behavior of the natural troop of Japanese monkeys on Koshima Island. *Primates* 6:1–30.

Kawakami, M., E. Terasawa, and T. Ibuki. 1970. Changes in multiple unit activity of the brain during the estrous cycle. *Neuroendocrinology* 6:30–48.

Kayser, C. 1965. Hibernation. In *Physiological Mammalogy,* vol. 3, ed. W. Mayer and R. W. van Gelder. New York: Academic Press.

Keane, B. 1990. The effect of relatedness on reproductive success and mate choice in the white-footed mouse, *Peromyscus leucopus. Anim. Behav.* 39:264–73.

Keeton, W. T. 1969. Orientation by pigeons: Is the sun necessary? *Science* 165:922–28.

———. 1970. Orientation by pigeons. *Science* 168:153.

———. 1971. Magnets interfere with pigeon homing. *Proc. Nat. Acad. Sci. USA* 68:102–6.

———. 1974. The orientation and navigational basis of homing in birds. *Adv. Stud. Behav.* 5:47–132.

Keeton, W. T., and A. I. Brown. 1976. Homing behavior of pigeons not disturbed by application of an olfactory stimulus. *J. Comp. Physiol.* 105:259–66.

Keeton, W. T., M. L. Kreithen, and K. L. Hermayer. 1976. Orientation by pigeons deprived of olfaction by nasal tubes. *J. Comp. Physiol.* 114:289–99.

Keller, R. 1975. Das Spielverhalten der Keas (*Nestor notabilis* Gould) des Zürcher Zoos. *Z. Tierpsychol.* 38:393–408.

———. 1976. Beitrag zur Biologie und Ethologie der Keas (*Nestor notabilis*) des Zürcher Zoos. *Zool. Beitr.* 22:111–56.

Kelley, D. B., J. I. Morrell, and D. W. Pfaff. 1975. Autoradiographic localization of hormone-concentrating cells in the brain of an amphibian, *Xenopus laevis.* I. Testosterone. *J. Comp. Neurol.* 164:47–62.

Kelley, D. B., and D. W. Pfaff. 1976. Hormone effects on male sex behavior in adult South African clawed frogs, *Xenopus laevis. Horm. Behav.* 7:159–82.

Kelly, R., J. W. Deutsch, S. S. Carlson, and J. A. Wagner. 1979. Biochemistry of neurotransmitter release. *Ann. Rev. Neurosci.* 2:399–446.

Kenagy, G. J., and S. C. Trombulak. 1986. Size and function of mammalian testes in relation to body size. *J. Mamm.* 67:1–22.

Kennedy, M., H. G. Spencer, and R. D. Gray. 1996. Hop, step and gape: Do the social displays of the Pelecaniformes reflect phylogeny? *Anim. Behav.* 51:273–91.

Kennleyside, M. H., and F. T. Yamamoto. 1962. Territorial behaviour of juvenile Atlantic salmon (*Salmo salar*). *Behaviour* 19:139–69.

Kerfoot, W. C. 1985. Adaptive value of vertical migration: Comments on the predation hypothesis and some alternatives. Pp. 91–113 in *Migration: Mechanisms and Adaptive Significance* (M. A. Rankin, ed.). Port Aransas, TX: Marine Science Institute.

Kessel, E. L. 1955. The mating activities of balloon flies. *Syst. Zool.* 4:97–104.

Ketterson, E. D. 1979. Aggressive behavior in wintering dark-eyed juncos: Determinants of dominance and their possible relation to geographic variation in sex ratio. *Wilson Bull.* 91:371–83.

Kettlewell, H. B. D. 1965. Insect survival and selection for pattern. *Science* 148:1290–96.

Kiesecker, J., D. Skelly, K. Beard, and E. Preisser. 1999. Behavioral reduction of infection risk. *Proc. Nat. Acad. Sci. USA* 96(16): 9165–68.

Kim, Y-K, and L. Ehrman. 1998. Developmental isolation and subsequent adult behavior of *Drosophila paulistorum* IV. Courtship. *Behav. Genet.* 28:57–66.

Kimura, D. 1995. Estrogen replacement therapy may protect against intellectual decline in postmenopausal women. *Horm. Behav.* 29:312–21.

King, A. P., and M. J. West. 1983a. Epigenesis of cowbird song—A joint endeavour of males and females. *Nature* 305:704–6.

———. 1983b. Dissecting cowbird song potency: assessing a song's geographic identity and relative appeal. *Z. Tierpsychol.* 63:37–50.

———. 1988. Searching for the functional origins of song in eastern brown-headed cowbirds, *Molothrus ater ater. Anim. Behav.* 36:1575–88.

King, J. A. 1968. Species specificity and early experience. In *Early Experience and Behavior*, ed. G. Newton and S. Levine. Springfield, IL: Thomas.

———. 1969. A comparison of longitudinal and cross-sectional groups in the development of behavior of deer mice. *An. N.Y. Acad. Sci.* 159:696–709.

King, J. A. 1955. Social behavior, social organization, and population dynamics in a black-tailed prairie dog town in the Black Hills of South Dakota. Contributions from the Laboratory of Vertebrate Zoology, University of Michigan, 67:1–123.

Klein, S. L., J. E. Hairston, A. C. DeVries, and R. J. Nelson. 1997. Social environment and steroid hormones affect species and sex differences in immune function among voles. *Horm. Behav.* 32:30–39.

Klinowska, M. 1985. Cetacean stranding sites relate to geomagnetic topography. *Aquatic Mammals* 1:27–32.

Kitchell, J. A. 1986. The evolution of predator-prey behavior: Naticid gastropods and their molluskan prey. In *Evolution of Animal Behavior. Paleontological and Field Approaches*, ed. M. H. Nitecki and J. A. Kitchell. New York: Oxford Univ. Press.

Kitching, J. A., and F. J. Ebling. 1967. Ecological studies at Lough Ine. *Adv. Ecol. Res.* 4:197–291.

Kleiman, D. G., and J. F. Eisenberg. 1973. Comparisons of canid and felid social systems from an evolutionary perspective. *Anim. Behav.* 21:637–59.

Kleinholz, L. H. 1970. A progress report on the separation and purification of crustacean neurosecretory pigmentary-effector hormones. *Gen. Comp. Endocr.* 14:578–88.

Klopfer, P. H. 1957. An experiment with empathic learning in ducks. *Amer. Nat.* 91:61–63.

———. 1963. Behavioral aspects of habitat selection: The role of early experience. *Wilson Bull.* 75:15–22.

Klopfer, P. H., D. K. Adams, and M. S. Klopfer. 1964. Maternal imprinting in goats. *Proc. Nat. Acad. Sci. USA* 52:911–14.

Klopfer, P. H., and J. U. Ganzhorn. 1985. Habitat selection: Behavioral aspects. In *Habitat selection in birds*, ed. M. L. Cody. New York: Academic Press.

Klopfer, P. H., and J. P. Hailman. 1965. Habitat selection in birds. *Adv. Stud. Behav.* 1:279–303.

Klopfer, P. H., and M. S. Klopfer. 1968. Maternal "imprinting" in goats: Fostering of alien young. *Zeit. Tierpsychol.* 25:862–66.

Klukowski, M., and C. E. Nelson. 1998. The challenge hypothesis and seasonal changes in aggression and steroids in male northern fence lizards (*Sceloporus undulatus hyacinthinus*). *Horm. Behav.* 33:197–204.

Knapp, R., J. C. Wingfield, and A. Bass. 1999. Steroid hormones and paternal care in the plainfin midshipman fish (*Porichthys notatus*). *Horm. Behav.* 35:81–89.

Knussman, R., K. Christiansen, and C. Couwenbergs. 1986. Relations between sex hormone levels and sexual behavior in men. *Arch. Sex Behav.* 15:429–45.

Kodric-Brown, A., and J. H. Brown. 1984. Truth in advertising: The kinds of traits favored by sexual selection. *Amer. Nat.* 124:309–23.

Koenig, W. D., R. L. Mumme, and F. A. Pitelka. 1984. The breeding system of the acorn woodpecker in central coastal California. *Z. Tierpsychol.* 65:289–308.

Koenig, W., and R. Mumme. 1987. Population ecology of the cooperative breeding acorn woodpecker. *Monograph in Population Biology,* no. 24. Princeton: Princeton University Press.

Kogure, M. 1933. The influence of light and temperature on certain characters of the silkworm, *Bombyx mori. J. Dept. Agr. Kyushu Univ.* 4:1–93.

Köhler, W. 1925. *The Mentality of Apes.* New York: Harcourt, Brace.

Kollack-Walker, S., and S. W. Newman. 1995. Mating and agonistic behavior produce different patterns of fos immunolabeling in the male Syrian hamster brain. *Neuroscience* 66:721–36.

Komdeur, J. 1997. Extreme adaptive modification in sex ration of the Seychelles warbler's eggs. *Nature* 385:522–25.

Komdeur, J. 1996. Facultative sex ratio bias in the offspring of Seychelles warblers. *Proc. Roy. Soc. Lond. B.* 263:661–66.

Komdeur, J. 1994. Experimental evidence for helping and hindering by previous offspring in the cooperative breeding Seychelles warbler (*Acrocephalus sechellensis*). *Behav. Ecol. Sociobiol.* 34:31–42.

Komdeur, J. 1992. Importance of habitat saturation and territory quality for evolution of cooperative breeding in the Seychelles warbler. *Nature* 358:493–95.

Komdeur, J., A. Huffstadt, W. Prast, G. Castle, R. Miltero, and J. Wattle. 1995. Transfer experiments of Seychelles warblers to new islands: Changes in dispersal and helping behaviour. *Anim. Behav.* 49:695–708.

Komisaruk, B. I. 1978. The nature of the neural substrate of female sexual behaviour in mammals and its hormonal sensitivity: Review and speculations. In *Biological Determinants of Sexual Behaviour,* ed. J. B. Hutchison. Chichester, England: Wiley.

Konishi, M., and E. Akutagawa. 1985. Neuronal growth, atrophy and death in a sexually dimorphic song nucleus in the zebra finch brain. *Nature* 315:145–47.

Konopka, R. J., and S. Benzer. 1971. Clock mutants of *Drosophila melanogaster. Proc. Nat. Acad. Sci. USA* 68:2112–16.

Konopka, R. J. 1979. Genetic dissection of the *Drosophila* circadian system. *Fed. Proc.* 38:2602–5.

Konopka, R. J., C. P. Kyriacous, J. C. Hall. 1996. Mosaic analysis in the *Drosophila* CNS of circadian and courtship-song rhythms affected by a period clock mutation. *J. Neurogenetics* 11:117.

Kramer, G. 1949. Über Richtungstendenzen bei der nächtlichen Zugunruhe gekäfigten Vögel. In *Ornithologie als biologische Wissenschaft,* ed. E. Mayr and E. Schüz. Heidelberg: Winter.

———. 1950. Orientierte Zugaktivitätgekäfiter Sing-vögel. *Naturwissenschaften* 37:188.

———. 1951. Eine neue Methode zur Erforschung der Zugorientierung und die bisher damit erzielten Ergebnisse. *Proc. Xth Inter. Ornithol. Congr.,* Uppsala. 271–80.

Krasne, F. B. 1973. Learning in Crustacea. In *Invertebrate Learning,* vol. 2., ed. W. C. Corning, J. A. Dyal, and A. O. D. Willows. New York: Plenum.

Krause, J. 1993. The effect of "Schreckstoff" on the shoaling behaviour of the minnow: A test of Hamilton's selfish herd theory. *Anim. Behav.* 45:1019–24.

Kravitz, E. A. 1988. Hormonal control of behavior: Amines and the biasing of behavioral output in lobsters. *Science* 241:1775–81.

Krebs, C. J. 1964. The lemming cycle at Baker Lake, Northwest Territories, during 1959–1962. Arctic Institute of North America Technical Paper No. 15.

Krebs, C. J. 1985. *Ecology: The Experimental Analysis of Distribution and Abundance.* 3rd ed. New York: Harper & Row.

Krebs, J. R. 1970. Regulation of numbers in the great tit (Aves: Passeriformes). *J. Zool. Lond.* 162:317–33.

———. 1971. Territory and breeding density in the great tit, *Parus major* L. *Ecology* 52:2–22.

Krebs, J. R., and N. B. Davies. 1987. *An Introduction to Behavioral Ecology.* Sunderland, MA: Sinauer Assoc.

Krebs, J. R., and R. Dawkins. 1984. Animal signals: Mindreading and manipulation. In *Behavioural Ecology: An Evolutionary Approach,* 2d ed., ed. J. R. Krebs and N. B. Davies. Oxford, England: Blackwell Scientific Publications, Ltd.

Krebs, J. R., A. Kacelnik, and P. Taylor. 1978. Test of optimal sampling by foraging great tits. *Nature* 275:27–31.

Krebs, J. R., J. T. Erichsen, M. I. Webber, and E. L. Charnov. 1977. Optimal prey-selection by the great tit (*Parus major*). *Anim. Behav.* 25:30–38.

Krebs, J. R., and N. B. Davies. 1993. *An Introduction of Behavioural Ecology,* 3rd ed. Oxford, England: Blackwell Science.

Krebs, J. R., and N. B. Davies, eds. 1997. *Behavioural Ecology: An Evolutionary Approach.* Oxford: Blackwell Science.

Krebs, J. R., D. F. Sherry, S. D. Healy, V. H. Perry, and A. L. Vaccarino. 1989. Hippocampal specialization of food storing birds. *Proc. Nat. Acad. Sci. USA* 86:1388–92.

Kreithen, M. L., and W. T. Keeton. 1974a. Detection of changes in atmospheric pressure by the homing pigeon, *Columba livia. J. Comp. Physiol.* 89:73–82.

———. 1974b. Attempts to condition homing pigeons to magnetic stimuli. *J. Comp. Physiol.* 91:355–62.

Krieger, D. T. 1980. Ventromedial hypothalamic lesions abolish food-shifted circadian adrenal and temperature rhythmicity. *Endocrinology* 106:649–54.

Kroodsma, D. E. 1978. Aspects of learning in the ontogeny of bird song: Where, from whom, when, how many, which, and how accurately? In *Development of Behavior,* ed. G. Burghardt and M. Bekoff. New York: Garland STPM Press.

———. 1981. Ontogeny of bird song. In *Behavioral Development,* ed. K. Immelmann et al. New York: Cambridge University Press.

Kroodsma, D. E., and R. Pickert. 1980. Environmentally dependent sensitive periods for avian vocal learning. *Nature* 288:477–79.

———. 1984. Repertoire size, auditory templates and selective vocal learning in songbirds. *Anim. Behav.* 32:395–99.

Kroodsma, D. E. 1989. Two North American song populations of the marsh wren reach distributional limits in the central Great Plains. *Condor* 91:332–40.

Kroodsma, D. E., and R. A. Canady. 1985. Differences in repertoire size, singing behavior, and associated neuroanatomy among marsh wren populations have a genetic basis. *Auk* 102:439–46.

Kroodsma, D. E., and B. E. Byers. 1998. Songbird song repertoires: An ethological approach to studying cognition. In *Animal Cognition in Nature* (R. P. Balda, I. M. Pepperberg, and A. C. Kamil, eds.) San Diego: Academic Press.

Kroodsma, D. E., and J. Verner. 1997. The marsh wren (*Cistothorus palustris*). In *The Birds of North America,* No. 308 (A. Poole and F. Gill eds.) Washington, D.C.: The American Ornithologists' Union.

Kruuk, H. 1972. *The Spotted Hyena: A Study of Predation and Social Behavior.* Chicago: University of Chicago Press.

Kummer, H. 1971. *Primate Societies: Group Techniques of Ecological Adaptation.* Chicago: Aldine-Atherton.

Kung, C., S. Y. Chang, Y. Satow, J. van Houten, and H. Hansma. 1975. Genetic dissection of behavior in *Paramecium. Science* 188:898–904.

Kuo, Z. Y. 1967. *Dynamics of Behavior Development.* New York: Random House.

Lack, D. 1933. Habitat selection in birds with special reference to the effects of afforestation on the Breckland avifauna. *J. Anim. Ecol.* 2:239–62.

———. 1961. *Darwin's Finches: An Essay on the General Biological Theory of Evolution* (New York: Harper & Bros.) First published in 1947 by Cambridge University Press, Cambridge, Eng.

Lageaux, M., P. Harry, and J. A. Hoffmann. 1981. Ecdysteroids are bound to vitellin in newly laid eggs of *Locusta. Mol. Cell Endocrinol.* 24:325–38.

Lagerspetz, K., and T. Heino. 1970. Changes in social reactions resulting from early experience with another species. *Psychol. Rep.* 27:255–62.

Laird, L. M. 1978. Marking fish. In *Animal Marking,* ed. B. Stonehouse, 95–101. Baltimore: University Park Press.

Landau, I. T., and W. G. Holmes. 1988. Mating of captive thirteen-lined ground squirrels and the annual timing of estrus. *Horm. Behav.* 22:474–87.

Lane-Petter, W. 1978. Identification of laboratory animals. In *Animal Marking,* ed. B. Stonehouse, 35–40. Baltimore: University Park Press.

Larkin, T. S., and W. T. Keeton. 1976. Bar magnets mask the effect of normal magnetic disturbances on pigeon orientation. *J. Comp. Physiol.* 110:227–31.

Lawick, H. V., and J. V. Lawick-Goodall. 1971. *Innocent Killers.* Boston: Houghton Mifflin.

Lawick-Goodall, J. 1970. Tool-using in primates and other vertebrates. Pp. 195–249. in *Advances in the Study of Behaviour,* Vol. 3., ed. D. S. Lehrman, R. A. Hinde, and E. Shaw. New York: Academic Press.

Le Boeuf, B. J. 1974. Male-male competition and reproductive success in elephant seals. *Amer. Zool.* 14:163–76.

Le Boeuf, B. J., and J. Reiter. 1988. Lifetime reproductive success in northern elephant seals. *Reproductive Success,* ed. T. H. Clutton-Brock. Chicago: University of Chicago Press.

Lee, R. B. 1980. Lactation, ovulation, infanticide, and women's work: A study of hunter-gatherer population regulation. In *Biosocial Mechanisms of Population Regulation,* ed. M. N. Cohen, R. S. Malpass, and H. G. Klein. New Haven: Yale University Press.

Lehner, P. N. 1979. *Handbook of Ethological Methods.* New York: Garland STPM Press.

Lehner, P. 1996. *Handbook of Ethological Methods.* New York: Cambridge University Press.

Lehrman, D. S. 1953. A critique of Konrad Lorenz's theory of instinctive behavior. *Quart. Rev. Biol.* 28:337–69.

———. 1955. The physiological basis of parental feeding behavior in ring doves (*Streptopelia risoria*). *Behaviour* 7:241–86.

———. 1958a. Effect of female sex hormones on incubation behavior in the ring dove (*Streptopelia risoria*). *J. Comp. Physiol. Psychol.* 51:142–45.

———. 1958b. Induction of broodiness by participation in courtship and nest-building in the ring dove (*Streptopelia risoria*). *J. Comp. Physiol. Psychol.* 51:32–36.

———. 1959. Hormonal responses to external stimuli in birds. *Ibis* 101:478–96.

———. 1961. Hormonal regulation of parental behavior in birds and infrahuman mammals. In *Sex and Internal Secretions,* ed. W. C. Young. Baltimore: Williams & Wilkins.

———. 1964. The reproductive behavior of ring doves. *Sci. Amer.* 211:48–54.

———. 1965. Interaction between internal and external environments in the regulation of the reproductive cycle of the ring dove. In *Sex and Behavior,* ed. F. A. Beach. New York: Wiley.

———. 1970. Semantic and conceptual issues in the nature-nurture problem. In *Development and Evolution of Behavior,* ed. L. R. Aronson et al. San Francisco: Freeman.

Lehrman, D. S., P. N. Brody, and R. P. Wortis. 1961. The presence of the mate and of nesting material as stimuli for the development of incubation behavior and for gonadotrophin secretion in the ring dove (*Streptopelia risoria*). *Endocrinology* 68:507–16.

Lent, P. C. 1966. Calving and related social behavior in the barren-ground caribou. *Zeit. Tierpsychol.* 23:701–56.

Leshner, A. I. 1975. A model of hormones and agonistic behavior. *Physiol. Behav.* 15:225–35.

Levey, D. J., and F. G. Stiles. 1992. Evolutionary precursors of long-distance migration: Resource availability and movement patterns in Neotropical landbirds. *Am. Nat.* 140:447–76.

Levi, R., and J. M. Camhi. 1996. Producing directed behaviour: Muscle activity patterns of the cockroach escape response. *J. Exp. Biol.* 199:563–68.

Levine, S. 1967. Maternal and environmental influences on the adrenocortical responses to stress in weanling rats. *Science* 156:258–60.

———. 1968. Hormones and conditioning. *Nebr. Symp. Motiv.* 16:85–101.

Lewis, E. R., and P. M. Narins. 1985. Do frogs communicate with seismic signals? *Science* 227:187–89.

Lewontin, R. C., and L. C. Dunn. 1960. The evolutionary dynamics of a polymorphism in the house mouse. *Genetics* 45:705–22.

Ley, W. 1968. *Dawn of Zoology.* Englewood Cliffs, NJ: Prentice-Hall.

Libersat, F., G. Haspel, J. Casagrand, K. Fouad. 1999. Localization of the site of effect of a wasp's venom in the cockroach escape circuitry. *J. Comp. Physiol. A.* 184:(3)333–45.

Licht, P. 1973. Influence of temperature and photoperiod on the annual ovarian cycle in the lizard *Anolis carolinensis. Copeia* 1973:465–72.

Liebig, J. 1847. *Chemistry Application to Agriculture and Physiology.* 4th ed. London: Taylor and Walton.

Lillie, F. R. 1916. The theory of the free-martin. *Science* 43:611–13.

Lima, S. L., and L. M. Dill. 1990. Behavioral decisions made under the risk of predation: A review and prospectus. *Can. J. Zool.* 68:619–40.

Lima, S. L. 1995. Back to the basics of anti-predatory vigilance: The group-size effect. *Anim. Behav.* 49:11–20.

Lima, S. L., and L. M. Dill. 1990. Behavioral decisions made under the risk of predation: A review and prospectus. *Can. J. Zool.* 68:619–40.

Lincoln, F. C. 1950. Migration of birds. *U.S. Fish and Wild. Serv. Washington, D.C. Circ.* 16:1–102.

Lindauer, M. 1961. *Communication among Social Bees.* Cambridge: Harvard University Press.

Lindsley, D. L., and G. G. Zimm. 1992. The genome of *Drosophila melanogaster.* San Diego: Academic Press.

Lindström, K. 1998. Energetic constraints on mating performance in the sand goby. *Behav. Ecol.* 9:297–300.

Lindzey, G., H. D. Winston, and M. Manosevitz. 1963. Early experience, genotype, and temperament in *Mus musculus J. Comp. Physiol. Psychol.* 56:622–29.

Lindzey, J., and D. Crews. 1988. Psychobiology of sexual behavior in a whiptail lizard, *Cnemidophorus inornatus. Horm. Behav.* 22:279–93.

Linn, I. J. 1978. Radioactive techniques for small mammal marking. In *Animal Marking,* ed. B. Stonehouse, 177–91. Baltimore: University Park Press.

Lissmann, H. W. 1958. On the function and evolution of electric organs in fish. *J. Exp. Biol.* 35:156–91.

Littlejohn, M. J. 1965. Premating isolation in the *Hyla ewingi* complex (Anura: Hylidae). *Evolution* 19:234–43.

Littlejohn, M. J., and J. J. Loftus-Hills. 1968. An experimental evaluation of premating isolation in the *Hyla ewingi* complex. *Evolution* 22:659–63.

Liu, C., D. R. Weaver, S. H. Strogatz, and S. M. Reppert. 1997. Cellular construction of a circadian clock: Period determination in the suprachiasmatic nuclei. *Cell* 94:855–60.

Lively, C. M. 1987. Evidence from a New Zealand snail for the maintenance of sex by parasitism. *Nature* 328:519–21.

Lloyd, J. E. 1965. Aggressive mimicry in *Photuris:* Firefly *femmes fatales. Science* 149:653–54.

———. 1980. Male *Photuris* fireflies mimic sexual signals of their females' prey. *Science* 210:669–71.

Lockard, R. B. 1971. Reflections on the fall of comparative psychology: Is there a message for us all? *Amer. Psychol.* 26:168–79.

Lockyer, C. H., and S. G. Brown. 1981. The migration of whales. In *Animal Migration,* ed. D. J. Aidley, 105–37. New York: Cambridge University Press.

Loeb, J. 1918. *Forced Movements, Tropisms and Animal Conduct.* Philadelphia: Lippincott.

Lohmann, K. J., and S. Johnsen. 2000. The neurobiology of magnetoreception in vertebrate animals. *Trends in Neurosciences* 23:153–59.

Lohrer, W. 1961. The chemical acceleration of the maturation process and its hormonal control in the male of the desert locust. *Proc. Roy. Soc. Lond.* (ser. B) 153:380–97.

Lohrer, W., and F. Huber. 1966. Nervous and endocrine control of sexual behavior in a grasshopper (*Comphocerus rufus* L., Acridinae). *Soc. Exp. Biol. Symp.* 20:381–400.

Loizos, C. 1967. Play behavior in high primates: A review. In *Primate Ethology,* ed. D. Morris. Chicago: Aldine.

Lombardi, J. R., and J. G. Vandenbergh. 1977. Pheromonally induced sexual maturation in females. Regulation by the social environment of the male. *Science* 196:545–46.

Longhurst, A. R. 1976. Vertical migration. In *Ecology of the Seas,* ed. D. H. Cushing and J. J. Walsh. Philadelphia: Saunders.

Lorenz, K. 1935. Der Kumpan in der Umwelt des Vogels. *J. Ornith.* 83:137–213, 289–413.

———. 1937. Über die Bildung des Instinktbegriffes. *Die Naturwissenschaften* 19:289–300.

———. 1966. *On Aggression.* New York: Harcourt, Brace & World.

Losey, G. S. 1979. The fish cleaning symbiosis: Proximate causes of host behaviour. *Anim. Behav.* 27:669–85.

Losos, J. B. 1999. Uncertainty in the reconstruction of ancestral character states and limitations on the use of phylogenetic comparative methods. *Anim. Behav.* 58:1319–24.

Lott, D. F. 1962. The role of progesterone in the maternal behavior of rodents. *J. Comp. Physiol. Psychol.* 55:610–13.

Lott, D. F., and J. S. Rosenblatt. 1969. Development of maternal responsiveness during pregnancy in the rat. In *Determinants of Infant Behavior,* vol. 4, ed. B. M. Foss. London: Methuen.

Low, B. S. 1978. Environmental uncertainty and the parental strategies of marsupials and placentals. *Amer. Nat.* 112:197–213.

Low, B. S. 1990. Marriage systems and pathogen stress in human societies. *Am. Zool.* 30:325–39.

Lowery, G. H. 1946. Evidence of trans-Gulf migration. *Auk* 63:175–211.

Lubin, M., M. Leon, H. Moltz, and M. Numan. 1972. Hormones and maternal behavior in the male rat. *Horm. Behav.* 3:369–74.

Luiselli, L. 1996. Individual success in mating balls of the grass snake, *Natrix natrix:* Size is important. *J. Zool.* 239: 731–40.

Luttge, W. G., and R. E. Whalen. 1970. Dihydrotestosterone, androstenodione, testosterone: Comparative effectiveness in masculinizing and defeminizing reproductive systems in male and female rats. *Horm. Behav.* 1:265–81.

Lutz, P. E., and C. E. Jenner. 1964. Life-history and photoperiodic responses of nymphs of *Tetragoneura cynosura* (Say). *Biol. Bull.* 127:304–16.

Lydon, J. P., F. J. DeMayo, C. R. Funk, S. K. Mani, A. R. Hughes, C. A. Montgomery, G. Shyamala, O. M. Conneely, and B. W. O'Malley. 1995. Mice lacking progesterone receptor exhibit pleiotropic reproductive abnormalities. *Genes Dev.* 9:2266–78.

Lyman, C. P. 1982. Who is among the hibernators? In *Hibernation and Torpor in Birds and Mammals,* ed. C. P. Lyman, J. S. Willis, A. Malan, and L. C. H. Wang. New York: Academic Press.

Lyman, C. P., J. S. Willis, A. Malan, and L. C. H. Wang. 1982. *Hibernation and Torpor in Mammals and Birds.* New York: Academic Press.

Lynch, M., and L. Deng. 1994. Genetic slippage in response to sex. *Amer. Nat.* 144:242–61.

MacArthur, R. H. 1965. Patterns of species diversity. *Biol. Rev.* 40:410–533.

MacArthur, R. H., and E. O. Wilson. 1967. *Theory of Island Biogeography.* Princeton: Princeton University Press.

MacArthur, R. H., and E. R. Pianka. 1966. On the optimal use of a patchy environment. *Am. Nat.* 100:603–9.

Machin, K. E., and H. W. Lissman. 1960. The mode of operation of the electric receptors in *Gymnarchus niloticus. J. Exp. Biol.* 37:801–11.

Mackintosh, N. J. 1975. A theory of attention: Variations in the associability of stimuli and reinforcement. *Psychol. Rev.* 82:276–98.

Mackintosh, N. J., and A. Cauty. 1971. Spatial reversal learning in rats, pigeons, and goldfish. *Psychonom. Sci.* 22:281–82.

Mackintosh, N. J., and V. Holgate. 1969. Spatial reversal training and nonreversal shift learning. *J. Physiol. Comp. Psychol.* 67:89–93.

Mackintosh, N. J., B. Wilson, and R. A. Boakes. 1985. Differences in mechanisms of intelligence among vertebrates. *Phil. Trans. Royal Soc. Lond.* B. 308:53–65.

Macphail, E. M. 1982. *Brain and Intelligence in Vertebrates.* Oxford: Clarendon Press.

———. 1985. Vertebrate intelligence: The null hypothesis. *Phil. Trans. Royal Soc. Lond.* B. 308:37–51.

———. 1987. The comparative psychology of intelligence. *Behav. Brain Sci.* 10:645–95.

Madison, D. M. 1985. Activity rhythms and spacing. In *Biology of the New World Microtus,* ed. R. H. Tamarin, 373–419. Special Publication of the American Society of Mammalogists #8.

Maguire, E., R. Frackowiak, C. Frith. 1997. Recalling routes around London: Activation of the right hippocampus in taxi drivers. *J. Neuroscience* 17:7103–10.

Magurran, A. E. 1999. The causes and consequences of geographic variation in antipredator behavior: Perspectives from fish populations. In *Geographic Variation in Behavior: Perspective on Evolutionary Mechanisms* (S. A. Foster, and J. A. Endler, eds.). New York: Oxford University Press.

Maher, C. R., and D. F. Lott. 2000. A review of ecological determinants of territoriality within vertebrate species. *Am. Mid. Nat.* 143:1–29.

Manning, A. 1966. Corpus allatum and sexual receptivity in female *Drosophila melanogaster. Nature* 211:1321–22.

———. 1967. The control of sexual receptivity in female *Drosophilia. Anim. Behav.* 15:239–50.

———. 1971. Evolution of behavior. In *Psychobiology: Behavior from a Biological Perspective,* ed. J. L. McGaugh. New York: Academic Press.

Manning, A., and T. E. McGill. 1974. Early androgen and sexual behavior in female house mice. *Horm. Behav.* 5:19–31.

Manning, J. T., and A. T. Chamberlain. 1993. Fluctuating asymmetry, sexual selection and canine teeth in primates. *Proc. Roy. Soc. London B.* 251:83–87.

Manning, J. T., and A. T. Chamberlain. 1994. Fluctuating asymmetry in gorilla canines: A sensitive indicator of environmental stress. *Proc. Roy. Soc. London B.* 255:189–93.

Mappes, J., A. Kaitala, and R. Alatalo. 1995. Joint brood guarding in parent bugs—an experiment on defence against predation. *Behav. Ecol. Sociobiol.* 36:343–47.

Margoliash, D. 1983. Acoustic parameters underlying the responses of song-specific neurons in the white-crowned sparrow. *J. Neurosci.* 3:1039–57.

———. 1986. Preference for autogenous song by auditory neurons in a song system nucleus of the white-crowned sparrow. *J. Neurosci.* 6:1643–61.

———. 1987. Neural plasticity in birdsong learning. In *Imprinting and Cortical Plasticity,* ed. J. P. Rauschecker and P. Marler, 23–54. New York: Wiley.

Margraf, R. R., and G. R. Lynch. 1993. Melantonin injections affect circadian behavior and SCN neurophysiology in Djungarian hamsters. *Am. J. Physiol.* 265:R615–R621.

Marler, P. 1967. Animal communication signals. *Science* 157:769–74.

———. 1968. Aggregation and dispersal: Two functions in primate communication. In *Primates: Studies in Adaptation and Variability,* ed. P. C. Jay. New York: Holt, Rinehart and Winston.

———. 1970. A comparative approach to vocal learning: Song development in white-crowned sparrows. *J. Comp. Physiol. Psychol.* 7:1–25.

———. 1973. A comparison of vocalizations of red tailed monkeys and blue monkeys, *Ceropithecus ascanius* and *C. Mitis,* in Uganda. *Zeit. Tierpsychol.* 33:223–47.

———. 1976. On animal aggression: The roles of strangeness and familiarity. *Amer. Psychol.* 31:239–46.

———. 1983. Some ethological implications for neuroethology: The ontogeny of birdsong. In *Advances in Neuroethology,* ed. J. P. Ewert, R. R. Capranica, and D. J. Ingle. New York: Plenum.

Marler, P., and S. Peters. 1977. Selective vocal learning in a sparrow. *Science* 198:519–21.

———. 1981. Sparrows learn adult song and more from memory. *Science* 213:780–82.

———. 1982. Structural changes in song ontogeny in the swamp sparrow *Melospiza georgiana. Auk* 99:446–58.

Marler, P., and M. Tamura. 1962. Song variation in three populations of white-crowned sparrows. *Condor* 64:368–77.

Marler, C. A., and M. C. Moore. 1989. Time and energy costs of aggression in testosterone implanted free-living male mountain spiny lizards (*Sceloporus jarrovi*). *Physiol. Zool.* 62:1334–50.

Marler, C. A., G. Walsberg, M. L. White, and M. Moore. 1995. Increased energy expenditure due to increased territorial defense in male lizards after phenotypic manipulation. *Behav. Ecol. Sociobiol.* 37:225–31.

Marler, P. 1990. Song learning: The interface between behaviour and neuroethology. *Trans. Roy. Soc. London B* 329:109–14.

Marsden, H. M. 1968. Agonistic behavior of young rhesus monkeys after changes induced in social rank of their mothers. *Anim. Behav.* 16:38–44.

Marshall, A. J. 1960. Annual periodicity in the migration and reproduction of birds. *Cold Spr. Harb. Symp. Quant. Biol.* 25:499–505.

Martan, J., and B. A. Shepherd. 1976. The role of the copulatory plug in reproduction of the guinea pig. *J. Exp. Zool.* 196:79–84.

Martin, G., O. Sorokine, M. Moniatte, and A. Vandorsselaer. 1998. The androgenic hormone of the crustacean isopod *Armadillidium vulgare. Ann. N.Y. Acad. Sci.* 282:97–98.

Martin, N. L., E. O. Price, S. J. R. Wallach, and M. R. Dally. 1987. Fostering lambs by odor transfer: The add-on experiment. *J. Anim. Sci.* 64:1378–83.

Martin, R. D., L. A. Willner, and A. Dettling. 1994. The evolution of sexual size dimorphism in primates. Pp. 159–202 in *The Differences Between the Sexes* (R. V. Short and E. Balaban eds.). Cambridge: Cambridge University Press.

Marx, J. L. 1980. Ape-language controversy flares up. *Science* 207:1330–33.

Marzluff, J. M., B. Heinrich, and C. S. Marzluff. 1996. Raven roosts are mobile information centres. *Anim. Behav.* 51:89–103.

Marzluff, J. M., and R. P. Balda. 1992. *The Pinyon Jay.* London: T. and A. D. Poyser.

Mason, W. A. 1960. The effects of social restriction on the behavior of rhesus monkeys. I. Free social behavior. *J. Comp. Physiol. Psychol.* 53:582–89.

———. 1961a. The effects of social restriction on the behavior of rhesus monkeys. II. Tests of gregariousness. *J. Comp. Physiol. Psychol.* 54:287–90.

———. 1961b. The effects of social restriction on the behavior of rhesus monkeys. III. Dominance tests. *J. Comp. Physiol. Psychol.* 54:694–99.

Mason, W. A., D. D. Long, and S. P. Mendoza. 1993. Temperament and mother-infant conflict in macaques: A transactional analysis. Pp. 205–27 in *Primate Social Conflict* (W. A. Mason and S. P. Mendoza, eds.). Albany, NY: SUNY Press.

Massey, A., and J. G. Vandenbergh. 1980. Puberty delay by a urinary cue from female house mice in feral populations. *Science* 209:821–22.

———. 1981. Puberty acceleration by a urinary cue from male mice in feral populations. *Biol. Reprod.* 24:523–27.

Masson, G. M. 1948. Effects of estradiol and progesterone on lactation. *Anat. Rec.* 102:513–21.

Mast, S. O. 1938. Factors involved in the process of orientation of lower organisms in light. *Biol. Rev.* 17:68–90.

Masterton, R. B., M. E. Bitterman, C. B. G. Campbell, and N. Hotton. 1976. *Evolution of Brain and Behavior in Vertebrates.* Hillsdale, NJ: Erlbaum.

Mateo, J. M., and R. E. Johnston. 2000. Kin recognition and the "armpit effect": Evidence of self-referent phenotype matching. *Proc. R. Soc. London B* 267:695–700.

Mather, M. H., and B. D. Roitberg. 1987. A sheep in wolf's clothing: Tephritid flies mimic spider predators. *Science* 236:308–10.

Mathews, G. V. T. 1968. *Bird Navigation,* 2d ed. Cambridge: Cambridge University Press.

Mathis, A. 1990. Territoriality in a terrestrial salamander: The influence of resource quality and body size. *Behaviour* 112:162–75.

Mathis, A., D. W. Schmidt, and K. A. Medley. 2000. The influence of residency status on agonistic behavior of male and female Ozark zigzag salamanders *Plethodon angusticlavius. Am. Mid. Nat.* 143:245–49.

Matsumoto, S. I., J. Basil, A. E. Jetton, M. N. Lehman, and E. L. Bittman. 1998. Regulation of the phase and period of circadian rhythms restored by suprachiasmatic transplants. *J. Biol. Rhythms* 11:145–62.

Maxson, S. C. 1981. The genetics of aggression in vertebrates. Pp. 69–104 in *The Biology of Aggression* (P. F. Brain and D. Benton, eds.). Rockville, MD: Sijthoff and Noordhoff.

Maxson, S. C. 1996. Searching for candidate genes with effects on an agonistic behavior, in mice. *Behav. Genet.* 26:471–76.

Maylon, C., and S. Healy. 1994. Fluctuating asymmetry in antlers of fallow deer, *Dama dama,* indicates dominance. *Anim. Behav.* 48:248–50.

Maynard Smith, J. 1974. The theory of games and the evolution of animal conflict. *J. Theoret. Biol.* 47:209–21.

———. 1976. Evolution and the theory of games. *Amer. Sci.* 64:41–45.

Maynard Smith, J., and G. R. Price. 1973. The logic of animal conflict. *Nature* 246:15–18.

Mayr, E. 1963. *Animal Species and Evolution.* Cambridge: Harvard University Press.

———. 1970. *Populations, Species and Evolution.* Cambridge: Harvard University Press.

———. 1977. Darwin and natural selection. *Amer. Sci.* 65:321–27.

Mayr, E. 1983. How to carry out the adaptationist program? *Amer. Nat.* 121:324–34.

Mazur, A. 1983. Hormones, aggression, and dominance in humans. In *Hormones and Aggressive Behavior,* ed. B. B. Svare. New York: Plenum.

McArthur, C., A. Hagerman, and C. T. Robbins. 1991. Physiological strategies of mammalian herbivores against plant defenses. Pp. 103–14 in *Plant Defenses Against Mammalian Herbivory* (R. T. Palo and C. T. Robbins, eds.). Boca Raton, FL: CRC.

McCabe, B. J., and A. U. Nicol. 1999. The recognition memory of imprinting: Biochemistry and lectrophysiology. *Behav. Brain Res.* 98:253–60.

McCarty, R., and C. H. Southwick. 1977. Cross-species fostering: Effects on the olfactory preference of *Onychomys torridus* and *Peromyscus leucopus. Behav. Biol.* 19:255–60.

McCaul, K. D., B. A. Gladue, and M. Joppa. 1992. Winning, losing, mood, and testosterone. *Horm. Behav.* 26:486–504.

McCauley, D. E., and M. J. Wade. 1980. Group selection: The genotypic and demographic basis for the phenotypic differentiation of small populations of *Triboleum castaneum. Evolution* 34:813–21.

McCleave, J. D., and R. C. Kleckner. 1985. Oceanic migrations of Atlantic eels (*Anguilla* spp.): Adults and their offspring. Pp. 316–37 in *Migration: Mechanisms and Adaptive Significance* (M. A. Rankin, ed.). Port Aransas, TX: Marine Science Institute.

McConnell, J. V., A. L. Jacobson, and D. P. Kimble. 1959. The effects of regeneration upon retention of a conditioned response in the planarian. *J. Comp. Physiol. Psychol.* 52:1–5.

McCullough, D. R. 1985. Long range movements of large terrestrial mammals. Pp. 444–65 in *Migration: Mechanisms and Adaptive Significance* (M. A. Rankin, ed.). Port Aransas, TX: Marine Science Institute.

McDonald, D. B., and W. K. Potts. 1994. Cooperative display and relatedness among males in a lek-mating bird. *Science* 266:1030–32.

McDonald, D. L., and L. G. Forslund. 1978. The development of social preferences in the voles, *Microtus montanus* and *Microtus canicaudus:* Effects of cross-fostering. *Behav. Biol.* 22:457–508.

McEachron, D. L., C. L. Lauchlan, and D. E. Midgley. 1993. Effects of thyroxine and thyroparathyroidectomy on circadian wheel-running in rats. *Pharmacology, Biochemistry and Behavior* 46:243–49.

McEwen, B. S., P. G. Davis, B. Parsons, and D. W. Pfaff. 1979. The brain as a target for steroid hormone action. *Ann. Rev. Neurosci.* 2:65–112.

McFarland, D. 1981. *The Oxford Companion to Animal Behaviour.* Oxford, England: Oxford University Press.

McGuire, T. R., and T. Tully. 1986. Food-search behavior and its relation to the central excitatory state in the genetic analysis of the blowfly (*Phormia regina*). *J. Comp. Psychol.* 100:52–58.

McKenna, J. J. 1981. Primate infant caregiving behavior. In *Parental Care in Mammals,* ed. D. J. Gubernick and P. H. Klopfer, 389–416. New York: Plenum.

McLaren, I. A. 1963. Effects of temperature on growth of zooplankton and the adaptive value of vertical migration. *J. Fish. Res. Bd. Can.* 20:685–727.

McNab, B. K. 1963. Bioenergetics and the determination of home range size. *Amer. Nat.* 97:133–40.

McNair, J. N. 1982. Optimal giving-up times and the marginal value theorem. *Amer. Nat.* 119:511–29.

McQuade, D. B., E. H. Williams, and H. B. Eichenbaum. 1986. Cues used for localizing food by the grey squirrel (*Sciurus carolinensis*). *Ethology* 72:22–30.

Meany, M. J., J. Stewart, and W. W. Beatty. 1982. The influence of glucocorticoids during the neonatal period on the development of play-fighting in Norway rat pups. *Horm. Behav.* 16:475–91.

Mech, L. D. 1970. *The Wolf: The Ecology and Behavior of an Endangered Species.* Garden City, NY: Natural History Press.

Medvin, M. B., P. K. Stoddard, and M. D. Beecher. 1993. Signals for parent-offspring recognition: A comparative analysis of the begging calls of cliff swallows and barn swallows. *Anim. Behav.* 45:841–50.

Meier, A. H. 1973. Daily hormone rhythms in the white-throated sparrow. *Amer. Sci.* 61:184–87.

Meier, G. W. 1965. Other data on the effects of social isolation during rearing upon adult reproductive behavior in the rhesus monkey (*Macaca mulatta*). *Anim. Behav.* 13:228–31.

Meijer, J. H., and W. J. Rietveld. 1989. Neurophysiology of the suprachiasmatic circadian pacemaker in rodents. *Physiol. Rev.* 69:671–707.

Meikle, D. B., B. L. Tilford, and S. H. Vessey. 1984. Dominance rank, secondary sex ratio and reproduction of offspring in polygynous primates. *Amer. Nat.* 124:173–88.

Meikle, D. B., and S. H. Vessey. 1981. Nepotism among rhesus monkey brothers. *Nature* 294:160–61.

Menaker, M., J. S. Takahashi, and A. Eskin. 1978. The physiology of circadian pacemakers. *Ann. Rev. Physiol.* 40:501–26.

Menaker, M., and J. S. Takahashi. 1995. Genetic analysis of the circadian system in mammals: Properties and prospects. *Sem. Neurosci.* 7:61–70.

Menaker, M., and N. Zimmerman. 1976. Role of the pineal in the circadian system of birds. *Amer. Zool.* 16:45–55.

Menzel, E. W. 1971. Communication about the environment in a group of young chimpanzees. *Folia Primat.* 15:220–32.

Merilae, J., and J. Sorjonen. 1994. Seasonal and diurnal patterns of singing and song-flight activity in bluethroats (*Luscinia svecica*). *Auk* 111:556–62.

Merilaita, S. 1998. Crypsis through disruptive coloration in an isopod. *Proc. Roy. Soc. London B.* 265:1059–64.

Merle, J. 1969. Fonctionnement ovarien et réceptivité sexuelle de *Drosophila melanogaster* après implantation de fragments de l'appareil génitale mâle. *J. Insect Physiol.* 14:1159–68.

Mettler, L. E., T. G. Gregg, and H. E. Schaffer. 1988. *Population Genetics and Evolution.* Englewood Cliffs, NJ: Prentice Hall.

Metzgar, L. H. 1967. An experimental comparison of screech owl predation on resident and transient white-footed mice (*Peromyscus leucopus*). *J. Mammal.* 48:387–91.

———. 1971. Behavioral population regulation in the woodmouse, *Peromyscus leucopus. Amer. Mid. Nat.* 86:434–48.

Metzger, M., S. Jiang, and K. Braun. 1998. Organization of the dorsocaudal neostriatal complex: A retrograde and anterograde tracing study in the domestic chick with special emphasis on pathways relevant to imprinting. *J. Comp. Neurol.* 395:380–404.

Meyer, D. R., and H. F. Harlow. 1952. Effects of multiple variables on delayed response performance by monkeys. *J. Genet. Psychol.* 81:53–61.

Meyer, D. R., F. R. Treichler, and P. M. Meyer. 1965. Discrete-trial training techniques and stimulus variables. In *Behavior of Nonhuman Primates,* vol. 1., ed. A. M. Schrier, H. F. Harlow, and F. Stollnitz. New York: Academic Press.

Michael, R. P., and R. W. Bonsall. 1977. Chemical signals and primate behavior. In *Chemical Signals in Vertebrates,* ed. D. Muller-Schwartze and M. M. Mozell. New York: Plenum.

Michel, R. 1972. Étude expérimentale de l'influence des glandes prothoraciques sur l'activité de vol du Criquet Pèlerin *Schistocerca gregaria. Gen. Comp. Endocr.* 19:96–101.

Michelsen, A., B. B. Andersen, J. Storm, W. H. Kirchner, and M. Lindauer. 1992. How honeybees perceive communication dances, studied by means of a mechanical model. *Behav. Ecol. Sociobiol.* 30:143–50.

Michelson, A., B. B. Andersen, W. H. Kirchner, and M. Lindauer. 1989. Honeybees can be recruited by a mechanical model of a dancing bee. *Naturwissenschaften* 76:277–80.

Michener, C. D. 1974. *The Social Behavior of the Bees: A Comparative Study.* Cambridge, MA: Belknap Press.

Miklósi, A., V. Csányi, and R. Gerlai. 1997. Antipredator behavior in paradise fish (*Macropodus opercularis*) larvae: The role of genetic factors and paternal influence. *Behav. Genet.* 27:191–200.

Miles, L. 1990. The cognitive foundations for reference in a signalling orangutan. In *Language and Intelligence in Monkeys and Apes,* ed. S. T. Parker and K. R. Gibson. New York: Columbia University Press.

Milinski, M. 1979. Can an experienced predator overcome the confusion of swarming prey more easily? *Anim. Behav.* 27:1122–26.

———. 1984. Parasites determine a predator's optimal feeding strategy. *Behav. Ecol. Sociobiol.* 15:35–38.

Miller, D. B. 1980. Maternal vocal control of behavioral inhibition in mallard ducklings (*Anas platyrhynchos*). *J. Comp. Physiol. Psychol.* 94:606–23.

———. 1982. Alarm call responsivity of mallard ducklings: I. The acoustical boundary between behavioral inhibition and excitation. *Develop. Psychobiol.* 16:185–94.

———. 1985. Methodological issues in the ecological study of learning. In *Issues in the Ecological Study of Learning,* ed. T. D. Johnston and A. T. Pietrewicz, 73–95. Hillsdale, NJ: Lawrence Erlbaum Associates.

———. 1988. Development of instinctive behavior. In *Handbook of Behavioral Neurobiology,* vol. 9, *Developmental Psychobiology and Behavioral Ecology,* ed. E. Blass, 415–44. New York: Plenum.

———. 1994. Social context affects the ontogeny of instinctive behavior. *Anim. Behav.* 48:627–34.

Miller, D. B., and C. F. Blaich. 1987. Alarm call responsivity of mallard ducklings: V. Age-related changes in repetition rate specificity and behavioral inhibition. *Devel. Psychobiol.* 20:571–86.

Miller, D. B., and G. Gottlieb. 1978. Maternal vocalizations of mallard ducks (*Anas platyrhynchos*). *Anim. Behav.* 26:1178–94.

Miller, D. B., G. Hicinbothom, and C. F. Blaich. 1990. Alarm call responsivity of mallard ducklings: Multiple pathways in behavioural development. *Anim. Behav.* 39:1207–12.

Miller, F. L., F. W. Anderka, C. Vithayasai, and R. L. McClure. 1975. Distribution, movements, and socialization of barren-ground caribou ratio-tracked on their calving and post-calving areas. In *First International Reindeer and Caribou Symposium,* eds. J. R. Luick, C. Lent, D. R. Klein, and P. G. White, 423–35. Fairbanks, AK: University of Alaska.

Miller, J. W. 1975. Much ado about starlings. *Natur. Hist.* 84 (7):38–45.

Miller, L. A., and J. T. Emlen, Jr. 1975. Individual chick recognition and family integrity in the ring-billed gull. *Behaviour* 52:124–44.

Miller, M. N., and J. A. Byers. 1990. Energetic cost of locomotor play in pronghorn fawns. *Anim. Behav.* 41:1007–13.

Miller, N. E. 1957. Experiments on motivation studies combining psychological, physiological, and pharmacological techniques. *Science* 126:1271–78.

Miller, R. C. 1922. The significance of the gregarious habit. *Ecology* 3:122–26.

Millesi, E., S. Huber, J. Dittami, I. Hoffmann, and S. Daan. 1998. Parameters of mating effort and success in male European ground squirrels, *Spermophilus citellus. Ethology* 104:298–313.

Mills, M. G. L., and M. L. Gorman. 1997. Factors affecting the density and distribution of wild dogs in the Kruger National Park. *Cons. Biol.* 11:1397–1406.

Milton, K. 1985. Mating patterns of woolly spider monkeys, *Brachyteles arachnoides:* Implications for female choice. *Behav. Ecol. Sociobiol.* 17:53–59.

Minituni, L., A. Innocenti, C. Bertolucci, and A. Foa. 1995. Circadian organization in the ruin lizard *Podacris sicula:* The role of the suprachiasmatic nuclei of the hypothalamus. *J. Comp. Physiol. A* 176:281–88.

Mitchell, G. 1970. Abnormal behavior in primates. In *Primate Behavior: Developments in Field and Laboratory Research,* vol. 1, ed. L. Rosenblum. New York: Academic Press.

Mitchell, G., and E. M. Brandt. 1970. Behavioral differences related to experience of mother and sex of infant in the rhesus monkey. *Dev. Psychol.* 3:149.

Mock, D. W. 1984. Siblicidal aggression and resource monopolization in birds. *Science* 225:731–33.

Mock, D. W., and G. W. Parker. 1997. The evolution of sibling rivalry. New York: Oxford University Press.

Mock, D. W., H. Drummond, and C. H. Stinson. 1990. Avian siblicide. *Am. Sci.* 78:438–49.

Mock, D. W. 1984. Siblicidal aggression and resource monopolization in birds. *Science* 225:731–33.

Mock, D. W., and B. J. Ploger. 1987. Parental manipulation of optimal hatch asynchrony in cattle egrets: An experimental study. *Anim. Behav.* 35:150–60.

Moehlman, P. D. 1979. Jackal helpers and pup survival. *Nature* 277:382–83.

Moehlman, P. D. 1983. Socioecology of silverbacked and golden jackals, *Canis mesomelas* and *C. aureus*. Pp. 423–53 in *Recent Advances in the Study of Mammalian Behavior* (J. F. Eisenberg and D. G. Kleiman, eds.). *Special Pub. No. 7, Am. Soc. Mammal.*

Moehlman, P. D. 1986. Ecology of cooperation in canids. Pp. 64–86 in *Ecological Aspects of Social Evolution* (D. I. Rubenstein and R. W. Wrangham, eds.). Princeton: Princeton University Press.

Moen, C. A., and R. J. Gutierrez. 1997. California spotted owl habitat selection in the central Sierra Nevada. *J. Wildl. Manage.* 61:1281–87.

Møffat, A. J. M., and R. R. Capranica. 1978. Middle ear sensitivity in anurans measured by light scattering spectroscopy. *J. Comp. Physiol.* 127:97–107.

Møller, A. P. 1988. Ejaculate quality, testes size, and sperm competition in primates. *J. Hum. Evol.* 17:479–88.

Møller, P. 1976. Electric signals and schooling behavior in a weakly electric fish, *Marcusenius cyprinoides* L. (Mormyriformes). *Science* 193:697–99.

Møller, A. P. 1990. Effects of a haematophagous mite on the barn swallow (*Hirundo rustica*): A test of the Hamilton and Zuk hypothesis. *Evolution* 44:771–84.

Møller, A. P. 1992. Female swallow preference for symmetrical male sexual ornaments. *Nature* 357:238–40.

Møller, A. P., J. J. Cuervo, J. J. Soler, and C. Zamora-Muñoz. 1996. Horn asymmetry and fitness in gemsbok, *Oryx g. gazella*. *Behav. Ecol.* 7:247–53.

Møller, A. P., P. Christe, and E. Lux. 1999. Parasitism, host immune function, and sexual selection. *Quart. Rev. Biol.* 74:3–20.

Møller, A. P., P. Y. Henry, and J. Erritzøe. 2000. The evolution of song repertoires and immune defense in birds. *Proc. Roy. Soc. Lond. B* 267:165–69.

Moltz, H. 1965. Contemporary instinct theory and the fixed action pattern. *Psychol. Rev.* 72:27–47.

Moltz, H., D. Robbins, and M. Parks. 1966. Caesarian delivery and the maternal behavior of primiparous and multiparous rats. *J. Comp. Physiol. Psychol.* 61:455–60.

Monaghan, E. P., and S. E. Glickman. 1992. Hormones and aggressive behavior. In *Behavioral Endocrinology,* ed. J. B. Becker, S. Marc Breedlove, and D. Crews. Cambridge, MA: MIT Press.

Monahan, E. J., and S. C. Maxson. 1998. Y chromosome, urinary chemosignals, and an agonistic behavior (offense) of mice. *Physiol. & Behav.* 64:123–32.

Money, J. 1977. Human hermaphroditism. In *Human Sexuality in Four Perspectives,* ed. F. A. Beach. Baltimore: Johns Hopkins University Press.

Money, J., and A. A. Ehrhardt. 1972. *Man and Woman, Boy and Girl: Differentiation and Dimmorphism of Gender Identity form Conception to Maturity.* Baltimore: Johns Hopkins University Press.

Moore, F., and P. Kerlinger. 1987. Stopover and fat deposition by North American wood warblers (Parulinae) following spring migration over the Gulf of Mexico. *Oecologia* 74:47–54.

Moore, J., and R. Ali. 1984. Are dispersal and inbreeding avoidance related? *Anim. Behav.* 32:94–112.

Moore, J. A. 1984. Science as a way of knowing—evolutionary biology. *Amer. Zool.* 24:467–534.

Moore, R. Y. 1980. Suprachiasmatic nucleus, secondary synchronizing stimuli and the central neural control of circadian rhythms. *Brain Res.* 183:13–28.

Moore, R. Y., and V. B. Eichler. 1972. Loss of a circadian adrenal corticosterone rhythm following suprachiasmatic lesions in the rat. *Brain Res.* 42:201–6.

Moore, R. Y., and N. J. Lenn. 1972. A retinohypothalamic projection in the rat. *J. Comp. Neurol.* 146:1–14.

Moore-Ede, M. C., F. M. Sulzman, and C. A. Fuller. 1982. *The Clocks That Time Us.* Cambridge: Harvard University Press.

Moore, D., J. E. Angel, I. M. Cheeseman, S. E. Fahrbach, and G. E. Robinson. 1998. Timekeeping in the honeybee colony: Integration of circadian rhythms and division of labor. *Behav. Ecol. Sociobiol.* 43:147–60.

Moore, M. C., D. K. Hews, and R. Knapp. 1998. Hormonal control and evolution of alternative male phenotypes: Generalizations of models for sexual differentiation. *Amer. Zool.* 38:133–51.

Mooring, M. S., and B. L. Hart. 1992. Animal grouping for protection from parasites: Selfish herd and encounter-dilution effects. *Behaviour* 123:173–93.

Morey, D. F. 1994. The early evolution of the domestic dog. *Amer. Scientist* 82:336–47.

Morgan, C. L. 1896. *Introduction to Comparative Psychology.* London: Scott.

Morin, P. A., J. J. Moore, R. Chakraborty, L. Jin, J. Goodall, and D. S. Woodruff. 1994. Kin selection, social structure, gene flow and the evolution of chimpanzees. *Science* 265:1193–1201.

Morrell, J. I., D. B. Kelley, and D. W. Pfaff. 1975. Sex steroid binding in the brains of vertebrates: Studies with light microscopic autoradiography. In *The Ventricular System in Neuroendocrine Mechanisms,* ed. K. M. Knigge et al. Basel, Switzerland: Karger.

Morrell, J. I., and D. W. Pfaff. 1981. Autoradiographic technique for steroid hormone localization. In *Neuroendocrinology of Reproduction,* ed. N. Adler. New York: Plenum.

Morris, D. 1956. The feather postures of birds and the problem of the origin of social signals. *Behaviour* 9:75–113.

Morris, M. R., and K. Casey. 1998. Female swordtail fish prefer symmetrical sexual signal. *Anim. Behav.* 55:33–39.

Morse, D. H. 1980. *Behavioral Mechanisms in Ecology.* Cambridge, MA: Harvard University Press.

Morton, M. L., and L. R. Mewaldt. 1962. Some effects of castration on migratory sparrows (*Zonotrichia atricapilla*). *Physiol. Zool.* 35:237–47.

Moyer, K. E. 1976. *Psychobiology of Aggression.* New York: Harper & Row.

Moynihan, M. 1956. Notes on the behaviour of some North American gulls. 1. Aerial hostile behaviour. *Behaviour* 10:126–78.

Mueller, H. 1966. Homing and distance-orientation in bats. *Zeit. Tierpsychol.* 23:403–21.

Mueller, H. 1971. Oddity and specific searching image more important than conspicuousness in prey selection. *Nature* 233:345–46.

Mueller, H., and J. T. Emlen. 1957. Homing in bats. *Science* 126:307–8.

Muller, H. J. 1964. The relation of recombination to mutational advance. *Mutation Res.* 1:2–9.

Muller-Schwarze, D. 1971. Pheromones in black-tailed deer. *Anim. Behav.* 19:141–52.

Murie, O. J. 1935. Alaska-Yukon caribou. *N. Amer. Fauna* (U.S. Dept. Agric.) 54:1–93.

Murie, O. J., and A. Murie. 1931. Travels of *Peromyscus*. *J. Mamm.* 12:200–9.

Murphy, M. R. 1980. Sexual preferences of male hamsters: Importance of preweaning and adult experience, vaginal secretion and olfactory or vomeronasal sensation. *Behav. Neural Biol.* 30:323–40.

Nadel, L. 1991. The hippocampus and space revisited. *Hippocampus* 1:221–29.

Nagamine, C. M., and A. W. Knight. 1980. Development, motivation and function of some sexually dimorphic structures of the Malaysian prawn *Macrobrachium rosenbergi. Crustaceana* 39:141–52.

Naisse, J. 1996a. Contróle endocrinen de la différenciation sexuelle chez l'insecte *Lampyris noctiluca.* I. Role androgéne des testicules. *Arch. Biol.* 77:139–201.

Naisse, J. 1996b. Contróle endocrinen de la différenciation sexuelle chez l'insecte *Lampyris noctiluca.* II. Phénoménes neurosécrétoires et endocrines au cours de développement postembryonnaire chez le mále et la femelle. *Gen. Comp. Endocrinol.* 7:85–104.

Naisse, J. 1996c. Contróle endocrinen de la différenciation sexuelle chez l'insecte *Lampyris noctiluca.* III. Influence des hormones de la pars intercerebralis. *Gen. Comp. Endocrinol.* 7:105–10.

Nams, V. O. 1997. Density-dependent predation by skunks using olfactory search images. *Oecologia* 110:440–48.

Narayanan, C. H., M. W. Fox, and V. Hamburger. 1971. Prenatal development of spontaneous and evoked activity in the rat. *Behaviour* 40:100–34.

National Institutes of Health. 1985. *Guide for the Care and Use of Laboratory Animals.* Washington, DC: U.S. Department of Health and Human Services.

National Institutes of Health. 1996. Guide for the Care and Use of Laboratory Animals. Washington, D.C.: National Academy Press.

Nelson, R. A., and K. A. Young. 1998. Behavior in mice with targeted disruption of single genes. *Neurosci. Biobehav. Rev.* 22:453–62.

Nelson, R. J., G. E. Demas, P. L. Huang, M. C. Fishman, V. L. Dawson, T. M. Dawson, and S. H. Snyder. 1995. Behavioral abnormalities in male mice lacking neuronal nitric oxide synthase. *Nature* 378:383–86.

Nelson, R. J. 1997. The use of genetic "knockout" mice in behavioral endocrinology research. *Horm. Behav.* 31:188–96.

Neumann, D. 1966. Die lunare and tägliche Schlüpfperiodik der Mücke *Clunio-* Steuerung und Abstimmung auf die Gezeitenperiodik. *Ziet. Vergleich. Physiol.* 53:1–61.

Nevitt, G., R. Veit, and P. Kareiva. 1995. Dimethyl sulphide as a foraging cue for Antarctic procellariiform seabirds. *Nature* 376:680–82.

Nice, M. M. 1941. The role of territory in bird life. *Amer. Mid. Nat.* 26:441–87.

Nicoll, C. S., and J. Meites. 1959. Prolongation of lactation in the rat by litter replacement. *Proc. Soc. Exp. Biol. Med.* 101:81–82.

Noakes, D. L. G. 1978. Ontogeny of behavior in fishes: A survey and suggestions. In *Development of Behavior,* ed. G. Burghardt and M. Bekoff. New York: Garland STPM Press.

Noirot, E. 1972. Ultrasounds and maternal behavior in small mammals. *Dev. Psychobiol.* 5:371–87.

Noor, M. A. F. 1999. Reinforcement and other consequences of sympatry. *Heredity* 83(5):503–8.

Nordeen, E. J., and K. W. Nordeen. 1988. Sex and regional differences in the incorporation of neurons born during song learning in zebra finches. *J. Neurosci.* 8:2869–74.

Nordenskiöld, E. 1928. *History of Biology.* New York: Knopf.

Norell, M. A., J. M. Clark, L. M. Chiappe, and D. Dashzeveg. 1995. A nesting dinosaur. *Nature* 378:774–76.

Norman, A., G. Jones, and A. Raphael. 1999. Noctuid moths show neural and behavioural responses to sounds made by some bat-marking rings. *Anim. Behav.* 57:829–35.

Nottebohm, F. 1975. Vocal behavior in birds. In *Avian Biology,* vol. 5, ed. J. R. King and D. S. Farner. New York: Academic Press.

———. 1980. Brain pathways for vocal learning in birds: A review of the first 10 years. In *Progress in Psychobiology and Physiology Psychology,* ed. J. M. Sprague and A. N. Epstein. 9:85–124.

———. 1981. A brain for all seasons: Cyclical anatomical changes in song control nuclei in the canary brain. *Science* 214:1368–70.

Nottebohm, F., and M. Nottebohm. 1976. Left hypoglossal dominance in the control of canary and white-crowned sparrow song. *J. Comp. Physiol.* 108:171–92.

Nottebohm, F., M. E. Nottebohm, and L. Crane. 1986. Developmental and seasonal changes in canary song and their relation to changes in the anatomy of song-control nuclei. *Behav. Neural Biol.* 46:445–71.

Novotny, M., B. Jemiolo, S. Harvey, D. Wiesler, and A. Marchlewska-Koj. 1986. Adrenal-mediated endogenous metabolites inhibit puberty in female mice. *Science* 231:722–25.

Nowicki, S., S. Peters, and J. Podos. 1998. Song learning, early nutrition and sexual selection in songbirds. *Amer. Zool.* 38:179–90.

Nyby, J., J. A. Matochik, and R. J. Barfield. 1992. Intracranial androgenic and estrogenic stimulation of male-typical behaviors in house mice. (*Mus domesticus*). *Horm. Behav.* 26:24–45.

O'Brien, S. J. 1994. A role for molecular-genetics in biological conservation. *Proc. Nat. Acad. Sci.* 91:5748–55.

O'Connell, M. E., C. Reboulleau, H. H. Feder, and R. Silver. 1981a. Social interactions and androgen levels in birds. I. Female characteristics associated with increased plasma androgen levels in the male ring dove (*Streptopelia risoria*). *Gen. Comp. Endocrinol.* 44:454–63.

———. 1981b. Social interactions and androgen levels in birds. II. Social factors associated with a decline in plasma androgen levels in male ring doves (*Streptopelia risoria*). *Gen. Comp. Endocrinol.* 44:464–69.

O'Keefe, J., and L. Nadel. 1978. The hippocampus as a cognitive map. Oxford, UK: Clarendon Press.

Oldham, R. S., and H. C. Gerhardt. 1975. Behavioral isolation of the treefrogs *Hyla cinerea* and *Hyla gratiosa. Copeia* 1975:223–31.

Oliveira, R. F., and V. C. Almada. 1998. Androgenization of dominant males in a cichlid fish: Androgens mediate the social modulation of sexually dimorphic traits. *Ethology* 104:841–58.

Oliverio, A. 1983. Genes and behavior: An evolutionary perspective. *Adv. Study Behav.* 13:191–217.

Olivier, B., J. Mos, R. Van Oorschot, and R. Hen. 1995. Serotonin receptors and animal models of aggressive behavior. *Pharmacopsychiatry* 28:80–90.

Olton, D. S. 1979. Mazes, maps, and memory. *Am. Psychol.* 34:583–96.

———. 1985. The temporal context of spatial memory. *Phil. Trans. Royal Soc. Lond.* B. 308:79–86.

Olton, D. S., C. Collison, and M. A. Werz. 1977. Spatial memory and radial arm maze performance of rats. *Learn. Motiv.* 8:289–314.

Olton, D. S., and R. J. Samuelson. 1976. Remembrance of places passed: Spatial memory in rats. *J. Exp. Psychol.: Anim. Behav. Process.* 2:97–116.

Oppenheim, R. W. 1970. Some aspects of embryonic behavior in the duck (*Anas platyrhynchos*). *Anim. Behav.* 18:335–52.

———. 1972. Pre-hatching and hatching behavior in birds: A comparative study of altricial and precocial species. *Anim. Behav.* 20:644–55.

———. 1982. Preformation and epigenesis in the origins of the nervous system and behavior: Issues, concepts, and their history. *Perspectives in Ethology* 5:1–100.

Oppenheim, R. W., W. Chu-Wang, and J. L. Maderut. 1978. Cell death of motorneurons in the chick embryo and spinal cord. III. *J. Comp. Neurol.* 177:87–112.

Oppenheim, R. W., J. L. Maderdrut, and D. J. Wells. 1982. Reduction of naturally occurring cell death in the thoraco-lumbar preganglionic cell column of the chick embryo by nerve growth factor and hemicholinium-3. *Dev. Brain Res.* 3:134–39.

Oppenheim, R. W., and R. Nunez. 1982. Electrical stimulation of hindlimb increases neuronal cell death in chick embryos. *Nature* 295:57–59.

Oppenheim, R. W., L. M. Schwartz, and C. J. Shatz. 1992. Neuronal death, a tradition of dying. *J. Neurobiol.* 23:1111–16.

Orians, G. H. 1960. Autumnal breeding in the tricolored blackbird. *Auk* 77:379–98.

———. 1961. The ecology of blackbird (*Agelaius*) social systems. *Ecol. Monogr.* 31:285–312.

———. 1969. On the evolution of mating systems in birds and mammals. *Amer. Nat.* 103:589–603.

Orians, G. H., and G. Collier. 1963. Competition and blackbird social systems. *Evolution* 17:449–59.

Orians, G. H., and N. E. Pearson. 1979. On the theory of central place foraging. Pp. 1254–77 in *Analysis of Ecological Systems* (D. J. Horn, R. D. Mitchell, and G. R. Stairs, eds.). Columbus: Ohio State University Press.

Oring, L. W., and D. B. Lank. 1982. Sexual selection, arrival times, philopatry and site fidelity in the polyandrous spotted sandpiper. *Behav. Ecol. Sociobiol.* 10:185–91.

Orr, R. T. 1970. *Animals in Migration.* New York: Macmillan.

Otronen, M., and M. T. Siva-Jothy. 1991. The effect of postcopulatory male behaviour on ejaculate distribution within the female sperm storage organs of the fly, *Dryomyza anilis* (Diptera: Dryomyzidae). *Behav. Ecol. Sociobiol.* 29:33–37.

Packer, C. 1979. Inter-troop transfer and inbreeding avoidance in *Papio anubis. Anim. Behav.* 27:1–36.

———. 1986. The ecology of sociality in felids. In *Ecological Aspects of Social Evolution,* ed. D. I. Rubenstein and R. W. Wrangham. Princeton: Princeton University Press.

Packer, C., D. Scheel, and A. E. Pusey. 1990. Why lions form groups: Food is not enough. *Amer. Nat.* 136:1–19.

Packer, C., D. A. Collins, A. Sindimwo, and J. Goodall. 1995. Reproductive constraints on aggressive competition in female baboons. *Nature* 373:60–63.

Page, R. E. Jr., K. D. Waddington, G. J. Hunt, and M. K. Fondrk. 1995. Genetic determinants of honey bee foraging behaviour. *Anim. Behav.* 50:1617–25.

Page, T. L. 1982. Transplantation of the cockroach circadian pacemaker. *Science* 216:73–75.

Page, T. L., P. C. Caldarola, and C. S. Pittendrigh. 1977. Mutual entrainment of bilaterally distributed circadian pacemakers. *Proc. Nat. Acad. Sci. USA* 74:1277–81.

Palanza, P., S. Parmigiani, and F. S. vom Saal. 1995. Urine marking and maternal aggression of wild female mice in relation to anogenital distance at birth. *Physiol. & Behav.* 58:827–35.

Palmer, J. D. 1990. The rhythmic life of crabs. *Bioscience* 40:352–58.

Panhuis, T. M., and G. S. Wilkinson. 1999. Exaggerated male eye span influences contest outcome in stalk-eyed flies (Diopsidae). *Behav. Ecol. Sociobiol.* 46:221–27.

Papaj, D. R. 1994. Oviposition site guarding by male walnut flies and its possible consequences for mating success. *Behav. Ecol. Sociobiol.* 34:187–95.

Papi, F. 1982. Olfaction and homing in pigeons: Ten years of experiments. In *Avian Navigation,* ed. F. Papi and H. G. Walraff, 149–59. Berlin: Springer-Verlag.

Papi, F., V. Fiaschi, S. Benvenuti, and N. E. Baldaccini. 1972. Olfaction and homing in pigeons. *Monit. Zool. Ital.* (n.s.) 6:85–95.

Papi, F., and L. Pardi. 1953. Ricerche sull'orientamento di *Talitrus saltator* Montagu (Crustacea-Amphipoda). *Z. Vergl. Physiol.* 35:490–518.

Papi, F. 1990. Olfactory navigation in birds. *Experientia* 46:352–63.

Papi, F. 1991. Olfactory navigation. Pp. 52–85 in *Orientation in Birds* (P. Berthold, ed.). Basel: Birkhauser Verlag.

Parker, G. A. 1970a. The reproductive behaviour and the nature of sexual selection in *Scatophaga stercoraria* L. (Diptera: Scatophagidae) IV. Epigamic recognition and competition between males for the possession of females. *Behaviour* 37:113–39.

———. 1970b. Sperm competition and its evolutionary consequences in the insects. *Biol. Rev.* 45:525–68.

———. 1974. The reproductive behavior and the nature of sexual selection in *Scatophaga stercoraria* L. IX. Spatial distribution of fertilization rates and evolution of male search strategy within the reproductive area. *Evolution* 28:93–108.

———. 1978. Searching for mates. In *Behavioural Ecology: An Evolutionary Approach,* ed. J. R. Krebs and N. B. Davies. Oxford, England: Blackwell Scientific Publications, Ltd.

———. 1984. Evolutionary stable strategies. In *Behavioural Ecology: An Evolutionary Approach,* 2d ed., eds. J. R. Krebs and N. B. Davies. Oxford, England: Blackwell Scientific Publications, Ltd.

Parker, G. A., R. R. Baker, and V. G. F. Smith. 1972. The origin and evolution of gamete dimorphism and the male-female phenomenon. *J. Theoret. Biol.* 36:529–53.

Parmagiano, S., and F. S. vom Saal. 1994. Infanticide and parental care. Langhorne, PA: Harwood Academic Publishers.

Parsons, L. M., and C. R. Terman. 1978. Influence of vision and olfaction on the homing ability of the white-footed mouse (*Peromyscus leucopus noveboracensis*). *J. Mamm.* 59:761–71.

Partridge, L. 1974. Habitat selection in titmice. *Nature* 247:573–74.

———. 1976. Field and laboratory observations on the foraging and feeding techniques of blue tits (*Parus caeruleus*) and coal tits (*Parus ater*) in relation to their habitats. *Anim. Behav.* 24:534–44.

———. 1978. Habitat selection. In *Behavioural Ecology: An Evolutionary Approach,* ed. J. R. Krebs and N. B. Davies. Oxford, England: Blackwell Scientific Publications, Ltd.

Paterson, A., G. P. Wallis, R. D. Gray. 1995. Penguins, petrels, and parsimony: Does cladistic analysis of behavior reflect seabird phylogeny? *Evolution* 49:974–89.

Paterson, H. 1985. The recognition concept of species. Pp. 21–29 in *Species and Speciation* (E. S. Vrba, ed.). Pretoria: Transvaal Mus. Monogr. No. 4.

Paton, J. A., and F. Nottebohm. 1984. Neurons generated in the adult brain are recruited into functional circuits. *Science* 225:1046–48.

Patterson, D. J. 1973. Habituation in a protozoan *Vorticella convallaria. Behaviour* 45:304–11.

Patterson, F. 1978. The gestures of a gorilla: Sign language acquisition in another pongid species. *Brain Lang.* 5:72–97.

Patterson, I. J. 1978. Tags and other distant-recognition markers for birds. In *Animal Marking,* ed. B. Stonehouse, 54–62. Baltimore: University Park Press.

Patterson, M. A., and S. H. Vessey. 1973. Tapeworm (*Hymenolepis nana*) infection in male albino mice: Effect of fighting among the hosts. *J. Mammal.* 54:784–86.

Pavey, C. R., and A. K. Smyth. 1998. Effects of avian mobbing on roost use and diet of powerful owls, *Ninox strenua. Anim. Behav.* 55:313–18.

Payne, K. B., and R. S. Payne. 1985. Large scale changes over 19 years in songs of humpback whales in Bermuda. *Zeit. für Tierpsychol.* 68:89–114.

Payne, K. B., W. R. Langbauer, Jr., and E. M. Thomas. 1986. Infrasonic calls of the Asian elephant (*Elephas maximus*). *Behav. Ecol. Sociobiol.* 18:297–301.

Payne, R. S., and S. McVay. 1971. Songs of humpback whales. *Science* 173:585–97.

Pearson, K. 1993. Common principles of motor control in vertebrates and invertebrates. *Ann. Rev. Neurosci.* 16:265–97.

Pellis, S. M., and V. C. Pellis. 1998. Structure-function interface in the analysis of play fighting. Pp. 115–40 in *Animal Play* (M. Bekoff and J. A. Byers, eds.). New York: Cambridge University Press.

Pener, M. P. 1965. On the influence of corpora allata on maturation and sexual behavior of *Schistocera gregaria. J. Zool. Lond.* 147:119–36.

Pengelley, E. T., and S. J. Asmundson. 1974. Circannural rhythmicity in hibernating animals. In *Circannual Clocks,* ed. E. T. Pengelley. New York: Academic Press.

Penn, D. J., and W. K. Potts. 1999. The evolution of mating preferences and major histocompatibility complex genes. *Am. Nat.* 153:145–64.

Pennycuik, C. J. 1978. Identification using natural markings. In *Animal Marking,* ed. B. Stonehouse, 147–58. Baltimore: University Press.

Pennycuik, C. J., and J. Rudnai. 1970. A method of identifying individual lions *Panther leo* with an analysis of the reliability of identification. *J. Zool. Lond.* 160:497–508.

Pepperberg, I. M. 1987a. Acquisition of the same/different concept by an African grey parrot (*Psittacus erithacus*): Learning with respect to categories of color, shape, and material. *Animal Learning & Behavior* 15:423–32.

———. 1987b. Evidence for conceptual quantitative abilities in the African grey parrot: Labeling of cardinal sets. *Ethology* 75:37–61.

———. 1990. Cognition in an African grey parrot (*Psittacus erithacus*): Further evidence for comprehension of categories and labels. *J. Comp. Psychol.* 104:41–52.

———. 1994a. Numerical competence in an African grey parrot (*Psittacus erithacus*). *J. Comp. Psychol.* 108:36–44.

———. 1994b. Vocal learning in grey parrots (*Psittacus erithacus*): Effects of social interaction, reference, and context. *Auk* 111:300–13.

Pepperberg, I. M., and M. V. Brezinsky. 1991. Acquisition of a relative class concept in an African grey parrot (*Psittacus erithacus*): Discriminations based on relative size. *J. Comp. Psychol.* 105:286–94.

Pepperberg, I. M., and F. A. Kozak. 1986. Object permanence in the African grey parrot (*Psittacus erithacus*). *Anim. Learn. Behav.* 14:322–30.

Pepperberg, I. M. 1998a. The African grey parrot: How cognitive processing might affect allospecific vocal learning. Pp. 381–410 in R. P. Balda, I. M. Pepperberg, and A. C. Kamil (eds.), *Animal Cognition in Nature.* New York: Academic Press.

Pepperberg, I. M. 1998b. Stimulus class formation by an African grey parrot. In *Stimulus Class Formation in Humans and Animals* (T. R. Zentall and P. Smeets, eds.). Amsterdam: North Holland Publishing.

Pepperberg, I. M. 2000. *The Alex Studies.* Cambridge: Harvard University Press.

Pepperberg, I. M., J. R. Naughton, and P. A. Banta. 1998. Allospecific vocal learning by grey parrots (*Psittacus erithacus*): A failure of videotaped instruction under certain conditions. *Behav. Process.* 42:139–58.

Pereira, H. S., and M. B. Sokolowski. 1993. Mutations in the larval foraging gene affect adult locomotory behavior after feeding in *Drosophila melanogaster. Proc. Nat. Acad. Sci.* 90:5044–46.

Peters, S., W. A. Searcy, and P. Marler. 1980. Species song discrimination in choice experiments with territorial male swamp and song sparrows. *Anim. Behav.* 28:393–404.

Petersen, J. C. B. 1972. An identification system for zebra (*Equus burchelli,* Gray). *E. Afr. Wildl. J.* 10:59–63.

Petersen, K., and D. F. Sherry. 1996. No sex difference occurs in the hippocampus, food-storing, or memory for food caches in black-capped chickadees. *Behav. Brain Res.* 79:15–22.

Pettigrew, J. D. 1999. Electoreception in monotremes. *J. Exp. Biol.* 202:1447–54.

Pfaff, D. W. 1981. Electrophysiological effects of steroid hormones in brain tissue. In *Neuroendocrinology of Reproduction,* ed. N. Adler. New York: Plenum.

Pfaff, D. W., and M. Keiner. 1973. Atlas of estradiol-concentrating cells in the central nervous system of the female rat. *J. Comp. Neurol.* 151:121–30.

Pfennig, D. W., H. K. Reeve, and P. W. Sherman. 1993. Kin recognition and cannibalism in spadefoot toad tadpoles. *Anim. Behav.* 46:87–94.

Philips, M., and S. N. Austad. 1990. Animal communication and social evolution. In *Interpretation and Explanation in the Study of Animal Behavior,* ed. M. Bekoff and D. Jamieson. Boulder, CO: Westview Press.

Phillips, J. B. and S. C. Borland. 1992. Behavioural evidence for use of a light-dependent magnetoreception mechanism by a vertebrate. *Nature* 359:142–144.

Phoenix, C. H. 1974. Prenatal testosterone in the nonhuman primate and its consequences for behavior. In *Sex Differences in Behavior,* eds. R. C. Friedman, R. N. Richart and R. L. Van de Wiele. New York: Wiley.

———. 1977. Induction of sexual behavior in ovariectomized rhesus females with 19-hydroxytestosterone. *Horm. Behav.* 8:356–62.

Phoenix, C. H., R. W. Goy, A. A. Gerrall, and W. C. Young. 1959. Organizing action of prenatally administered testosterone propionate on the tissues mediating mating behavior in the female guinea pig. *Endocrinology* 65:369–82.

Pietrewicz, A. T., and A. C. Kamil. 1979. Search image formation in the blue jay (*Cyanocitta cristata*). *Science* 204:1332–33.

———. 1981. Search images and the detection of cryptic prey: An operant approach. In *Foraging Behavior: Ecological, Ethological, and Psychological Approaches,* eds. A. C. Kamil and T. D. Sargent. New York: Garland STPM Press.

Pietsch, T. W., and D. B. Grobecker. 1978. The compleat angler: Aggressive mimicry in an Antennariid anglerfish. *Science* 201:369–70.

Pinsker, H., I. Kupfermann, V. Castellucci, and E. R. Kandel. 1970. Habituation and dishabituation of the gill-withdrawal reflex in *Aplysia. Science* 167:1740–42.

Pittendrigh, C. S. 1954. On temperature independence in the clock system controlling emergence time in *Drosophila. Proc. Nat. Acad. Sci. USA* 40:1018–29.

———. 1960. Circadian rhythms and the circadian organization of living systems. *Cold Spr. Harb. Symp. Quant. Biol.* 25:159–84.

———. 1993. Temporal organization: Reflections of a Darwinian clock-watcher. *Ann. Rev. Physiol.* 55:17–54.

Pittman, R., and R. W. Oppenheim. 1979. Cell death of motorneurons in the chick embryo spinal cord. IV. *J. Comp. Neurol.* 187:425–46.

Pleim, E. T., and and R. J. Barfield. 1988. Progesterone versus estrogen facilitationof female sexual behavior in intracranial administation to female rats. *Horm. Behav.* 22:150–59.

Plomin, R. J. C. DeFries, and J. C. Loehlin. 1977. Genotype-Environment interaction and correlation in the analysis of human behavior. *Psych. Bull.* 84:309–22.

Plomin, R. 1994. *Genetics and Experience: The Interplay Between Nature and Nurture.* Thousand Oaks, CA: Sage Publications.

Plomin, R., and I. Craig. 1997. Human behavioural genetics of cognitive abilities and disabilities. *BioEssays* 19:1117–24.

Plomin, R., J. D. DeFries, G. E. McClearn, and M. Rutter. 1997. *Behavioral Genetics,* 3rd edition. New York: W. H. Freeman and Company.

Podos, J., S. Peters, T. Rudnicky, P. Marler, and S. Nowicki. 1992. The organization of song repertoires of song sparrows: Themes and variations. *Ethology* 90:89–106.

Pompanon, F., P. Fouillet, and M. Bouletreau. 1995. Emergence rhythms and protandry in relation to daily patterns of locomotor activity in *Trichogramma* species. *Evol. Ecol.* 9:467–77.

Poole, J. H., K. Payne, W. R. Langbauer, Jr., and C. J. Moss. 1988. The social contexts of some very low frequency calls of African elephants. *Behav. Ecol. Sociobiol.* 22:385–92.

Potts, W. K., C. J. Manning, and E. K. Wakeland. 1991a. The evolution of MHC-based mating preferences in Mus. In *Molecular Evolution of the Major Histocompatibility Complex* (J. Klein and D. Klein, eds.). Berlin: Springer-Verlag.

Potts, W. K., C. J. Manning, and E. K. Wakeland. 1991b. Mating patterns in seminatural populations of mice influenced by MHC genotype. *Nature* 352:619–21.

Poulsen, H. R., and D. Chiszar. 1975. Interaction of predation and intraspecific aggression in bluegill sunfish *Lepomis macrochirus*. *Behaviour* 55:268–86.

Powers, D. R., and T. McKee. 1994. The effect of food availability on time and energy expenditures of territorial and nonterritorial hummingbirds. *Condor* 96:1064–75.

Pratt, C. L., and G. P. Sackett. 1967. Selection of social partners as a function of peer contact during rearing. *Science* 155:1133–35.

Prechtl, H. F. R. 1984. Continuity and change in early neural development. In *Continuity of Neural Functions from Prenatal to Postnatal Life,* ed. H. F. R. Prechtl. Philadelphia: J. B. Lippincott.

Premak, D. 1971. Language in chimpanzee? *Science* 172:808–22.

Price, E. O. 1972. Domestication and early experience effects on escape conditioning in the Norway rat. *J. Comp. Physiol. Psychol.* 79:51–55.

———. 1984. Behavioral aspects of animal domestication. *Quart. Rev. Biol.* 59:1–32.

Price, E. O., and P. L. Belanger. 1977. Maternal behavior of wild and domestic Norway rats. *Behav. Biol.* 20:60–69.

Price, E. O., and U. W. Huck. 1976. Open-field behavior of wild and domestic Norway rats. *Anim. Learn. Behav.* 4:125–30.

Price, M. V., N. M. Waser, and T. A. Bass. 1984. Effects of moonlight on microhabitat use by desert rodents. *J. Mamm.* 65:353–56.

Prince, G. J., and P. A. Parsons. 1977. Adaptive behaviour of *Drosophila* adults in relation to temperature and humidity. *Aust. J. Zool.* 25:285–90.

Prinz, K., and W. Wiltschko. 1992. Migratory orientation of pied flycatchers: interaction of stellar and magnetic information during ontogeny. *Anim. Behav.* 44:539–45.

Proctor, H. C. 1991. Courtship in the water mite *Neumania papillator:* Males capitalize on female adaptations for predation. *Anim. Behav.* 42:589–98.

Pugesek, B. 1981. Increased reproductive effort with age in the California gull (*Larus californicus*). *Science* 212:822–23.

———. 1983. The relationship between parental age and reproductive effort in the California gull (*Larus californicus*). *Behav. Ecol. Sociobiol.* 13:161–71.

Pulliam, H. R., and B. J. Danielson. 1991. Sources, sinks, and habitat selection: A landscape perspective on population dynamics. *Am. Nat.* 137:S50–S66.

Purcell, J. E. 1980. Influence of siphonophore behavior upon their natural diets: Evidence for aggressive mimicry. *Science* 209:1045–47.

Pusey, A. E., and C. Packer. 1987. The evolution of sex-biased dispersal in lions. *Behaviour* 101:275–310.

Pyke, G. H. 1979a. The economics of territory size and time budget in the golden-winged sunbird. *Amer. Nat.* 114:131–45.

Quadagno, D. M., and E. M. Banks. 1970. The effect of reciprocal cross-fostering on the behavior of two species of rodents, *Mus musculus* and *Baiomys taylori ater. Anim. Behav.* 18:379–90.

Quartermus, C., and J. A. Ward. 1969. Development and significance of two motor patterns used in contacting parents by young orange chromides (*Etroplus maculatus*). *Anim. Behav.* 17:624–35.

Queller, D. 1989. The evolution of eusociality: Reproductive head starts of workers. *Proc. Nat. Acad. Sci. USA* 86:3224–26.

Quiatt, D. 1979. Aunts and mothers: Adaptive implications of allomaternal behavior in nonhuman primates. *Amer. Anthropol.* 81:310–19.

Rainey, R. C. 1959. Some new methods for the study of flight and migration. *Proc. 15th Inter. Congr. Zool.,* London. 866–70.

———. 1962. The mechanisms of desert locust swarm movements and the migration of insects. *Proc. 11th Inter. Congr. Ent.* 3:47–49.

Ralls, K. 1971. Mammalian scent marking. *Science* 171:443–49.

Ralls, K., K. Brugger, and J. Ballou. 1979. Inbreeding and juvenile mortality in small populations of ungulates. *Science* 206:1101–03.

Ralph, M. R., R. G. Foster, F. C. Davis, and M. Menaker. 1990. Transplanted suprachiasmatic nucleus determines circadian period. *Science* 247:975–78.

Ralph, M. R., and M. Menaker. 1988. A mutation of the circadian system in golden hamsters. *Science* 241:1225–27.

Randall, J. A. 1984. Territorial defense and advertisement by footdrumming in bannertail kangaroo rats (*Dipodomys spectabilis*) at high and low population densities. *Behav. Ecol. Sociobiol.* 16:11–20.

Randall, J. A. 1994. Discrimination of footdrumming signatures by kangaroo rats. *Anim. Behav.* 47:45–54.

Randall, J. A., and M. D. Matocq. 1997. Why do kangaroo rats (*Dipodomys spectabilis*) footdrum at snakes? *Behav. Ecol.* 8:404–13.

Rasika, S., B. Alvarez, and F. Nottebohm. 1999. BDNF mediates the effects of testosterone on the survival of new neurons in an adult brain. *Neuron* 22:53–62.

Ratner, S. C., and A. R. Gilpin. 1974. Habituation and retention of habituation of responses to air puff of normal and decerebrate earthworms. *J. Comp. Physiol. Psychol.* 86:911–18.

Ratner, S. C., and K. R. Miller. 1959. Classical conditioning in earthworms, *Lumbricus terrestris. J. Comp. Physiol. Psychol.* 52:102–5.

Rayor, L. S., and G. W. Uetz. 1990. Trade-offs in foraging success and predation risk with spatial position in colonial spiders. *Behav. Ecol. Sociobiol.* 27:77–85.

Rayor, L. S., and G. W. Uetz. 1993. Ontogenetic shifts within the selfish herd: Predation risk and foraging trade-offs change with age in colonial web-building spiders. *Oecologia* 95:1–8.

Read, A. F., and P. H. Harvey. 1989. Reassessment of comparative evidence for Hamilton and Zuk theory on the evolution of secondary sexual characteristics. *Nature* 339:618–20.

Real, L. A., ed. 1994. *Behavioral Mechanisms in Evolutionary Ecology.* Chicago: University of Chicago Press.

Redfield, R. J. 1994. Male mutation rates and the cost of sex for females. *Nature* 369:145–47.

Redfield, R. J. 1994. Male mutation rates and the cost of sex for females. *Nature* 369:145–47.

Reffinetti, R., and S. J. Susalka. 1997. Circadian rhythm of temperature selection in a nocturnal lizard. *Physiol. Behav.* 62:331–36.

Reinisch, J. M. 1981. Prenatal exposure to synthetic progestins increases potential for aggression in humans. *Science* 211:1171–73.

Reiter, R. J. 1974a. Pineal regulation of the hypothalamic-pituitary axis: Gonadotrophins. In *Handbook of Physiology, Endocrinology,* vol. 4, part 2, ed. E. Knobil and W. H. Sawyer. Washington, DC: American Physiological Society.

———. 1974b. Circannual reproductive rhythms in mammals related to photoperiod and pineal function: A review. *Chronobiology* 1:365–95.

———. 1980. The pineal gland: A regulator of regulators. *Prog. In Psychobiol. And Physiol. Psych.* 9:323–56.

Renn, S. C. P., J. H. Park, M. Rosbash, J. C. Hall, and P. H. Taghert. 1999. A pdf neuropeptide gene mutation and ablation of PDF neurons each cause severe abnormalities of behavioral circadian rhythms in *Drosophila. Cell* 99:791–802.

Renner, M. 1959. Über ein weiteres Versetzungsexperiment zur Analyse des Zeitsinnes und der Sonnenorientierung der Honigbiene. *Zeit. Vergleich. Physiol.* 42:449–83.

———. 1960. The contribution of the honey bees to the study of time-sense and astronomical orientation. *Cold Spr. Harb. Symp. Quant. Biol.* 25:361–67.

Renner, M. J. 1987. Experience-dependent changes in exploratory behavior in the adult rat (*Rattus norvegicus*): Overall activity level and interactions with objects. *J. Comp. Psychol.* 101:94–100.

————. 1990. Neglected aspects of exploratory and investigatory behavior. *Psychobiology* 18:16–22.

Renner, M. J. and P. J. Pierre. 1991. Patterns of development in the exploratory behavior of the Norway rat (*Rattus norvegicus*): Locomotion and interactions with objects. Manuscript submitted for publication.

Renner, M. J., and M. R. Rosenzweig. 1986. Social interactions among rats housed in grouped and enriched conditions. *Develop. Phychobiol.* 19:303–13.

Reppert, S. M., D. R. Weaver, S. A. Rivkees, and E. G. Stopa. 1988. Putative melatonin receptors in a human biological clock. *Science* 242:78–81.

Rescorla, R. A. 1968. Probability of shock in the presence and absence of CS in fear conditioning. *J. Comp. Physiol. Psychol.* 66:1–5.

————. 1988a. Behavioral studies of Pavlovian conditioning. *Ann. Rev. Neurosci.* 11:329–52.

————. 1988b. Pavlovian conditioning. It's not what you think it is. *Amer. Psychol.* 43:151–60.

Rescorla, R. A., and A. R. Wagner. 1972. A theory of Pavlovian conditioning: Variations in the effectiveness of reinforcement and nonreinforcement. In *Classical Conditioning II: Current Research and Theory,* ed. A. H. Black and W. F. Prokasy. New York: Appleton-Century-Crofts.

Ressler, R. H., R. B. Cialdini, M. L. Ghoca, and S. M. Kleist. 1968. Alarm pheromone in the earthworm *Lumbricus terrestris. Science* 161:597–99.

Rettenmeyer, C. W. 1963. Behavioral studies of army ants. *Kansas University Science Bulletin* 44:281–465.

Reyer, H. 1990. Pied kingfishers: Ecological causes and reproductive consequences of cooperative breeding. Pp. 527–58 in *Cooperative Breeding in Birds: Long-Term Studies of Ecology and Behavior* (P. Stacey and W. Koening, eds.). Cambridge: Cambridge University Press.

Ribbands, C. R. 1953. *Behaviour and Social Life of Honey Bees.* London: Bee Research Assoc.

Rice, W. R. 1987. Speciation via habitat specialization: The evolution of reproductive isolation as a correlated character. *Evol. Ecol.* 1:301–14.

Richards, M. P. 1967. Maternal behavior in rodents and lagomorphs. In *Advances in Reproductive Physiology,* vol. 2, ed. A. McLaren. New York: Academic Press.

Richardson, W. J. 1971. Spring migration and weather in eastern Canada: A radar study. *Amer. Birds* 25:684–90.

————. 1972. Autumn migration and weather in eastern Canada. *Amer. Birds* 26:10–17.

Richter, C. P. 1922. A behavioristic study of the activity of the rat. *Comp. Psychol. Monogr.* 1:1–55.

Riechert, S. E. 1984. Games spiders play III: Cues underlying context-associated changes in agonistic behaviour. *Anim. Behav.* 32:1–15.

Riechert, S. E. 1998. Game theory and animal contests. Pp. 64–93 in *Game Theory and Animal Behavior* (L. A. Dugatkin and H. K. Reeve, eds.). New York: Oxford University Press.

Riechert, S. E. 1999. The use of behavioral ecotypes in the study of evolutionary processes. Pp. 3–32 in *Geographic Variation in Behavior: Perspective on Evolutionary Mechanisms* (S. A. Foster and J. A. Endler, eds.). New York: Oxford University Press.

Riechert, S. E., and A. V. Hedrick. 1990. Levels of predation and genetically based anti-predator behaviour in the spider, *Agelenopsis aperta. Anim. Behav.* 40:679–87.

Ringo, J. M. 1976. A communal display in Hawaiian *Drosophila* (Diptera: Drosophilidae). *An Entomol. Soc. Am.* 69:209–14.

————. 1978. The development of behavior in *Drosophila.* In *Development of Behavior,* ed. G. Burghardt and M. Bekoff. New York: Garland STPM Press.

Rivier, C., J. Rivier, and W. Vale. 1986. Stress-induced inhibition of reproductive functions: Role of endogenous corticotropin-releasing factor. *Science* 231:607–9.

Roberts, E. P., Jr., and P. D. Weigl. 1984. Habitat preference in the dark-eyed junco (*Junco hyemalis*): The role of photoperiod and dominance. *Anim. Behav.* 32:709–14.

Roberts, S. K. 1966. Circadian activity rhythms in cockroaches. III. The role of endocrine and neural factors. *J. Cell Physiol.* 67:473–86.

————. 1974. Circadian rhythms in cockroaches. Effects of the optic lobe lesions. *J. Comp. Physiol.* 88:21–30.

Roberts, W. A. 1981. Retroactive inhibition in rat spatial memory. *Anim. Learn. Behav.* 9:566–74.

Roberts, G. 1996. Why individual vigilance declines as group size increases. *Anim. Behav.* 51:1077–86.

Roberts, G., and T. Sherratt. 1998. Development of cooperative relationships through increasing investment. *Nature* 394:175–79.

Robertson, I., W. Robertson, and B. Roitberg. 1998. A model of mutual tolerance and the origin of communal associations between unrelated females. *J. Insect Behav.* 11:265–86.

Robinson, B. W., M. Alexander, and G. Browne. 1969. Dominance reversal resulting from aggressive responses evoked by brain telestimulation. *Physiol. Behav.* 4:749–52.

Robinson, D., ed. 1998. *Neurobiology.* Berlin: Springer.

Roeder, K. D. 1967. *Nerve Cells and Insect Behavior.* Cambridge: Harvard University Press.

Roeder, K. D. 1970. Episodes in insect brains. *Amer. Scientist* 58:378–892.

Roeder, K. D., and A. E. Treat. 1961. The detection and evasion of bats by moths. *Amer. Scientist* 49:135–48.

Roitblat, H. R. 1987. Introduction to Comparative Cognition. New York: W. H. Freeman.

Roitblat, H. L., and J. A. Meyer, eds. 1995. *Comparative Approaches to Cognitive Science.* Cambridge, MA: MIT Press.

Romero, L. M., K. K. Soma, K. M. O'Reilly, R. Suydam, and J. C. Wingfield. 1998. Hormones and territorial behavior during breeding in snow buntings (*Plectrophenax nivalis*): An arctic-breeding songbird. *Horm. Behav.* 33:40–47.

Rood, J. P. 1983. The social system of the dwarf mongoose. In *Recent Advances in the Study of Mammalian Behavior,* ed. J. F. Eisenberg and D. G. Kleiman. Special Pub. No. 7, Amer. Soc. Mammal.

Rose, G. J., R. Zelick, and S. Rand. 1988. Auditory processing of temporal information in a neotropical frog is independent of signal intensity. *Ethology* 77:330–36.

Rose, R. N., I. S. Bernstein, and T. P. Gorden. 1975. Consequences of social conflict on plasma testosterone levels in rhesus monkeys. *Psychosomatic Med.* 37:50–61.

Roseler, P. F., I. Roseler, and A. Strambi. 1986. Roles of ovaries and ecdysteroid in dominance hierarchy establishment among foundresses of the primitively social wasp, *Polistes gallicus. Behav. Ecol. Sociobiol.* 18:9–13.

Rosenblatt, J. S. 1967. Nonhormonal basis for maternal behavior in the rat. *Science* 156:1512–14.

————. 1970. Views on the onset and maintenance of maternal behavior in the rat. In *Development and Evolution of Behavior,* ed. L. R. Aronson et al. San Francisco: Freeman.

Rosenblatt, J. S., and L. R. Aronson. 1958. The decline in sexual behavior of male cats after castration with special reference to the role of prior sexual experience. *Behaviour* 12:285–338.

Rosenblatt, J. S., and D. S. Lehrman. 1963. Maternal behavior of the laboratory rat. In *Maternal Behavior in Mammals,* ed. H. L. Rheingold. New York: Wiley.

Rosenblatt, J. S., H. I. Siegel, and A. D. Mayer. 1979. Progress in the study of maternal behavior in the rat: Hormonal, nonhormonal, sensory, and developmental aspects. *Adv. Stud. Behav.* 10:226–311.

Rosenblatt, J. S., and K. Ceus. 1998. Estrogen implants in the medial preoptic area stimulate maternal behavior in male rats. *Horm. Behav.* 33:23–30.

Rosenblum, L. A., and I. C. Kaufman. 1967. Laboratory observations of early mother-infant relations in pigtail and bonnet macaques. In *Social Communication among Primates,* ed. S. A. Altmann. Chicago: University of Chicago Press.

———. 1968. Variations in infant development and response to maternal loss in monkeys. *Amer. J. Orthopsych.* 38:418–26.

Rosenheim, J. A. 1993. Comparative and experimental approaches to understanding insect learning. Pp. 273–307 in *Insect Learning: Ecological and Evolutionary Perspectives* (D. R. Papaj and A. C. Lewis, eds.). New York: Chapman and Hall.

Rowan, W. 1925. Relation of light to bird migration and development changes. *Nature* 115:494–95.

Rosenwasser, A. M., and N. T. Adler. 1986. Structure and function in circadian timing systems: Evidence for multiple coupled circadian oscillators. *Neurosci. Biobehav. Rev.* 10:431–48.

Rosenzweig, M. L. 1985. Some theoretical aspects of habitat selection. In *Habitat Selection in Birds,* ed. M. L. Cody. New York: Academic Press.

———. 1990. Do animals choose habitats? In *Interpretation and Explanation in the Study of Animal Behavior,* ed. M. Bekoff and D. Jamieson. Boulder, CO: Westview Press.

———. 1991. Habitat selection and population interactions: The search for mechanism. *Amer. Nat.* 137:S5–S28.

Ross, D. M. 1965. Complex and modifiable behavior patterns in *Calliactis* and *Stomphia. Amer. Zool.* 5:573–80.

Ross, R. M., G. S. Losey, and M. Diamond. 1983. Sex change in a coral reef fish: Dependence of stimulation and inhibition on relative size. *Science* 221:574–75.

Rothenbuler, W. C. 1964a. Behaviour genetics of nest cleaning in honeybees. I. Responses of four in-bred lines to disease-killed brood. *Anim. Behav.* 12:578–83.

———. 1964b. Behaviour genetics of nest cleaning in honeybees. IV. Responses of F-1 and backcross generations to disease-killed brood. *Amer. Zool.* 4:111–23.

Rottman, S. J., and C. T. Snowdon. 1972. Demonstration and analysis of an alarm pheromone in mice. *J. Comp. Physiol. Psychol.* 81:483–90.

Rowell, T. I. 1974. The concept of social dominance. *Behav. Biol.* 11:131–54.

Rozin, P. 1968. Specific aversions and neophobia resulting from vitamin deficiency or poisoning in half wild and domestic rats. *J. Comp. Physiol. Psychol.* 66:82–88.

Rubenstein, D., and M. Hohmann. 1989. Parasites and social behavior of island feral horses. *Oikos* 55:312–20.

Rumbaugh, D. M. 1968. The learning and sensory capacities of the squirrel monkey in phylogenetic perspective. In *The Squirrel Monkey,* ed. L. A. Rosenblum and R. W. Cooper. New York: Academic Press.

Rumbaugh, D. M. Ed. 1977. *Language Learning by a Chimpanzee. The Lana Project.* New York: Academic Press.

Rumbaugh, D. M., and T. V. Gill. 1976. The mastery of language-type skills by the chimpanzee (Pan). In *Origins and Evolution of Language and Speech,* eds. S. R. Harnard. H. D. Steklis, and J. Lancaster. Annals of the New York Academy of Sciences. 280:562–78.

Rusak, B., and I. Zucker. 1979. Neural regulation of circadian rhythms. *Physiol. Rev.* 59:449–526.

Rushforth, N. D. 1973. Behavioral modifications in coelenterates. In *Invertebrate Learning,* vol. 1., ed. W. C. Corning, J. A. Dyal, and A. O. D. Willows. New York: Plenum.

Russell, P. F., and T. R. Rao. 1942. On relation of mechanical obstruction and shade to ovipositing of *Anopheles culifacies. J. Exp. Zool.* 99:303–29.

Rust, C. C. 1965. Hormonal control of pelage cycles in the short-tailed weasel (*Mustela erminea bangsi*). *Gen. Comp. Endocr.* 5:222–31.

Rust, C. C., and R. K. Meyer. 1969. Coat color, molt, and testis size in male short-tailed weasels treated with melatonin. *Science* 165:921–22.

Ryan, M. J., and W. Wilczynski. 1988. Coevolution of sender and receiver: Effect on local mate preference in cricket frogs. *Science* 240:1786–88.

Sade, D. S. 1965. Some aspects of parent-offspring and sibling relations in a group of rhesus monkeys, with a discussion of grooming. *Amer. J. Phys. Anthrop.* 23:1–17.

———. 1967. Determinants of dominance in a group of free-ranging rhesus monkeys. In *Social Communication among Primates,* ed. S. Altman. Chicago: University of Chicago Press.

Sade, D. S., K. Cushing, P. Cushing, J. Dunaif, A. Figueroa, J. R. Kaplan, C. Laver, D. Rhodes, and J. Schneider. 1976. Population dynamics in relation to social structure on Cayo Santiago. *Yearbook Phys. Anthrop.* 20:253–62.

Sadowski, J. A., A. J. Moore, and E. D. Brodie. 1999. The evolution of nuptial gifts in a dance fly, *Empis snoddyi* (Diptera: Empididae): Bigger isn't always better. *Behav. Ecol. Sociobiol.* 45:161–66.

Saigusa, M., and P. Kawagoye. 1997. Circatidal rhythm in an intertidal crab, *Hemigraspus sanguineus:* Synchrony with unequal tide height and involvement of a light-response mechanism. *Mar. Biol.* 129:87–96.

Sale, P. F. 1970. A suggested mechanism for habitat selection by the juvenile manini *Acanthurus triostegus sandvicensis* Streets. *Behavior* 35:27–44.

Salzen, E. A., and C. C. Meyer. 1967. Imprinting: Reversal of a preference established during the critical period. *Nature* 215:785–86.

Salzen, E. A., and W. Sluckin. 1959. The incidence of the following response and the duration of responsiveness in domestic fowl. *Anim. Behav.* 7:172–79.

Sanchez-Vazquez, F. J., J. A. Madrid, S. Zamora, and M. Tabata. 1997. Daily cycles in plasma and ocular melatonin in demond-ped sea bass *Dicentrarchus labrax* L. *J. Comp. Physiol. A* 181:121–32.

Sanders, G. D. 1973. The cephalopods. In *Invertebrate Learning,* vol. 3., ed. W. C. Corning, J. A. Dyal, and A. O. D. Willows. New York: Plenum.

Santschi, F. 1911. Le mechanisme d'orientation chez les fourmis. *Rev. Suisse Zool.* 19:117–34.

Sapolsky, R. M. 1990. Adrenocortical function, social rank, and personality among wild baboons. *Biol. Psychiatry* 28:862–85.

———. 1991. Testicular function, social rank, and personality among wild baboons. *Psychoneuroendocrinology* 16:281–93.

———. 1992. Cortisol concentrations and the social significance of rank instability among wild baboons. *Psychoneuroendocrinology* 17:701–9.

Sarnat, H. B., and M. G. Netsky. 1974. *Evolution of the Nervous System.* New York: Oxford University Press.

Sato, T., and H. Kawamura. 1984. Effects of bilateral suprachiasmatic nucleus lesions on the circadian rhythm of a diurnal rodent, the Siberian chipmunk (*Eutamias sibiricus*). *J. Comp. Physiol. A.* 155:745–52.

Saudou, F., D. A. Amara, A. Diericht, M. LeMeurru, S. Ramboz, L. Segu, M. C. Buhot, and R. C. L. Hen. 1994. Enhanced aggressive behavior in mice lacking 5-HT-1β receptor. *Science* 265:1875–78.

Sauer, E. G. F. 1957. Die Sternenorientierung nächtlich ziehender Grasmücken (*Sylvia atricapilla, borin* und *curruca*). *Zeit. Tierpsychol.* 14:29–70.

———. 1963. Migration habits of golden plovers. *Proc. Inter. Ornithol. Congr., 13th.* 454–67.

Saunders. D. S. 1978. Internal and external coincidence and the apparent diversity of photoperiodic clocks in the insects. *J. Comp. Physiol. A* 127:197–207.

Savage-Rumbaugh, E. S. 1986. *Ape Language: From Conditioned Response to Symbol.* New York: Columbia University Press.

———. 1987. A new look at ape language: Comprehension of vocal speech and syntax. *Nebr. Symp. Motiv.* 35:201–55.

Savage-Rumbaugh, E. S., and K. E. Brakke. 1990. Animal language: Methodological and interpretive issues. In *Interpretation and Explanation in the Study of Animal Behavior,* ed. M. Bekoff and D. Jamieson. Boulder, CO: Westview Press.

Savage-Rumbaugh, E. S., K. McDonald, R. A. Sevcik, W. D. Hopkins, and E. Rubert. 1986. Spontaneous symbol acquisition and communicative use by two pygmy chimpanzees. *J. Exp. Psychol.: General* 115:211–35.

Savage-Rumbaugh, E. S., D. M. Rumbaugh, and S. Boysem. 1980. Do apes use language? *Amer. Scientist* 68:49–61.

Sawyer, T. G., K. V. Miller, and R. L. Marchinton. 1994. Patterns of urination and rub-urination in female white-tailed deer. *J. Mammal.* 74:477–79.

Sayigh, L. S., P. L. Tyack, R. S. Wells, A. R. Solow, M. D. Scott, and A. B. Irvine. 1998. Individual recognition in wild bottlenose dolphins: A field test using playback experiments. *Anim. Behav.* 57:41–50.

Scapini, F. 1986. Inheritance of direction finding in sandhoppers. In *Orientation in Space,* ed. G. Beugnon, 111–19. Toulouse, France: Editions Pivat.

Schiavi, R. C., A. Theilgaard, D. R. Owen, and D. White. 1984. Sex chromosome anomalies, hormones, and aggressivity. *Arch. Gen. Psychiatr.* 41:93–99.

Schaller, G. B. 1972. *The Serengeti Lion: A Study of Predator-Prey Relations.* Chicago: University of Chicago Press.

Schein, M. W., and E. B. Hale. 1959. The effect of early social experience on male sexual behavior of androgen injected turkeys. *Anim. Behav.* 7:189–200.

Schenkel, R. 1956. Zur Deutung der Balzleisturgen einiger Phasianiden und Tetraoniden. *Ornithologische Beobachter* 53:182–201.

Schiller, P. H. 1996. On the specificity of neurons and visual areas. *Behav. Brain Res.* 76:21–35.

Schjelderup-Ebbe, T. 1922. Beitrage zur Sozialpsychologie des Haushuhns. *Zeit. Psychol.* 88:225–52.

Schleidt, W. 1974. "How fixed is the fixed action pattern?" *Zeit. Tierpsychol.* 36:184–211.

Schmacher, M., J. C. Hendrick, and J. Balthazart. 1989. Sexual differentiation in quail: Critical period and hormonal specificity. *Horm. Behav.* 23:130–49.

Schmidt-Koenig, K., and J. B. Phillips. 1978. Local anethesia of the olfactory membrane and homing in pigeons. In *Animal Migration, Navigation, and Homing,* ed. K. Schmidt-Koenig and W. T. Keeton. New York: Springer-Verlag.

Schmidt-Koenig, K., and H. J. Schlichte. 1972. Homing in pigeons with reduced vision. *Proc. Nat. Acad. Sci. USA* 69:2446–47.

Schmidt-Koenig, K., and C. Walcott. 1973. Flugwege und Verbleib von Brieftauben mit getrübten Haftschalen. *Naturwissenschaften* 60:108–9.

Schneider, D. 1974. The sex-attractant receptor of moths. *Sci. Amer.* 231:28–35.

Schneiderman, H. A. 1972. Insect hormones and insect control. In *Insect Juvenile Hormones,* eds. J. Menn and M. Beroza. New York: Academic Press.

Schnierla, T. C. 1950. The relationship between observation and experimentation in the field study of behavior. *Ann. N.Y. Acad. Sci.* 51:1022–44.

Schnierla. T. C., and G. Piel. 1948. The army ant. *Sci. Amer.* 178:16–23.

Schoech, S. J., R. L. Mumme, and M. C. Moore. 1991. Reproductive endocrinology and mechanisms of breeding inhibition in cooperatively breeding Florida USA scrub jays *Aphelocoma coerulescens. Condor* 93:354–64.

Schoener, T. W., and C. A. Toft. 1983. Spider populations: Extraordinarily high densities on islands without top predators. *Science* 219:1353–55.

Schoener, T. W. 1971. Theory of feeding strategies. *Ann. Rev. Ecol. Syst.* 2:369–404.

Schoener, T. W. 1979. Generality of the size-distance relation in models of optimal feeding. *Am. Nat.* 114:902–14.

Schröder, J. H., and M. Sund. 1984. Inheritance of water-escape performance and water-escape learning in mice. *Behav. Genet.* 14:221–34.

Schuler, W., and E. Hesse. 1985. On the function of warning coloration: A black and yellow pattern inhibits prey attack by naive domestic chicks. *Behav. Ecol. Sociobiol.* 16:249–56.

Schultz, F. 1965. Sexuelle Prägung bei Anatiden. *Zeit. Tierpsychol.* 22:50–103.

Schumacher, M., J. C. Hendrick, and J. Balthazart. 1989. Sexual differentiation in quail: Critical period and hormonal specificity. *Horm. Behav.* 23:130–49.

Schusterman, R. J., and K. Krieger. 1986. Artificial language comprehension and size transposition by a California sea lion (*Tursiops californianus*). *J. Comp. Psychol.* 100:348–55.

Schusterman, R. J., and D. Kastak. 1998. Functional equivalence in a California sea lion: Relevance to animal and social communicative interaction. *Anim. Behav.* 55:1087–95.

Schusterman, R. J., C. J. Reichmuth, and D. Kastak. 2000. How animals classify friends and foes. *Curr. Dir. Psychol. Sci.* 9:1–6.

Schwagmeyer, P. L. 1988. Scramble-competition polygyny in an asocial mammal: Male mobility and mating success. *Amer. Nat.* 131:885–92.

Schwartz, J., and H. Gerhardt. 1998. The neuroethology of frequency preferences in the spring peeper. *Anim. Behav.* 56:55–69.

Scott, D. 1978. Identification of individual Bewick's swans by bill patterns. In *Animal Marking,* ed. B. Stonehouse, 160–68. Baltimore: University Press.

Scott, J. P. 1966. Agonistic behavior of mice and rats: A review. *Amer. Zool.* 6:683–701.

———. 1972. *Animal Behavior.* 2d ed. Chicago: University of Chicago Press.

———. 1975. Violence and the disaggregated society. *Aggressive Behav.* 1:235–60.

———. 1976. The control of violence: Human and nonhuman societies compared. In *Violence in Animal and Human Societies,* ed. A. Neal. Chicago: Nelson-Hall.

Scott, J. P., and E. Fredericson. 1951. The causes of fighting in mice and rats. *Physiol. Zool.* 24:273–309.

Scott, J. P., and J. L. Fuller. 1965. *Genetics and the Social Behavior of the Dog.* Chicago: University of Chicago Press.

Scott, M. P. 1990. Brood guarding and the evolution of male parental care in burying beetles. *Behav. Ecol. Sociobiol.* 26:31–40.

Scott, M. 1997. Reproductive dominance and differential ovicide in the communally breeding burying beetle *Nicrophorus tomentosus. Behav. Ecol. Sociobiol.* 40:313–20.

Searcy, W. A. 1984. Song repertoire size and female preferences in song sparrows. *Behav. Ecol. Sociobiol.* 14:281–86.

Seehausen, O., J. van Alphen, F. Witte. 1997. Cichlid fish diversity threatened by eutrophication that curbs sexual selection. *Science* 277:1808–11.

Seeley, T. 1983. The ecology of temperate and tropical honeybees. *Am. Sci.* 71:264–72.

Sehgal, A. 1995. Genetic dissection of the circadian clock: A timeless story. *Sem. Neurosci.* 7:27–35.

Seligman, M. E. P. 1970. On the generality of the laws of learning. *Psychol. Rev.* 77:406–18.

Selye, H. 1950. *Stress.* Montreal: Acta.

Semm, P. and C. Demaine. 1986. Neurophysiological properties of magnetic cells in the pigeon's visual system. *J. Comp. Physiol.* 59:619–25.

Seyfarth, R. M., D. L. Cheyney, and P. Marler. 1980. Monkey responses to three different alarm calls: Evidence of predator classification and semantic communication. *Science* 210:801–3.

Shaikh, M. B., and A. Siegel. 1997. The role of substance P receptors in amygdaloid modulation of aggressive behavior in the cat. Pp. 29–50 in *Aggression: Biological, Developmental, and Social Perspectives* (S. Feshbach and J. Zagrodzka, eds.). New York: Plenum Press.

Shannon, C. E., and W. Weaver. 1949. *Mathematical Theory of Communication.* Urbana: University of Illinois Press.

Shapiro, D. Y. 1979. Social behavior, group structure, and the control of sex reversal in hermaphroditic fish. In *Advances in the Study of Behavior,* eds. J. S. Rosenblatt, R. A. Hinde, and C. Beer. New York: Academic Press.

Sharma, V. K., and M. K. Chandrashekaran. 1997. Rapid phase resetting of a mammalian circadian rhythm by brief light pulses. *Chronobiol. Int.* 14:537–48.

Sharp, F. R., J. S. Kauer, and G. M. Shepherd. 1975. Local sites of activity-related glucose metabolism in rat olfactory bulb during olfactory stimulation. *Brain. Res.* 98:596–600.

Sharp, P. E., H. T. Blair, D. Etkin, and D. B. Tzanetos. 1995. Influences of vestibular and visual motion information on the spatial firing patterns of hippocampal place cells. *J. Neuroscience* 15:173–89.

Shell, W. F., and A. J. Riopelle. 1957. Multiple discrimination learning in raccoons. *J. Comp. Physiol. Psychol.* 50:585–87.

Shelly, T. E., and T. S. Whittier. 1997. Lek behavior of insects. Pp. 273–93 in *The Evolution of Mating Systems in Insects and Arachnids* (J. C. Choe and B. J. Crespi, eds.). Cambridge: Cambridge University Press.

Sheridan, M., and R. H. Tamarin. 1988. Space use, longevity, and reproductive success in meadow voles. *Behav. Ecol. Sociobiol.* 22:85–90.

Sherman, P. W. 1977. Nepotism and the evolution of alarm calls. *Science* 197:1246–53.

———. 1981. Kinship, demography, and Belding's ground squirrel nepotism. *Behav. Ecol. Sociobiol.* 8:251–59.

Sherman, P. W., J. U. M. Jarvis, and R. D. Alexander, eds. 1991. *The Biology of the Naked Mole-rat.* Princeton, NJ: Princeton University Press.

Sherry, D. F., and A. L. Vaccarino. 1989. The hippocampus and memory for food caches in blackcapped chickadees. *Behav. Neurosci.* 103:308–18.

Sherry, D. F., A. L. Vaccarino, K. Buckenham, and R. S. Herz. 1989. The hippocampal complex of food-storing birds. *Brain Behav. Evol.* 34:308–17.

Sherry, D. F. 1998. The ecology and neurobiology of spatial memory. Pp. 261–96 in *Cognitive Ecology* (R. Dukas, ed.). Chicago: University of Chicago Press.

Shettleworth, S. J. 1972. Constraints on learning. *Adv. Stud. Behav.* 4:1–68.

———. 1984. Learning and behavioural ecology. In *Behavioural Ecology: An Evolutionary Approach,* 2d ed., eds. J. R. Krebs and N. B. Davies. Oxford: Blackwell Scientific Publications. Ltd.

Shettleworth, S. J. 1995. Comparative studies of memory in food storing birds: From the field to the Skinner box. In *Behavioral Brain Research in Naturalistic and Semi-Naturalistic Settings.* (E. Alleva, A. Fasolo, H. P. Lipp, L. Nadel, and L. Ricceri, eds.). Dordrecht, Netherlands: Kluwer Academic.

Shettleworth, S. J. 1998. *Cognition, Evolution, and Behavior.* New York: Oxford University Press.

Shettleworth, S. J., and R. R. Hampton. 1998. Adaptive specializations of spatial cognition in food-storing birds? Approaches to testing a comparative hypothesis. Pp. 65–98 in *Animal Cognition in Nature* (R. P. Balda, I. M. Pepperberg, and A. C. Kamil, eds.). San Diego: Academic Press.

Shuster, S. M. 1992. The reproductive behavior of alpha male, beta male, and gamma male morphs in *Paracerceis sculpta,* a marine isopod crustacean. *Behaviour* 121:231–58.

Shuster, S. M., and C. Sassaman. 1997. Genetic interaction between male mating strategy and sex ratio in a marine isopod. *Nature* 388:373–77.

Shuster, S. M., and M. J. Wade. 1991. Equal mating success among male reproductive strategies in a marine isopod. *Nature* 350:608–10.

Shields, W. M. 1982. *Philopatry, Inbreeding, and the Evolution of Sex.* Albany: State University of New York Press.

Shockley, W. B. 1969. Human quality problems and research taboos. In *New Concepts and Direction in Education,* ed. J. A. Pintus. Greenwich, CT: Educational Records Bureau.

Sigurdson, J. E. 1981a. Automated discrete-trials techniques of appetitive conditioning in honeybees. *Behav. Res. Meth. Instr.* 13:1–10.

———. 1981b. Measurement of consummatory behavior in honeybees. *Behav. Res. Meth. Instr.* 13:308–10.

Silberglied, R. E., A. Aiello, and D. M. Windsor. 1980. Disruptive coloration in butterflies: Lack of support in *Anartia fatima. Science* 209:617–19.

Silberglied, R. E., J. G. Shepherd, and J. L. Dickinson. 1984. Eunuchs: The role of apyrene sperm in Lepidoptera? *Amer. Nat.* 123:255–65.

Silver, R. 1978. The parental behavior of ring doves. *Amer. Sci.* 66:209–15.

Simmons, L. W. 1994. Courtship role reversal in bush crickets: Another role for parasites? *Behav. Ecol.* 5:259–66.

Simonds, P. E. 1965. The bonnet macaque in South India. In *Primate Behavior: Field Studies of Monkeys and Apes,* ed. I. DeVore. New York: Holt, Rinehart and Winston.

Simons, E. L. 1972. *Primate Evolution.* New York: Macmillan.

Simpson, J., I. B. M. Riedel, and N. Wilding. 1968. Invertase in the hypopharyngeal glands of the honeybee. *J. Apicult. Res.* 7:29–36.

Simpson, G. G. 1944. *Tempo and Mode in Evolution.* New York: Columbia University Press.

Singh, M., E. M. Meyer, W. J. Millard, and J. W. Simpkins. 1994. Ovarian steroid deprivation results in a reversible learning impairment and compromised cholinergic function in female Sprague-Dawley rats. *Brain Res.* 644:305–12.

Skinner, B. F. 1938. *Behavior of Organisms: An Experimental Analysis.* New York: Appleton-Century-Crofts.

———. 1953. *Science and Human Behavior.* New York: Macmillan.

Sklorz, B., and T. Volz. 1990. Innate and learned preference in food selection of the chicken (*Gallus domesticus*). *J. für Ornithol.* 131:157–60.

Skutelsky, O. 1996. Predation risk and state-dependent foraging in scorpions: Effects of moonlight on foraging in the scorpion *Buthus occitanus. Anim. Behav.* 49:57–.

Slater, P. J. B. 1983. The study of communication. In *Animal Behaviour,* ed. T. R. Halliday and J. B. Slater. New York: Freeman.

———. 1989. Bird song learning: Causes and consequences. *Ethol. Ecol. Evol.* 1:19–46.

Smale, L., K. E. Holekamp, M. Weldele, L. G. Frank, and S. E. Glickman. 1995. Competition and cooperation between littermates in the spotted hyaena, *Crocuta crocuta. Anim. Behav.* 50:671–82.

Smetzer, B. 1969. Night of the palolo. *Nat. Hist.* 78:64–71.

Smith, A. 1776. *Wealth of Nations.* New York: Modern Library. 1937.

Smith, C. C. 1968. The adaptive nature of social organization in the genus of tree squirrels *Tamiasciurus. Ecol. Monogr.* 38:31–63.

Smith, D. G. 1976. An experimental analysis of the function of red-winged blackbird song. *Behaviour* 56:136–56.

———. 1981. The association between rank and reproductive success of male rhesus monkeys. *Amer. J. Primat.* 1:83–90.

Smith, J. E. 1950. Some observations on the nervous mechanisms underlying the behavior of starfish. *Symp. Soc. Exp. Biol. Med.* 4:196–220.

Smith, M., M. Manlove, and J. Joule. 1978. Spatial and temporal dynamics of the genetic organization of small mammal populations. In *Populations of Small Mammals under Natural Conditions,* ed. D. Snyder. Pittsburgh: University of Pittsburgh.

Smith, N. G. 1966. Evolution of some arctic gulls (*Larus*): An experimental study of isolating mechanisms. *Ornith. Monogr.* 4:1–99.

Smith, R. E., and R. J. Hock. 1963. Brown fat: Thermogenic effector of arousal in hibernators. *Science* 149:199–200.

Smith, S. M. 1975. Innate recognition of coral snake pattern by a possible avian predator. *Science* 187:759–60.

Smith, W. J. 1984. *Behavior of Communicating.* 2d ed. Cambridge: Harvard University Press.

Smith, R. L. 1997. Evolution of parental care in giant water bugs (Heteroptera: Belostomatidae). Pp. 116–49 in *Competition and Cooperation Among Insects and Arachnids. II. Evolution of Sociality* (J. C. Choe and B. J. Crespi, eds.). Cambridge: Cambridge University Press.

Smith, R. S., C. Guilleminault, and B. Efron. 1997. Circadian rhythms and enhanced athletic performance in the National Football League. *Sleep* 20:362–65.

Smotherman, W. P., L. S. Richards, and S. R. Robinson. 1984. Techniques for observing fetal behavior *in utero:* A comparison of chemomyelotomy and spinal transection. *Devel. Psychobiol.* 17:661–74.

Smythe, N. 1977. The function of mammalian alarm advertising: Social signals or pursuit invitation? *Amer. Nat.* 111:191–94.

Sneddon, L. U., A. C. Taylor, and F. A. Huntingford. 1999. Metabolic consequences of agonistic behaviour: Crab fights in declining oxygen tensions. *Anim. Behav.* 57:353–63.

Snodgrass, R. E. 1956. *Anatomy of the Honeybee.* Ithaca, NY: Cornell University Press.

Snyder, N., and H. Snyder. 1970. Alarm response of *Diadema antillarium. Science* 168:276–78.

Sober, E., and D. S. Wilson. 1998. *Unto Others: The Evolution and Psychology of Unselfish Behavior.* Cambridge: Harvard University Press.

Sokoloff L., M. Reivich, C. Kennedy, M. Des Rosiers, C. Patlak, K. Pettigrew, O. Sakurada, and M. Shinohara. 1972. The [^{14}C]-deoxyglucose method for the measurement of local cerebral glucose utilization: Theory, procedure and normal values in the conscious and anesthetized rat. *J. Neurochem.* 28:897–916.

Sokolove, P. G. 1975. Localization of the cockroach optic lobe circadian pacemaker with microlesions. *Brain Res.* 87:13–21.

Sontag, L. W., and R. F. Wallace. 1935. The movement response of the human fetus to sound stimuli. *Child Dev.* 6:253–58.

Sordahl, T. A. 1981. Sleight of wing. *Nat. Hist.* 90(8):42–49.

Southern, W. E. 1969. Orientation behavior of ring-billed gull chicks and fledglings. *Condor* 71:418–25.

———. 1972. Magnets disrupt the orientation of juvenile ring-billed gulls. *BioScience* 22:476–79.

Southwick, C. H. 1955. The population dynamics of confined house mice supplied with unlimited food. *Ecology* 36:212–25.

———. 1967. An experimental study of intragroup agonistic behavior in rhesus monkeys, *Macaca mulatta. Behaviour* 28:182–209.

Southwick, C. H., and L. H. Clark. 1968. Interstrain differences in aggressive and exploratory activity of inbred mice, Part A. *Comm. Behav. Biol.* 1:49–59.

Southwick, C. H., M. F. Siddiqi, M. Y. Farooqui, and B. C. Pal. 1974. Xenophobia among freeranging rhesus groups in India. In *Primate Aggression Territoriality and Xenophobia,* ed. R. L. Halloway. New York: Academic Press.

Southwood, T. R. E. 1978. Marking invertebrates. In *Animal Marking,* ed. B. Stonehouse, 102–5. Baltimore: University Park Press.

Spalding, D. A. 1873. Instinct with original observations on young animals. *Macmillan's* 27:282–93. Reprinted in *Brit. J. Anim. Behav.* 2:2–11.

Spencer, R. 1978. Ringing and related durable methods of marking birds. In *Animal Marking,* ed. B. Stonehouse, 43–53. Baltimore: University Park Press.

Sperry, R. A., J. S. Stamm, and N. Miner. 1956. Relearning tests for interoccular transfer following division of optic chiasma and corpus callosum in rats. *J. Comp. Physiol. Psychol.* 49:529–33.

Spieth, H. T. 1966. Courtship behavior of endemic Hawaiian *Drosophila. Univ. Texas Publ.* 6615:245–313.

Stacey, P., and J. Ligon. 1987. Territory quality and dispersal options in the Acorn woodpecker, and a challenge to the habitat-saturation model of cooperative breeding. *Am. Nat.* 130:654–76.

Staddon, J. E. R. 1983. *Adaptive Behavior and Learning.* New York: Cambridge University Press.

Stamps, J. A. 1983. Territoriality and the defence of predator-refuges in juvenile lizards. *Anim. Behav.* 31:857–70.

Stamps, J. 1995. Motor learning and the value of familiar space. *Am. Nat.* 146:41–58.

Stanley, S. M. 1981. *The New Evolutionary Timetable.* New York: Basic Books.

Stanley, S. M. 1993. *Earth and Life Through Time,* 2nd ed. New York: W. H. Freeman.

Starkey, S. J., M. P. Walker, I. J. M. Beresford, and R. M. Hagan. 1995. Modulation of the rat suprachiasmatic circadian clock by melatonin in vitro. *Neuroreport* 14:1947–51.

Staub, N. L. 1993. Intraspecific agonistic behavior of the salamander *Aneides flavipunctatus* (Amphibia: Plethodontidae) with comparisons to other plethodontid species. *Herpetologica* 49:271–82.

Stephan, F. K., J. M. Swann, and C. L. Sisk. 1979a. Anticipation of 24-hour feeding schedules in rats with lesions of the suprachiasmatic nucleus. *Behav. Neural Biol.* 25:346–63.

———. 1979b. Entrainment of circadian rhythms by feeding schedules in rats with suprachiasmatic lesions. *Behav. Neural Biol.* 25:545–54.

Stephan, F. K., and I. Zucker. 1972. Circadian rhythms in drinking behavior and locomotor activity of rats are eliminated by hypothalamic lesions. *Proc. Nat. Acad. Sci. USA* 69:1583–86.

Stephens, D. W., and J. R. Krebs. 1986. *Foraging Theory.* Princeton, NJ: Princeton University Press.

Stephens, P., and W. Sutherland. 1999. Consequences of the Allee effect for behaviour, ecology and conservation. *TREE* 14:401–4.

Stewart, M. M., and F. H. Pough. 1983. Population density of tropical forest frogs: Relation to retreat sites. *Science* 221:570–72.

Stewart, B. S., and R. L. DeLong. 1995. Double migrations of the northern elephant seal, *Mirounga angustirostris. J. Mamm.* 76:196–205.

Stierle, I. E., M. Getman, and C. M. Comer. 1994. Multisensory control of escape in the cockroach *Periplaneta americana. J. Comp. Physiol.* 174:1–11.

Stoddard, P. K. 1999. Predation enhances complexity in the evolution of electric fish signals. *Nature* 400: 254–56.

Stoddard, P. K., M. D. Beecher, and M. S. Willis. 1988. Response of territorial male song sparrows to song types and variations. *Behav. Ecol. Sociobiol.* 22:125–30.

Stoddard, P. K., M. D. Beecher, C. L. Horning, and M. S. Willis. 1990. Strong neighbor-stranger discrimination in song sparrows. *Condor* 92:1051–56.

Stoddard, P. K., M. D. Beecher, C. L. Horning, and S. E. Campbell. 1991. Recognition of individual neighbors by song in the song sparrow, a species with song repertoires. *Behav. Ecol. Sociobiol.* 29:211–15.

Stonehouse, B., ed. 1978. *Animal Marking.* Baltimore: University Park Press.

Strassman, J. E., Y. Zhu, and D. Queller. 2000. Altruism and social cheating in the social amoeba *Dictyostelium discoideum. Nature* 408:965–67.

Strickberger, M. W. 1996. *Evolution,* 2nd edition. Sudbury, MA: Jones and Bartlett Publishers.

Strier, K. B. 1992. *Faces in the Forest.* New York: Oxford University Press.

Strier, K. B., T. E. Ziegler, and D. J. Wittwer. 1999. Seasonal and social correlates of fecal testosterone and cortisol levels in wild male muriquis (*Brachyteles arachnoides*). *Horm. Behav.* 35:125–34.

Struhsaker, T. T. 1967. Auditory communication among vervet monkeys (*Cercopithecus aethiops*). In *Social Communication among Primates*, ed. S. A. Altmann. Chicago: University of Chicago Press.

Sturmbauer, C., J. S. Levinton, and J. Christy. 1996. Molecular phylogeny analysis of fiddler crabs: Test of the hypothesis of increasing behavioral complexity in evolution. *Proc. Nat. Acad. Sci.* 93:10855–57.

Suarez, S. D., and G. G. Gallup, Jr. 1981. Predatory overtones of open-field testing in chickens. *Anim. Learn. Behav.* 9:153–63.

Sullivan, K. A. 1984. The advantages of social foraging in downy woodpeckers. *Anim. Behav.* 32:16–22.

Sun, L., and S. Muller. 1998a. Anal gland secretion codes for family membership in the beaver. *Behav. Ecol. Sociobiol.* 44:199–208.

Sun, L., and S. Muller. 1998b. Anal gland secretion codes for relatedness in the beaver, *Castor canadensis. Ethology* 104:917–27.

Suomi, S. J. 1973. Surrogate rehabilitation of monkeys reared in total social isolation. *J. Child. Psychol. Psychiat.* 14:71–77.

Suomi, S. J., and H. F. Harlow. 1977. Early separation and behavioral maturation. In *Genetics, Environment and Intelligence*, ed. A. Oliverio. Elsevier: North-Holland.

Suomi, S. J., H. F. Harlow, and M. A. Novak. 1974. Reversal of social deficits produced by isolation rearing in monkeys. *J. Hum. Evol.* 3:527–34.

Suthers, R. A. 1966. Optomotor responses by echolocating bats. *Science* 152:1102–4.

Swaddle, J. P. 1997. Developmental stability and predation success in an insect predator-prey system. *Behav. Ecol.* 8:433–36.

Sweeney, B., and J. W. Hastings. 1960. Effects of temperature upon diurnal rhythms. *Cold Spr. Harb. Symp. Quant. Biol.* 25:87–104.

Switzer, P. V., and P. K. Eason. 2000. Proximate constraints on intruder detection in the dragonfly *Perithemis tenera* (Odonata: Libellulidae): Effects of angle of approach and background. *Ann. Entomol. Soc. Amer.* 93:333–39.

Taitt, M. J. 1981. The effect of extra food on small rodent populations: I. Deermice (*Peromyscus maniclatus*). *J. Anim. Ecol.* 50:11–24.

Takahashi, J. S., L. H. Pinto, and M. H. Vitaterna. 1994. Forward and reverse genetic approches to behavior in the mouse. *Science* 264:1724-33.

Takahashi. J. S., and M. Zatz. 1982. Regulation of circadian rhythicity. *Science* 217:1104–11.

Takumi, T., C. Matsubara, Y. Shigeoshi, K. Taguchi, Y. Yagita, Y. Maebayashi, Y. Sakakida, K. Okumura, N. Takashima, and H. Okamura. 1998. A new mammalian *period* gene predominantly expressed in the suprachiasmatic nucleus. *Genes to Cells* 3:167–76.

Tamarin, R. H. 1980. Dispersal and population regulation in rodent. In *Biosocial Mechanisims of Population Regulation*, ed. M. N. Cohen, R. S. Malpass, and H. G. Klein. New Haven: Yale University Press.

Tamarkin, L., S. Brown, and B. Goldman. 1975. Neuroendocrine regulation of seasonal reproductive cycles in the hampster. *Abstr. 5th Ann. Mtg. Soc. Neurosci,* 458.

Tamm, G. R., and J. S. Cobb. 1978. Behavior and the crustacean molt cycle: Changes in aggression in *Homarus americanus. Science* 200:79–81.

Tamura, N. 1989. Snake-directed mobbing by the Formosan squirrel *Callosciurus erythraeus thaiwanensis. Behav. Ecol. Sociobiol.* 24:175–80.

Taylor, O. R., S. Perez, and R. Jander. 1997. A sun compass in monarch butterflies. *Nature* 387:29.

Taylor, O. R., S. Perez, and R. Jander. 1998. Monarch butterflies (*Danaus plexippus*) are disoriented by a strong magnetic pulse. *Naturwissenschafen* 86(3):140.

Templeton, A. R. 1989. The meaning of species and speciation: A genetic perspective. In *Speciation and Its Consequences* (D. Otte and J. Endler, eds.). Sunderland, MA: Sinauer Associates.

Terasawa, E., and C. H. Sawyer. 1969. Changes in electrical activity in the rat hypothalamus related to electrochemical stimulation of adenohypophyseal function. *Endocrinology* 85:143–51.

Terkel, J., and J. S. Rosenblatt. 1968. Maternal behavior induced by maternal blood plasma injection into virgin rats. *J. Comp. Physiol. Psychol.* 65:479–82.

———. 1971. Aspects of nonhormonal maternal behavior in the rat. *Horm. Behav.* 2:161–71.

———. 1972. Humoral factors underlying maternal behavior at parturition: Cross transfusion between freely moving rats. *J. Comp. Physiol. Psychol.* 80:365–71.

Terman, C. R. 1987. Intrinsic behavioral and physiological differences among laboratory populations of prairie deermice. *Amer. Zool.* 27:853–66.

Terrace, H. S., L. A. Petitto, R. J. Sanders, and T. G. Bever. 1979. Can an ape create a sentence? *Science* 206:891–902.

Thomas, D. W. 1983. The annual migrations of three species of West African fruit bats (Chiroptera: Pteropodidae). *Can. J. Zool.* 61:2266–72.

Thompson, R., and J. V. McConnell. 1955. Classical conditioning in the planarian, *Dugesia dorotocephala. J. Comp. Physiol. Psychol.* 48:65–68.

Thompson, W. R. 1957. Influence of prenatal maternal anxiety on emotionality in young rats. *Science* 125:698–99.

Thompson, K. 1998. Self assessment in juvenile play. In *Animal Play* (M. Bekoff and J. A. Byers, eds.). New York: Cambridge University Press.

Thorne, B. 1997. Evolution of eusociality in termites. *Ann. Rev. Ecol. Syst.* 28:27–54.

Thornhill, R. 1976. Sexual selection and paternal investment in insects. *Amer. Nat.* 110:153–63.

———. 1979. Adaptive female-mimicking behavior in a scorpionfly. *Science* 205:412–14.

Thornhill, R. 1992a. Female preference for the pheromone of males with low fluctuating asymmetry in the Japanese scorpionfly (*Panorpa japonica:* Mecoptera). *Behav. Ecol.* 3:277–83.

Thornhill, R. 1992b. Fluctuating asymmetry and the mating system of the Japanese scorpionfly, *Panorpa japonica. Anim. Behav.* 44:867–79.

Thornhill, R., and J. Alcock, eds. 1983. *The Evolution of Insect Mating Systems.* Cambridge, MA: Harvard University Press.

Thorpe, W. H. 1945. The evolutionary significance of habitat selection. *J. Anim. Ecol.* 14:67–70.

———. 1956. *Learning and Instinct in Animals.* Cambridge: Harvard University Press.

———. 1963. *Learning and Instinct in Animals,* 2d ed. London: Methuen.

Thorpe, W. H., and F. G. W. Jones. 1937. Olfactory conditioning in a parasitic insect and its relation to the problem of host selection. *Proc. Roy. Soc. London,* (Series B) 124:56–81.

Tinbergen, N. 1963. On aims and methods of ethology. *Zeitschrift für Tierpsychologie* 20:410–29.

Todt, D., and H. Hultsch. 1998. Hierarchical learning, development and representation of song. In *Animal Cognition in Nature* (R. P. Balda, I. M. Pepperberg, and A. C. Kamil, eds.). San Diego: Academic Press.

Tomioka, K., M. Nakamichi, and M. Yukizane. 1994. Optic lobe circadian pacemaker sends its information to the contralateral optic lobe in the cricket *Gryllus bimaculatus. J. Comp. Physiol. A* 175:381–88.

Timberlake, W. 1990. Natural learning in laboratory rodents. In *Contemporary Issues in Comparative Psychology,* ed. D. A. Dewsbury. Sunderland, MA: Sinauer Assoc.

Timberlake, W., and G. A. Lucas. 1989. Behavior systems and learning: From misbehavior to general principles. In *Contemporary Learning Theories, Instrumental Conditioning Theory and the*

Impact of Biological Constraints on Learning, eds. S. B. Klein and R. R. Mowrer. Hillsdale, NJ: L. Erlbaum Assoc.

Tinbergen, N. 1948. Social releasers and the experimental method required for their study. *Wilson Bull.* 60:6–52.

———. 1951. *The Study of Instinct.* Oxford: Clarenden Press.

———. 1953. *The Herring Gull's World.* London: Collins.

———. 1958. *Curious Naturalists.* New York: American Museum of Natural History.

———. 1960. *The Herring Gull's World.* Garden City, NY: Doubleday.

———. 1963a. On aims and methods of ethology. *Zeitschrift für Tierpsychologie.* 20:410–33.

———. 1963b. The shell menace. *Nat. Hist.* 72:28–35.

Tinbergen, N., G. J. Broekhuysen, F. Feekes, J. C. W. Houghton, H. Kruuk, and E. Szulc. 1962. Egg shell removal by the black-headed gull *Larus ridibundus* L.: A behavior component of camouflage. *Behaviour* 19:74–118.

Tinbergen, N., and W. Kruyt. 1938. Uber die Orientierung des Bienenwolfes (*Philanthus triangulum* Fabr): III Die Bevorzugung bestimmter Wegmarken. *Zeit. Vergl. Physiol.* 25:292–334.

Tinbergen, N., and A. C. Perdeck. 1950. On the stimulus situation releasing the begging response in the newly hatched herring gull chick (*Larus a argentatus* Ponstopp). *Behaviour* 3:1–38.

Tkadlec, F., and R. Gattermann. 1993. Circadian changes in susceptibility of voles and golden hamsters to acute rodenticides. *J. Interdiscipl. Cycle Res.* 24:153–61.

Toates, F. M. 1986. *Motivational Systems.* Cambridge, England: Cambridge University Press.

Tollman, J., and J. A. King. 1956. The effects of testosterone propionate on aggression in male and female C57BL/10 mice. *Anim. Behav.* 4:147–49.

Tombak, D. 1980. How nutcrackers find their seed stores. *Condor* 82:10–19.

Toran-Allerand, C. D. 1978. Gonadal hormones and brain development: Cellular aspects of sexual differentiation. *Amer. Zool.* 18:553–65.

Toyoda, F., M. Ito, S. Tanaka, and S. Kikuyama. 1992. Hormonal induction of male courtship behavior in the Japanese newt, *Cynops Pyrrhogaster. Horm. Behav.* 27:511–22.

Tramontin, A. D., and E. A. Brenowitz. 2000. Seasonal plasticity in the adult brain. *Trends in Neuroscience* 23:51–258.

Travis-Neideffer, M. N., J. D. Niedeffer, and S. F. Davis. 1982. Free operant single and double alternation in the albino rat: A demonstration. *Bull. Psychonom. Soc.* 19:287–90.

Trivers, R. L. 1971. The evolution of reciprocal altruism. *Quart. Rev. Biol.* 46:35–57.

———. 1972. Parental investment and sexual selection. In *Sexual Selection and the Descent of Man 1871–1971,* ed. B. Campbell. Chicago: Aldine.

———. 1974. Parent-offspring conflict. *Amer. Zool.* 14:249–64.

———. 1985. *Social Evolution.* Reading, MA: Benjamin Cummings.

Trivers, R. L., and H. Hare. 1976. Haplodiploidy and the evolution of the social insects. *Science* 191:249–63.

Trivers, R. L., and D. E. Willard. 1973. Natural selection of parental ability to vary the sex ratio of offspring. *Science* 179:90–92.

Truman, J. W. 1971. Physiology of insect ecdysis. I. The eclosion behavior of saturniid moths and its hormonal release. *J. Exp. Biol.* 54:805–14.

Truman, J. W., and O. S. Dominick. 1983. Endocrine mechanisms organizing invertebrate behavior. *BioScience* 33:546–51.

Truman, J. W., A. M. Fallon, and G. R. Wyatt. 1976. Hormonal release of programmed behavior in silk moths. *Science* 194:1432–33.

Truman, J. W., and L. M. Riddiford. 1970. Neuroendocrine control of ecdysis in silkmoths. *Science* 167:1624–26.

Trumbo, S. 1995. Nesting failure in burying beetles and the origin of communal associations. *Evol. Ecol.* 9:125–30.

Trumbo, S., and A. Fiore. 1994. Interspecific competition and the evolution of communal breeding in burying beetles. *Am. Midl. Nat.* 131:169–74.

Trut, L. N. 1999. Early canid domestication: The farm-fox experiment. *Amer. Sci.* 87:160–69.

Tsien, J. 2000. Building a brainier mouse. *Scientific American* 282:62–68.

Tsukuda, Y., T. Kanamatsu, and H. Takahara. 1999. Neurotransmitter release from the medial hyperstriatum ventrale of the chick forebrain accompanying filial imprinting behavior, measured by in vivo microdialysis. *Neurochem. Res.* 24:315–20.

Turner, C. D., and J. T. Bagnara. 1976. *General Endocrinology.* Philadelphia: Saunders.

Tyack, P. 1981. Interactions between singing Hawaiian humpback whales and conspecifics nearby. *Behav. Ecol. Sociobiol.* 8:105–16.

Tyack, P. L. 1997. Development and social functions of signature whistles in bottlenose dolphins *Tursiops truncatus. Bioacoustics* 8:21–46.

Uetz, G., and C. Hieber. 1997. Colonial web-building spiders: Balancing the costs and benefits of group-living. In *The Evolution of Social Behavior in Insects and Arachnids* (J. Choe and B. Crespi, eds.) Cambridge: Cambridge University Press.

Uetz, G. W. 1989. The "ricochet effect" and prey capture in colonial spiders. *Oecologia* 81:154–59.

Uetz, G. W. 1996. Risk-sensitvitiy and the paradox of colonial web-building in spiders. *Amer. Zool.* 36:459–70.

Ukegbu, A. A., and F. A. Huntingford. 1988. Brood value and life expectancy as determinants of parental investment in male three-spined sticklebacks (*Gasterosteus aculeatus*). *Ethology* 78:72–86.

Ulrich, R. E., R. R. Hutchinson, and N. H. Azrin. 1965. Pain-elicited aggression. *Psych. Record* 15:111–26.

Ulrich, R. E. 1966. Pain as a cause of aggression. *Amer. Zool.* 6:643–62.

Uvarov. B. 1966. *Grasshoppers and Locusts.* New York: Cambridge University Press.

Valenstein, P., and D. Crews. 1977. Mating-induced termination of behavioral estrus in the female lizard *Anolis carolinensis. Horm. Behav.* 9:362–70.

Van Hemel, P. E., and J. S. Myer. 1970. Satiation of mouse killing by rats in an operant situation. *Psychon. Sci.* 21:129–30.

van Lawick, H., and J. van Lawick-Goodall. 1970. *Innocent Killers.* Boston: Houghton Mifflin.

van Lawick-Goodall, J. 1968. A preliminary report on expressive movements and communication in the Gombe Stream chimpanzees. In *Primates: Studies in Adaptation and Variability,* ed. P. Jay. New York: Holt, Rinehart and Winston.

van Lawick-Goodall, J. 1970. Tool-using in primates and other vertebrates. *Adv. Study Behav.* 3:195–249.

van Schaik, C. P., and S. Blaffer Hrdy. 1991. Intensity of local resource competition shapes the relationship between maternal rank and sex ratios in cercopithecine primates. *Amer. Nat.* 138:1555–62.

Van Tets, G. F. 1965. A comparative study of some social communication patterns in the Pelecaniformes. *Ornith. Monogr.* 2:1–88.

van Tyne, J., and A. J. Berger, 1959. *Fundamentals of Ornithology.* New York: Wiley.

Van Valen, L. 1973. A new evolutionary law. *Evol. Theory* 1:1–30.

Vandenbergh, J. G. 1965. Hormonal basis of sex skin in male rhesus monkeys. *Gen. Comp. Endocr.* 5:31–34.

———. 1967. Effect of the presence of a male on the sexual maturation of female mice. *Endocrinology* 81:345–48.

———. 1969a. Endocrine coordination in monkeys: Male sexual responses to the female. *Physiol. Behav.* 4:261–64.

———. 1969b. Male odor accelerates female sexual maturation in mice. *Endocrinology* 84:658.

———. 1971. The effects of gonadal hormones on the aggressive behavior of adult golden hamsters. *Anim. Behav.* 19:589–94.

Vandenbergh, J. G., and D. M. Coppola. 1986. The physiology and ecology of puberty modulation by primer pheromones. *Adv. Stud. Behav.* 16:71–108.

Vandenbergh, J. G., and L. C. Drickamer. 1974. Reproductive coordination among free-ranging rhesus monkeys. *Physiol. Behav.* 13:373–76.

Vandenbergh, J. G., L. C. Drickamer, and D. R. Colby. 1972. Social and dietary factors in the sexual maturation of female mice. *J. Reprod. Fertil.* 28:397–405.

Vandenbergh, J. G., and S. Vessey. 1968. Seasonal breeding of free-ranging rhesus monkeys and related ecological factors. *J. Reprod. Fert.* 15:71–79.

Vander Wall, S. B., and R. P. Balda. 1981. Ecology and evolution of food storage behavior in conifer-seed-caching corvids. *Z. Tierpsychol.* 56:217–42.

Vander Wall, S. B. 1990. *Food Hoarding in Animals.* Chicago: University of Chicago Press.

Vaughan, T. A., and R. M. Hansen. 1964. Experiments on interspecific competition between two species of pocket gophers. *Amer. Mid. Nat.* 72:444–52.

Vehernecamp, S. L. 1977. Relative fecundity and parental effort in communally nesting anis, *Crotophaga sulcirostris. Science* 197:403–5.

Verme, L. J., and J. J. Ozoga. 1981. Sex ratio of white-tailed deer and the estrus cycle. *J. Wildl. Manage.* 45:710–15.

Verner, J. 1976. Complex song repertoire of male long-billed marsh wrens in eastern Washington. *Living Bird* 14:263–300.

Vernikos-Danellis, J., and C. D. Winget. 1979. The importance of light, postural, and social cues in the regulation of the plasma cortical rhythms in man. In *Chronopharmacology,* eds. A. Reinberg, and F. Halbert. New York: Pergamon.

Vessey, S. H. 1967. Effects of chlorpromazine on aggression in laboratory populations of wild house mice. *Ecology* 48:367–76.

———. 1971. Free-ranging rhesus monkeys: Behavioural effects of removal, separation, and reintroduction of group members. *Behaviour* 40:216–27.

———. 1973. Night observations of free ranging rhesus monkeys. *Amer. J. Phys. Anthrop.* 38:613–20.

———. 1987. Long-term population trends in white-footed mice and the impact of supplemental food and shelter. *Amer. Zool.* 27:879–90.

Vessey, S. H., and H. M. Marsden. 1975. Oviduct ligation in rhesus monkeys causes maladaptive epimeletic (care-giving) behavior. In *Contemporary Primatology,* ed. S. Kondo, M. Kawai, and A. Ehara. Basel: S. Karger.

Vessey, S. H., and D. B. Meikle. 1984. Free-ranging rhesus monkeys: Adult male interactions with infants and juveniles. In *Paternal Behavior,* ed. D. Taub. New York: Van Nostrand Reinhold.

Vessey, S. H. 1968. Interactions between free-ranging groups of rhesus monkeys. *Folia Primatologica* 8:228–39.

Viemerö, V. 1996. Factors in childhood that predict later criminal behavior. *Aggressive Behav.* 22:87–97.

Vilá, C., J. Maldonado, and R. Wayne. 1999. Phylogenetic relationships, evolution, and genetic diversity of the domestic dog. *J. Heredity* 90:71–77.

Vilá, C., P. Savolainen, J. E. Maldonado, I. R. Amorim, J. E. Rice, R. L. Honeycutt, K. A. Crandall, J. Lundeberg, and R. K. Wayne. 1997. Multiple and ancient origins of the domestic dog. *Science* 276:1687–89.

Villars, T. A. 1983. Hormones and aggressive behavior in teleost fish. In *Hormones and Aggressive Behavior,* ed. B. B. Svare. New York: Plenum.

Vince, M. A. 1964. Social facilitation of hatching in bobwhite quail. *Anim. Behav.* 12:531–34.

———. 1966. Artificial acceleration of hatching in quail embryos. *Anim. Behav.* 14:389–94.

———. 1969. Embryonic communication, respiration, and the synchronization of hatching. In *Bird Vocalizations: Their Relations to Current Problems in Biology and Psychology: Essays Presented to W. H. Thorpe,* ed. R. A. Hinde. Cambridge: Cambridge University Press.

Vince, M. A. 1973. Effects of external stimulation on the onset of lung ventilation and the time of hatching in the fowl, duck, and goose. *Brit. J. Poult. Sci.* 14:389–401.

Vinogradova, Y. B. 1965. An experimental study of the factors regulating induction of imaginal diapause in the mosquito *Aedes togoi* Theob. *Entomol. Rev.* 44:309–15.

Virgin, C. E., Jr., and R. M. Sapolsky. 1997. Styles of male social behavior and their endocrine correlates among low-ranking baboons. *Am. J. Primatol.* 42:25–39.

Vivien-Roels, B., J. Arendt, and J. Bradtke. 1979. Circadian and circannual fluctuations of pineal indoleamines (serotonin and melatonin) in *Testudo hermanni* Gmelin (Reptilia, Chelonia). I. Under natural conditions of photoperiod and temperature. *Gen. Comp. Endocr.* 37:197–210.

Vogt, J. L. 1984. Interactions between adult males and infants in prosimians and New World monkeys. Pp. 346–76 in *Primate Paternalism* (D. M. Taub, ed.). New York: Van Nostrand Reinhold.

vom Saal, F. S. 1983. Models of early hormonal effects on intrasex aggression in mice. In *Hormones and Aggressive Behavior,* ed. B. B. Svare. New York: Plenum.

———. 1989. Sexual differentiation in litter-bearing mammals: Influence of sex of adjacent fetuses in utero. *J. Anim. Sci.* 67:1824–40.

vom Saal, F. S., and F. H. Bronson. 1978. In utero proximity of female mouse fetuses to males: Effect on reproductive performance during later life. *Biol. Reprod.* 19:842–53.

———. 1980a. Variation in length of the estrous cycle in mice due to former intrauterine proximity to male fetuses. *Biol. Reprod.* 22:777–80.

———. 1980b. Sexual characteristics of adult female mice are correlated with their blood testosterone levels during prenatal development. *Science* 208:597–99.

vom Saal, F. S., M. M. Montano, and M. H. Wang. 1992. Sexual differentiation in mammals. In *Chemically Induced Alterations in Sexual and Functional Development: The Wildlife-Human Connection* (T. Colburn and C. Clement, eds.). Princeton, NJ: Princeton Scientific.

von der Emde, G. 1999. Active electrolocation of objects in weakly electric fish. *J. Exp. Biol.* 202:1205–15.

von Frisch, K. 1914. Der Farbensinn und Formensinn der Biene. *Zool. Jahr.* 35:1–188.

———. 1967a. *Bees, Their Vision, Chemical Senses and Language.* Ithaca, NY: Cornell University Press.

———. 1967b. *Dance Language and Orientation of Bees.* Cambridge: Harvard University Press.

Waage, J. K. 1979. Dual function of the damselfly penis: Sperm removal and transfer. *Science* 203:916–18.

Waage, J. K. 1997. Parental investment—Minding the kids or keeping control? In *Feminism and Evolutionary Biology: Boundaries, Interactions, and Frontiers* (P. A. Gowaty, ed.). New York: Chapman and Hall.

Wade, M. J. 1976. Group selection among laboratory populations of *Tribolium. Proc. Nat. Acad. Sci.* 73:4604–7.

Wade, M. J. 1979. The primary characteristics of *Tribolium* populations group selected for increased and decreased population size. *Evolution* 33:749–64.

Wade, M. J. 1985. Soft selection, hard selection, kin selection, and group. *Amer. Nat.* 155:61–73.

Walcott, C. 1969. A spider's vibration receptor: Its anatomy and physiology. *Amer. Zool.* 9:133–44.

———. 1972. Bird navigation. *Natural History* 81(June):32–43.

———. 1977. Magnetic fields and orientation of homing pigeons under sun. *J. Exp. Biol.* 70:105–23.

Walcott, C., J. L. Gould, and J. L. Kirschvink. 1979. Pigeons have magnets. *Science* 205:1027–29.

Waldman, B. 1982. Sibling association among schooling toad tadpoles: Field evidence and implications. *Anim. Behav.* 30:700–13.

Walker, B. W. 1949. Periodicity of spawning of the grunion, *Leuresthes tenuis.* Ph.D. thesis, University of California, Los Angeles.

Walker, M. M., J. L. Kirschvink, G. Ahmed, and A. E. Dizon. 1992. Evidence that fin whales respond to the geomagnetic field during migration. *J. Exp. Biol.* 171:67–78.

Walker, T. J. 1991. Butterfly migration from and to peninsular Florida (USA). *Ecol. Entomol.* 16:241–52.

Wallraf, H. G. 1991. Conceptual approaches to avian orientation systems. Pp. 128–65 in *Orientation in Birds* (P. Berthold, ed.). Basel: Birkhauser Verlag.

Walls, S. C., and A. R. Blaustein. 1995. Larval marbled salamanders, *Ambystoma opacum,* eat their kin. *Anim. Behav.* 50:537–45.

Waloff, Z. 1958. The behaviour of locusts in migrating swarms. *Proc. 10th Inter. Congr. Ent.,* Montreal 2:567–70.

Walraff, H. G. 1986. Relevance of olfaction and atmospheric odours to pigeon homing. In *Orientation in Space,* ed. G. Beugnon, 71–80. Toulouse, France: Editions Privat.

Walsh, J. 1983. Wide world of reports. *Science* 220:804–5.

Walters, J. 1990. Red-cockaded woodpeckers: A 'primitive' cooperative breeder. In *Cooperative Breeding in Birds: Long-Term Studies of Ecology and Behavior* (P. Stacey and W. Koenig, eds.). Cambridge: Cambridge University Press.

Wang, L. C. H. 1978. Energetic and field aspects of mammalian torpor: The Richardson's ground squirrel. In *Strategies in Cold: Natural Torpidity and Thermogenesis,* ed. L. C. H. Wang and J. W. Hudson. New York: Academic Press.

Ward, P. 1971. The migration patterns of *Quelea quelea* in Africa. *Ibis* 113:275–97.

Ward, P., and A. Zahavi. 1973. The importance of certain assemblages of birds as "information centres" for food-finding. *Ibis* 115:517–34.

Warman, C. G., and E. Naylor. 1995. Evidence for multiple, cue-specific circatidal clocks in the shore crab *Carcinus maenas. J. Exp. Mar. Bio. Ecol.* 189:93–101.

Warren, J. M. 1965. Learning in vertebrates. In *Behavior of Nonhuman Primates: Modern Research,* ed. A. M. Schrier, H. F. Harlow, and F. Stollnitz. New York: Academic Press.

———. 1973. Learning in vertebrates. In *Comparative Psychology: A Modern Survey,* ed. D. A. Dewsbury and D. A. Rethlingshafer. New York: McGraw-Hill.

———. 1983. Learning in vertebrates. In *Comparative Psychology: A Modern Survey,* ed. D. A. Dewsbury and D. A. Rethlingshafer. New York: McGraw-Hill.

Warren, J. M., and A. Barron. 1956. The formation of learning sets by cats. *J. Comp. Physiol. Psychol.* 49:227–31.

Waterman, T. H. 1966. Systems analysis and the visual orientation of animals. *Amer. Sci.* 54:15–45.

Waterman, T. H., and H. Hashimoto. 1974. E-vector discrimination by the goldfish optic tectum. *J. Comp. Physiol.* 95:1–12.

Waters, D. A. 1996. The peripheral auditory characteristics of noctuid moths: Information encoding and endogenous noise. *J. Exp. Biol.* 199:857–68.

Waters, D. A., and G. Jones. 1994. Wingbeat-generated ultrasound in noctuid moths increases the discharge rate of the bat-detecting A1 cell. *Proc. Roy. Soc. Lond. B* 258:41–46.

Watson, J. B. 1930, *Behaviorism.* New York: Norton.

Watt, W. B., P. A. Carter, and K. Donohue. 1986. Females' choice of "good genotypes" as mates is promoted by an insect mating system. *Science* 233:1187–90.

Watt, P. J., S. F. Nottingham, and S. Young. 1997. Toad tadpole aggregation behaviour: Evidence for a predator avoidance function. *Anim. Behav.* 54:865–72.

Weaver, N. 1966. Physiology of caste determination. *Ann. Rev. Entomol.* 11:79–102.

Wecker, S. C. 1963. The role of early experience in habitat selection by the prairie deer mouse, *Peromyscus maniculatus bairdi. Ecol. Monogr.* 33:307–25.

Weis, C. S., and J. P. Coughlin. 1979. Maintained aggressive behavior in gonadectomized male Siamese fighting fish (*Betta splendens*). *Physiol. Behav.* 23:173–77.

Welch, D. M., and M. Meselson. 2000. Evidence for the evolution of bdelloid rotifers without sexual reproduction or genetic exchange. *Science* 288:1211–15.

Welker, W. I. 1957. "Free" versus "forced" exploration of a novel situation by rats. *Psychol. Rep.* 3:95–108.

Weller, S. J., N. L. Jacobson, and W. E. Conner. 1999. The evolution of chemical defences and mating systems in tiger moths (Lepidoptera: Arctiidae). *Biol. J. Linn. Soc.* 68:557–78.

Wells, P. H. 1973. Honey bees. In *Invertebrate Learning,* vol. 2., ed. W. C. Corning, J. A. Dyal, and A. O. D. Willows. New York: Plenum.

Wells, M. J, and J. Wells. 1959. Hormonal control of sexual maturity in *Octopus. J. Exp. Biol.* 36:1–33.Wells, M., and C. Henry. 1992. Behavioural responses of green lacewings (*Neuroptera: Chrysopidae: Chrysoperla*) to synthetic mating songs. *Anim. Behav.* 44:641–52.

Wells, M. M., and C. S. Henry. 1998. In *Endless Forms: Species and Speciation* (D. J. Howard and S. H. Berlocher, eds.). Oxford: Oxford University Press.

Wenner, A. M. 1967. Honeybees: Do they use the distance information contained in their dance maneuver? *Science* 155:847–49.

———. 1971. *Bee Language Controversy.* Boulder, CO: Educational Programs Improvement Corp.

———. 1974. Information transfer in honeybees: A population approach. In *Nonverbal Communication,* vol. 1, ed. L. Krames, P. Pliner, and T. Alloway. New York: Plenum.

Werren, J. H. 1983. Sex ratio evolution under local mate competition. *Evolution* 37:116–24.

West, M. J., and A. P. King. 1985. Social guidance of vocal learning by female cowbirds: Validating its functional significance. *Z. Tierpsychol.* 70:225–35.

———. 1986. Song repertoire development in male cowbirds (*Molothrus ater*): Its relation to female assessment of song potency. *J. Comp. Psychol.* 100:296–303.

———. 1987. Settling nature and nurture into an ontogenetic niche. *Develop. Psychobiol.* 20:549–62.

West Eberhard, M. J. 1969. The social biology of Polistine wasps. *Misc. Pub. Mus. Zool.,* (University of Michigan, Ann Arbor) 140:1–101.

Wickler, W. 1968. *Mimicry in Plants and Animals.* London: World University Library.

———. 1969. Zur Sociologie des Brabanthbuntbarshes, *Tropheus moorei* (Pisces, cichlidae) *Zeit. Für Tierpsychol.* 26:967–87.

———. 1972. *The Sexual Code: The Social Behavior of Animals and Men.* Garden City, NY: Doubleday.

Wiklund, C., and T. Jarvi. 1982. Survival of distasteful insects after being attacked by naive birds: A reappraisal of the theory of aposematic coloration evolving through individual selection. *Evolution* 36(5):998–1002.

Wilcox, R. S. 1972. Communication by surface waves: Mating behavior of a water strider (Gerridae). *J. Comp. Physiol.* 80:255–66.

———. 1979. Sex discrimination in *Gerris remigis:* Role of a surface wave signal. *Science* 206:1325–27.

Wilcox, R. S. 1995. Ripple communication in aquatic and semi-aquatic insects. *Ecoscience* 2:109–15.

Wildhaber, M. L., R. F. Green, and L. B. Crowder. 1994. Bluegills continuously update patch giving-up times based on foraging experience. *Anim. Behav.* 47:318–26.

Wildt, D. E., M. Bush, K. L. Goodrowe, C. Packer, A. E. Pusey, J. L. Brown, P. Joslin, and S. J. O'Brien. 1987. Reproductive and genetic consequences of founding isolated lion populations. *Nature* 329:328–31.

Wiley, R. H. 1973. Territoriality and nonrandom mating in sage grouse, *Centrocerus urophasianus. Anim. Behav. Monogr.* 6:85–169.

Wiley, R. H. 1983. The evolution of communication: Information and manipulation. Pp. 156–89 in *Animal Behavior,* Vol. 2: *Communication* (T. R. Halliday and P. J. B. Slater, eds.). New York: W. H. Freeman and Co.

Wilkinson, G. S. 1984. Reciprocal food sharing in the vampire bat. *Nature* 308:181–84.

Wilkinson, D. 1984. Reciprocal food sharing in vampire bats. *Nature* 309:181–84.

Wilkinson, D. 1986. Social grooming in the common vampire bat, *Desmodus rotundus. Anim. Behav.* 34:1880–89.

Wilkinson, G. S. 1987. Equilibrium analysis of sexual selection in *Drosophila melanogaster. Evolution* 41:11–21.

Wilkinson, G. S. 1999. Male eye span in stalk-eyed flies indicates genetic quality by meiotic drive suppression. *Nature* 391:276–79.

Wille, A., and E. Orozco. 1970. The life cycle and behavior of the social bee *Lasioglossum (Dialictus) umbripenne. Rev. Biol. Trop.* (Costa Rica): 17:199–245.

Williams, G. C. 1966. *Adaptation and Natural Selection: A Critique of Some Current Evolutionary Thought.* Princeton: Princeton University Press.

———. 1975. *Sex and Evolution.* Princeton: Princeton University Press.

———. 1979. The question of adaptive sex ratio in outcrossed vertebrates. *Proc. Roy. Soc. Lond.* B 205:567–80.

Williams, T. C., and J. M. Williams. 1967. Radio tracking of homing bats. *Science* 155:1435–36.

———. 1970. Radio tracking of homing and feeding flights of a neotropical bat, *Phyllostomus hastatus. Anim. Behav.* 18:302–9.

Williams, T. C., J. M. Williams, and D. R. Griffin. 1966. The homing ability of the neotropical bats, *Phyllostomus hastatus,* with evidence for visual orientation. *Anim. Behav.* 14:468–73.

Williams, C. L. 1998. Estrogen effects on cognition across the lifespan. *Horm. Behav.* 34:80–84.

Wilsbacher, L. D. and J. S. Takahashi. 1998. Circadian rhythms: Molecular basis of the clock. *Curr. Opin. Genet. Dev.* 8:595–602.

Wilson, B., N. J. Mackintosh, and R. A. Boakes. 1985. Transfer of relationship rules in matching and oddity learning by pigeons and corvids. *Quart. J. Exp. Psychol.* 37B:313–32.

Wilson, D. S. 1980. *The Natural Selection of Populations and Communities.* Menlo Park, CA: Benjamin/Cummings.

Wilson, E. O. 1971. *Insect Societies.* Cambridge: Harvard University Press.

———. 1973. Group selection and its significance for ecology *BioScience* 23:631–38.

———. 1975. *Sociobiology: The New Synthesis* Cambridge: Harvard University Press.

Wilson, D. M. 1960. The central nervous control of flight in a locust. *J. Exp. Biol.* 38:471–90.

Wilson, D. S. 1998. Game theory and human behavior. In *Game Theory and Animal Behavior* (L. Dugatkin, and H. Reeve, eds.). Oxford: Oxford University Press.

Wiltschko, W. 1972. The influence of magnetic total intensity and inclination on directions preferred by migrating European robins (*Erithacus rubecula*). *NASA Spec. Publ.* NASA SP-262:569–78.

Wiltschko, W., and R. Wiltschko. 1972. The magnetic compass of European robins, *Erithacus rubecula. Science* 176:62–64.

Wiltschko, W., and R. Wiltschko, 1988. Magnetic orientation in birds. *Current Ornithology* 5:67–121.

Winberg, S., and O. Lepage. 1998. Elevation of brain 5-HT activity, POMC expression, and plasma cortisol in socially subordinate rainbow trout. *Am. J. Physiol.* 274:R645–R654.

Wingfield, J. C., G. F. Ball, A. M. Dufty, R. E. Hegner, and M. Ramenofsky. 1987. Testosterone and aggression in birds. *Amer. Sci.* 75:602–8.

Winkler, D. W., and F. H. Sheldon. 1993. Evolution of nest construction in swallows (Hirundinidae): A molecular phylogenetic perspective. *Proc. Nat. Acad. Sci.* 90:5705.

Witham, T. G. 1980. The theory of habitat selection: Examined and extended using *Pemphigus* aphids. *Amer. Nat.* 115:449–66.

Wittenberger, J. F. 1981. *Animal Social Behavior.* Boston, MA: Duxbury Press.

Wodinsky, J. 1977. Hormonal inhibition of feeding and death in *Octopus:* Control by optic gland secretion. *Science* 198:948–51.

Wolff, J. O., K. I. Lundy, and R. Baccus. 1988. Dispersal, inbreeding avoidance, and reproductive success in white-footed mice. *Anim. Behav.* 36:456–65.

Wolff, J. R. 1981. Some morphogenetic aspects of the development of the central nervous system. In *Behavioral Development,* ed. K. Immelmann et al. New York: Cambridge University Press.

Wolfson, A. 1948. Bird migration and the concept of continental drift. *Science* 108:23–30.

Woolfenden, G. E. 1975. Florida scrub jay helpers at the nest. *Auk* 92:1–15.

Woolfenden, G. E., and J. W. Fitzpatrick. 1984. *The Florida Scrub Jay: Demography of a Cooperative-Breeding Bird.* Princeton: Princeton University Press.

Wourms, M. K., and F. E. Wasserman. 1985. Butterfly wing markings are more advantageous during handling than during the initial strike of an avian predator. *Evolution* 39:845–51.

Wright, W. G., and A. L. Shanks. 1993. Previous experience determines territorial behavior in an archaeogastropod limpet. *J. Exp. Marine Biol. Ecol.* 166:217–29.

Wursig, B., and M. Wursig. 1977. The photographic determination of group size, composition, and stability of coastal porpoises (*Tursiops truncatus*). *Science* 198:755–56.

Wyman, R. L., and J. A. Ward. 1973. The development of behavior in the cichlid fish *Etroplus maculatus* Bloch. *Z. Tierpsychol.* 33:461–91.

Wynne-Edwards, V. C. 1962. *Animal Dispersion in Relation to Social Behavior.* Edinburgh: Oliver and Boyd.

Wyttenbach, R. A., M. L. May, and R. R. Hoy. 1996. Categorical perception of sound frequency by crickets. *Science* 273:1542–44.

Xiaowei, J., L. P. Shearman, D. R. Weaver, M. J. Zylka, G. J. De Vries, and S. M. Reppert. 1999. A molecular mechanism regulating rhythmic output from the suprachiasmatic circadian clock. *Cell* 96:57–68.

Yamazaki, K., E. A. Boyse, V. Mike, H. T. Thaler, B. J. Mathieson, J. Abbott, J. Boyse, Z. A. Zayas. 1976. Control of mating preferences in mice by genes in the major histocompatibility complex. *J. Exp. Med.* 144, 1324–1335.

Yamasaki, K., E. A. Boyse, V. Mike, H. T. Thaler, B. J. Mathieson, J. Abbot, J. Boyse, Z. A. Zayas, and L. Thomas. 1976. Control of mating preferences in mice by genes in the major histocompatibility complex. *J. Exp. Med.* 144:1324–35.

Ydenberg, R., and P. Hurd. 1998. Simple models of feeding with time and energy constraints. *Behav. Ecol.* 9:43–48.

Yerkes, R. M. 1907. *The Dancing Mouse.* New York: Macmillan.

Yoerg, S. 1991. Social feeding reverses learned flavor aversions in spotted hyenas (*Crocuta crocuta*). *J. Comp. Psych.* 105:185–89.

Young, W. C., R. W. Goy, and C. H. Phoenix. 1964. Hormones and sexual behavior. *Science* 143:212–18.

Zahavi, A. 1975. Mate selection—A selection for a handicap. *J. Theor. Biol.* 53:205–14.

Zahavi, A., and A. Zahavi. 1997. *The Handicap Principle.* New York: Oxford University Press.

Zahl, P. A. 1963. The mystery of the monarch butterfly. *Nat. Geogr.* 123:588–98.

Zarrow, M. X. 1961. Gestation. In *Sex and Internal Secretions,* ed. W. C. Young. Baltimore: Williams & Wilkins.

Zentall, T. R., and D. E. Hogan. 1974. Abstract concept learning in the pigeon. *Journal of Experimental Psychology* 102:393–98.

Zeveloff, S. I., and M. S. Boyce. 1980. Parental investment and mating systems in mammals. *Evolution* 34:973–82.

Zigmond, R. E., F. Nottebohm, and D. W. Pfaff. 1973. Androgen-concentrating cells in the midbrain of a songbird. *Science* 179:1005–6.

Zimen, E. 1972. *Vergleichende Verhaltungsbeobachtungen an Wölfen und Königspudeln.* Munich: Piper.

Zucker, I., B. Rusak, and R. G. King. 1976. Neural bases for circadian rhythms in rodent behavior. *Adv. Psychobiol.* 3:36–74.

Zuk, M. 1984. A charming resistance to parasites. *Nat. Hist.* 4 (84):28–34.

Zuk, M. 1990. Reproductive strategies and sex differences is disease susceptibility: An evolutionary viewpoint. *Parasitology Today* 6:231–33.

Zuk, M. 1994. Immunology and the evolution of behavior. In *Behavioral Mechanisms in Evolutionary Ecology.* (L. A. Real, ed.). Chicago: University of Chicago Press.

Zuk, M., K. Johnson, R. Thornhill, and J. D. Ligon. 1990. Parasites and male ornaments in free-ranging and captive red jungle fowl. Behaviour 114:232–48.

Zumpe, D., and R. P. Michael. 1968. The clutching reaction and orgasm in the female rhesus monkey (*Macaca mulatta*). *J. Endocr.* 40:117–23.

GLOSSARY

acquisition Gaining a response in a learning paradigm.

action potential A self-propagating wave of depolarization that moves along the membrane of a nerve or muscle.

activational effects Effects of hormones that act as triggering influences on the expression of particular behavior patterns; response is rapid, in hours or days. Contrast with organizational effects.

active avoidance learning A form of operant conditioning where the animal has to act in order to avoid some noxious consequence (e.g., shock).

adaptation In evolutionary biology, any structure, physiological process, or behavioral trait that makes an animal better able to survive and reproduce compared to conspecifics. Also used to describe the process of evolutionary change leading to the formation of such a trait.

adaptive behavior Behavior patterns that make an organism more fit to survive and reproduce in comparison with other members of the same species.

adaptive value A numerical value that expresses the potential for survival and reproductive success of a particular genotype relative to other genotypes.

adrenal glands Paired endocrine glands, located next to the kidneys in the abdomen. The adrenal cortex produces steroid hormones involved in water balance, glucose metabolism, and electrolyte balance. The adrenal medulla produces adrenaline and noradrenaline, which are involved in glucose metabolism, heart rate, and blood pressure.

aggression Behavior that appears to be intended to inflict noxious stimulation or destruction on another organism.

aggressive mimicry A technique for capturing prey in which the predator uses lures or other means to misinform the prey.

agonistic behavior A suite of behavior patterns used during conflict with a conspecific, usually indicating whether an individual is going to submit to the other animal or fight if the other does not submit.

Allee effect The unfavorable consequences of low numbers of individuals, as when populations fail to breed when numbers are below some critical density.

allele A particular form of a gene (locus) that can be distinguished from other forms (alleles) of that gene.

alliance A relationship between two or more individuals, in which they support each other in competitive interactions. Often used for dolphins and primates in reference to long-term coalitions.

allopatric speciation The formation of new species by a process involving geographic barriers.

allopatry Populations or species with nonoverlapping geographic distributions. Contrast with sympatry.

altricial young Young that are born or hatched in a relatively immature or helpless condition. Compare with precocial.

altruism The situation when one individual acts to increase the fitness of another to the cost of its own fitness.

amplitude The maximum absolute value of a periodically varying quantity, as the peak in activity in a circadian or circannual cycle.

analogy Similarities due to convergent evolution.

anisogamy The condition in which the female gamete (ovum) is larger than the male gamete (sperm).

anosmic Lacking the sense of smell.

antagonism The condition of being an opposing principle, force, or factor, as when two hormones have opposite effects on target tissues.

aposematic coloration Conspicuous markings of noxious animals that are easily recognizable and avoidable by a potential predator.

appetitive behavior The flexible introductory phase of a behavior sequence during which the organism is searching to obtain something to meet a need, as in seeking food, a mate, stimulation of a specific type, etc. *See* consummatory behavior.

applied animal behavior A subdiscipline concerned with all aspects of the behavioral biology of pets, domestic animals, animals in zoos and circuses, and animals in aquaria.

arrhythmic activity Activity that does not exhibit any clear cyclical pattern.

Aschoff's Rule Pertaining to circadian rhythms, the rule stating that when nocturnal animals are held in constant darkness, their free-running period becomes slightly shorter each day and when a diurnal animal is held in constant darkness, the free-running period becomes slightly longer each day.

axon In a nerve cell, the extension that carries information away from the cell body.

basilar membrane A flat band of tissue that bears the auditory hair cells in the cochlea of the vertebrate ear; the membrane vibrates in response to pressure waves.

Bateman gradient The degree to which the reproductive success of males increases with the number of mates they obtain.

Batesian mimics Palatable species that evolve morphologies and behaviors similar to those of unpalatable species.

behavioral ecology A subdiscipline within animal behavior that deals with the ways in which animals interact with their environment. It emphasizes the effect of behavior on survival and reproductive success.

behavioral isolating mechanisms Differences in behavior (usually courtship) that prevent genetic exchange between members of different populations or species.

behavioral genetics The study of the role that genes play in controlling behavior.

behaviorism A view of the actions of animals that postulates that behavior can be analyzed functionally in terms of stimulus and response combinations. Thus, in this view most of behavior is a function of the experience of the organism.

biological clock An internal timing mechanism that involves both an internal self-sustaining pacemaker and cyclic environmental synchronizers.

biological communication An action by one organism (or cell) that causes a reaction from another organism (or cell) in a fashion adaptive to either one or both of the participants.

biological rhythm A cyclical pattern of behavior, occurring at some regular period (e.g., diurnal). *See* biological clock.

Bruce effect In mice, the effect of a strange male, or his odor, that causes females to abort and become receptive.

cannibalism The killing and eating of members of the same species.

cannulation The use of a hollow electrode or a fine tube to introduce a substance to a tissue (e.g., an area of the brain) or to sample chemicals that are released in the tissue.

caste Physiologically morphologically, and/or behaviorally different forms of individuals that perform specialized labor in colonies; especially well known in social insects.

central nervous system (CNS) That part of the nervous system that is condensed and centrally located; for example, the brain and spinal cord of vertebrates and the brain and ganglia of insects.

central pattern generator A set of neurons in the central nervous system that controls repetitive movement without sensory input.

central place foragers Animals that collect more than one food item at a time and bring the items to a central location for storage or feeding to offspring.

cephalization The evolutionary trend of the concentration of sensory and other nervous system tissue in the head region of animals.

cerebellum Part of the hindbrain in vertebrates; important in coordination.

cerebrum Part of the forebrain in vertebrates; includes the paleocortex and the neocortex.

character A trait used in constructing a phylogeny.

character displacement The process by which two closely related species interact so as to cause one or both of them to diverge evolutionarily in one or more traits.

character state The value of a particular taxonomic character (e.g., present or absent, or a particular measurement).

chemical synapse The type of synapse where communication between neurons is through a chemical intermediary.

chemoreceptor A sensory receptor that responds to a specific molecule or molecules.

chromosome A long, complex molecule of DNA, portions of which are the different units of heredity (genes).

circadian rhythm A biological rhythm of about a day in length or period.

circannual rhythm A biological rhythm of about a year in length or period.

cladistics A method of reconstructing phylogenies based on the identification of monophyletic groups.

classical (Pavlovian) conditioning A type of learning whereby an unconditioned stimulus (US) that elicits a specific response (unconditioned response [UCR]) is paired with a neutral stimulus. (After it becomes associated with the US, the neutral stimulus, becomes the conditioned stimulus (CS), and results in a conditioned response (CR).

cochlea A portion of the inner ear of vertebrates that is coiled like a snail's shell and that contains the sensory structures for detecting sound.

coefficient of relationship The fraction of genes identical by common descent shared between two individuals.

coevolution The change in gene frequencies resulting from two species acting as strong selective forces on one another.

cognition The processes in the minds of animals or their general mental functions, including perception, representation, and memory.

commonality In learning, investigating whether species share similar types of learning and using this to draw conclusions about the evolution of learning.

communal nesting More than one female in a nest raising the young of more than one female.

comparative method A comparison of the behavior of two or more species for the purpose of either elucidating some common aspects of the ecology and evolution of behavior or exploring the mechanisms underlying behavior.

comparative psychology A branch of psychology involving the study of animals. Some comparative psychologists are more concerned with learning, cognition, and intelligence in human and nonhuman animals, while others are indistinguishable from other animal behaviorists in that they explore the causation, development, evolution, and functional aspects of behavior in a broad range of species.

competition The attempt of two or more organisms to utilize the same resource.

composite signal A signal formed by combining two or more simpler signals.

compound eye The multiunit arthropod eye that is made up of many ommatidia.

conditioned response (CR) The behavior pattern that becomes conditioned during classical conditioning.

conditioned stimulus (CS) The previously neutral stimulus that, through classical conditioning, now elicits the conditioned response.

cone A vertebrate photoreceptor cell with a tapered outer segment; unlike rod cells, cone cells are usually sensitive to color.

confusion effect The reduced capture efficiency experienced by a predator caused by high densities or swarms of prey.

conspecifics Animals belonging to the same species.

constraints In an optimality model, the limits, both internal and external, that define the range of an animal's behavior.

consummatory behavior Actions of an animal completing a behavior sequence, as in consuming food, mating, etc. *See* appetitive behavior.

context In communication, stimuli other than the signal that are impinging on the receiver that might alter the meaning of a signal.

contiguity The association of events in time, especially as used in classical conditioning.

continuous variable One that can assume any value whatsoever between certain limits, versus a discrete variable that can assume only specific values (usually integers).

control group An observed, unmanipulated set of test subjects. In an experimental manipulation, we must also perform the actions involved in that manipulation on a second test group of organisms in order to ascertain whether any effects of the experimental treatment are due to that treatment and not to the actions involved in providing the treatment.

cooperation Two or more individuals behaving to achieve a common purpose.

cooperative breeding A breeding system in which nonparents share in rearing the young.

core area The area of heaviest use within the home range.

corpora allata Paired invertebrate endocrine glands, located in the thoracic region, which secrete juvenile hormone that prevents maturation.

corpora cardiaca Paired invertebrate endocrine glands, located near the heart, which are involved in regulation of metabolism. They receive neurosecretions from the brain and release endocrine products into the hemocoel.

cost of meiosis The reduction in numbers of alleles passed on to subsequent generations as a result of reproducing sexually compared to asexually.

coursers Predators that chase their prey over long distances.

crepuscular Daily cycles with peak activity around dusk and/or dawn.

critical period A discrete portion of a sensitive period during development, during which the probability for forming and reinforcing the behavior is greatest.

cryptic female choice Choice of mates by females that takes place during or after copulation in cases where females mate with more than one male. Sperm of particular males is favored in fertilizing eggs.

crystallized song The final stage in song development, when the song variation (repertoire) has decreased and become more fixed.

cupula A small inverted cup that covers another structure; in vertebrate lateral line and equilibrium organs, the cupula covers hair cells in a gelatinous matrix.

currency The resource that is being maximized in an optimality model; usually assumed to be energy in an optimal foraging model.

cycle Repeating units that make up the pattern of biological rhythms.

decisions In an optimality model, the set of strategies that is available to an animal.

delayed response problem An operant conditioning procedure in which the test animal sees the experimenter set up the test situation, including what will be the correct response, and then must wait until the experimenter provides the animal with an opportunity to respond.

dendrites The part of the nerve cell that conveys information toward the cell body; most neurons have several.

dependent variable What may change (e.g., aggression) in response to manipulation of an independent variable (e.g., food supply) by the experimenter.

depolarized The reduction in the difference in charge (potential) between the outside and inside of a membrane.

deprivation experiments Experimental manipulations involving removal of particular types of stimuli (e.g., social, sensory, motor) to ascertain the effects later in development.

despot A dominant individual that controls resources.

diapause phase A period of dormancy, common in insect species, which occurs during the more rigorous portions of the annual climatic cycle.

diet selection model A type of optimality model that concerns the sort of food an animal should attempt to acquire and eat.

dilution effect Safety in numbers; when predators can attack only one prey at a time, animals in a group are less likely to be eaten than solitary animals.

dimorphic Having more than one form, size, or appearance. Sexual dimorphism refers to the difference between conspecific males and females.

diploid Having both members of all chromosome pairs.

direct fitness A measure of an individual's potential to contribute genes to future generations via personal reproduction.

discrete signal A form of communication that is digital, or all or none; usually given at the same intensity each time, such as an alarm call.

discrete variable A variable that can take on only certain values (usually integers), as opposed to a continuous variable.

discrimination The ability to detect differences between two or more stimuli and make choices based on those differences in a learning situation.

dispersal A more or less permanent movement of an individual from an area, such as movement of a juvenile away from its place of birth.

displacement activity An innate or stereotypic response to a stimulus that seems inappropriate or irrelevant to the situation.

display Any behavior pattern especially adapted in physical form or frequency to function as a social signal in communication.

diurnal An animal with an activity period during the light portion of the daily cycle.

dizygotic Twins that arise from two different zygotes, hence, individuals no more closely related than two different-aged siblings.

dominant trait A trait whose phenotype is determined by a single allele at a particular locus.

ecdysone A hormone secreted by the prothoracic gland, which controls molting in insects.

eclosion The process whereby the adult form of an insect emerges from the pupa.

ecological niche The range of ecological variables (e.g., temperature, moisture, etc.) in which a species can exist and reproduce.

economic defendability Defense of a resource that yields benefits that outweigh the costs of defending it.

effector A tissue or organ that responds to an action potential or a hormone.

electrical synapse The type of synapse where communication between neurons is through electrical transmission rather than a chemical intermediary.

electric organ discharges (EODs) Produced by some species of fish; may be used in defense, prey capture, and as a channel of intraspecific communication.

electrophoresis The separation of different proteins or nucleic acids within a gel matrix that is subjected to an electric field; separation is based on size and/or charge of the molecule(s).

electroreceptor Sensory receptors that detect electric fields.

endocrine glands A series of ductless glands in both invertebrates and vertebrates that release hormones into the body through the blood or lymph.

endogenous Processes within the animal; used here with particular reference to the internal, genetically based components of biological rhythms.

endogenous clock mechanism Any internal processes that are genetically based and that play a role in setting or regulating biological rhythms.

enrichment experiments Experimental manipulations that involve providing organisms with particular types of stimuli (e.g., social, sensory, motor) to ascertain the effects later in development.

entrainment The process by which a biological clock is set or reset by synchronizing with the period of some external, environmental stimulus.

epicycle See ultradian rhythm.

epigenesis The integrated, interactive process of behavior development involving the genetics of the organism and its environment, including all of its experiences.

epistasis One or more alleles at one locus that influence the expression of alleles at another locus.

equation With respect to comparative learning, the attempts to match situations and procedures for examining learning behavior in different species.

estrous cycle The period of behavioral and physiological changes from one ovulatory event (estrus) to another.

estrus The period of sexual receptivity at the time of ovulation.

ethogram An inventory of all of the behavior patterns of a species.

ethology The study of patterns of animal behavior in natural environments, stressing the analysis of adaptation and the evolution of patterns.

eusocial A social system involving reproductive division of labor (i.e., castes) and cooperative rearing of young by members of previous generations.

evolution A change in the frequency of alleles in a population over generations. The change is caused by natural selection and/or genetic drift.

evolutionarily stable strategy (ESS) A strategy that when employed by most individuals in a population cannot be out-competed by some alternative strategy.

experience All of the interactions between an organism and its environment, beginning at conception and including both external and internal influences.

experimental design The prescription of the treatment groups, dependent and independent variables, and sample sizes for the testing of a particular hypothesis.

experimental hypotheses The biological ideas, both broad and specific, developed from knowledge about previous investigations and our own imagination and creativity, that lead to tests of statistical null hypotheses using experimental designs.

exploitative A type of competition in which organisms passively use up limited resources; also called scramble competition. Contrast with interference or contest competition, in which organisms defend or otherwise control resources.

exploratory behavior The spontaneous search for and active investigation of objects, situations, or other organisms in the absence of any homeostatic need.

extinction In learning, the decrease of response rate or magnitude of response with lack of reinforcement in a learning situation.

fatigue Loss of efficiency in the performance of a motor act when that act is repeated in rapid succession.

feedback loops A sequence of events in which the level of a hormone or related endocrine product circulating in the blood leads to alterations in the rate of production and release of other hormones from one or more endocrine glands. Often, feedback loops involve the hypothalamic portion of the brain where specialized sensory cells monitor circulating blood levels of numerous compounds, including both hormones and products from hormone actions at various body locations.

female defense polygyny Males controlling access to females directly by competing with other males.

filial imprinting The process by which young animals form a social attachment for a particular stimulus, often the mother.

fitness The potential for an individual to contribute genes to future generations as a function of its adaptive traits.

fixed action pattern (FAP) An innate behavior pattern that is stereotyped, spontaneous, and independent of immediate control, genetically encoded, and independent of individual learning.

flagging behavior Alarm signaling, as with the use of the tail.

flehmen A retraction of the upper lip exhibited by certain mammals soon after sniffing the anogenital region of another or while investigating freshly voided urine.

fluctuating asymmetry Random deviations from bilateral symmetry in paired traits.

focal-animal sampling A technique for recording the behavior of individual animals in which the observer concentrates on the activities of one individual for a set time period and then switches to watch another animal for a set time period.

foci Groups of cells in an embryo that differentiate into specific structures and organs.

founder effect The establishment of a genetically distinct population based on a random event in which the founders of the population represent only a subset of the genetic variability in the parent population.

free-running rhythm The activity cycle that an animal exhibits when placed in a constant environment; its period is different from any known cyclic environmental variable.

functional neuroanatomy The study of the size, structure, and arrangement of cells within the nervous system, particularly the brain.

functionalists People who conduct studies involving attempts to discern how the mind works, as opposed to studies of its structure.

fundamental niche The multidimensional space that an animal would occupy under optimal conditions in the absence of competitors; also called preferred niche. Contrast with realized niche.

gametes Mature haploid cells (sperm and ova) that fuse to form a zygote.

ganglion A distinct cluster of nerve cell bodies.

gene A sequence of nucleotides on a DNA molecule that is the basic unit of heredity.

gene flow The movement of genes from one population to another via migration or interbreeding.

generalists Species that occupy a wide range of habitats and eat a variety of food types. Contrast with specialists.

generalization The phenomenon in which stimuli similar to those used in training also elicit a response.

generation time The length of time between when an organism is born and when it first reproduces.

genetic drift The occurrence of random changes in the gene frequency in a population due to chance rather than selection, gene flow, or mutation.

genetic screening The use of a number of mutant forms of a species to test for a specific type of change or deficit, as with respect to biological clocks.

genotype The genetic constitution of an individual.

glial cells Supportive cells that are closely associated with neurons.

gonads Glands responsible for the production of gametes and where certain gonadal hormones are produced. These consist of the ovaries in females and the testes in males.

graded signal A form of communication that is analog, or varying in intensity or frequency, providing quantitative information about the strength of the stimulus.

green beard effect A mechanism for recognizing kin that does not require previous experience. Recognition genes enable an organism to identify and behave altruistically toward other organisms bearing that same gene.

group selection Selection that operates on two or more genetic lineages (groups) as units; broadly defined, this includes kin and interdemic selection.

gustatory Having to do with the sensation of taste, the chemoreception of molecules in solution by specialized epithelial receptor cells.

habitat imprinting The tendency of an animal to prefer a place to live based on early experience in that habitat.

habitat selection The choosing of a place to live.

habituation The relatively persistent waning of a response that results from repeated presentations not followed by any form of reinforcement.

hair cell An epithelial cell that bears modified cilia on one end; when the cilia are deflected, they change the receptor cell's potential.

handicap hypothesis The hypothesis that apparently deleterious sexual "ornaments" possessed by males in many species are attractive to females because they indicate that the males have such vigor (good genes) that they can survive even with the handicap.

handling time The time it takes to process a food item for consumption.

haplodiploidy The mode of sex determination by which males develop from haploid (unfertilized) eggs and females from diploid (fertilized) eggs; characteristic of the insect order Hymenoptera.

haploid Containing one member of each pair of chromosomes, or half the chromosomes of a diploid cell.

Hardy-Weinberg equilibrium The application of the binomial expansion to determine the frequency of two alleles in the absence of any evolutionary forces.

helpers-at-the-nest Individuals other than the parents that forego reproduction to assist rearing the young of others.

heritability in the broad sense The amount of the variance in a trait in a population that is due to genetic factors; defined as the ratio of the variance due to genetic factors to total phenotypic variance.

heritability in the narrow sense A measure of the extent to which genes inherited from parents determine phenotypes; defined as the ratio of the additive genetic variance to total phenotypic variance; also called realized heritability.

heterozygous A condition where an individual has two different alleles of the same gene on homologous chromosomes.

hibernation A condition of deep sleep and reduced metabolic activity observed in some animals, particularly during the winter months.

hippocampus Part of the vertebrate brain important in learning and spatial memory.

homeostasis A tendency toward a stable or equilibrium state with respect to the internal physiological conditions of an animal.

home range The area in which an animal spends most of its time.

homing The ability of an animal to return to its home site after being displaced.

homologous chromosomes (homologue) Chromosomes that exist in pairs; each homologue possesses the same genes or loci, but the homologues may have different alleles at the same locus; one member of each pair comes from each parent.

homology A similarity between two structures that is due to inheritance from a common ancestor.

homozygous A condition where an individual has two copies of the same allele at the same locus on homologous chromosomes.

hormones Chemical products of ductless glands and neurosecretory cells that are carried by the circulatory system and that influence various physiological processes in the body.

hyperpolarized A description for a membrane whose polarity is greater than its typical resting potential.

ideal free distribution The distribution of individuals among resource patches of different quality that equalizes the net rate of gain of each individual. Assumes that organisms are free to move and have ideal knowledge about patch quality.

imitation An act that occurs when an animal immediately copies the actions of another animal while they are in each other's presence.

immunocompetence A measure of the capacity of an organism's immune system to respond to challenges in the form of pathogens such as bacteria or viruses.

imprinting A process that occurs when an animal learns to make a particular response to only one type of animal or object. The sensory modes used for establishing such a connection can be visual, auditory, olfactory, or some combination of these, depending upon the animal.

inbreeding depression Reduction in offspring fitness resulting from mating among kin; caused in part by the accumulation of deleterious recessives.

inclusive fitness The sum of an individual's direct and indirect fitness. *See* direct and indirect fitness.

independent data points Many statistical tests require that data points not depend on each other in anyway. For example, data points from the same individual measured twice could not be considered independent.

independent variables The variables that define the treatment conditions in an experiment; literally, the variables that we manipulate to test their effects on behavior.

indicator traits In reference to sexual selection, traits that provide information to members of the opposite sex about the health or fitness of the bearer.

indirect fitness A measure of the potential for an individual to contribute copies of its genes to future generations through its influence on the reproduction of nondescendant relatives. Only the portion of the nondescendant kins' reproduction due to helping by the relative is included in the measure of its inclusive fitness, and it is devalued in proportion to the degree of relatedness.

individual distance The fixed minimal distance an animal keeps between itself and other members of its species.

induced ovulators Species in which copulation is necessary to effect ovulation.

industrial melanism The increase in the prevalence of darker-colored morphs as a result of environmental changes caused by industrial pollution.

information In communication, a measure of the uncertainty of the behavioral response of an individual. In learning, refers to the fact that for successful classical conditioning to occur there must be some intuitive knowledge of the relationship between the CS and US.

inhibition In the context of modern classical conditioning, the learning of negative relationships between the CS and US.

innate Behavior that has either a fixed genetic basis or a high degree of genetic preprogramming.

innate releasing mechanism (IRM) A neural process, preprogrammed for receiving a particular sign stimulus, that mediates specific behavioral responses.

instar The name given to the form of an insect or arthropod between molting episodes.

instrumental conditioning *See* operant conditioning.

intelligence A collection of mental capacities including imagination, problem-solving ability, memory, the ability to use information gained from past experiences, perceptiveness, and behavioral flexibility. The processes by which animals obtain information about their environment, and retain and use the information to make decisions during the course of their behavioral activities.

intention movement The preparatory movements that an animal makes prior to a complete behavior pattern, as with wing movements before taking flight.

interference A form of competition in which organisms defend or otherwise control limited resources; also called contest competition. Contrast with exploitative competition.

intermittent reinforcement Schedules of reinforcement that involve providing rewards, not on every trial, but on usually one of four other types of schedules: fixed ratio, fixed interval, variable ratio, or variable interval.

interneuron A neuron that connects two or more other neurons.

interobserver reliability A method for checking on the agreement between two or more people watching the same animals behave. This becomes important both as a check on the definitions developed for particular behavior patterns and when we have two or more individuals gathering the data in a particular investigation and we wish to know that they are taking down their records of the data in as nearly similar a manner as possible.

intersexual selection Selection on characteristics of one sex (usually males) based on mate choices made by members of the other sex (usually females).

intrasexual selection Selection on characteristics of the sexes based on competition among them (usually males) for access to members of the other sex (usually females).

irritability The ability of a cell to undergo a change in membrane potential.

isogamy A condition in which gametes are equal in size.

isolating mechanisms Any physical or behavioral characteristic that prevents successful exchange of genes between members of different species or populations.

iteroparity The production of offspring by an organism in successive bouts. Contrast with semelparity.

kinesis Random locomotion patterns in which there is no orientation of the organism's body axis to the source of stimulation.

kin selection The selection of genes due to individuals assisting the survival and reproduction of nondescendant relatives who share genes by common descent.

knockout gene A genetically engineered mutant gene that has been targeted for disruption so that it no longer functions normally.

K selection Selection favoring characteristics that are adapted to stable, predictable environments.

latent learning Associations made with neither immediate reinforcement or reward nor with particular behavior evident at the time of learning. The animal may store and use such information in appropriate situations at a later time.

learning The relatively permanent change in behavior or potential for behavior that results from experience.

learning curve A graphical presentation of the response measures in a learning situation. This may be depicted as correct or incorrect responses, or as a proportion of the trials given to the animal on which it gave correct or incorrect responses.

learning set The acquisition of a learning strategy by the animal; the ability to transfer learning between problems of the same type.

lek An area used, usually consistently, for communal courtship displays.

lesion An area of tissue that has been destroyed by an agent such as electric current or a chemical.

life range The larger geographic area that an animal utilizes over the course of its life.

limbic system A group of nuclei and brain regions in vertebrates that control emotion, and sex and feeding drives.

local mate competition Competition for mates that may arise among siblings depending on the population size and operational sex ratio and that may favor the evolution of skewed offspring sex ratios.

local resource competition model In female philopatric primates, females giving birth to a disproportionate number of males that will disperse and not compete with each other, or high-ranking females giving birth to a disproportionate number of daughters that are able to compete for resources effectively.

locomotion-fear dichotomy A process during the imprinting process in young birds whereby they initially are more likely to locomote and follow a stimulus object and later develop a fear toward objects; thus, imprinting occurs when the locomotor tendency is high and fear of objects is low.

locus (pl. loci) The position of a particular gene on a chromosome.

lymphocyte A class of white blood cells consisting of a variety of types that function, in the presence of foreign materials, to immobilize, disable, and destroy the pathogen.

macroevolution Evolutionary events occurring above the species level.

macrophage White blood cells that phagocytize (engulf and destroy) foreign materials such as bacteria or viruses.

male dominance polygyny A mating system in which males compete and acquire dominance ranks that influence their access to females, with higher-ranking males often having greater mating activity.

marginal value theorem A model that predicts when an organism should cease foraging in one patch and travel to another.

maternal condition model In polygynous mammals, females that are high ranking, well nourished, or otherwise in good condition giving birth to a greater proportion of sons since their sons will outreproduce their daughters and vice versa for females that are in poor condition.

mating system The species-typical pattern of mate finding, reproduction, and parenting of offspring.

mean The central tendency of a series of measurements or observations, calculated by summing all values in a distribution and dividing the total by the number of values.

mechanoreceptor A sensory cell that is stimulated by some mechanical force or pressure.

meiosis The process in which haploid cells (gametes) are formed from diploid cells by the separation of homologous pairs of chromosomes.

menses The period of shedding of the lining (endometrium) of the uterus and associated fluids if an ovum is not fertilized, most notably in primates.

menstrual cycle The period from the end of one ovulatory cycle, as demarcated by menstrual flow, to the end of the next cycle in female primates.

metacommunication Communication about communication, where one signal changes the meaning of signals that follow.

microevolution Changes in gene frequencies or traits that occur in small increments.

migration A seasonal movement from one location to another.

modal action pattern (MAP) A spatiotemporal behavior pattern that is common to members of a species; different individuals perform the pattern in a recognizably similar fashion.

molt A process in arthropods, such as crustaceans and insects, as well as in some vertebrates, such as snakes, whereby the organism sheds its outer exoskeleton or skin periodically as it grows. Also refers to changing the pelage or feathers in a mammal or bird.

monestrous A species in which the female is receptive for only a few days once each year.

monogamy A mating system in which a male and female bond for some period of time and share in the rearing of offspring.

monophyletic origin A group of organisms that evolved from a single ancestral type.

monozygotic Twins that arise from a single zygote, hence, two genetically identical individuals.

mosaic An organism whose tissues are made up of two or more genetically different types.

motivation Internal processes that arouse and direct behavior. Today this refers more to examination of the consequences of behavior and feedback to the animal concerning the consequences of those actions.

motor neuron A neuron that synapses with a muscle membrane.

Müllerian mimics Noxious species that resemble each other.

Muller's ratchet The steady accumulation of mutations in a population of asexual organisms over time.

mutation A change in the sequence of nucleotides in a gene.

mutational analysis An analysis that examines the effects of specific mutants for the effects on particular phenomena with respect to morphology, physiology, or behavior.

mutualism Interaction between two organisms of the same or different species from which both benefit.

natural selection The disproportionate survival and reproductive success of organisms that possess certain alleles, as a result of the influence of those alleles.

navigation The process by which an animal uses various cues to determine its position in reference to a goal.

neoteny The persistence of juvenile traits in the adult form.

nerve cell body The largest part of a neuron that typically contains the nucleus.

nerve net A diffuse, noncentralized network of nerve cells.

neuromodulation A change in the response of a neuron to further input.

neuron A nerve cell.

neurosecretions (neurosecretory cells) Hormonelike chemical substances that are produced by specialized neurons that affect various physiological processes in the body.

neurotransmitter A chemical released by the presynaptic membrane of a synapse that attaches to receptor molecules on the postsynaptic membrane and causes a change in the permeability of that membrane.

nocturnal Animals whose primary activity occurs during the dark portion of the daily cycle.

null hypothesis In a statistical sense, the statement that there will not be any difference between effects of experimental treatments.

null model A hypothesis that tells us what to expect if no special forces are at work. An example is the Hardy-Weinberg equilibrium.

observability Not all of the members of a group are always seen the same proportion of time. Animals of different ages, sexes, or of different dominance status may be seen more or less than other animals.

observational learning The tendency to perform an appropriate action or response as the result of having observed another animal's performance of similar actions in the same situation.

observational method In an historical context, the notion, developed by C. Lloyd Morgan, that only data gathered by direct experiment and watching the animals under study could provide the basis for generalizations. This was in contrast to Romanes' notion that much about animal behavior could be gained by inference from what was seen. In the modern sense, this refers to any of a number of techniques used to watch and record the actions and interactions of animals.

olfactory Having to do with the sense of smell, the chemoreception of molecules suspended in air.

ommatidia The functional units of the invertebrate compound eye, each one of which has a lens, focusing cone, photoreceptive cells, and forms an image.

on-center The inner circle of cells of a receptive field that stimulate ganglion cells if they are stimulated by light.

ontogenetic niche The multitude of inherited ecological and social traits that are, in effect, passed on from generation to generation and that play integral roles, in concert with genetic inheritance, influencing behavior development.

operant (instrumental) conditioning A type of learning that involves the animal's forming a variety of associations during which it learns to associate its behavior with the consequences of that behavior. That is, the sequence of events is dependent upon the behavior of the animal.

operational sex ratio The number of sexually mature males and females in a population.

optimality modeling A model that allows us to predict which decision an animal should make under a given set of circumstances in order to maximize its fitness.

organizational effects Refers to the effects of exposure to certain hormones during critical periods in development that affect the central nervous system and other structures, producing morphological and behavioral changes in adulthood. Contrast with activational effects.

orientation The way an organism positions itself in relation to external cues.

oscillator The internal mechanism that is the clock in a biological rhythm.

outbreeding depression A reduction in the fitness of offspring resulting from matings among individuals from different populations or demes, possibly caused by the breaking up of coadapted gene complexes.

pancreas A gland located in the abdomen that produces both digestive enzymes (exocrine pancreas) and hormones (endocrine pancreas). Key hormones produced by the pancreas are insulin and glucagon, which play roles in regulating blood glucose levels.

parental investment Any investment in an offspring that increases its chances of survival and reproduction at the expense of the parents' ability to invest in other offspring.

parental manipulation The selective providing of care to some offspring at the expense of other offspring so as to maximize the parents' reproductive success.

passive avoidance learning A form of operant conditioning, in which the animal does not make an overt response, but rather is trained to avoid making a particular response by being exposed to some type of noxious stimulus each time the response occurs.

patches Regions of localized concentrations of resources.

patch model A type of optimal foraging model that assumes that prey occur in discrete clumps and that seeks to predict where and when organisms will forage for these prey.

path integration A process by which an organism uses internal spatial localization to return from an outward-bound trip. Also called idiothetic orientation or dead reckoning.

Pavlovian conditioning *See* classical conditioning.

payoff matrix The outcome of a game, in which the costs and benefits resulting from different strategies are listed for each player.

pelage The body covering of hair in a mammal.

perception The analysis and interpretation of sensory information.

perceptual world The manner in which an organism's brain analyzes and interprets all incoming stimuli. Each species has a different view of the world, its *Umwelt.*

period The duration of one cycle of a biological rhythm.

peripheral nervous system All nerves that run to and from central nervous system.

phase A specified, recognizable portion of the activity cycle.

phenotype The observable characteristics of an organism that result from the influence of both the organism's genotype and environmental factors.

phenotype matching A mechanism by which kin may recognize one another; individuals use kin as a referent whose phenotypes are learned by association.

pheromone A species-specific odor cue released by animals that influences the behavior and/or physiology of conspecifics.

philopatric Remaining near the place of birth after sexual maturation.

phonotaxis Orientation with respect to sound.

photopigment A molecule in visual receptor cells (e.g., rhodopsin) that responds to light energy.

photoreceptors Sensory cells that contain photopigments and respond specifically to light energy.

phototaxis Orientation with respect to light.

phylogeny The evolutionary history of a group of organisms.

piloting The use of familiar landmarks to find a direction or area.

pineal gland An endocrine gland located near the midline of the brain that produces melatonin, a hormone involved in biological rhythms, particularly in annual cycles.

pituitary gland The master gland (along with the hypothalamus) of the endocrine system of vertebrate animals. Located directly below the hypothalamus of the brain, the pituitary produces or releases a variety of hormones that target other endocrine glands of the body.

plastic song One of the stages of song development in birds during which the bird has begun to sing using the species-typical pattern, but with a certain degree of variation.

play behavior Certain locomotor, social, and manipulative behavior patterns exhibited by young and some adult mammals and birds.

pleiotropy A description for a gene or set of genes that influences the phenotype of more than one characteristic.

polarized A description for a membrane that has a potential difference due to an unequal distribution of ions across the membrane.

polyandry A mating system in which some females mate with more than one male and males usually invest heavily in offspring.

polygamy A mating system in which both males and females mate with several members of the opposite sex.

polygenic trait Traits that are influenced by many genes and that are usually continuously distributed within a population.

polygyny A mating system in which some males mate with more than one female and females usually invest heavily in offspring.

polygyny threshold The point at which a female will benefit more by joining an already mated male possessing a good territory rather than an unmated male on a poor territory.

polymorphic A locus that contains two or more alleles within a population.

polyphyletic Describes a group of taxa that arose from multiple ancestors.

postzygotic isolating mechanisms Isolating mechanisms (barriers to interbreeding) that occur after the formation of zygotes.

precocial young Young that are born or hatched at a relatively advanced state of development and are capable of a more independent existence beginning at birth or hatching. Compare to altricial.

preparedness The genetically based predisposition to learn. Depending upon the learning situation, animals of a particular species may be prepared, unprepared, or contraprepared to learn.

prezygotic isolating mechanisms Isolating mechanisms (barriers to interbreeding) that occur before the formation of zygotes.

priming pheromone A chemical signal that alters the physiology of another organism, eventually causing a change in its behavior. Contrast with releasing pheromone.

profitability The ratio of energy gained to the handling time of a type of food.

promiscuity A mating system in which there is no prolonged association between the sexes and multiple mating by at least one sex.

proprioreceptor Sensory receptor that relays information about the position and movement of the body; located primarily in muscles, tendons, and the inner ear.

protandry A sequentially hermaphroditic species in which individuals change from males to females.

prothoracic gland An invertebrate endocrine gland located in the prothorax region directly behind the head, which secretes ecdysone, a steroid hormone that promotes molting.

protogyny A sequentially hermaphroditic species in which individuals change from females to males.

proximate causation (factors) Mechanistic explanations for how behavior occurs, including, in particular, hormones, the nervous system, and behavior development.

pseudoconditioning The strengthening of a response to a previously neutral stimulus by repeatedly eliciting the response with another stimulus without pairing the presentation of the two stimuli.

pseudoreplication Artificially inflating sample size by incorrectly counting units as independent when they are not. Can result from either repeated testing of the same animal without proper statistical correction using a repeated measures design, or counting members of an interacting social group as separate data points. In both cases, problems with independence of data points arises with pseudoreplication.

psychobiology The study of the mechanism and function of the central nervous system from both psychological and biological perspectives.

psychopharmacology The study of the brain's behavior in terms of chemical, physiological, and psychological parameters.

puberty The age at which an organism can first reproduce.

punctuated equilibria The hypothesis that evolution occurs in relatively rapid bursts, interspersed with long periods of stasis.

---

quantitative genetics The study of continuously varying traits.

quantitative trait locus (QTL) The set of genes that governs the quantity of a trait that is not completely determined by any one gene acting alone.

---

realized heritability *See* heritability in the narrow sense.

realized niche The multidimensional space that an animal actually occupies in the presence of competitors, predators, pathogens, and limited food. Contrast with fundamental niche.

receiver bias The increased sensitivity of perceptual systems to certain stimuli as a result of natural selection. These biases may then influence the evolution of communicative behavior.

receptive field The region of a sensory surface that causes the change in activity of a neuron.

recessive trait A trait caused by an allele, the phenotype of which is suppressed when it occurs with the dominant allele. Hence, two recessive alleles are necessary for the recessive phenotype to be displayed.

reciprocal altruism The trading of altruistic acts by individuals at different times, i.e., the payback to the altruist occurs some time after the receipt of the act.

recombination The formation of new combinations of alleles through exchange of sections of homologous chromosomes during the process of meiosis; sometimes refers to the combinations of new alleles that result from fertilization.

reconciliation Social interactions, such as appeasement, that take place between two or more individuals following agonistic encounters.

Red Queen hypothesis The hypothesis that sexual reproduction has evolved because the genetic variation that results is adaptive in the evolutionary "arms race" between hosts and their pathogens and parasites as well as between predators and prey.

reflex A simple, stereotyped behavior in response to a specific stimulus.

refractory period The time following a nerve impulse during which another nerve impulse cannot occur.

regulatory gene A gene that turns other genes on and off.

reinforcement In learning, anything that alters the probability of behavior. See behavioral isolating mechanism.

reinforcement of prezygotic isolation The concept that prezygotic reproductive isolating mechanisms may be under direct selection.

releasing factors Hormones or neurosecretions from the hypothalamus that travel either via the hypothalamic-pituitary portal system or along axons between the hypothalamus and the pituitary where they exert their effect in terms of production and release of hormones.

replication (1) The sample size or number of animals in each particular treatment in an experiment. (2) The ability of other investigators to use your description of the methods that you have employed in order to repeat exactly what you have done in your experiment.

reproductive effort The energy expended and risk taken to reproduce, measured in terms of the decrease in ability of the organism to reproduce at a later time.

reproductive success The number of progeny born, or surviving progeny produced by an organism.

resource defense polygyny Males control access to females indirectly by monopolizing critical resources.

response to selection (R) The difference in the mean phenotypic value between a group of progeny and the mean value for the parents of those progeny.

resting potential The unequal distribution of charges on either side of a neuron membrane that is not undergoing an action potential.

risk-sensitive foraging A model that incorporates variation in prey encounter rates to predict where predators forage. Depending on their internal state and other factors, predators may be risk-prone, foraging where encounter rates are variable, or risk-averse, foraging where encounter rates are relatively constant.

ritualization Evolutionary process by which behavior patterns become modified to serve as communication signals.

rod A class of vertebrate photoreceptors that, unlike cone cells, is usually insensitive to color.

r selection Selection favoring rapid rates of reproduction and growth, especially among species that specialize in colonizing short-lived habitats.

runaway selection Selection for male ornaments that happens due to the genetic correlation, and resulting positive feedback relationship, between the trait and the preference for the trait.

salience In the context of classical conditioning, the fact that animals can learn to make certain associations, depending upon which stimuli are involved, more rapidly than other associations.

scan sampling A method for recording behavior observations that involves looking briefly at each animal at prescribed time intervals (e.g., every minute or every five minutes) to record their activity at the time of the sample.

scramble polygyny A mating system in which males actively search for mates without overt competition.

search image The hypothetical mental image of a prey species that a predator forms as it improves in its ability to capture that particular species.

self-propagating A description of the events occurring during an action potential, with each regional depolarization by sodium voltage-gated channels causing a similar event at an adjacent area downstream.

semantic communication Of or relating to meaning of signals; specifically used to denote the use of different alarm signals to warn about different predators.

semelparity The production of offspring by an organism once in its life.

sensation The process of transducing environmental stimuli or energy into action potentials.

sensilla A chitinous, hollow, hairlike projection of the arthropod exoskeleton that serves as an auxiliary structure for sensory neurons.

sensitive period The time interval when an animal can develop an imprinting attachment. More broadly, in behavior development, this refers to time intervals when particular events must occur for proper ontogenetic sequencing.

sensitization (1) Enhanced responsiveness to a repeated stimulus. (2) Strengthening of a response that was initially produced via a CS resulting from a pairing with a US and UR. (3) A stimulus priming the animal to pay particular attention to what follows.

sensory adaptation A process that occurs at the level of the sensory receptors and that consists of a slowing down or cessation of nerve impulses transmitted to the central nervous system.

sensory deprivation Withholding all or a specified portion of the sensory input that an animal would normally be receiving.

sensory filter Neural circuits that selectively transmit some features of a sensory input and ignore other features.

sensory neuron A neuron that is modified to respond to a particular set of stimuli.

sexual imprinting The process by which young of many species establish an attachment to opposite sex conspecifics; the effect manifests itself later in life during the process of finding mates.

sexual selection Selection in relation to mating. It is composed of intrasexual competition among members of one sex (usually males) for access to the other sex and intersexual choice of members of one sex by members of the other sex (usually females).

shaping The procedure used in operant conditioning wherein the experimenter can reinforce particular responses or behavioral actions by the subject and not others, thus enhancing the frequency of some actions and extinguishing other actions.

siblicide The killing of one sibling by another.

signal Any behavior pattern that conveys information. Patterns modified through evolution to convey information are called displays.

signaling pheromone A chemical signal that is quickly perceived and causes an immediate response. Also called releasing pheromone. Contrast with priming pheromone.

signs *See* sign stimuli.

sign stimuli Specific external stimuli that trigger stereotyped responses from conspecifics, usually called fixed action patterns.

social deprivation Withholding or removal of contact with any form of stimulation from conspecifics.

social organization The species-typical pattern of relationships among all members of a group. This would include spatial distribution patterns, interindividual relationships involving dominance hierarchies or territoriality, mating systems, parenting, and dispersal.

society A group of individuals belonging to the same species and organized in a cooperative manner. Usually assumed to extend beyond sexual behavior and parental care of offspring.

sociobiology A study that involves the application of the principles of evolution to the study of the social behavior and social systems of animals.

somatic Having to do with body cells, i.e., those that do not produce gametes.

sound window The use of frequencies for communication that are transmitted through the environment with little loss of strength (attenuation).

specialists Species that occupy a narrow range of habitats and eat a narrow range of foods. Contrast with generalists.

speciation The process of two populations that share a common descent evolving in different ways so that they ultimately do not interbreed and are therefore different species.

species The basic (natural) unit of classification, consisting of groups of actually or potentially interbreeding individuals that are reproductively isolated from other such groups.

species-typical behavior Actions and displays that are broadly characteristic of a species.

sperm competition A situation in which one male's sperm fertilize a disproportionate number of eggs when a female copulates with more than one male.

spontaneous ovulators Species in which females release eggs whether they have copulated or not.

spontaneous recovery A process in which a conditioned response has been extinguished and then is followed by some time interval (generally one minute up to one day or more depending upon the species and experimental conditions), after which the animal may immediately exhibit a nearly normal correct response rate upon reintroduction to the test situation.

stalk-and-rush Predators that approach prey as closely as possible, then close with a sudden burst of speed. Contrast with coursers.

standard deviation A statistical measure of the degree of variation from the mean value among the individual measurements in a series of values.

standard error of the mean (SEM) A statistical measure of variation most properly restricted to use with a group of means, though often reported as a measure of variation around the mean value in a series of individual measurements.

stereocilia The short, modified cilia at the apex of a hair cell.

strategy Alternative course of action.

stretch receptor Sensory receptor that responds to stretch; found in muscle tissue, lungs, and other organs that undergo changes in position or size.

stridulation The production of sound by an insect rubbing one body part against another (e.g., in male crickets).

structural gene A gene that determines the amino acid sequence of a protein.

sun compass The mechanism in which animals use the sun and an internal clock that allows them to adjust for movement in the sun as they navigate.

suprachiasmtic nucleus (SCN) A brain nucleus involved in the visual pathway that has been clearly identified for its involvement in biological rhythms mediated by photoperiod.

sympatry Populations or species with overlapping geographic distributions. Contrast with allopatry.

synapse A junction between two neurons or between a neuron and a muscle fiber in which action potentials along the presynaptic membrane will influence the postsynaptic membrane.

synaptic processes The threadlike extensions at the end of an axon.

synergism The interaction of two or more agents or forces so that their combined effect is greater than the sum of their individual effects, as when two hormones combine to affect target tissues.

syntax Information provided by the sequence in which signals are transmitted.

syrinx In birds, a part of the respiratory tract specialized for sound production.

systematic variation In learning, a procedure that involves changing the conditions under which animals are tested in a systematic way in order to better compare species from different taxonomic groups.

taxis (pl. taxes) Directed reactions to a stimulus involving an orientation of the long axis of the body in line with the stimulus source.

temperature-compensated rhythm The relative insensitivity of biological rhythms to the effects of temperature; this contrasts with the fact that many chemical reactions double in rate for every $10°C$ increase in temperature.

territory An area occupied exclusively by an animal or group of animals that is defended.

thermoreceptors Sensory cells that are sensitive to changes in temperature.

threshold The minimum stimulus necessary to initiate an all-or-none response.

thyroid gland An endocrine gland, located in the throat region, whose major endocrine products are protein hormones involved in regulation of cellular metabolism.

tradition A behavior pattern that is passed from one generation to the next by learning.

transection The severing of a nerve to determine its specific function(s).

translocation Moving animals from one location to another, for instance to determine whether and how soon they shift their activity cycle to match the photoperiod and/or other features in their new location.

transplantation The movement of neural or hormonal tissue from one area of an animal to another or from one animal to another. Also, the movement of individuals or colonies from one location to another.

trophic hormones (trophic neurosecretions) Hormonal or neurosecretory products from endocrine glands or neurosecretory cells that influence the production and release of other hormone products from endocrine glands.

true language Communication that includes the use of symbols to represent abstract objects or ideas, and syntax, where those symbols convey different messages depending on their sequence.

true navigation The ability to maintain or establish reference to a goal without the use of landmarks.

ultimate causation (factors) Those aspects of behavior that are concerned with why the behavior evolved and its functional significance in an ecological context.

ultradian rhythm A cyclical rhythm of less than 24 hours.

Umwelt The sensory and perceptual world of the animal, dependent upon the types of sensory receptors that it possesses and the internal nervous system processes for receiving and interpreting those stimuli.

unconditioned response (UCR) In classical conditioning, the animal's response that is initially given to the unconditioned stimulus.

unconditioned stimulus (US) In classical conditioning, the stimulus that produces what is initially the unconditioned response.

utility In foraging theory, the value that a resource has to an animal.

variance A statistical measure of the amount of variation in a series of measurements.

voltage-gated channels Ion-specific channels that open only in response to a specific polarity across the cell membrane.

warning coloration *See* aposematic coloration.

Wisconsin General Test Apparatus (WGTA) An apparatus designed originally at the University of Wisconsin Primate Center for use with various species of primates in a series of learning problems. Modifications of this apparatus have been used for a variety of animal species.

Zeitgeber Any entraining agent that plays a role in setting or resetting an internal biological clock. Examples include sunrise or sunset.

Zugunruhe Restlessness that an animal displays at the time of its usual migration.

CREDITS

Chapter 3

Figure 3.3 From "Conditioned Discrimination of Airborne Odorants by Garter Snakes . . ." by D. Begunn, et al., 1988, *Journal of Comparative Psychology, 102,* pp. 35-43. Copyright 1988 by the American Psychological Association. Reprinted by permission.

Chapter 4

Figure 4.1 Data from Donald E. Kroodsma, "The North American Song Populations of the March Wren Reach Distributional Limits in the Central Great Plains," in *The Condor,* 91:332-340, 1989.

Figure 4.5 From E. Mayr, "Darwin and Natural Selection," in *American Scientist,* 65:321-327. Copyright © 1977. Reprinted by permission of *American Scientist,* Journal of Sigma Xi, The Scientific Research Society.

Chapter 5

Figure 5.5 Modified from Hugh Dingle in *Popular Biology: Ecological and Evolutionary Viewpoints,* K. Wohrman and S. Jain, eds. Copyright © 1990, Springer-Verlag.

Figure 5.6 In C. R. B. Boake, ed. *Quantitative Genetic Studies of Behavioral Evolution.* University of Chicago Press, Chicago. Used with permission.

Chapter 6

Text Art 6.1 From R. E. Irwin, "The Evolutionary Importance of Behavioural Development," *Animal Behaviour,* 36:814-824, 1988.

Figure 6.10 From J. Rosenheim, "Comparative and Experimental Approaches to Understanding Insect Learning," in D. Papaj and A. Lewis, *Insect Learning.* Used with permission by Kluwer Academic Publishers.

Figure 6.13 From M. J. Littlejohn, "Premating Isolation in the *Hyla ewingi* Complex *(Anura: Hylidae),* in *Evolution,* 19:234-243. Copyright © 1965. Reprinted by permission.

Chapter 7

Figure 7.10 From: *Animal Physiology* by Roger Eckert, David Randall, and George Augustine © 1988 by W. H. Freeman and Company. Used with permission.

Figure 7.11 From: *Animal Physiology* by Roger Eckert, David Randall, and George Augustine © 1988 by W. H. Freeman and Company. Used with permission.

Figure 7.19 Source: Frye & Olberg, "Visual Receptive Field Properties of Feature Detecting Neurons in the Dragonfly," in *Journal of Comparative Physiology A* 177:569-578, 1995. Used with permission by Springer-Verlag.

Figure 7.22 From R. R. Capranica and A. J. M. Moffat, "Neurobehavioral Correlates of Sound Communication in Anurans," in *Advances in Vertebrate Neuroethology,* p. 706. Copyright © 1983 Plenum Press, New York, NY. Reprinted by permission.

Figure 7.23 From F. Nottebohm and M. Nottebohm, "Left Hypoglossal Dominance in the Control of Canary and White-Crowned Sparrow Song," in *Journal of Comparative Physiology,* 108:171-172. Copyright © 1976 Springer-Verlag, Heidelberg, Germany. Reprinted by permission.

Figure 7.24 Source: Maguire, et al, "Navigation-related structural change in the hippocampal of taxi drivers," *PNAS,* 97:4398-4403.

Chapter 8

Figure 8.7 From E. Adkins-Regan, "Sexual Differentiation in Birds," *Trends in Neuroscience,* 10:518. Copyright © 1987 Elsevier Science Publications Ltd., Cambridge. Used by permission.

Figure 8.10 From K. D. McCaul, et al., "Winning, Losing, Mood and Testosterone," *Hormones and Behavior,* vol. 26, pp. 486-504. Copyright © 1992 by Academic Press. Reproduced by permission of the publisher.

Figure 8.17 From R. E. Hegner and J. C. Wingfield, "Effects of Experimental Manipulation of Testosterone Levels on Parental Investment and Breeding Success in Male House Sparrows," *Auk,* 104:464. Copyright © 1987 American Ornithologists Union. Reprinted by permission.

Chapter 10

Figure 10.3 From Wolfgang Schleidt, "How fixed is the Fixed Action Pattern?" in *Zeitschrift für Tierpsychologie,* 36:194. Copyright 1974 Paul Parey Verlagsbuchhandlung, Hamburg. Reprinted with permission of the author.

Figure 10.7 From R. W. Openheim, "Some Aspects of Embryonic Behavior in the Duck," *Animal Behaviour,* 18:335-352, 1970. Reproduced with the kind permission of Bailliere Tindall, London, England.

Figure 10.11 From D. B. Miller, "Social Context Affects the Ontogeny of Instinctive Behavior," in *Animal Behaviour,* 48:627-34. Used by permission of Academic Press Ltd., London.

Figure 10.17 From P. Merler and S. Peters, "Structural Changes in Song Ontogeny in the Swamp Sparrow *Melospiza georgiana,* in *Auk,* 99:448. Copyright © 1982 by the American Ornithologists' Union. Reprinted by permission.

Figure 10.18 From Nowicki, Peters, and Podos, *American Zoologist,* 38:184, 1998.

Figure 10.19 Used with permission of Garland Publishing, from *The Development of Behavior.* Copyright © 1978. All rights reserved.

Figure 10.20 From *Behaviour and Social Life of Honeybees* by C. R. Ribbands. Published by International Bee Research Association, 18 North Road, Cardiff, CF1 3DY, UK. Reprinted by permission.

Figure 10.21 From R. L. Wyman and J. A. Ward, "The Development of Behavior in the Cichlid Fish *Etroplus maculatus* (Bloch)," *Zeitschrift für Tierpsychologie,* 33:461-191, 1973 Paul Parey Verlagsbuchhandlung, Berlin.

Chapter 11

Text Art 11.1 From "Social Factors in the Poison Avoidance and Feeding Behavior of Wild and Domesticated Rat Pups," 1971, *Journal of Comparative and Physiological Psychology, 75,* p. 352. Copyright 1971 by the American Psychological Association. Reprinted by permission.

Figure 11.10 From G. D. Sanders, "The Cephalopode," *Invertebrate Learning,* Vol. III, 1975, ed. by W. C. Corning, J. A. Dyal, and A. O. D. Willows. Copyright © 1975 by Plenum Publishing Corporation, New York, NY. Reprinted by permission.

Figure 11.11 From J. E. Sigurdson, "Automated Discrete-trials Techniques of Appetitive Conditioning in Honeybees," *Behavior Research Methods and Instrumentation,* vol. 13, p. 7. Used by permission.

Figure 11.15 From "Remembrance of Places Passed: Spatial Memory in Rats" by David S. Olton and Robert J. Samuelson, *Journal of Experimental Psychology: Animal Behavior Processes, 1976,* 2, 97-116. Copyright 1975 by the American Psychological Association, Inc. Reprinted by permission.

Figure 11.17 From "Visual Classes and Natural Categories in the Pigeon," by J. Cerella, 1979, *Journal of Experimental Psychology: Human Perception and Performance,* 5, p. 70. Copyright 1979 by the American Psychological Association. Adapted by permission.

Figure 11.19 From P. K. Stoddard, et al., *Behavioral Ecology and Sociobiology,* 29:211-215, p. 214. Copyright © 1991 Springer-Verlag, Heidelberg, Germany. Reprinted by permission.

Chapter 12

Figure 12.1 Source: After Trumler, E. 1959. Das "Rossigkeitsgesicht" und ähnliches Ausdrucksverhalten bei Einhufern. Zeit. für Tierpsychol, 16:478-488.

Figure 12.4 Reprinted with permission from M. C. Baker, et al., "Early Experience Determines Song Dialect Responsiveness of Female Sparrows" in *Science,* 214:819-821 (13 Nov 1981). Copyright 1981 by American Association for the Advancement of Science.

Figure 12.6 Reprinted by permission from: Stephen T. Emlen, "An Experimental Analysis of the Parameters of Bird Song Eliciting Species Recognition," in *Behavior,* 41 (1972) pp. 130-171 by E. J. Brill, Leiden.

Figure 12.11 Reprinted with permission from H. Kruuk, *The Spotted Hyena: A Study of Predation and Social Behavior,* p. 259. © 1972 by The University of Chicago. All rights reserved.

Figure 12.12 Reproduced with the kind permission of Bailliere Tindall from C. H. Brown and P. M. Waser, "Hearing and Communication in Blue Monkeys (*Cercopithecus mitis*)" in *Animal Behaviour*, 32:66-75, 1984.

Figure 12.14 Source: Redrawn from H. C. Proctor, "Courtship in the water mite *Meumania papillator*, in *Animal Behaviour*, 42:589-598, © 1991, Academic Press. Reprinted with permission of the publisher.

Figure 12.21 From J. L. Gould, "The Dance-Language Controversy," in *Quarterly Review of Biology*, 51:211-244, copyright 1976 by Stony Brook Foundation, Inc., Stony Brook, NY. Used with permission of the *Quarterly Review of Biology*.

Chapter 13

Figure 13.5 Redrawn after A. Carr, "Adaptive Aspects of the Scheduled Travel of Chelonia," in *Animal Orientation and Navigation,* p. 43, ed. by R. M. Storm. Copyright © 1967 Oregon State University Press, Corvallis, OR. Reprinted by permission.

Figure 13.10 From R. Wehner, "Arthropods," pp. 45-144, in *Animal Homing* (F. Papi, ed.), 1992. Used with the kind permission from Kluwer Academic Publishers.

Figure 13.11 From S. T. Emlen, "Migration, Orientation and Navigation" in *Avian Biology*, vol. 5, fig. 5, p. 152, © 1975 Academic Press, based on data from G. Kramer (1951), as presented by S. T. Emlen in *Proceedings of the X International Ornithologists Congress*, pp. 169-280. Used by permission.

Figure 13.13 From S. T. Emlen, "Migrating Orientation in the Indigo Bunting. I. Evidence for Use of Celestial Cues," *Auk*, 84:309-342. Copyright 1967 by the American Ornithologists' Union. Reprinted by permission.

Figure 13.14 From S. T. Emlen, "Migratory Orientation in the Indigo Bunting, *Passerina cayanea*, in *Auk*, 84:463-489. Copyright © 1967 by the American Ornithologists' Union. Reprinted by permission.

Figure 13.16 From K. Prinz and W. Wiltschko, "Migratory Orientation of Pied Flycatchers: Interaction of Stellar and Magnetic Information During Ontogeny," *Animal Behaviour*, 44:539-545. Copyright 1992 The Association for the Study of Animal Behaviour. Used by permission of Academic Press Ltd, London.

Chapter 14

Figure 14.4 After J. M. Diamond, "Ecological Consequences of Island Colonization by Southwest Pacific Birds" in *Proceedings of the National Academy of Sciences*, 67:529-536, 1970. Used by permission of the author.

Figure 14.5 Source: J. H. Connell, "The influence of interspecific competition and other factors on the distribution of the barnacle *Chthamalus stellatus*, in *Ecology*, 42:710-723. Copyright © 1961 by the Ecological Society of America.

Figure 14.10 Reprinted by permission from: Peter F. Sale from "A Suggested Mechanism for Habitat Selection by the Juvenile Manini *Acanthurus triostegus sandvicensis* Streets" in *Behaviour*, 35 (1970), pp. 27-44 by E. J. Brill, Leiden.

Figure 14.11 From J. J. Hughes, et al., "Predation Risk and Competition Affect Habitat Selection and Activity of Namib Desert Gerbils," *Ecology*, 75:1397-1405. Copyright 1994 Ecological Society of America. Used by permission.

Figure 14.12 From R. L. Hutto, "Habitat Selection by Nonbreeding, Migratory Land Birds," *Habitat Selection in Birds*, M. L. Cody, ed., pp. 455-476. Copyright © 1985 by Academic Press. Reproduced by permission of the publisher.

Figure 14.13 From L. Partridge, "Habitat Selection," in *Behavioural Ecology: An Evolutionary Approach*, ed. by J. R. Krebs and N. B. Davies. Copyright 1978 by Blackwell Scientific Publications Ltd, Oxford, England. Used by permission.

Figure 14.15 Source: Redrawn from S. D. Fretwell, "Populations in a Seasonal Environment." Copyright © 1972 by Princeton University Press, Princeton, NJ. Reprinted by permission of Princeton University Press.

Chapter 15

Figure 15.1 From Bruce A. Colvin, "Owl Foraging Behavior and Secondary Poisoning Hazard from Rodenticide Use on Farms," Ph.D. dissertation, Bowling Green State University, 1984. Reprinted by permission.

Figure 15.3 Redrawn from J. R. Krebs and R. H. McCleery, "Optimization in Ecology," in *Behavioural Ecology: An Evolutionary Approach*, 2d ed., ed. by J. R. Krebs and N. B. Davies. Copyright 1984 by Blackwell Scientific Publications Ltd, Oxford. Used by permission.

Figure 15.4 Reprinted by permission from *Nature*, 268:137-139. Copyright © 1977 Macmillan Magazines Ltd.

Figure 15.6 From D. W. Stephens and J. R. Krebs, *Foraging Theory*. Copyright © 1986 Princeton University Press, Princeton, NJ. Reprinted by permission.

Figure 15.7 Source: D. Stephens and J. Krebs, *Foraging Theory*. Copyright © 1986 by Princeton University Press, Princeton, NJ. Reprinted by permission of Princeton University Press.

Figure 15.8 From N. B. Davies and A. I. Houston, "Territory Economics," in *Behavioural Ecology: An Evolutionary Approach*, 2d edition, ed. by J. R. Krebs and N. B. Davies. Copyright © 1984 by Blackwell Scientific Publications Ltd, Oxford. Reprinted by permission.

Figure 15.11 From R. F. Foelix, *Biology of Spiders*, 2d edition, pp. 124-125. Copyright © 1996, Oxford University Press.

Figure 15.19 Reprinted with permission from T. Caracao and L. L. Wolf, "Ecological Determinents of Group Sizes of Foraging Lions," *American Naturalist*, 109:343-352. © 1975 by The University of Chicago. All rights reserved.

Figure 15.21 From *The Wolf* by L. David Mech. Copyright © 1970 by L. David Mech. Used by permission of Doubleday, a division of Random House, Inc.

Figure 15.29 Source: V. Heiber, *Behavioral Ecology*, 5:326-333. Copyright © 1994 Oxford University Press. Used by permission of the publisher.

Chapter 16

Figure 16.2 From D. G. C. Harper, "Competitive Foraging in Mallards: Ideal Free Ducks" in *Animal Behaviour*, 30:575-584. Copyright © 1982. Reproduced with the kind permission of Bailliere Tindall, London, England.

Figure 16.3 Figure 2.7 from *Primate Behavior: Field Studies of Monkeys and Apes* by Irven DeVore, copyright © 1965 by Holt, Rinehart and Winston, Inc. Reprinted by permission of the publisher.

Figure 16.5 Source: J. R. Krebs, "Territory and breeding density in the great tit, *Parus major* L. in *Ecology*, 52:2-22. Copyright © 1971 by the Ecological Society of America.

Figure 16.8 Source: C. A. Marler and M. C. Moore, "Time and Energy Costs of Aggression in Testosterone Implanted Free-Living Male mountain Spiny Lizards (*Sceloporus jarrovi*), in *Physiological Zoology*, 62:1334-1350, 1989.

Figure 16.11 Reprinted from *Psychoneuroendocrinology*, Volume 17, R. M. Sapolsky, "Cortisol Concentrations and the Social Significance of Rank Instability Among Wild Baboons," pp. 701-709. Copyright 1992 with kind permission from Elsevier Science Ltd, The Boulevard, Langford Lane, Kidlington OX5 1GB, UK.

Figure 16.12 Reprinted from *Psychoneuroendocrinology*, Volume 16, R. M. Sapolsky, "Testicular Function, Social Rank and Personality Among Wild Baboons," pp. 281-293. Copyright 1990, with kind permission from Elsevier Science Ltd, The Boulevard, Langford Lane, Kidlington OX5 1GB, UK.

Figure 16.16b Source: R. Huber et al., "Serotonin and aggressive motivation in crustaceans: Altering the decision to retreat, *Proc. Natl. Acad. Sci.*, 94:5939-5942, 1997.

Chapter 17

Figure 17.3 Reprinted with permission from T. H. Clutton-Brock, *Reproductive Success*. © 1988 by The University of Chicago. All rights reserved.

Figure 17.6 Reprinted with permission from R. L. Trivers, "Parental Investment and Sexual Selection," in *Sexual Selection and The Descent of Man*, edited by Bernard Campbell. Copyright © 1972 by Aldine de Gruyter, New York, revised 2001.

Figure 17.9 Reprinted by permission from *Nature*, 299:818-820. Copyright © 1982 Macmillan Magazines Ltd.

Figure 17.10 Reprinted by permission from *Nature*, 357:238-240. Copyright © 1992 Macmillan Journals Ltd.

Figure 17.15 Reprinted by permission from *Nature*, 299:818-820. Copyright © 1982 Macmillan Magazines Ltd.

Chapter 18

Figure 18.1 From John R. Krebs and Nicholas B. Davies, *An Introduction to Behavioural Ecology*, 2d ed. Copyright © 1987 by Blackwell Scientific Publications Ltd., Oxford, England. Reprinted by permission.

Figure 18.6 From G. H. Orians, "The Ecology of Blackbird (*Agelaius*) Social Systems," in *Ecological Monographs,* 31:285-312. Copyright © 1961 by the Ecological Society of America, Tempe, AZ. Reprinted by permission.

Figure 18.8 Source: M. R. Gross, "Salmon breeding behavior and life history evolution in changing environments," in *Ecology,* 72:1180-1186. Copyright © 1991 by the Ecological Society of America.

Figure 18.9 Source: S. M. Shuster, "The reproductive behavior of alpha male, beta male, and gamma male morphs in *Paracercis sculpta,* a marine isopod crustacean," in *Bahaviour,* 121:231-258, 1992.

Figure 18.12 Reprinted with permission from R. L. Trivers, "Parental Investment and Sexual Selection," in *Sexual Selection and The Descent of Man,* edited by Bernard Campbell. Copyright © 1972 by Aldine de Gruyter, New York, revised 2001.

Chapter 19

Figure 19.7 Reprinted by permission from *Nature,* 308:181-184. Copyright © 1984 Macmillan Magazines Ltd.

Figure 19.12 Source: F. de Waal and M. Berger, "Payment for Labour in Monkeys," in *Nature,* 404:563. Copyright © 2000, Macmillan Magazines Ltd. Reprinted by permission.

Figure 19.15 Source: H. Reyer in P. B Stacey and W. D. Koenig, eds., *Cooperative Breeding in Birds.* Copyright 1990 Cambridge University Press. Reprinted with the permission of Cambridge University Press.

Figure 19.18 Redrawn with permission from R. L. Honeycutt, "Naked Mole-Rats," *American Scientist,* 80:43-53. Copyright © 1992 Sigma Xi, the Scientific Research Society.

SUBJECT INDEX